Spectral Methods for Uncertainty Quantification

Scientific Computation

For other titles published in this series, go to
www.springer.com/series/718

O.P. Le Maître · O.M. Knio

Spectral Methods for Uncertainty Quantification

With Applications to Computational Fluid Dynamics

Springer

Prof. Dr. O.P. Le Maître
LIMSI-CNRS
Université Paris-Sud XI
91403 Orsay cedex
France
olm@limsi.fr

Prof. Dr. O.M. Knio
Department of Mechanical Engineering
The Johns Hopkins University
3400 North Charles Street 223 Latrobe Hall
Baltimore MD 21218-2682
USA
knio@jhu.edu

ISBN 978-90-481-3519-6 e-ISBN 978-90-481-3520-2
DOI 10.1007/978-90-481-3520-2
Springer Dordrecht Heidelberg London New York

Library of Congress Control Number: 2010921813

Cover design: eStudio Calamar S.L.

Printed on acid-free paper

Springer is part of Springer Science+Business Media (www.springer.com)

To the Ladies, certainly,
Marie-Christine & May

Preface

This book deals with the application of spectral methods to problems of uncertainty propagation and quantification in model-based computations. It specifically focuses on computational and algorithmic features of these methods which are most useful in dealing with models based on partial differential equations, with special attention to models arising in simulations of fluid flows. Implementations are illustrated through applications to elementary problems, as well as more elaborate examples selected from the authors' interests in incompressible vortex-dominated flows and compressible flows at low Mach numbers.

Spectral stochastic methods are probabilistic in nature, and are consequently rooted in the rich mathematical foundation associated with probability and measure spaces. Despite the authors' fascination with this foundation, the discussion only alludes to those theoretical aspects needed to set the stage for subsequent applications. The book is authored by practitioners, and is primarily intended for researchers or graduate students in computational mathematics, physics, or fluid dynamics. The book assumes familiarity with elementary methods for the numerical solution of time-dependent, partial differential equations; prior experience with spectral methods is naturally helpful though not essential. Full appreciation of elaborate examples in computational fluid dynamics (CFD) would require familiarity with key, and in some cases delicate, features of the associated numerical methods. Besides these shortcomings, our aim is to treat algorithmic and computational aspects of spectral stochastic methods with details sufficient to address and reconstruct all but those highly elaborate examples.

This book is composed of 10 chapters. Chapter 1 discusses the relevance and (ever increasing) role of uncertainty propagation and quantification in model-based predictions. This is followed with brief comments on various approaches used to deal with model data uncertainties, focusing in particular on a probabilistic framework that forms the foundation for subsequent discussion. The remaining nine chapters are divided into two parts.

Part I (Chaps. 2–6) focuses on basic formulations and mechanics, providing diverse illustrations based on elementary examples. Chapter 2 discusses fundamentals of spectral expansions of random parameters and processes. Treated in detail are the classical concepts underlying Karhunen-Loève (KL) expansions, homogeneous chaos, and polynomial chaos (PC). An outline is also provided of the application of these concepts to the representation of uncertain model data, and to the representation of the corresponding uncertain model outputs. Chapter 3 discusses so-called non-intrusive spectral methods of uncertainty propagation. These resemble collocation methods used in the numerical solution of PDEs, and are termed non-intrusive since they generally do not require modification of existing or legacy simulation codes. The discussion covers several approaches falling within this class of spectral methods, including stochastic quadratures, as well as cubature and regression methods. In Chap. 4, we discuss Galerkin (intrusive) approaches to uncertainty propagation, focusing in particular on weak formulations of stochastic problems involving data uncertainty. Stochastic basis function expansions are introduced, and the setup of the resulting stochastic problem is discussed in detail. Special attention is paid to the estimation of nonlinearities, and a brief outline of solution methods is provided. Chapter 5 provides detailed illustration of the implementation of PC methods to simple problems, namely through application to transient diffusion equations in two space dimensions, and to the steady Burgers equation in one space dimension. Chapter 6 then provides several examples illustrating the application of various approaches introduced in Chaps. 3 and 4 to flows governed by the time-dependent Navier-Stokes equations. Examples include incompressible flows, variable-density flows at low-Mach-number, and electrokinetically driven flows.

Part II (Chaps. 7–10) focuses exclusively on Galerkin methods, and deals with more advanced topics, more recent developments, or more elaborate applications. Chapter 7 discusses the application of specialized solution methods that are of general interest in stochastic flow computations. These include methods for finding stochastic stationary flow solutions, stochastic multigrid solvers, and a brief discussion of pre-conditioning and Krylov methods for the resolution of large systems of linear equations arising in Galerkin projections. Chapter 8 deals with generalized spectral representation concepts, particularly wavelet and multiwavelet representations, as well as multi-resolution analysis of stochastic problems. The applicability of these schemes to problems exhibiting discontinuous dependence on model data is emphasized, and is illustrated using applications to simple dynamical problems and to flow computations. Chapter 9 deals with adaptive representations, stochastic domain decomposition techniques, stochastic error estimation and refinement, and reduced basis approximations. New challenges, open questions, and closing remarks are mentioned in Chap. 10.

Orsay, France — O.P. Le Maître
Baltimore, Maryland — O.M. Knio

Acknowledgements

We wish to thank Prof. Roger Ghanem for his persistence in conveying his passion in the current subject matter. OMK, in particular, discovered that he had already learned quite a bit from Prof. Ghanem even before deliberately charging into the "uncertain," by osmosis and random collisions that have spanned multiple years. Much of our initial work took place within the framework of two focused projects that have brought us together a number of colleagues and collaborators, including Prof. Ghanem of the University of Southern California, and Drs. Habib Najm, Bert Debusschere, and Matthew Reagan of Sandia National Laboratories. Interactions and exchanges with these colleagues had made tremendous contributions to our appreciation of the subject matter, as well as developments outlined in this monograph. These exchanges have been made possible through the support of the Defense Advanced Research Projects Agency (DARPA) and Air Force Research Laboratory, Air Force Materiel Command, USAF, under Agreement F30602-00-2-0612, and by the Laboratory Directed Research and Development Program at Sandia National Laboratory, funded by the US Department of Energy.
OLM wishes to acknowledge the support of the two institutions that hosted him along the past years when working on stochastic spectral methods: the Laboratoire de Mécanique et d'Energétique at the Université d'Evry Val d'Essonne (LMEE) and the Laboratoire d'Informatique pour la Mécanique et les Sciences de l'Ingénieur (LIMSI) of the Centre National de la Recherche Scientifique (CNRS). The directors of these two institutions, Olivier Daube (LMEE) and Patrick Le Quéré (LIMSI), deserve special thanks for having provided OLM with the best possible working conditions and the necessary freedom to start new adventurous researches in the uncertainty world. OLM is also grateful to the Johns Hopkins University who supported him on many occasions over the last decade while visiting OMK: a large part of the materials presented in this monograph was initiated, and sometime performed, during stays at the Johns Hopkins University. Different financial supports from the French office for nuclear energy (CEA), funding agencies ANR (JCJC-080022) and Digiteo, and research network MoMaS were also benefical to OLM. Working on these projects and others, OLM was involved in collaborations with French colleagues; he wishes to particularly acknowledge numerous and fruitful

discussions with Drs. Lionel Mathelin (LIMSI), Jean-Marc Martinez (CEA) and Profs. Anthony Nouy (Université de Nantes), Christian Soize (Université de Paris Est), Alexandre Ern (Ecole des Ponts) and Serge Huberson (Université de Poitiers).
OMK wishes to express his gratitude to Prof. Rupert Klein of the Free University of Berlin for helpful contributions to his recent work on uncertainty. Exchanges with Prof. Klein have been supported by the Humboldt Foundation under a Friechrich Wilhelm Bessel research award. He also wishes to acknowledge support from the US Department of Energy under Awards DE-SC0001980 and DE-SC0002506. These collaborative efforts involving Prof. Roger Ghanem, Prof. Youssef Marzouk of Massachusetts Institute of Technology, Prof. Kevin Long of Texas Tech University, and Dr. Habib Najm, Dr. Bert Debusschere and Dr. Helgi Adalsteinsson of Sandia National Laboratories have inspired some of the material presented in Part II and many of ideas outlined in the Epilogue. He finally wishes to articulate his indebtedness to Prof. Serge Huberson of the Université de Poitiers for connecting him with OLM, and for his unwavering support.
We are grateful to Prof. Pierre Sagaut for suggesting the preparation of this monograph. We are also grateful to Dr. Ramon Khanna, Mr. Tobias Schwaibold and the Springer staff for their encouragement and assistance during this project. During the initial conception stages, we had anticipated delivering about a 300-page manuscript in April 2009. Consequently, we also wish to express our gratitude to the Springer editors and staff for their patience and persistence, along with our commitment to incorporate experiences and knowledge gained during this project into in future endeavors.

Contents

Chapter 1
Introduction: Uncertainty Quantification and Propagation

1.1 Introduction

Numerical modeling and simulation of complex systems is continuously developing at an incredible rate in many fields of engineering and sciences. This development has been made possible thanks to the constant evolution of numerical techniques and increasing availability of computational resources. Nowadays, simulations are essential tools for engineers throughout the design process. The simulations minimize the need for costly physical experiments that may be even impossible during early design stages. However, numerical simulations have to be carefully designed, performed, and verified, to yield useful and reliable information regarding the system being studied. In fact, the confidence one has in a computation is a key aspect when interpreting and analyzing simulations results. Indeed, simulations inherently involve errors, understanding and quantification of which is critical to assess the differences between the numerical predictions and actual system behavior. Classically, the errors leading to discrepancies between simulations and real world systems are grouped into three distinct families [215]:

- **Model error:** Simulations rely on the resolution of mathematical models accounting for the essential characteristics of the systems being studied. The mathematical models express essential principles (*e.g.* conservation laws, thermodynamics laws,...) and are generally supplemented with appropriate modeling of the physical characteristics of the systems (*e.g.* constitutive equations, state equations,...). Often, simplifications of the mathematical model are performed, essentially to facilitate its resolution, based on the analysis of the problem characteristics and on some assumptions, with the direct effect of modeling some sort of ideal system different from the actual one. In computational fluid dynamics for instance, incompressible flow or inviscid fluid models are convenient approximations of the exact model (*e.g.* the compressible Navier-Stokes equations) describing an actual fluid flow. In other circumstances, a two-dimensional approximation may be found suitable, though the flow takes place in a three-dimensional world. Also, physical phenomena may be simply disregarded when they are deemed to

O.P. Le Maître, O.M. Knio, *Spectral Methods for Uncertainty Quantification*, Scientific Computation,
DOI 10.1007/978-90-481-3520-2_1, © Springer Science+Business Media B.V. 2010

have a negligible contribution, such as radiative transfer in many natural convection models. Clearly, the resulting mathematical model will not be able to exactly reproduce the behavior of the real system, but one expects that the predictions based on the simplified model will remain sufficiently accurate to conduct a suitable analysis. In fact, the validity of the assumptions has to be carefully verified *a posteriori*.

- **Numerical errors:** The mathematical model selected will have to be solved using a numerical method. The numerical resolution of the model will then introduce some numerical error in the prediction, because numerical methods usually provide approximations of the exact model solution. Indeed, mathematical models consist in sets of equations (differential, integral, algebraic,...) whose resolution calls for appropriate discretization techniques and algorithms. The numerical errors can be controlled and reduced to an arbitrary low level, at least theoretically, by using finer discretizations (say finer spatial meshes and smaller time steps) and more computational resources (for instance to lower convergence criteria in iterative algorithms). This is made possible by the design of numerical methods that specifically incorporate specific measures, based for instance on notions of convergence, consistency and stability, to ensure a small numerical error, which is however always nonzero due to the finite representation of numbers in computers.
- **Data errors:** The mathematical model also needs to be complemented with data and parameters that specify physical characteristics of the simulated system among the class of systems spanned by the model. These data may concern the system's geometry, the boundary and initial conditions, the external forcings. Parameters may be physical or model constants prescribing the constitutive laws of the system. In many situations, the data cannot be exactly specified, because of some limitations in experimental data available (for instance in the measurement or identification of model constants), in the knowledge of the system (for instance at early design stage where forcing and boundary conditions may not be precisely defined yet), or because of inherent variability of the systems studied (for instance due to dimensional tolerances in fabrication and assembly processes, variability in operating conditions,...). Using data which partially reflect the nature of the exact system induces additional errors, called data errors, on the prediction.

The sources of error and uncertainty have distinct origins, which may be rooted in different disciplines. On the one hand, the choice of the physical model depends primarily on the experience of the modeler or specialist, and the subject of modeling of the physical system is outside the scope of this book. Nonetheless, we note that in the presence of uncertainty in input data, if the intrinsic predictive capability of the model is not satisfactory, it may not always be possible or desirable to augment the model complexity, *e.g.* by relaxing simplifying assumptions. In some cases, it may be preferable to consider the problem and model as uncertain, and to apply a non-parametric probabilistic analysis [216]. Such non-parametric approaches are quite recent, and have been primarily applied to models governed by linear elasticity theory; their application to more complex systems and situations, *e.g.* nonlinear models, are yet to be explored. From the numerical simulation point of view, we note that numerous approaches exist for discretizing mathematical formulations, and various methods are available for estimating and minimizing the associated discretization errors. These aspects of the problem, namely numerical error control and reduction, are also outside the scope of this book.

The last category of uncertainties, namely those associated with data describing the physical system, constitute the central theme of this volume. Specifically, the methods and analyses that we shall describe aim at characterizing the effects of variability in the data, or in other words the impact of imprecise knowledge in the input data used to specify the system on the predicted response of the associated model.

1.1.1 Simulation Framework

Once a suitable mathematical model of a physical system is formulated, the numerical simulation task typically involves several elementary steps:

1. Case specification. In this step, the input parameters are specified. The types of data that are fixed generally depend on the physical problem that is being simulated, and on the model used to represent it. Generally, one needs to specify the geometry associated with the system, and particularly the computational domain. Boundary conditions on the model solutions are also imposed, which may also depend on the model and the mathematical formulation utilized. In the case of transient systems, initial conditions are also provided and, when present, external forcing functions that are applied to the system. Finally, physical constants are specified that describe the properties of the system, as well as modeling or calibration data for representing phenomena that are not explicitly resolved by the model.
2. Having defined the input data, one then proceeds to the simulation stage. It is often necessary to define a computational grid on which the model solution is first discretized. This involves selection of a discretization approach, and of associated discretization parameters. Additional parameters related to time integration, whenever relevant, are also specified. Numerical solution of the resulting discrete analogue of the mathematical model can then be performed. Here, we shall restrict ourselves to deterministic numerical models having a well-posed mathematical formulation. In particular, we shall assume that: (i) for the input data fixed during the specification step, the original mathematical model admits a **unique solution**, (ii) provided that discretization approaches and numerical parameters are judiciously selected, the discrete model analogue admits a unique solution that converges to the model solution, and (iii) sufficiently small discretization errors can be achieved.
3. The final step concerns analysis of the computed solution. Typically, this involves visualization as well as various post-treatments which aim at extracting and representing quantities of interest, metrics, and information facilitating subsequent decision support.

This methodology outlined above is illustrated in Fig. 1.1, which depicts schematically the process flow and the links between various steps of a simulation.

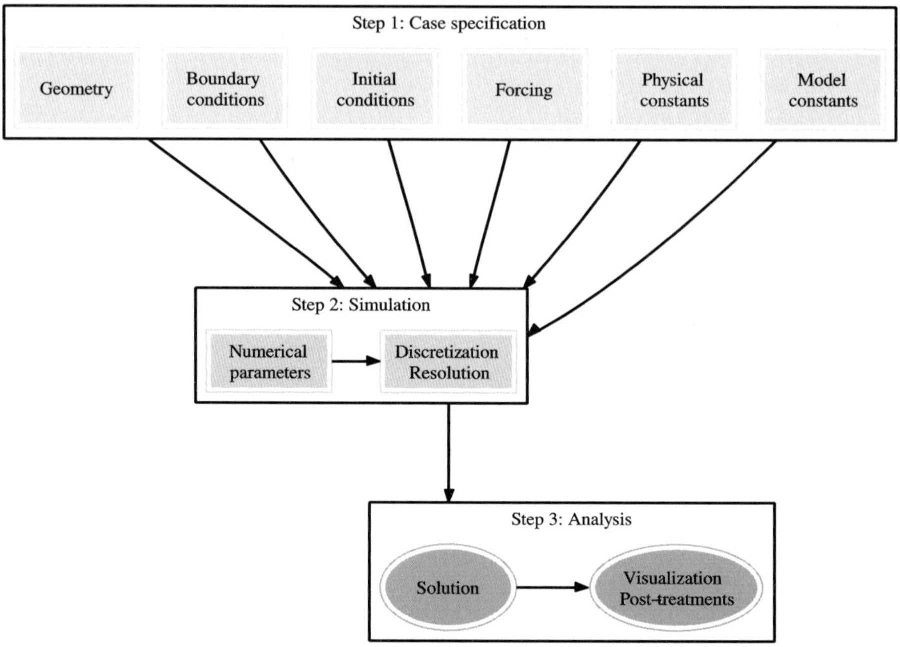

Fig. 1.1 Flow chart illustrating the various steps of a numerical simulation, including case specification, numerical solution, and post-treatment

1.1.2 Uncertainties

The simulation methodology above reflects an idealized situation that may not be always achieved in practice. In many cases, the input data set may not be completely specified, for instance due to incomplete knowledge of the real system or due to intrinsic variability. The associated uncertainties may have different origins, and in many cases relate to a subset of the input data. For example, it may not be possible to determine precisely the boundary conditions of the system, or the forcing that it is subjected to. Furthermore, the physical properties of the system may not be exactly known. Also arising frequently are parametrization uncertainties, which may affect constants that one can bound but cannot determine exactly *a priori*, possibly because direct measurements are not practical.

Thus, though model equations may be deterministic, it may not be possible to rely on a single deterministic simulation because the input data are not precisely known, or are known to admit intrinsic variabilities. Consequently, one must associate with the simulation results an uncertainty resulting from incomplete knowledge of the input data.

Admittedly, in situations involving detailed fundamental studies of simplified problems, the idealized nature of deterministic numerical models may offer an advantage, as it enables analysis of relevant settings that are impossible to address

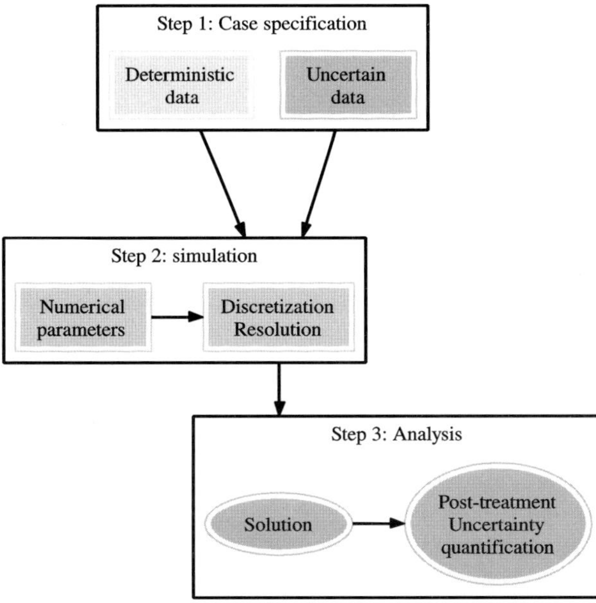

Fig. 1.2 Schematic illustration of model-based simulation in the presence of uncertain data. Though model equations are deterministic, the solution is uncertain, and the associated uncertain levels must be quantified

experimentally, *e.g.* due to limitations in controlling experiments or due to experimental imperfections that may not be eliminated. However, in general, the idealized nature of deterministic simulations presents a severe limitation, since one generally wishes to characterize and quantify the impact of uncertainties in model data on numerical predictions. To do so, one must generalize the previous framework to accommodate propagation of data uncertainty, as schematically illustrated in Fig. 1.2.

1.2 Uncertainty Propagation and Quantification

1.2.1 Objectives

The principal objectives of uncertainty propagation and quantification in model-based simulations are briefly addressed through the following partial list:

- **Validation:** simulations must be validated against measurements performed on real systems. Note, however, that physical measurements are inherently affected by uncertainties, due both to measurement errors as well as system imperfections. Measurement uncertainties are typically represented using error bars, which are indicative of their range. Clearly, the validation task must carefully take into consideration both experimental and computational uncertainty ranges.

- **Variance analysis:** the variation of the system response around its mean (or nominal) value provides important information that is relevant to design and optimization, as well as decision support. It characterizes the robustness of the prediction, the controllability of the system, and provides a confidence measure in computed predictions.
- **Risk analysis:** based on the probability laws of the input data, it is often desired to determine the probabilities of the system exceeding certain critical values, or operation thresholds. In turn, these probabilities can be used to conduct reliability or risk assessment analyses.
- **Uncertainty management:** in cases where the system is subject to multiple (distinct) sources of uncertainty, a key question concerns their relative impacts on the response of the system. This is required in order to establish effective strategies, including priorities, for managing, observing and eventually reducing dominant sources of uncertainty.

The objectives above are quite general in nature, and may take different incarnations depending on the nature of the problem, the disciplines to which it belongs, and the methodologies used to characterize it.

1.2.2 Probabilistic Framework

A probabilistic framework appears to be well suited for the pursuit of the objectives stated above. Since the input data cannot be defined exactly, it is legitimate to consider these as random quantities. Later, we shall often describe the random input data in terms of a stochastic vector, $\mathbf{d}$, belonging to a probability space (Ω, σ, μ), whose existence will be implicitly assumed without being systematically mentioned. We shall also assume that the probability law of $\mathbf{d}$ is known. For a detailed treatment of probability theory, see [137].

1.3 Data Uncertainty

A schematic representation of the probabilistic framework defined here is shown in Fig. 1.3. As illustrated in the figure, the input data follow a known probability law. The spectral uncertainty quantification (UQ) methods that are the central theme of this book are essentially based on a parametrization of the uncertain input data using a set of independent random variables (RVs) that is often called the germ. Several methods are at our disposal for constructing such parametrizations, and their selection may depend on the nature of the components of $\mathbf{d}$ [57]. As discussed in the following chapter, these include Karhunen-Loève (KL) decompositions of stochastic processes, or more generally Polynomial Chaos (PC) decompositions. The germ that parameterizes the random data follows a probability law that is not necessarily the same as that of the random data itself, particularly when parametrization of the

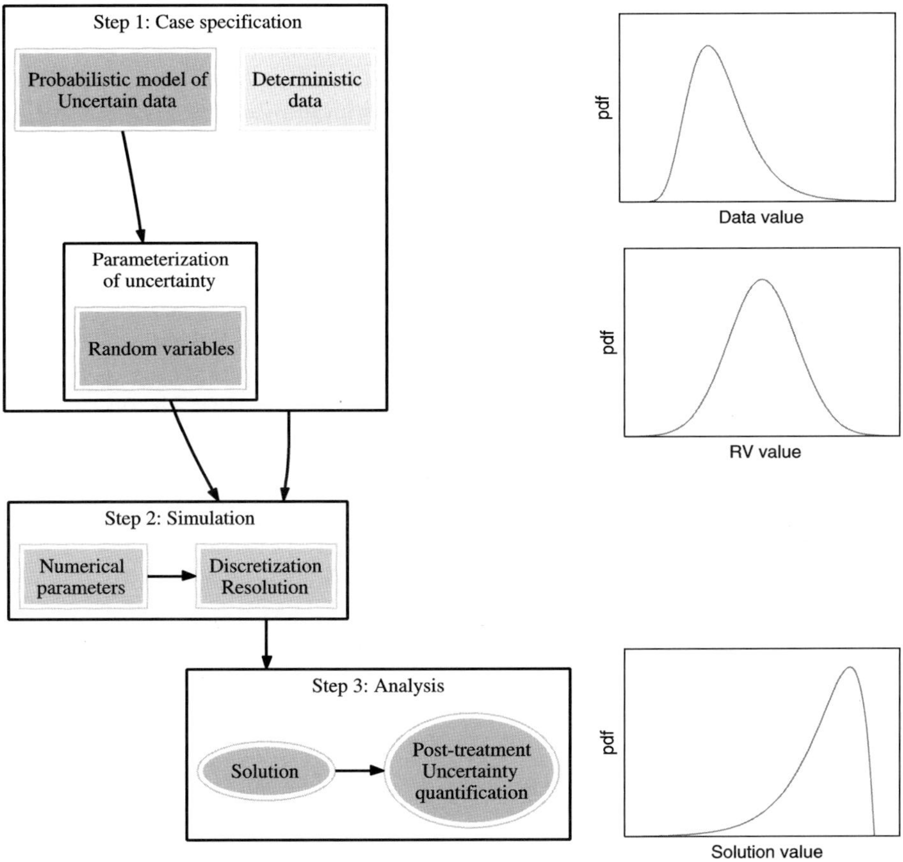

Fig. 1.3 Schematic view of the various stages of uncertainty propagation using spectral methods. Based on a known probability law of uncertain data **d** (*top*), one constructs a parametrization based on a set of random variables also having known probability laws (*middle*). Uncertainty propagation consists in determining the probability law of the solution that is induced by the germ (*bottom*)

data involves nonlinear functionals. Figure 1.3 identifies links between key steps in the application of spectral methods to uncertainty propagation, starting with probabilistic representation of the random data, data parametrization using independent RVs, and propagation of data uncertainty to model solution. Within this framework, the propagation step can be thought of as determining the functional dependence of the solution on the RVs that parametrize the data.

1.4 Approach to UQ

Regardless of the nature of the model considered, an underlying assumption in our approach to the probabilistic characterization of the system response to uncertain data is the existence of a numerical tool for the prediction of the deterministic system. The deterministic model may be simple or elaborate, depending both on the nature of the system and of level of fidelity that one desires to achieve. Thus, in some cases, application of the deterministic model itself may itself require substantial computational resources.

The numerical methods used by the model may be quite diverse, including finite-element, finite-difference, finite-volume, spectral, particle, or an hybridization of these discretization methods. Uncertainty propagation using the associated models should aim to the extent possible to minimize the overheads necessitated by the stochastic representation, while at the same time strive to keep a sufficient degree of generality so as to facilitate applications to different models and various discretizations.

Suppose that one seeks to determine the response, $\mathbf{s}$, of a system governed by an operator $\mathcal{M}$ (the model), which we shall abstractly denote as follows:

$$\mathcal{M}(\mathbf{s}, \mathbf{d}) = 0.$$

Formally, we seek the probability law of the solution, $\mathbf{s}$, which is induced by the random data, $\mathbf{d}$. Following the discussion above, the data is parametrized using, $\boldsymbol{\xi} = \{\xi_1, \xi_2, \ldots\}$, a vector of independent RVs, and the dependence of the data on the germ will be formally expressed according to:

$$\mathbf{d} \equiv \mathbf{d}(\boldsymbol{\xi}).$$

This leads us to seek the expression of $\mathbf{s}(\boldsymbol{\xi})$. Knowledge of the probability law of $\boldsymbol{\xi}$ will then yield the probability law of $\mathbf{s}$.

Below, we shall briefly outline the application of spectral methods for extracting the probabilistic content of $\mathbf{s}(\boldsymbol{\xi})$. To provide a brief sense of perspective, brief comments are first provided concerning classical Monte-Carlo approaches.

1.4.1 Monte Carlo Methods

These methods are certainly quite popular, and also the simplest to implement. The fundamental idea on which Monte-Carlo (MC) methods rely is a pseudo-random sampling of the germ $\boldsymbol{\xi}$ in order to construct a set of realization of the input data, $\{\mathbf{d}^1, \mathbf{d}^2, \ldots\}$. To each of these realizations corresponds a unique solution of the model, which is denoted by $\mathbf{s}^i \equiv \mathbf{s}(\boldsymbol{\xi}^i)$, $i = 1, 2, \ldots$. The collection of $\{\mathbf{s}^1, \mathbf{s}^2, \ldots\}$ is called the sample solution set. Based on the latter, it is possible to apply sampling methods to estimate the statistics of $\mathbf{s}$, the statistics of a particular observable $h(\mathbf{s})$, the correlations between components of the solution, probability laws, etc. For instance, the mathematical expectation of $\mathbf{s}$ can be estimated according to:

$$\langle \mathbf{s} \rangle = \lim_{M \to \infty} \frac{1}{M} \sum_{i=1}^{M} \mathbf{s}^i \partial_i, \quad \sum_{i=1}^{M} \partial_i = M,$$

where M is the total number of realizations, and $\eth_i$ is the relative weight associated with realization i. (For a non-biased sampling, $\eth_i \equiv 1$.) One of the advantages of MC methods is that it is sufficient to resolve the deterministic model, i.e. to determine **s** for a particular realization of **d**. Thus, the effort needed to propagate the uncertainty in **d** essentially amounts to obtaining a (generally large) number of individual deterministic model realizations. The MC approach is also quite robust, since its implementation does not necessitate any hypothesis or condition on the variance of **d**, which may be quite large, nor on the regularity of $\mathbf{s}(\boldsymbol{\xi})$, nor on the form of the model. The convergence of MC methods can be assessed based on indicators that relate directly to computed solutions, without the need for intervention into the underlying model. Furthermore, the convergence is independent of the dimensionality of the germ, which may be advantageous for problems involving a large number of independent RVs.

One of the principal limitations of MC methods concerns their convergence rate with the number, M, of realizations. In fact, the convergence of variance estimates behaves as $M^{-1/2}$, which is relatively low compared to the convergence of spectral methods. Numerous sampling methods have been proposed in order to accelerate the statistical convergence of estimators (importance sampling, variance reduction, Latin hypercube,... [94, 134, 153]) but these are generally insufficient to provide accurate characterization of uncertain systems that motivate the present development, for which the application of MC methods would be prohibitively expensive.

1.4.2 Spectral Methods

As outlined above, MC methods are collocation methods: for a specific realization of **d**, one obtains a local information on the solution, and the domain of **d** must be sampled with sufficiently fine resolution to determine the variability induced on **s**. Consequently, one immediately realizes that the local nature of the information associated with each realization penalizes the problem of determining the global variability of the solution, both in terms of efficiency and the limited analytical capabilities afforded by the local representation.

In contrast, spectral methods are based on a radically different approach, namely one based on constructing (or reconstructing) the functional dependence of the solution on the germ. This functional dependence is typically expressed in terms of a series:

$$\mathbf{s}(\boldsymbol{\xi}) = \sum_{k=0}^{\infty} \mathbf{s}_k \Psi_k(\boldsymbol{\xi}), \tag{1.1}$$

where the Ψ_k's are suitably selected functionals of the RVs, and the $\mathbf{s}_k$'s are deterministic coefficients. Once available, the series development may be immediately exploited to determine the statistics of **s**, either analytically or via sampling of $\boldsymbol{\xi}$.

Determination of the development of the solution, **s**, in the series form given by (1.1), constitutes the central theme of the spectral methods described in this book.

Without going into details, we provide a short (partial) list of prior work that illustrates diverse areas of application of spectral UQ methods. We point to the work of Ghanem and Spanos [90] as being at the origin of the recent spreading of spectral UQ methods. This early work primarily aimed at elasticity problems, considering in particular uncertainty in mechanical properties and external forcing [89, 90]. These methods were subsequently refined to deal with problems of increasing complexity, see e.g. [21, 96, 149], and the review in [204]. In parallel, numerous applications to heat transfer have appeared, e.g. [101, 107, 135, 165, 210]), including theoretical studies of spectral UQ methods for associated elliptic problems [9, 50].

The first applications of spectral UQ methods to fluid flow considered Darcy flows in porous media [85, 87, 150]. Following these developments, the solution incompressible flows described by stochastic Navier-Stokes equations has been accomplished using spectral methods, particularly in [123, 128, 247], and were later extended to stochastic flows at low Mach number [127], and fully compressible flows [146, 148]. Spectral uncertainty propagation methods were also used in more complex settings, such as flow-structure interaction [249], and protein label into electro-chemical microchannel flow [51]. For a review of the application of spectral UQ methods in fluids, the reader is referred to the reviews in [115, 163].

1.5 Overview

This book consists of ten chapters, and comprises two main parts. Part I (Chaps. 2–6) discusses the underlying theory and construction of stochastic spectral methods. It also provides a detailed exposition of elementary examples, as well as an overview of selected applications in computational fluid dynamics (CFD). Part II (Chaps. 7–9) discusses selected advanced topics, namely concerning iterative solvers, multi-resolution approaches, and adaptive methods. Concluding remarks immediately follow in Chap. 10.

In Chap. 2, we introduce the fundamental concepts on which further developments are based. As further described throughout this monograph, we shall regard the solution of a model depending on random input as being an element of a suitable product of Hilbert spaces, namely an L^2 space describing the deterministic solution, and an L^2 probability space that adequately represents the random data. Thus, a particular realization of the random solution corresponds to fixing a specific value of the random inputs. Based on this fundamental concept, statistics and other transformations of the random solution can be obtained by appropriately exploiting the measures and inner products associated with these Hilbert spaces. Theoretical foundations are briefly alluded to, and relevant background material is relegated to Appendix A.

The bulk of Chap. 2 is devoted to classical spectral representation approaches of random processes. We introduce the classical Karhunen-Loève decomposition of a second-order random process, based on the spectral decomposition of its autocorrelation function. We then derive the spectral representation of the autocorrelation in terms of the eigenvalues and eigenfunctions of the associated eigenvalue problem. The properties of the KL expansion are then discussed, and the spectral decomposition is extended to approximate the random process itself. The corresponding approximation error is briefly analyzed, and then examined in detail in light of a practical example where analytical solution of the eigenvalue problem is available. Numerical alternatives are briefly discussed for situations where analytical methods are not available. In particular, we outline a Galerkin formulation that is suitable for this purpose. Classical Polynomial Chaos decompositions are discussed next. We start by defining the space of polynomials in Gaussian random variables, and in particular recall the definitions of Homogeneous Chaos and Polynomial Chaos (PC) of order p. Based on these definitions, we outline formal expansions of a second-order random variable in terms of the Polynomial Chaos, and define it as its PC decomposition. The construction of the PC system is examined for the case of Gaussian random variables. We consider both the one-dimensional system, where classical Hermite polynomials are recovered, as well as its multi-dimensional systems, where the PC basis is defined in terms a partial tensorization of 1D polynomials. We then consider the truncation of PC expansions at finite order, and discuss errors associated with truncated expansions.

Following the basic outline above, generalized PC decompositions are addressed. The generalization accommodates random variables that are not necessarily Gaussian. In Chap. 2, we limit the discussion to polynomials of non-Gaussian variables, and thus outline a straightforward extension of the Hermite chaos.[1] On the other hand, brief remarks are provided for the case of dependent random inputs. We finally provide a brief discussion regarding the application of PC representations, and thus set the stage for subsequent developments.

In Chap. 3, we provide a brief overview of so-called "non-intrusive" uncertainty propagation methods. The fundamental concept behind these methods essentially consists in the (repeated) application of a deterministic solver in order to determine the unknown expansion coefficients appearing in the spectral expansion of the solution. This approach is called non-intrusive, because (existing or legacy) deterministic solvers can be immediately applied without modification. Within this broad framework, we explore different strategies for obtaining the spectral coefficients. We start with classical sampling approaches, and discuss in particular the application of Gauss quadrature methods and cubature formulas. Both the 1D and multi-dimensional cases are considered in the discussion. Due to their relevance to a wide class of computational approaches, a more detailed discussion of quadrature formulas is provided in Appendix B. We then turn our attention to regression-based approaches, and conclude with a discussion of key features of deterministic and stochastic sampling methods.

[1] This restriction is later extended in Chap. 7 where discontinuous or localized polynomials are used in the context of wavelet and multiwavelet representations.

Chapter 4 is devoted to spectral Galerkin methods. Unlike non-intrusive methods, Galerkin methods are inherently "intrusive", because they are based on the solution of the system of governing equations for the spectral coefficients in the PC representation of the solution. Thus, careful adaptation of deterministic solver is at a minimum required to address the resulting task. In Chap. 4, an abstract description is provided of the setup of stochastic Galerkin problem, starting with the statement of the deterministic problem and its generalization to account for random inputs. Following the framework introduced in Chap. 2, probabilistic representations of the random data and of the stochastic solution are adopted. Basis function expansions in the appropriate function spaces are used for this purpose. We then derive the weak form of the stochastic problem, and construct discrete parametrizations of both the random data and the model solutions. Using these discretized representations, a weighted residual formalism is used to define the so-called "spectral problem," which governs the behavior of the unknown solution coefficients. The structure of the spectral problem is analyzed in detail for the special case of linear operators, and suitable solution methods are briefly outlined. Approaches for estimating nonlinear terms are then addressed, and their application is illustrated for selected examples that frequently arise in practical applications. In many cases, these approaches rely on PC product and moment formulas, whose implementation is further discussed in Appendix C.

Chapter 5 provides a detailed treatment of elementary examples using intrusive PC expansions. Attention is focused on the steady 2D heat equation, and the steady Burgers equation. The discussion covers the setup of the deterministic and stochastic problems, stochastic and spatial discretizations, parameter selection, and analysis of computed results. For both model problems, we focus on a finite element methodology in the product space spanned by the spatial and stochastic basis functions. For the heat equation, a variational formulation of the problem is adopted, which is coupled with a Galerkin formulation along stochastic dimensions. For the Burgers equation, we start from the weak form, and rely on Galerkin projections along both the stochastic and spatial dimensions. Computed results are used to provide detailed illustrations of the application of intrusive PC expansions, the dependence of the results on spatial and stochastic discretization parameters, as well as utilization of stochastic representations to quantify solution uncertainty and extract relevant statistics.

In Chap. 6, we provide detailed examples of the application of intrusive and non-intrusive PC expansions to fluid flows governed by the transient Navier-Stokes equations. We start with the development of an incompressible solver based on a pressure projection formalism, and discuss its application to 2D internal flow. The resulting stochastic projection method is then extended to Boussinesq flows, and later generalized to compressible flows in the zero-Mach-number limit. We then outline the construction of a stochastic particle method, and illustrate its application to buoyancy-driven flow at high Reynolds and Rayleigh numbers. Finally, an example is provided of the application of the stochastic projection scheme to a multiphysics problem, namely the analysis of protein labeling reactions in electro-chemical microchannel flow.

Chapter 7 discusses the development of specialized solvers for equation systems that arise frequently in PC applications. We first focus on iterative methods for linear problems, and address in particular the implementation of Krylov methods and preconditioning. We then turn our attention to the application of multigrid methods to systems governed by the stochastic Poisson equation. Finally, a specialized solver is presented that is suitable for the simulation of the steady Navier-Stokes equations in the presence of random data.

Chapter 8 deals with the application of multi-resolution analysis (MRA) schemes to intrusive PC computations. We start by developing PC expansions based on Haar wavelets, and generalize this approach to multiwavelet (MW) basis functions. Application of the resulting MRA schemes is then illustrated based on simplified examples of dynamical systems, and in more elaborate examples involving buoyancy-dominated flow and a simplified Lorenz system. One of the interesting features of these developments is that they enable the treatment of problems involving steep or discontinuous dependence of the solution on random inputs, phenomena which are shown to cause major difficulties when global polynomial representations are used.

Chapter 9 explores four approaches for the construction of adaptive PC methods. Attention is focused primarily on adaptivity along the random dimensions, and consequently on strategies for refinement of the stochastic representation. We start by outlining the development of adaptive multiwavelet expansions, which are based on refinement of the MW basis itself. An alternative approach is then presented based on an adaptive partitioning of the space of random data. The third approach considered relies on a refinement strategy based on a posteriori error estimates. Finally, a generalized spectral decomposition approach is presented which is based on constructing an "optimal" set of eigenfunctions, which are later used as a "reduced" basis in the PC decomposition. The implementation of each of these four adaptive strategies is illustrated through practical examples, which are in particular used to quantify the effectiveness of the corresponding techniques.

In Chap. 10, we provide a brief discussion of open questions and selected topics of ongoing research. Specifically, we outline specific areas where further developments and improvements of the concepts and methods outlined in earlier chapters may be possible. We also briefly address topics that lie outside the scope of this monograph, but may likely yield substantial benefits to the present uncertainty quantification and management capabilities.

Part I
Basic Formulations

Chapter 2
Spectral Expansions

In this chapter, we discuss fundamental and practical aspects of spectral expansions of random model data and of model solutions. We focus on a specific class of random process in L^2 (see Appendix A) and seek Fourier-like expansions that are convergent with respect to the norm associated with the corresponding inner product. To clarify the discussion, a brief introduction of notation adopted is first provided; see Appendix A for additional details.

Let (Θ, Σ, P) be a probability space and θ a random event belonging to Θ. We denote $L_2(\Theta, P)$ the space of second-order random variables defined on (Θ, Σ, P) equipped with the inner product $\langle \cdot, \cdot \rangle$ and associated norm $\| \cdot \|_\Theta$:

$$\langle U, V \rangle = \int_\Theta U(\theta) V(\theta) \, \mathrm{d}P(\theta) = \mathrm{E}[UV] \quad \forall U, V \in L_2(\Theta, P),$$
$$U \in L_2(\Theta, P) \rightarrow \langle U, U \rangle = \|U\|_\Theta^2 < \infty, \tag{2.1}$$

where $\mathrm{E}[\cdot]$ is the expectation operator.

We consider $\mathbb{R}$-valued stochastic processes, indexed by $\mathbf{x} \in \Omega \subseteq \mathbb{R}^d$, $d \geq 1$:

$$U : (\mathbf{x}, \theta) \in \Omega \times \Theta \mapsto U(\mathbf{x}, \theta) \in \mathbb{R},$$

where for any fixed $\mathbf{x} \in \Omega$, the function $U(\mathbf{x}, \cdot)$ is a random variable. We shall consider second-order stochastic processes:

$$U(\mathbf{x}, \cdot) \in L_2(\Theta, P) \quad \forall \mathbf{x} \in \Omega. \tag{2.2}$$

Conversely, for a fixed event θ, the function $U(\cdot, \theta)$ is called a realization of the stochastic process. We will assume that the realizations $U(\cdot, \theta)$ are almost surely in the Hilbert space $L_2(\Omega)$. We denote $(\cdot, \cdot)$ and $\| \cdot \|_\Omega$ the inner product and norm on this space; specifically,

$$(u, v) \equiv \int_\Omega u(\mathbf{x}) v(\mathbf{x}) \, \mathrm{d}\mathbf{x} \quad \forall u, v \in L^2(\Omega),$$
$$u \in L_2(\Omega) \rightarrow \|u\|_\Omega^2 = (u, u) < \infty. \tag{2.3}$$

O.P. Le Maître, O.M. Knio, *Spectral Methods for Uncertainty Quantification*,
Scientific Computation,
DOI 10.1007/978-90-481-3520-2_2, © Springer Science+Business Media B.V. 2010

The assumption $U \in L_2(\Omega)$ almost surely means that

$$P(\{\theta : (U(\cdot,\theta), U(\cdot,\theta)) < \infty\}) = 1. \tag{2.4}$$

In Sect. 2.1, we focus our attention on the Karhunen-Loève (KL) representation of the stochastic process U. This classical approach essentially amounts to a bi-orthogonal decomposition based on the eigenfunctions obtained through analysis of its correlation function. Basic results pertaining to these decompositions are first stated; implementation of KL decompositions is then outlined through specific examples. Section 2.2 discusses Polynomial Chaos (PC) representations of random variables. We start by introducing the classical concepts of Homogeneous Chaos, and the associated one-dimensional and multi-dimensional Hermite bases expansions. The convergence of these expansions is briefly discussed. These concepts are later extended in Sect. 2.3, which deals with generalized PC expansions based on different families of orthogonal polynomials, as well as situations involving dependent random variables. Finally these expansions are extended to random vectors and stochastic processes in Sect. 2.4. An elementary road map for the applications of spectral representations is then provided in Sect. 2.5.

2.1 Karhunen-Loève Expansion

The Karhunen-Loève (KL) decomposition is well-known, and its use is widespread in many disciplines, including mechanics, medicine, signal analysis, biology, physics, and finance. It is also known as proper orthogonal decomposition (POD), principal component analysis (PCA) in the finite dimensional case. It has been proposed independently by different authors in the 1940's [106, 108, 136]. As further discussed below, the KL decomposition essentially involves the representation of a stochastic process according to a spectral decomposition of its correlation function.

2.1.1 Problem Formulation

Consider a stochastic process U,

$$(U : \Omega \times \Theta \mapsto \mathbb{R}) \in L_2(\Omega) \times L_2(\Theta, P), \tag{2.5}$$

for bounded Ω. Without loss of generality we restrict ourselves to centered stochastic processes,

$$\mathrm{E}[U(\mathbf{x},\cdot)] = \int_\Theta U(\mathbf{x},\theta)\,\mathrm{d}P(\theta) = 0 \quad \forall x \in \Omega, \tag{2.6}$$

and assume that U is continuous in the mean square sense:

$$\lim_{\mathbf{x}' \to \mathbf{x}} \|U(\mathbf{x},\cdot) - U(\mathbf{x}',\cdot)\|_\Theta^2 = 0 \quad \forall \mathbf{x} \in \Omega. \tag{2.7}$$

The autocorrelation function of the process,

$$C_{UU} : \Omega \times \Omega \mapsto \mathbb{R}, \tag{2.8}$$

has for expression

$$C_{UU}(\mathbf{x}, \mathbf{x}') = \mathrm{E}\big[U(\mathbf{x}, \cdot)U(\mathbf{x}', \cdot)\big] \quad \forall \mathbf{x}, \mathbf{x}' \in \Omega. \tag{2.9}$$

Under the assumptions above, it can be shown the C_{UU} is continuous on $\Omega \times \Omega$ and

$$\int_\Omega \int_\Omega C_{UU}(\mathbf{x}, \mathbf{x}')\, d\mathbf{x}\, d\mathbf{x}' < +\infty. \tag{2.10}$$

Therefore, the linear operator, based on the so-called correlation kernel K, defined by

$$(Kv, w) = \int_\Omega \int_\Omega C_{UU}(\mathbf{x}, \mathbf{x}')v(\mathbf{x})w(\mathbf{x}')\, d\mathbf{x}\, d\mathbf{x}', \tag{2.11}$$

is a symmetric semi-positive Hilbert-Schmidt operator on $H = L_1^2(\Omega, \mathbb{R})$ equipped with inner product $(\cdot, \cdot)$. We then have the following relevant results [13, 43, 214]:

- the kernel K has real eigenvalues λ_i;
- the eigenvalues λ_i are non-negative and can be arranged in decreasing order,

$$\lambda_1 \geq \lambda_2 \geq \cdots \to 0;$$

- the eigenvalues of K are countable, and are such that

$$\sum_{j \geq 1} {\lambda_j}^2 < +\infty;$$

- for each eigenvalue, there exists a finite number of linearly independent eigenvectors;
- the collection of eigenvectors $\{u_i, i \geq 1\}$ constitutes an orthogonal basis of H. Furthermore, the eigenvectors may be normalized so that

$$(u_i, u_j) = \delta_{ij}.$$

The kernel K has thus the spectral decomposition

$$K(\mathbf{x}, \mathbf{x}') = \sum_{i \geq 1} \lambda_i u_i(\mathbf{x}) u_i(\mathbf{x}'). \tag{2.12}$$

In fact, the eigenvalues and eigenvectors of K are the solutions of the Fredholm equation of the second-kind:

$$\int_\Omega K(\mathbf{x}, \mathbf{x}')u_i(\mathbf{x}')\, d\mathbf{x}' = \lambda_i u_i(\mathbf{x}). \tag{2.13}$$

The KL decomposition of the stochastic process U is consequently given by:

$$U(\mathbf{x},\theta)=\sum_{i\geq 1}\sqrt{\lambda_i}u_i(\mathbf{x})\eta_i(\theta), \tag{2.14}$$

where the random variables $\eta_i(\theta)$ are given by:

$$\eta_i(\theta)=\frac{1}{\sqrt{\lambda_i}}(U(\mathbf{x},\theta),u_i(\mathbf{x})). \tag{2.15}$$

It is immediate to demonstrate that the random variables η_i have zero mean, unit variance, and are mutually uncorrelated:

$$\mathrm{E}[\eta_i]=0, \qquad \mathrm{E}[\eta_i\eta_j]=\delta_{ij}.$$

So the random variables are orthogonal; however, they are generally not independent (except in particular for the case of Gaussian processes).

As for the correlation kernel, the correlation function can also be expressed in terms of its eigenvalues and eigenfunctions, namely

$$\begin{aligned}C_{UU}(\mathbf{x},\mathbf{x}')=\mathrm{E}\left[U(\mathbf{x},\cdot)U(\mathbf{x}',\cdot)\right]&=\mathrm{E}\left[\left(\sum_{i\geq 1}\sqrt{\lambda_i}u_i(\mathbf{x})\eta_i\right)\left(\sum_{j\geq 1}\sqrt{\lambda_j}u_j(\mathbf{x}')\eta_j\right)\right]\\&=\sum_{i\geq 1}\sum_{j\geq 1}\sqrt{\lambda_i\lambda_j}u_i(\mathbf{x})u_j(\mathbf{x}')\mathrm{E}[\eta_i\eta_j]\\&=\sum_{i\geq 1}\lambda_i u_i(\mathbf{x})u_i(\mathbf{x}').\end{aligned}$$

2.1.2 Properties of KL Expansions

We will assume in the following that the eigenvalues are arranged in decreasing order, *i.e.*

$$\lambda_1\geq\lambda_2\geq\cdots.$$

The KL expansion is optimal in the mean square sense; that is, when truncated after a finite number, N_{KL}, of terms, the resulting approximation $\hat{U}$ minimizes the mean square error [136, 138]

$$\begin{aligned}\epsilon^2_{N_{KL}}\equiv\mathrm{E}\left[\|U(\mathbf{x},\cdot)-\hat{U}(\mathbf{x},\cdot)\|^2_\Omega\right]&=\sum_{i,j>N_{KL}}\sqrt{\lambda_i\lambda_j}(u_i,u_j)\mathrm{E}[\eta_i\eta_j]\\&=\sum_{i,j>N_{KL}}\sqrt{\lambda_i\lambda_j}\delta_{ij}\delta_{ij}=\sum_{i>N_{KL}}\lambda_i\end{aligned}$$

In other words, no other approximation of U in a series of N_{KL} terms results in a smaller mean square error.

Formally, the KL decomposition provides an optimal representation of a process U satisfying the assumptions in Sect. 2.1.1, using a series expansion involving a finite number of random variables:

$$U(\mathbf{x},\theta) \approx \hat{U}(\mathbf{x},\theta) = \sum_{i=1}^{N_{KL}} \sqrt{\lambda_i} u_i(\mathbf{x}) \eta_i(\theta). \tag{2.16}$$

The mean square truncation error decreases monotonically with the number of terms retained in the expansion, at a rate that depends on the decay of the spectrum of K. The higher the rate of spectral decay, the smaller the number of terms needed in the expansion. Specifically, the number of terms to achieve a specified error threshold depends on the correlation function of the process. The more correlated the process, the smaller the number of terms needed to achieve the desired threshold. Conversely, if the process is poorly correlated, a higher number of terms is needed. In the limit where U corresponds to a white noise, i.e. $C_{UU}(\mathbf{x},\mathbf{x}') \sim \delta_0(\|\mathbf{x}-\mathbf{x}'\|)$, an infinite number of terms would be necessary, as further discussed below. Finally, one can also show (see for instance [90]) that the KL decomposition of K based on its eigenfunctions is the only expansion that results in orthogonal random variables.

2.1.3 *Practical Determination*

We now provide examples of the solution of (2.13) for $\Omega = [0, 1]$.

2.1.3.1 Rational Spectra

Suppose that the process U has a known rational spectrum of the form:

$$S(f) = \frac{N(f^2)}{D(f^2)}, \tag{2.17}$$

where N and D are polynomial functions of the frequency, f. In the case of a stationary process, the Fredholm's equation becomes:

$$\int_{\Omega} u_j(x') \int_{-\infty}^{\infty} \exp^{-i|x-x'|f} \frac{N(f^2)}{D(f^2)} \, \mathrm{d}f \, \mathrm{d}x' = \lambda_j u_j(x), \tag{2.18}$$

where in this equation $i^2 = -1$. Equation (2.18) can be recast as a second-order differential equation for the eigenfunction $u_i(x)$ [90]:

$$\lambda_i D\left[\frac{\mathrm{d}^2}{\mathrm{d}x^2}\right] u_i(x) = N\left[\frac{\mathrm{d}^2}{\mathrm{d}x^2}\right] u_i(x), \tag{2.19}$$

which must be solved for each of the eigenvalues, λ_i.

There exist analytical solutions for (2.19) for some classical spectra [251]. One of the known solutions concerns the exponential kernel:

$$K(x,x') = \sigma_U^2 \exp^{-|x-x'|/b}, \tag{2.20}$$

which features in the study of first-order Markov processes. The (dimensional) parameter $b > 0$ refers to the correlation length or correlation time, and σ_U^2 is the process variance. In this case, N and D are given by [221]:

$$N(f) = 2b, \qquad D(f) = 1 + b^2 f^2, \tag{2.21}$$

and the analytical solution of (2.19) is given by [90]:

$$u_i(x) = \begin{cases} \dfrac{\cos[\omega_i(x-1/2)]}{\sqrt{\frac{1}{2}+\frac{\sin(\omega_i)}{2\omega_i}}} & \text{if } i \text{ is even,} \\[2ex] \dfrac{\sin[\omega_i(x-1/2)]}{\sqrt{\frac{1}{2}-\frac{\sin(\omega_i)}{2\omega_i}}} & \text{if } i \text{ is odd} \end{cases} \tag{2.22}$$

and

$$\lambda_i = \sigma_U^2 \frac{2b}{1+(\omega_i b)^2}, \tag{2.23}$$

where ω_i are the (ordered) positive roots of the characteristic equation:

$$\left[1 - b\,\omega \tan(\omega/2)\right]\left[b\,\omega + \tan(\omega/2)\right] = 0. \tag{2.24}$$

In Fig. 2.1, we plot the first 10 eigenvalues and eigenfunctions for the exponential kernel with $b = 1$, $\sigma_U = 1$ and $\Omega = [0, 1]$. One observes that the eigenfunctions $u_i(x)$ exhibit oscillations whose frequencies increase with increasing index i.

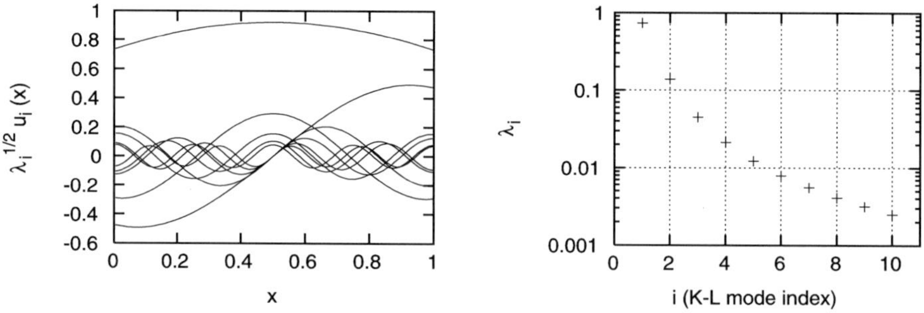

Fig. 2.1 Karhunen-Loève decomposition of the exponential kernel on [0, 1]. The correlation length $b = 1$. *Left*: first ten eigenfunctions, $\sqrt{\lambda_i}u_i(x)$. *Right*: spectrum of eigenvalues, λ_i. Adapted from [128]

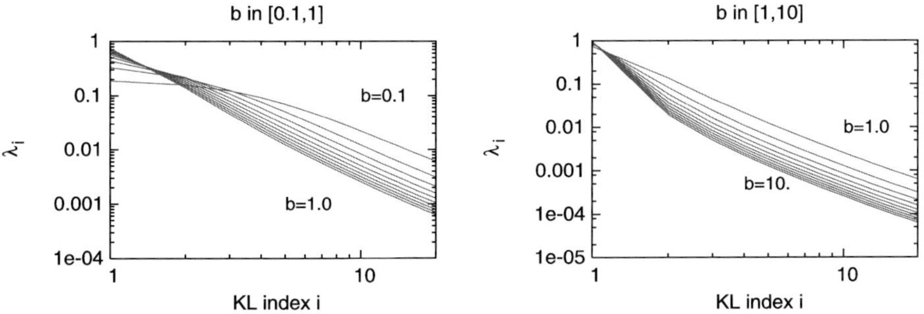

Fig. 2.2 The spectrum of the exponential kernel for different values of the correlation length, b. *Left*: $b \in [0.1, 1]$. *Right*: $b \in [1, 10]$. Note the logarithmic scale in the plots

Table 2.1 $E^2_{\sigma_U}$ and $E^\infty_{\sigma_U}$ for different values of N_{KL}, with $b = 1/2$, 1, and 2

		$N_{KL} = 4$	$N_{KL} = 6$	$N_{KL} = 10$	$N_{KL} = 20$	$N_{KL} = 40$
$E^2_{\sigma_U}(N_{KL})$	$b = 1/2$	0.588E–1	0.375E–1	0.216E–1	0.104E–1	0.513E–2
$E^2_{\sigma_U}(N_{KL})$	$b = 1$	0.295E–1	0.187E–1	0.108E–1	0.521E–2	0.256E–2
$E^2_{\sigma_U}(N_{KL})$	$b = 2$	0.147E–1	0.934E–2	0.538E–2	0.260E–2	0.128E–2
$E^\infty_{\sigma_U}(N_{KL})$	$b = 1/2$	0.108E–0	0.659E–1	0.345E–1	0.130E–1	0.559E–2
$E^\infty_{\sigma_U}(N_{KL})$	$b = 1$	0.535E–1	0.325E–1	0.170E–1	0.643E–2	0.279E–2
$E^\infty_{\sigma_U}(N_{KL})$	$b = 2$	0.266E–1	0.161E–1	0.846E–2	0.320E–2	0.139E–2

Table 2.2 E^2_K and E^∞_K for different values of N_{KL}, with $b = 1/2$, 1, and 2

		$N_{KL} = 4$	$N_{KL} = 6$	$N_{KL} = 10$	$N_{KL} = 20$	$N_{KL} = 40$
$E^2_K(N_{KL})$	$b = 1/2$	0.337E–1	0.179E–1	0.818E–2	0.299E–2	0.125E–2
$E^2_K(N_{KL})$	$b = 1$	0.174E–1	0.908E–2	0.411E–2	0.150E–2	0.625E–3
$E^2_K(N_{KL})$	$b = 2$	0.879E–2	0.456E–2	0.206E–2	0.788E–3	0.312E–3
$E^\infty_K(N_{KL})$	$b = 1/2$	0.219E–0	0.144E–0	0.846E–1	0.415E–1	0.205E–1
$E^\infty_K(N_{KL})$	$b = 1$	0.113E–0	0.728E–1	0.425E–1	0.207E–1	0.103E–1
$E^\infty_K(N_{KL})$	$b = 2$	0.570E–1	0.366E–1	0.213E–1	0.104E–1	0.513E–2

As previously mentioned, the number of terms in the KL expansion needed for adequate approximation of the stochastic process will be large if the spectrum decays slowly. In order to illustrate this effect, Fig. 2.2 depicts the dependence of the first 20 eigenvalues on the correlation length, b.

In practice, the KL expansion must be truncated and the truncation error must be carefully evaluated. Tables 2.1 and 2.2 provide the L^2 and L^∞ norms of the errors incurred to the approximation of the correlation kernel, and to the local standard deviation σ_U for $x \in \Omega$. These norms are defined according to:

$$E_K^p(N_{KL}) = \left[\int_0^1 \int_0^1 \left| K(x,x') - \sum_{i=1}^{N_{KL}} \lambda_i u_i(x) u_i(x') \right|^p \mathrm{d}x \, \mathrm{d}x' \right]^{1/p}, \quad (2.25)$$

$$E_{\sigma_U}^p(N_{KL}) = \left[\int_0^1 \left| \sigma_U - \sqrt{\sum_{i=1}^{N_{KL}} \lambda_i u_i^2(x)} \right|^p \mathrm{d}x \right]^{1/p}. \quad (2.26)$$

In the present example, $\sigma_U^2 = C_{UU}(x,x) = 1$. Table 2.1 provides $E_{\sigma_U}^2$ and $E_{\sigma_U}^\infty$ for different values of N_{KL}, with $b = 1/2$, 1 and 2. Table 2.2 provides the corresponding values of E_K^2 and E_K^∞. The results show that with N_{KL} fixed, the truncation error behaves approximately as $1/b$. With a fixed value of b, $E_{\sigma_U}^2$ and E_K^∞ decay as N_{KL}^{-1}, whereas a more rapid decay is observed for $E_{\sigma_U}^\infty$ and E_K^2.

To better appreciate the effect of truncation, we plot in Fig. 2.3 the correlation function $C_{\hat{U}\hat{U}}(x,x')$ of $\hat{U}$ in $[0,1] \times [0,1]$, for $b = 1$ and using $N_{KL} = 6$, as well as the difference with the exact correlation function C_{UU}. One observes that the error peaks for $x \approx x'$, and that it has an oscillatory decay as one gets away from the axis $x = x'$. One can thus conclude that the truncation essentially affects small-scale correlations. This kind of behavior is not unique to rational spectra, and is generally observed for other types of kernels as well. Consequently, one can generalize the present observation by noting that if the autocorrelation function does not rapidly decay with respect to the size of the domain Ω, then the KL expansion would rapidly converge, and a small number of terms need be retained in the series without incurring significant error.

2.1.3.2 Non-rational Spectra

In the case of non-rational spectra, we do not dispose of a systematic means to decompose analytically the correlation kernel. For certain kernels, however, it is still possible to transform the Fredholm integral equation into a differential equation by means of differentiation, but analytical solutions are available only for particular cases. Notable examples are the triangular kernel [90] and band-limited white noise [208]. This approach is difficult to generalize to multidimensional domains Ω, in particular when the latter does not have a simple geometry [209]. In these situations, it is necessary to rely on numerical solution methods, which have a wider scope of applicability, including multidimensional domains, non-stationary processes, etc.

2.1.3.3 Numerical Resolution

Numerical methods for the solution of the eigenvalue problem generally fall into two broad categories, namely those relying on variational formulation of (2.13) or on Galerkin-type approximations. We shall focus on the latter approach.

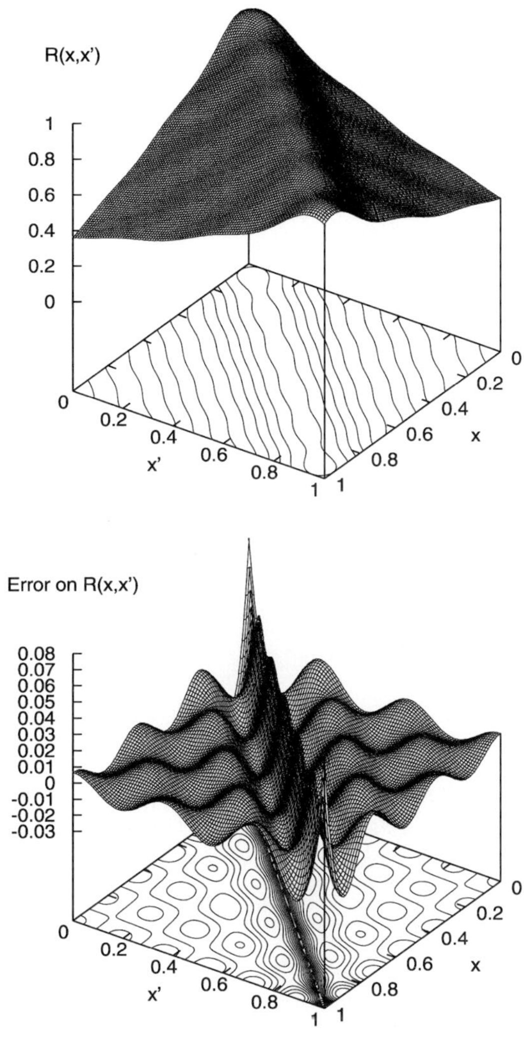

Fig. 2.3 *Top*: approximated correlation function corresponding to $\hat{U}$, using $N_{KL} = 6$. *Bottom*: difference $C_{UU}(x, x') - C_{\hat{U}\hat{U}}(x, x')$. The correlation scale $b = 1$ and the process variance $\sigma_U^2 = 1$

Let $h_i(\mathbf{x})$, $i = 1, \ldots, \mathrm{N_x}$ be a set of basis functions of a Hilbert space $\mathcal{V} \subset L_2(\Omega)$. In this space, the eigenfunction $u_i(\mathbf{x})$ can be approximated as:

$$u_i(\mathbf{x}) \approx \sum_{k=1}^{\mathrm{N_x}} d_k^{(i)} h_k(\mathbf{x}).$$

Substituting this approximation into (2.13) we get:

$$\epsilon_{\mathrm{N_x}}(\mathbf{x}) = \sum_{k=1}^{\mathrm{N_x}} d_k^{(i)} \left[\int_\Omega K(\mathbf{x}, \mathbf{x}') h_k(\mathbf{x}')\, \mathrm{d}\mathbf{x}' - \lambda_i h_k(\mathbf{x}) \right]. \tag{2.27}$$

We now seek the coefficients $d_k^{(i)}$ such that the approximation error $\epsilon_{N_x}(\mathbf{x})$ is orthogonal to the subspace spanned by the h_k's. In other words, $(h_k, \epsilon_{N_x}) = 0$, $\forall k \in \{1, \ldots, N_x\}$. Enforcement of this constraint results in a system of equations of the form:

$$\sum_{k=1}^{N_x} d_k^{(i)} \left[\int_\Omega \int_\Omega K(\mathbf{x}, \mathbf{x}') h_k(\mathbf{x}') h_j(\mathbf{x})\, d\mathbf{x}'\, d\mathbf{x} - \lambda_i \int_\Omega h_k(\mathbf{x}) h_j(\mathbf{x})\, d\mathbf{x} \right] = 0, \quad (2.28)$$

for $j = 1, 2, \ldots, N_x$. This system may be expressed in a matrix form; omitting the index, i, of the eigenfunction, we have:

$$\left([K]_{jk} - \lambda [M]_{jk}\right) d_k = 0. \quad (2.29)$$

Thus, we need to solve a generalized eigenvalue problem with stiffness matrix,

$$[K]_{kl} = \int_\Omega \int_\Omega K(\mathbf{x}, \mathbf{x}') h_k(\mathbf{x}') h_j(\mathbf{x})\, d\mathbf{x}'\, d\mathbf{x} \in \mathbb{R}^{N_x \times N_x}, \quad (2.30)$$

and mass matrix

$$[M]_{kl} = \int_\Omega h_k(\mathbf{x}) h_j(\mathbf{x})\, d\mathbf{x} \in \mathbb{R}^{N_x \times N_x}. \quad (2.31)$$

Note that these matrices inherit the properties of the correlation kernel; in particular, $[K]$ and $[M]$ are symmetric and positive semi-definite. Efficient numerical libraries can be used to solve such eigenvalue problems; examples in the open domain include LAPACK[1] and ARPACK[2].

One of the challenges to the numerical solution of the eigenvalue problem concerns the cost of the computations. In the case where the correlation function decays rapidly with $\|\mathbf{x} - \mathbf{x}'\|$, the eigenfunctions are generally highly oscillatory, and a large amount of basis functions $h_i(\mathbf{x})$ may be consequently needed to properly represent these functions on Ω. Accordingly, the dimension of the matrices $[K]$ and $[M]$ is also large, which directly contributes to the cost of the computations. If on the other hand, the decay of the correlation function is slow (strongly-correlated process), then a small basis may be used but the stiffness and mass matrices generally tend to be full. This limits the efficiency of iterative solvers, which is generally higher when the system is sparse. Not surprisingly, the development of efficient eigenvalue solvers (such as multipole expansions) continues to be the subject of focused research efforts, particularly for 3D domains Ω.

---

[1] http://www.netlib.org/lapack/.

[2] http://www.caam.rice.edu/software/ARPACK/.

2.1.4 Gaussian Processes

As discussed above, the KL decomposition is in fact a representation of a stochastic process in terms of the eigenfunctions of its correlation kernel. Note that there are infinitely many second-order processes sharing the same correlation kernel, and consequently admit expansions on the same set of eigenfunctions. What distinguishes processes having the same autocorrelation kernel is the joint probability of the random variables η_i. By construction, these random variables have zero mean, are mutually orthogonal, and have unit variance. In many cases, the random process is (or can be assumed to be) Gaussian, leading to significant simplifications. Indeed, the KL expansion of a Gaussian process involves random variables which not only are uncorrelated but independent. In particular, a steady Gaussian process U has for expansion

$$\hat{U}(\mathbf{x},\theta) = \sum_{i=1}^{N_{KL}} \sqrt{\lambda_i} u_i(\mathbf{x}) \xi_i(\theta), \tag{2.32}$$

where the ξ_i are independent, centered normalized, Gaussian random variables:

$$\begin{cases} \mathrm{E}[\xi_i] = 0, \\ \mathrm{E}[\xi_i^2] = 1, \\ \mathrm{E}[\xi_i \xi_{j\neq i}] = 0, \end{cases} \tag{2.33}$$

In addition, the joint probability density of the ξ_i factorizes to

$$p_{\boldsymbol{\xi}}(y_1, \dots, y_{N_{KL}}) = \prod_{i=1}^{N_{KL}} \frac{1}{\sqrt{2\pi}} \exp -y_i^2/2. \tag{2.34}$$

The factorized form of the joint probability density of the ξ_i greatly simplifies its sampling for the simulation of the process.

It should be emphasized that, although $\hat{U}(\mathbf{x},\cdot) \in L_2(\Theta, P)$, Gaussian processes are not bounded. In some cases, this is problematic, particularly when KL expansions are used to represent bounded physical processes, for instance temperature or diffusivity fields. In these situations, one wishes to construct approximations $\hat{U}$ of U that are physically meaningful, for any truncation level. Several studies are being directed towards the construction of admissible truncated KL expansion; see [218] for recent results. For brevity, these methods are not discussed further in this monograph; we simply mention that this can be achieved through the approximation of dependent η_i by means of polynomial chaos expansions.

Remark Let us first point out that in order to implement the KL decomposition one must know or be able to determine the correlation function of the process to be represented. Therefore, KL expansions are particularly useful when one is representing random model data, which may generally be amenable to analysis and/or subjected to experimental observations and diagnostics. In predictive contexts, on the other hand, the KL formalism may not be as powerful, since the probability laws of the random process, not yet determined, are generally not known *a priori*. Specifically, the correlation function of the solution of a stochastic model problem is generally one of the quantities that one seeks to determine, and consequently not known or specified. Thus, alternative means of representing stochastic processes are needed, though the KL formalism may still be used to guide the definition of essential ingredients of these alternatives representations.

Essentially, the KL decomposition is a series expansion involving **deterministic functions** (namely the eigenfunctions, u_i) and **random variables** (the random variables, η_i). The deterministic functions are fixed by the form of the autocorrelation kernel, whereas the joint probability law of the η_i's remains unknown in the absence of information other than the second-order properties of the process. Specifically, one can only ascertain that the random variables have zero mean, unit variance, and are mutually orthogonal.

One can envision modifying the structure of the KL expansion by relaxing its bi-orthogonal character, so offering more flexibility. In particular, one can envision prescribing *a priori* the functional form of the random coefficients in the expansion, for instance as polynomials of independent random variables with given distribution, and then seek the (not necessarily) deterministic functions that minimize a given error norm, much in the same fashion that spectral approximations are used in the solution of deterministic PDEs. Such approach corresponds precisely to PC decompositions which we shall discuss next.

2.2 Polynomial Chaos Expansion

In this section, we discuss the PC expansion of a random variable. Let us first introduce some definitions and notations. We consider $\mathbb{R}$-valued random variables U defined on a probability space (Θ, Σ, P),

$$U : \Theta \mapsto \mathbb{R}, \tag{2.35}$$

and denote $L_2(\Theta, P)$ the set of second-order random variables.

Let $\{\xi_i\}_{i=1}^{\infty}$ be a sequence of centered, normalized, mutually orthogonal, Gaussian variables. Let $\hat{\Gamma}_p$ denote the space of polynomials in $\{\xi_i\}_{i=1}^{\infty}$ having degree less or equal to p; Γ_p denotes the set of polynomials that belong to $\hat{\Gamma}_p$ and orthogonal to $\hat{\Gamma}_{p-1}$; and $\tilde{\Gamma}_p$ we denote the space spanned by Γ_p. We have:

$$\hat{\Gamma}_p = \hat{\Gamma}_{p-1} \oplus \tilde{\Gamma}_p, \qquad L_2(\Theta, P) = \bigoplus_{i=0}^{\infty} \tilde{\Gamma}_i. \tag{2.36}$$

The subspace $\tilde{\Gamma}_p$ of $L_2(\Theta, P)$ is called the p-th Homogeneous Chaos, whereas Γ_p is called the Polynomial Chaos of order p. Thus, the Polynomial Chaos of order

p consists of all polynomials of order p, involving all possible combinations of the random variables $\{\xi_i\}_{i=1}^{\infty}$. Note that since random variables are functions, the polynomial chaoses are functions of functions, and are thus regarded as functionals.

Each second-order, random variable $U \in L_2(\Theta, P)$ admits a PC representation of the form [25]:

$$\begin{aligned} U(\theta) = u_0\Gamma_0 &+ \sum_{i_1=1}^{\infty} u_{i_1}\Gamma_1(\xi_{i_1}(\theta)) \\ &+ \sum_{i_1=1}^{\infty}\sum_{i_2=1}^{i_1} u_{i_1 i_2}\Gamma_2(\xi_{i_1}(\theta), \xi_{i_2}(\theta)) \\ &+ \sum_{i_1=1}^{\infty}\sum_{i_2=1}^{i_1}\sum_{i_3=1}^{i_2} u_{i_1 i_2 i_3}\Gamma_3(\xi_{i_1}(\theta), \xi_{i_2}(\theta), \xi_{i_3}(\theta)) \\ &+ \sum_{i_1=1}^{\infty}\sum_{i_2=1}^{i_1}\sum_{i_3=1}^{i_2}\sum_{i_4=1}^{i_3} u_{i_1 i_2 i_3 i_4}\Gamma_4(\xi_{i_1}(\theta), \xi_{i_2}(\theta), \xi_{i_3}(\theta), \xi_{i_4}(\theta)) \\ &+ \cdots . \end{aligned} \tag{2.37}$$

This representation is convergent in the mean-square sense:

$$\lim_{p\to\infty} \mathrm{E}\left[\left(u_0\Gamma_0 + \cdots + \sum_{i1=1}^{\infty} \cdots \sum_{i_p=1}^{i_{p-1}} u_{i_1 \ldots u_p}\Gamma_p(\xi_{i_1}, \ldots, \xi_{i_p}) - U\right)^2\right] = 0. \tag{2.38}$$

By construction, chaos polynomials whose orders are greater than $p = 0$ have vanishing expectation:

$$\mathrm{E}[\Gamma_{p>0}] - 0. \tag{2.39}$$

Also, all polynomials are mutually orthogonal with regards to the Gaussian measure associated to the random variables in $\{\xi_i\}_{i=1}^{\infty}$. In fact, one can express the expectation of U either in the original space,

$$\mathrm{E}[U] = \int_{\Theta} U(\theta)\, \mathrm{d}P(\theta), \tag{2.40}$$

or in the Gaussian space spanned by $\{\xi_i\}_{i=1}^{\infty}$,

$$\mathrm{E}[U] = \int_{\Theta} U(\boldsymbol{\xi}(\theta))\, \mathrm{d}P(\theta) = \int \cdots \int U(\mathbf{y}) p_{\boldsymbol{\xi}}(\mathbf{y})\, \mathrm{d}\mathbf{y} \equiv \langle U \rangle, \tag{2.41}$$

where $U(\boldsymbol{\xi})$ is understood as the PC representation of $U(\theta)$ and $p_{\boldsymbol{\xi}}$ stands for the Gaussian probability density function:

$$p_{\boldsymbol{\xi}}(\mathbf{y}) = \prod_{i=1}^{\infty} \frac{1}{\sqrt{2\pi}} \exp(-y_i^2/2). \tag{2.42}$$

In the following, we use the brackets $\langle \cdot \rangle$ to make clear that the expectation is measured with regard to the probability distribution of the random variables used in the expansion.[3]

Classically, in order to facilitate the manipulation of the PC expansion, we rely on an univocal relation between the PC $\Gamma()$ and new functionals $\Psi()$. It results in a more compact expression of the random variable expansion:

$$U(\boldsymbol{\xi}) = \sum_{k=0}^{\infty} u_k \Psi_k(\boldsymbol{\xi}), \quad \boldsymbol{\xi} = \{\xi_1, \xi_2, \dots\}, \tag{2.43}$$

where the deterministic expansion coefficients u_k are simply called PC coefficients. The convention that $\Psi_0 = \Gamma_0$ is often adopted, and it is further assumed that the univocal relation is set such that Ψ_i's are ordered with increasing polynomial order.

2.2.1 Polynomial Chaos System

The construction outlined above involves an infinite collection, ξ_i, of normalized, uncorrelated, random Gaussian variables. In practice, particularly for computational purposes, it is necessary to restrict the representation to a finite number of random variables, which leads to PC expansions of finite dimension. Specifically, the PC of dimension N and order p is the subspace of $\tilde{\Gamma}_p$ generated by the elements of Γ_p that only involve N random variables, $\xi_1, \dots, \xi_{\mathrm{N}}$. For finite dimensions, the infinite sums in (2.37) are replaced by finite sums over N dimensions. For instance, for an expansion with two dimensions (N = 2), (2.37) becomes:

$$\begin{aligned} U = u_0\Gamma_0 &+ \sum_{i_1=1}^{2} u_{i_1}\Gamma_1(\xi_{i_1}) + \sum_{i_1=1}^{2}\sum_{i_2=1}^{i_1} u_{i_1 i_2}\Gamma_2(\xi_{i_1}, \xi_{i_2}) \\ &+ \sum_{i_1=1}^{2}\sum_{i_2=1}^{i_1}\sum_{i_3=1}^{i_2} u_{i_1 i_2 i_3}\Gamma_3(\xi_{i_1}, \xi_{i_2}, \xi_{i_3}) \\ &+ \sum_{i_1=1}^{2}\sum_{i_2=1}^{i_1}\sum_{i_3=1}^{i_2}\sum_{i_4=1}^{i_3} u_{i_1 i_2 i_3 i_4}\Gamma_4(\xi_{i_1}, \xi_{i_2}, \xi_{i_3}, \xi_{i_4}) + \cdots \end{aligned} \tag{2.44}$$

[3] We will see later that these random variables are not necessarily Gaussian.

or alternatively,

$$\begin{aligned} U &= u_0\Gamma_0 + u_1\Gamma_1(\xi_1) + u_2\Gamma_2(\xi_2) \\ &\quad + u_{11}\Gamma_2(\xi_1,\xi_1) + u_{21}\Gamma_2(\xi_2,\xi_1) + u_{22}\Gamma_2(\xi_2,\xi_2) \\ &\quad + u_{111}\Gamma_3(\xi_1,\xi_1,\xi_1) + u_{211}\Gamma_3(\xi_2,\xi_1,\xi_1) + u_{221}\Gamma_3(\xi_2,\xi_2,\xi_1) \\ &\quad + u_{222}\Gamma_3(\xi_2,\xi_2,\xi_2) + u_{1111}\Gamma_4(\xi_1,\xi_1,\xi_1,\xi_1) + \cdots. \end{aligned} \tag{2.45}$$

2.2.2 *One Dimensional PC Basis*

One simple fashion of constructing the N-dimensional PC is to follow a partial tensorization of the 1D polynomials. Thus, we first focus on polynomials of a single random variable, ξ. Recall that the chaos polynomials are orthogonal, and that the probability density of ξ is given by:

$$p_\xi(y) = \frac{1}{\sqrt{2\pi}} \exp\left[-y^2/2\right]. \tag{2.46}$$

By $\psi_p(\xi)$ we denote the 1D chaos of order p. Following the same convention, the polynomial of degree 0 is $\psi_0(\xi) = 1$. The orthogonality condition can be expressed as:

$$\begin{aligned} \mathrm{E}[\psi_i\psi_j] &= \int_\Theta \psi_i(\xi(\theta))\psi_j(\xi(\theta))\,\mathrm{d}P(\theta) = \langle\psi_i,\psi_j\rangle \\ &= \int_{\mathbb{R}} \psi_i(y)\psi_j(y)p_\xi(y)\,\mathrm{d}y = \delta_{ij}\langle\psi_i^2\rangle, \end{aligned}$$

with the polynomials (conventionally) normalized so that $\langle\psi_k^2\rangle = k!$. The 1D polynomials thus defined, which are mutually orthogonal with respect to the Gaussian measure, constitute a well-known family, namely the Hermite polynomials [1]. The first seven Hermite polynomials, $\psi_0, \dots, \psi_6$, are given in (B.22–B.28); they are plotted in Fig. 2.4 for $\xi \in [-3, 3]$.

2.2.3 *Multidimensional PC Basis*

We know proceed to the N-dimensional case, and seek to construct Γ_p starting from the 1D Hermite polynomials, ψ_q. We will denote $\boldsymbol{\xi} = \{\xi_1, \dots, \xi_\mathrm{N}\}$. Since these random variables are independent, the probability density of $\boldsymbol{\xi}$ is given by:

$$p_{\boldsymbol{\xi}}(\mathbf{y}) = \prod_{i=1}^{\mathrm{N}} p_\xi(y_i). \tag{2.47}$$

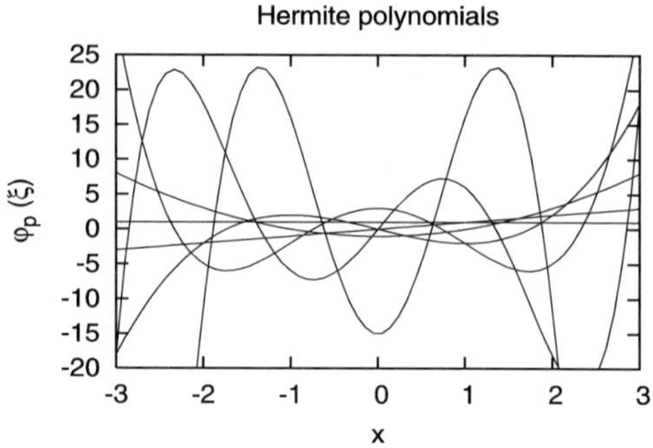

Fig. 2.4 One-dimensional Hermite polynomials, $\psi_p(\xi)$, for $p = 0, \ldots, 6$

Let γ denote the multi-index, $\gamma = \{\gamma_1, \ldots, \gamma_N\}$ and let $\lambda(p)$ denote the following set of multi-indices:

$$\lambda(p) = \left\{ \gamma : \sum_{i=1}^{N} \gamma_i = p \right\}. \tag{2.48}$$

Following these definitions, one constructs the p-th order polynomial chaos according to:

$$\Gamma_p = \left\{ \bigcup_{\gamma \in \lambda(p)} \prod_{\gamma_1}^{\gamma_N} \psi_{\gamma_i}(\xi_i) \right\}. \tag{2.49}$$

Thus, for the 2D case, the Hermite expansion can be expressed as

$$\begin{aligned} U &= u_0\psi_0 + u_1\psi_1(\xi_1) + u_2\psi_1(\xi_2) \\ &\quad + u_{11}\psi_2(\xi_1) + u_{21}\psi_1(\xi_2)\psi_1(\xi_1) + u_{22}\psi_2(\xi_2) \\ &\quad + u_{111}\psi_3(\xi_1) + u_{211}\psi_1(\xi_2)\psi_2(\xi_1) + u_{221}\psi_2(\xi_2)\psi_1(\xi_1) \\ &\quad + u_{222}\psi_3(\xi_2) + u_{1111}\psi_4(\xi_1) + \cdots. \end{aligned} \tag{2.50}$$

The above expression can be recast in the following, more compact, form:

$$U = \sum_{k=0}^{\infty} u_k \Psi_k(\xi_1, \xi_2). \tag{2.51}$$

The first 2D polynomial chaoses are plotted in Figs. 2.5 and 2.6.

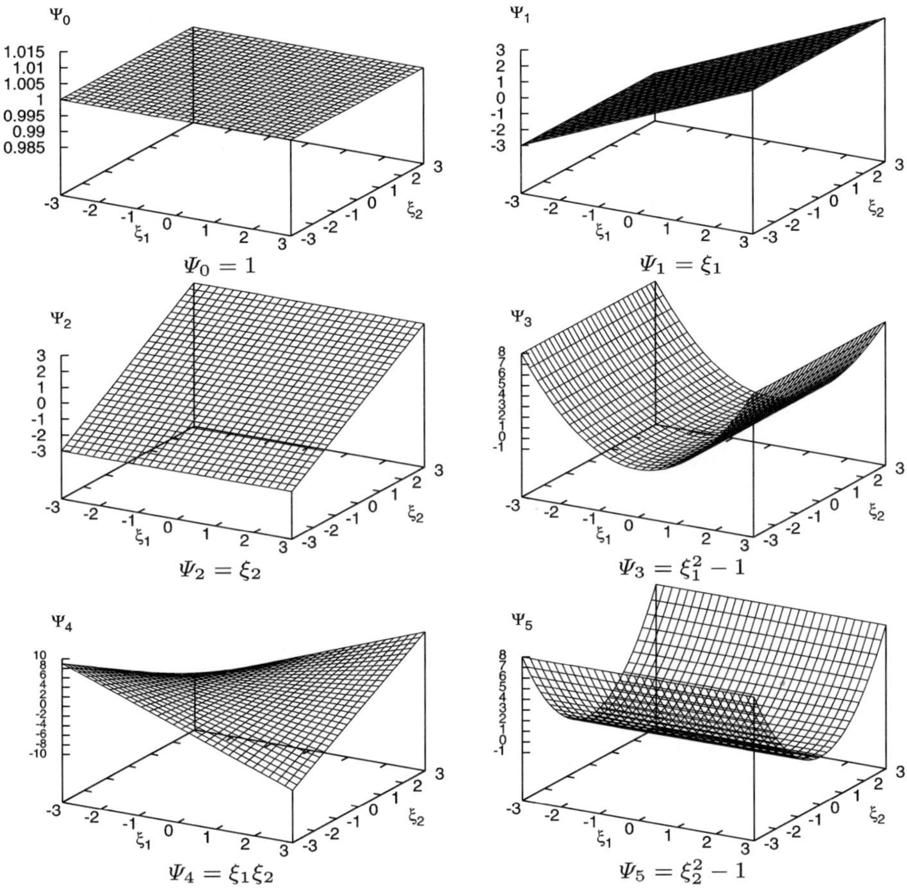

Fig. 2.5 Two-dimensional (N = 2) Hermite chaoses of order 0, 1, and 2

2.2.4 Truncated PC Expansion

In the following, we will adopt a condensed notation of the expansion of the random variable U, specifically

$$U = \sum_{k=0}^{\infty} u_k \Psi_k(\boldsymbol{\xi}). \tag{2.52}$$

As mentioned earlier, it is necessary to conduct computations with a finite number, N, of random variables, ξ_i, $i = 1, \ldots, \mathrm{N}$. One also needs to truncate the PC expansion to order p, so that the expansion is also finite. The number of terms retained in the expansion, after the double truncation at N dimensions and order p, is given

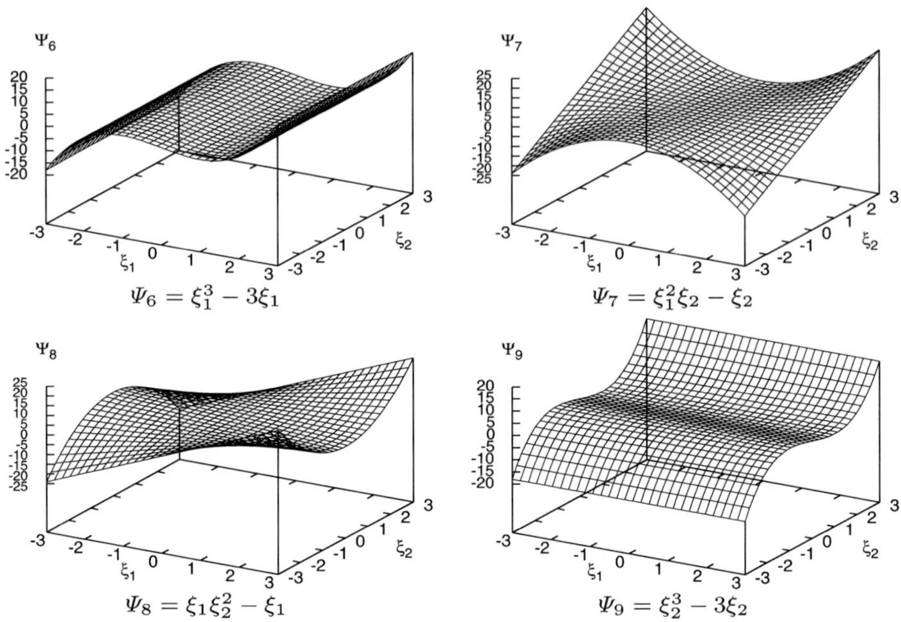

Fig. 2.6 Two-dimensional (N = 2) Hermite chaoses of order 3

Table 2.3 Number of terms (P + 1) in the N-dimensional PC expansion truncated at order p

p/N	1	2	3	4	5	6	p/N	1	2	3	4	5	6
1	2	3	4	5	6	7	4	5	15	35	70	126	210
2	3	6	10	15	21	28	5	6	21	56	126	252	462
3	4	10	20	35	56	84	6	7	28	84	210	462	924

by [90]:

$$\mathrm{P} + 1 = \frac{(\mathrm{N} + p)!}{\mathrm{N}!p!}. \tag{2.53}$$

The dependence of (P + 1) on N and p is illustrated in Fig. 2.7. Table 2.3 provides values (P + 1) for p and N in the interval [1, 6].

The truncated expansion of a random variable U can consequently be expressed as:

$$U = \sum_{k=0}^{\mathrm{P}} u_k \Psi_k(\boldsymbol{\xi}) + \epsilon(\mathrm{N}, p). \tag{2.54}$$

where the truncation error depends on both N and p. This error is itself a random variable. The truncated expansion converges in the mean square sense as N *and* p

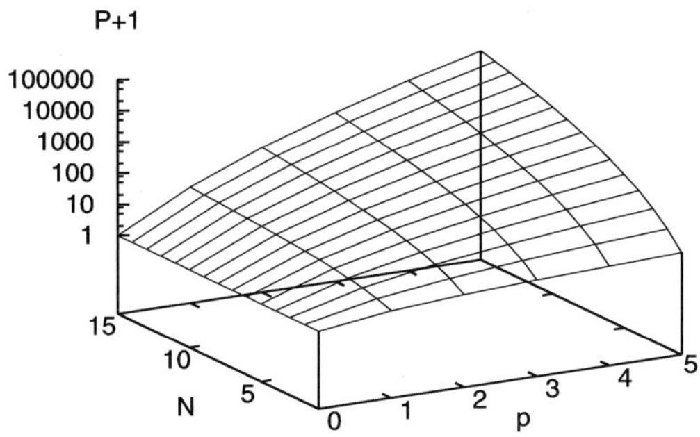

Fig. 2.7 Number of terms in the PC expansion plotted against the order, p, and the number of dimensions, N

go to infinity [25], *i.e.*

$$\lim_{\mathrm{N},p\to\infty} \left\langle \epsilon^2(\mathrm{N},p) \right\rangle = 0. \tag{2.55}$$

In light of the dependence of P on the order and the number of random variables, the PC representation will be computationally efficient when small values of N and p are sufficient for an accurate representation of U, in other words when $\langle \epsilon^2_{(\mathrm{N},p)} \rangle \to 0$ rapidly with N and p. As we shall see later, in practice N is governed by the number and structure of the sources of uncertainty. Meanwhile, in order to achieve a given error threshold, the order p needed is governed by the random variable U that one seeks to represent, particularly by its probability law.

2.3 Generalized Polynomial Chaos

As further discussed in Sect. 2.5, the number N of random variables ξ_i in the PC expansion is fixed by the parametrization of the uncertain model under scrutiny. Therefore, we will be essentially concerned here with the convergence of the expansion with the PC order p.

2.3.1 Independent Random Variables

We temporarily restrict the discussion to the case of expansions involving a single random variable $\xi_i = \xi$.

As we have just observed, the rate of convergence of the expansion of U with p depends on the distribution of the random variable that one seeks to represent. It is thus natural to explore whether there exists other families of orthogonal polynomials that lead to a smaller representation error for the same number of terms in the expansion. One can immediately note that if $U(\theta)$ is a Gaussian random variable, then the Hermite PC basis is optimal, since an expansion with $p = 1$ provides an exact representation. This remark can in fact be generalized, namely to assert that the

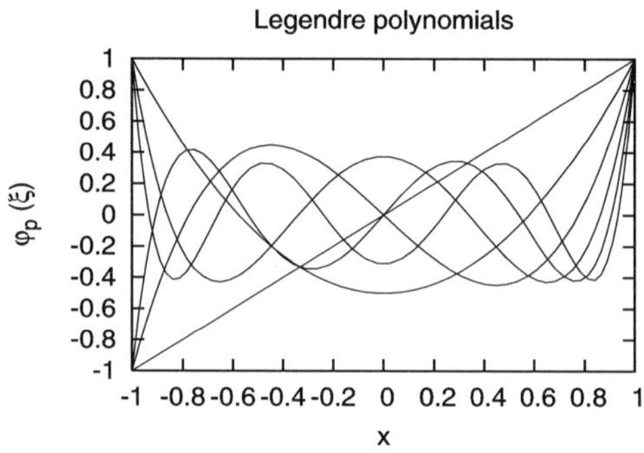

Fig. 2.8 One-dimensional Legendre polynomials of order $p = 0, \ldots, 6$

optimal polynomial expansion is that constructed using the measure corresponding to the probability law of the random variable that we seek to represent. However, for the class of problems of interest here, particularly for model-based predictions, the probability law of the model solution to be determined is generally *not* known *a priori*. It is therefore not possible to construct, *a priori*, the optimal orthogonal family. Nonetheless, since one knows *a priori* the probability law of the data uncertainties that one wishes to propagate, it may be useful to utilize, if possible, the measure associated with these uncertainties in order to generate the PC basis. Thus, the resulting basis will at least be optimal with respect to the uncertain model data representation. However, there is generally no guaranty of the optimality concerning the uncertain model solution, except possibly in isolated situations such as the highly idealized case of Gaussian input data and a model in which the output is linear with regards to the input data and Gaussian input data.

Following the remarks above, when the uncertain data corresponds to a random variable uniformly distributed over a given finite interval, we choose the basis generated for random variables ξ_i that are uniformly distributed on $[-1, 1]$, which leads to the Legendre polynomials [1] (see Appendix B). The first seven Legendre polynomials are plotted in Fig. 2.8. From a broader perspective, one notes that the Legendre polynomials are members of the Jacobi family of polynomials which cover the set of type β probability laws [246]. Xiu and Karniadakis [246] have in fact shown that for a large number of common probability laws, the corresponding families of polynomials are determined using the Askey scheme [6]. Selected probability laws (measures) and the corresponding orthogonal polynomial sets are given in Table 2.4. Note that the PC decomposition can also be accomplished on the basis of discrete random variables (RVs). Furthermore, in the case of measures for which one does not readily dispose of an orthogonal family of polynomials, it is generally possible to rely on a numerical construction of the PC basis, following a Gram-Schmidt orthogonalization process [220].

We also note that the expression of the multidimensional PCs in terms of products of one-dimensional polynomials, thanks for instance to multi-index constructions (see Appendix C), offers the possibility of natural extension to the case where the random variables ξ_i are associated with different measures. Such constructions may be particularly useful for the propagation of multiple uncertainties in complex models, where the various ξ_i's may be associated with different sources of uncer-

Table 2.4 Families of probability laws and corresponding families of orthogonal polynomials

	Distribution ξ	Polynomials $\psi_k(\xi)$	Support
Continuous RV	Gaussian	Hermite	$(-\infty, \infty)$
	γ	Laguerre	$[0, \infty)$
	β	Jacobi	$[a, b]$
	Uniform	Legendre	$[a, b]$
Discrete RV	Poisson	Charlier	$\{0, 1, 2, \dots\}$
	Binomial	Krawtchouk	$\{0, 1, 2, \dots, n\}$
	Negative binomial	Meixner	$\{0, 1, 2, \dots\}$
	Hypergeometric	Hahn	$\{0, 1, 2, \dots, n\}$

tainty having different probability laws. In addition, this approach lends itself to a modular implementation and to automatic basis construction.

2.3.2 Chaos Expansions

Although the original (Hermite) PC expansion, and the generalized PC, rely on global polynomials, the approach is not limited to polynomial bases. In fact, we will see later in Chap. 8 that using piecewise polynomial functions can greatly improve the convergence properties of the stochastic expansion in certain complex situations.

Strictly speaking, the term of Polynomial Chaos expansion should be used when (2.43) involves polynomials functionals Ψ_k, and Chaos expansion for other types of functionals. In this monograph, we shall often rely on the terminology of Wiener-Hermite, Wiener-Legendre, etc..., for PC expansions using Hermite, Legendre, ...polynomials, and use the generic term stochastic spectral expansion to designate any type of expansion (including piecewise polynomials) of the form in (2.43).

2.3.3 Dependent Random Variables

So far, we have restricted ourselves to situations where the random variables ξ_i are independent, such that their joint density has product form. We have seen that for such structure of the probability law, the N-dimensional PC basis can be constructed by tensorizing one-dimensional bases.

In some situations, expansion in terms of independent random variables may not be possible. This may be the case when considering the representation of complex model data. The general situation corresponds to a set $\boldsymbol{\xi} = \{\xi_1, \dots, \xi_N\}$ of N random variables with given joint density $p_{\boldsymbol{\xi}}$. We denote Ξ the support of the joint density, *i.e.* $p_{\boldsymbol{\xi}}(\mathbf{y}) = 0$ for $\mathbf{y} \notin \Xi$. A construction method of orthonormal Chaos bases for general probability laws was introduced in [218]. The construction is performed in two steps.

First, a set of N one-dimensional generalized PC bases is determined in relation with the *marginal* densities of the ξ_i. Let us denote p_i the marginal density of the i-th random variable. We recall (see Appendix A) that the marginal density p_i is obtained by integration of the joint density along all its dimensions but the i-th one:

$$p_i(y) = \int \mathrm{d}y_1 \cdots \int \mathrm{d}y_{i-1} \int \mathrm{d}y_{i+1} \cdots \int \mathrm{d}y_\mathrm{N}\, p_\mathbf{y}(y_1, \ldots, y_\mathrm{N}). \tag{2.56}$$

For convenience, we assume orthonormal (Hilbert) bases with regard to the p_i, and denote $\{\phi_p^{(i)}(\xi)\}$ the corresponding sets of polynomials satisfying

$$\left\langle \phi_p^{(i)}, \phi_{p'}^{(i)} \right\rangle_{p_i} \equiv \int \phi_p^{(i)}(y)\phi_{p'}^{(i)}(y) p_i(y)\, \mathrm{d}y = \delta_{pp'}. \tag{2.57}$$

Second, the N dimensional Chaos basis is constructed. It consists in the set of random functionals $\Psi_\gamma(\boldsymbol{\xi})$, where $\gamma = \{\gamma_1, \ldots, \gamma_\mathrm{N}\}$ is a multi-index of $\mathbb{N}^\mathrm{N}$. The functionals Ψ_γ are defined for $\boldsymbol{\xi} \in \Xi$ as

$$\Psi_\gamma(\boldsymbol{\xi}) = \left[\frac{p_1(\xi_1) \cdots p_\mathrm{N}(\xi_\mathrm{N})}{p_{\boldsymbol{\xi}}(\boldsymbol{\xi})} \right]^{1/2} \phi_{\gamma_1}^{(1)}(\xi_1) \ldots \phi_{\gamma_\mathrm{N}}^{(\mathrm{N})}(\xi_\mathrm{N}). \tag{2.58}$$

It is immediate to show that $\{\Phi_\gamma\}$ is an orthonormal set for the measure $p_{\boldsymbol{\xi}}$. Indeed,

$$\begin{aligned}
\left\langle \Psi_\gamma, \Psi_\beta \right\rangle &= \int_\Xi \Psi_\gamma(\mathbf{y}) \Psi_\beta(\mathbf{y}) p_{\boldsymbol{\xi}}(\mathbf{y})\, \mathbf{dy} \\
&= \int_\Xi \frac{p_1(y_1) \ldots p_\mathrm{N}(y_\mathrm{N})}{p_{\boldsymbol{\xi}}(\mathbf{y})} \left(\phi_{\gamma_1}^{(1)}(y_1) \ldots \phi_{\gamma_\mathrm{N}}^{(\mathrm{N})}(y_\mathrm{N}) \right) \\
&\quad \times \left(\phi_{\beta_1}^{(1)}(y_1) \ldots \phi_{\beta_\mathrm{N}}^{(\mathrm{N})}(y_\mathrm{N}) \right) p_{\boldsymbol{\xi}}(\mathbf{y})\, \mathbf{dy} \\
&= \int_\Xi \left(\phi_{\gamma_1}^{(1)}(y_1) \phi_{\beta_1}^{(1)}(y_1) p_1(y_1) \right) \cdots \\
&\quad \times \left(\phi_{\gamma_\mathrm{N}}^{(\mathrm{N})}(y_\mathrm{N}) \phi_{\beta_1}^{(\mathrm{N})}(y_\mathrm{N}) p_\mathrm{N}(y_\mathrm{N}) \right) \mathrm{d}y_1 \cdots \mathrm{d}y_\mathrm{N} \\
&= \prod_{i=1}^{\mathrm{N}} \left\langle \phi_{\gamma_i}^{(i)}, \phi_{\beta_i}^{(i)} \right\rangle_{p_i} = \delta_{\gamma\beta}.
\end{aligned} \tag{2.59}$$

We observe that the definition in (2.58) yields bases which are not polynomials for general joint densities, although the one-dimensional functionals $\phi_j^{(i)}$ may be. As a result, analytical manipulation of the functionals are quite complex or even impossible, and appropriate numerical procedures are needed.

Furthermore, construction of Chaos bases for general probability laws requires the complete description of $p_{\boldsymbol{\xi}}$ for any point in Ξ. In fact, in many situations the explicit form of the joint density is unknown. This is the case when the ξ_i are related to physical quantities as for the parametrization of stochastic model data. In this case, one may only have a sample set of realizations of the data, from physical measurements for instance, and it is necessary to estimate the probability law of $\boldsymbol{\xi}$ through identification or optimization procedures.

2.4 Spectral Expansions of Stochastic Quantities

2.4.1 Random Variable

Let U be a second-order $\mathbb{R}$-valued random variable defined on a probability space $(\Theta, \Sigma, \mathrm{P})$ and

$$U = \sum_{k=0}^{\infty} u_k \Psi_k(\boldsymbol{\xi}), \tag{2.60}$$

its expansion on the orthogonal PC basis $\{\Psi_0, \Psi_1, \ldots\}$, where we assume an indexation such that $\Psi_0 = 1$. We see immediately that the expectation of U is given by:

$$\mathrm{E}[U] = \left\langle U(\boldsymbol{\xi})\right\rangle = \left\langle \Psi_0 U(\boldsymbol{\xi})\right\rangle = \sum_{k=0}^{\mathrm{P}} u_k \left\langle \Psi_0, \Psi_k\right\rangle = u_0, \tag{2.61}$$

by virtue of orthogonality of the basis. Therefore, from the indexation convention the coefficient u_0 is in fact the mean of the random variable U. Further, from the definition of the random variable variance, σ_U^2, we obtain:

$$\begin{aligned}\sigma_U^2 &= \mathrm{E}\left[(U - \mathrm{E}[U])^2\right] = \mathrm{E}\left[\left(\sum_{k=1}^{\mathrm{P}} u_k \Psi_k\right)^2\right] \\ &= \sum_{k,l=1}^{\mathrm{P}} u_k u_l \left\langle \Psi_k, \Psi_l\right\rangle = \sum_{k=1}^{\mathrm{P}} u_k^2 \left\langle \Psi_k^2\right\rangle.\end{aligned} \tag{2.62}$$

In other words, the variance of U is given as a weighted sum of its squared PC coefficients. Similar expressions can be derived for the higher order moments of U in terms of its PC coefficients; however, higher order moments do not have an expression as simple as that for the first and second-order ones. Alternatively, the statistics of the random variable can be estimated by means of sampling strategies: realizations $U(\theta)$ can be obtained by a sampling of $\boldsymbol{\xi}$ following its density $p_{\boldsymbol{\xi}}$ followed by the evaluation of the PC series at the sample points $\boldsymbol{\xi}(\theta)$. We shall rely heavily on such sampling procedure to estimate densities, cumulative density functions, probabilities, etc.... Note that in the context of the analysis of model data uncertainty, the sampling scheme for a model output known from its PC expansion is substantially simpler and more efficient than the full evaluation of the model realizations such as in MC methods. Also in the same context, if the (independent) random variables ξ_i used for the expansion of U can be related to physical sources of uncertainty, one has immediate access to the second-order characterization of the impact of different uncertainty sources, for instance through the ANOVA (analysis of variance) of U. An example of the ANOVA for an stochastic elliptic model is presented in Chap. 5.

2.4.2 Random Vectors

The PC expansion of a random variable can be immediately extended to the representation of second-order $\mathbb{R}^d$-random vectors:

$$\mathbf{U} : \Theta \mapsto \mathbb{R}^d. \tag{2.63}$$

Denoting U_i the i-th component of the random vector, its PC expansion on the truncated basis is

$$U_i \approx \sum_{k=0}^{\mathrm{P}} (u_i)_k \, \Psi_k(\boldsymbol{\xi}). \tag{2.64}$$

The random vector expansion can be recast in the vector form

$$\mathbf{U} = \sum_{k=0}^{\mathrm{P}} \mathbf{u}_k \Psi_k(\boldsymbol{\xi}), \tag{2.65}$$

where $\mathbf{u}_k = ((u_1)_k \cdots (u_d)_k)^t \in \mathbb{R}^d$ contains the coefficients of the k-th PC coefficients of the random vector components. The vector $\mathbf{u}_k$ will be called the k-th stochastic mode of the random vector $\mathbf{U}$. Clearly, $\mathbf{u}_0$ is the mean of the random vector. Further, two components U_i and U_j are orthogonal if and only if

$$\sum_{k=0}^{\mathrm{P}} (u_i)_k (u_j)_k \langle \Psi_k^2 \rangle = 0, \tag{2.66}$$

with the same condition for uncorrelated components but with the sum index starting at $k = 1$. In fact, the correlation and covariance matrices of vector $\mathbf{U}$ can be respectively expressed as:

$$\mathbf{r} = \sum_{k=0}^{\mathrm{P}} \mathbf{u}_k \mathbf{u}_k^t \langle \Psi_k^2 \rangle, \qquad \mathbf{c} = \sum_{k=1}^{\mathrm{P}} \mathbf{u}_k \mathbf{u}_k^t \langle \Psi_k^2 \rangle. \tag{2.67}$$

Note that a simple construction of PC expansions for random vectors with independent components consists in using different random variables for the independent components.

2.4.3 Stochastic Processes

The PC expansion can be immediately extended to a second-order stochastic processes U,

$$U : \Omega \times \Theta \mapsto \mathbb{R}, \tag{2.68}$$

by letting the deterministic coefficients to depend on the index $\mathbf{x} \in \Omega$, namely

$$U(\mathbf{x}, \boldsymbol{\xi}) \simeq \sum_{k=0}^{\mathrm{P}} u_k(\mathbf{x}) \Psi_k(\boldsymbol{\xi}). \tag{2.69}$$

Consistently with the case of random vectors, the deterministic functions $u_k(\mathbf{x})$ will be called the stochastic modes of the process. The expansion (2.69) is obtained by considering $U(\mathbf{x}, \cdot)$ as a random variable of $L_2(\Theta, P)$ for any $\mathbf{x} \in \Omega$, such that its expansion coefficients on the PC basis are also indexed with $\mathbf{x}$. Again, from the convention $\Psi_0 = 1$ we have $u_0(x) = \mathrm{E}[U(\mathbf{x}, \cdot)]$. Also, due to orthogonality of the PC basis, the k-th stochastic mode in the expansion of U is given by:

$$\begin{aligned} \langle U(\mathbf{x}, \cdot), \Psi_k \rangle &= \left\langle \left(\sum_{l=0}^{\mathrm{P}} u_l(\mathbf{x}) \Psi_l \right), \Psi_k \right\rangle \\ &= \sum_{l=0}^{\mathrm{P}} u_l(\mathbf{x}) \, \langle \Psi_l, \Psi_k \rangle = u_k(\mathbf{x}) \big\langle \Psi_k^2 \big\rangle. \end{aligned} \tag{2.70}$$

It shows that mode $u_k(\mathbf{x})$ is, up to a suitable normalization factor, given by the correlation between U and Ψ_k.

Due to the orthogonality of the PC basis, one immediately observes that the correlation function of U can be expressed in terms of its expansion modes, according to:

$$\begin{aligned} R_{UU}(\mathbf{x}, \mathbf{x}') = \big\langle U(\mathbf{x}, \cdot) U(\mathbf{x}', \cdot) \big\rangle &= \left\langle \left(\sum_{k=0}^{\mathrm{P}} u_k(\mathbf{x}) \Psi_k \right) \left(\sum_{k=0}^{\mathrm{P}} u_k(\mathbf{x}') \Psi_k \right) \right\rangle \\ &= \sum_{k=0}^{\mathrm{P}} \sum_{l=0}^{\mathrm{P}} u_k(\mathbf{x}) u_l(\mathbf{x}') \, \langle \Psi_k \Psi_l \rangle \\ &= \sum_{k=0}^{\mathrm{P}} u_k(\mathbf{x}) u_k(\mathbf{x}') \big\langle \Psi_k^2 \big\rangle. \end{aligned} \tag{2.71}$$

Note that the knowledge of the correlation function is not sufficient to uniquely determine the set of coefficients $u_k(\mathbf{x})$ appearing in the expansion of U. This illustrates the limitation of the characterization of U based only on its second-order properties, and indicates that additional information is needed to define the stochastic modes.

Conversely, this reflects the larger amount of information contained in a PC expansion, that clearly transcends the second-order characteristics.

Comparison of the KL expansion of a stochastic process in (2.14) with the PC expansion in (2.69) leads to additional observations. First, whereas the stochastic coefficients (η_k and Ψ_k) are in both cases orthogonal, the stochastic modes, u_k, of the PC expansion of U are not orthogonal, unlike their counterpart in the KL decomposition. Indeed, in the PC expansion (2.69), the modes $u_k(\mathbf{x})$ are generally not orthogonal, so we cannot determine the random functionals Ψ_k through $(U(\cdot, \boldsymbol{\xi}), u_k)$. Nonetheless, the two expansions can still be related. To this end, we write the KL expansion as:

$$U(\mathbf{x}, \theta) = \sum_l u_l^{(KL)} \sqrt{\lambda_l} \eta_l(\theta), \qquad \left(u_k^{(KL)}, u_l^{(KL)}\right) = \delta_{kl}, \qquad \mathrm{E}[\eta_l \eta_{l'}] = \delta_{ll'}, \tag{2.72}$$

where superscript (KL) has been added to the KL-modes to avoid confusion with the PC modes of U. Now, because the random coefficients in the KL expansion are second-order random variables, they have a convergent PC expansion:

$$\eta_l(\theta) = \sum_k (\eta_l)_k \Psi_k(\boldsymbol{\xi}(\theta)). \tag{2.73}$$

Inserting these expansions into the KL decomposition, and rearranging the terms results in:

$$U(\mathbf{x}, \theta) = \sum_k \left[\sum_l \sqrt{\lambda_l} (\eta_l)_k u_l^{(KL)}(\mathbf{x}) \right] \Psi_k(\boldsymbol{\xi}(\theta)), \tag{2.74}$$

which is equivalent to (2.69) if we define

$$u_k(\mathbf{x}) = \left[\sum_l \sqrt{\lambda_l} (\eta_l)_k u_l^{(KL)}(\mathbf{x}) \right]. \tag{2.75}$$

Further, this simple manipulation shows that

$$(u_k, u_{k'}) = \sum_l \sum_{l'} (\eta_l)_k (\eta_{l'})_{k'} \left(u_l^{(KL)}, u_{l'}^{(KL)}\right) = \sum_l (\eta_l)_k (\eta_l)_{k'}, \tag{2.76}$$

which is generally non-zero for $k \neq k'$. A noticeable exception occurs for the case of Gaussian processes for which $\eta_l \sim \mathrm{N}(0, 1)$. In this case, the η_l's being independent, we can construct a straightforward first-order Wiener-Hermite expansion for them simply by using $\xi_l = \eta_l$. For appropriate indexation of the PC basis, we have $\eta_l(\theta) = \xi_l(\theta) = \Psi_l(\boldsymbol{\xi})$ where $\boldsymbol{\xi} = \{\xi_1, \xi_2, \ldots\}$. For this particular expansion of the η_l, the stochastic modes of the PC expansion of U and the KL modes are simply related by

$$u_k(\mathbf{x}) = \sqrt{\lambda_k} u_k^{(KL)}(\mathbf{x}). \tag{2.77}$$

Such simple relation between stochastic PC modes and KL modes does not exist for general (non-Gaussian) processes, for which the second-order properties of U, that determine the KL modes $u^{(KL)}$, do not suffice.

2.5 Application to Uncertainty Quantification Problems

The stochastic spectral expansion of a random quantity (random variable, vector or stochastic process) provides a convenient representation in view of its characterization (extraction of mean, moments, analysis of correlations, density estimates, measure of the probability of events, local and global sensitivity analysis, ...). All this information is available at low computational cost provided that the coefficients of the representation of relevant quantities are known. Therefore, stochastic expansions will have practical utility if one disposes of methods allowing efficient determination of the associated coefficients—scalars, vectors, functions, or fields, as the case may be.

We distinguish here between two types of problems.

- In the first case, one disposes of information regarding a random quantity, and seeks to construct a spectral expansion to represent it. The type of information available may greatly differ from one application to another; this may include the complete probability law (for instance its density, joint-density or full set of finite dimensional distributions, etc.) or simply a (sometimes coarse) sample set realizations. To construct the spectral representation, one thus needs to establish an optimization problem to determine the expansion coefficients. The actual form of the optimization problem depends on the information available. For example, one approach may be based on minimizing the distance between the characteristics of some actual quantities (*e.g.* moments, densities, etc.) and those of the corresponding approximation. The definition of the distance may also differ from one approach to another, leading to optimization problems of very different nature.

  A central question regarding the spectral representation of random quantities concerns the rate of convergence, which in the context of PC expansions depends both on the expansion order p and the number N of random variables ξ_i. This is particularly delicate when one has only a sample set of realizations, since in this case it is not possible to relate an individual realization of the random quantity to a specific value of the random vector $\boldsymbol{\xi}$. Prescribing this relation *a priori* may be an option, but over-fitting issues may occur when one refines the expansion bases. Clearly, optimization techniques with regularization properties are needed here. Let us just mention algorithms based on Bayesian inference and maximum entropy principles that appear to offer the required properties in terms of robustness and convergence [45, 55, 56, 217, 219]. These aspects of stochastic spectral approximation are at the center of many current investigations, and significant advances are expected in the coming years.
- The second type of problem, which will be of central concern in this monograph, consists in the propagation of uncertainty in some model input data, D, which

are assumed to be already parametrized using a finite set of random variables $\boldsymbol{\xi}$. Knowing the density of the random vector $\boldsymbol{\xi}$, the goal of the uncertainty propagation problem is to determine the stochastic expansion of the model solution, say $U(\boldsymbol{\xi})$, induced by the stochastic data $D(\boldsymbol{\xi})$. In the following chapters, we shall discuss two broad classes of methods that can be used to address this goal: non-intrusive methods and spectral Galerkin methods.

Chapter 3
Non-intrusive Methods

In this chapter, we focus our attention on non-intrusive methods for the approximation of an output of a model involving random data, parametrized by a finite set of independent random parameters $\boldsymbol{\xi}(\theta)$ defined on a probability space (Θ, Σ, P), and probability density $p_{\boldsymbol{\xi}}(\boldsymbol{\xi})$. We denote Ξ the support of the density $p_{\boldsymbol{\xi}}$. As discussed in Chap. 2, we are concerned with models having a unique solution for almost all realizations of the random parameters, so the model can be seen as a surjective mapping from the parameters domain to the image solution space. Because this mapping involves models which are generally complex to solve (for instance PDEs), a natural idea that has been used for a long time is to construct a much simpler mapping, or surrogate model, that approximates the actual complex model. Denoting s the model output of interest, we seek the mapping

$$s : \boldsymbol{\xi} \in \Xi \mapsto s(\boldsymbol{\xi}) \in \mathbb{R}. \tag{3.1}$$

From this perspective, stochastic spectral expansions can be seen as a particular form of response surface methods where the approximation of the mapping is sought in terms of a basis of (orthogonal) random functionals $\{\Psi_0, \Psi_1, \dots\}$ such that

$$s : \boldsymbol{\xi} \in \Xi \mapsto s(\boldsymbol{\xi}) = \sum_i s_i \Psi_i(\boldsymbol{\xi}). \tag{3.2}$$

The so-called non-intrusive methods rely on a set of *deterministic model resolutions*, corresponding to some specific values or realizations of $\boldsymbol{\xi}$, to construct approximations $s(\boldsymbol{\xi})$. Along this line, a deterministic simulation code can be used as a black-box, which associates to each realization of the parameters the corresponding model output.

The most attractive feature of non-intrusive methods comes with the approximation of $s(\boldsymbol{\xi})$ which requires a deterministic solver only, and so needs no particular adaptation of existing codes to generate the outputs. This feature has to be contrasted with the stochastic Galerkin projection techniques described in the following chapter, which require the resolution of the so-called spectral problem resulting from a reformulation of the problem. In addition, it is usually possible to plan the

O.P. Le Maître, O.M. Knio, *Spectral Methods for Uncertainty Quantification*, Scientific Computation,
DOI 10.1007/978-90-481-3520-2_3, © Springer Science+Business Media B.V. 2010

needed deterministic model simulations in non-intrusive approaches, so they can be distributed and performed in parallel. These characteristics make non-intrusive methods very attractive for parametric uncertainty propagation in complex models, industrial applications and situations where only deterministic (commercial or legacy) codes are available. Using the computer code as a black-box also presents the advantage of making non-intrusive methods applicable to models of virtually any complexity (multiphysics, coupled problems, highly nonlinear models,...).

However, the numerical cost of non-intrusive methods essentially scales with the number of deterministic model resolutions one has to perform to construct the approximation. This number of model resolutions can be large. In particular, non-intrusive methods suffer from the curse of dimensionality (the number of model resolutions increases exponentially with the number of independent random variables in the parametrization). This drawback makes non-intrusive methods computationally intensive and costly if the underlying deterministic model is expensive to solve. Consequently, the reduction of the complexity in non-intrusive methods is the focus of many ongoing efforts. These have yielded a great deal of advances, in particular following the recent introduction of adaptive sparse grid methods (see below).

While this monograph focuses primarily on stochastic Galerkin methods, this chapter aims at presenting an overview of the different non-intrusive strategies for the construction of stochastic spectral expansions of some model output. For simplicity, we shall restrict ourselves to the expansion of single real scalar model output ($s \in \mathbb{R}$), function of N independent real-valued random variables, $\boldsymbol{\xi}(\theta) = (\xi_1(\theta), \dots, \xi_{\mathrm{N}}(\theta)) \in \Xi \subseteq \mathbb{R}^{\mathrm{N}}$, with probability density function

$$p_{\boldsymbol{\xi}}(\boldsymbol{\xi}) = \prod_{i=1}^{\mathrm{N}} p_i(\xi_i).$$

Extension of the methodologies outlined to more general model outputs (vectors and fields) is straightforward. Problems with non-independent random parameters raise several additional issues that will not be discussed.

In Sect. 3.1, we provide an overview of the so-called non-intrusive spectral projection (NISP) which aims at computing the coefficients of the model output through its orthogonal projection on a prescribed stochastic basis. In Sect. 3.2 we discuss different pseudo-random sampling strategies for NISP. Section 3.3 introduces deterministic integration methods, or cubature rules; we focus on the quadrature formulas (Sect. 3.3.1) and on multidimensional cubature formulas based on the tensor product of quadrature formulas (Sect. 3.3.2). We complete our abbreviated outline of NISP methods with a discussion of the so-called sparse grid methods (Sect. 3.4), which provide means for constructing coarse cubature formulas.

Section 3.5 deals with an alternative formulation of the problem of estimating the expansion coefficients of the model output, through the resolution of a least squares minimization problem. In particular, we discussed the connection and differences between this non-intrusive approach and NISP.

Section 3.6 briefly reviews collocation methods. These are non-intrusive approaches in which the set of deterministic model outputs is used in conjunction with interpolation schemes in order to construct an approximation of the stochastic solution. Collocation techniques for stochastic problems have received considerable attention recently. Due to the fast evolution of the resulting algorithms, we shall limit ourselves to the essential ideas underlying these approaches, and provide recent references.

3.1 Non-intrusive Spectral Projection

3.1.1 Orthogonal Basis

The Non-Intrusive Spectral Projection (NISP) aims at computing the projection coefficients of random model output $s(\boldsymbol{\xi})$ on a finite dimensional stochastic subspace $\mathcal{S}^{\mathrm{P}}$ of $L_2(\Xi, p_{\boldsymbol{\xi}})$. The space $L_2(\Xi, p_{\boldsymbol{\xi}})$ equipped with the inner product $\langle , \rangle$,

$$\langle u, v\rangle = \int_{\Xi} u(\boldsymbol{\xi})v(\boldsymbol{\xi})p_{\boldsymbol{\xi}}(\boldsymbol{\xi})\mathrm{d}\boldsymbol{\xi}, \tag{3.3}$$

is a Hilbert space. We denote $(\mathrm{P}+1)$ the dimension of $\mathcal{S}^{\mathrm{P}}$, and let $\{\Psi_0(\boldsymbol{\xi}), \ldots, \Psi_{\mathrm{P}}(\boldsymbol{\xi})\}$ be an orthogonal basis of $\mathcal{S}^{\mathrm{P}}$, *i.e.*

$$\left\langle \Psi_i, \Psi_j \right\rangle = \int \cdots \int \Psi_i(\boldsymbol{\xi})\Psi_j(\boldsymbol{\xi})p_{\boldsymbol{\xi}}(\boldsymbol{\xi})\, \mathrm{d}\xi_1 \cdots \mathrm{d}\xi_{\mathrm{N}} = \langle \Psi_i, \Psi_i\rangle\, \delta_{ij}. \tag{3.4}$$

Most often, the basis will consist of the set of Generalized Polynomial Chaoses truncated to order No, but other types of orthogonal basis (*e.g.* non-polynomial ones) may also be considered for NISP.

3.1.2 Orthogonal Projection

Assuming that the model output is a second-order random variable, *i.e.* $\langle s, s\rangle < \infty$, the orthogonal projection of s onto $\mathcal{S}^{\mathrm{P}}$, denoted $(s)^{\parallel}$, is

$$(s)^{\parallel}(\boldsymbol{\xi}) = \sum_{k=0}^{\mathrm{P}} s_k \Psi_k(\boldsymbol{\xi}) \in \mathcal{S}^{\mathrm{P}}. \tag{3.5}$$

Thanks to the orthogonality of the selected basis of $\mathcal{S}^{\mathrm{P}}$, the projection coefficients s_k are given by

$$s_k = \frac{\langle s, \Psi_k\rangle}{\langle \Psi_k, \Psi_k\rangle}, \qquad k = 0, 1, \ldots, \tag{3.6}$$

such that

$$s - s^{\parallel} \perp \mathcal{S}^{\mathrm{P}} \Leftrightarrow \left\langle s - s^{\parallel}, \Psi_k \right\rangle = 0 \quad \text{for } 0 \le k \le \mathrm{P}. \tag{3.7}$$

The normalization factor $\langle \Psi_k, \Psi_k \rangle$ depends only on the basis used. For GPC bases and classical probability density functions, the $\langle \Psi_k, \Psi_k \rangle$ are known analytically, such that only the determination of the numerator in the left-hand side of (3.6) is to be discussed. It is also important to observe that the expression of s_k does not depend on the dimension of the projection subspace, but only on the basis function Ψ_k. This implies that once the basis has been selected, the determination of the expansion coefficients s_k in NISP are independent from each other. This characteristic of NISP has to be contrasted with other approaches such as the stochastic Galerkin projection and other non-intrusive methods, where the determination of the expansion coefficients is generally coupled.

Remark Introducing the definition of the inner product in the expression of the projection coefficient results in:

$$\langle \Psi_k, \Psi_k \rangle \, s_k = \langle s, \Psi_k \rangle = \int_{\Xi} s(\xi_1, \ldots, \xi_{\mathrm{N}}) \Psi_k \, p_1(\xi_1) \ldots p_{\mathrm{N}}(\xi_{\mathrm{N}}) \, d\xi_1 \cdots d\xi_{\mathrm{N}}. \tag{3.8}$$

Equation (3.8) shows that the coefficients s_k are given, up to the normalization factor $\langle \Psi_k, \Psi_k \rangle$, as a N-dimensional integral over Ξ. In fact this integral is nothing but the correlation between the random model output $s(\boldsymbol{\xi})$ and the k-th basis function $\Psi_k(\boldsymbol{\xi})$. Both interpretations for the expression of s_k, in terms of integrals or correlations, are equally valid and result in a problem of similar complexity. In particular, they both face the issue of the curse of dimensionality as N increases.

Different techniques have been proposed to numerically estimate the right-hand side of (3.8). These techniques can be classified into simulation approaches, where one relies on pseudo-random sampling strategies, and cubature methods, which involve deterministic schemes to numerically estimate the integrals.

3.2 Simulation Approaches for NISP

In the simulation approach, the correlation between $s(\boldsymbol{\xi})$ and $\Psi_k(\boldsymbol{\xi})$ is estimated by means of a pseudo-random sampling of the parameter space Ξ. Different methods can be applied towards this end, as briefly outlined below.

3.2.1 Monte Carlo Method

The Monte Carlo method is the simplest simulation technique: a sample set of independent realizations of $\boldsymbol{\xi}$ is generated from $p_{\boldsymbol{\xi}}$ using a (pseudo) random number generator. Since the ξ_i's are independent, it is sufficient to draw independently values for the ξ_i from their respective distributions p_i. Denoting $\boldsymbol{\xi}^{(i)}$ the i-th element of the sample set, and $s^{(i)} \equiv s(\boldsymbol{\xi}^{(i)})$ the corresponding model output, the empirical correlation for s_k is given by

$$\langle s, \Psi_k \rangle = \frac{1}{\mathrm{M}} \sum_{i=1}^{\mathrm{M}} s^{(i)} \Psi_k(\boldsymbol{\xi}^{(i)}) + \epsilon_{\mathrm{M}}, \tag{3.9}$$

where M is the sample set dimension and ϵ_{M} is the sampling error.

We are thus concerned with the convergence of the sampling error ϵ_M as the sample set dimension increases. Since the sample set is randomly generated, ϵ_{M} is random. It can be easily shown that the empirical estimate in (3.9) is unbiased ($\mathrm{E}[\epsilon_{\mathrm{M}}] = 0$) and the variance of ϵ_{M} goes to zero asymptotically as $1/\sqrt{\mathrm{M}}$ for sufficiently large M, according to the law of large numbers. This low convergence rate of the sampling error is the main limitation in using a Monte Carlo method for NISP. Indeed, for most applications the computational cost is dominated by the resolution of the model for the parameter realizations $\boldsymbol{\xi}^{(i)}$, and so directly scales with M: higher convergence rates with M are needed if accurate predictions are to be efficiently obtained.

3.2.2 Improved Sampling Strategies

Lower sampling errors, for fixed M, can be achieved using more efficient sampling strategies, such as Latin Hypercube Sampling (LHS, see [153]) and Quasi Monte Carlo sampling (QMC, see [158]). The essential idea underlying these improved sampling schemes, in the case of uniform measures, is to force the sampler to draw points that cover the parameter domain in a more uniform way than for the standard MC.

For LHS, this is achieved by forcing the sampler to draw a realization within equiprobable bins in the parameter range. This strategy imposes that the sample set dimension is fixed a priori to define the bins. In QMC, a low discrepancy deterministic sequence of points is generated so as to maximize the uniformity of the sample points. For uniform measures, the minimal discrepancy corresponds to points distributed on a regular N-dimensional grid, but such a grid has a number of nodes which scales exponentially with N. To circumvent this scaling, sequences of points are used in place of grids. Different sequences have been proposed, *e.g.* Halton's [98] and Sobol's [213] for which public domain codes are available. For both algorithms, transformations of random variables to uniformly distributed ones can be used.

Examples of sample sets, with respective dimensions $\mathrm{M} = 128$, 256 and 512, in the unit square ($\mathrm{N} = 2$) with uniform probability distribution are shown in Fig. 3.1. Compared are sample sets obtained for the MC, LHS and QMC (using Halton's sequence) samplers. It is seen that the MC sample sets exhibit clusters of points

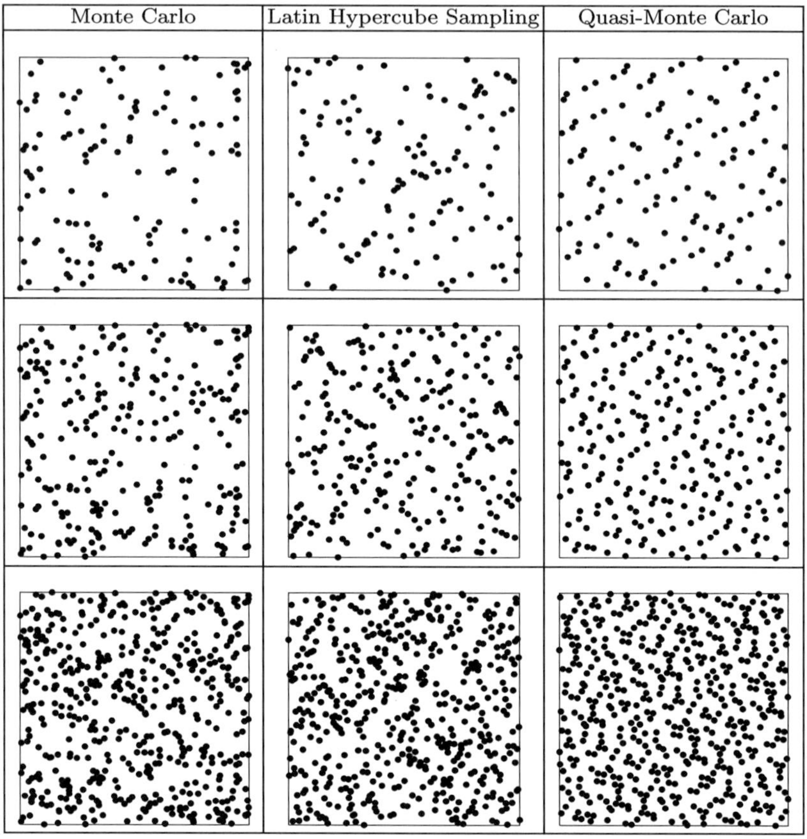

Fig. 3.1 Illustration of pseudo-random samples sets over the unit square, with increasing dimension $M = 128$, 256 and 512 (*from top to bottom*), using Monte Carlo, LHS and QMC methods. The QMC sampling uses paired Halton's sequences with respective primes 3 and 5 along the first and second directions (see [158])

and areas with low density of points. The LHS performs slightly better than MC in terms of sample distribution, while the QMC samples are clearly more uniformly distributed. In practice MC and LHS samplings lead to the same asymptotic convergence rate in $\mathcal{O}(1/\sqrt{M})$, while QMC offers improvement of the convergence rate, up to $\mathcal{O}(\ln M^N/M)$.

Remark The primary advantage of simulation approaches resides in their convergence rate independent of the dimensionality N of the parameter space and of the smoothness of $s(\xi)\Psi_k(\xi)$. This characteristic suggests a robust behavior of pseudo-random sampling schemes in the case of high dimensionality of the parameter space, and of solutions $s(\xi)$ that are weakly regular or non-smooth. These situations are challenging for the alternative integration methods discussed below, which are most efficient for low to moderate dimensionality of the parameter space, smooth model output and smooth basis functions. Furthermore, one disposes of well-established error estimators for the sampling errors. Nonetheless, the slow convergence rate of sampling errors justifies consideration of alternative deterministic methods which in some situations prove to be more practical.

3.3 Deterministic Integration Approach for NISP

The definition of the projection coefficients s_k in (3.8) shows that their evaluation amounts to the computation of a multidimensional integral of a function $f(\boldsymbol{\xi}) \equiv s(\boldsymbol{\xi})\Psi_k(\boldsymbol{\xi})$ over Ξ with a non-negative weight $p_{\boldsymbol{\xi}}$ having a product form. In fact, one has to compute a set of integrals

$$\mathcal{I}f = \int_{\Xi} f(\boldsymbol{\xi}) p_{\boldsymbol{\xi}}(\boldsymbol{\xi})\, \mathrm{d}\boldsymbol{\xi}. \qquad (3.10)$$

When the dimensionality N of the integration is not too large, as is the case for many problems in practice, deterministic cubatures can be effectively used. A cubature is an approximation of the multidimensional integral $\mathcal{I}f$ as a discrete sum:

$$\mathcal{I}f \approx \sum_{i=1}^{\mathrm{N_Q}} f\left(\boldsymbol{\xi}^{(i)}\right) W^{(i)}, \qquad (3.11)$$

where $\boldsymbol{\xi}^{(i)} \in \Xi$ and $W^{(i)} \in \mathbb{R}$, $i = 1, \ldots, \mathrm{N_Q}$, are the nodes and weights of the $\mathrm{N_Q}$-node cubature. Cubatures can be substantially more economical than the simulation approach, in terms of the number of model resolutions $\mathrm{N_Q}$, to meet a prescribed error threshold and provided that N is not to large.

Cubatures are generally constructed on the basis of 1D integration rules, or quadrature formulas, which we now discuss.

3.3.1 *Quadrature Formulas*

We denote $\mathcal{I}^{(1)} f$ the univariate integration and $Q^{(1)} f$ its quadrature approximation:

$$\mathcal{I}^{(1)} f = \int_a^b f(\xi) p(\xi)\, \mathrm{d}\xi \approx Q^{(1)} f \equiv \sum_{i=1}^{n} f\left(\xi^{(i)}\right) w^{(i)}, \qquad (3.12)$$

where the integration bounds a and b may be infinite.

3.3.1.1 Gauss Quadratures

In the context of NISP, the integration weights are the product of probability density function p_i of the random parameters, which will be assumed strictly positive over their respective domains. Then, $p(\xi) > 0$ for $\xi \in (a, b)$ in (3.12). Consequently, $\mathcal{I}^{(1)} f$ can be approximated using Gauss quadrature formulas (see for instance [220, 222]), which are high-order approximations for integrals such as in (3.12). The n-node Gauss quadrature can be expressed as:

$$Q^{(1)} f = \sum_{i=1}^{n} f^{(i)} w^{(i)}, \quad f^{(i)} \equiv f\left(\xi^{(i)}\right), \tag{3.13}$$

where the nodes location $a < \xi^{(i)} < b$ and the weights $w^{(i)} > 0$ depend on the domain of integration and on the weight function $p(\xi)$. The Gauss formula is closely related to the orthogonal polynomial family for the weight function $p(\xi)$, which is also the basis for Generalized Polynomial Chaos expansions. It then appears natural to consider Gauss formulas for NISP. In fact, the nodes $\xi^{(i)}$ of the quadrature rule are the zeros of the n-th order orthogonal polynomial with regard to the measure $p(\xi)$. Gauss quadratures yield the highest possible degree of exactness: they exactly integrate polynomial functions of degree $\leq 2n - 1$. For given $n \geq 1$ the nodes and weights can be computed by solving an eigenvalue problem. Below we address references to classical weight functions.

- **Gauss-Legendre quadrature.** In the case of a uniform measure on the interval $[-1, 1]$, we have:

$$\mathcal{I}^{(1)} f = \int_{-1}^{1} f(\xi) \frac{d\xi}{2} \approx \sum_{i=1}^{n} f\left(\xi^{(i)}\right) w^{(i)}. \tag{3.14}$$

  The nodes and weights of the Gauss-Legendre can be obtained from Table B.1 for different values of n.
- **Gauss-Hermite quadrature.** In the case of a standard Gaussian measure, we have:

$$\mathcal{I}^{(1)} f = \frac{1}{\sqrt{2\pi}} \int_{-\infty}^{\infty} f(\xi) \exp\left(-\xi^2/2\right) d\xi \approx \sum_{i=1}^{n} f\left(\xi^{(i)}\right) w^{(i)}. \tag{3.15}$$

  The Gauss-Hermite quadrature points and weights are given in Table B.2, for few different values of n.
- Analogous Gauss' formulas are available for other measures, in particular the γ and β distributions [1]. Open-source is available for the calculation of the nodes and weights of Gauss quadratures corresponding to various weight functions.[1]

Although yielding an optimal degree of exactness, Gauss integration formulas present the drawback of having a node distribution that depends on n. This dependence makes Gauss quadratures unsuited for the development of composite integration strategies, where the convergence of the numerical integration is controlled by

[1] See for instance http://www.netlib.org.

progressively increasing the number of nodes: generally the model evaluations at the nodes ξ_i for the n-points formula cannot be re-used in another formula involving a different number of nodes. In addition, in common problems, although the stochastic basis may be chosen to be polynomial, the model solution rarely is, since the integrand is not polynomial. Consequently, achieving the maximum polynomial degree of exactness may not be as essential as one could initial envision. In fact, lower-order quadratures may actually perform as well as Gauss quadrature (see for instance [227]) for the numerical integration of practical functions. This observation calls for adopting more flexible quadratures, particular nested formulas, where the nodes of a given quadrature also feature as nodes of higher-order formulas.

3.3.1.2 Nested Quadratures

To distinguish formulas with different orders (levels) of accuracy, we assign a level index $l \geq 1$ to the quadrature approximation of (3.12); we write

$$\mathcal{I}^{(1)} f \approx Q_l^{(1)} f = \sum_{i=1}^{n_l} f\left(\xi_l^{(i)}\right) w_l^{(i)}, \tag{3.16}$$

where n_l is the number of nodes for the l-level quadrature formula. A quadrature rule is nested, if nodes at a given level l appears in the formulas with level $l' \geq l$. Nested formulas usually involve sets of nodes whose dimensions double each time the level l is incremented, *i.e.*

$$n_l \sim 2^l. \tag{3.17}$$

To simplify the exposition, we consider the case of uniform measure over [0, 1]. For NISP, a natural transformation to reduce the integration to the [0, 1] interval is the iso-probabilistic transformation which consists in the mapping from the random parameter space to the cumulative distribution function (CDF) of ξ:

$$x : \xi \in (a, b) \mapsto x(\xi) \equiv F(\xi) = \int_a^{\min(\xi, b)} p(\xi')\, d\xi' \in [0, 1]. \tag{3.18}$$

This mapping is bijective provided that $p(\xi) > 0$ for all ξ in (a, b). We shall assume in the following a one-to-one correspondence between the mapped variable x and the random parameter ξ. With this transformation of variable, (3.12) is equivalent to

$$\mathcal{I}^{(1)} f = \tilde{\mathcal{I}}^{(1)} g = \int_0^1 g(x)\, dx, \quad g(x) \equiv f\left(F^{-1}(x)\right). \tag{3.19}$$

Thus, one ends up with an integration over [0, 1], with unity weight function. Denoting $\tilde{Q}_l^{(1)} g$ the level l quadrature rule in the transformed domain,

$$\tilde{Q}_l^{(1)} g = \sum_{i=1}^{n_l} g\left(x_l^{(i)}\right) \tilde{w}_l^{(i)}, \tag{3.20}$$

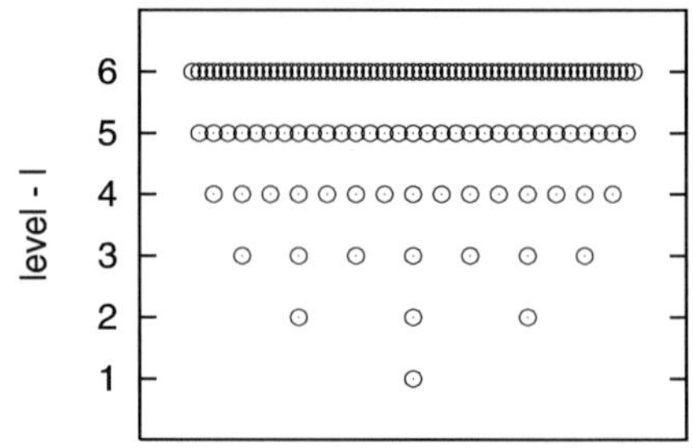

Fig. 3.2 Nodes of the trapezoidal rule at levels $1 \leq l \leq 6$

the corresponding quadrature for f is expressed as:

$$Q_l^{(1)} f = \sum_{i=1}^{n_l} f\left(\xi_l^{(i)}\right) w_l^{(i)}, \tag{3.21}$$

with $\xi_l^{(i)} \equiv F^{-1}(x_l^{(i)})$ and $w_l^{(i)} = \tilde{w}_l^{(i)}$. We now provide examples of classical nested quadrature formulas.

- **Trapezoidal rule.** A simple nested quadrature formula is the trapezoidal rule for integration on [0, 1] with unit weight. It corresponds to

$$\begin{aligned} n_l &= 2^l - 1, \qquad x_l^{(i)} = \frac{i}{n_l + 1}, \\ \tilde{Q}_l^{(1)} g &= \frac{1}{n_l + 1}\left[\frac{3}{2}\left(g\left(x_l^{(1)}\right) + g\left(x_l^{(n_l)}\right)\right) + \sum_{i_l=2}^{n_l-1} g\left(x_l^{(i)}\right)\right]. \end{aligned} \tag{3.22}$$

  Figure 3.2 shows the modes at levels $1 \geq l \geq 6$ of the nested trapezoidal rule. It is seen that the nodes at given level are equidistant, contrary to the Gauss points. In addition, no node is on the boundary of the integration domain, a characteristic which makes the trapezoidal rule suited for NISP in case of unbounded random parameters. However, the trapezoidal rule has a slow convergence rate with the number of nodes involved in the integration, due to the underlying piecewise linear approximation of $g(x)$. Higher order Newton-Cotes composite formulas [220] can be used to improve the convergence rate with the number of nodes.
- **Clenshaw-Curtis [36] and Fejèr rules.** These quadratures approximates $\tilde{\mathcal{I}}^{(1)} g$ by the exact integral of the Chebychev polynomial expansion of g. The nodes of these quadratures are the maximum of the Chebychev polynomials, including (Clenshaw-Curtis) or excluding (Fejèr) the two boundary nodes. Efficient strategies can be used to compute the nodes and weights for a given level l [81, 82, 173, 233]. Again, using $n_l = (2^l) \pm 1$ results in nested sets of nodes. The Clenshaw-Curtis and Fejèr quadrature nodes are plotted in Fig. 3.3 for $1 \leq l \leq 6$.

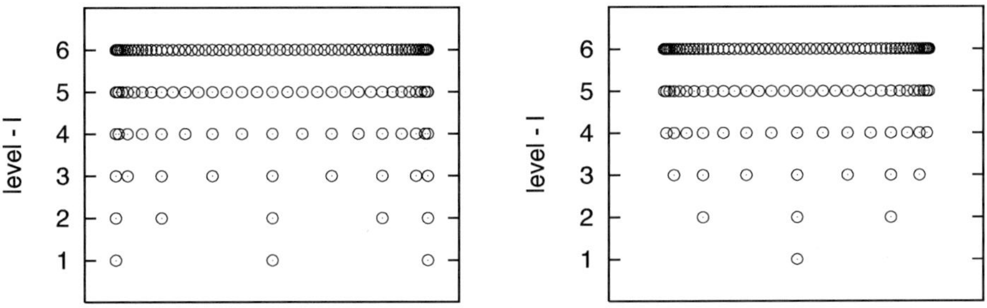

Fig. 3.3 Nodes of the Clenshaw-Curtis (*left*) and Fejèr (*right*) rules for levels $1 \le l \le 6$

3.3.2 Tensor Product Formulas

We now return to the N-dimensional numerical integration in (3.11). From the 1D quadrature formula, say $Q_l^{(1)}$, N-dimensional cubature rules can be constructed by tensorization. For instance, if the random parameters are identically distributed, such that the same quadrature $Q_l^{(1)}$ can be used along all directions, we obtain

$$\mathcal{I} f \approx Q^{(\mathrm{N})} f = \left(Q_l^{(1)} \otimes \cdots \otimes Q_l^{(1)}\right) f. \tag{3.23}$$

The previous definition can be extended to situations where the independent random parameters have different distributions, simply by using different quadratures rules along the different integration directions. Different levels in each direction can also be used. The latter case results in formulas of the form

$$\mathcal{I} f \approx \left(Q_{l_1}^{(1)} \otimes \cdots \otimes Q_{l_\mathrm{N}}^{(1)}\right) f. \tag{3.24}$$

It is important to recognize that the tensor product results in summations over all possible combinations of the indices:

$$\left(Q_{l_1}^{(1)} \otimes \cdots \otimes Q_{l_\mathrm{N}}^{(1)}\right) f = \sum_{i_1=1}^{n_{l_1}} \cdots \sum_{i_\mathrm{N}=1}^{n_{l_\mathrm{N}}} f\left(\xi_1^{(i_1)}, \ldots, \xi_\mathrm{N}^{(i_\mathrm{N})}\right) w_{l_1}^{(i_1)} \cdots w_{l_\mathrm{N}}^{(i_\mathrm{N})}. \tag{3.25}$$

The telescopic sum can be recast to a unique sum as in (3.11), with appropriate indexation of the cubature points $\boldsymbol{\xi}^{(i)}$ and weights $W^{(i)}$. Product formulas are regular grids of integration points, as illustrated in Fig. 3.4 for the Fejèr quadrature with $\mathrm{N} = 2$. We see that the integrand f has to be evaluated at a set of points lying on a structured grid in the integration domain Ξ.

We further observe that for the product formulas in (3.23), the total number $\mathrm{N_Q}$ of cubature points is

$$\mathrm{N_Q} = (n_l)^\mathrm{N}, \tag{3.26}$$

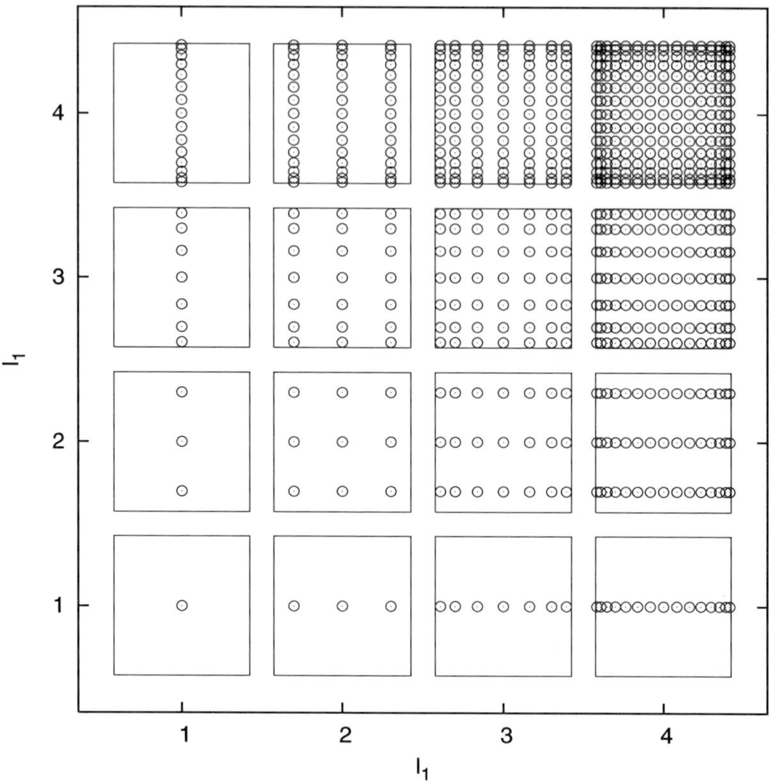

Fig. 3.4 Illustration of cubature rules constructed by products of nested Fejèr quadratures: plotted are the 2D grids of integration nodes from (3.25) for different values of the levels l_1 and l_2 along the integration dimensions. Grids on the diagonal plots correspond to the definition (3.23) of the cubature

while for (3.25) it is $N_Q = \prod_i n_{l_i}$. In both cases, N_Q exhibits an exponential increase with N. This result, known as the curse of dimensionality, shows that even if optimal quadrature formulas are used (Gauss type), the tensored form for cubature formula can be practical only when expanding low-dimensional model outputs needing low to moderate order stochastic basis functions, in which case a low level formula is sufficient.

3.4 Sparse Grid Cubatures for NISP

The sparse tensorization of quadrature formulas constitutes an efficient way to temper the curse of dimensionality of cubature rules. It is first observed that cubatures resulting from full tensorization are non-optimal in the sense that their degree of exactness could actually be achieved using a lower number of nodes. However, no general method is available to construct optimal cubature rules for arbitrary number of dimensions. Although usually non-optimal, the sparse grid cubatures discussed in this section offer an appreciable reduction in the number of cubature nodes for given degree of exactness compared to tensored product formulas. The sparse grid algorithms discussed below also present the advantage of being applicable to any dimensionality N.

3.4.1 Sparse Grid Construction

The first sparse grid method was proposed by Smolyak [212] in the context of multidimensional quadrature and interpolation. It consists in a general algorithm based on sparse tensor product construction. The construction is outlined as follows.

Consider a family of quadrature rules $Q_l^{(1)} f$, as in (3.21), and define the difference formulas:

$$\begin{aligned} \Delta_l^{(1)} &\equiv \left(Q_l^{(1)} - Q_{l-1}^{(1)}\right) f, \\ Q_0^{(1)} f &\equiv 0. \end{aligned} \tag{3.27}$$

Note that $\Delta_l^{(1)} f$ is also a quadrature rule. For nested formulas, $\Delta_l^{(1)} f$ contains the set of nodes of $Q_l^{(1)} f$ with weights equal to the difference of weights between level l and $l-1$.

To build the sparse cubature, we introduce the multi-index $\mathbf{l} = (l_1, \dots, l_{\mathrm{N}}) \in \mathbb{N}^{\mathrm{N}}$ and define

$$|\mathbf{l}| \equiv \sum_{i=1}^{\mathrm{N}} l_i. \tag{3.28}$$

Using this multi-index, the sparse cubature formula at level l is expressed as

$$Q_l^{(\mathrm{N})} f = \sum_{|\mathbf{l}| \le l+\mathrm{N}-1} \left(\Delta_{l_1}^{(1)} \otimes \cdots \otimes \Delta_{l_{\mathrm{N}}}^{(1)}\right) f. \tag{3.29}$$

This definition results in a coarser tensorization compared to the product form in (3.23), which has the following difference-formula expression:

$$Q_l^{(\mathrm{N})} f \equiv \sum_{\max \mathbf{l} \le l} \left(\Delta_{l_1}^{(1)} \otimes \cdots \otimes \Delta_{l_{\mathrm{N}}}^{(1)}\right) f, \quad \max \mathbf{l} \equiv \max\{l_1, \dots, l_{\mathrm{N}}\}. \tag{3.30}$$

Comparison of (3.30) and (3.29) shows that the product cubature consists in a summation over a hypercube in the multi-index space, whereas the sparse cubature reduces the summation over the simplex $|\mathbf{l}| \le l$. This is illustrated for the two-dimensional case ($\mathrm{N} = 2$) and $l = 4$ in Fig. 3.5. Also shown are the product and sparse grids of nested Fejèr nodes.

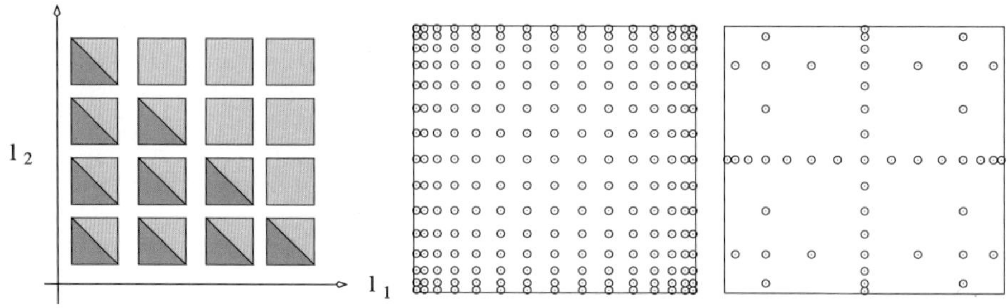

Fig. 3.5 Comparison of product and sparse tensorizations in the construction of cubature formulas of level $l = 4$, for the numerical integration in N = 2 dimensions. The *left plot* shows the indexes of the summation of difference formulas $\Delta_k^{(N)}$ for the product form in (3.30) (*squares*) and Smolyak's algorithm in (3.29) (*triangle*). The resulting grids for the Fejèr nested quadrature rule are shown in the *middle* (product form) and *right* (sparse grid) *plots*

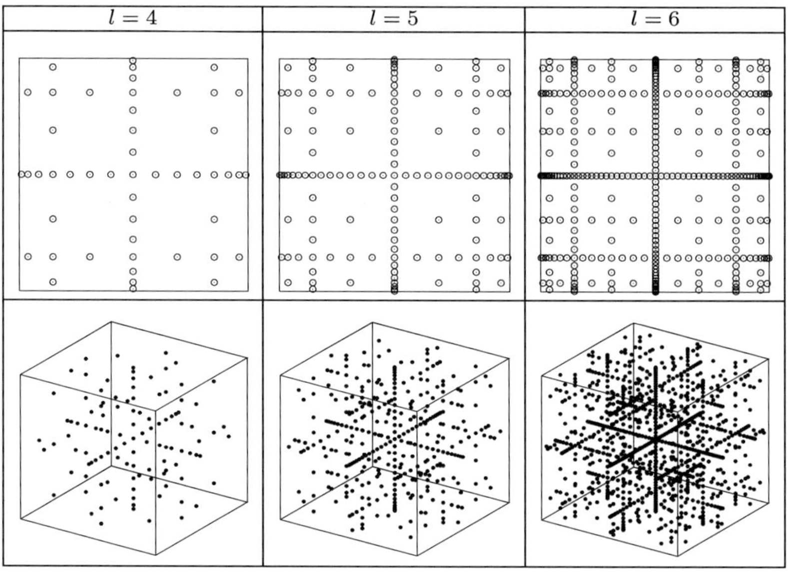

Fig. 3.6 Illustration of the sparse grid cubature nodes in N = 2 and N = 3 dimensions for the Smolyak's method and nested Fejèr quadrature formulas. Different levels l are considered as indicated

It is seen that the sparse cubature involves a significantly reduced number of points. In higher dimension, the ratio of number of nodes for product and sparse cubature increases. More information concerning the construction of sparse cubature formulas, their properties and discussion on efficient implementation strategies can be found in [83, 84, 173, 186, 187]. We provide in Fig. 3.6 examples of sparse grids based on the nested Fejèr quadrature, in N = 2 and 3 dimensions and different levels l in the Smolyak construction.

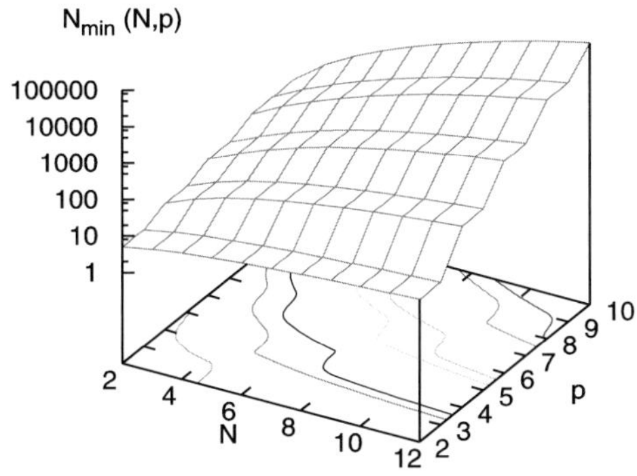

Fig. 3.7 Minimum number of nodes $N_{\min}$ for the Smolyak's sparse cubature for exact integration of polynomial integrands with degree $\leq p$ over hypercubes with uniform weight (nested Clenshaw-Curtis rules and Smolyak's sparse tensorization)

In Fig. 3.7, we plot the minimal number of nodes for the exact integration of polynomial with degree $\leq p$ as a function of the number N of dimensions. Shown is the case of a constant weight and sparse grid based on Clenshaw-Curtis rules. For $\mathrm{N} = 12$, exact integration up to the sixth order requires a few thousand of cubature nodes (and consequently as many individual deterministic model solutions); this is significantly less than 3^{12} (roughly half a million nodes!) the number of resolutions yielding the same accuracy for the product Gauss formula. Application of sparse grid methods for NISP is highly attractive, due to its reduced computational complexity, compared to product form cubature formulas and Monte Carlo sampling strategies, for moderate number of random parameters. The first use of sparse grid techniques was proposed in [109, 110], for the projection of elliptic equation solutions with random coefficients. The Smolyak sparse tensorization was used, relying on a package written by Knut Petras [186, 187] for the computation of the integration nodes and weights. However, although drastically improving the computational complexity of NISP, it was soon acknowledged that it remains too expensive to be applied for numerically demanding deterministic models and/or stochastic problems involving a larger number of random variables. This observation sets the starting point of researches toward adaptive sparse grid techniques.

3.4.2 Adaptive Sparse Grids

The definition of the multi-index set for the construction in (3.29) can be modified to yield more general sparse grid cubatures. The general form is

$$Q_l^{(\mathrm{N})} f = \sum_{\mathbf{l} \in \mathcal{I}(l)} \left(\Delta_{l_1}^{(1)} \otimes \cdots \otimes \Delta_{l_\mathrm{N}}^{(1)} \right) f, \tag{3.31}$$

where the multi-index set $\mathcal{I}(l)$ is function of the level l. The Smolyak's construction corresponds to the definition

$$\mathcal{I}(l) = \left\{ \mathbf{l} \in \mathbb{N}^{\mathrm{N}} : \sum_{i=1}^{\mathrm{N}} l_i \leq l + \mathrm{N} - 1 \right\}, \tag{3.32}$$

whereas the product cubature in (3.23) has the following multi-index set

$$\mathcal{I}(l) = \{\mathbf{l} \in \mathbb{N}^{\mathrm{N}} : l_i \leq l,\ i = 1, \dots, \mathrm{N}\}. \tag{3.33}$$

3.4.2.1 Dimension-Adaptive Sparse Grid

Changing the definition of the multi-index set allows grid adaptation. A straightforward sparse grid adaptation consists in distorting the simplex of the multi-indices in the summation of (3.29), simply by affecting different weights along each dimension. This can be easily implemented by considering a weights vector $\mathbf{a} \in \mathbb{R}_{+}^{\mathrm{N}}$ and defining the multi-index set as [84]:

$$\mathcal{I}(l) = \left\{ \mathbf{l} \in \mathbb{N}^{\mathrm{N}} : \mathbf{l} \cdot \mathbf{a} = \sum_{i=1}^{\mathrm{N}} a_i l_i \leq l + \mathrm{N} - 1 \right\}. \tag{3.34}$$

For this multi-index set one obtains anisotropic sparse cubature formulas, with variable order of accuracy along the integration dimensions. Such sparse grids are illustrated in Fig. 3.8 for the integration over a square. In Fig. 3.8, the weights vectors are selected such as to maintain a constant range for the second indices l_2 while the range of the first one varies depending on a_1.

The main issue with this dimension-adaptivity strategy is the prescription of appropriate weights $\mathbf{a}$, particularly since we have no prior knowledge on the characteristics of the model output. Only for few specific models is one able to determine *a priori* a relevant vector $\mathbf{a}$ for the problem at hand (see [73] for the case of an elliptic equations with random coefficients). Moreover, although dimension adaptive strategies can effectively coarsen the set of sparse grid nodes, and consequently reduce the computational complexity, these methods cannot result in a general or arbitrary coarsening of the grid, because they are naturally restricted to specific structures of the index set $\mathcal{I}$. This limitation, together with the difficulty of prescribing appropriate vectors $\mathbf{a}$, has motivated the introduction of more flexible adaptive strategies for the construction of sparse cubatures.

3.4.2.2 General Adaptive Sparse Grid Method

An adaptive sparse grid construction was proposed in [84]. The adaptation is based on a sequential construction of the multi-index set, which is progressively enriched starting from $\mathcal{I} = \{\mathbf{1} = (1, \dots, 1)\}$. To detail the construction, the notion of admissible multi-index set has to be introduced first. An index set $\mathcal{I}$ is admissible if for each multi-index $\mathbf{l} \in \mathcal{I}$ we have

$$\mathbf{l} - \mathbf{e}_j \in \mathcal{I}, \quad \text{for } 1 \leq j \leq \mathrm{N},\ l_j > 1, \tag{3.35}$$

where $\mathbf{e}_j$ is j-th unit vector ($(\mathbf{e}_j)_i = \delta_{ij}$). In words, this condition means that any element of $\mathcal{I}$ has a predecessor in all directions. It ensures the validity of

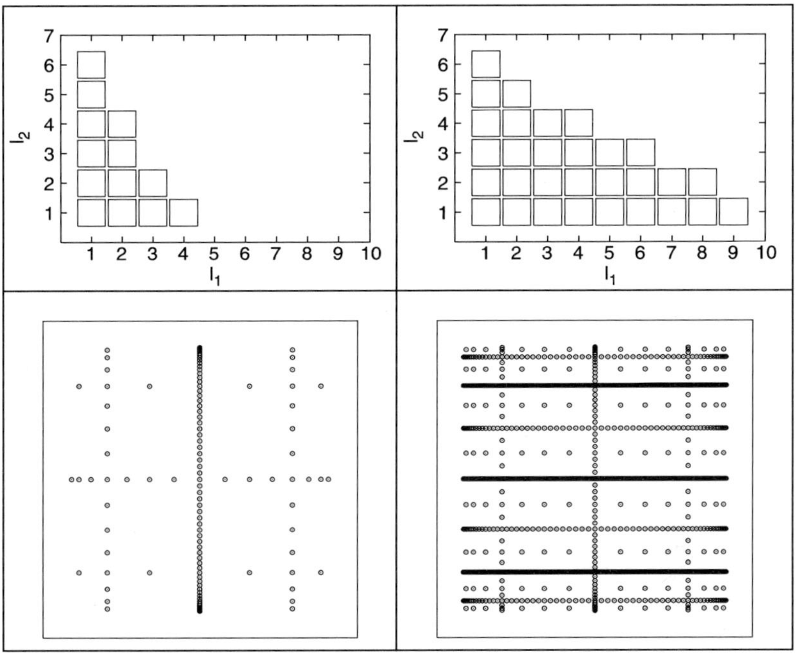

Fig. 3.8 Example of two-dimensional cubatures constructed with the dimension-adaptive strategy using $a_1 = 1.5$ (*left*) and $a_1 = 0.6$ (*right*) and $a_2 = 1 + (1 - a_1)/l$, $l = 6$. Plotted are the respective multi-index sets (*top row*, see (3.34)) and the corresponding sparse grids (*bottom row*, Fejèr nested nodes)

the telescope sum expansion in terms of difference rules, see (3.31), when defining a sparse grid cubature. Examples of admissible and non-admissible multi-index sets are shown in Fig. 3.9 for the case of $\mathrm{N} = 2$. The adaptive sparse grid method requires that the progressive enrichment of the multi-index set maintains the admissibility. In addition, the enrichment should reduce the integration error in the most efficient way. To this end, an indicator is used to determine which multi-index should be added to $\mathcal{I}$. Following [84], we denote $g_{\mathbf{l}}$ the error indicator associated to a given multi-index $\mathbf{l}$. The indicator $g_{\mathbf{l}}$ combines information from the associated difference term, $\Delta_{\mathbf{l}}^{(\mathrm{N})} f$, with the computational complexity involved in its estimation. The latter is measured by $n_{\mathbf{l}}$ defined as the number of cubature nodes in the evaluation of $\Delta_{\mathbf{l}}^{(\mathrm{N})} f$. A convenient form for $g_{\mathbf{l}}$ was proposed in [84]:

$$g_{\mathbf{l}} \equiv \max \left\{ \alpha \left| \frac{\Delta_{\mathbf{l}}^{(\mathrm{N})} f}{\Delta_{\mathbf{1}}^{(\mathrm{N})} f} \right|, (1 - \alpha) \frac{n_{\mathbf{1}}}{n_{\mathbf{l}}} \right\}, \tag{3.36}$$

where $0 \leq \alpha \leq 1$ weights the difference contribution and computational cost.

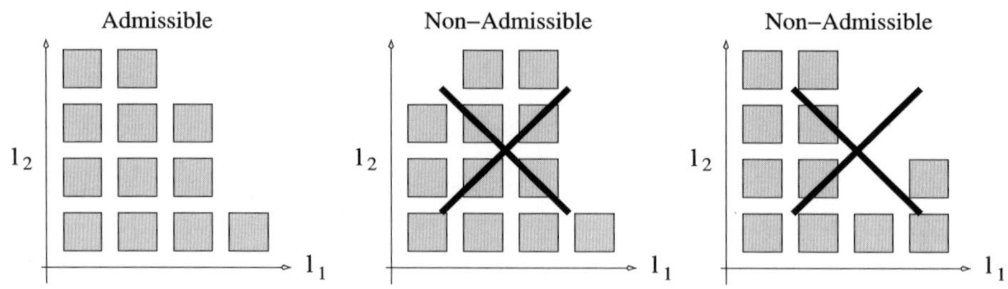

Fig. 3.9 Examples of admissible and non-admissible multi-index sets in two dimensions

With this indicator, the enrichment of $\mathcal{I}$ can proceed. Assume that at a given step of the adaptation we have constructed an admissible multi-index set $\mathcal{I}$, and define for $\mathbf{l} \in \mathcal{I}$ its forward neighborhood $\mathcal{F}_{\mathbf{l}}$ as

$$\mathcal{F}_{\mathbf{l}} \equiv \{\mathbf{l} + \mathbf{e}_j, 1 \le j \le \mathrm{N}\}. \tag{3.37}$$

The next multi-index to be included in $\mathcal{I}$, denoted $\mathbf{k}$, is to be selected such that $g_{\mathbf{k}}$ is the largest and

$$\begin{aligned} &\mathbf{k} \notin \mathcal{I}, && (a)\\ &\mathbf{k} \in \bigcup_{\mathbf{l}\in\mathcal{I}} \mathcal{F}_{\mathbf{l}}, && (b)\\ &\mathcal{I} \cup \{\mathbf{k}\} \quad \text{is admissible.} && (c) \end{aligned}$$

These conditions state that $\mathbf{k}$ should be a new multi-index (a), taken in the forward neighborhood of $\mathcal{I}$ (b), whose inclusion leaves $\mathcal{I}$ admissible (c).

The procedure is most efficiently implemented considering two subsets $\mathcal{O}$ and $\mathcal{A}$. The set $\mathcal{O}$ contains the "old" multi-indexes which need not be tested anymore, while $\mathcal{A}$ contains those who are candidates for inclusion in $\mathcal{I}$. The set $\mathcal{O}$ is initialized to $\{\mathbf{1}\}$ and $\mathcal{A}$ to $\mathcal{F}_{\mathbf{1}}$. We then select the multi-index in $\mathcal{A}$, say $\mathbf{l}$, having the highest indicator. The multi-index $\mathbf{l}$ is removed form $\mathcal{A}$ and added to $\mathcal{O}$; $\mathcal{A}$ is then completed by the multi-indexes in the forward neighborhood $\mathcal{F}_{\mathbf{l}}$ that maintain $\mathcal{I} = \mathcal{O} \cup \mathcal{A}$ admissible, and the error indicators $g_{\mathbf{k}\in\mathcal{F}_{\mathbf{l}}}$ of the new multi-indexes in $\mathcal{A}$ are computed. The procedure is repeated as long as the global error indicator η, defined as

$$\eta \equiv \sum_{\mathbf{l}\in\mathcal{A}} g_{\mathbf{l}}, \tag{3.38}$$

is greater than a prescribed error tolerance ϵ. The algorithm can be summarized as follows:

Initialization:

set $\mathcal{O} = \{\mathbf{1}\}$

set $\mathcal{A} = \mathcal{F}_{\mathbf{1}}$
set $r = \sum_{\mathbf{l} \in \mathcal{A} \cup \mathcal{O}} \Delta_{\mathbf{l}}^{(\mathrm{N})} f$
set $\eta = \sum_{\mathbf{l} \in \mathcal{A}} g_{\mathbf{l}}$

while $(\eta > \epsilon)$ **do**

select $\mathbf{l} \in \mathcal{A}$ with largest $g_{\mathbf{l}}$
$\mathcal{O} \leftarrow \mathcal{O} \cup \{\mathbf{l}\}$
$\mathcal{A} \leftarrow \mathcal{A}/\{\mathbf{l}\}$
$\eta \leftarrow \eta - g_{\mathbf{l}}$
For $\mathbf{k} \in \mathcal{F}_{\mathbf{l}}$ such that $(\mathbf{k} - \mathbf{e}_j) \in \mathcal{O}$ for $j = 1, \ldots, \mathrm{N}$ **do**
$\mathcal{A} \leftarrow \mathcal{A} \cup \{\mathbf{k}\}$
$r \leftarrow r + \Delta_{\mathbf{k}}^{(\mathrm{N})} f$
$\eta \leftarrow \eta + g_{\mathbf{k}}$
end for

end while
return r

The stopping criterion may also involve a limit in the computational effort (namely the number of model evaluations). Note that in the previous algorithms, the result r involves all multi-indices in $\mathcal{I} = \mathcal{A} \cup \mathcal{O}$: all the difference terms evaluated are exploited. In Fig. 3.10, we provide an example of the evolution of the multi-index sets during an adaptive procedure for $\mathrm{N} = 2$. The corresponding sparse grids are also shown.

3.5 Least Squares Fit

Least squares estimation is a popular tool in statistics where one seeks to determine model parameters from a set of noisy measurements or observations. Here, the idea is to consider the model output s as a variable dependent on the model parameters $\boldsymbol{\xi}$. The true model is replaced by its surrogate $\hat{s}(\boldsymbol{\xi})$, consisting in its stochastic spectral expansion,

$$\hat{s}(\boldsymbol{\xi}) = \sum_{k=0}^{\mathrm{P}} \hat{s}_k \Psi_k(\boldsymbol{\xi}), \tag{3.39}$$

and the $(\mathrm{P}+1)$ expansion coefficients are to be determined on the basis of a set of measurements (computations or observations). Consistently with the previous notation, we denote $(\boldsymbol{\xi}^{(i)}, s^{(i)})$, $i = 1, \ldots, m$, the set of observations, where as previously $s^{(i)} \equiv s(\boldsymbol{\xi}^{(i)})$ is the model output for the parameters given by $\boldsymbol{\xi}^{(i)}$.

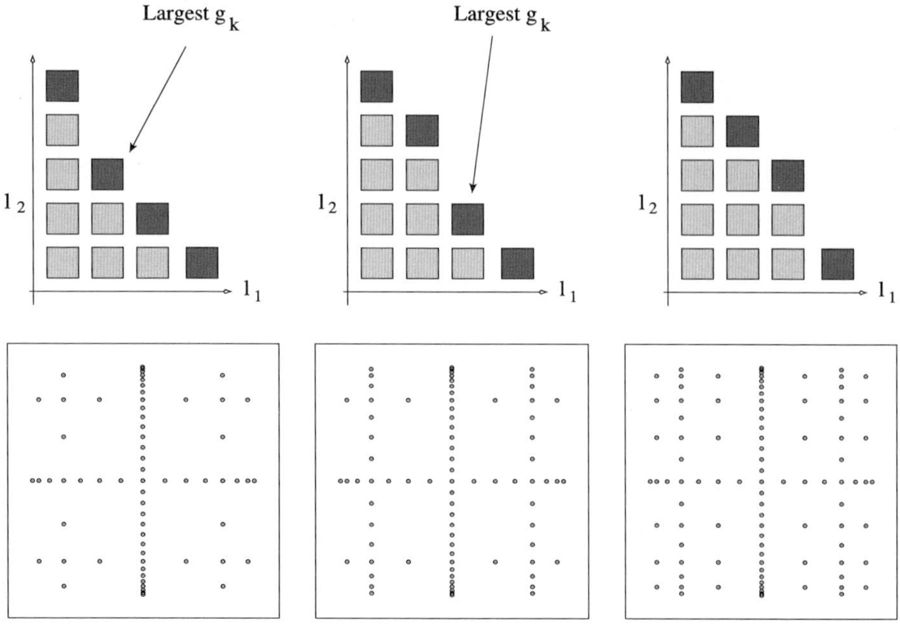

Fig. 3.10 Illustration of the adaptive sparse grid procedure for N = 2. The plots on the *top row* show the evolution of the multi-index set $\mathcal{I}$, distinguishing the sets of old multi-indexes $\mathcal{O}$ (*light gray squares*) and active multi-indexes $\mathcal{A}$ (*dark gray squares*). The corresponding sparse grids are plotted in the *bottom row*

3.5.1 Least Squares Minimization Problem

The expansion coefficients in (3.39) can be defined as the solution of an optimization problem for the sum of the squares of the residuals R,

$$R(\hat{s}) \equiv \sum_{i=1}^{m} \left(r^{(i)}\right)^2 = \sum_{i=1}^{m} \left(s^{(i)} - \hat{s}\left(\boldsymbol{\xi}^{(i)}\right)\right)^2, \tag{3.40}$$

where the residuals $r^{(i)}$ are simply the distances between the observations and the predictions of the surrogate model:

$$r^{(i)} \equiv s^{(i)} - \hat{s}\left(\boldsymbol{\xi}^{(i)}\right). \tag{3.41}$$

The coefficients $\hat{s}_k$ are then sought to minimize R.

Expressing the stationarity of R with regard to the expansion coefficients $\hat{s}_k$, *i.e.* $\partial R / \partial \hat{s}_k$, we end up with a set of $(\mathrm{P}+1)$ linear equations:

$$\frac{\partial R}{\partial \hat{s}_k} = -2 \sum_{i=1}^{m} \left(s^{(i)} - \hat{s}\left(\boldsymbol{\xi}^{(i)}\right)\right) \Psi_k\left(\boldsymbol{\xi}^{(i)}\right) = 0, \quad 0 \le k \le \mathrm{P}. \tag{3.42}$$

This system can be conveniently rewritten in terms of the matrix Z,

$$Z \equiv \begin{pmatrix} Z_{10} & \dots & Z_{1\mathrm{P}} \\ \vdots & \ddots & \vdots \\ Z_{m0} & \dots & Z_{m\mathrm{P}} \end{pmatrix}, \qquad Z_{ij} = \Psi_j\left(\boldsymbol{\xi}^{(i)}\right). \tag{3.43}$$

Denoting $\hat{\mathbf{s}} \equiv (\hat{s}_0 \ \dots \ \hat{s}_{\mathrm{P}})^t$ the vector of unknown expansion coefficients, $\mathbf{s} \equiv (s^{(1)} \ \dots \ s^{(m)})$ the vector of model output, the solution of the minimization problem satisfies

$$\left(Z^t Z\right) \hat{\mathbf{s}} = Z^t \mathbf{s}. \tag{3.44}$$

Provided that the information matrix $(Z^t Z)$ is non-singular, (3.44) has for solution $\hat{\mathbf{s}} = (Z^t Z)^{-1} Z^t \mathbf{s}$

The matrix Z plays a central role in the least squares definition of the expansion coefficients. Indeed, it defines an orthogonal projection operator Π from $\mathbb{R}^m$ to the subspace spanned by the $(\mathrm{P}+1)$ columns of Z. To show this, let $\Pi = Z(Z^t Z)^{-1} Z^t$; clearly, Π is symmetric, idempotent ($\Pi\Pi = \Pi$), and columns of Z are Π-stable ($\Pi Z = Z$). Therefore, the solution of (3.44) lives in the subspace spanned by the $(\mathrm{P}+1)$ columns of Z. As direct consequence, in general the model residual,

$$r(\boldsymbol{\xi}) \equiv s(\boldsymbol{\xi}) - \sum_{k=0}^{\mathrm{P}} \hat{s}_k \Psi_k(\boldsymbol{\xi}), \tag{3.45}$$

will be orthogonal to $\mathcal{S}^{\mathrm{P}}$ only in the limit of $m \to \infty$ and appropriate selection of the minimization points. Thus, the least squares fit is an orthogonal projection on $\mathcal{S}^{\mathrm{P}}$ only for particular sets of minimization points.

3.5.2 Selection of the Minimization Points

The minimization problem is well posed if $(Z^t Z)$ has full rank, a necessary (and sufficient) condition to make the problem solvable. Besides this necessary condition, the discussion on the projection properties of Π raises the question of the selection of the minimization points $\boldsymbol{\xi}^{(i)}$.

A reasonable approach consists in selecting points at random, following the probability law of $\boldsymbol{\xi}$. This point of view leads to Monte Carlo sampling strategies, as in the NISP method of Sect. 3.2.1. In fact, adopting this sampling strategy for the minimization points results in methods that are asymptotically equivalent to NISP: in the limit $m \to \infty$, Π convergences to the projector on $\mathcal{S}^{\mathrm{P}}$. Indeed, denoting $\overline{s}_k$ the MC-NISP definition of the expansion coefficients of $s(\boldsymbol{\xi})$,

$$\overline{s}_k \left\langle \Psi_k^2 \right\rangle = \lim_{m\to\infty} \frac{1}{m} \sum_{i=1}^{m} s^{(i)} \Psi_k(\boldsymbol{\xi}^{(i)}), \tag{3.46}$$

it can be easily shown that they solve the limiting least squares problem

$$\begin{aligned}
&\lim_{m\to\infty}\sum_{i=1}^{m}\left(s^{(i)}-\sum_{l=0}^{\mathrm{P}}\overline{s}_l\Psi_l\left(\boldsymbol{\xi}^{(i)}\right)\right)\Psi_k\left(\boldsymbol{\xi}^{(i)}\right)\\
&=\lim_{m\to\infty}\left(\sum_{i=1}^{m}s^{(i)}\Psi_k\left(\boldsymbol{\xi}^{(i)}\right)-\sum_{i=1}^{m}\sum_{l=0}^{\mathrm{P}}\sum_{j=1}^{m}\frac{s^{(j)}\Psi_l(\boldsymbol{\xi}^{(j)})}{m\langle\Psi_l^2\rangle}\Psi_l\left(\boldsymbol{\xi}^{(i)}\right)\Psi_k\left(\boldsymbol{\xi}^{(i)}\right)\right)\\
&=\lim_{m\to\infty}\left(\sum_{i=1}^{m}s^{(i)}\Psi_k\left(\boldsymbol{\xi}^{(i)}\right)-\sum_{j=1}^{m}s^{(j)}\sum_{l=0}^{\mathrm{P}}\Psi_l\left(\boldsymbol{\xi}^{(j)}\right)\sum_{i=1}^{m}\frac{\Psi_l(\boldsymbol{\xi}^{(i)})\Psi_k(\boldsymbol{\xi}^{(i)})}{m\langle\Psi_l^2\rangle}\right)\\
&=\lim_{m\to\infty}\left(\sum_{i=1}^{m}s^{(i)}\Psi_k\left(\boldsymbol{\xi}^{(i)}\right)-\sum_{j=1}^{m}s^{(j)}\Psi_k\left(\boldsymbol{\xi}^{(j)}\right)\right)=0.
\end{aligned}\tag{3.47}$$

In the derivation above, we have used the assumed orthogonality of the stochastic basis, specifically in

$$\lim_{m\to\infty}\frac{1}{m}\sum_{i=1}^{m}\Psi_k\left(\boldsymbol{\xi}^{(i)}\right)\Psi_l\left(\boldsymbol{\xi}^{(i)}\right)=\delta_{kl}\left\langle\Psi_l^2\right\rangle.\tag{3.48}$$

However, with finite m the two methods generally yield different expansions, except if the minimization points satisfy the empirical orthogonality condition

$$\sum_{i=1}^{m}\Psi_l\left(\boldsymbol{\xi}^{(i)}\right)\Psi_k\left(\boldsymbol{\xi}^{(i)}\right)=Z_l\cdot Z_k=m\left\langle\Psi_l^2\right\rangle\delta_{kl},\quad 0\le l,k\le\mathrm{P}\tag{3.49}$$

where Z_k is the k-th column of the matrix Z.

Equation (3.49) highlights the crucial role of Z^tZ in the characteristics of the least squares solution. In fact, design of experiment (DOE) or optimal design methods are research areas focusing on the construction of sample sets to provide optimal properties to least squares problems (see for instance [193]). Interestingly enough, the space filling methods of DOE share similarities with the QMC sampling discussed in Sect. 3.2.2. Advanced techniques in DOE are based on the optimization of the spectral properties of the information matrix Z^tZ, or directly of the projection operator Π [99, 193]. Classical optimal design methods and their respective objectives are listed in Table 3.1.

The analysis of these techniques and the algorithms that actually determine optimal set of points go far beyond the scope of the present monograph and will not be discussed here. We simply mention the existence of software tools[2] to construct optimal sample sets based on various optimality criteria. One should be aware that the

[2] See for instance http://www.research.att.com/~njas/gosset/index.html.

Table 3.1 Classical approaches for optimal design

Name	Objective	Object
A-optimality	minimize the trace	$(Z^t Z)^{-1}$
D-optimality	maximize determinant	$Z^t Z$
E-optimality	maximize lower singular value	$Z^t Z$
G-optimality	minimize largest diagonal term	Π

statistical theories underlying these sample set constructions assume model observations with random noise, a situation which is quite different from the framework considered here.

3.5.3 Weighted Least Squares Problem

In [20, 23], algorithms were introduced based on *ad hoc* selection of minimization points, consisting of tensored Gauss points, or sparse cubature nodes. Such approaches could be motivated by the use of weighted sums of the squares of local residuals. Consider for instance the weighted sum of squares of residuals:

$$R_w(\hat{s}) \equiv \sum_{i=1}^{m} w^{(i)} \left(s^{(i)} - \hat{s}\left(\xi^{(i)}\right)\right)^2, \quad w^{(i)} > 0. \tag{3.50}$$

The solution of this weighted least squares problem is given by

$$\sum_{l=0}^{\mathrm{P}} \hat{s}_k \left(\sum_{i=1}^{m} w^{(i)} \Psi_l \left(\xi^{(i)}\right) \Psi_k \left(\xi^{(i)}\right)\right) = \sum_{i=1}^{m} w^{(i)} s^{(i)} \Psi_k \left(\xi^{(i)}\right), \tag{3.51}$$

or in matrix form

$$\left(Z^t W Z\right) \hat{\mathbf{s}} = Z^t W \mathbf{s}, \quad W = \begin{pmatrix} w^{(1)} & 0 & \dots & 0 \\ 0 & \ddots & \ddots & \vdots \\ \vdots & \ddots & \ddots & 0 \\ 0 & \dots & 0 & w^{(m)} \end{pmatrix}. \tag{3.52}$$

For weights satisfying

$$\sum_{i=1}^{m} w^{(i)} \Psi_l \left(\xi^{(i)}\right) \Psi_k \left(\xi^{(i)}\right) = \left\langle \Psi_l^2 \right\rangle \delta_{kl}, \quad 0 \le k, l \le \mathrm{P}, \tag{3.53}$$

(3.51) reduces to

$$\hat{s}_k \left\langle \Psi_k^2 \right\rangle = \sum_{i=1}^{m} w^{(i)} s^{(i)} \Psi_k \left(\xi^{(i)}\right). \tag{3.54}$$

Equation (3.54) is formally equivalent to a cubature approximation of the NISP definition of $\hat{s}_k$; see (3.6). But then (3.53) shows that for any two random variables of $\mathcal{S}^{\mathrm{P}}$ we have

$$\int_{\Xi} u(\boldsymbol{\xi})v(\boldsymbol{\xi})p_{\boldsymbol{\xi}}(\boldsymbol{\xi})\,\mathrm{d}\boldsymbol{\xi} = \sum_{i=1}^{m} u(\boldsymbol{\xi}^{(i)})v(\boldsymbol{\xi}^{(i)})w^{(i)}, \quad \forall u, v \in \mathcal{S}^{\mathrm{P}}, \tag{3.55}$$

so the set of minimization points and weights can be thought as a cubature rule.

We can conclude that when the unweighted (resp. weighted) least squares problem uses points (resp. points and weights) satisfying condition (3.49) (resp. (3.53)) it reduces to a NISP method. In fact, the orthogonality of the stochastic basis functions is not naturally taken into account in the non-intrusive least squares formulations whereas it is implicit in NISP. This is not a surprise because any polynomial basis spanning $\mathcal{S}^{\mathrm{P}}$ can actually be used in the expansion of the least squares solution: the orthogonality of the basis is not exploited and an additional condition has to be introduced to recover the L_2-projection definition of the solution.

If one expects that the least squares solution $\hat{s}$ should really correspond to the projection of s on $\mathcal{S}^{\mathrm{P}}$, in the limit of $m \to \infty$, then it should be clear that a random sampling or a cubature rule of sufficient accuracy is needed to generate the minimization points (and eventually the least squares weights). Therefore, the NISP and least squares approaches are of comparable complexities and the apparent flexibility of the least squares problem may be misleading: one cannot consider arbitrary sets of minimization points. It is also interesting to note that the discrete orthogonality conditions in (3.49) and (3.53) lead to diagonal matrices $Z^t Z$ and $Z^t W Z$. In this case, the individual modes are decoupled, and this is related to the NISP definition of the projection coefficients (3.6), in which the orthogonality of the Ψ_k is exploited.

3.6 Collocation Methods

3.6.1 Approximation Problem

Unlike non-intrusive methods discussed above, collocation methods do not aim at determining the projection of the model output $s(\boldsymbol{\xi})$ on a pre-defined stochastic subspace, but instead rely on interpolation. This brings an essential difference with previous methods: the approximation *and* the approximation space are implicitly prescribed by the selected points $\boldsymbol{\xi}^{(i)}$. As the number of points increases, the space over which the solution is sought becomes accordingly larger.

In collocation methods, one specifically seeks an approximation $\tilde{s}(\boldsymbol{\xi})$ of $s(\boldsymbol{\xi})$ such that

$$\tilde{s}\left(\boldsymbol{\xi}^{(i)}\right) = s\left(\boldsymbol{\xi}^{(i)}\right), \quad 1 \le i \le m, \tag{3.56}$$

for given set of interpolation points $\boldsymbol{\xi}^{(i)}$. The approximation is exact at the m collocation points. Because this requirement is imposed on a finite set of points the method is said to be collocative.

Assuming that the collocation points are distinct, we can construct an approximation in a vector space of dimension at least equal to m, to satisfy in general the m constraints in (3.56). In the following, we shall consider approximation spaces, denoted $\tilde{\mathcal{S}}_m$, with dimension m and bases $\{\Phi_i(\boldsymbol{\xi})\}_{i=1}^{i=m}$. The subscript m on $\tilde{\mathcal{S}}_m$ stresses the dependence of the approximation space on the number of collocation points. The approximation on $\tilde{\mathcal{S}}_m$ is given in terms of the expansion

$$\tilde{s}(\boldsymbol{\xi}) = \sum_{i=1}^{m} \tilde{s}_i \Phi_i(\boldsymbol{\xi}), \tag{3.57}$$

where the expansion coefficients $\tilde{s}_i$ are to be determined from the constraint in (3.56). A straightforward way to enforce the constraints is to consider vector spaces spanned by bases functions having the properties

$$\Phi_i(\boldsymbol{\xi}^j) = \begin{cases} 1, & \text{if } i = j, \\ 0, & \text{otherwise}, \end{cases} \qquad 1 \le i, j \le m. \tag{3.58}$$

Then, (3.56) leads to

$$\tilde{s}_i = s^{(i)} = s\left(\boldsymbol{\xi}^{(i)}\right), \qquad 1 \le i \le m. \tag{3.59}$$

Note that the basis will be orthogonal only for particular sets of collocation points, and that (3.57) is generally not an orthogonal expansion with respect to the inner product of $L_2(\Xi, p_{\boldsymbol{\xi}})$.

In the following, we shall consider constructions of the form given by (3.57)–(3.58). It then remains to define and construct the basis functions Φ_i, called hereafter the interpolation functions associated to the set of collocation nodes. Different options are available, the most popular choice corresponding to the polynomial interpolation that we discuss below. We shall further restrict ourselves to global interpolation methods, where the support of the functions Φ_i is the entire domain Ξ, *i.e.* we consider spectral interpolation methods. Other polynomial interpolation methods, such as spline approximations, will not be discussed here, even though they possess interesting features.

3.6.2 Polynomial Interpolation

One-dimensional interpolation: Consider a function $s(\xi)$ and a set of m_1 distinct points $\xi^{(i)}$. We seek a polynomial of $\mathbb{P}_{m_1}$ that interpolates $s(\xi)$, where $\mathbb{P}_{m_1}$ is the set of polynomial with degree less than m_1. We denote this polynomial by $q(\xi)$. It satisfies:

$$q(\xi^{(i)}) = s\left(\xi^{(i)}\right) = s^{(i)}, \qquad 1 \le i \le m_1. \tag{3.60}$$

The determination of $q(\xi)$ is a classical problem in approximation theory. It can be shown that $q(\xi)$ always exists and is unique. A stable and robust method to construct $q(\xi)$ is to express it using the Lagrange polynomials associated to the set of interpolation points $\xi^{(i)}$. The i-th one-dimensional Lagrange polynomial, $\mathcal{L}_i^{(m_1)}$ is defined as

$$\mathcal{L}_i^{(m_1)}(\xi) = \prod_{\substack{j=1 \\ j\neq i}}^{m} \frac{\xi - \xi_j}{\xi_i - \xi_j}. \tag{3.61}$$

It is easily verified that $\mathcal{L}_i^{(m_1)} \in \mathbb{P}_{m_1}$, and

$$\mathcal{L}_i^{(m_1)}(\xi_j) = \begin{cases} 0, & i \neq j, \\ 1, & i = j, \end{cases} \qquad 1 \leq i, j, \leq m_1. \tag{3.62}$$

Therefore, the polynomial q can be expressed as:

$$q(\xi) = \sum_{i=1}^{m} s^{(i)} \mathcal{L}_i^{(m_1)}(\xi) \in \mathbb{P}_{m_1}. \tag{3.63}$$

In the following, we denote

$$q = \mathcal{C}_{m_1}^{(1)} s, \tag{3.64}$$

the 1D polynomial interpolation of s, based on m_1 interpolation points. Many results can be found in the literature for the convergence properties of the polynomial interpolation (see for instance [2, 26]). We only mention here the importance of the interpolation points used on the quality of the approximation and its convergence properties as m increases. In particular, the use of equidistant points may not be suitable for global interpolations, as it can result in highly oscillating approximations. Instead, Gauss or Chebychev points offer better properties and are generally preferred whenever possible. Chebychev points also present the advantage of being nested sets, a particularly attractive feature for adaptive multi-dimensional interpolation.

Multidimensional interpolation: Extension of 1D interpolation formulas to the N-dimensional case is immediate using tensored grids, in a similar way as for integration. In tensored constructions, both the grid of interpolation points and interpolation polynomials are products of the associated 1D objects. The N-dimensional interpolation formula can be expressed as:

$$\tilde{s}(\xi_1, \ldots, \xi_N) = \sum_{i_1=1}^{m_1} \cdots \sum_{i_N=1}^{m_N} s\left(\xi_1^{(i_1)}, \ldots, \xi_N^{(i_N)}\right) \mathcal{L}_{i_1}^{(m_1)}(\xi_1) \cdots \mathcal{L}_{i_N}^{(m_N)}(\xi_N), \tag{3.65}$$

or in the tensor product form

$$\tilde{s}(\boldsymbol{\xi}) = \mathcal{C}^{(N)} s = \left(\mathcal{C}_{m_1}^{(1)} \otimes \cdots \otimes \mathcal{C}_{m_N}^{(1)}\right) s. \tag{3.66}$$

In this expression, different numbers of interpolation points have been considered along the N-dimensions. This results in a regular grid pattern with a total number of points $m = m_1 \times \cdots \times m_N$. Again, appropriate indexation can be used to reduce the telescopic sum as in (3.57). Further, the interpolation space $\tilde{\mathcal{S}}_m$ is spanned by functions Φ_i which are product of the 1D Lagrange polynomials: this structure highlights the dependence of the approximation space with regard to the constructed grid, and the importance of the underlying 1D interpolation formula.

3.6.3 Sparse Collocation Method

We see that the complexity of the collocation methods, based on tensored 1D polynomial interpolations, scales with the number of interpolation points. Formulas such as (3.66) quickly become expensive, and sparse methods can once again be used to limit the increase of m with the number of stochastic dimensions, N. In fact, the sparse grid techniques, such as the Smolyak algorithm outlined in Sect. 3.4, can be readily reused. For instance, considering a set of 1D interpolation formulas, parametrized by an index level l, one can reuse the general form of (3.31), replacing the 1D quadrature rules $Q^{(1)}$ by the 1D interpolation formulas $\mathcal{C}^{(1)}$ and defining consistently difference formulas between interpolations at consecutive levels. A large family of sparse interpolation methods can thus be obtained, including adaptive interpolations schemes in the spirit of the techniques discussed in Sect. 3.4.2.

Examples of sparse grid collocation methods for stochastic problems can be found in [7, 73, 145, 239, 244]; adaptive techniques were recently considered in [72, 77, 139, 166].

3.7 Closing Remarks

For NISP methods, it is natural to question whether one mode of sampling is preferable over the other depending on the nature of problem or of the model. We have already seen that the stochastic sampling methods, by virtue of having a convergence rate that is independent of the number N of stochastic dimensions, appear to be more attractive when N is large. Conversely, for fixed level of accuracy, quadrature and cubature approaches require a number of realizations that increases substantially with N, which may limit their application to situations where the stochastic dimension is small or moderate. Furthermore, by the nature of their construction, quadrature and cubature formulas aim at achieving the highest order of precision, with a minimal number of points in higher dimensions. But the solution that one seeks to project is not necessarily polynomial in $\boldsymbol{\xi}$, so that straightforward application of these formulas may not always prove to be efficient in practice.

The comparison of sampling and deterministic strategies is even more difficult after the emergence of sparse grid methods, which have greatly improved the computational efficiency of NISP, and consequently increased its domain of applicability. It is evident that current researches on the development of adaptive sparse methods will sustain this trend in the near future.

In addition to rate of convergence and computational complexity considerations, other aspects of stochastic solutions schemes may also need to be accounted for. In particular, the robustness of the deterministic code, used to compute the individual realizations, may constitute an important factor. In fact, when the random inputs are not bounded, quadrature and cubature formulas involve realizations corresponding to extreme values of these inputs. Specifically, when the variance of the random inputs is large, and the stability or performance of the code is sensitive to extreme values of the input, it may prove difficult to evaluate the solution for all the quadrature or cubature nodes. In contrast, when a non-biased stochastic sampling is used, extreme values of the input have a very low probability of being selected; it is often the practice in stochastic sampling strategies to reject extreme values leading to model resolution issues, as long as the rejection rate remains small.

Another relevant aspect pertaining to non-intrusive spectral projection techniques concerns the truncation order of the stochastic expansion. It is clear that the choice of the order of quadrature and cubature formulas, and hence the number of points N_Q, is strongly dependent on the order of truncation of the expansion. In practice, the dependence of the stochastic solution with regard to the uncertain inputs is rarely polynomial, and hence one should generally expect a non-vanishing projection error. For many problems, this projection error is small, and decreases rapidly as the order of the expansion (and accordingly the number of node points) increases. For some problems, however, the convergence may be very slow and projection errors may become critical. In such situations, a stochastic sampling may be preferable. In addition, all the integration techniques described in this chapter, rely on a global refinement of the set of integration points, with possibly a different treatment along the integration directions in adaptive sparse methods. This may be sub-optimal, particularly in situations where one would rather concentrate the computational resources in the areas of the parameter space where the integrand requires more efforts than in other places. This local adaptivity of the cubature formulas, in contrast to the dimension adaptivity, remains to be explored although numerical procedures are already available (see for instance [38]). Strategies of adaptive subdivision of the integration domain appear to be well suited when considering non-global basis functions expansions, as discussed later in Chap. 8. The same remark applies to collocation methods, for which the solution space is refined as nodes are locally added.

In fact, a general observation regarding non-intrusive methods and related algorithms, even for the more advanced ones, is the lack of theoretically well grounded error estimators and theoretical convergence rate for the approximation of general (nonlinear) models. Consequently, adaptive strategies are still based on *ad-hoc* or heuristic rules to guide the adaptive procedure or to decide that the results are sufficiently converged. Note that this observation applies also to Galerkin projection methods addressed in the following chapters, even though some attempts have been made toward rigorous (*a posteriori*) error estimation, as in Sect. 9.3.

Chapter 4
Galerkin Methods

Unlike non-intrusive approaches which rely on individual realizations to determine the stochastic model response to random inputs, Galerkin methods are based on a weighted residual formalism to form systems of governing equations for the solution's PC coefficients. This chapter is devoted to the formulation of Galerkin methods of UQ using PC expansions.

In Sect. 4.1, we start with an abstract representation of the deterministic problem, and its stochastic extension. We restrict our attention to the case of random model data, and introduce appropriate representations of both the random data and the model solution. We finally outline an abstract representation of the weak form of the stochastic problem.

Based on this restricted framework, spectral stochastic representation is discussed in Sect. 4.2. Generic basis functions expansions are introduced, and are formally applied to the representation of the random data and model solution. Using these expansions, the Galerkin formalism is outlined in Sect. 4.3. Stochastic residuals are in particular defined, and projections onto orthogonal bases are used to derive governing equations for the unknown PC coefficients.

The structure of the stochastic problem is further analyzed in Sect. 4.4 for the special case of linear problems. The relationship to the underlying deterministic problem is explored, and solution methods are briefly highlighted.

In Sect. 4.5 we address the issue of model nonlinearities. We avoid the discussion of solution methods and concentrate instead on approaches suitable for the construction of stochastic model equations. In particular, methods are outlined for estimating various nonlinear transformations of PC expansions.

We conclude in Sect. 4.6 with brief remarks concerning the construction and implementation of Galerkin methods, and potential cost/benefit consideration vis-à-vis non-intrusive methods.

O.P. Le Maître, O.M. Knio, *Spectral Methods for Uncertainty Quantification*,
Scientific Computation,
DOI 10.1007/978-90-481-3520-2_4, © Springer Science+Business Media B.V. 2010

4.1 Stochastic Problem Formulation

4.1.1 Model Equations and Notations

4.1.1.1 Deterministic Problem

We consider a mathematical model $\mathcal{M}$ of a physical system. Among the class of all the physical systems that the mathematical model can generate, the actual system to be modeled is selected by providing additional information. We globally designate all the information required to specify the system among the class spanned by the mathematical model using the loose term of set of data or simply data. Depending on the context and the type of system studied, the information needed to fully characterize the system varies. Classically, the set of data prescribes the geometry, some physical and modeling constants, boundary and initial conditions, forcing terms and any other relevant characteristic needed to completely specify the system and to make the mathematical model solvable.

Then, denoting u the solution of the mathematical model, and d the set of data, one has to solve for u a well defined mathematical problem. To denote the relation between the model solution u and the data d, we write

$$\mathcal{M}(u; d) = 0, \tag{4.1}$$

to abstractly represent the mathematical problem satisfied by u given d. This notation is purely formal, and models of various types may thus be represented, including an ordinary differential equation (ODE), systems of ODEs, partial differential equations, integral equations, algebraic equations, or more generally mixed models. The actual type of equation is not important at the conceptual level considered here, though it will obviously affect the form of the resulting spectral problem and so dictate the numerical strategy to be used for its resolution. In addition, the notation $\mathcal{M}$ involves the full set of equations satisfied by the solution u, including, if relevant, boundary conditions, initial conditions, constitutive equations, source terms and any other constraint satisfied by the solution. In a similar way, the abstract notation u for the solution stands for *all* the variables (scalars, vectors, fields) involved in the model formulation, for instance the velocity, pressure, temperature and density fields in a fluid flow problem. It is further assumed that the mathematical problem is well-posed.

4.1.1.2 Stochastic Problem

To propagate and quantify the impact of uncertainty in d on the solution, u, we introduce an abstract probability space (Θ, Σ, P), where Θ is the set of outcomes, Σ the σ-algebra of events and P the associated probability measure. We write $d(\theta)$ to stress the dependence of the data on the outcome $\theta \in \Theta$. In most situations, not all the information prescribing the system is random but only a subset of the data. For instance, the geometry of system may be perfectly known and only a subset of the model parameters may be subject to uncertainty. Still, to avoid the need of making a distinction between the deterministic and random subsets of the data, we simply write $d(\theta)$.

Clearly, the data being random, the model solution u is also random. Our objective is therefore to determine $u(\theta)$ satisfying

$$\mathcal{M}(u(\theta); d(\theta)) = 0. \tag{4.2}$$

Equation (4.2) for the random solution $u(\theta)$ is referred to as the stochastic problem. It makes explicit the functional dependence between the model solution and the data. It is also remarked that the random data have to be such that the stochastic problem is well posed. Specifically, it will be assumed that the probability law of the data are such that the mathematical problem $\mathcal{M}(\cdot; d(\theta))$ has almost surely a unique solution in a suitable Hilbert space. One should also note that the equality in (4.2) should be interpreted in a probabilistic sense. The Galerkin approach adopted below corresponds to one specific interpretation, though many approaches may be formulated, based for instance on concepts of mean square convergence, convergence in the sense of distributions, or convergence in measure, see Appendix A.

4.1.2 Functional Spaces

To solve problem (4.2) we need to define some functional spaces to work with.

We denote $\mathcal{V}$ a suitable Hilbert space for the deterministic solution u of (4.1). It will be assumed in the following that $\mathcal{V}$ is independent of the data, or more precisely of the event $\theta \in \Theta$, so we call $\mathcal{V}$ the deterministic space.

A suitable space for real-valued random variables is also required. We denote by $\mathcal{S}$ this function space, or refer to it as the stochastic space. In the following, we assume that all random quantities appearing in the problem are second order ones, and we take $\mathcal{S} = L_2(\Theta, P)$ as the stochastic space.

With the deterministic and stochastic spaces, the appropriate functional space for the solution u being sought is the tensor product space $\mathcal{V} \otimes \mathcal{S}$.

Remark 4.1 The assumption of the independence of the deterministic space with regards to the outcome together with the tensor product form of the solution space is not free of important consequences, especially at the discrete level. It first implies that the same deterministic space is valid for all realizations of the data. This restriction may cause severe difficulties if the data uncertainty level can lead to changes in the nature of the model equations (*e.g.* degeneracy of operators, transition from parabolic to hyperbolic character,...).

Secondly, since one needs to rely on a discretization of the deterministic space, it is clear that the tensor product form implies that the deterministic discretization should be somehow fine enough to capture all scales (spatial and temporal) that the data can yield.[1] It is thus clear that the tensor product construction is consequently non-optimal in the sense one may save computational resources by using discretizations adapted to the random outcome. Such adaptation makes however the problem much more complex and will be discussed later in the second part of this book.

Lagrangian particle methods for the simulation of vortex dominated flows is an example of mathematical model where the independence of the deterministic space with regard to the outcome does not hold at the discrete level. Indeed for these methods, the discrete deterministic basis $\{\phi_i\}_{i=1}^{m}$ spanning $\mathcal{V}^h$ is implicitly related to the solution, by moving the elements with the flow velocity: whenever the flow field is random, one has to consider approximation bases that depend on the outcome. An appropriate treatment for the Galerkin projection of particle methods is proposed in [121].

Finally, there is a large class of problems for which the dependence of the deterministic space on the outcome is inherent. This arises, for example, when the physical domain geometry is uncertain, or when the uncertainty on the data leads to uncertainty on the physical domain, as for instance in a fluid-structure interaction problem. In such situations, one has to solve the mathematical problem on a deterministic space $\mathcal{V}$ dependent on the outcome, unless it is possible to reformulate the problem in a reference domain, independent of the outcome, using a random mapping. A generalization of the stochastic Galerkin projection to situations where $\mathcal{V}$ depends on the outcome can be found in [172], which considers elasticity problems on random geometries.

4.1.3 Case of Discrete Deterministic Problems

It is observed that the developments given above have for starting point the continuous formulation of the mathematical problem $\mathcal{M}$. However, in view of applying Galerkin methods to existing deterministic codes, starting from the discrete problem can be a fruitful and time-saving strategy. Let us denote $\mathcal{V}^h \subset \mathcal{V}$ the discrete deterministic approximation space defined as

$$\mathcal{V}^h = \text{span}\{\phi_1, \ldots, \phi_m\}, \tag{4.3}$$

where the set $\{\phi_i\}_{i=1}^{i=m}$ forms a basis of $\mathcal{V}^h$. The deterministic approximate solution u_h of u on $\mathcal{V}^h$ is written as

$$u_h = \sum_{i=1}^{m} u_i \phi_i = \Phi \cdot \mathbf{U}^T, \tag{4.4}$$

[1] In fact, experience shows that the weak formulation introduced below significantly weakens this requirement.

where we have denoted $\Phi = (\phi_1 \cdots \phi_m)$ and $\mathbf{U} = (u_1 \cdots u_m) \in \mathbb{R}^m$. The deterministic solution satisfies the discretized version of (4.1), namely

$$\mathcal{M}_h(\mathbf{U}; d) = 0. \tag{4.5}$$

For random data $d(\theta)$, because the deterministic space is independent of the outcome, the discrete solution has random coefficients $\mathbf{U}(\theta) \in \mathbb{R}^m \otimes \mathcal{S}$ satisfying:

$$\mathcal{M}_h(\mathbf{U}(\theta); d(\theta)) = 0. \tag{4.6}$$

It is seen that in order to compute the discrete random solution one has to determine a number of random variables equal to the dimension of the deterministic approximation space $\mathcal{V}^h$, namely m.

4.1.4 Weak Form

At this point, we have to seek a stochastic solution $u \in \mathcal{V} \otimes \mathcal{S}$ (resp. $u_h \in \mathcal{V}^h \otimes \mathcal{S}$ in the discrete case) that satisfies (4.2) (resp. (4.6)) in a probabilistic sense, yet to be defined.

The weak form of the continuous problem is:
Find $u \in \mathcal{V} \otimes \mathcal{S}$ such that

$$\mathrm{E}[\mathcal{M}(u; d)\beta] = 0 \quad \forall \beta \in \mathcal{S}, \tag{4.7}$$

where $\mathrm{E}[\cdot]$ is the mathematical expectation:

$$\mathrm{E}[f] := \int_\Theta f(\theta)\,\mathrm{d}P(\theta). \tag{4.8}$$

It is seen from (4.7) that the weak form of the stochastic problem in fact corresponds to an interpretation of the stochastic problem in the *mean sense*.

Remark 4.2 For the discrete case, the weak form of (4.6) is:
Find $\mathbf{U} \in \mathbb{R}^m \otimes \mathcal{S}$ such that

$$\int_\Theta \mathcal{M}_h(\mathbf{U}(\theta); d(\theta))\beta(\theta)\,\mathrm{d}P(\theta) = \mathrm{E}[\mathcal{M}_h(\mathbf{U}; d)\beta] = 0 \quad \forall \beta \in \mathcal{S}. \tag{4.9}$$

Considering the weak form (4.9) of the discrete equations or proceeding to the discretization of the weak form (4.7) of the continuous equation is generally equivalent.

4.2 Stochastic Discretization

The second step in the derivation of the spectral problem is to perform a stochastic discretization of the weak form (4.7), for a problem already discretized at the deterministic level.

4.2.1 Stochastic Basis

At this point, we introduce the parametrization of the data. Let $\boldsymbol{\xi} = \{\xi_1, \ldots, \xi_\mathrm{N}\}$ be a finite set of N real-valued random variables, with known distribution. Let Ξ denote the range of $\boldsymbol{\xi}$, and $(\Xi, \mathcal{B}_\Xi, P_\Xi)$ the associated probability space. The stochastic solution is then sought in the image probability space $(\Xi, \mathcal{B}_\Xi, P_\Xi)$ in place of the abstract probability (Θ, Σ, P), *i.e.* we will seek to determine the random solution $u(\boldsymbol{\xi})$. For sake of simplicity, we shall focus on the case where $\boldsymbol{\xi}$ is a vector of independent random variables, so the probability measure P_Ξ can be expressed in product form as:

$$\mathrm{d}P_\Xi(\mathbf{y}) = \prod_{i=1}^{\mathrm{N}} p_{\xi_i}(y_i)\,\mathrm{d}\mathbf{y}, \tag{4.10}$$

where p_{ξ_i} is the density function of ξ_i. Furthermore, the expectation operator has the following expression in the image probability space:

$$\mathrm{E}[f] = \int_\Theta f(\boldsymbol{\xi}(\theta))\,\mathrm{d}P(\theta) = \int_\Xi f(\mathbf{y})\,\mathrm{d}P_\Xi(\mathbf{y}) := \langle f \rangle. \tag{4.11}$$

We will use brackets $\langle\cdot\rangle$ to make the distinction between the expectation operators in the abstract space (Θ, Σ, P) and expectation in $(\Xi, \mathcal{B}_\Xi, P_\Xi)$. It is clear from (4.11) that $f \in L_2(\Theta, P)$ implies that $f \in L_2(\Xi, P_\Xi)$, the space of second-order random variables spanned by $\boldsymbol{\xi}$.

To proceed with the determination of the stochastic solution, we consider a finite dimensional stochastic approximation space $\mathcal{S}^\mathrm{P} \subset L_2(\Xi, P_\Xi)$. Different discretizations may be sought at the stochastic level. These include polynomial expansions, expansions using piecewise polynomials, and multiwavelet decompositions. For the sake of simplicity, we shall restrict ourselves in this chapter to the PC and GPC bases introduced in Chap. 2. However, the developments presented hereafter extend immediately to any other orthogonal basis.

Let $\{\Psi_i\}_{i=0}^{\infty}$ be a basis of $\mathcal{S} = \mathrm{L}_2(\Xi, P_\Xi)$, where $\Psi_i(\boldsymbol{\xi})$ are mutually orthogonal polynomials in $\boldsymbol{\xi}$. Let us denote No the order of truncation, so the stochastic approximation basis is $\{\Psi_i\}_{i=0}^{\mathrm{P}}$, where P is given by

$$\mathrm{P} + 1 = \frac{(\mathrm{N} + \mathrm{No})!}{\mathrm{N!No!}}. \tag{4.12}$$

Recall that due to orthogonality of the polynomials we have:

$$\left\langle \Psi_i, \Psi_j \right\rangle = \int_\Xi \Psi_i(\mathbf{y})\Psi_j(\mathbf{y})\,\mathrm{d}P_\Xi(\mathbf{y}) = \left\langle \Psi_i^2 \right\rangle \delta_{ij}. \tag{4.13}$$

We denote $\mathcal{S}^\mathrm{P} \subset \mathcal{S}$ the stochastic approximation space

$$\mathcal{S}^\mathrm{P} = \mathrm{span}\{\Psi_0, \ldots, \Psi_\mathrm{P}\}. \tag{4.14}$$

The dimension of the stochastic approximation space is therefore given by

$$\dim(\mathcal{S}^{\mathrm{P}}) = \mathrm{P} + 1, \tag{4.15}$$

and any random variable $\beta \in \mathcal{S}^{\mathrm{P}}$ can be expressed as:

$$\beta(\boldsymbol{\xi}) = \sum_{i=0}^{\mathrm{P}} \beta_i \Psi_i(\boldsymbol{\xi}). \tag{4.16}$$

Provided that $\beta(\boldsymbol{\xi})$ is known, the spectral coefficients β_i in (4.16) can be obtained from:

$$\beta_i = \frac{\langle \beta \Psi_i \rangle}{\langle \Psi_i^2 \rangle}. \tag{4.17}$$

4.2.2 Data Parametrization and Solution Expansion

We now consider the parametrization of the data using the random vector $\boldsymbol{\xi}$:

$$d(\theta) = d(\boldsymbol{\xi}(\theta)). \tag{4.18}$$

Because the mathematical models considered involve no randomness other than in the data, the stochastic solution u is dependent on the same outcome $\boldsymbol{\xi}(\theta)$ as the data. Consequently, it is defined on the same probability space as d, and the solution will be sought in terms of an expansion of the form:

$$\mathcal{V} \otimes \mathcal{S}^{\mathrm{P}} \ni u(\boldsymbol{\xi}) = \sum_{i=0}^{\mathrm{P}} u_i \Psi_i(\boldsymbol{\xi}), \tag{4.19}$$

where $u_i \in \mathcal{V}$ are the stochastic modes of the solution.

Remark 4.3 For clarity, we have not relied explicitly on the expansion of the data on the stochastic basis of $\mathcal{S}^{\mathrm{P}}$. However, this may be interesting in view of an efficient numerical implementation. The expansion of the data on the stochastic approximation space $\mathcal{S}^{\mathrm{P}}$ is

$$d(\boldsymbol{\xi}(\theta)) = \sum_{k=0}^{\mathrm{P}} d_k \Psi_k(\boldsymbol{\xi}(\theta)). \tag{4.20}$$

Equation (4.20) is understood in a Galerkin sense, namely

$$\left\langle \beta, \left(d(\boldsymbol{\xi}) - \sum_{k=0}^{\mathrm{P}} d_k \Psi_k(\boldsymbol{\xi}) \right) \right\rangle = 0 \quad \forall \beta \in \mathcal{S}^{\mathrm{P}}. \tag{4.21}$$

Since the parametrization of the data using the random vector $\boldsymbol{\xi}$ is given, (4.17) can be used to compute the spectral coefficients $\{d_i\}_{i=0}^{i=\mathrm{P}}$ of the data.

4.3 Spectral Problem

Galerkin projection is a classical tool for solving spectral problems (see [24] for an introduction to spectral problems and relevant chapters in [26] for deterministic spectral methods in fluid flows). In the stochastic context, it was proposed as a computational method to determine the PC expansion of the solution to stochastic linear equations by Ghanem and Spanos [90]. The Galerkin projection method involves two basic steps: (a) introducing the solution expansions into the formulation of the stochastic problem and, (b) projecting the resulting stochastic equation onto the expansion basis to yield a set of equations that the expansion coefficients must satisfy.

4.3.1 Stochastic Residual

At this point, we have to solve equation (4.7) or its semi-discrete form (4.9). Since derivations for the continuous and semi-discrete forms are essentially the same, we focus on the continuous case.

The next step is now to introduce the solution expansion (4.19) in the stochastic problem (4.2). One obtains:

$$\mathcal{M}\left(\sum_{i=0}^{\mathrm{P}} u_i \Psi_i(\boldsymbol{\xi}); d(\boldsymbol{\xi})\right) = \mathcal{R}^{\mathrm{P}}(u). \tag{4.22}$$

Generally, the right-hand side of (4.22) is not zero, in any sense, unless $\mathrm{P} \to \infty$. In fact, provided that the random residual $\mathcal{R}^{\mathrm{P}}(u)$ is a second-order random quantity, it can be formally expanded as:

$$\mathcal{R}^{P}(u) = \sum_{i=0}^{\infty} \mathcal{R}_i \Psi_i(\boldsymbol{\xi}). \tag{4.23}$$

Thus, the weak form (4.7) would consist in finding the spectral modes $u_i \in \mathcal{V}$ of the solution such that

$$\left\langle \beta, \mathcal{R}^{\mathrm{P}}\left(\sum_{i=0}^{P} u_i \Psi_i\right)\right\rangle = 0 \quad \forall \beta \in \mathcal{S}. \tag{4.24}$$

However, it is generally impossible to find a set of spectral modes $\{u_i\}_{i=0}^{i=\mathrm{P}}$ satisfying (4.24), for any $\beta \in \mathcal{S}$, due to the infinite dimensionality of $\mathcal{S}$. Consequently, (4.24) has to be solved in a finite-dimensional stochastic approximation space.

4.3.2 Galerkin Method

The final step in the derivation of the spectral problem consists in selecting an appropriate stochastic approximation space $\mathcal{S}_\beta$ for the test random variables β in equation (4.24), in order to form an approximate solution of the stochastic problem.

The Galerkin method is precisely based on using the *same* stochastic approximation space for the solution and test random variables, $\mathcal{S}_\beta = \mathcal{S}^{\mathrm{P}}$. For this choice, the test random variables β have for expansion (4.16), and the weak form of the stochastic problem is

$$\left\langle \left(\sum_{i=0}^{\mathrm{P}} \beta_i \Psi_i\right), \left(\sum_{j=0}^{\mathrm{P}} u_j \Psi_j; d\right)\right\rangle = \sum_{i=0}^{\mathrm{P}} \beta_i \left\langle \Psi_i, \mathcal{M}\left(\sum_{j=0}^{\mathrm{P}} u_j \Psi_j; d\right)\right\rangle$$
$$\forall (\beta_0, \ldots, \beta_{\mathrm{P}}) \in \mathbb{R}^{\mathrm{P}+1}. \qquad (4.25)$$

This is seen to be equivalent to the following set of coupled problems, which is referred to as the spectral problem, for the modes $u_i \in \mathcal{V}$ of the solution:

$$\left\langle \mathcal{M}\left(\sum_{i=0}^{\mathrm{P}} u_i \Psi_i(\xi); d(\xi)\right), \Psi_k(\xi)\right\rangle = 0, \quad \forall k \in \{1, \ldots, \mathrm{P}\}. \qquad (4.26)$$

Remark 4.4 For the spatially-discretized deterministic model $\mathcal{M}_h$, the spectral problem corresponding to equation (4.9) is:
Find $\mathbf{U} \in \mathbb{R}^m \otimes \mathcal{S}^{\mathrm{P}}$ such that

$$\left\langle \mathcal{M}_h\left(\sum_{i=0}^{\mathrm{P}} \mathbf{U}_i \Psi_i; \sum_{j=0}^{\mathrm{P}} d_j \Psi_j\right), \Psi_k\right\rangle = 0, \quad \forall k \in \{1, \ldots, \mathrm{P}\}, \qquad (4.27)$$

where $\mathbf{U}_i \in \mathbb{R}^m$ are the stochastic modes of $\mathbf{U}(\xi)$ on the stochastic basis of $\mathcal{S}^{\mathrm{P}}$, i.e.

$$\mathbf{U}(\xi) = \sum_{i=0}^{\mathrm{P}} \mathbf{U}_i \Psi_i(\xi) \in \mathbb{R}^m \otimes \mathcal{S}^{\mathrm{P}}. \qquad (4.28)$$

4.3.3 Comments

The derivation of the spectral problem has been conducted in an abstract way that in fact hides most the difficulties faced when one actually applies the stochastic Galerkin method to complex mathematical models and large-scale problems. There are in fact two major sources of difficulties inherent to the Galerkin methodology.

One type of difficulty is related to the dimension of the spectral problem: the tensor product form of the discrete solution space $\mathcal{V}^h \otimes \mathcal{S}^{\mathrm{P}}$ already indicates that compared to the deterministic problem, the size of the set of equations to be solved will be $P+1$ times larger than that of the deterministic problem. Thus, the development of efficient numerical strategies for the resolution of the spectral problem is a central concern for Galerkin methods. Different approaches can be envisioned to maintain a computational cost as low as possible. In a first class of strategies, one tries, whenever possible, to derive formulations that decouple the resolution of the spectral modes u_i (or U_i) of the solution. Thus, in this situation one has to solve a set of $\dim(\mathcal{S}^{P})$ problems, each with dimension equal to that of the deterministic problem, m. Since it is usually more efficient to solve a set of $(P+1)$ independent problems of size m than solving a single problem of size $(P+1)\times m$, decoupling strategies often yield significant savings. Many examples of decoupling strategies can be found in examples provided in this book, for instance in the applications tackled in Chaps. 5 and 6, and also during the development of concepts of generalized spectral decomposition in Chap. 9.

Regardless of the computational strategy used to reduce the complexity of the spectral problem, the systems considered in this book frequently result in large systems of equations, so that direct solution methods, even for linear problems, may not be practical. In these situations, iterative techniques, carefully tailored to the spectral problems considered, may provide a more suitable alternative. Some of these are addressed in some detail in Chap. 7.

A second source of difficulties arising form the Galerkin procedure concerns the treatment of nonlinearities. This is a significant challenge for Galerkin methods, and Sect. 4.5 is specifically dedicated to this aspect. However, before considering the issues related to the nonlinearities, and proposing some appropriate treatments, we provide a detailed analysis of the spectral problem resulting for the Galerkin projection of linear models.

4.4 Linear Problems

Linear problems are of practical importance in computational science, whether as stand-alone mathematical problems or as ingredients of numerical methods (*e.g.* iteration techniques for the resolution of nonlinear problems). In this section, we analyze the structure of the spectral problem arising from linear systems, and briefly examine implications regarding suitable solution strategies.

4.4.1 General Formulation

We shall consider in this section linear problems being discretized at the deterministic level. The deterministic problem corresponds to (4.6) which, owing to the assumed linearity of $\mathcal{M}_h$ with regard to $\mathbf{U}$, can be recast in the matrix form

$$[A]\mathbf{U}=\mathbf{B}, \tag{4.29}$$

where $[A]$ is a $(m \times m)$ real matrix, and $\mathbf{U}$ and $\mathbf{B}$ are vectors of $\mathbb{R}^m$. In presence of random data $d(\boldsymbol{\xi})$ affecting both the matrix $[A]$ and right-hand-side, $\mathbf{B}$, of (4.29) becomes:

$$[A](\boldsymbol{\xi})\mathbf{U}(\boldsymbol{\xi}) = \mathbf{B}(\boldsymbol{\xi}). \tag{4.30}$$

Seeking the solution $\mathbf{U}(\boldsymbol{\xi})$ in a subspace $\mathbb{R}^m \otimes \mathcal{S}^{\mathrm{P}}$ of $\mathbb{R}^m \otimes L_2(\Xi, P_\Xi)$, the Galerkin projection of (4.30) is expressed as:

$$\sum_{i=0}^{\mathrm{P}} \langle \Psi_k, [A]\Psi_i \rangle \mathbf{U}_i = \langle \Psi_k, \mathbf{B} \rangle, \quad k \in \{0, \dots, \mathrm{P}\}. \tag{4.31}$$

This equation can be eventually recast into a larger (block) system of linear equations, namely

$$\begin{bmatrix} [A]_{00} & \dots & [A]_{0\mathrm{P}} \\ \vdots & \ddots & \vdots \\ [A]_{\mathrm{P}0} & \dots & [A]_{\mathrm{PP}} \end{bmatrix} \begin{pmatrix} \mathbf{U}_0 \\ \vdots \\ \mathbf{U}_{\mathrm{P}} \end{pmatrix} = \begin{pmatrix} \mathbf{B}_0 \\ \vdots \\ \mathbf{B}_{\mathrm{P}} \end{pmatrix}, \tag{4.32}$$

where $[A]_{ij}$ is the $(m \times m)$ matrix given by:

$$[A]_{ij} \equiv \left\langle \Psi_i, [A]\Psi_j \right\rangle, \tag{4.33}$$

and $\mathbf{B}_i$ the m-dimensional real vector defined as

$$\mathbf{B}_i \equiv \langle \Psi_i, \mathbf{B} \rangle. \tag{4.34}$$

4.4.2 *Structure of Linear Spectral Problems*

Equation (4.32) illustrates how the Galerkin projection will generally result in a linear spectral problem coupling all the stochastic modes $\mathbf{U}_i$ of the stochastic solution. By mode coupling, we mean that it is not possible to compute independently the components $\mathbf{U}_i$. It is also seen that the size of the spectral problem to be solved will be large, since the dimension of the discrete solution is $m \times \dim \mathcal{S}^{\mathrm{P}} = m(\mathrm{P}+1)$. As most problems of interest will usually have large m, the direct inversion of system (4.32) may not be feasible, and a better understanding of the block structured system is instructive to design and apply well-suited numerical methods.

4.4.2.1 Case of Deterministic Operator

A first particular case occurs when the random data have no impact on the linear operator $[A]$ but only on the right-hand-side $\mathbf{B}$. This is a common situation for linear systems with deterministic properties but uncertain boundary conditions and/or forcing terms. In this case, due to the orthogonality of the stochastic basis we have

$$\left\langle \Psi_i, [A]\Psi_j \right\rangle = [A]\left\langle \Psi_i, \Psi_j \right\rangle = [A]\delta_{ij}\left\langle \Psi_i^2 \right\rangle, \tag{4.35}$$

and (4.32) becomes

$$\begin{bmatrix} [A] & [0] & & \dots & [0] \\ [0] & [A] & \ddots & & \vdots \\ \vdots & \ddots & \ddots & \ddots & \vdots \\ \vdots & & \ddots & [A] & [0] \\ [0] & \dots & \dots & [0] & [A] \end{bmatrix} \begin{pmatrix} \mathbf{U}_0 \\ \mathbf{U}_1 \\ \vdots \\ \mathbf{U}_{P-1} \\ \mathbf{U}_P \end{pmatrix} = \begin{pmatrix} \overline{\mathbf{B}}_0 \\ \overline{\mathbf{B}}_1 \\ \vdots \\ \overline{\mathbf{B}}_{P-1} \\ \overline{\mathbf{B}}_P \end{pmatrix}, \tag{4.36}$$

where $\overline{\mathbf{B}}_i$ is defined according to:

$$\overline{\mathbf{B}}_i \equiv \frac{\langle \Psi_i \mathbf{B} \rangle}{\langle \Psi_i^2 \rangle}. \tag{4.37}$$

Thus, the modes $\mathbf{U}_i$ satisfy

$$\mathbf{U}_i = [A]^{-1}\overline{\mathbf{B}}_i, \tag{4.38}$$

and can be computed independently. In this case, the resolution of the spectral problem consists in solving a set of $(P+1)$ linear systems of size $(m \times m)$. Since the system matrices are all the same, this amounts to solving the same linear system $[A]$ for $(P+1)$ different right-hand sides.

4.4.2.2 General Case

In general, the matrix $[A]$ will not be deterministic, but will have a stochastic expansion given by:

$$[A](\boldsymbol{\xi}) = \sum_{i=0}^{P} [A]_i \Psi_i(\boldsymbol{\xi}). \tag{4.39}$$

Using (4.39), one immediately obtains:

$$[A]_{ij} = \left\langle \Psi_i, [A]\Psi_j \right\rangle = \sum_{k=0}^{P} [A]_k \left\langle \Psi_i, \Psi_j \Psi_k \right\rangle, \tag{4.40}$$

so (4.32) can be conveniently recast as

$$\begin{bmatrix} [\overline{A}]_{00} & \dots & [\overline{A}]_{0P} \\ \vdots & \ddots & \vdots \\ [\overline{A}]_{P0} & \dots & [\overline{A}]_{PP} \end{bmatrix} \begin{pmatrix} \mathbf{U}_0 \\ \vdots \\ \mathbf{U}_P \end{pmatrix} = \begin{pmatrix} \overline{\mathbf{B}}_0 \\ \vdots \\ \overline{\mathbf{B}}_P \end{pmatrix}, \tag{4.41}$$

where $\overline{\mathbf{B}}_i$ are given by (4.37),

$$[\overline{A}]_{ij} \equiv \sum_{k=0}^{\mathrm{P}} [A]_k C_{kji}, \tag{4.42}$$

and

$$C_{ijk} \equiv \frac{\langle \Psi_i \Psi_j \Psi_k \rangle}{\langle \Psi_k \Psi_k \rangle}. \tag{4.43}$$

The third order tensor C_{ijk} plays a fundamental role in stochastic Galerkin methods, especially in nonlinear problems as we will see later in this chapter. Due to its fundamental importance, the tensor C_{ijk} will be referred to as the "multiplication tensor" throughout this book. It is first remarked that this tensor is symmetric with regard to the two first indices, $C_{ijk} = C_{jik}$. This highlights some symmetry in the structure of the linear spectral problem (4.41), namely $[\overline{A}]_{ij} = [\overline{A}]_{ji}$, as one may readily notice from (4.32). Furthermore, many of the $(\mathrm{P}+1)^3$ entries of the multiplication tensor are zero, thanks to the orthogonality of the stochastic basis. This can already be seen from the definition of the first block of system (4.41), which in fact reduces to:

$$[\overline{A}]_{00} = \sum_{k=0}^{\mathrm{P}} [A]_k C_{k00} = [A]_0, \tag{4.44}$$

since by convention $\Psi_0 = 1$ so $\langle \Psi_k \Psi_0 \Psi_0 \rangle = \langle \Psi_k \rangle = \delta_{k0}$. Similarly, the sum for the upper-right block (and lower-left block) actually reduces to a unique term: $[\overline{A}]_{0\mathrm{P}} = [A]_{\mathrm{P}} / \langle \Psi_{\mathrm{P}}^2 \rangle$. Many other, sometimes less obvious, simplifications occur. Since the multiplication tensor depends only on the stochastic basis, it is recommended that in practice one computes and stores its non-zero entries in a pre-processing step. To this end, an efficient computational strategy is provided in Appendix C.

The sparse structure of the multiplication tensor depends essentially on the number, N, of random variables in $\boldsymbol{\xi}$ and on the expansion order, No, of the PC or GPC basis. It not only allows for a drastic reduction of the sums in the definition of the blocks $[\overline{A}]_{ij}$, it also leads to a significant number of null blocks. In other words, the block system (4.41) is sparse. In order to illustrate the sparsity of the linear spectral problems, we provide in Figs. 4.1 and 4.2 the resulting block structure of (4.42), for stochastic bases corresponding to different dimensions N, and expansion orders, No. The plots in Figs. 4.1 and 4.2 depict all the blocks $[\overline{A}]_{ij}$ that are generally non-zero; in each case, the plots also provide the dimension of the basis and a measure of the sparsity over the system. The latter is quantified using the ratio, S, of the number of non-zero blocks over the total number of block $(\dim \mathcal{S}^{\mathrm{P}})^2$. It can be observed from these figures that the block structure becomes sparser as N increases with No fixed, whereas for fixed N the value of S varies slightly as No increases. Note that individual blocks usually have a sparse structure as well.

N = 4-dim $\mathscr{S}^{\mathrm{P}} = 35$-$S = 0.58$ N = 6-dim $\mathscr{S}^{\mathrm{P}} = 84$-$S = 0.41$

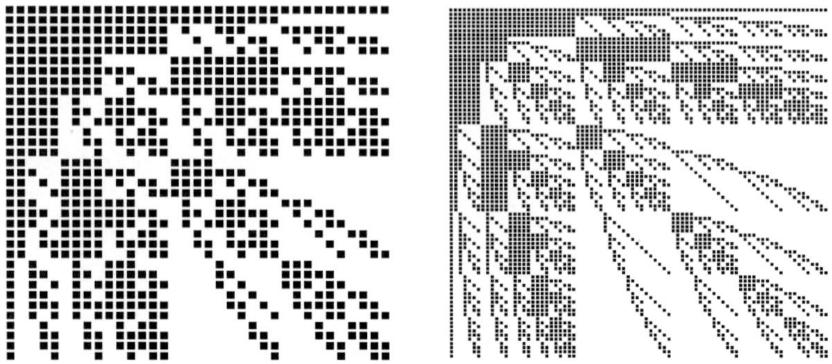

N = 8-dim $\mathscr{S}^{\mathrm{P}} = 165$-$S = 0.31$ N = 10-dim $\mathscr{S}^{\mathrm{P}} = 286$-$S = 0.23$

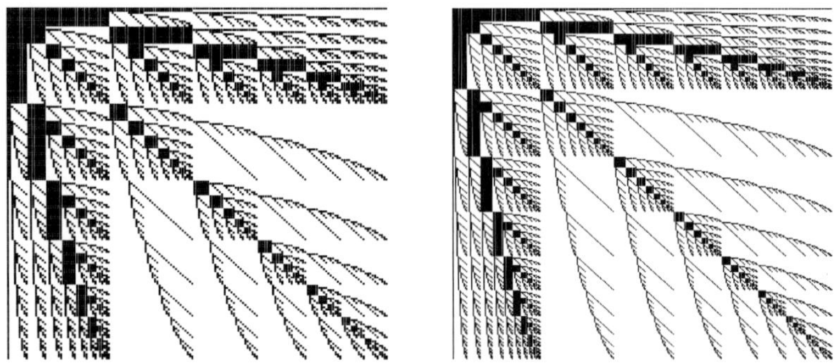

Fig. 4.1 Illustration of the sparse structure of the matrices of the linear spectral problem for different dimensions, N, with No = 3. Matrix blocks $[\overline{A}]_{ij}$ that are generally non-zero appear as *black squares*

In the previous analysis, it was assumed that the random matrix $[A](\boldsymbol{\xi})$ had a full spectrum on the stochastic basis. However, in many situations the stochastic basis is selected to provide an exact representation of the random linear operator with a low expansion order, often a first-order expansion. When $[A](\boldsymbol{\xi})$ has a first-order expansion, the block structure of the linear spectral problem becomes even sparser, as illustrated in Fig. 4.3, and this should be contrasted with the block structures shown in Fig. 4.2 where higher orders were considered. It is seen that for a linear stochastic matrix having a first-order expansion, the block structure of the discrete spectral problem is extremely sparse. This behavior motivates the selection, whenever possible, of an approximation based on a first-order operator.

No $= 2$-dim $\mathscr{S}^{\mathrm{P}} = 21$-$S = 0.52$	No $= 3$-dim $\mathscr{S}^{\mathrm{P}} = 56$-$S = 0.49$

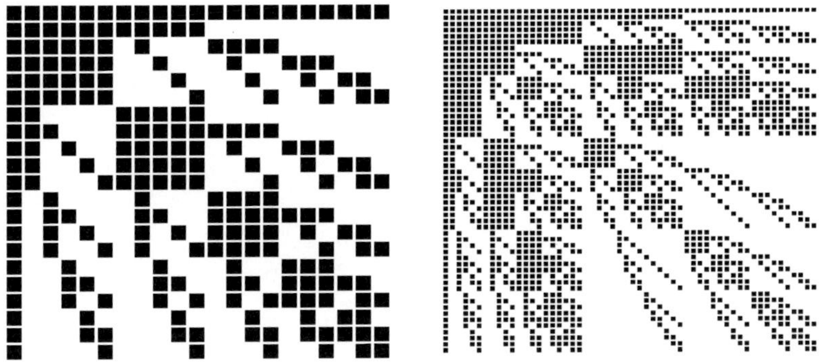

No $= 4$-dim $\mathscr{S}^{\mathrm{P}} = 126$-$S = 0.54$	No $= 5$-dim $\mathscr{S}^{\mathrm{P}} = 252$-$S = 0.55$

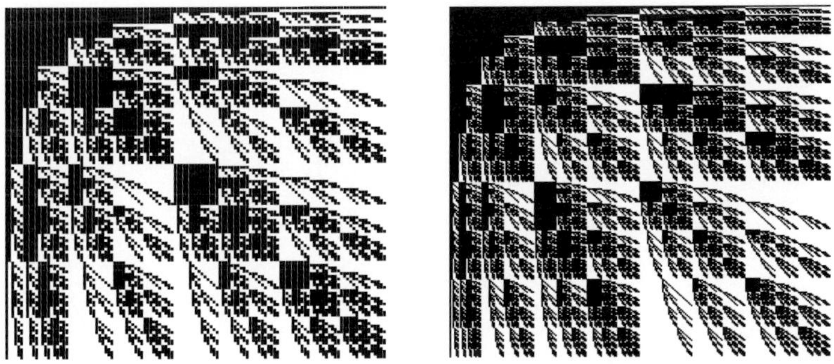

Fig. 4.2 Illustration of the sparse structure of the matrices of the linear spectral problem for different expansion orders No, with $\mathrm{N} = 5$. Matrix blocks $[\overline{A}]_{ij}$ that are generally non-zero appear as *black squares*

4.4.3 Solution Methods for Linear Spectral Problems

As stated previously, one of the main issues in solving discrete linear spectral problems concerns the size of the system of equations. In fact, the structure of the linear spectral problem, and its associated sparsity discussed above, suggests the adoption of iterative solution strategies. In many situations, one can re-use the iterative solver developed for the deterministic problem (*e.g.* conjugate gradient techniques for symmetric systems, and Krylov subspace methods [92]). However, the efficiency of iterative techniques often depends on the availability of an appropriate preconditioner, and such preconditioner needs be adapted to the spectral problem. The block-structure of the linear spectral problem provides us with some clues to construct preconditioners. It is seen, most clearly in Fig. 4.3, that diagonal blocks $[\overline{A}]_{ii}$

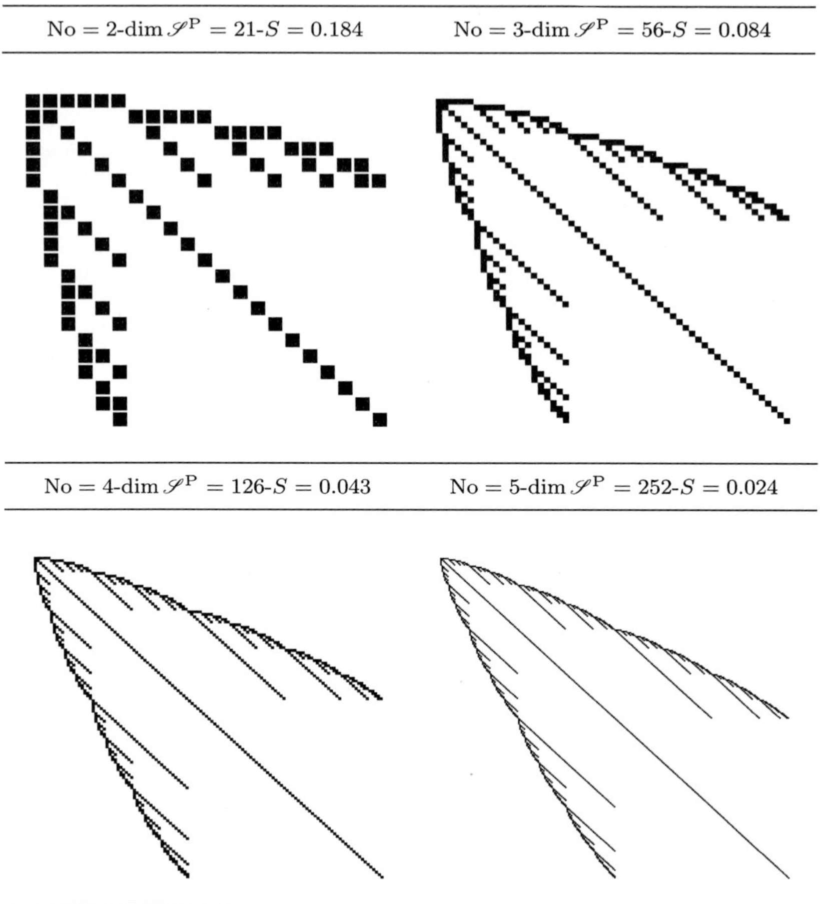

Fig. 4.3 Same as in Fig. 4.2 but for a linear stochastic operator $[A](\boldsymbol{\xi})$ having a first-order expansion

generally do not vanish. In fact, in view of (4.42) we have

$$[\overline{A}]_{ii} = \sum_{k=0}^{\mathrm{P}} [A]_k \frac{\langle \Psi_k \Psi_i \Psi_i \rangle}{\langle \Psi_i \Psi_i \rangle}, \tag{4.45}$$

so the sum contains at least one term, namely $[A]_0$. Since $[A]_0$ is the mean linear operator, *i.e.* the deterministic operator of the mean mathematical problem (roughly speaking the problem for the averaged data if the model is linear in the data), it is expected to be dominant in the expression of $[\overline{A}]_{ii}$ at least for reasonable uncertainty level. In other words, one can consider the mean operator $[A]_0$ as representative of the random operator $[A](\boldsymbol{\xi})$. Therefore, an appropriate preconditioner $[P]$ can be constructed by considering only the diagonal blocks as the full matrix in system (4.41) and approximating them using $[A]_0$, *i.e.*

$$[P] = \begin{bmatrix} [A]_0 & 0 & \dots & 0 \\ 0 & \ddots & \ddots & \vdots \\ \vdots & \ddots & \ddots & 0 \\ 0 & \dots & 0 & [A]_0 \end{bmatrix}. \tag{4.46}$$

Due to the structure of $[P]$, its inversion results in the inversion of $[A]_0$ whose dimension is much less than that of the spectral problem. Indeed, we have

$$[P]^{-1} = \begin{bmatrix} ([A]_0)^{-1} & 0 & \dots & 0 \\ 0 & \ddots & \ddots & \vdots \\ \vdots & \ddots & \ddots & 0 \\ 0 & \dots & 0 & ([A]_0)^{-1} \end{bmatrix}. \tag{4.47}$$

The preconditioned spectral problem can now be expressed as:

$$[P]^{-1} \begin{bmatrix} [\overline{A}]_{00} & \dots & [\overline{A}]_{0P} \\ \vdots & \ddots & \vdots \\ [\overline{A}]_{P0} & \dots & [\overline{A}]_{PP} \end{bmatrix} \begin{pmatrix} \mathbf{U}_0 \\ \vdots \\ \vdots \\ \mathbf{U}_P \end{pmatrix} = [P]^{-1} \begin{pmatrix} \mathbf{B}_0 \\ \vdots \\ \vdots \\ \mathbf{B}_P \end{pmatrix}. \tag{4.48}$$

Iteration strategies and preconditioning are further discussed in Chap. 7.

4.5 Nonlinearities

Many models involve nonlinearities of various types. Their treatment is critical in the context of stochastic Galerkin methods, which require the projection of these nonlinearities on the expansion basis.

The general situation can be cast as follows. Let $\{\Psi_k(\boldsymbol{\xi})\}_{k=0}^{P}$, $\boldsymbol{\xi} \in \Xi \subset \mathbb{R}^N$, be an orthogonal basis of $\mathcal{S}^P \subset L_2(\Xi, P_\Xi)$, and consider a nonlinear functional f with arguments $u, v, \dots$:

$$u, v, \dots \in \mathbb{R} \mapsto f(u, v, \dots) \in \mathbb{R}. \tag{4.49}$$

If we extend the definition of the functional to random arguments,

$$u(\boldsymbol{\xi}), v(\boldsymbol{\xi}), \dots \in \mathbb{R} \otimes \mathcal{S}^P,$$

we generally have $f(u, v, \dots) = f(\boldsymbol{\xi}) \notin \mathbb{R} \otimes \mathcal{S}^P$. However, if $f(\boldsymbol{\xi}) \in \mathbb{R} \otimes L_2(\Xi, P_\Xi)$, it has an orthogonal projection onto $\mathcal{S}^P$ which converges as P increases,

$$f(\boldsymbol{\xi}) \approx \hat{f} = \sum_{k=0}^{P} f_k \Psi_k, \quad f_k = \frac{\langle f(u, v, \dots), \Psi_k \rangle}{\langle \Psi_k^2 \rangle}. \tag{4.50}$$

The problem is therefore to derive efficient strategies to compute the expansion coefficients f_k of $f(\boldsymbol{\xi})$ from the expansion coefficients of its arguments $u(\boldsymbol{\xi}), v(\boldsymbol{\xi}), \ldots$. In the remainder of this chapter, we discuss different strategies for the projection of nonlinear functionals. We start with the simple case involving products of stochastic quantities, and then address more elaborate transformations.

4.5.1 Polynomial Nonlinearities

4.5.1.1 Galerkin Product

The product of two quantities appears in many models; it corresponds to the case $f(\boldsymbol{\xi}) = w(\boldsymbol{\xi}) = u(\boldsymbol{\xi})v(\boldsymbol{\xi})$, for $u, v \in \mathcal{S}^{\mathrm{P}}$ with known expansions, *i.e.*

$$u(\boldsymbol{\xi}) = \sum_{k=0}^{\mathrm{P}} u_k \Psi_k(\boldsymbol{\xi}), \qquad v(\boldsymbol{\xi}) = \sum_{k=0}^{\mathrm{P}} v_k \Psi_k(\boldsymbol{\xi}). \tag{4.51}$$

Clearly,

$$w(\boldsymbol{\xi}) = \sum_{i=0}^{\mathrm{P}} \sum_{j=0}^{\mathrm{P}} u_i v_j \Psi_i(\boldsymbol{\xi}) \Psi_j(\boldsymbol{\xi}) \tag{4.52}$$

and in general $w(\boldsymbol{\xi}) \notin \mathcal{S}^{\mathrm{P}}$ though it is always in $L_2(\Xi, P_\Xi)$. Therefore, $\hat{w}$, the orthogonal projection of w on $\mathcal{S}^{\mathrm{P}}$, has expansion coefficients

$$w_k = \frac{\langle w, \Psi_k \rangle}{\langle \Psi_k^2 \rangle} = \sum_{i=0}^{\mathrm{P}} \sum_{j=0}^{\mathrm{P}} u_i v_j C_{ijk}, \tag{4.53}$$

where C_{ijk} is the Galerkin multiplication tensor defined in (4.43). The result of the orthogonal projection of $w = uv$ on $\mathcal{S}^{\mathrm{P}}$ is called the Galerkin product of u and v. As a shorthand notation, we write the Galerkin product of u and v as $u * v$, *i.e.*

$$\hat{w} = u * v \equiv \sum_{k=0}^{\mathrm{P}} \left(\sum_{i=0}^{\mathrm{P}} \sum_{j=0}^{\mathrm{P}} u_i v_j C_{ijk} \right) \Psi_k. \tag{4.54}$$

The Galerkin product is exact in the sense that it corresponds to the exact projection of uv on $\mathcal{S}^{\mathrm{P}}$. It however introduces a truncation error by disregarding the components of uv which are orthogonal to $\mathcal{S}^{\mathrm{P}}$. Some interesting properties of the Galerkin product are:
$\forall u, v, w \in \mathcal{S}^{\mathrm{P}}$ and $\alpha, \beta \in \mathbb{R}$:

$$u * v = v * u,$$

$$(\alpha u) * (\beta v) = \alpha\beta(u * v),$$

$$(u + v) * w = u * w + v * w.$$

4.5.1.2 Higher-Order Polynomial Nonlinearity

Higher-order polynomial nonlinearities are also frequent in physical models (for instance in chemical systems). This requires the Galerkin projection of products involving three or more stochastic quantities. Consider first the triple product $f(\xi) = u(\xi)v(\xi)w(\xi)$, with u, v and w having known expansions on $\mathcal{S}^{\mathrm{P}}$:

$$u = \sum_{k=0}^{\mathrm{P}} u_k \Psi_k, \qquad v = \sum_{k=0}^{\mathrm{P}} v_k \Psi_k, \qquad w = \sum_{k=0}^{\mathrm{P}} w_k \Psi_k. \tag{4.55}$$

One can again perform an exact Galerkin projection of the triple product as we have done previously:

$$\widehat{uvw} := \sum_{m=0}^{\mathrm{P}} \widehat{uvw}_m \Psi_m = \sum_{m=0}^{\mathrm{P}} \Psi_m \left(\sum_{j,k,l=0}^{\mathrm{P}} T_{jklm} u_j v_k w_l \right),$$
$$T_{jklm} \equiv \frac{\langle \Psi_j \Psi_k \Psi_l \Psi_m \rangle}{\langle \Psi_m \Psi_m \rangle}. \tag{4.56}$$

This exact Galerkin projection of the triple product involves the fourth-order tensor T_{jklm}. Although this tensor has a sparse structure, owing to the orthogonality of the basis, and many symmetries, computation and storage of T_{jklm} becomes quickly prohibitive as the dimension $\mathrm{P}+1$ of the stochastic space increases. In addition, the exact Galerkin projection can hardly be extended further to higher-order polynomial nonlinearities. An alternative is necessary, and it is often preferred to rely on approximations for polynomial nonlinearity of order larger than 2. For the triple product, an immediate approximation is

$$\widehat{uvw} \approx u * (v * w) = \widehat{u\widehat{vw}}, \tag{4.57}$$

i.e. we first apply the Galerkin product (4.54) to compute $v * w = \widehat{vw}$, whose result is subsequently multiplied by u. This strategy can be extended to higher polynomial nonlinearities by using successive Galerkin projections. For instance,

$$\widehat{abc\cdots d} \approx a * (b * (c * (\cdots * d))). \tag{4.58}$$

It is clear that this procedure does not provide the exact Galerkin projection, since every intermediate product disregards the part orthogonal to $\mathcal{S}^{\mathrm{P}}$ so that the subsequent Galerkin products will not account for the truncated part. In fact, even for the triple product it is remarked that, in general

$$u * (v * w) \neq (u * v) * w \neq (u * w) * v. \tag{4.59}$$

In other words, the order in which the successive Galerkin products are applied affects the result. To see this, let us develop the approximate triple product for $(u * v) * w$:

$$
\begin{aligned}
(u * v) * w &= \sum_{m=0}^{P} \Psi_m \left(\sum_{k,l=0}^{P} C_{klm} (u * v)_k w_l \right) \\
&= \sum_{m=0}^{P} \Psi_m \left(\sum_{k,l=0}^{P} C_{klm} \left(\sum_{i,j=0}^{P} C_{ijk} u_i v_j \right) w_l \right) \\
&= \sum_{m=0}^{P} \Psi_m \left(\sum_{i,j,k,l=0}^{P} C_{ijk} C_{klm} u_i v_j w_l \right) \\
&= \sum_{m=0}^{P} \Psi_m \left(\sum_{i,j,l=0}^{P} C^{(2)}_{ijlm} u_i v_j w_l \right),
\end{aligned}
$$

where

$$
C^{(2)}_{ijlm} \equiv \sum_{k=0}^{P} C_{ijk} C_{klm} = \sum_{k=0}^{P} \frac{\langle \Psi_i \Psi_j \Psi_k \rangle \langle \Psi_k \Psi_l \Psi_m \rangle}{\langle \Psi_k \Psi_k \rangle \langle \Psi_m \Psi_m \rangle}. \tag{4.60}
$$

This calculation shows that, eventhough $C^{(2)}_{ijlm} = C^{(2)}_{jilm}$ (expressing the fact that $(u * v) * w = (v * u) * w$), in general $C^{(2)}_{ijlm} \neq C^{(2)}_{ljim} \neq C^{(2)}_{iljm}$ expressing a lack of commutativity of the operator. In contrast, the triple product tensor T_{iljm} is invariant for any permutation of the first 3 indices.

Even though inexact, the approximations of triple products (and higher-order polynomial nonlinearities) by successive Galerkin products are extensively used, and also applied in the Galerkin projection of more general nonlinear transformations (see below). This is justified when the truncation error associated with each Galerkin product is negligible in well resolved computations. (One should, however, verify that this is in fact the case!) In addition, it should be stressed that relying on successive applications of the Galerkin product does not reduce the computational complexity (*i.e.* the operation count) of the polynomial nonlinearity evaluations, but essentially bypasses the need to compute and store high-order tensors. In fact, the evaluation of high-order products remains a computationally demanding task when the dimension of the stochastic space is large.

4.5.2 Galerkin Inversion and Division

Inversion and division also feature in many physical models. For the inversion in the stochastic context, one has to determine the expansion coefficients of the stochastic inverse $u^{-1}(\boldsymbol{\xi})$ of a random variable $u(\boldsymbol{\xi})$ with known expansion on $\mathcal{S}^{\mathrm{P}}$:

$$u^{-1}(\boldsymbol{\xi}) = \frac{1}{u(\boldsymbol{\xi})} = \left(\sum_{k=0}^{\mathrm{P}} u_k \Psi_k(\boldsymbol{\xi})\right)^{-1}, \tag{4.61}$$

such that

$$u^{-1}(\boldsymbol{\xi}) u(\boldsymbol{\xi}) = 1 \quad \text{a.s.} \tag{4.62}$$

Since the expansion of u^{-1} is sought in the finite dimensional space $\mathcal{S}^{\mathrm{P}}$, (4.62) needs to be interpreted in a weak sense. Using the previous notation, the weak interpretation corresponds to

$$\widehat{u^{-1}} * u = \Psi_0, \tag{4.63}$$

where we have made use of the usual convention $\Psi_0(\boldsymbol{\xi}) = 1$. Therefore, using (4.54) one ends up with a system of linear equations for the expansion coefficients of u^{-1} on $\mathcal{S}^{\mathrm{P}}$, which is expressed in matrix form as:

$$\begin{pmatrix} \sum_{j=0}^{\mathrm{P}} C_{j00} u_j & \dots & \sum_{j=0}^{\mathrm{P}} C_{j\mathrm{P}0} u_j \\ \vdots & \ddots & \vdots \\ \sum_{j=0}^{\mathrm{P}} C_{j0\mathrm{P}} u_j & \dots & \sum_{j=0}^{\mathrm{P}} C_{j\mathrm{P}\mathrm{P}} u_j \end{pmatrix} \begin{pmatrix} u_0^{-1} \\ \vdots \\ u_{\mathrm{P}}^{-1} \end{pmatrix} = \begin{pmatrix} 1 \\ \vdots \\ 0 \end{pmatrix}. \tag{4.64}$$

This system of equations has a unique solution provided that the determinant of its matrix is non-zero. In fact, it can be shown that the conditioning of system (4.64) degrades as the density $\boldsymbol{\xi}$ such that $u(\boldsymbol{\xi}) = 0$ increases. This behavior reflects the fact that u^{-1} is not defined for $u = 0$ and that u^{-1} may not be in $L_2(\Xi, P_\Xi)$ if the density of u does not vanish at 0. Still, for finite P, the approximation of u^{-1} on the truncated space $\mathcal{S}^{\mathrm{P}}$ may exist, provided the density of u at 0 it is not too large, but great care must be taken in such a situation as the convergence of $u^{-1} \in \mathcal{S}^{\mathrm{P}}$ as $\mathrm{P} \to +\infty$ is not to be expected in general. These effects are illustrated in Fig. 4.4, which depicts the computed pseudo-spectral inverses of $u(\xi) = 1 + \alpha\xi$, $\xi \sim \mathrm{N}(0, 1)$, on the Hermite polynomials basis with increasing expansion orders No and $\alpha = 1/5$, $1/4$ and $1/3$. The convergence issue as No increases can be seen from Fig. 4.5 which shows the standard deviations of the approximate $\widehat{u^{-1}}$ as a function of No for different α. It is seen that when the expansion order of the pseudo-spectral approximation increases, its expansion does not converge, denoting the fact that in this example u^{-1} is not square-integrable. This seems to prevent us from using this approach for the determination of inverses, especially for stochastic bases involving unbounded domains as for the Wiener-Hermite expansions. In practice however, one needs to invert some physical quantities $u(\boldsymbol{\xi})$ (like density, temperature, ...) which must not vanish and one is still prompted the use the same expansion order for both the quantity and its inverse. As a result, increasing the expansion order in a simulation usually ensures the convergence $u(\boldsymbol{\xi})$ which in turn has a well-defined inverse $\widehat{u^{-1}}(\boldsymbol{\xi})$.

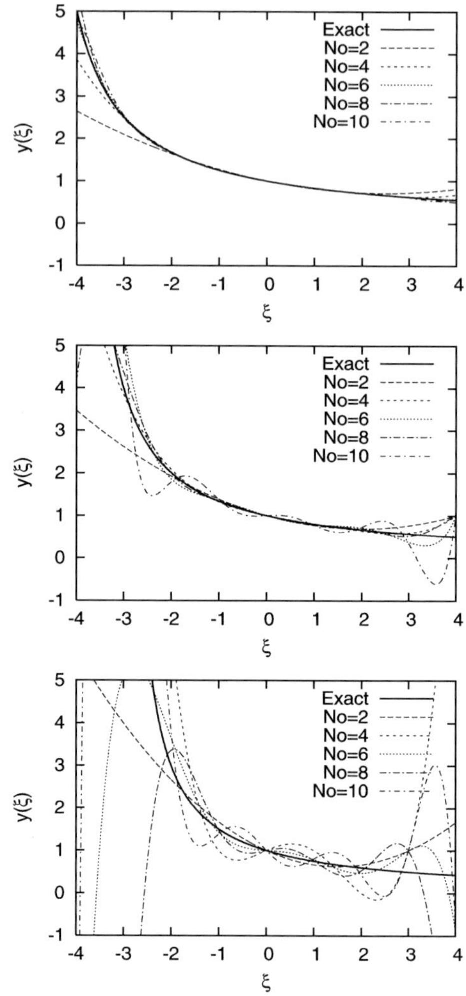

Fig. 4.4 Pseudo-spectral approximation at different orders of the inverse $y(\xi) = \widehat{u^{-1}}(\xi)$ of $u(\xi) = 1 + \alpha\xi$ with $\xi \sim \mathrm{N}(0, 1)$: $\alpha = 1/5$ (*top*), $1/4$ (*middle*) and $1/3$ (*bottom*). Wiener-Hermite expansions are used

The procedure can also be used to perform division by stochastic quantities. For instance, let us assume the expansions of u and v known on $\mathcal{S}^{\mathrm{P}}$ and define w such that

$$w = \frac{u}{v}. \tag{4.65}$$

This equation can be interpreted as $wv = u$, which in a weak sense gives $\widehat{w} * v = u$, leading to the set of linear equations for the expansion coefficients of w on $\mathcal{S}^{\mathrm{P}}$:

$$\begin{pmatrix} \sum_{j=0}^{\mathrm{P}} C_{j00} v_j & \cdots & \sum_{j=0}^{\mathrm{P}} C_{j\mathrm{P}0} v_j \\ \vdots & \ddots & \vdots \\ \sum_{j=0}^{\mathrm{P}} C_{j0\mathrm{P}} v_j & \cdots & \sum_{j=0}^{\mathrm{P}} C_{j\mathrm{PP}} v_j \end{pmatrix} \begin{pmatrix} w_0 \\ \vdots \\ w_{\mathrm{P}} \end{pmatrix} = \begin{pmatrix} u_0 \\ \vdots \\ u_{\mathrm{P}} \end{pmatrix}. \tag{4.66}$$

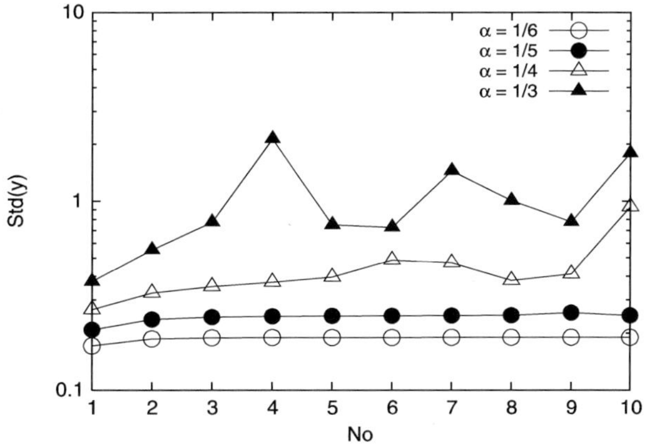

Fig. 4.5 Standard deviations for different expansion orders No of the pseudo-spectral inverse $y(\xi)$ of $x(\xi) = 1 + \alpha\xi$ with $\xi \sim N(0, 1)$ and α as indicated. Wiener-Hermite expansions are used

It is remarked that if the division has to be performed for various $u(\boldsymbol{\xi})$, it is more advantageous to first compute $\widehat{v^{-1}}(\boldsymbol{\xi})$, by inversion of $v(\boldsymbol{\xi})$ as described above, and then use $\hat{w} = u * \widehat{v^{-1}}$ for the different u.

4.5.3 Square Root

The definition of the spectral product in (4.54) can also be used to find the stochastic expansion of square roots. Indeed, for given $u(\boldsymbol{\xi})$ we have almost surely

$$\sqrt{u}(\boldsymbol{\xi})\sqrt{u}(\boldsymbol{\xi}) = u(\boldsymbol{\xi}). \tag{4.67}$$

Again, we can rely on the Galerkin product to yield a set of nonlinear equations for the expansion coefficients of $\sqrt{u}(\boldsymbol{\xi})$ on $\mathcal{S}^{\mathrm{P}}$. This results in:

$$\begin{pmatrix} \sum_{j=0}^{\mathrm{P}} C_{j00}\sqrt{u}_j & \cdots & \sum_{j=0}^{\mathrm{P}} C_{j\mathrm{P}0}\sqrt{u}_j \\ \vdots & \ddots & \vdots \\ \sum_{j=0}^{\mathrm{P}} C_{j0\mathrm{P}}\sqrt{u}_j & \cdots & \sum_{j=0}^{\mathrm{P}} C_{j\mathrm{PP}}\sqrt{u}_j \end{pmatrix} \begin{pmatrix} \sqrt{u}_0 \\ \vdots \\ \sqrt{u}_{\mathrm{P}} \end{pmatrix} = \begin{pmatrix} u_0 \\ \vdots \\ u_{\mathrm{P}} \end{pmatrix}. \tag{4.68}$$

This system can be solved using a standard technique such as Newton-Raphson iterations (see for instance [220]). Choosing for the initial guess $\widehat{\sqrt{u}}(\boldsymbol{\xi}) = \pm\sqrt{u_0}$ allows for the selection of the positive or negative square root of $u(\boldsymbol{\xi})$. As for the determination of pseudo-spectral inverses, some conditions on $u(\boldsymbol{\xi})$ must hold to ensure the existence of a solution to system (4.68). Specifically, as $\mathrm{P} \to \infty$, $u(\boldsymbol{\xi})$ should be almost surely positive.

4.5.4 Absolute Values

The absolute value is another type of common nonlinearity. Given the stochastic expansion of a random variable $u(\boldsymbol{\xi})$, a pseudo-spectral approximation of the projection of $|u|(\boldsymbol{\xi})$ on $\mathcal{S}^{\mathrm{P}}$ can be constructed by identifying it with the positive square root of $\widehat{u^2} = u * u$ using the techniques described above.

Here, the main difficulty comes from the lack of differentiability of the absolute value function at $u = 0$. The approximation of $|u|(\boldsymbol{\xi})$ on the polynomial basis exhibits a slow convergence rate whenever $u(\boldsymbol{\xi})$ vanishes for some $\boldsymbol{\xi}$. This is illustrated in Fig. 4.6 where pseudo-spectral approximations of $|u|(\xi)$ are plotted for various random variables $u(\xi)$ expanded on Wiener-Hermite and Wiener-Legendre bases.

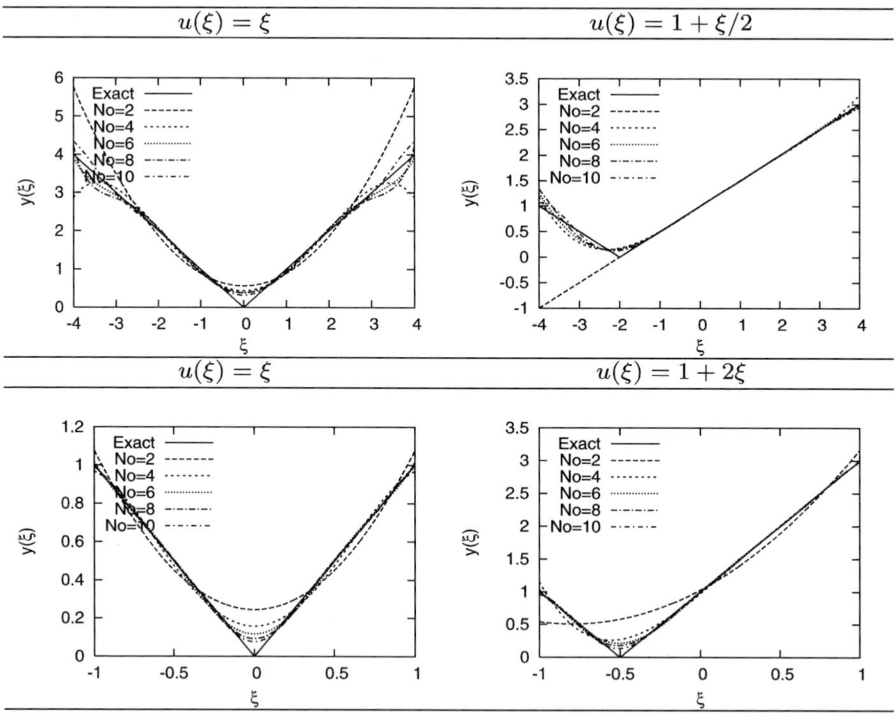

Fig. 4.6 Convergence with the expansion order No of $y(\xi)$, the pseudo-spectral approximation on $\mathcal{S}^{\mathrm{No}}$ of $y(\xi) = |u(\xi)|$ for different $u(\xi)$. *Top plots*: $\xi \sim \mathrm{N}(0,1)$ and Wiener-Hermite expansions are used. *Bottom plots*: $\xi \sim \mathcal{U}(-1,1)$ and Wiener-Legendre expansions are used

4.5.5 Min and Max Operators

Deterministic simulations often require the evaluation of Min and Max operators:

$$u, v \in \mathbb{R} \mapsto \mathrm{Min}(u, v) = \begin{cases} u, & u \leq v, \\ v, & u > v, \end{cases} \tag{4.69}$$

$$u, v \in \mathbb{R} \mapsto \mathrm{Max}(u, v) = \begin{cases} u, & u \geq v, \\ v, & u < v. \end{cases} \tag{4.70}$$

Extension of these operators to stochastic arguments is thus needed, together with a methodology to approximate the expansion coefficients of $\mathrm{Min}(u(\mathbf{x}), v(\boldsymbol{\xi}))$ and $\mathrm{Max}(u(\mathbf{x}), v(\boldsymbol{\xi}))$. Since $\mathrm{Min}(u, v) = -\mathrm{Max}(-u, -v)$ we only consider the case of the Max operator.

The idea we explore below is to construct a stochastic sequence $\{x^n\} \in \mathcal{S}^{\mathrm{P}}$ that converges to $\mathrm{Max}(u, v)$ in a weak sense. To this end, we first consider the deterministic case.

Deterministic sequence: For $u, v \in \mathbb{R}$ given, let us consider the functional $g(\cdot; u, v)$ defined as:

$$x \in \mathbb{R} \mapsto g(x; u, v) = -(x-u)(x-v)(x-w) \in \mathbb{R}, \qquad w \equiv \frac{u+v}{2}. \tag{4.71}$$

Clearly u and v are zeros of $g(\cdot; u, v)$, so $\mathrm{Max}(u, v)$ is the largest zero of g, the third zero being w which lies between $\mathrm{Min}(u, v)$ and $\mathrm{Max}(u, v)$. Considering Newton-Raphson iterations for the determination of the zero of g, one constructs the sequence $\{x^n\}$ using the iterations:

$$x^{n+1} = h(x^n) \equiv x^n - \frac{g(x^n; u, v)}{g'(x^n, u, v)} = x^n + \frac{(x^n-u)(x^n-v)(x^n-w)}{3\,(x^n)^2 - 6wx^n + uv + uw + vw}. \tag{4.72}$$

It is easy to show that starting from $x^0 > \mathrm{Max}(u, v)$ the sequence will converge to the largest zero of g, i.e. $\lim_{n\to\infty} x^n = \mathrm{Max}(u, v)$.

Stochastic sequence: This iterative technique can now be extended to the stochastic case. The iteration becomes

$$h(x(\boldsymbol{\xi})) = x(\boldsymbol{\xi}) + \frac{(x(\boldsymbol{\xi}) - u(\boldsymbol{\xi}))(x(\boldsymbol{\xi}) - v(\boldsymbol{\xi}))(x(\boldsymbol{\xi}) - w(\boldsymbol{\xi}))}{3x(\boldsymbol{\xi})x(\boldsymbol{\xi}) - 6w(\boldsymbol{\xi})x(\boldsymbol{\xi}) + u(\boldsymbol{\xi})v(\boldsymbol{\xi}) + u(\boldsymbol{\xi})w(\boldsymbol{\xi}) + v(\boldsymbol{\xi})w(\boldsymbol{\xi})}, \tag{4.73}$$

and we need to compute the expansion coefficients of $h(x(\boldsymbol{\xi}))$. These can be approximated by applying, step by step, the pseudo-spectral techniques described above.

First, using the Galerkin product, we compute the expansion coefficients on $\mathcal{S}^{\mathrm{P}}$ of g':

$$\widehat{g'} = \sum_{k=0}^{\mathrm{P}} g'_k \Psi_k = 3x * x - 6x * w + u * w + u * v + v * w. \tag{4.74}$$

Then, the pseudo-spectral inverse of $\widehat{g'}$ is determined, giving the expansion coefficients of $\widehat{\widehat{g'}^{-1}}(\boldsymbol{\xi}) = \sum_{k=0}^{\mathrm{P}} \widehat{g'}_k^{-1} \Psi_k(\boldsymbol{\xi})$. Successive Galerkin products are then applied to approximate g/g',

$$\widehat{g/g'}(\boldsymbol{\xi}) = -(x-u) * (x-v) * (x-w) * \widehat{\widehat{g'}^{-1}} = \sum_{k=0}^{\mathrm{P}} (g/g')_k \Psi_k(\boldsymbol{\xi}), \tag{4.75}$$

and the result is finally subtracted from $x(\boldsymbol{\xi})$ yielding the subsequent term of the stochastic sequence. Operations count shows a total of 4 Galerkin products and one pseudo-spectral inversion per iteration. The iterations are stopped when the difference between two successive terms in the sequence, $\Delta x^{n+1} = x^{n+1} - x^n$, falls below a prescribed tolerance. The magnitude of Δx^n is measured using its stochastic norm $\|\Delta x^n\|^2_{P_\Xi}$:

$$\|\Delta x^n\|^2_{P_\Xi} = \mathrm{E}[(\Delta x^n)^2] = \left\langle \left(\Delta x^{n+1}\right)^2 \right\rangle = \sum_{k=0}^{\mathrm{P}} \left(\Delta x_k^n\right)^2 \left\langle \Psi_k^2 \right\rangle. \tag{4.76}$$

To expect convergence of the sequence $\{x^n\}$ to $\mathrm{Max}(u, v)$, it has to be initialized such that $x^0(\boldsymbol{\xi}) > \mathrm{Max}(u(\boldsymbol{\xi}), v(\boldsymbol{\xi}))$. A possible initialization consists in setting the initial term to a large deterministic value, $x^0(\boldsymbol{\xi}) = X^0$, where X^0 may be taken as

$$X^0 = \alpha \sqrt{\|u\|^2_{P_\Xi} + \|v\|^2_{P_\Xi}}, \tag{4.77}$$

where $\alpha > 1$ is a prescribed constant.

Example: Figures 4.7 and 4.8 illustrate a sequence $\{x^n\}$ approaching $\mathrm{Max}(u, v)$ for a two-dimensional stochastic space: $\boldsymbol{\xi} = \{\xi_1, \xi_2\}$. The random variables ξ_1 and ξ_2 are independent and uniformly distributed on $[-1, 1]^2$, and the stochastic discretization uses Wiener-Legendre expansions truncated to $\mathrm{No} = 5$. The left plot in Fig. 4.7 shows $u(\xi_1, \xi_2)$ and $v(\xi_1, \xi_2)$: u is quadratic in the two random variables, while v is constant and equal to zero. The right plot shows the convergence of the algorithm as measured by the norm of the successive increments Δx^n of the sequence. We observe a first stage where the convergence exhibits a behavior typical of the Newton-Raphson iterations with a fast decay rate of the increments norm. The first five terms of the sequence are plotted in Fig. 4.8 to illustrate the convergence. In a second stage, the convergence rate slows down as the effects of pseudo-spectral approximations of g/g' become larger. In fact, when using high-order expansions,

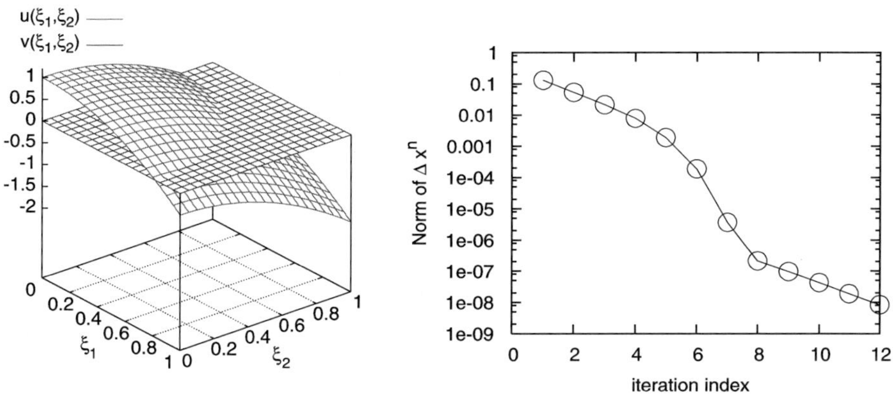

Fig. 4.7 *Left*: $u(\xi_1, \xi_2)$ and $v(\xi_1, \xi_2)$ for which $\mathrm{Max}(u, v)$ is sought. Only a portion of the stochastic domain Ξ is shown for clarity. *Right*: convergence of the sequence $\{x^n\}$ measured by the stochastic norm of $\Delta x^n = x^n - x^{n-1}$ approximating $\mathrm{Max}(u, v)$

it may be necessary to rely on under-relaxed iterations, modifying the sequence to

$$x^{n+1} = \tilde{h}(x^n) = x^n - \omega \frac{g(x^n; u, v)}{g'(x^n; u, v)} \tag{4.78}$$

for some $0 < \omega \leq 1$, in order to ensure the convergence of the algorithm when truncation effects become significant.

As for the absolute value, $\mathrm{Max}(u, v)(\boldsymbol{\xi})$ is generally not differentiable and the stochastic projection on continuous polynomial spaces will generally exhibit a slow convergence with the expansion order. In Fig. 4.9 we show an example of the convergence of the spectral approximation $w(\boldsymbol{\xi}) = \mathrm{Max}(u, v)(\boldsymbol{\xi})$, for two-dimensional Wiener-Legendre bases with increasing order No. It is seen that though u and v have a first-order expansion, a significantly larger order is needed to accurately represent $w = \mathrm{Max}(u, v)$.

4.5.6 Integration Approach

An integration approach was proposed in [51] for the projection of a class of differentiable nonlinearities. Consider again the problem of approximating the expansion on $\mathcal{S}^{\mathrm{P}}$ of a functional $f(u(\boldsymbol{\xi}))$, given the expansion coefficients u_k of $u(\boldsymbol{\xi})$. If $f(\cdot)$ is analytical with derivative $f'(\cdot)$, f can be defined as some integral of f'. However, the integration path has to be deterministic. To this end, consider the following

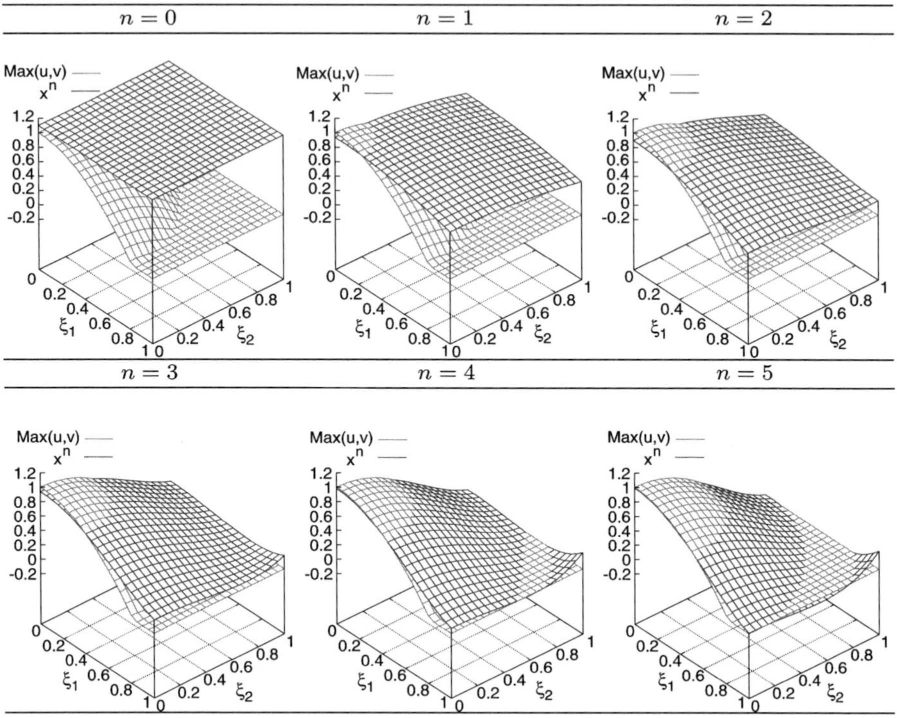

Fig. 4.8 First terms of the sequence $\{x^n\}$ in the pseudo-spectral approximation of $\text{Max}(u, v)$ on a two dimensional stochastic space. Only a portion of the stochastic domain Ξ is shown for clarity. Wiener-Legendre expansions with $\text{No} = 5$ are used

stochastic processes of $L^2(\Xi, P_\Xi)$ indexed by $s \in \mathbb{R}$:

$$y = y(s, \boldsymbol{\xi}) = \sum_{k=0}^{\mathrm{P}} y_k(s)\Psi_k(\boldsymbol{\xi}), \qquad f = f(s, \boldsymbol{\xi}) = \sum_{k=0}^{\mathrm{P}} f_k(s)\Psi_k(\boldsymbol{\xi}),$$
$$f' = f'(s, \boldsymbol{\xi}) = \sum_{k=0}^{\mathrm{P}} f'_k(s)\Psi_k(\boldsymbol{\xi}). \tag{4.79}$$

Therefore, we have

$$\int_{s_1}^{s_2} \frac{\partial f}{\partial s} \, \mathrm{d}s = \int_{s_1}^{s_2} f' \frac{\partial y}{\partial s} \, \mathrm{d}s. \tag{4.80}$$

Introducing the expansion of f in the previous equation leads to:

$$\sum_{k=0}^{\mathrm{P}} \Psi_k \int_{s_1}^{s_2} \frac{\mathrm{d} f_k}{\mathrm{d}s} \, \mathrm{d}s = \sum_{k=0}^{\mathrm{P}} \Psi_k \left[f_k(s_2) - f_k(s_1) \right]$$

$$= \sum_{i=0}^{\mathrm{P}} \sum_{j=0}^{\mathrm{P}} \Psi_i \Psi_j \int_{s_1}^{s_2} f_i'(s) \frac{\mathrm{d}y_j}{\mathrm{d}s} \, \mathrm{d}s. \tag{4.81}$$

If we now set the integration path such that for all $k = 0, \ldots, \mathrm{P}$

$$y_k(s_1, \boldsymbol{\xi}) = \hat{u}_k, \qquad y_k(s_2) = u_k, \tag{4.82}$$

we obtain after projection on the stochastic basis the expression of the expansion coefficients of $f(u(\boldsymbol{\xi}))$

$$f_k = \left(f(\hat{u})\right)_k + \sum_{i=0}^{\mathrm{P}} \sum_{j=0}^{\mathrm{P}} C_{ijk} \int_{\hat{u}_j}^{u_j} f_i' \, \mathrm{d}y_j, \quad \forall k = 0, \ldots, \mathrm{P}. \tag{4.83}$$

As remarked in [51], (4.83) will be useful provided two conditions hold. The first is that the expansion coefficients of $f(\hat{u})$ are known. This can be easily achieved, for instance by selecting an integration path starting from a deterministic $\hat{u}$, *i.e.* using $\hat{u}_k = 0$ for $k = 1, \ldots, \mathrm{P}$. A classical choice is $\hat{u} = \langle u \rangle$. The second is that the expansion of $f'(y(s, \boldsymbol{\xi}))$ can be easily determined along the integration path. This is

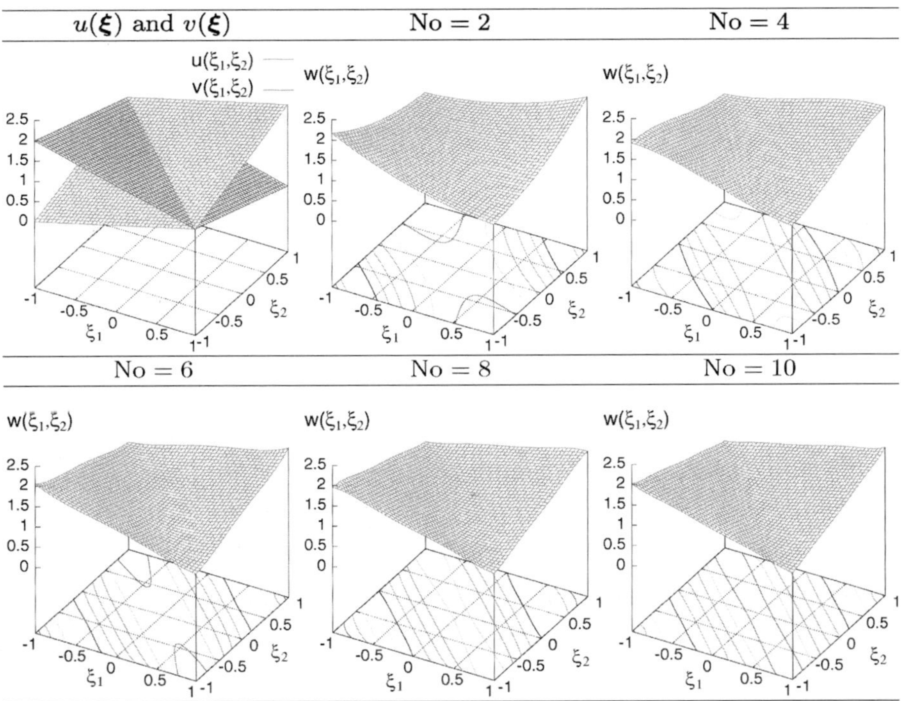

Fig. 4.9 Convergence with truncation order No of the stochastic spectral expansion of $w(\boldsymbol{\xi})$ approximating $\mathrm{Max}(u, v)(\boldsymbol{\xi})$ on the two dimensional stochastic space of uniformly distributed random variables. The random variables $u(\boldsymbol{\xi})$ and $v(\boldsymbol{\xi})$ are linear in ξ_1 and ξ_2 as depicted in the *top-left plot*

a more stringent condition which directly limits the applicability of the integration method to nonlinearities whose derivative can be expanded on the stochastic basis using one the techniques described above. Examples of nonlinearities amenable to integration techniques include exponentials, square roots, logarithms and powers [51].

We now provide details concerning the integration technique for the exponential, $f(u) = \exp(u)$. From $u(\boldsymbol{\xi}) = \sum_{k=0}^{\mathrm{P}} u_k \Psi_k(\boldsymbol{\xi})$, we set

$$y(s, \boldsymbol{\xi}) = s \left(\sum_{k=0}^{\mathrm{P}} u_k \Psi_k(\boldsymbol{\xi}) \right),$$

so that $y(s = 0, \boldsymbol{\xi}) = 0$ and $y(s = 1, \boldsymbol{\xi}) = u(\boldsymbol{\xi})$. Because $f'(y) = f(y)$ we have

$$\int_0^1 \frac{\partial f(s, \boldsymbol{\xi})}{\partial s} \,\mathrm{d}s = \int_0^1 f'(s, \boldsymbol{\xi}) \frac{\partial y}{\partial s} \,\mathrm{d}s = \int_0^1 f(s, \boldsymbol{\xi}) u(\boldsymbol{\xi}) \,\mathrm{d}s. \tag{4.84}$$

The PC expansions of $f(s, \boldsymbol{\xi})$ and $u(\boldsymbol{\xi})$ are then introduced in the previous expansions, which after projection on Ψ_k reduce to:

$$f_k(s = 1) - f_k(s = 0) = \sum_{i=0}^{\mathrm{P}} \sum_{j=0}^{\mathrm{P}} \int_0^1 C_{ijk} f_i(s) u_j \,\mathrm{d}s, \quad k = 0, \dots, \mathrm{P}. \tag{4.85}$$

In fact, since $f_k(s = 0) = \langle \exp(y(s = 0, \boldsymbol{\xi})) \Psi_k(\boldsymbol{\xi}) \rangle = \exp(0) \delta_{0,k}$, the evaluation of the expansion of $f(\boldsymbol{\xi}) = \exp(u(\boldsymbol{\xi}))$ amounts to the integration up to $s = 1$ of the set of $\mathrm{P} + 1$ ODEs

$$\frac{\mathrm{d} f_k}{\mathrm{d}s} = \sum_{i=0}^{\mathrm{P}} \sum_{j=0}^{\mathrm{P}} C_{ijk} u_i f_k, \quad k = 0, \dots, \mathrm{P}, \tag{4.86}$$

starting from the initial condition (at $s = 0$)

$$f_0 = 1, \qquad f_k = 0 \quad \text{for } k = 1, \dots, \mathrm{P}. \tag{4.87}$$

A standard integration technique (*e.g.* Euler, Runge-Kutta, ...) can be used for this purpose. It is clear however, that the integration strategy provides only approximate coefficients of $f(\boldsymbol{\xi})$, due to integration errors and stochastic truncation errors. The integration error can be controlled by selecting an appropriate integration scheme with small enough integration step Δs, while the truncation error essentially depends on the order of the expansion which has to be sufficiently large. It is however difficult to know *a priori* the expansion order needed to maintain low truncation errors and one should carefully verify the convergence of the approximation with regard to No.

4.5.7 Other Types of Nonlinearities

The previous projection techniques cover a limited class of nonlinearities. More general treatments are needed. The development of general methodologies is the object of ongoing research and we outline in this section additional approaches that can be applied toward this goal.

4.5.7.1 Taylor Expansion

From a deterministic functional $f : x \in \mathbb{R} \mapsto f(x) \in \mathbb{R}$, we want to estimate the expansion coefficients of the random functional $f(u)$ for random arguments $u(\boldsymbol{\xi}) \in \mathcal{S}^{\mathrm{P}}$. To this end, let us rewrite $u(\boldsymbol{\xi})$ as the sum of its expectation $\langle u \rangle = u_0$ and zero mean deviation $u' = \sum_{k=1}^{\mathrm{P}} u'_k \Psi_k$. If the functional is infinitely differentiable, the Taylor expansion of $f(u)$ about $\langle u \rangle$ is

$$f(u(\boldsymbol{\xi})) = f(\langle u \rangle) + u'(\boldsymbol{\xi}) \left.\frac{\mathrm{d}f}{\mathrm{d}x}\right|_{x=\langle x \rangle} + \frac{(u'(\boldsymbol{\xi}))^2}{2} \left.\frac{\mathrm{d}^2 f}{\mathrm{d}x^2}\right|_{x=\langle u \rangle} + \frac{(u'(\boldsymbol{\xi}))^3}{3!} \left.\frac{\mathrm{d}^3 f}{\mathrm{d}x^3}\right|_{x=\langle u \rangle} + \cdots + \frac{(u'(\boldsymbol{\xi}))^n}{n!} \left.\frac{\mathrm{d}^n f}{\mathrm{d}x^n}\right|_{x=\langle u \rangle} + \cdots. \tag{4.88}$$

This series is convergent provided that $u'(\boldsymbol{\xi})$ is a.s. within the radius of convergence of the Taylor expansion of f about $\langle u \rangle$. Also, since as $\mathrm{d}^n f / \mathrm{d}x^n(\langle u \rangle)$ is deterministic, applying (4.88) essentially amounts to the evaluation of the stochastic expansion of the successive powers of $u'(\boldsymbol{\xi})$. These in turn can be approximated using the recursive pseudo-spectral approximation described above:

$$\left(u'(\boldsymbol{\xi})\right)^n = u'(\boldsymbol{\xi}) * \left(u'(\boldsymbol{\xi})\right)^{n-1} = u'(\boldsymbol{\xi}) * \left(u'(\boldsymbol{\xi}) * \left(\cdots * \left(u'\boldsymbol{\xi}\right)\right)\right).$$

Clearly, the projection of an infinitely differentiable functional f, through its Taylor expansion, will be feasible only for random argument u having low variability, in order to ensure a fast convergence of the Taylor series. Otherwise, spectral aliasing and truncation errors arising from the systematic projection of the successive powers of u' would compromise the accuracy of the resulting approximation, or may even prevent convergence of the series. In fact, though the Taylor series approach provides a general method for expanding the functional, in practice it has to be applied carefully, in particular with a control of the series convergence.

4.5.7.2 Non-intrusive Projection

One can rely on one of the non-intrusive projection techniques described in Chap. 3 to obtain the expansion coefficients of a nonlinear random functional $f : u(\boldsymbol{\xi}) \in$

$\mathcal{S}^{\mathrm{P}} \mapsto f(u(\boldsymbol{\xi})) \in L^2(\Xi, P_\Xi)$. For instance, using a cubature formula with N_c points $\{\boldsymbol{\xi}^l\}$ and associated weights $\{\omega^l\}$, the projection of $f(u)$ on $\mathcal{S}^{\mathrm{P}}$ is

$$f(u(\boldsymbol{\xi})) \approx \sum_{k=0}^{\mathrm{P}} f_k \Psi_k(\boldsymbol{\xi}), \tag{4.89}$$

where

$$f_k = \frac{\langle f(u(\boldsymbol{\xi})), \Psi_k(\boldsymbol{\xi})\rangle}{\langle \Psi_k^2 \rangle} \approx \frac{1}{\langle \Psi_k^2 \rangle} \sum_{l=1}^{N_c} f(u(\boldsymbol{\xi}^l))\Psi_k(\boldsymbol{\xi}^l)\omega^l. \tag{4.90}$$

Compared to the Taylor series, the non-intrusive projection is not limited to infinitely differentiable functionals, but the accuracy of the quadrature used must be verified to ensure that the expansion coefficients are correctly estimated. Moreover, if one needs to rely on such a non-intrusive sampling, it is legitimate to question whether a fully non-intrusive method would be better suited. In fact, the answer to this question essentially depends on the number of times one has to perform such non intrusive projections in the Galerkin context, and in practice it should be limited to the projection of some equation coefficients, ideally during a preprocessing stage.

4.6 Closing Remarks

This chapter has provided an abstract outline of a general approach to intrusive PC formulations of stochastic problems. The presentation focused on the underlying weighted residual formalism, but also included a brief analysis of the structure of linear problems as well as a discussion of pseudo-spectral approaches for the estimation of nonlinear transformations.

The methods resulting from the intrusive approaches outlined above naturally inherit well-known properties that one usually associates with Galerkin methods. These include the fact that the stochastic representation error is orthogonal to the finite subspace spanned by the truncated basis, so that the representation is optimal in the mean-square norm. With a judicious choice of basis functions, one also obtains representations exhibiting exponential convergence. These two key properties provide a strong motivation for the development and implementation of intrusive methods.

Despite its fairly abstract nature, it is evident from the discussion above that the intrusive PC formalism results in a larger system of equations than in the deterministic case. In addition, the stochastic system may generally exhibit a different structure than its deterministic counterpart, which implies that one may require a new class of solvers and computational codes to simulate the discretized stochastic system. Consequently, the merits of the intrusive approach will depend on the tradeoffs between the additional computing overheads associated with the numerical solution of the stochastic PC system, and the accuracy benefits due to the spectral convergence of the PC representation.

In the following two chapters, examples are provided which demonstrate that, when endowed with efficient solution schemes, intrusive Galerkin approaches provide substantial advantages over non-intrusive solvers. These examples include both simplified settings, which are discussed in great detail, as well as CFD applications which include elaborate models. Thus, investment in intrusive approaches may be appropriate not only in idealized settings but also for large-scale applications.

Chapter 5
Detailed Elementary Applications

This chapter deals with elementary applications of PC methods in idealized settings, specifically involving the heat equation in 2D and the 1D Burgers equation. Attention is focused on the intrusive formalism. Furthermore, we restrict ourselves to steady problems, the extension of the methods to transient situations being relatively straightforward. Within this restricted framework, we strive to provide a detailed outline of the formulation of the stochastic problem, the discretization schemes employed, and solution methods for the resulting discrete stochastic systems.

In Sect. 5.1, we address the steady 2D heat equation. We start with the deterministic case, and present the problem in its original strong form, as well as its variational formulation. By defining the solution space as a tensor product of the spatial FE basis and the stochastic PC basis, Sect. 5.1.2 extends the variational formulation to the stochastic case. A general approach is adopted which generally accommodates random diffusivity and/or boundary conditions. The spatial and stochastic discretizations of the variational problem are discussed in detail, and solution methods for the discrete spectral problem are briefly outlined.

Application of the concepts developed in Sect. 5.1.2 is illustrated in Sects. 5.1.3–5.1.6. In Sect. 5.1.3, we analyze the case of random conductivity, and show that the intrusive approach leads to coupled systems for the stochastic modes. Experiences indicate that for the problem considered accurate predictions can be obtained using low-order PC expansions. In Sect. 5.1.4 a slightly more elaborate problem is addressed, with independent stochastic conductivities on two sub-domains. Computations are used to demonstrate the efficiency of the PC representation, and its ability to accurately capture not only the solution first moments but also its pdf. In Sect. 5.1.5 the case of uncertain boundary conditions is considered, and the computations are applied to a problem involving multiple sources of uncertainty, including both material properties and boundary conditions. This same setting is further examined in Sect. 5.1.6, where a variance analysis is conducted in order to demonstrate capabilities afforded by the PC representation, particularly in assessing the impact of different sources of uncertainty.

Section 5.2 deals with the steady 1D viscous Burgers equation. Similarly to the analysis of Sect. 5.1, we start in Sect. 5.2.1 with a detailed description of the deterministic problem, in both its strong form and its weak form. We rely on the

O.P. Le Maître, O.M. Knio, *Spectral Methods for Uncertainty Quantification*,
Scientific Computation,
DOI 10.1007/978-90-481-3520-2_5,

weak form in the formulation of the discrete spatial problem, the solution being represented in terms of a high-order basis function expansion based on Lagrange interpolants. This results in a system of nonlinear equations in terms of the nodal unknowns, which is solved using a standard Newton solver.

In Sect. 5.2.2, a stochastic variant of the steady viscous Burgers problem is presented. We specifically consider the case of a random diffusivity, parametrized in terms of an N-dimensional stochastic vector $\boldsymbol{\xi}$. The stochastic solution is then sought in the tensor product space of the discrete space spanned by the spatial Lagrange interpolants, and the PC of order No. A Galerkin approach is used to construct the nonlinear system governing the unknown expansion coefficients. As for the deterministic system, a Newton solver is used to determine the discrete stochastic solution.

Application of the resulting stochastic Galerkin scheme is illustrated in Sect. 5.2.3 for the case of a log-normal diffusivity, and the convergence of the solution mean and standard deviation is analyzed for different orders of the PC expansion. Further analysis of the predictions is presented in Sects. 5.2.4 and 5.2.5, which contrast Galerkin predictions to results obtained using NISP and MC sampling, respectively. A percentile analysis is finally presented that illustrates the efficiency afforded by the PC expansion in sampling the random solution.

5.1 Heat Equation

5.1.1 Deterministic Problem

In this section, the stochastic Galerkin projection methodology is detailed for the linear steady heat equation in an isotropic two-dimensional domain Ω, with boundary $\partial\Omega$. For the deterministic problem, one has to compute the temperature field $u : \mathbf{x} \in \Omega \mapsto u(x) \in \mathbb{R}$, satisfying the heat equation:

$$\nabla \cdot (\nu(\mathbf{x}) \nabla u(\mathbf{x})) = -f(\mathbf{x}), \tag{5.1}$$

where $\nu > 0$ is the thermal conductivity and $f \in L_2(\Omega)$ is a given source term. The heat equation (5.1) needs be complemented with boundary conditions; we consider homogeneous Dirichlet and Neumann conditions over the respective portions Γ_d and Γ_n of the domain boundary $\partial\Omega = \Gamma_d \cup \Gamma_n$, i.e.

$$u(\mathbf{x}) = 0, \quad \mathbf{x} \in \Gamma_d, \tag{5.2}$$

and

$$\frac{\partial u}{\partial n} = 0, \quad \mathbf{x} \in \Gamma_n. \tag{5.3}$$

The computational domain and the Dirichlet and Neumann boundaries are depicted in the left plot of Fig. 5.1.

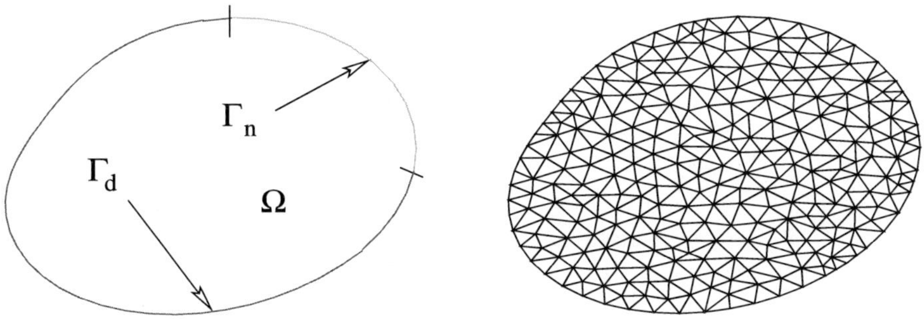

Fig. 5.1 *Left*: sketch of the domain Ω and decomposition of the boundary $\partial\Omega$ into Dirichlet Γ_d and Neumann Γ_n regions. *Right*: example of a finite-element mesh with 508 elements and 284 nodes

5.1.1.1 Variational Formulation

Let $\mathcal{V}$ be the set of functionals on Ω such that:

$$\mathcal{V} = \{u \in H^1(\Omega) : u = 0 \text{ on } \Gamma_d\}, \tag{5.4}$$

where $H^1(\Omega)$ is the Sobolev space of square integrable functionals whose first-order derivatives are also square integrable. When $\Gamma_n = \emptyset$, $\mathcal{V}$ is often denoted $H_0^1(\Omega)$. The heat equation (5.1) together with the boundary conditions in (5.2)–(5.3) can be expressed in the following variational form:
Find $u \in \mathcal{V}$ such that

$$a(u, v) = b(v) \quad \forall v \in \mathcal{V}, \tag{5.5}$$

where $a(u, v)$ and $b(v)$ are bilinear and linear forms respectively defined as:

$$a(u, v) \equiv \int_\Omega \nu(\mathbf{x}) \nabla u(\mathbf{x}) \cdot \nabla v(\mathbf{x})\, d\mathbf{x}, \tag{5.6}$$

and

$$b(v) \equiv \int_\Omega f(\mathbf{x}) v(\mathbf{x})\, d\mathbf{x}. \tag{5.7}$$

5.1.1.2 Finite Element Approximation

For the resolution of (5.5) we rely of a finite element discretization involving classical P_1 finite elements. Let $\mathcal{T} = \{\Sigma_1, \ldots, \Sigma_{ne}\}$ be a triangulation of Ω with ne non-overlapping triangular elements Σ_i, such that the intersection of two distinct elements is either empty, a common vertex, or a common edge. An example of a finite element mesh of Ω is depicted in the right plot of Fig. 5.1.

Locally on each Σ_i, the finite element approximation v^h of a functional $v \in \mathcal{V}$ is linear. Moreover, conformity of the approximation is imposed by enforcing the continuity of v^h at the inter-element boundaries. The global degrees of freedom of v^h are then its values at the mesh nodes, and the finite-element approximation can be recast as:

$$v^h(\mathbf{x}) = \sum_{i \in \mathcal{N}} v_i^h \Phi_i(\mathbf{x}), \tag{5.8}$$

where $\mathcal{N}$ is the set of nodes of the finite-element mesh which are not lying on Γ_d and the $\Phi_i(\mathbf{x})$ are the shape functions associated to these nodes. We denote $\mathcal{V}^h$ the corresponding finite-element approximation space:

$$\mathcal{V}^h = \text{span}\{\Phi_i\}_{i \in \mathcal{N}}. \tag{5.9}$$

Plugging the finite-element approximations of u and v into the variational form (5.5), and accounting for the linearity of a and b, results in:

$$\sum_{i \in \mathcal{N}} \sum_{j \in \mathcal{N}} a_{i,j} u_i v_j = \sum_{j \in \mathcal{N}} b_j v_j, \tag{5.10}$$

where

$$a_{i,j} = \int_\Omega \nu(\mathbf{x}) \nabla \Phi_i(\mathbf{x}) \cdot \nabla \Phi_j(\mathbf{x}) \, d\mathbf{x}, \tag{5.11}$$

and

$$b_i = \int_\Omega f(\mathbf{x}) \Phi_i(\mathbf{x}) \, d\mathbf{x}. \tag{5.12}$$

Using an appropriate indexing of the nodes in $\mathcal{N}$, the finite-element approximation of the variational equation can be recast as a set of linear equations for the set of nodal values u_i of u^h, according to:

$$\begin{pmatrix} a_{1,1} & \dots & a_{1,n} \\ \vdots & \ddots & \vdots \\ a_{n,1} & \dots & a_{n,n} \end{pmatrix} \begin{pmatrix} u_1 \\ \vdots \\ u_n \end{pmatrix} = \begin{pmatrix} b_1 \\ \vdots \\ b_n \end{pmatrix}, \tag{5.13}$$

where $n = \text{Card}(\mathcal{N})$ is the number of unknowns or degrees of freedom of the finite-element problem and the system coefficients $a_{i,j}$ and right-handside components b_i still given by (5.11) and (5.12), respectively. In fact, the matrix $[a]$ is sparse; it is also symmetric, as can be readily deduced from the expression of the $a_{i,j}$'s in (5.11). It can be further shown that the system (5.13) has a unique solution for the finite-element discretization used here [70]. A classical method for large sparse linear systems, such as the conjugate gradient method, can be employed for the resolution of (5.13).

5.1.2 Stochastic Problem

Consider now the situation where the heat equation (5.1) involves a random conductivity and a random source term, *i.e.* that ν and f are functions of a random event θ

of an abstract probability space (Θ, Σ, P):

$$\nu = \nu(\mathbf{x}, \theta), \qquad f = f(\mathbf{x}, \theta). \tag{5.14}$$

Then, the solution is also random and satisfies almost surely the stochastic problem

$$\begin{cases} \nabla \cdot (\nu(\mathbf{x}, \theta)\nabla u(\mathbf{x}, \theta)) = -f(\mathbf{x}, \theta), & \mathbf{x} \in \Omega, \\ u(\mathbf{x}, \theta) = 0, & \mathbf{x} \in \Gamma_d, \\ \dfrac{\partial u(\mathbf{x}, \theta)}{\partial n} = 0, & \mathbf{x} \in \Gamma_n. \end{cases} \tag{5.15}$$

5.1.2.1 Stochastic Variational Formulation

The deterministic space for the random solution is still $\mathcal{V}$, and the complete functional space for $u(\mathbf{x}, \theta)$ will be $\mathcal{V} \otimes L_2(\Theta, P)$. In other words,

$$u(\cdot, \theta) \in \mathcal{V}, \qquad u(\mathbf{x}, \cdot) \in L_2(\Theta, P), \tag{5.16}$$

or: for a given event θ, the solution $u \in \mathcal{V}$, while for fixed $\mathbf{x} \in \Omega$, u is a second-order random variable. The variational form of the problem then becomes:
Find $u \in \mathcal{V} \otimes L_2(\Theta, P)$ such that

$$A(u, v) = B(v) \quad \forall v \in \mathcal{V} \otimes L_2(\Theta, P), \tag{5.17}$$

where

$$A(u, v) \equiv \mathrm{E}[a(u, v)] = \int_\Theta \left[\int_\Omega \nu(\mathbf{x}, \theta)\nabla u(\mathbf{x}, \theta) \cdot \nabla v(\mathbf{x}, \theta)\, \mathrm{d}\mathbf{x} \right] \mathrm{d}P(\theta) \tag{5.18}$$

and

$$B(v) \equiv \mathrm{E}[b(v)] = \int_\Theta \left[\int_\Omega f(\mathbf{x}, \theta) v(\mathbf{x}, \theta)\, \mathrm{d}\mathbf{x} \right] \mathrm{d}P(\theta). \tag{5.19}$$

5.1.2.2 Deterministic Discretization

Introducing the deterministic finite-element space $\mathcal{V}^h$, defined previously, the semi-discrete stochastic solution u^h is expressed as:

$$u^h(\mathbf{x}, \theta) = \sum_{i \in \mathcal{N}} u_i(\theta)\Phi_i(\mathbf{x}) \in \left(\mathcal{V}^h \otimes L_2(\Theta, P) \right). \tag{5.20}$$

This expression shows that, following deterministic discretization, one must determine a set of $n = \mathrm{Card}(\mathcal{N})$ random variables $u_i(\theta)$, such that

$$\sum_{i \in \mathcal{N}} \sum_{j \in \mathcal{N}} \mathrm{E}\left[A_{i,j}(\theta) u_i(\theta) v_j(\theta)\right] = \sum_{i \in \mathcal{N}} \mathrm{E}[B_i(\theta) v_i(\theta)],$$
$$\forall v_i(\theta) \in L_2(\Theta, P),\ i \in \mathcal{N}, \tag{5.21}$$

where

$$A_{i,j}(\theta) = \int_\Omega \nu(\mathbf{x},\theta)\nabla\Phi_i(\mathbf{x})\cdot\nabla\Phi_j(\mathbf{x})\,\mathrm{d}\mathbf{x}, \tag{5.22}$$

and

$$B_i(\theta) = \int_\Omega f(\mathbf{x},\theta)\Phi_i(\mathbf{x})\,\mathrm{d}\mathbf{x}. \tag{5.23}$$

5.1.2.3 Stochastic Discretization

At this point, we introduce the stochastic discretization of $L_2(\Theta, P)$ to represent the random solution. We assume that the random conductivity and source fields are parameterized using a set of N independent normalized Gaussian random variables $\boldsymbol{\xi} = (\xi_1 \cdots \xi_{\mathrm{N}})$ defined on (Θ, Σ, P):

$$\nu(\mathbf{x},\theta) = \nu(\mathbf{x},\boldsymbol{\xi}(\theta)), \qquad f(\mathbf{x},\theta) = f(\mathbf{x},\boldsymbol{\xi}(\theta)). \tag{5.24}$$

If the random fields ν and f are independent, their respective parameterizations in fact involve distinct subsets of $\boldsymbol{\xi}$, but to simplify the notation we do not make this distinction; this point will be clarified in the examples below. The joint probability density function of $\boldsymbol{\xi}(\theta)$ is

$$p_{\boldsymbol{\xi}}(\boldsymbol{\eta} = (\eta_1 \ldots \eta_{\mathrm{N}})) = \frac{1}{(2\pi)^{\mathrm{N}/2}} \prod_{i=1}^{\mathrm{N}} \exp\left(-\frac{\eta_i^2}{2}\right). \tag{5.25}$$

The image probability space is $(\Xi, \mathcal{B}^{\mathrm{N}}, P_\Xi)$, where $\Xi = \mathbb{R}^{\mathrm{N}}$ and the associate probability measure

$$\mathrm{d}P_\Xi(\boldsymbol{\eta}) = p_{\boldsymbol{\xi}}(\boldsymbol{\eta})\,\mathrm{d}\boldsymbol{\eta}. \tag{5.26}$$

The expectation of a random quantity $g(\boldsymbol{\xi})$, denoted using the brackets $\langle\cdot\rangle$, has for expression:

$$\langle g(\boldsymbol{\xi})\rangle \equiv \mathrm{E}\left[g(\boldsymbol{\xi})\right] = \int_\Theta g(\boldsymbol{\xi}(\theta))\,\mathrm{d}P(\theta) = \int_\Xi g(\boldsymbol{\eta})p_{\boldsymbol{\xi}}(\boldsymbol{\eta})\,\mathrm{d}\boldsymbol{\eta}. \tag{5.27}$$

The space of second-order random functionals in $\boldsymbol{\xi}$ is spanned by the Polynomial Chaos basis:

$$\mathcal{S} = \mathrm{span}\{\Psi_k(\boldsymbol{\xi})\}_{k=0}^{k=\infty} = L_2(\mathbb{R}^2, P_\Xi), \tag{5.28}$$

where the Ψ_i's form the set of orthogonal multidimensional Hermite polynomials in $\boldsymbol{\xi}$:

$$\left\langle\Psi_i, \Psi_j\right\rangle = \int_\Xi \Psi_i(\boldsymbol{\eta})\Psi_j(\boldsymbol{\eta})p_{\boldsymbol{\xi}}(\boldsymbol{\eta})\,\mathrm{d}\boldsymbol{\eta} = \delta_{ij}\left\langle\Psi_i^2\right\rangle. \tag{5.29}$$

Therefore, provided that the conductivity and source fields are second-order quantities, they have orthogonal representations:

$$\nu(\mathbf{x},\boldsymbol{\xi}) = \sum_{k=0}^{\infty} \nu_k(\mathbf{x})\Psi_k(\boldsymbol{\xi}), \qquad f(\mathbf{x},\boldsymbol{\xi}) = \sum_{k=0}^{\infty} f_k(\mathbf{x})\Psi_k(\boldsymbol{\xi}). \tag{5.30}$$

Similarly, the expansion of the solution discrete solution u^h is

$$u^h(\mathbf{x},\boldsymbol{\xi}) = \sum_{i\in\mathcal{N}} \left(\sum_{k=0}^{\infty} u_{i,k}\Psi_k(\boldsymbol{\xi})\right)\Phi_i(\mathbf{x}). \tag{5.31}$$

For practical computations, the stochastic expansions are truncated to finite polynomial order. Different orders of truncation may be considered for the conductivity, source and solution expansions. For simplicity, we shall use in the following the same truncation order, No, for all stochastic expansions. Consequently, all expansions involve the same number $\mathrm{P}+1$ of terms with

$$\mathrm{P}+1 = \frac{(\mathrm{No}+\mathrm{N})!}{\mathrm{No}!\mathrm{N}!}. \tag{5.32}$$

This truncation corresponds to a stochastic approximation space

$$\mathcal{S}^{\mathrm{P}} \equiv \mathrm{span}\{\Psi_0,\ldots,\Psi_{\mathrm{P}}\} \subset \mathcal{S}. \tag{5.33}$$

5.1.2.4 Spectral Problem

We are now able to derive the set of equations for the stochastic expansion coefficients of u^h. To this end, the expansions of ν, f, u^h and test functions $v \in \mathcal{V}^h \otimes \mathcal{S}^{\mathrm{P}}$ are introduced into the variational form of the semi discrete stochastic problem (5.21). This results in:
Find u_i^k, $i \in \mathcal{N}$ and $k = 0,\ldots,\mathrm{P}$, such that

$$\sum_{i,j\in\mathcal{N}} \sum_{k,l,m=0}^{\mathrm{P}} \langle \Psi_k\Psi_l\Psi_m\rangle A_{i,j}^k u_{i,l} v_{j,m} = \sum_{i\in\mathcal{N}} \sum_{k=0}^{\mathrm{P}} B_i^k v_{i,k},$$
$$\forall v_{i,k},\ i\in\mathcal{N},\ k=0,\ldots,\mathrm{P} \tag{5.34}$$

where

$$A_{i,j}^k \equiv \int_\Omega \nu_k(\mathbf{x})\nabla\Phi_i(\mathbf{x})\cdot\nabla\Phi_j(\mathbf{x})\,d\mathbf{x}, \qquad B_i^k \equiv \left\langle \Psi_k^2\right\rangle \int_\Omega f_k(\mathbf{x})\Phi_i(\mathbf{x})\,d\mathbf{x}. \tag{5.35}$$

It is observed that (5.34) involves deterministic quantities only. It is also noted that the integrals involved in the definition of $A_{i,j}^k$ simplify due to the orthogonality of the stochastic expansion basis; see (5.29). To better appreciate the structure of the system, let us denote $\mathbf{u}_k \equiv (u_{1,k} \dots u_{n,k})^t$ the vector of nodal values of the k-th stochastic mode of the solution, where $n = \text{Card}(\mathcal{N})$ as in the deterministic case, so that

$$\frac{\langle u^h(\mathbf{x}, \boldsymbol{\xi}) \Psi_k(\boldsymbol{\xi}) \rangle}{\langle \Psi_k^2 \rangle} = (\Phi_1(\mathbf{x}) \cdots \Phi_n(\mathbf{x}))\, \mathbf{u}_k. \tag{5.36}$$

With this notation, the spectral problem becomes: find $\mathbf{u}_0, \dots, \mathbf{u}_{\mathrm{P}}$ such that for all $k = 0, \dots, \mathrm{P}$

$$\sum_{l=0}^{\mathrm{P}} \sum_{m=0}^{\mathrm{P}} \langle \Psi_k \Psi_l \Psi_m \rangle \left[A^l \right] \mathbf{u}_m = \mathbf{B}^k, \tag{5.37}$$

where the matrix $[A^l]$ has for coefficients $A^l_{i,j}$ and the vector $\mathbf{B}^k = (B^k_1 \dots B^k_n)^t$. This set of systems can be formally expressed as a single system

$$[\mathbf{A}]\mathbf{u} = \mathbf{B},$$

where the global system matrix $[\mathbf{A}]$ has the block structure corresponding to:

$$\begin{pmatrix} \mathbf{A}_{0,0} & \dots & \mathbf{A}_{0,\mathrm{P}} \\ \vdots & \ddots & \vdots \\ \mathbf{A}_{\mathrm{P},0} & \dots & \mathbf{A}_{\mathrm{P},\mathrm{P}} \end{pmatrix} \begin{pmatrix} \mathbf{u}_0 \\ \vdots \\ \mathbf{u}_{\mathrm{P}} \end{pmatrix} = \begin{pmatrix} \mathbf{B}_0 \\ \vdots \\ \mathbf{B}_{\mathrm{P}} \end{pmatrix}. \tag{5.38}$$

The matrix blocks are given by:

$$\mathbf{A}_{i,j} = \sum_{m=0}^{\mathrm{P}} [A^m] \langle \Psi_i \Psi_j \Psi_m \rangle, \quad 0 \le i, j \le P.$$

The system $[\mathbf{A}]\mathbf{u} = \mathbf{B}$ is called the spectral problem, and its solution yields the expansion coefficients of u^h on $\mathcal{V}^h \otimes \mathcal{S}^{\mathrm{P}}$. It will be also referred to as the Galerkin problem as it can also be obtained through the Galerkin projection of the semi-discrete stochastic equations on the Ψ_k as shown next.

Alternative derivation of the spectral problem: Another, perhaps more direct, way to derive the spectral problem consists in starting from the discretized deterministic system (5.13), and formally expressing the dependence of the matrix coefficients $a_{i,j}$, solution $(u_1 \dots u_n)^t$ and right-hand-side $(b_1 \dots b_n)^t$ on $\boldsymbol{\xi}$. One obtains:

$$\begin{pmatrix} a_{1,1}(\boldsymbol{\xi}) & \dots & a_{1,n}(\boldsymbol{\xi}) \\ \vdots & \ddots & \vdots \\ a_{n,1}(\boldsymbol{\xi}) & \dots & a_{n,n}(\boldsymbol{\xi}) \end{pmatrix} \begin{pmatrix} u_1(\boldsymbol{\xi}) \\ \vdots \\ u_n(\boldsymbol{\xi}) \end{pmatrix} = \begin{pmatrix} b_1(\boldsymbol{\xi}) \\ \vdots \\ b_n(\boldsymbol{\xi}) \end{pmatrix}. \tag{5.39}$$

Next, the truncated PC expansion of the random matrix, solution and right-hand-side vectors,

$$\begin{pmatrix} a_{1,1}(\boldsymbol{\xi}) & \dots & a_{1,n}(\boldsymbol{\xi}) \\ \vdots & \ddots & \vdots \\ a_{n,1}(\boldsymbol{\xi}) & \dots & a_{n,n}(\boldsymbol{\xi}) \end{pmatrix} \approx \sum_{k=0}^{\mathrm{P}} \left[A^k\right] \Psi_k(\boldsymbol{\xi}), \tag{5.40}$$

$$\begin{pmatrix} u_1(\boldsymbol{\xi}) \\ \vdots \\ u_n(\boldsymbol{\xi}) \end{pmatrix} \approx \sum_{k=0}^{\mathrm{P}} \mathbf{u}_k \Psi_k(\boldsymbol{\xi}), \qquad \begin{pmatrix} b_1(\boldsymbol{\xi}) \\ \vdots \\ b_n(\boldsymbol{\xi}) \end{pmatrix} \approx \sum_{k=0}^{\mathrm{P}} \mathbf{b}_k \Psi_k(\boldsymbol{\xi}), \tag{5.41}$$

are substituted into the system (5.39). In general, this equation cannot be satisfied. Indeed, the left-hand-side has a polynomial degree equal or less than 2No while the right-hand-side has a polynomial degree less or equal to No. As a result, one obtains:

$$\left(\sum_{l=0}^{\mathrm{P}} \left[A^l\right] \Psi_l(\boldsymbol{\xi})\right) \left(\sum_{m=0}^{\mathrm{P}} \mathbf{u}_m \Psi_m(\boldsymbol{\xi})\right) = \sum_{k=0}^{\mathrm{P}} \mathbf{b}_k \Psi_k(\boldsymbol{\xi}) + \mathbf{R}(\boldsymbol{\xi}), \tag{5.42}$$

where $\mathbf{R}(\boldsymbol{\xi}) \in \mathbb{R}^n \otimes \mathcal{S}$ is the stochastic residual. A weak solution of the previous system is obtained by seeking the $\mathbf{u}_k$'s such that the stochastic residual is orthogonal to $\mathbb{R}^n \otimes \mathcal{S}^{\mathrm{P}}$. It amounts to the resolution of the set of $\mathrm{P}+1$ coupled linear systems: For $k = 0, \dots, \mathrm{P}$,

$$\sum_{l=0}^{\mathrm{P}} \sum_{m=0}^{\mathrm{P}} \langle \Psi_k \Psi_l \Psi_m \rangle \left[A^l\right] \mathbf{u}_m = \left\langle \Psi_k^2 \right\rangle \mathbf{b}_k = \mathbf{B}^k, \tag{5.43}$$

which is identical to (5.37).

Solution method: The matrix **[A]** of the spectral problem (5.38) has a block symmetric structure, $\mathbf{A}_{i,j} = \mathbf{A}_{j,i}$, since $\langle \Psi_i \Psi_j \Psi_m \rangle = \langle \Psi_j \Psi_i \Psi_m \rangle$. In addition, the blocks are themselves symmetric matrices because $A^k_{i,j} = A^k_{j,i}$ (see (5.35)) so the matrix **[A]** is symmetric and solution techniques for symmetric linear systems can be used for the resolution of (5.38). In fact, due to the size of the spectral problem, which is $n \times (\mathrm{P}+1)$, the full assembly of matrix **[A]** is generally to be avoided. Instead, it is advantageous to rely on a sparse storage method for **[A]**, since many of the blocks $\mathbf{A}_{i,j}$ are actually zero (see examples below), or even better on a matrix-free approach where only the result of **[A]u** (the so-called matrix-vector product) has to be computed for a given vector **u**. Efficient strategies for the resolution of large linear systems arising from the stochastic Galerkin projection are further discussed in Chap. 7.

Existence and uniqueness of the solution: An important issue that needs to be addressed is the existence and uniqueness of the solution **u** satisfying system (5.38). This essential aspect has been the focus of many works (see *e.g.* [8, 9, 75, 151]).

For Dirichlet boundary conditions, it has been proved that the Galerkin system for stochastic elliptic problems has a unique solution provided that the random conductivity field satisfies some probabilistic (sufficient) conditions. For the deterministic discretization with P_1 finite-elements as used below, these probabilistic conditions of the random conductivity field reduce to

$$\frac{1}{\nu(\cdot, \boldsymbol{\xi})} \in L_2(\Xi, P_\Xi), \tag{5.44}$$

i.e.: for any point in the computational domain, the inverse of the conductivity must be a second-order random variable.

As for the deterministic problem, when Neumann boundary conditions are applied all over the boundary of Ω there is an infinity of solutions: $u(\mathbf{x}, \boldsymbol{\xi})$ is defined up to an arbitrary random variable. In addition, an integral constraint on the source term is necessary. Specifically, when homogeneous Neumann boundary conditions are considered on $\partial\Omega$, the source term must satisfy the solvability condition

$$\int_\Omega f(\mathbf{x}, \boldsymbol{\xi})\, \mathrm{d}\mathbf{x} = F(\boldsymbol{\xi}) = 0 \quad \text{a.s.} \tag{5.45}$$

5.1.3 Example 1: Uniform Conductivity

We consider the simple case where the domain Ω consists of the unit square, $\Omega = [0, 1]^2$, with Dirichlet boundary conditions over 3 edges and Neumann conditions over the left edge $x = 1$. The domain and boundary conditions, together with a typical finite element mesh are shown in Fig. 5.2.

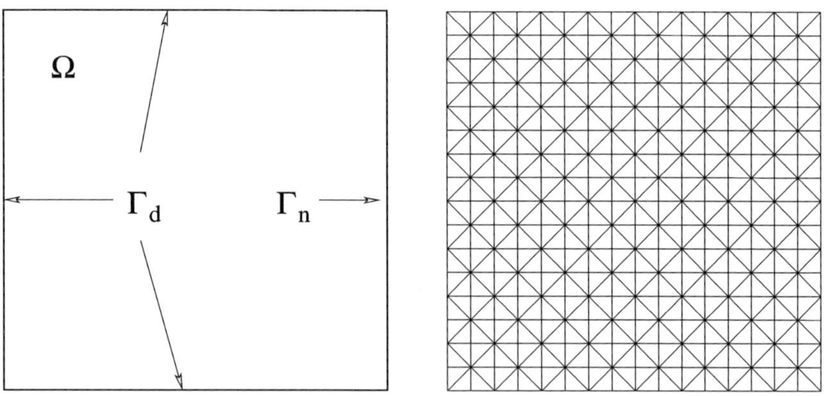

Fig. 5.2 *Left*: computational domain Ω and decomposition of the boundary $\partial\Omega$ into Dirichlet Γ_d and Neumann Γ_n parts for the examples section. *Right*: typical finite-element triangulation of Ω using 512 elements and 289 nodes

5.1.3.1 Trivial Cases

Before considering situations requiring the resolution of the spectral problem (5.38), let us analyze a few trivial situations for which the computation of the stochastic modes u_k^h can in fact be decoupled.

We start with the simplest situation where the conductivity is deterministic, *i.e.* independent of the random event: $\nu(\mathbf{x}, \boldsymbol{\xi}(\theta)) = \nu(\mathbf{x})$. The source term is however stochastic: $f = f(\mathbf{x}, \boldsymbol{\xi})$. In this case, recalling that by convention $\Psi_0(\boldsymbol{\xi}) = 1$, we have $\nu(\mathbf{x}) = \nu_0(\mathbf{x})\Psi_0(\boldsymbol{\xi})$, so that the coefficients $A_{i,j}^k$ in (5.35) reduce to

$$A_{i,j}^0 = a_{i,j} = \int_\Omega \nu(\mathbf{x})\nabla\Phi_i(\mathbf{x}) \cdot \nabla\Phi_j(\mathbf{x})\,d\mathbf{x}, \qquad A_{i,j}^{k>0} = 0. \tag{5.46}$$

Furthermore, the blocks $\mathbf{A}_{i,j}$ of $[\mathbf{A}]$ are given by:

$$\mathbf{A}_{i,j} = \sum_{m=0}^{\mathrm{P}} [A^m]\langle\Psi_i\Psi_j\Psi_m\rangle = [A^0]\langle\Psi_i\Psi_j\rangle = [A^0]\left\langle\Psi_i^2\right\rangle\delta_{i,j}, \tag{5.47}$$

showing that the spectral problem is block-diagonal:

$$\begin{pmatrix} \mathbf{A}_{0,0} & [0] & \dots & [0] \\ [0] & \ddots & \ddots & \vdots \\ \vdots & \ddots & \ddots & [0] \\ [0] & \dots & [0] & \mathbf{A}_{\mathrm{P,P}} \end{pmatrix} \begin{pmatrix} \mathbf{u}_0 \\ \vdots \\ \vdots \\ \mathbf{u}_{\mathrm{P}} \end{pmatrix} = \begin{pmatrix} \mathbf{B}_0 \\ \vdots \\ \vdots \\ \mathbf{B}_{\mathrm{P}} \end{pmatrix}. \tag{5.48}$$

In fact, each stochastic mode $\mathbf{u}_k$ is solution of an independent problem:

$$\mathbf{A}_{k,k}\mathbf{u}_k = \left\langle\Psi_k^2\right\rangle[A^0]\mathbf{u}_k = \mathbf{B}_k = \left\langle\Psi_k^2\right\rangle\mathbf{b}_k. \tag{5.49}$$

Dividing by $\langle\Psi_k^2\rangle$, we get:

$$[A^0]\mathbf{u}_k = \mathbf{B}_k = \left\langle\Psi_k^2\right\rangle\mathbf{b}_k, \tag{5.50}$$

which indicates that the computation of the stochastic modes requires the resolution of $(\mathrm{P}+1)$ identical deterministic problems, but for different right-hand sides $\mathbf{b}_k$.

If in addition the source term has a deterministic spatial shape but a stochastic magnitude, *i.e.* it can be decomposed as

$$f(\mathbf{x}, \boldsymbol{\xi}) = \overline{f}(\mathbf{x})\alpha(\boldsymbol{\xi}), \tag{5.51}$$

the stochastic solution on $\mathcal{V}^h \otimes \mathcal{S}^{\mathrm{P}}$ can be expressed as:

$$u^h(\mathbf{x}, \boldsymbol{\xi}) = \overline{u}^h(\mathbf{x}) \sum_{k=0}^{\mathrm{P}} \alpha_k\Psi_k(\boldsymbol{\xi}), \quad \alpha_k = \frac{\langle\alpha\Psi_k\rangle}{\langle\Psi_k^2\rangle}, \tag{5.52}$$

where we have denoted $\overline{u}^h$ the solution of the deterministic problem for the forcing $\overline{f}(\mathbf{x})$. In this trivial situation we have to solve a single deterministic problem only!

Another situation leading to great simplification occurs when the conductivity is random but has a decomposition similar to (5.51), *i.e.* it can be written as

$$\nu(\mathbf{x}, \boldsymbol{\xi}) = \overline{\nu}(\mathbf{x})\beta(\boldsymbol{\xi}), \tag{5.53}$$

with $\overline{\nu}(\mathbf{x}) > 0\ \forall \mathbf{x} \in \Omega$. In this case, (5.42) becomes

$$\beta(\boldsymbol{\xi})\left[\overline{A}\right]\left(\sum_{l=0}^{\mathrm{P}} \mathbf{u}_l \Psi_l(\boldsymbol{\xi})\right) = \sum_{k=0}^{\mathrm{P}} \mathbf{b}_k \Psi_k(\boldsymbol{\xi}) + \mathbf{R}(\boldsymbol{\xi}), \tag{5.54}$$

with

$$\overline{A}_{i,j} = \int_{\Omega} \overline{\nu}(\mathbf{x}) \nabla \Phi_i(\mathbf{x}) \cdot \nabla \Phi_j(\mathbf{x})\, d\mathbf{x}. \tag{5.55}$$

Multiplying both sides of (5.54) by $1/\beta(\boldsymbol{\xi})$ and proceeding further with the stochastic Galerkin projection, we get:

$$\left[\overline{A}\right]\mathbf{u}_k = \tilde{\mathbf{b}} = \sum_{i=0}^{\mathrm{P}} \sum_{j=0}^{\mathrm{P}} C_{ijk} \left(\frac{1}{\beta}\right)_i \mathbf{b}_j, \quad k = 0, \ldots, P \tag{5.56}$$

where C_{ijk} is the Galerkin product tensor (see Sect. 4.5.1.1) and the expansion coefficients $(1/\beta)_k$ of the inverse of $\beta(\boldsymbol{\xi})$ can be computed using the technique described in Sect. 4.5.2. Note that this manipulation (multiplication by $1/\beta$ and projection to yield (5.56)) makes sense only if $(1/\beta)(\boldsymbol{\xi})$ is a second-order quantity, which supposes in turn that $\nu(\mathbf{x}, \boldsymbol{\xi})$ satisfies the condition (5.44).

It is seen that for this particular expression of the conductivity field, it is possible to decouple the resolution of the stochastic modes of u^h through an appropriate redefinition of the system right-hand-side. If in addition, the stochastic source term can be expressed as in (5.51), the stochastic solution $u^h \in \mathcal{V} \otimes \mathcal{S}^{\mathrm{P}}$ can be expressed as

$$u^h(\mathbf{x}, \boldsymbol{\xi}) = \overline{u}^h(\mathbf{x}) \sum_{k=0}^{\mathrm{P}} \left(\frac{\alpha}{\beta}\right)_k \Psi_k(\boldsymbol{\xi}), \tag{5.57}$$

where the expansion coefficients of α/β on $\mathcal{S}^{\mathrm{P}}$ can be computed by the method outlined in Sect. 4.5.2, and $\overline{u}^h$ is the deterministic solution of the discrete problem for the conductivity and source fields $\overline{\nu}$ and $\overline{f}$.

5.1.3.2 Validation

We use this latter trivial case for the validation of our numerical code. We consider a deterministic constant source term and a random uniform conductivity field:

$$f(\mathbf{x}, \theta) = f(\mathbf{x}) = 1, \qquad \nu(\mathbf{x}, \theta) = \beta(\theta). \tag{5.58}$$

The random conductivity β is assumed to be log-normal, with unit median value $\overline{\beta} = 1$ and a coefficient of variation $C \geq 1$. This indicates that the probability of $\beta \in [\overline{\beta}/C, \overline{\beta}C]$ is equal to 99.9%. In the following, we set $C = \sqrt{10}$ so roughly speaking, the variability of the conductivity ranges over a decade. Furthermore, the parametrization of β uses a unique normalized Gaussian variable $\xi_1(\theta)$ so that the PC expansion of the problem has $\mathrm{N} = 1$, $\boldsymbol{\xi} = (\xi_1)$ and the PC basis is made of the one-dimensional Hermite polynomials. The parametrization of the conductivity is:

$$\beta(\xi_1) = \exp\left(\mu_\beta + \sigma_\beta \xi_1\right), \quad \mu_\beta = \log\left(\overline{\beta}\right) \text{ and } \sigma_\beta = \frac{\log C}{2.85}. \tag{5.59}$$

Its PC expansion coefficients have a closed form expression [86]:

$$\beta(\xi_1) = \sum_{k=0}^{\infty} \beta_k \Psi_k(\xi_1), \quad \beta_k = \exp\left(\mu_\beta + \sigma_\beta^2/2\right) \frac{\sigma_\beta^k}{\langle \Psi_k^2 \rangle}. \tag{5.60}$$

The discrete solution is found by solving system (5.38) which for the present 1D case with a broad conductivity spectrum, has a full block structure.

Solution modes: Figure 5.3 shows the computed stochastic modes of u^h. The solution uses the finite-element mesh shown in Fig. 5.2 and an expansion order $\mathrm{No} = \mathrm{P} = 4$. As expected, the modes $u_k^h(\mathbf{x})$ exhibit a common spatial shape, which is scaled by a factor whose magnitude decreases with the mode index k, as suggested by (5.57). Also provided in the bottom-right plot is the resulting standard deviation of the solution which has obviously the same spatial structure.

Convergence with expansion order: Since β is log-normal, so is its inverse, and the expansion of $1/\beta$ is consequently given by:

$$\left(\frac{1}{\beta}\right)(\xi_1) = \exp\left(-\mu_\beta - \sigma_\beta \xi_1\right) = \sum_{k=0}^{\infty} \left(\frac{1}{\beta}\right)_k \Psi_k(\xi_1), \tag{5.61}$$

where

$$\left(\frac{1}{\beta}\right)_k = \exp\left(-\mu_\beta + \sigma_\beta^2/2\right) \frac{(-\sigma_\beta)^k}{\langle \Psi_k^2 \rangle}. \tag{5.62}$$

One thus expects the spectrum of the numerical solution to decay as $|\sigma_\beta|^k/k!$. Note that the expansion of $1/\beta$ provided above explains the negative values of u_k^h for odd $k > 0$. The spectral decay of the solution, *i.e.* u_k^h/u_0^h as a function of k, is verified in Fig. 5.4 for the point $\mathbf{x} = (1, 0.5)$: compared is the theoretical spectrum $|\sigma_\beta|^k/k!$ with numerical predictions for expansion orders $\mathrm{No} = 4, \ldots, 10$. It is seen that for a given expansion order No, the scaling of the modes is in excellent agreement with the theory, except for the last mode $k = \mathrm{P} = \mathrm{No}$, for which a small deviation from the theoretical value is reported. This deviation can be attributed to the approximate Galerkin product which neglects terms of order greater than No.

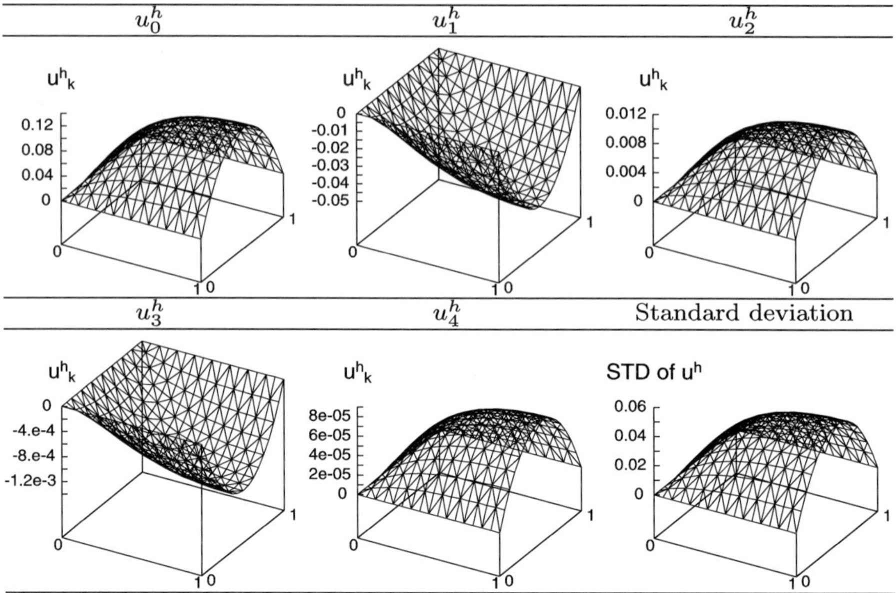

Fig. 5.3 Solution modes u_k^h for $k = 0, \ldots, 4$ computed with No $= 4$. The *bottom right plot* depicts the standard deviation. Case of a deterministic uniform source term ($f = 1$) and random uniform log-normal conductivity with median value 1 and coefficient of variation $C = \sqrt{10}$

Fig. 5.4 Normalized spectra of the random solution u_k^h at node $\mathbf{x} = (1, 0.5)$ as computed using different expansion orders No. Also plotted is the theoretical spectrum (*solid line*). Case of a deterministic uniform source ($f = 1$) and random uniform log-normal conductivity with median value 1 and coefficient of variation $\sqrt{10}$

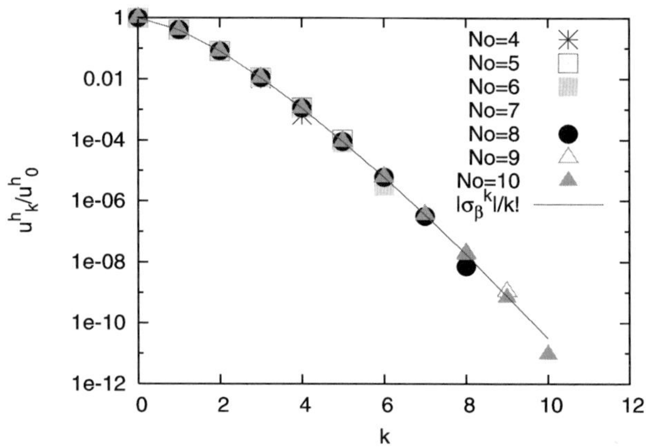

Convergence of probability density functions: For a better appreciation of the convergence of the stochastic solution as No increases, we provide in Fig. 5.5 the pdfs of u^h at node $\mathbf{x} = (1, 0.5)$ computed for No $= 1, \ldots, 6$. The computation of these pdfs relies on direct sampling of the Gaussian variable ξ_1. A pseudo-random

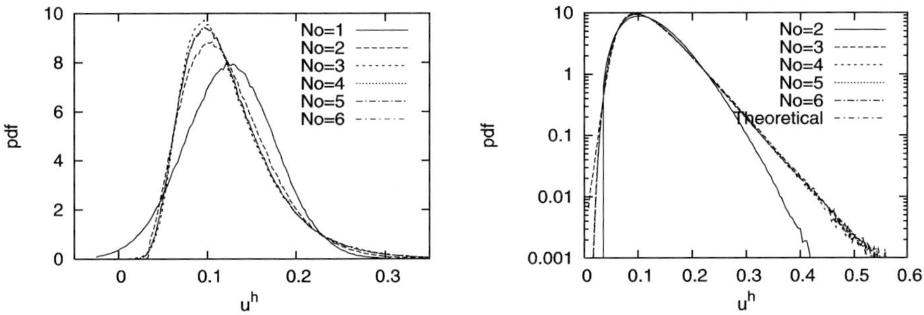

Fig. 5.5 Computed probability density functions of u^h at $\mathbf{x} = (1, 0.5)$ for different expansion orders No as indicated. *Left plot*: $\text{No} = 1, \dots, 6$. *Right plot*: same pdfs in log scale for $\text{No} = 2, \dots, 6$ together with the theoretical pdf from (5.63). Case of a deterministic uniform source ($f = 1$) and random uniform log-normal conductivity with median value 1 and coefficient of variation $\sqrt{10}$

number generator[1] is used for this purpose. For each element of the random sample set, the expansion series of u^h is evaluated; each amounts to the evaluation of a set of P polynomials, which requires only few operations, so that a large sample set for u^h can be generated at a low computational cost. From the resulting realizations of u^h, the pdfs are estimated through a discretization of the range of u_h into bins of equal size; the relative frequency of finding u^h in the bins is used to plot the pdfs after an appropriate rescaling by the bin size. Here, a set containing 5,000,000 samples was used for the analysis. It takes only seconds to generate the sample set and to conduct the analysis. In contrast, performing resolutions of the deterministic heat equation for the corresponding 5,000,000 realizations of the conductivity would take hours.

The left plot in Fig. 5.5, which uses a linear scale for the pdfs, shows the fast convergence with No for high probability events: on the linear scale, pdfs of solutions for $\text{No} \geq 4$ are essentially not discernible. One may observe that for $\text{No} = 1$, which corresponds to the Gaussian approximation, the probability of having $u^h < 0$ is significant, which is unphysical for the deterministic source term considered here: even though the heat equation is linear, a first-order expansion is not enough to achieve a meaningful solution. The right plot in Fig. 5.5 uses a logarithmic scale for the pdfs in order to better appreciate the convergence of the PC solution for low probability events. It compares the tails of the pdfs for $\text{No} = 2, \dots, 6$ with the pdf of the "exact" random solution

$$u^{ex}(\mathbf{x}, \xi_1) = \left(\frac{1}{\beta}\right)(\xi_1)\overline{u}^h(\mathbf{x}), \tag{5.63}$$

where $\overline{u}^h$ is the discrete deterministic solution for $\nu = 1$ and $f = 1$. $1/\beta$ is given by (5.61) and so its pdf is known analytically. We observe from the plot that $\text{No} = 2$ leads to a significant error for both tails of the pdf of u^h. For $\text{No} = 3$, only the left-tail of the pdf is plagued by errors, though only for events with quite low probabilities.

[1] Specifically, the generator from the `zufall` package, written by P.W. Petersen, and available from http://www.netlib.org.

For $\text{No} \geq 4$, no significant differences occur in the tails of the pdfs, and these are in excellent agreement with the theoretical pdf discussed above. It is emphasized that this agreement between computed pdfs and the theoretical pdf ranges over roughly 4 decades. In fact, small differences can be observed for the highest values of u^h but they are essentially masked by the sampling noise: an adapted sampling strategy would be necessary for a better assessment of the low probability events contained in the stochastic expansion. At any rate, it is remarkable that the stochastic expansion of the solution already captures accurately the statistics of these low probability events with a very limited number of modes. In fact, the high efficiency of the representation may have already been guessed from the spectra plotted in Fig. 5.4. It is precisely this essential feature of the spectral expansions that makes them very attractive and powerful, as they provide means to obtain very accurate solutions through the computation of a limited number of modes.

Remark We have shown that for the present example, the solution of the stochastic Galerkin problem quickly converges as the expansion order No increases, with a rate consistent with the theoretical analysis of the problem. It is noted that the theoretical analysis was conducted for a fixed deterministic discretization, since we expressed the stochastic solution on $\mathcal{V}^h \otimes \mathcal{S}$ as a function of some deterministic solution $\overline{u}^h \in \mathcal{V}^h$. In addition to the stochastic convergence as $\mathcal{S}^{\mathrm{P}} \to \mathcal{S}$ one should also be concerned with the accuracy of the solution with regards to the deterministic discretization space, *i.e.* the convergence of the solution as $\mathcal{V}^h \to \mathcal{V}$.

5.1.4 Example 2: Nonuniform Conductivity

5.1.4.1 Setup

Let us now reconsider the previous example with a random conductivity field defined as follows:

$$\nu(\mathbf{x}, \theta) = \begin{cases} \nu^1(\theta), & x \leq 0.5, \\ \nu^2(\theta), & x > 0.5 \end{cases} \tag{5.64}$$

where ν^1 and ν^2 are two independent log-normal random variables with respective medians $\overline{\nu}^1$ and $\overline{\nu}^2$, and coefficients of variation C^1 and C^2. Two normalized Gaussian variables ξ_1 and ξ_2 are then needed to parametrize the conductivity of the two sub-domains, and the problem can be regarded as two bodies with perfect thermal contact. The stochastic dimension is therefore $\mathrm{N} = 2$, $\boldsymbol{\xi} = (\xi_1, \xi_2)$, and the stochastic basis will now be the set of two dimensional Hermite polynomials.

The conductivity $\nu(\mathbf{x}, \boldsymbol{\xi})$ cannot be reduced to a form as in (5.53) so that no trivial solution is available here. Since ν^1 and ν^2 are independent, we adopt the following parameterization; for $i = 1, 2$:

$$\nu^i(\boldsymbol{\xi}) = \exp[\mu_i + \sigma_i \xi_i], \quad \mu_i = \log \overline{\nu}^i \text{ and } \sigma_i = \frac{\log C^i}{2.85}. \tag{5.65}$$

To show how non-trivial effects can emerge from simple stochastic settings, let us examine the stochastic solution for $\overline{\nu}^1 = \overline{\nu}^2 = 1$ and $C^1 = C^2 = \sqrt{10}$. For these settings, the probability law of $\nu(\cdot, \boldsymbol{\xi})$ is the same as the previous example, thus one may naively believe that the solution should remain essentially the same. This is however not the case as shown below. Before doing so, we make a few remarks.

First, though the conductivity field has locally the same probability distribution as in the previous example, its covariance function,

$$c_\nu(\mathbf{x}, \mathbf{y}) \equiv \langle (\nu(\mathbf{x}, \boldsymbol{\xi}) - \langle \nu(\mathbf{x}, \boldsymbol{\xi}) \rangle) (\nu(\mathbf{y}, \boldsymbol{\xi}) - \langle \nu(\mathbf{y}, \boldsymbol{\xi}) \rangle) \rangle, \tag{5.66}$$

exhibits a completely different structure. Indeed, whereas the constant conductivity field of the previous example leads to a constant covariance function equal to the variance of the conductivity $\sigma^2(\nu)$, in the present case it is given by:

$$c_\nu(\mathbf{x}, \mathbf{y}) = \sigma^2(\nu) \left(\mathbb{I}_{x \le 0.5}(\mathbf{x}) \mathbb{I}_{x \le 0.5}(\mathbf{y}) + \mathbb{I}_{x > 0.5}(\mathbf{x}) \mathbb{I}_{x > 0.5}(\mathbf{y}) \right), \tag{5.67}$$

where $\mathbb{I}$ is the indicator function.

We also observe that the conductivity field in (5.64) is discontinuous along the internal boundary $x = 0.5$. This character calls for an appropriate finite-element mesh to adequately represent this feature: we enforce the triangulation to be fitted to with the boundary $x = 0.5$, *i.e.* that no element intersects it. As a result, the conductivity over an element is a random variable and exhibits no spatial dependency. Specific treatments would be necessary for finite element meshes non-fitted to the discontinuous conductivity field; this has been considered in [170, 172], together with the even more complex situation of internal discontinuities of material properties with random locations.

Second, the expansion on $\mathcal{S}^{\mathrm{P}}$ of the random conductivity field,

$$\nu(\mathbf{x}, \boldsymbol{\xi}) = \sum_{k=0}^{\mathrm{P}} \nu_k(\mathbf{x}) \Psi_k(\boldsymbol{\xi}), \tag{5.68}$$

has non-zero modes $\nu_k(\mathbf{x})$ only for indices k corresponding to $\Psi_k(\boldsymbol{\xi})$ being actually a one-dimensional polynomial; in the present case where $\mathrm{N} = 2$, it means that $\nu_k(\mathbf{x}) = 0$ if $\Psi_k(\boldsymbol{\xi}) = \psi_{k_1}(\xi_1)\psi_{k_2}(\xi_2)$ is such that k_1 and k_2 are both ≥ 0. Consequently, some elementary matrices $[A^l]$ (see (5.35)) are zero, a fact resulting in a sparse block structure for the Galerkin system (5.38). The sparsity of the full Galerkin matrix system $[\mathbf{A}]$ is illustrated in Fig. 5.6 for $\mathrm{No} = 4, \ldots, 6$.

5.1.4.2 Mean and Standard Deviation

The Galerkin system is solved using $\mathrm{No} = 5$ so the 2D basis has dimension $\dim(\mathcal{S}^{\mathrm{P}}) = 21$. In Fig. 5.7 we compare the mean and standard deviation of the solution u^h with those of Example 1 in Sect. 5.1.3. To facilitate the comparison, the same iso-contours are used to represent the two cases and the plots for the case of a

Fig. 5.6 Block structure of the Galerkin system matrix [**A**] for the example of Sect. 5.1.4 for No = 4 (*left*), No = 5 (*center*) and No = 6 (*right*). The dimensions of the stochastic spaces $\mathcal{S}^{\mathrm{P}}$ are $\mathrm{P}+1 = 15$, 21 and 28, respectively

constant random conductivity have been transposed. We observe that for the discontinuous random conductivity a slightly lower expected temperature u^h is obtained with flatter profiles along the y direction. Differences in the standard deviation fields are even more pronounced. Specifically, the standard deviation field has no longer the same spatial structure as the expectation field: the effect of the discontinuity in the random conductivity field, along the inner boundary $x = 0.5$, is also clearly seen in the spatial distribution of the standard deviation, with significant deformations of the iso-contours compared to the case of the constant conductivity in Example 1. In addition, one observes that allowing for the conductivities to take independently random values depending on x smaller or greater than 0.5, though from the same probabilistic distribution, leads overall to a lower uncertainty level. This can be better appreciated from Fig. 5.8 where the uncertainty in u^h is depicted in terms of bars extending one standard deviation above and below the mean. The reduction in the resulting variability can be roughly explained by first noting that, from the physical standpoint, a higher conductivity means a lower temperature since the heat flows more easily. Now consider an event (or realization) such that the conductivity of the left domain is higher than its median value. Since the two conductivities are independent, knowing ν^1 provides no information on the value of ν^2. However, we can assess that ν^2 is more likely to be lower than ν^1 (remember that ν^1 and ν^2 have the same probability distribution), so on average ν^2 will be less than it would have been for the case of Example 1, where $\nu^1 = \nu^2$ with probability 1. This leads to a decrease in the variability of the solution.

5.1.4.3 Analysis of the Solution Modes

The standard deviation field just shown indicates a more complex structure for the stochastic modes as compared to the case of Example 1 in Sect. 5.1.3. The modes u_k^h, $k = 0, \ldots, 9$ corresponding to components Ψ_k with degree less or equal to 3 are reported in Fig. 5.9. Inspection of the spatial structure of the modes shows that they cannot be deduced from each other by means of a simple scaling as in Example 1. Even for the two modes associated to degree 1 polynomials, the asymmetry of the

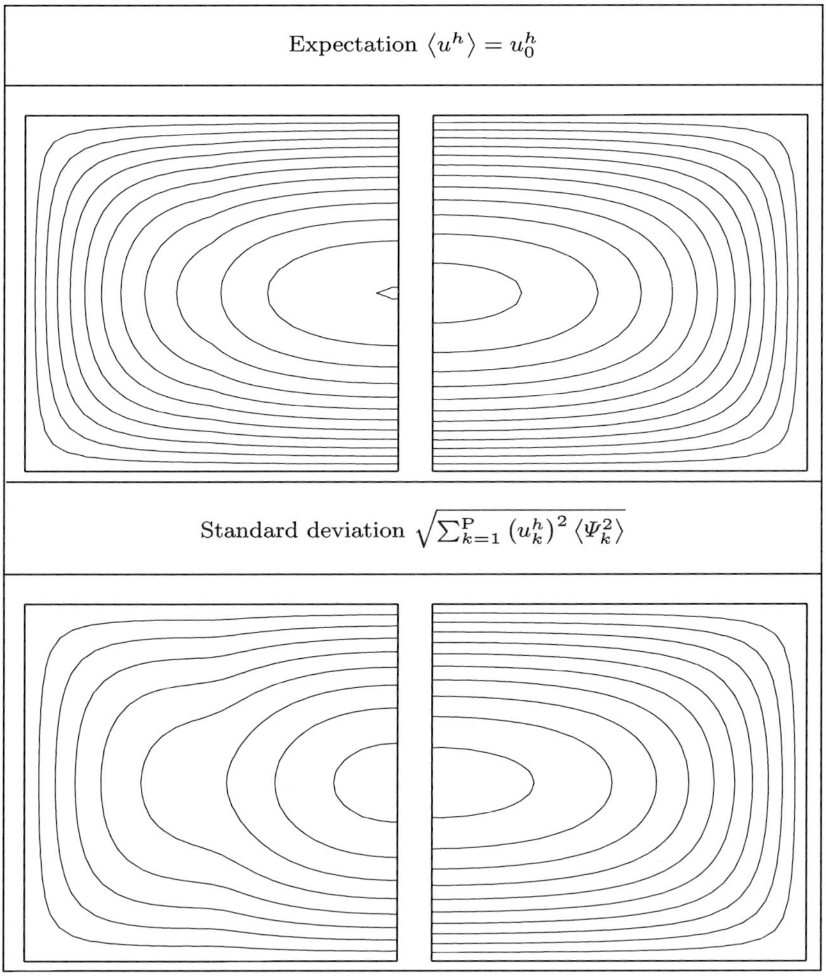

Fig. 5.7 Expectations (*top*) and standard deviations (*bottom*) of u^h for No = 5. *Left*: random conductivities (N = 2, P = 20) for the case of Sect. 5.1.4. *Right*: random conductivity (N = 1, P = 5) for the case of Sect. 5.1.3. To facilitate the comparison, the plots have been transposed into the x direction. The *plots* use the same iso-contours starting from 0 with constant increment 0.01 and 0.005 for the expectation and standard deviation fields respectively. Conductivities have log-normal distributions with median 1 and coefficient of variation $\sqrt{10}$

dependence of the solution with regard to ξ_1 and ξ_2, caused by the non-symmetric boundary conditions, results in clearly unsymmetrical modes. Note that the modes standing on the far left (resp. far right) of the lines correspond to components involving variations of ξ_1 (resp. ξ_2) only, while modes in between accounts for the mixed contributions of variations of ξ_1 and ξ_2.

Though it cannot be readily appreciated from the plots where scales for u_k^h are not shown, the magnitude of the modes decreases with the degree of the correspond-

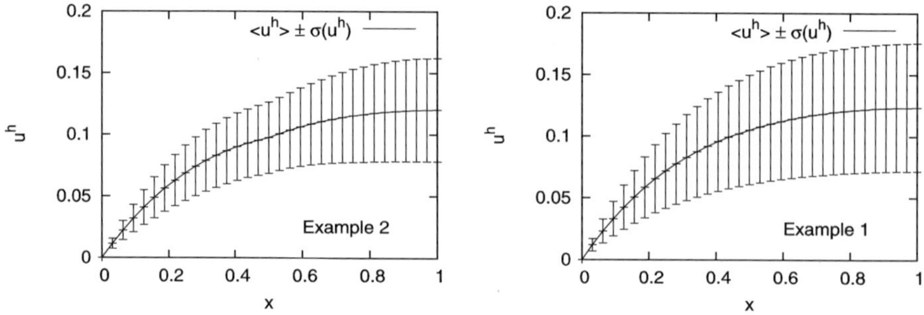

Fig. 5.8 Uncertainty in u^h along the line $y = 0.5$ represented in terms of bars extending one standard deviation above and below the mean. *Left*: case of Example 2 in Sect. 5.1.4. *Right*: Case of Example 1 in Sect. 5.1.3. In both cases $\nu(\mathbf{x}, \theta)$ is log-normal with median 1 and coefficient of variation $C = \sqrt{10}$. Solutions are computed with $\mathrm{No} = 5$

ing polynomial as for Example 1. This again demonstrates the convergence of the Galerkin projection as No increases.

5.1.4.4 Probability Density Functions

In Fig. 5.10 we compare the pdfs of the temperature at $\mathbf{x} = (1, 0.5)$ for Examples 1 and 2. It is observed that the maximum of the pdf for Example 2 is slightly shifted toward the highest temperature compared to Example 1. This is not in contradiction with our previous observation of a lower temperature expectation for Example 1, since for skewed pdfs the maximum of the pdf does not coincide with the mean. In fact, the main difference between the pdfs of the two examples is the significant faster decay of the pdf tails for Example 2. This is emphasized in the right plot where the two pdfs are plotted on a logarithmic scale.

5.1.5 Example 3: Uncertain Boundary Conditions

We now consider the case of uncertainty in the boundary conditions. An immediate consequence of this extension is that boundary conditions can no longer be considered homogeneous, even though they may still vanish in the mean. This is the case because there exists a set of random events with non-zero probability corresponding to non-vanishing boundary values.

5.1.5.1 Treatment of Uncertain Boundary Conditions

Treatment of non-homogeneous Neumann boundary conditions in the finite element method amounts to a modification of the form $b(v)$ in (5.12). Specifically, assuming

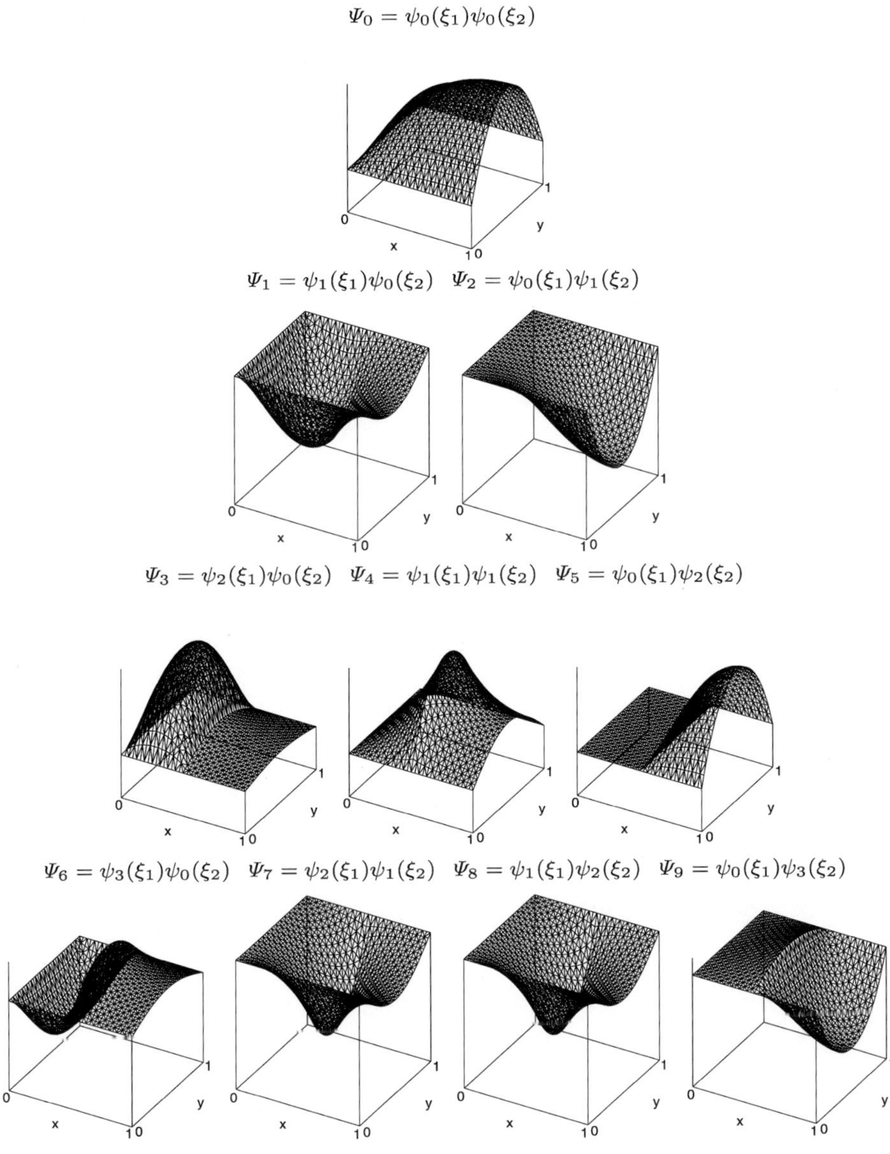

Fig. 5.9 Modes $u_k(\mathbf{x})$ of the stochastic solution for the non-uniform conductivity problem. The mode index is indicated. The *plots* are arranged in rows, in which the degree of the associated polynomial chaos, $\Psi_k(\boldsymbol{\xi})$, is the same. The latter increases *from top to bottom*, starting with 0 for the *top row*

a random Neumann boundary condition,

$$\nu(\mathbf{x}, \theta)\frac{\partial u}{\partial n} = \Phi^n(\mathbf{x}, \theta), \quad \mathbf{x} \in \Gamma_n, \tag{5.69}$$

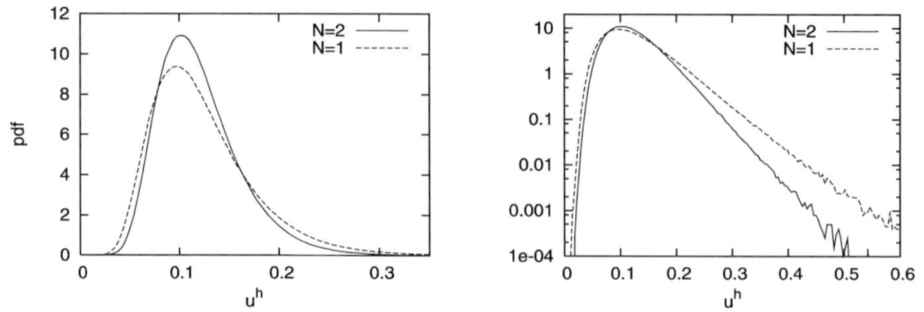

Fig. 5.10 Comparison of the solution pdfs for the random conductivity (N = 1) of Sect. 5.1.3 and random conductivities (N = 2) of Sect. 5.1.4. *Left*: linear scale; *right*: logarithmic scale. All conductivities have log-normal distribution with unit median value and coefficient of variation $C = \sqrt{10}$

leads to a stochastic form for $v \in \mathcal{V} \otimes L_2(\Theta, P)$:

$$b^n(v) = \int_\Omega f(\mathbf{x}, \theta) v(\mathbf{x}, \theta)\, d\mathbf{x} + \int_{\Gamma_n} \Phi^n(\mathbf{x}, \theta) v(\mathbf{x}, \theta)\, d\mathbf{x}. \tag{5.70}$$

Non-homogeneous Dirichlet boundary conditions can be handled as follows. Let us denote $\mathcal{U} \supset \mathcal{V}$ the deterministic space of continuous functionals with square integrable derivatives on Ω. Assuming the Dirichlet boundary condition to be

$$u(\mathbf{x}, \theta) = U^d(\mathbf{x}, \theta) \quad \mathbf{x} \in \Gamma_d, \tag{5.71}$$

the solution of the heat equation with non-homogeneous Dirichlet boundary conditions can be expressed as

$$u(\mathbf{x}, \theta) = u^0(\mathbf{x}, \theta) + u^d(\mathbf{x}, \theta), \tag{5.72}$$

where $u^0 \in \mathcal{V} \otimes L_2(\Theta, P)$ and u^d is an arbitrary function of $\mathcal{U} \otimes \mathrm{L}_2(\Theta, P)$ such that $u^d(\mathbf{x}, \theta) = U^d(\mathbf{x}, \theta)$ for $\mathbf{x} \in \Gamma_d$. It is then observed that, thanks to the linearity of the stochastic form,

$$A(u^0 + u^d, v) = A(u^0, v) + A(u^d, v), \tag{5.73}$$

u^0 satisfies the weak form of the stochastic problem (5.17) with the slight modification of $B(v)$, according to:

$$B(v) = \mathrm{E}\left[b^d(v)\right], \quad b^d(v) \equiv \int_\Omega f(\mathbf{x}, \theta) v(\mathbf{x}, \theta)\, d\mathbf{x} - a(u^d, v). \tag{5.74}$$

Therefore, the general situation of both uncertain Neumann and Dirichlet boundary problems can be cast as:
Find $u \in \mathcal{V} \otimes L_2(\Theta, P)$ such that

$$\mathrm{E}[a(u, v)] = \mathrm{E}[b(v)] \quad \forall v \in \mathcal{V} \otimes L_2(\Theta, P), \tag{5.75}$$

where

$$a(u, v) = \int_{\Omega} \nu(\mathbf{x}) \nabla u(\mathbf{x}, \theta) \cdot \nabla v(\mathbf{x}, \theta) \, d\mathbf{x}, \tag{5.76}$$

as previously defined, b having the generalized form:

$$b(v) = \int_{\Omega} f(\mathbf{x}, \theta) v(\mathbf{x}, \theta) \, d\mathbf{x} - a(u^d, v) + \int_{\Gamma_n} \Phi^n(\mathbf{x}, \theta) v(\mathbf{x}, \theta) \, d\mathbf{x}. \tag{5.77}$$

$u^d \in \mathcal{U} \otimes L_2(\Theta, P)$ is a smooth field that satisfies the Dirichlet solution, and the full solution is $u + u^d$. Except from this redefinition of the form $b(v)$, the derivation of the Galerkin problem remains essentially unchanged. The solution method can be readily applied, since the generalization only involves a modification of the right-hand side of the spectral problem. Note that for the finite-element discretization used here, the solution space $\mathcal{U}^h$ is obtained by extending $\mathcal{V}^h$ using the shape functions Φ_i of nodes belonging to the Dirichlet boundary:

$$\mathcal{U}^h = \mathcal{V}^h \oplus \operatorname{span}\{\Phi_i\}_{i \in \mathcal{N}_d}, \tag{5.78}$$

where $\mathcal{N}_d$ is the set of Dirichlet nodes.

5.1.5.2 Test Case

We consider an idealized two-dimensional heat exchanger, which consists of a rectangular piece of material with uniform but random conductivity $\nu^1(\theta)$, perforated by two symmetric circular holes. The first hole contains devices that inject heat at a random rate $W_n(\theta)$ into the exchanger. This is modeled as a uniform flux entering the computational domain through the hole's boundary, Γ_n, according to:

$$\Phi^n(\mathbf{x}, \theta) = \frac{W_n(\theta)}{\operatorname{meas}(\Gamma_n)}, \quad \mathbf{x} \in \Gamma_n,$$

where $\operatorname{meas}(\Gamma_n)$ is the perimeter of the hole. In the second hole, a cooling fluid is circulating to extract the heat. The temperature of the cooling fluid is a random variable, $T_c(\theta)$, the random Dirichlet condition is applied at the boundary, Γ_c, of the second hole, i.e.

$$U^d(\mathbf{x}, \theta) = T_c(\theta), \quad \mathbf{x} \in \Gamma_c.$$

Finally, a layer of a second material with a low random conductivity $\nu^2(\theta)$ is positioned around the exchanger thermally isolate it from its surroundings. The external boundary of the casing, denoted Γ_e, has a random temperature $T_e(\theta)$ so the following Dirichlet boundary condition is applied:

$$U^d(\mathbf{x}, \theta) = T_e(\theta), \quad \mathbf{x} \in \Gamma_e.$$

The system is depicted in Fig. 5.11.

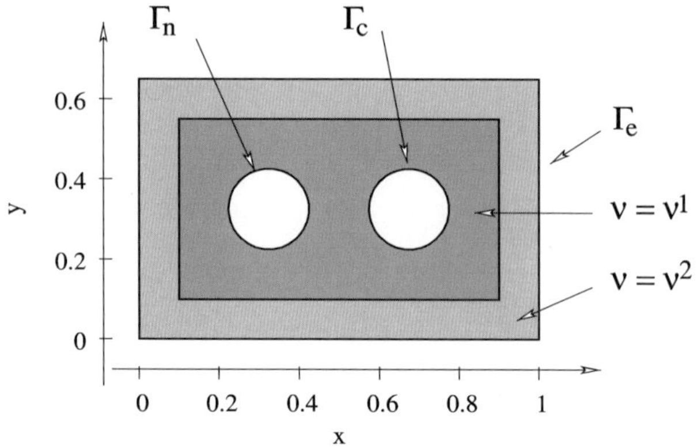

Fig. 5.11 Computational domain for the heat exchanger problem of Sect. 5.1.5. Γ_n is a Neumann boundary with flux Φ^n; Γ_c and Γ_e are Dirichlet boundaries with temperatures T_c and T_n, respectively. The conductivity of the inner domain is ν^1 while the conductivity of the casing is ν^2

Table 5.1 Distributions of the random quantities for the problem of Sect. 5.1.5, and associated random variables

Log normal	Median value	COV	Gaussian variable
ν^1	1	$\sqrt{5}$	ξ_1
ν^2	10^{-2}	$\sqrt{5}$	ξ_2
W_n	125	$\sqrt{2}$	ξ_3
Uniform	**Lower bound**	**Upper bound**	**Uniform variable**
T_c	5	10	ξ_4
T_e	10	30	ξ_5

We assume ν^1, ν^2, W_n, T_c and T_e to be independent random variables. Therefore, N = 5 random variables ξ_i are required for the stochastic parametrization of the problem. We assume log-normal distributions for ν^1, ν^2 and W_n, whereas the temperatures T_c and T_e have uniform distributions. Characteristics of these distributions are summarized in Table 5.1. Normalized Gaussian variables ξ_1, ξ_2 and ξ_3 are used for the parametrization of ν^1, ν^2 and W_n respectively, whereas T_c and T_e are respectively parametrized using ξ_4 and ξ_5, which are uniformly distributed on $[-1, 1]$. Consequently, $\Xi = \mathbb{R}^3 \times [-1, 1]^2$ and the stochastic polynomials Ψ_k are products of Hermite and Legendre polynomials along dimensions $(1, 2, 3)$ and $(4, 5)$ respectively.

5.1.5.3 Simulations

The solution is computed using an expansion with No = 4, so the dimension of $\mathcal{S}^{\mathrm{P}}$ is $(\mathrm{P} + 1) = 126$. It is remarked that the matrix $[\mathbf{A}]$ of the Galerkin system has the sparse structure shown in Fig. 5.12. Indeed, it depends on ν^1 and ν^2 only, and the matrix $[A](\boldsymbol{\xi})$ has actually a 2D expansion (in terms of ξ_1 and ξ_2). It is

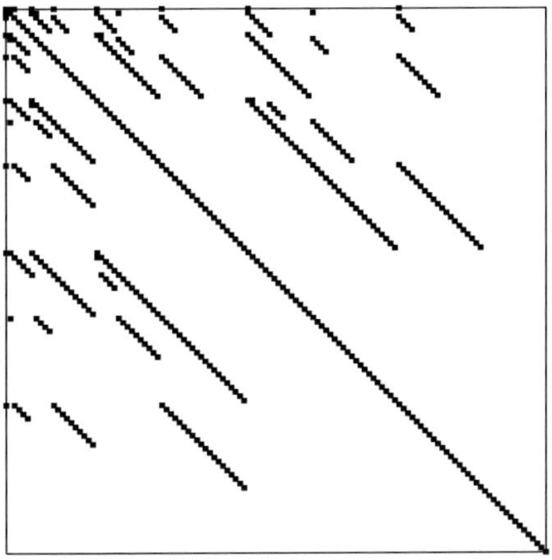

Fig. 5.12 Structure of the Galerkin system of problem in Sect. 5.1.5 for No = 4

further remarked that the lower part of the Galerkin problem matrix has only non-zero blocks on the diagonal, so the corresponding modes are decoupled. This is due to the fact that this part of the matrix corresponds to the equations for the modes $\mathbf{u}_k^h$ of the solution associated with the polynomials Ψ_k of total degree No (here No = 4) involving only the three last random variables, *i.e.* using the notation of Appendix C, polynomials of the form $\Psi_k = \psi_{\alpha_3^k}\psi_{\alpha_4^k}\psi_{\alpha_5^k}$ where $\alpha_3^k + \alpha_4^k + \alpha_5^k = \mathrm{No}$. On the other hand, projection of the stochastic heat equation on this Ψ_k gives:

$$\begin{aligned}
&\sum_{l=0}^{\mathrm{P}}\sum_{i=0}^{\mathrm{P}}[A^l]\mathbf{u}_i^h\,\langle\Psi_l\Psi_i\Psi_k\rangle\\
&\quad=\sum_{l=0}^{\mathrm{P}}\sum_{i=0}^{\mathrm{P}}[A^l]\mathbf{u}_i^h\left\langle\psi_{\alpha_1^l}\psi_{\alpha_2^l}\psi_{\alpha_1^i}\cdots\psi_{\alpha_{\mathrm{N}}^i}\psi_{\alpha_3^k}\psi_{\alpha_4^k}\psi_{\alpha_5^k}\right\rangle\\
&\quad=\sum_{l=0}^{\mathrm{P}}\sum_{i=0}^{\mathrm{P}}[A^l]\mathbf{u}_i^h\left\langle\psi_{\alpha_1^l}\psi_{\alpha_1^i}\right\rangle\left\langle\psi_{\alpha_2^l}\psi_{\alpha_2^i}\right\rangle\left\langle\psi_{\alpha_3^i}\psi_{\alpha_3^k}\right\rangle\cdots\left\langle\psi_{\alpha_5^i}\psi_{\alpha_5^k}\right\rangle\\
&\quad=\sum_{l=0}^{\mathrm{P}}\sum_{i=0}^{\mathrm{P}}[A^l]\mathbf{u}_l^h\,\delta_{\alpha_1^l,\alpha_1^i}\delta_{\alpha_2^l,\alpha_2^i}\delta_{\alpha_3^i,\alpha_3^k}\delta_{\alpha_4^i,\alpha_4^k}\delta_{\alpha_5^i,\alpha_5^k},
\end{aligned}\tag{5.79}$$

where we have used the fact that

$$[A](\boldsymbol{\xi}) = \sum_{l=0}^{\mathrm{P}}[A^l]\Psi_l(\boldsymbol{\xi}) = \sum_{l=0}^{\mathrm{P}}[A^l]\psi_{\alpha_1^l}(\xi_1)\psi_{\alpha_2^l}(\xi_2),$$

because, as previously mentioned, the random matrix has expansion along the first two dimensions only. Therefore, since $\alpha_1^i + \cdots + \alpha_5^i \le \mathrm{No}$, (5.79) reduces to

$$\sum_{l=0}^{\mathrm{P}}\sum_{i=0}^{\mathrm{P}}[A^l]\mathbf{u}_i^h\,\langle\Psi_l\Psi_i\Psi_k\rangle = [A^0]\mathbf{u}_k^h,\tag{5.80}$$

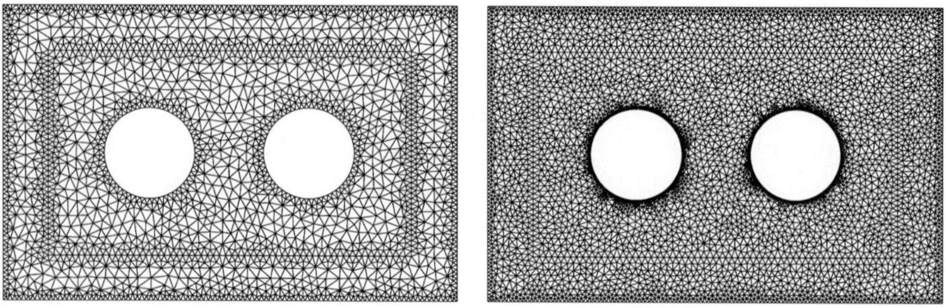

Fig. 5.13 Coarse (*left*) and fine (*right*) finite-element meshes used for the resolution of the problem of Sect. 5.1.5. The coarse mesh has 4,326 elements and 2,365 nodes, while the fine mesh has 9,140 elements and 4,947 nodes

for k such that $\alpha_3^k + \alpha_4^k + \alpha_5^k = \text{No}$.

For the spatial discretization, two finite-element meshes will be considered: a coarse mesh with 4,326 elements and a fine mesh with 9,140 elements. The two meshes are plotted in Fig. 5.13. These two meshes will be used to analyze the effect of the spatial discretization on the statistics of the solution.

Mean and standard deviation of the temperature field: Figure 5.14 shows the mean of the temperature field over the computational domain, computed using the fine finite-element mesh. The highest expected temperature is reported for a point on the boundary Γ_n, located along the domain mid-line and opposite to the hole containing the coolant fluid. This finding may have been anticipated from physical considerations, assuming that the heat flux Φ^n is large enough to heat the neighborhood of the first hole to a temperature greater than T_e (otherwise, the maximum temperature would be observed on Γ_e). The effect of the insulating casing is also seen from the large temperature gradient in the casing domain on the side of the first (hot) hole.

The standard deviation of the temperature field, shown in Fig. 5.15, has a spatial structure which resembles the mean field. Note that contrary to the previous example, the standard deviation is everywhere strictly greater than zero, since the Dirichlet boundary conditions on Γ_c and Γ_e are both stochastic. In fact, the lowest standard deviation is reported for the second hole boundary which has indeed a lower variability than that of the external boundary. It is further observed that the highest level of uncertainty (as measured by the standard deviation) is reported for the point of Γ_n opposite to the second hole, *i.e.* the point where the averaged temperature is also the highest. In the following, we denote this point $\mathbf{x}_{max}$.

Probability density functions: We start by analyzing the pdf of the temperature at the point $\mathbf{x}_{max}$ which is located on Γ_n. The pdf is plotted in Fig. 5.16 using both linear (left plot) and logarithmic scales (right plot). A skewed pdf is found with a longer tail toward the highest temperatures. A large variability is observed, with significant probability of events where the temperature exceeds twice its expected value. These plots present the two distributions computed for the fine and coarse

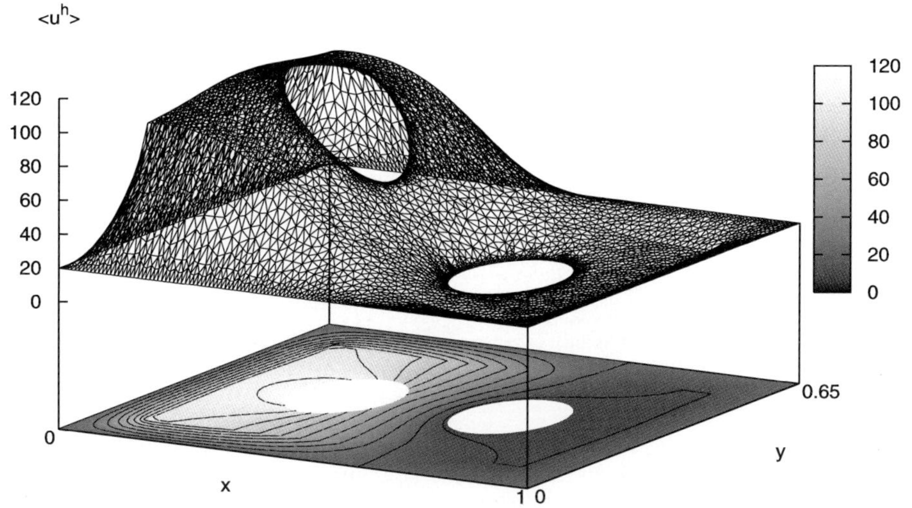

Fig. 5.14 Mean temperature field for the problem of Sect. 5.1.5. The computations are performed using No = 4 (126 spectral modes) and the fine finite-element mesh

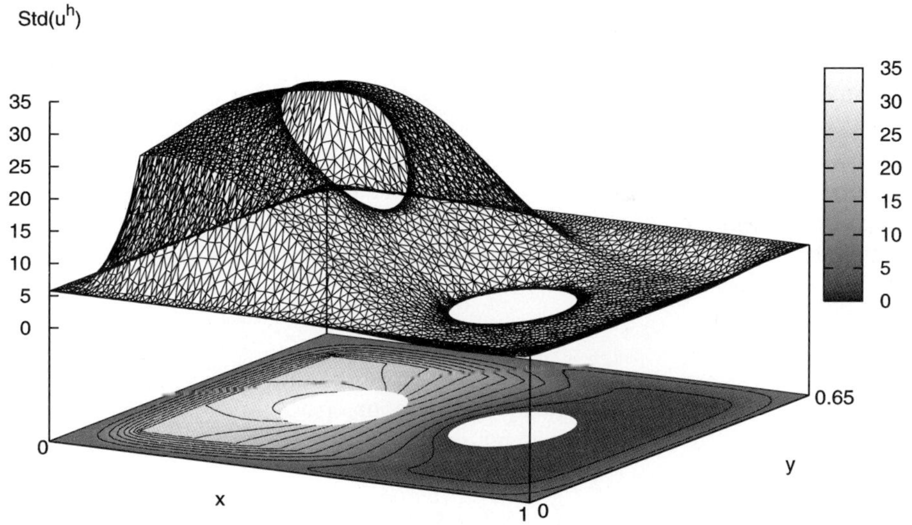

Fig. 5.15 Standard deviation of the temperature field for the problem of Sect. 5.1.5. The computations are performed using No = 4 (126 spectral modes) and the fine finite-element mesh

meshes. An excellent agreement is observed for the predictions for the two meshes, demonstrating the convergence of the computed temperatures with regard to the spatial mesh size. A similar verification was performed for the convergence with regard

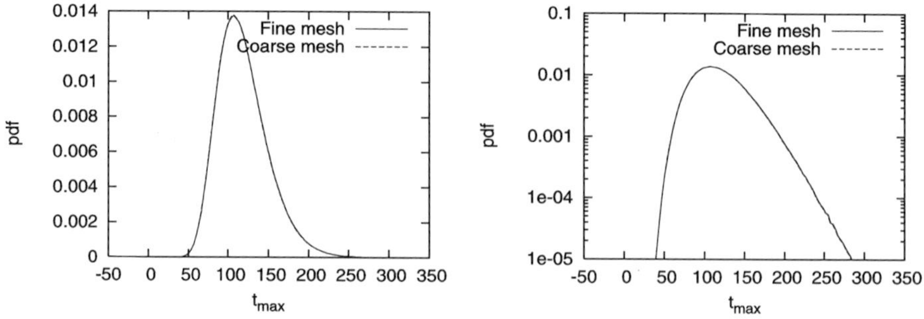

Fig. 5.16 Probability density function of temperature at $\mathbf{x}_{max}$; *left*: linear scale; *right*: logarithmic scale. Results are shown for the coarse and fine finite-element meshes. In both cases, an expansion with No = 4 expansion (126 spectral modes) is used

to the expansion order No and leads to the conclusion that No = 4 is sufficiently large.

The efficiency of the exchanger is analyzed based on the global fluxes across the various boundaries of the domain. We denote the total fluxes across Γ_c and Γ_e by W_c and W_e, respectively. These fluxes are given by:

$$W_c \equiv \int_{\Gamma_c} \nu^1 \frac{\partial u}{\partial n}\, d\mathbf{x}, \qquad W_e \equiv \int_{\Gamma_e} \nu^2 \frac{\partial u}{\partial n}\, d\mathbf{x}. \tag{5.81}$$

W_c and W_e are obtained from the discrete solution $u^h(\boldsymbol{\xi})$, by evaluating numerically the gradient of the finite-element solution on the elements lying along the boundary, which is constant over the elements, projecting it to the normal direction and integrating along the boundary edge of the elements. A similar treatment is applied to estimate the actual numerical heat flux along boundary Γ_n, denoted W_n^h, which is not exactly equal to W_n, due to the weak form of the problem solved. We denote W_n^h, W_c^h and W_e^h these discrete fluxes. From conservation of energy, we should have:

$$W_n^h + W_c^h + W_e^h = 0,$$

though for finite discretizations some conservation errors are expected as the finite element discretization used here is not conservative. We denote this conservation error

$$\epsilon_W^h \equiv W_n^h + W_c^h + W_e^h. \tag{5.82}$$

Note that in the present case all these quantities are random functionals of $\boldsymbol{\xi}$.

Plotted in Fig. 5.17 is the pdf of W_e, *i.e.* the total amount of heat leaving the casing to the exterior of the domain. It is seen that the flux is predominantly negative (system leaking heat to the exterior), though the logarithmic scale allows us to appreciate that events with $W_e > 0$ have small but finite probability. Indeed, they are events where the entering flux W_n is too low, so that heat is actually "pumped" by the cooling fluid towards the casing, whose boundary temperature T_e is almost

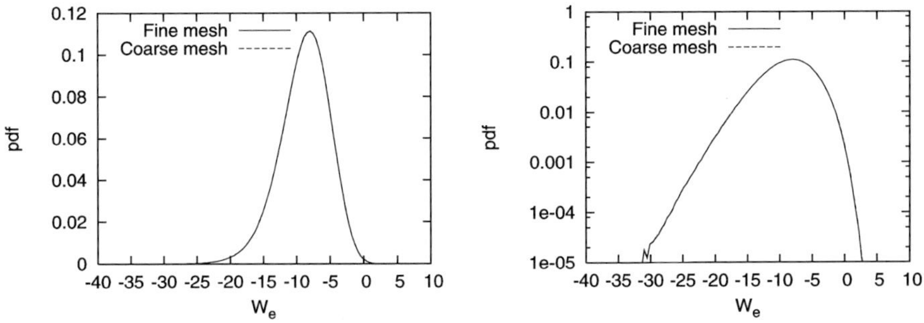

Fig. 5.17 Probability density functions of W_e^h across the external boundary, Γ_e, of the casing for the problem of Sect. 5.1.5; *left*: linear scale; *right*: logarithmic scale. Results are shown for the coarse and fine finite-element meshes. In both cases, an expansion with No $= 4$ expansion (126 spectral modes) is used

surely greater than T_c. Although not negligible, these events have low predicted probabilities. It is also seen from Fig. 5.17 that W_e^h appears well converged with regard to the spatial discretization.

The pdfs of W_c^h are shown in Fig. 5.18. It is seen that for the two meshes the fluxes W_c^h are always negative, denoting that the heat always flows from the exchanger to the cooling fluid, a result consistent with the problem settings, for which $T_c < T_e$ and $W_n \geq 0$ almost surely. However, in contrast with the results reported for the fluxes W_e^h in Fig. 5.17, significant differences are observed between the approximations on the two meshes. Specifically, compared to the coarse mesh predictions, the fine mesh results yield a distribution of the flux W_c^h which is shifted toward the lower values with a larger variability (broader distribution). The difference in the mean values of the fluxes is about 5 units. A deeper analysis reveals that these differences are essentially due to the slow convergence of the flux boundary condition W_n^h toward its exact value W_n, as $h \to 0$. In fact, for the finite-element method used here, one expects a convergence of the solution u^h as $\mathcal{O}(h^2)$, but a slower convergence in $\mathcal{O}(h)$ or less for the fluxes. Alternative formulations, such as a mixed one, would be better suited to achieve a more accurate approximation of the fluxes with larger convergence rates. It is however stressed that the slow convergence is due to the deterministic discretization techniques, and is in no way related to the stochastic spectral decomposition of the random solution.[2] The latter still exhibits an exponential convergence rate with the expansion order No for fixed h.

Conservation errors are analyzed in Fig. 5.19, which shows the pdf of error term defined in (5.82) for the two meshes. The conservation errors are seen to decay with the size h of the mesh, both in mean and variability, demonstrating the improvement of conservation properties of the solution as the mesh is refined.

[2]Examples of Chap. 6 use conservative discretizations at the deterministic level (finite-differences and finite-volumes) and it will be shown there that the Galerkin projection yields a conservative discrete spectral problem.

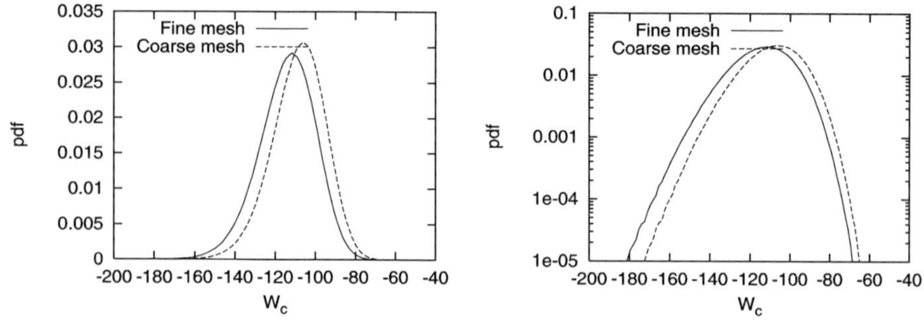

Fig. 5.18 Probability density functions of W_c across the cooling boundary Γ_2 for the problem of Sect. 5.1.5; *left*: linear scale, *right*: logarithmic scale. Results are shown for the coarse and fine finite-element meshes. In both cases, an expansion with No = 4 expansion (126 spectral modes) is used

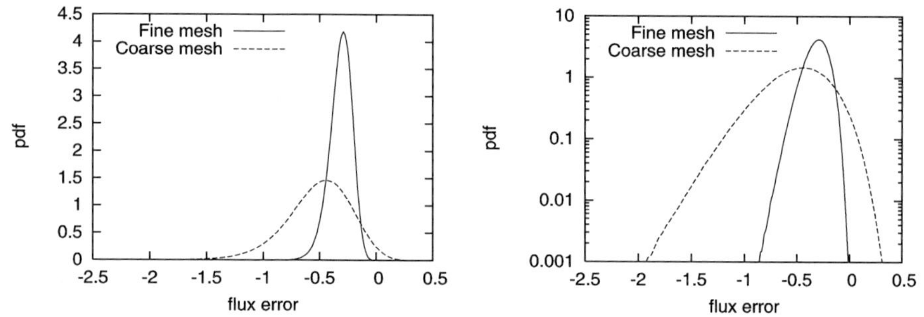

Fig. 5.19 Probability density functions of ϵ_W^h for the problem of Sect. 5.1.5; *left*: linear scale, *right*: logarithmic scale. Results are shown for the coarse and fine finite-element meshes. In both cases, an expansion with No = 4 expansion (126 spectral modes) is used

Finally, the efficiency of the heat exchanger is characterized using the ratio of heat extracted by the cooling fluid and the amount of heat W_n released in the system. Specifically, the efficiency is defined according to:

$$\eta = -\frac{W_c}{W_n} = 1 + \frac{W_e}{W_n}.$$

Due to the conservation errors, the two expressions of the efficiency will not be strictly equal for meshes with finite h, so we consider

$$\eta_c^h = -\frac{W_c^h}{W_n^h}, \qquad \eta_e^h = 1 - \frac{W_e^h}{W_n^h}. \tag{5.83}$$

From the spectral expansions of the random fluxes, we can determine the expansions of the efficiencies η_c^h and η_e^h using the methods described in Sect. 4.5.2. The probability density functions of the two efficiencies η_c^h and η_e^h are reported in Fig. 5.20 for the fine and coarse meshes. Again, we observe the convergence of the predictions as

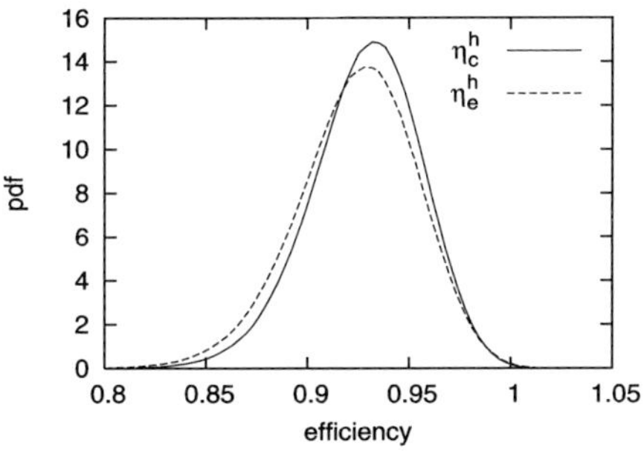

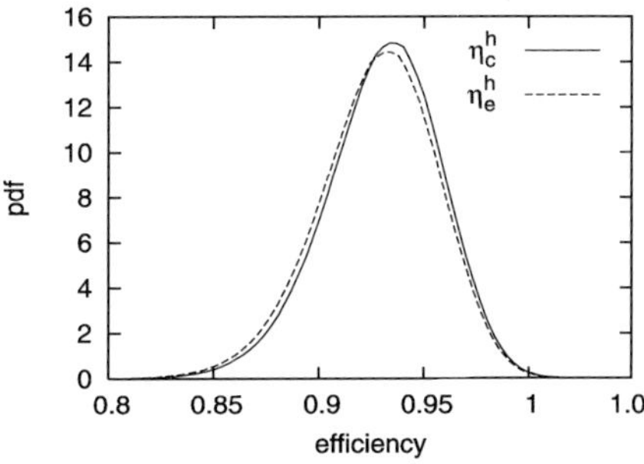

Fig. 5.20 Probability density functions of η_c^h and η_e^h of problem in Sect. 5.1.5 for the coarse (*left*) and fine (*right*) finite-element meshes. Solutions with No = 4 (126 spectral modes) are used

the mesh is refined and a better agreement between the two estimates η_c^h and η_e^h for the finer mesh. This again denotes the decay of the conservation error as the mesh is refined. Interestingly enough, we see that the random efficiency exceeds unity with some small but non-vanishing probability. This is consistent with our previous finding for the random flux W_e where it was found to be positive for a non-vanishing set of random event: the heat extracted by the cooling fluid may exceed the heat released at Γ_n resulting in an efficiency greater than one.

5.1.6 Variance Analysis

From a practical perspective, it is essential to characterize the variability of the system due to uncertainties in properties, forcing, or more generally input data. In the previous example, the predictions for the maximum temperature and efficiency densities provided useful information. However, in order to "manage" (or reduce) the corresponding variabilities, a more complete understanding of the respective impacts of the different uncertainty sources is needed. Indeed, one may need to know which of the uncertainties in the materials conductivities ν^1 and ν^2, temperatures T_c and T_e, and flux W_n, yield the largest variability on a given system observable, before taking any action to reduce this variability. For instance, if the prediction of the temperature at point $\mathbf{x}_{max}$ is considered critical, a relevant question may be: which of the uncertain system characteristics should one focus on to most effectively reduce the uncertainty in the temperature at $\mathbf{x}_{max}$? This is clearly not a trivial question and appropriate tools are needed. Below, we show that the stochastic expansion of the solution provides an immediate way to characterize variabilities induced by different sources of uncertainties, thanks to the orthogonal nature and structure of the PC bases which make explicit the dependence between the uncertain data and model solution.

5.1.6.1 Functional Decomposition

Any second-order functional $g(\xi_1,\ldots,\xi_\mathrm{N})$ has a unique orthogonal functional decomposition of the form:

$$g(\xi_1,\ldots,\xi_\mathrm{N}) = \sum_{s\subseteq\{1,\ldots,\mathrm{N}\}} g_s(\boldsymbol{\xi}_s),$$

where for $s=\{i_1,\ldots,i_p\}$ we denote $\boldsymbol{\xi}_s=\{\xi_{i_1},\ldots,\xi_{i_p}\}$ and $g_\emptyset \equiv \langle g\rangle$. This decomposition is often called the Hoeffding or Sobol decomposition of g in the literature. From the stochastic expansion of $g(\boldsymbol{\xi})$ we can easily identify to which component of the Sobol's decomposition a given mode g_k contributes. Indeed, if we write the k-th polynomial of the stochastic basis as the product of one-dimensional polynomials (see Appendix C),

$$\Psi_k(\boldsymbol{\xi}) = \prod_{i=1}^{\mathrm{N}} \psi_{\alpha_i^k}(\xi_i),$$

we see immediately that Ψ_k has a Sobol decomposition involving a single term:

$$\Psi_k(\boldsymbol{\xi}) = (\Psi_k)_{s=\{i_1,\ldots,i_p\}}(\xi_{i_1},\ldots,\xi_{i_p}), \quad \text{s.t.} \begin{cases} i\in s \Rightarrow \alpha_i^k \neq 0, \\ i\notin s \Rightarrow \alpha_i^k = 0. \end{cases} \tag{5.84}$$

Therefore, the Sobol's expansion of g can be obtained from its PC representation; for instance,

$$g_s(\boldsymbol{\xi}_u) = \sum_{k\in S_s} g_k\Psi_k(\boldsymbol{\xi}),$$

where S_s is a subset of stochastic modes indexes, with $S_s\cap S_{s'}=\emptyset$ for $s\neq s'$. It is seen, however, that truncating the stochastic expansion of g to order No < N will only provide an approximation of functional g_s with Card$(s)\leq$ No.

The main interest of Sobol's decomposition lies in its orthogonal nature,

$$\langle g_s g_{s'}\rangle = \left\langle g_s^2\right\rangle \delta_{s,s'}, \tag{5.85}$$

such that the variance of g as for expression:

$$V(g) = \left\langle (g-g_\emptyset)^2\right\rangle = \sum_{\substack{s\in\{1,\ldots\mathrm{N}\}\\ s\neq\emptyset}} \left\langle g_s^2\right\rangle. \tag{5.86}$$

This decomposition is often used for the analysis of uncertainty, since $V_s(g)\equiv\langle g_s\rangle$ is the contribution to the total variance of the interaction of the set of random parameters $\{\xi_i, i\in s\}$. For instance, $V_{\{i\}}(g)$ is the variance of $g(\boldsymbol{\xi})$ due to the random variable ξ_i *only*, while $V_{\{i,j\}}(g)$ is the variance due to the *combined* effects of ξ_i and ξ_j. The latter should be distinguished from the total variance due to ξ_i and ξ_j

which is $V_{\{i\}}(g) + V_{\{j\}}(g) + V_{\{i,j\}}(g)$. This definition allows for a comparison of the respective impacts of different sources of uncertainty, *i.e.* caused by different ξ_i. The comparison is generally made from the sensitivity indexes, which are defined as

$$\mathcal{S}_s = \frac{V_s(g)}{V(g)}, \quad \text{so} \quad \sum_{\substack{s \in \{1,\ldots,\mathrm{N}\} \\ s \neq \emptyset}} \mathcal{S}_s = 1$$

and allows for a direct assessment of the relative impacts. Because there are $2^{\mathrm{N}} - 1$ such sensitivity indexes, it often found convenient to summarize the impact of ξ_i using its so-called total sensitivity index, denoted $\mathcal{T}_{\{i\}}$, the sum of all the sensitivity indexes to which ξ_i contributes:

$$\mathcal{T}_{\{i\}} \equiv \sum_{\substack{s \in \{1,\ldots,\mathrm{N}\} \\ s \ni i}} \mathcal{S}_s.$$

Note that the sum $\sum_i \mathcal{T}_{\{i\}}$ is greater than one. This concept can be even extended to a group of variables, say $\{\xi_{i_1}, \ldots, \xi_{i_p}\}$ corresponding to a set $\bar{s} = \{i_1, \ldots, i_p\}$:

$$\mathcal{T}_{\bar{s}} = \sum_{\substack{s \in \{1,\ldots,\mathrm{N}\} \\ s \supset \bar{s}}} \mathcal{S}_s.$$

This can be particularly useful in situations where the group contains random variables parameterizing a single uncertainty source.

5.1.6.2 Application

We apply the analysis above to the fine-mesh results obtained for the example of Sect. 5.1.5. For simplicity, we denote $\mathcal{T}_{\nu^1}$, $\mathcal{T}_{\nu^2}$, $\mathcal{T}_{W_n}$, $\mathcal{T}_{T_c}$ and $\mathcal{T}_{T_e}$, the total sensitivity indexes related to the uncertain conductivities ν^1, ν^2, flux W_n, temperatures T_c and T_e respectively. The total sensitivity indexes are computed for different observables: fluxes W_n, W_c and W_e, temperature T_{max} at point $\mathbf{x}_{max}$, and the two efficiencies η_1 and η_2. Results are reported in Table 5.2

The first line of Table 5.2 shows the total sensitivity indexes of the numerical flux W_n. It is seen that all indexes are zero, excepts $\mathcal{T}_{W_n}$ which is equal to 1: this is expected as W_n is a *prescribed* stochastic boundary condition parameterized with the random variable ξ_3, so its variability does not depend on the other uncertainties which are parameterized with different, independent, random variables.[3]

[3]The flux W_n analyzed being actually the numerical flux, *i.e.* the integral of $\nu^1 \partial u/\partial n$ along Γ_n, having $\mathcal{T}_{W_n} = 1$, and other total indexes equal to zero, shows that the convergence of the W_n to the prescribed Neumann boundary condition depends only on the deterministic (spatial) discretization and not on the stochastic discretization.

Table 5.2 Total sensitivity indices of the random fluxes and efficiencies

Quantity	$\mathcal{T}_{\nu^1}$	$\mathcal{T}_{\nu^2}$	$\mathcal{T}_{W_n}$	$\mathcal{T}_{T_c}$	$\mathcal{T}_{T_e}$	Total
W_n	0.0000	0.0000	1.0000	0.0000	0.0000	1.0000
W_c	0.0254	0.0481	0.9185	0.0006	0.0114	1.0041
W_e	0.2729	0.5116	0.1298	0.0076	0.1216	1.0435
T_{max}	0.0132	0.8082	0.1910	0.0016	0.0008	1.0147
η_1	0.2980	0.5747	0.0121	0.0080	0.1437	1.0364
η_2	0.3137	0.6000	0.0128	0.0095	0.1510	1.0869

Analysis of the variability of the flux W_c (second line of Table 5.2) shows that it is essentially due to the variability in W_n ($\mathcal{T}_{W_n} \approx 0.92$), while uncertainty in the conductivities are seen to induce a low variability, though the uncertainty in the conductivity of the casing (ν^1) has a larger impact than the uncertainty in the conductivity of the exchanger (ν^2): $\mathcal{T}_{\nu^2} \approx 2\mathcal{T}_{\nu^1}$. Variabilities in the boundary temperatures have even weaker impact. Again, this finding could have been anticipated since most of the heat is effectively extracted through the boundary Γ_c, so that one could have reasonably expected the uncertainty on W_n to have the largest impact on W_c. However, assessment of the relative impacts of the temperatures and conductivities would have been difficult to anticipate.

More interesting is the analysis of the variability in the flux W_e (third line of Table 5.2) which, we recall, corresponds to the flux leaking from the casing to the external domain. It is seen that it is not the variability of the input flux W_n that yields most of W_e's variability (this effect is in fact ranked third in terms of total sensitivity indexes); instead, the conductivities of the exchanger $\mathcal{T}_{\nu^2} \approx 0.51$ and of the casing $\mathcal{T}_{\nu^1} \approx 0.27$ have the largest impact on W_e, while uncertainties in temperatures have weaker influence.

For the temperature at point $\mathbf{x}_{max}$ (fourth line of Table 5.2) it is the conductivity of the exchanger which is the predominant source of variability, followed by the heat flux W_n, while the conductivity of the casing and temperatures have minor effects.

Finally, the last two lines of Table 5.2 report the total sensitivity indexes for the efficiencies of the heat exchanger, as defined in (5.83). It is first observed that the two efficiencies lead to similar total sensitivity indexes. Next, it is seen that the uncertainty in W_n has a low impact on the variability of the efficiency of the exchanger. This may be explained by the fact that the fluxes W_c and W_e are essentially proportional to W_n with a relation of the form:

$$W_c(\theta) \propto W_n(\theta)\alpha(\nu^1(\theta), \nu^2(\theta), T_e(\theta), T_c(\theta)),$$

$$W_e(\theta) \propto W_n(\theta)\beta(\nu^1(\theta), \nu^2(\theta), T_e(\theta), T_c(\theta)),$$

so that efficiencies are essentially independent of W_n. Therefore, among the remaining sources of uncertainties, conductivity of the exchanger appears to have the largest impact on the efficiency $\mathcal{T}_{\nu^2} \simeq 0.6$, followed by that on the casing $\mathcal{T}_{\nu^1} \simeq 0.3$, while temperatures have lower, though non-negligible, impact.

Finally, the sum of the total sensitivity indices, reported on the far right column of Table 5.2, are all greater than one except for W_n. The larger from one is the deviation of the sum of total sensitivity indexes, the larger are the contributions to the variance of effects involving coupled uncertainty sources. Conversely, the closer to one is the sum of the total sensitivity indexes the more additive are the impact of the different uncertainty sources.

5.2 Stochastic Viscous Burgers Equation

We now extend the methodology detailed for the linear heat equation to the 1D steady viscous Burgers equation. The Burgers equation is often considered as a prototype of fluid flow models as it involves both linear diffusion and nonlinear convective terms. We restrict ourselves to the steady case as the extension to the unsteady case is immediate.

5.2.1 Deterministic Problem

We consider the 1D steady viscous Burgers equation on the spatial domain $\Omega = (-1,1)$,

$$w\frac{\partial w}{\partial x} - \nu\frac{\partial^2 w}{\partial x^2} = 0, \quad \forall x \in \Omega, \tag{5.87}$$

with boundary conditions

$$w(-1) = 1, \qquad w(1) = -1, \tag{5.88}$$

where $\nu > 0$ is the fluid viscosity. The solution is sought in the functional space

$$\mathcal{U} \equiv \{v \in H^1(\Omega);\, v(-1) = 1,\, v(1) = -1\}. \tag{5.89}$$

The space $\mathcal{U}$ is affine, and we denote $\mathcal{V}$ the corresponding vector space:

$$\mathcal{V} = \{v \in H^1(\Omega);\, v(-1) = 0,\, v(1) = 0\}. \tag{5.90}$$

The solution w is decomposed as

$$w(x) = \tilde{u}(x) + u(x), \quad \tilde{u} \in \mathcal{U}, \tag{5.91}$$

where $\tilde{u}$ is chosen arbitrarily while u is sought in $\mathcal{V}$. An obvious choice for $\tilde{u}$ is $\tilde{u}(x) = -x$. With this decomposition, Burgers equation becomes:

$$u\frac{\partial u}{\partial x} + \tilde{u}\frac{\partial u}{\partial x} + u\frac{\partial \tilde{u}}{\partial x} - \nu\frac{\partial^2 u}{\partial x^2} = -\tilde{u}\frac{\partial \tilde{u}}{\partial x} + \nu\frac{\partial^2 \tilde{u}}{\partial x^2}, \tag{5.92}$$

with boundary conditions

$$u(x=-1)=u(x=1)=0. \tag{5.93}$$

The variational form of (5.92) is:
Find $u \in \mathcal{V}$ such that for all $v \in \mathcal{V}$

$$n(u,u,v)+n(\tilde{u},u,v)+n(u,\tilde{u},v)+\nu a(u,v)=-n(\tilde{u},tu,v)-\nu a(\tilde{u},v), \tag{5.94}$$

where

$$n(u,v,w)\equiv\int_\Omega u\frac{\partial v}{\partial x}w\,\mathrm{d}x, \quad \text{and} \quad a(u,v)\equiv\int_\Omega \frac{\partial u}{\partial x}\frac{\partial v}{\partial x}\,\mathrm{d}x.$$

5.2.1.1 Spatial Discretization

We denote by $\mathbb{P}_{\mathrm{N_x}+1}(\Omega)$ the space of polynomials on Ω of degree $\leq \mathrm{N_x}+1$. We define the finite dimensional approximation space $\mathcal{V}^h$ as:

$$\mathcal{V}^h=\{v\in\mathbb{P}_{\mathrm{N_x}+1}(\Omega):v(-1)=0,v(1)=0\}\subset\mathcal{V}. \tag{5.95}$$

Let $x_{i\in\{0,\dots,\mathrm{N_x}+1\}}$ be the $\mathrm{N_x}+2$ Gauss-Lobatto points [1] of the interval $[-1,1]$, such that

$$x_0=-1<x_1<\cdots<x_{\mathrm{N_x}}<x_{\mathrm{N_x}+1}=1. \tag{5.96}$$

We denote by $L_i(x)\in\mathbb{P}_{\mathrm{N_x}+1}$, $i=0,\dots,\mathrm{N_x}+1$ the Lagrange polynomials constructed on the Gauss-Lobatto points,

$$L_i(x)=\prod_{\substack{j=0\\ j\neq i}}^{\mathrm{N_x}+1}\frac{x-x_j}{x_i-x_j}. \tag{5.97}$$

These polynomials are orthogonal and satisfy

$$L_i(x_j)=\begin{cases}0 & \text{if } i\neq j,\\ 1 & \text{if } i=j\end{cases} \qquad \forall j=0,\dots,\mathrm{N_x}+1. \tag{5.98}$$

A basis of $\mathcal{V}^h$ is

$$\mathcal{V}^h=\mathrm{span}\{L_i, i=1,\dots,\mathrm{N_x}\}. \tag{5.99}$$

Indeed, any $v\in\mathcal{V}^h$ can be expressed as:

$$v(x)=\sum_{i=1}^{\mathrm{N_x}}v^i L_i(x), \tag{5.100}$$

and the derivative of v is given by:

$$\frac{\partial v}{\partial x} = \sum_{i=1}^{\mathrm{N_x}} v^i L_i'(x), \qquad L_i' \equiv \frac{\partial L_i}{\partial x}. \tag{5.101}$$

5.2.1.2 Discrete Deterministic Problem

The forms a and n are discretized using the following quadrature formula over the Gauss-Lobatto points [26]. Specifically, for $u, v \in \mathcal{V}^h$, we have

$$\begin{aligned} a(u,v) &= \int_\Omega \frac{\partial u}{\partial x}\frac{\partial v}{\partial x}\,\mathrm{d}x = \int_\Omega \left(\sum_{i=1}^{\mathrm{N_x}} u^i L_i'\right)\left(\sum_{i=1}^{\mathrm{N_x}} v^i L_i'\right)\mathrm{d}x \\ &= \sum_{i,j=1}^{\mathrm{N_x}} u^i v^j \int_\Omega L_i'(x) L_j'(x)\,\mathrm{d}x = \sum_{k,i=1}^{\mathrm{N_x}} a_{k,i} u^i v^k, \end{aligned} \tag{5.102}$$

where

$$a_{k,i} \equiv \sum_{j=0}^{\mathrm{N_x}+1} L_i'(x_j) L_k'(x_j)\omega_j, \tag{5.103}$$

and ω_k, $k = 0, \ldots, \mathrm{N_x}+1$, are the Gauss-Lobatto quadrature weights [1]. Similarly, for $u, v \in \mathcal{V}$, we approximate $n(u,u,v)$ using:

$$n(u,u,v) \approx \sum_{k,i=1}^{\mathrm{N_x}} n_{k,i} u^k u^i v^k, \qquad n_{k,i} = L_i'(x_k)\omega_k. \tag{5.104}$$

The two previous discrete forms can be extended to arguments $\tilde{u} \in \mathcal{U}$ by defining $\tilde{u}^i = \tilde{u}(x_i)$, for $i = 0, \ldots, \mathrm{N_x}+1$, and appropriate modifications of the summation bounds in (5.102) and (5.104). Specifically, for $u, v \in \mathcal{V}^h$ and $\tilde{u} \in \mathcal{U}$ we use:

$$\begin{aligned} a(\tilde{u}, v) &\approx \sum_{k=1}^{\mathrm{N_x}} \sum_{i=0}^{\mathrm{N_x}+1} a_{k,i} \tilde{u}^i v^k, \\ n(\tilde{u}, u, v) &\approx \sum_{k,i=1}^{\mathrm{N_x}} n_{k,i} \tilde{u}^k u^i v^k, \\ n(u, \tilde{u}, , v) &\approx \sum_{k=1}^{\mathrm{N_x}} \sum_{i=0}^{\mathrm{N_x}+1} n_{k,i} u^k \tilde{u}^i v^k, \\ n(\tilde{u}, \tilde{u}, v) &\approx \sum_{k=1}^{\mathrm{N_x}} \sum_{i=0}^{\mathrm{N_x}+1} n_{k,i} \tilde{u}^k \tilde{u}^i v^k. \end{aligned}$$

Introducing these discrete forms into (5.94) leads to the following set of $\mathrm{N_x}$ nonlinear equations for $u^1, \ldots, u^{\mathrm{N_x}}$:

$$\sum_{i=1}^{\mathrm{N_x}} n_{k,i} u^k u^i + \sum_{i=1}^{\mathrm{N_x}} n_{k,i} \tilde{u}^k u^i + \left(\sum_{i=0}^{\mathrm{N_x}+1} n_{k,i} \tilde{u}^i \right) u^k + \nu \sum_{i=1}^{\mathrm{N_x}} a_{k,i} u^i$$
$$= \sum_{i=0}^{\mathrm{N_x}+1} n_{k,i} \tilde{u}^i \tilde{u}^k - \nu \sum_{i=0}^{\mathrm{N_x}+1} a_{k,i} \tilde{u}^i, \quad k = 1, \ldots, \mathrm{N_x}. \tag{5.105}$$

This system is solved using Newton iterations.

5.2.2 *Stochastic Problem*

We consider a stochastic variant of the steady Burgers equation, namely for a random viscosity $\nu(\theta)$ defined on an abstract probability space (Θ, Σ, P). Since the viscosity is stochastic, the solution w is also stochastic and depends on the random event, *i.e.* $w = w(x, \theta)$.

Assuming $\nu(\theta) \geq \alpha > 0$ almost surely to avoid degeneracy, the stochastic Burgers equation,

$$w \frac{\partial w}{\partial x} - \nu(\theta) \frac{\partial^2 w}{\partial x^2} = 0 \tag{5.106}$$

with the deterministic boundary conditions $w(x = \pm 1, \theta) = \pm 1$, has a unique continuous solution almost everywhere on Θ. Furthermore, if ν and $1/\nu$ are in $L_2(\Theta, P)$, the solution $w(x, \cdot) \in L_2(\Theta, P)$. In addition, due to the deterministic boundary condition, w can still be decomposed as

$$w(x, \theta) = \tilde{u} + u(x, \theta), \tag{5.107}$$

where $\tilde{u} \in \mathcal{U}$ and

$$u(x = \pm 1, \theta) = 0 \quad \text{almost surely.} \tag{5.108}$$

5.2.2.1 Stochastic Discretization

Similar to Sect. 5.1.2.3, the random viscosity $\nu(\theta)$ is parameterized using a set of N independent, real, second-order random variables, $\boldsymbol{\xi} = \{\xi_1, \ldots, \xi_{\mathrm{N}}\}$:

$$\nu(\theta) = \nu(\boldsymbol{\xi}(\theta)). \tag{5.109}$$

Again, we denote by Ξ the range of $\boldsymbol{\xi}$, by $p_{\boldsymbol{\xi}}$ the joint probability density function of $\boldsymbol{\xi}$, and by $(\Xi, \mathcal{B}_\Xi, P_\Xi)$ the associated probability space. The stochastic solution is sought in the image probability space $(\Xi, \mathcal{B}_\Xi, P_\Xi)$ instead of (Θ, Σ, P), *i.e.* we compute $w(x,\xi)$. We then construct the finite dimensional stochastic approximation space $\mathcal{S}^{\mathrm{P}} \subset L_2(\Xi, P_\Xi)$, spanned by the Generalized Polynomial Chaoses Ψ_k whose total degrees are less than or equal to No,

$$\mathcal{S}^{\mathrm{P}} = \mathrm{span}\{\Psi_0, \ldots, \Psi_{\mathrm{P}}\}, \qquad \mathrm{P}+1 = \dim(\mathcal{S}^{\mathrm{P}}) = \frac{(\mathrm{No}+\mathrm{N})!}{\mathrm{No}!\mathrm{N}!}. \tag{5.110}$$

The solution is then sought in $\mathcal{S}^{\mathrm{P}}$.

5.2.2.2 Stochastic Galerkin Projection

As for the heat equation, the Galerkin problem can be derived starting from the continuous Burgers equation (5.106), or starting from the discrete deterministic nonlinear system (5.105). The latter approach is adopted here, *i.e.* the stochastic solution will be sought as

$$w(x,\xi) \approx \tilde{u}(x) + u(x,\xi), \tag{5.111}$$

with $u(x,\xi) \in \mathcal{V}^h \otimes \mathcal{S}^{\mathrm{P}}$, so the stochastic unknown u has the spectral expansion:

$$u(x,\xi) = \sum_{i=1}^{\mathrm{N_x}} u^i(\xi) L_i(x) = \sum_{i=1}^{\mathrm{N_x}} \sum_{l=0}^{\mathrm{P}} u_l^i L_i(x) \Psi_l(\xi). \tag{5.112}$$

The system of equations for the $\mathrm{N_x} \times (\mathrm{P}+1)$ unknown deterministic coefficients of the stochastic solution are derived by inserting (5.112) into the discrete variational form (5.105), and requiring the residual of the resulting system to be orthogonal to the stochastic approximation space $\mathcal{S}^{\mathrm{P}}$. Doing so, one obtains

$$\begin{aligned}
&\sum_{l,m=0}^{\mathrm{P}} \left\langle \Psi_l \Psi_m \Psi_j \right\rangle \sum_{i=1}^{\mathrm{N_x}} n_{k,i} u_l^k u_m^i + \left\langle \Psi_j^2 \right\rangle \left[\sum_{i=1}^{\mathrm{N_x}} n_{k,i} \tilde{u}^k u_j^i + \left(\sum_{i=0}^{\mathrm{N_x}+1} n_{k,i} \tilde{u}^i \right) u_j^k \right] \\
&\quad + \sum_{l=0}^{\mathrm{P}} \left\langle \Psi_l \Psi_j \nu(\xi) \right\rangle \sum_{i=1}^{\mathrm{N_x}} a_{k,i} u_l^i \\
&= \left\langle \Psi_j \right\rangle \sum_{i=0}^{\mathrm{N_x}+1} n_{k,i} \tilde{u}^i \tilde{u}^k - \left\langle \Psi_j \nu(\xi) \right\rangle \sum_{i=0}^{\mathrm{N_x}+1} a_{k,i} \tilde{u}^i, \\
&\quad \text{for } k = 1, \ldots, \mathrm{N_x} \text{ and } j = 0, \ldots, \mathrm{P}.
\end{aligned} \tag{5.113}$$

Note that in deriving (5.113), we have made use of the orthogonality of the basis: $\langle \Psi_l \Psi_j \rangle = \langle \Psi_j^2 \rangle \delta_{l,j}$. It is seen that the stochastic Galerkin projection of the discrete system (5.105) results in a set of $\mathrm{N_x} \times (\mathrm{P}+1)$ nonlinear equations for the $\mathrm{N_x} \times (\mathrm{P}+1)$ unknown coefficients u_l^i of the solution. Unlike the case the of linear heat equation treated in Sect. 5.1, the Galerkin approximation of the stochastic Burgers equation not only couples the resolution of all modes through the diffusion terms associated to the random viscosity, but also involves a nonlinear interaction between the stochastic modes of u^i due to the quadratic nonlinearity of the convective term.

The system (5.113) can be solved using a standard iterative technique for nonlinear systems of equations. In the following, we use Newton iterations. More advanced algorithms can of course be considered, in particular relying on approximate tangent operators to reduce CPU times. Such techniques are not considered here but will be discussed later in Chap. 7.

5.2.3 Numerical Example

To illustrate the Galerkin methodology, we consider the problem of a shifted log-normal random viscosity expressed as

$$\nu(\theta) = \overline{\nu} + \nu'(\theta), \tag{5.114}$$

where $\overline{\nu} > 0$ is deterministic and $\nu'(\theta)$ has a log-normal distribution. These settings ensure the positivity of the viscosity for any random event. The random part ν' of the viscosity is expanded using a single, centered normalized Gaussian variable $\xi \sim N(0, 1)$. Denoting the median value of ν' by $\overline{\nu}'$, and the coefficient of variation by $C_{\nu'}$, we set

$$\nu(\xi) = \overline{\nu} + \exp\left[\mu + \sigma_\nu \xi\right], \quad \mu = \log\left(\overline{\nu}'\right) \text{ and } \sigma_\nu = \frac{\log C_{\nu'}}{2.85}. \tag{5.115}$$

Note that the actual median value of the viscosity is $\overline{\nu} + \overline{\nu}'$. Since the random variable ξ is Gaussian, we rely in the following computations on the Wiener-Hermite basis. The Wiener-Hermite expansion of the viscosity is given by:

$$\nu(\xi) = \overline{\nu}\Psi_0(\xi) + \exp(\mu + \sigma_\nu^2/2) \sum_{k \geq 0} \frac{\sigma_\nu^k}{\langle \Psi_k^2 \rangle} \Psi_k(\xi), \tag{5.116}$$

where Ψ_k is the k-th order Hermite polynomial. In the computations below, we use $\overline{\nu} = \overline{\nu}' = 0.05$, and a fairly large coefficient of variation, $C_{\nu'} = 10$. The probability density function of the viscosity is shown in Fig. 5.21.

Because the spatial discretization is independent of the random event (viscosity value) it has to accommodate all realizations of ν, and accordingly possible steep spatial gradients arising when the viscosity is small. To ensure a well converged solution for the deterministic problem (5.87) with $\nu = \overline{\nu}$, we set $N_x = 201$. In other words, the spatial grid is sufficiently fine to accurately capture the solution for the smallest possible viscosity value.

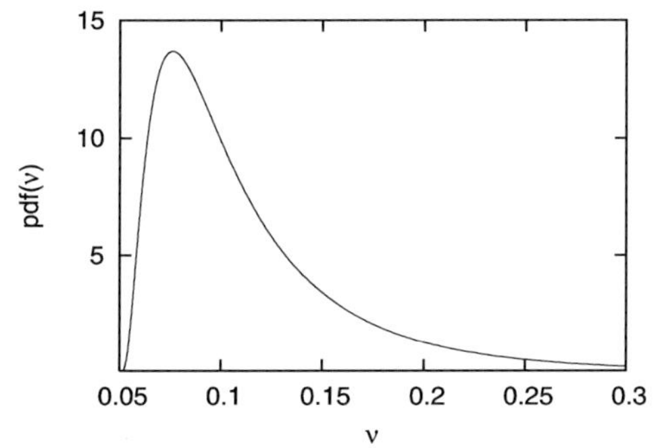

Fig. 5.21 Probability density function of the viscosity for the problem of Sect. 5.2.3.1

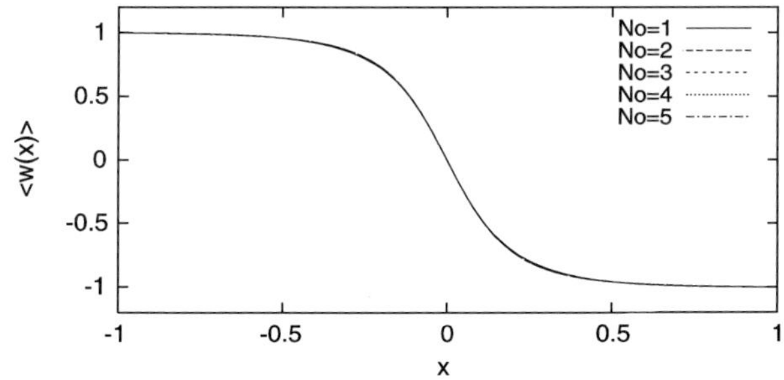

Fig. 5.22 Mean $w_0(x)$ of the Galerkin approximation of $w(x, \xi)$ for different expansion orders as indicated. Problem settings are given in Sect. 5.2.3.1

5.2.3.1 Convergence of the Stochastic Approximation

We start by inspecting the convergence of the solution to the Galerkin system (5.113) for increasing expansion order No.

Plotted in Fig. 5.22 is the expectation of the solution $w(x, \xi)$, for No $= 1, \dots, 5$. The results show that the mean $\langle w(x, \xi) \rangle = w_0(x)$ is well approximated for all $x \in [-1, 1]$, even with No $= 1$. The curves for the different expansion orders are in fact indistinguishable at the scale of the plot.

Unlike the mean, curves for the standard deviation, presented in Fig. 5.23, indicate that the first-order expansion significantly overestimates the variability in w in the neighborhood of $x = 0$, and underestimates it elsewhere. However, the standard deviation appears to converge when No increases, at different rates depending on the location in the spatial domain. Furthermore, one observes that the variability in w vanishes at the domain center, as expected from the symmetric setting of the problem. Indeed, at $x = 0$, $w = 0$ independently of the viscosity value.

Figure 5.24 depicts the stochastic modes of the Galerkin solution obtained with No $= 5$. The modes have been rescaled for clarity. They exhibit a non-trivial spatial

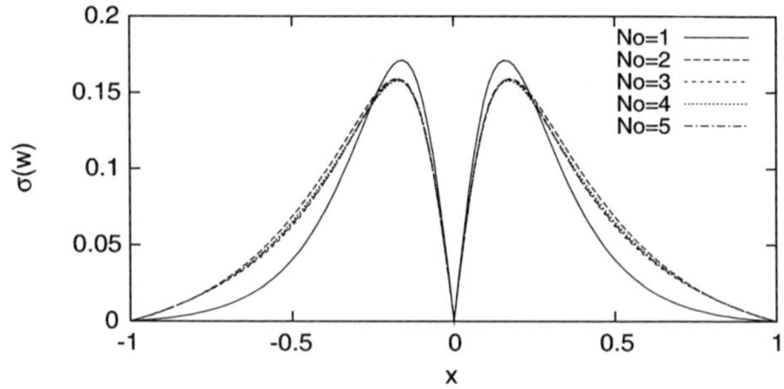

Fig. 5.23 Standard deviation of the Galerkin approximation of $w(x, \xi)$ for different expansion orders as indicated

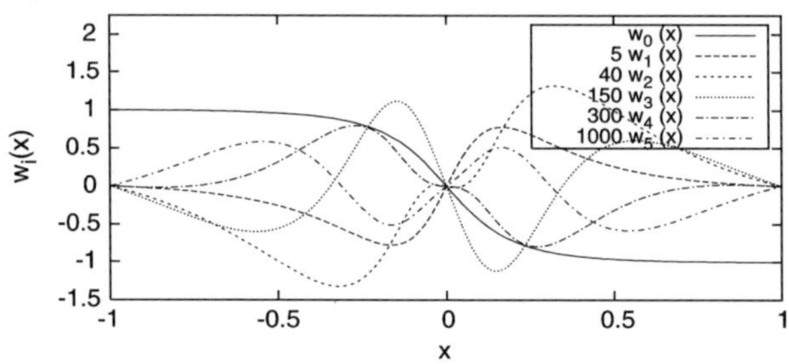

Fig. 5.24 First six stochastic modes $w_k(x)$ of the solution obtained with No = 5. Modes have been rescaled as indicated for clarity

structure that depends on the mode index. This highlights the dependence of the solution statistical distribution on the spatial variable x. Also note that the increasing scaling factor with the mode index further demonstrates the convergence of the expansion.

5.2.4 *Non-intrusive Spectral Projection*

We now take advantage of the simple (one-dimensional) parameterization of the viscosity, and compare the Galerkin projection results with predictions obtained with NISP (see Chap. 3).

5.2.4.1 Quadrature Formula

The NISP implementation below relies on a one-dimensional Gauss quadrature rule to estimate the integrals defining the stochastic modes $w_k(x)$. Using tildes to distinguish the NISP modes from their Galerkin counterparts, the former are defined as

$$\tilde{w}_k(x) = \frac{\langle w(x, \xi)\Psi_k(\xi)\rangle}{\langle \Psi_k^2(\xi)\rangle} \approx \frac{1}{\langle \Psi_k^2\rangle} \sum_{i=1}^{\mathrm{N_Q}} w^{(i)}(x)\Psi_k(\xi^{(i)})\omega^{(i)}, \tag{5.117}$$

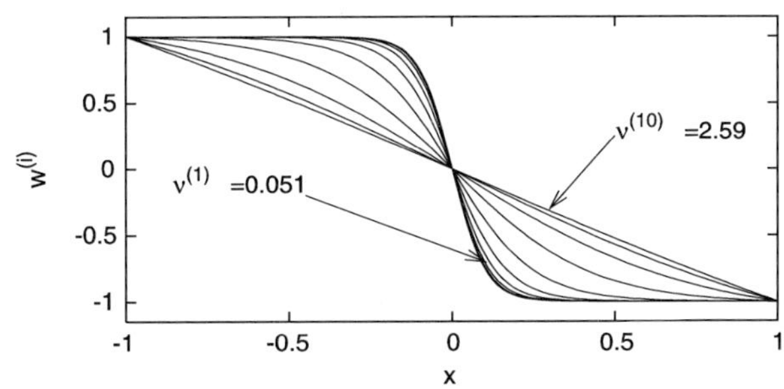

Fig. 5.25 Plot of the deterministic solutions $w^{(i)}(x)$ corresponding to the NISP quadrature points with $\mathrm{N_Q} = 10$

where $\xi^{(i)}$ and $\omega^{(i)}$ respectively denote the nodes and weights of the Gauss-Hermite quadrature formula (see Appendix B.2), whereas $w^{(i)}(x)$ denotes the solution of the deterministic Burgers equation (5.87) with $\nu = \nu^{(i)} \equiv \nu(\xi^{(i)})$. Therefore, the NISP amounts to the resolution of a set of $\mathrm{N_Q}$ deterministic Burgers equations, each for a viscosity value $\nu^{(i)}$, and then applying (5.117).

The selection of the number $\mathrm{N_Q}$ of quadrature points is an important issue in NISP. The quadrature formula is exact for polynomial integrand of degree less or equal to $2\mathrm{N_Q} - 1$. However, the solution of the Burgers equation at a given x is not necessarily a polynomial function of the viscosity, which itself depends exponentially on ξ. Thus, $w(\cdot, \xi)$ is not polynomial in ξ and there is no clear rule for the selection of $\mathrm{N_Q}$, except that one should generally use $\mathrm{N_Q} > \mathrm{No}$.

Since previous numerical tests for the Galerkin projection have shown that $\mathrm{No} = 5$ provides a well converged spectral expansion of the stochastic solution, we also apply NISP with $\mathrm{No} = 5$. Owing to the low computational cost of the deterministic Burgers equation resolution we use $\mathrm{N_Q} = 2\mathrm{No} = 10$, and we expect quadrature errors to be small compared to the truncation errors.

Plotted in Fig. 5.25 are the $\mathrm{N_Q} = 10$ deterministic solutions $w^{(i)}$ used for the projection. The Gauss points correspond to viscosity values in the range $[0.051, 2.59]$, with an equal number of values above and below the median viscosity (0.1), but a significant bias toward the highest viscosity values as expected for a shifted log-normal distribution. The figure clearly shows the boundedness of the stochastic solution. for $x \le 0$ (resp. $x > 0$), $-x \le w(x,\xi) \le w(x, 0.05)$ (resp. $w(x, 0.05) \le w(x,\xi) \le -x$), almost surely. The bounded character of the exact solution has to be contrasted with its truncated Wiener-Hermite expansion which is unbounded whenever $w_k(x) \ne 0$ for some $k \ge 1$. This point will be further discussed in Sect. 5.2.5.

5.2.4.2 Comparison with the Galerkin Projection

In Fig. 5.26, we present the first No + 1 stochastic modes $\tilde{w}_i(x)$ obtained after applying (5.117) to the set of $\mathrm{N_Q}$ deterministic solutions $w^{(i)}$ shown in Fig. 5.25.

The NISP modes in Fig. 5.26 are compared with the Galerkin modes presented in Fig. 5.24, and the same scaling factors have been used to facilitate the comparison. It is seen that the agreement between the Galerkin and NISP modes is excellent

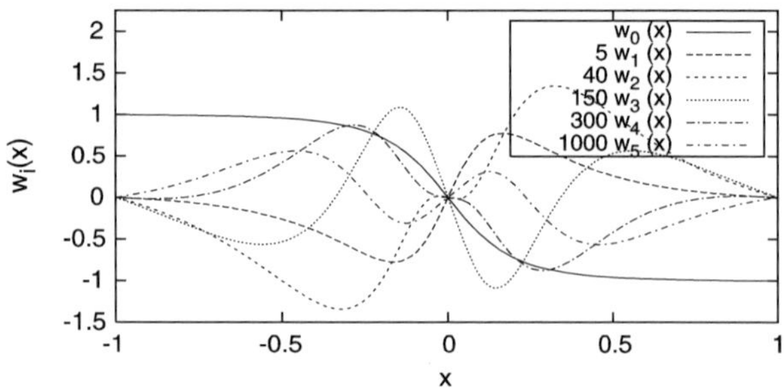

Fig. 5.26 First six stochastic modes $\tilde{w}_i(x)$ obtained using NISP with $N_Q = 10$. Modes have been rescaled as indicated for clarity

although a few differences are observed for the higher index modes. To better appreciate these differences, we plot in Fig. 5.27 for $i = 0, \dots, \text{No} = 5$ the Galerkin mode $w_i(x)$, the NISP mode $\tilde{w}_i(x)$ and the difference $\Delta_i = |w_i(x) - \tilde{w}_i(x)|$. The figure indicates that the discrepancy between the Galerkin and NISP modes increases with the mode index. Because of the decay of the modes with the index, the discrepancy in the second-order moment of the Galerkin and NISP solutions is however negligible.

Clearly, the numerical parameter governing the error on the NISP modes is the number of quadrature points in the quadrature formula. However, it is important to understand that contrary to the Galerkin projection method, NISP decouples the determination of the modes, *i.e.* the projection on Ψ_i is independent of the projection on Ψ_j for $i \neq j$. This feature is to be contrasted with the Galerkin projection which couples the resolution of all the modes. In fact, the integration error in NISP may be small for some low-order modes whereas it may be large for higher order ones. In the present case, numerical tests have shown that using $N_Q = 10$ is sufficient to accurately compute the modes up to order $\text{No} = 5$: increasing further N_Q does not reduce the discrepancy Δ_i between the Galerkin and NISP modes. Instead, this discrepancy appears to be essentially caused by the truncation errors in the Galerkin computation. To demonstrate this point, we show in Fig. 5.28 the evolution of the difference Δ_5 for increasing expansion order No in the Galerkin method. It is seen that the agreement between the NISP and Galerkin modes with index $i = 5$ improves with the order of the Galerkin computation. This numerical experiment stresses the fact already noted in Sect. 5.1 that, in Galerkin computations, the highest-order modes are generally the most affected by truncation errors.

5.2.5 Monte-Carlo Method

We finally consider application of Monte-Carlo simulation techniques. The objective is first to validate the Galerkin solution and to assess the efficiency of the Galerkin projection method for the stochastic Burgers equation. Secondly, in Sect. 5.2.5.3 we provide an illustration of a Monte-Carlo sampling strategy for the post-treatment of stochastic spectral expansions, specifically for the numerical estimation of the solution percentiles.

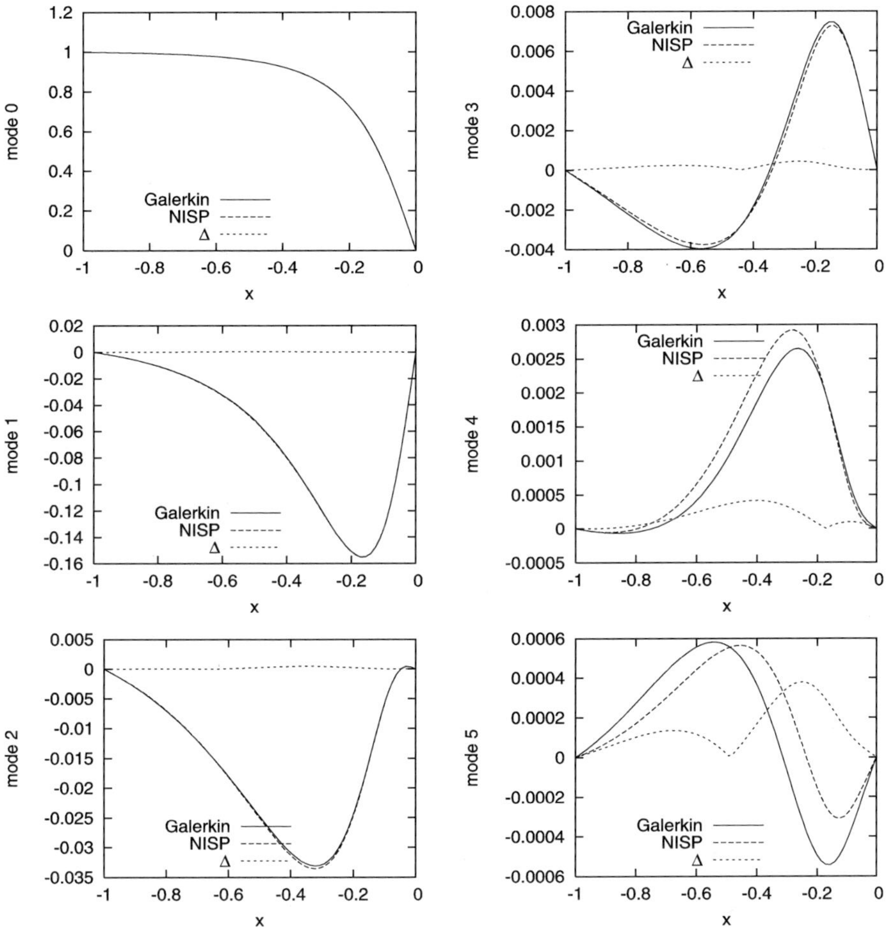

Fig. 5.27 Comparison of first six Galerkin and NISP modes. Galerkin projection uses No = 5 whereas NISP uses $N_Q = 10$ quadrature points. Only half of the spatial domain is shown due to the symmetry of the modes

5.2.5.1 Monte-Carlo Sampling

For simplicity, we rely on the simplest MC strategy based on a random unbiased sampling to construct a sample set of realizations for the stochastic Burgers solution.

Let $V = \{\nu^{(1)}, \dots, \nu^{(\mathrm{M})}\}$ be the sample set of the stochastic viscosity $\nu(\theta)$, where $\mathrm{M} = \dim S_{\mathrm{M}}$ is the sample set dimension. The elements $\nu^{(i)}$ are independent and randomly drawn from the shifted log-normal distribution of $\nu(\theta)$. Accordingly, let $W = \{w^{(1)}, \dots, w^{(\mathrm{M})}\}$ be the corresponding sample set of stochastic Burgers solutions $w(x,\theta)$, where $w^{(i)}(x)$ is the solution of (5.87) with $\nu = \nu^{(i)} \in V$.

We use a pseudo-random number generator to obtain M independent realizations of a normalized Gaussian random variable, which are subsequently transformed to

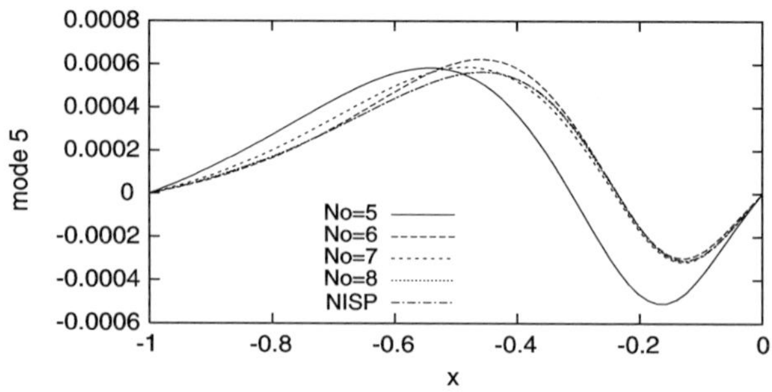

Fig. 5.28 Convergence with the order No in the Galerkin method of $w_5(x)$ toward its well resolved NISP estimation $\tilde{w}_5$ obtained with $N_Q = 10$

construct the realizations corresponding to the shifted log-normal distribution. After resolution of the corresponding deterministic problems, a realization of W is obtained. We can then proceed with the estimation of some statistics of w based on the realization of W. It is known that in MC methods the statistics based on a sample set with dimension M converge to the true values as $1/\sqrt{\mathrm{M}}$. The sampling error for finite M is conveniently estimated using the standard error (SE), defined as the standard deviation of the empirical estimator induced by the random sample W. Since the probability law of W is usually unknown, the SE has to be estimated. A popular technique is the bootstrap method [61] which relies on resampling of the MC sample set. The main advantage of the bootstrap technique is that it requires only weak assumptions on the distributions and needs no additional resolution of the deterministic model.

5.2.5.2 First- and Second-Order Estimates

Based on the MC sample set W, the empirical estimator of the expectation of $w(x, \theta)$, denoted $\hat{w}_W(x)$, is

$$\hat{w}_W(x) = \frac{1}{\mathrm{M}} \sum_{i=1}^{\mathrm{M}} w^{(i)}(x). \tag{5.118}$$

Similarly, the empirical estimator of the standard deviation $\hat{\sigma}_W(w)$ is

$$\hat{\sigma}_W(w)(x) = \sqrt{\frac{1}{\mathrm{M}} \sum_{i=1}^{\mathrm{M}} \left(w^{(i)}\right)^2(x) - \hat{w}_W^2(x)}. \tag{5.119}$$

In Fig. 5.29 we present the MC estimate $\hat{w}_W$ with ± 3 bootstrap standard error bounds, for two sample sets with respective dimensions $\mathrm{M} = 100$ and $\mathrm{M} = 1000$. Also shown for comparison is the expectation of the Galerkin solution. It is seen that for $\mathrm{M} = 100$, the MC estimate is in excellent agreement with the expectation for the Galerkin solution. In fact, increasing M to 1000 essentially improves the confidence in $\hat{w}_W$ by reducing the error bounds.

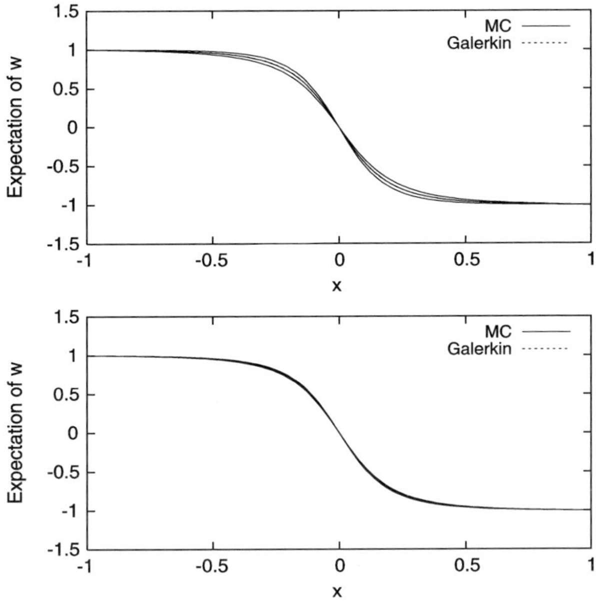

Fig. 5.29 Comparison of the MC estimate of the expectation of w with ± 3 standard error bounds and Galerkin solution w_0 computed with $\mathrm{No} = 5$; *top*: $\mathrm{M} = 100$, *bottom*: $\mathrm{M} = 1000$

The plots in Fig. 5.29 shows that in the present case the MC simulation provides an excellent estimate of the expectation of the stochastic solution for low sample set dimensions. This is contrasted with the MC determination of second moments of the stochastic solution, as illustrated by the estimation of the standard deviation reported in Fig. 5.30. Plotted are the MC estimate of $\sigma_W(w)(x)$ for $\mathrm{M} = 100$, 1000 and 10000. The estimate is compared with the standard deviation of the Galerkin solution computed with $\mathrm{No} = 5$. Significant discrepancies between MC and Galerkin predictions are obtained, which remain noticeable even for the largest sample set with $\mathrm{M} = 10000$. However, the ± 3 standard error bounds in the MC estimate, constructed from the bootstrap procedure, are seen to reduce with M, according to the expected asymptotic behavior, and to collapse to the Galerkin prediction. In addition to validating the Galerkin solution, this observation demonstrates that for this problem the Galerkin projection is much more efficient than MC sampling. Specifically, the Galerkin approach provides accurate and reliable estimates with a low-order expansion and for a much lower computational cost compared to MC. For $\mathrm{M} = 10000$, the ratio of computational Galerkin to MC computational times is about $\sim 1/100$. The ratio is even larger for the NISP shown in the previous section, as it scales roughly as $\mathrm{N_Q}/\mathrm{M}$.

The greater efficiency of the Galerkin projection and NISP, compared to MC, illustrates the potential of PC expansions and moregenerally that of spectral decompositions. Unfortunately this level of superiority cannot be always be expected or achieved, as the numerical cost of stochastic spectral methods highly depends on the problem considered, and particularly on the dimension of the stochastic basis needed. In contrast, MC methods have convergence rates which asymptotically depend on the sample set dimension M only, and not on the complexity of the problem.

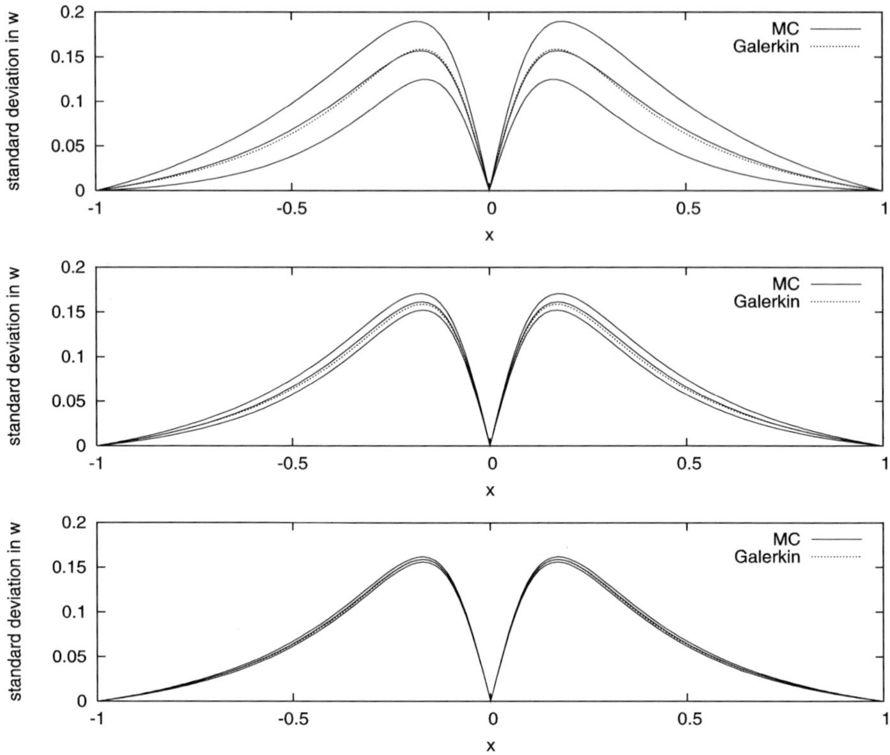

Fig. 5.30 Comparison of the MC estimate (with ±3 bootstrap standard error bounds) of the standard deviation of w with the corresponding Galerkin prediction computed with No = 5, for three sample sets of dimensions M = 100 (*top*) and M = 1000 (*middle*) and M = 10000 (*bottom*)

5.2.5.3 Determination of Percentiles

MC methods are often used to retrieve useful statistics from the spectral expansion of the solution. As an example, we present below percentile estimations for w. Percentiles are commonly used in reliability analysis and system safety. For $\alpha \in]0, 100[$, we define the percentile $w^{\alpha\%}$ as the value such the probability of $w(\xi) \leq w^{\alpha\%}$ is equal to α (in %):

$$\mathrm{P}(\mathrm{w} \leq \mathrm{w}^{\alpha\%}) = \alpha(\%). \tag{5.120}$$

$w^{\alpha\%}$ can be estimated from MC sampling as follows. From a sufficiently large sample set W, with dimension M, the percentile $w^{\alpha\%}$ can be obtained from the empirical cumulative distribution function of W. This amounts to sorting the elements $w^{(i)}$ in W in increasing order, and affecting them with a probability $1/\mathrm{M}$. For instance, the median value of w, which is by definition $w^{50\%}$, will be estimated as $w^{(\mathrm{M}/2)}$, assuming the elements of W sorted. Clearly, this MC estimation requires larger and larger sample sets as α goes to zero or 100, *i.e.* when inspecting tails of the distribution of w. In addition, as for the moments of w, one has to assess the error in the

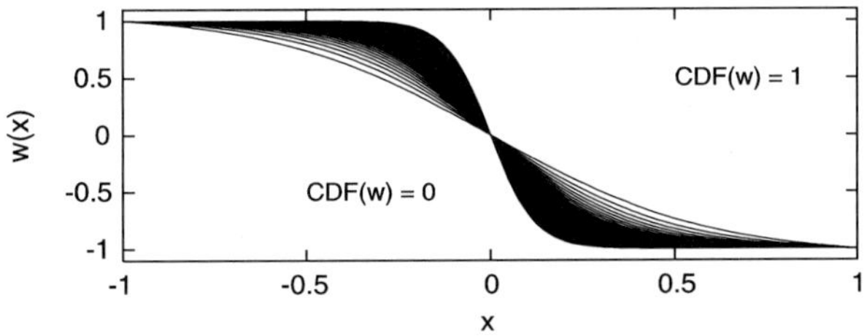

Fig. 5.31 Isolevels of the F of the exact solution of the stochastic Burgers equation

estimate of $w^{(\alpha\%)}$ due to finite sampling. The latter can be again characterized by the bootstrap standard error.

Below, we present the percentiles obtained from a MC sample set W, with dimension M, corresponding to the sampling of the Galerkin solution $w = \sum_k w_k \Psi_k(\xi)$. The sample set W is constructed by sampling randomly ξ in the range $]-\infty, \infty[$, according to the standard normal distribution, and then evaluating the solution expansion to obtain $w^{(i)}$. For comparison purposes, we take advantage of the monotonic character of the Burgers solution with regard to the viscosity value, to compute the exact percentile:

$$w_{ex}^{\alpha\%}(x) = \begin{cases} w(x|\nu^{\alpha\%}), & x \le 0, \\ w(x|\nu^{(100-\alpha)\%}), & x > 0 \end{cases} \tag{5.121}$$

where $w(x|\nu^{\alpha\%})$ is the solution of the deterministic Burgers equation (5.87) for the viscosity equal to its α-percentile. The latter is exactly known thanks to the explicit parametrization. In fact, the percentile $w^{\alpha\%}$ corresponds to an isolevel in the (x, w) plane of the cumulative distribution function $F(w(x))$. A few isolevels of the exact CDF $F(w(x))$ are shown in Fig. 5.31.

We set $\mathrm{M} = 1000$ and focus on the analysis of the 2-, 50- (median value) and 98-percentiles. The estimates of the percentiles are reported in Fig. 5.32 for $-1 \le x \le 0$ and different expansion orders. The bootstrap ± 3 standard error bounds on the estimated percentiles are also shown to assess the reliability of the MC computation, together with the exact percentiles plotted with symbols. We observe very different magnitudes of the standard errors depending on the percentiles: the standard error is large for the 2-percentile, low for the median estimate and nearly zero for the 98-percentile. This trend can be explained from the results in Fig. 5.31, which show that the distance between consecutive isolevels is larger for low values of the F (when $x < 0$) and so the empirical distribution is more sensitive to finite sampling errors. As a result, the discrepancy between the percentile estimate and its exact value improves as α increases. However, the discrepancies are not only due to the sampling errors, but also to truncation of the stochastic expansion. The latter can be appreciated from Fig. 5.32 which shows results corresponding to different expansion orders. As No increases from 3 to 7, the estimates come closer to the exact percentiles. In fact the agreement for the 50- and 98-percentiles is excellent for $\mathrm{No} \ge 5$.

For $w^{2\%}(x)$ and $\mathrm{No} = 7$, the difference between the MC estimate and the exact value, even though small, is still noticeable. Further tests (not shown) were con-

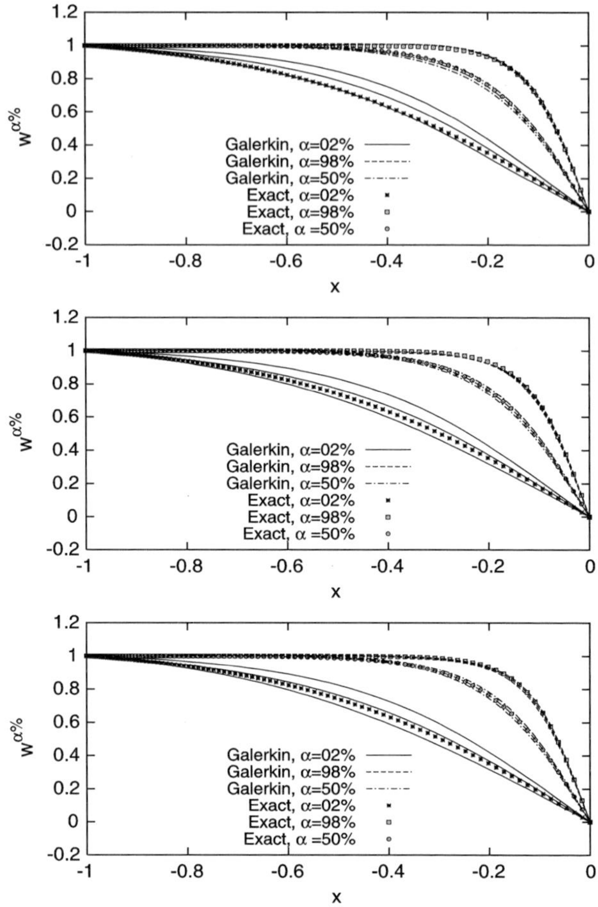

Fig. 5.32 Comparison of the MC estimates with ±3 bootstrap standard error bounds of the percentiles $w^{\alpha\%}$ for $\alpha = 2$, 50 and 98, obtained using the Galerkin projection with No = 3 (*top*), 5 (*middle*) and 7 (*bottom*). Also shown for comparison are the exact percentiles (*symbols*). The computations use MC sample sets with M = 1000

ducted with larger sample sets to reduce the sampling error. These tests indicate that in the present case the truncation error is essentially responsible for the discrepancy. A similar behavior is observed for $w^{98\%}(x)$ and $x > 0$ (not shown). This experience demonstrates the difficulty to capture the complete probabilistic distribution of the stochastic Burgers solution using Wiener-Hermite expansions, particularly since a very large expansion order is needed to accurately capture the tails of the distribution. This is not surprising since we know from the properties of the Burgers equation that the stochastic solution is bounded, a property that cannot be satisfied using projections onto polynomials of unbounded random variables.

Chapter 6
Application to Navier-Stokes Equations

This chapter is devoted to application of PC methods to fluid flows governed by the transient Navier-Stokes equations. Attention is primarily focused on the development and application of Galerkin (intrusive) solution methods. Non-intrusive approaches, including Monte Carlo sampling and Gauss quadratures, are used in a limited fashion, primarily for the purpose of verifying intrusive predictions and for analyzing the efficiency of the corresponding solvers.

In Sect. 6.1, we start with a general stochastic formulation for incompressible, constant-density, flow. We also consider the presence of a passive scalar, whose evolution is governed by an advection-diffusion equation. We rely on a weighted residual formalism and derive evolution equations for the velocity, pressure and scalar modes. A numerical scheme is then constructed for the solution of the system of equations governing the stochastic modes. The scheme is based on a fractional step update of the momentum equation, in which the convection and diffusion terms are accounted for in a first step, whereas the pressure term is accounted for in a second step. The latter involves the solution of a decoupled system of Poisson equations for the pressure modes, which generalize pressure projection step in the deterministic case. Implementation of the resulting stochastic projection method (SPM) is then illustrated in terms of simplified examples involving a single random variable.

In Sect. 6.2, we show that the SPM for incompressible flow extends in a natural fashion to variable-density flows in the Boussinesq limit. We also consider the case where model data is given by a random process, and thus deal with truncation of its (otherwise infinite-dimensional) representation. The behavior of the solver is analyzed in detail. Addressed are questions concerning effects of the truncation of the input data, the order of the PC representation, and the efficiency of the intrusive scheme. In order to address these questions, the analysis relies on non-intrusive computations, including results obtained with Monte-Carlo sampling and with Gauss quadratures.

In Sect. 6.3, we address the generalization of the SPM to low-Mach-number flows, particularly situations where the Boussinesq approximation no longer holds. The analysis shows that the generalization must be conducted carefully, particularly to ensure that the divergence constraints associated with the individual pressure modes are properly satisfied. The resulting low-Mach-number scheme is then

O.P. Le Maître, O.M. Knio, *Spectral Methods for Uncertainty Quantification*, Scientific Computation,
DOI 10.1007/978-90-481-3520-2_6, © Springer Science+Business Media B.V. 2010

applied to analyze the behavior of a heated cavity under stochastic boundary conditions. Predictions are first compared against deterministic results available from the literature, and the computations are then used to examine the response of the flow and heat transfer to the imposed random forcing.

In Sect. 6.4, we consider a particle-mesh scheme for the simulation of vorticity-form of the incompressible momentum equations in the Boussinesq limit, for unbounded flows at high Reynolds and Rayleigh numbers. In these situations, particle-based approximations provide significant advantages, owing to the fact that the schemes have minimal numerical diffusion, naturally handle far-field boundary conditions, and concentrate resources in regions of finite vorticity and temperature gradients. However, extension of these schemes to situations involving random data is problematic, since straightforward implementation of PC expansions leads to the definition of multiple sets of particles. This difficulty is addressed using a specialized construction in which a single set of particles is defined, the particles being transported with the mean flow velocity. This approach results in an efficient stochastic particle-mesh scheme, whose validity is first verified against analytical solutions for simple advection-dominated and diffusion-dominated flows. Application of the scheme is then illustrated for the case of a rising plume, and the computations are used to analyze the impact of uncertainty in the Rayleigh number.

In Sect. 6.5, a more elaborate example is provided of electrokinetically-driven flow in a microchannel. A detailed physical formulation is adopted which considers the fully coupled momentum, species transport, and electrostatic field equations, including a model for the dependence of the zeta potential on pH and buffer molarity. A mixed finite-rate, partial-equilibrium formulation is applied for the chemical reactions, where "fast" electrolyte reactions are described by equilibrium constraints, while the remaining "slow" protein labeling reactions are modeled with finite-rate kinetics. Implementation of the Galerkin formalism to the resulting systems leads to computational challenges, particularly concerning the enforcement of nonlinear constraints associated with fast chemical kinetics, and the solution of non-separable elliptic equations for the stochastic modes of the electric potential. Means for addressing these challenges are first discussed. Implementation of the stochastic scheme is then illustrated through applications to protein-labeling in a homogeneous buffer, as well as in two-dimensional electrochemical microchannel flow.

6.1 SPM for Incompressible Flow

In this section, we aim at numerical simulation of the incompressible Navier-Stokes equations, and particularly at the construction of a robust and efficient PC-based computational solver. As further described below, the present approach is based on stochastic extension of a conservative-difference, pressure projection scheme. We outline the construction of the solver for the case of advection and mixing in a 2D flow, starting from the deterministic system of governing equations provided in Sect. 6.1.1.

As discussed in Sect. 6.1.2, construction of the flow solver is based on representing random variables and field quantities in terms of suitable PC expansions, introducing these expansions into the governing equations and using a Galerkin formalism to determine the coefficients in the expansions. This approach results in a coupled system of advection-diffusion equations for the stochastic velocity (and temperature) fields, and a decoupled set of stochastic divergence constraints. This feature is exploited by constructing an efficient stochastic projection method (SPM) which provides a stochastic characterization of the solution process at a cost that is essentially proportional to the number of terms in the spectral expansion.

In order to demonstrate the application of the resulting scheme, we consider in Sect. 6.1.3 a set of simplified problems in which the uncertainty in the solution process arises due to random viscosity or due to the dependence of the viscosity on a random temperature. Thus, our attention is restricted to relatively simple conditions involving low-Reynolds-number flows. Thus, potential representation difficulties [31] that may arise in complex settings involving shocks or energy cascade are naturally avoided. We first start with a simplified setting in which the temperature is treated as a Gaussian random variable that is spatially uniform. Thus, the fluid viscosity is also uniform, but both linear and nonlinear viscosity laws are considered in the analysis. A more complicated setting is then considered, which consists of a double-inlet channel where the inlet temperature of one of the streams has a random Gaussian component. In this case, the uncertain boundary condition leads to stochastic velocity and temperature fields that are coupled by the temperature dependence of viscosity.

6.1.1 Governing Equations

We consider the two-dimensional flow, in the (x, y) plane, of an incompressible, uniform-density Newtonian fluid inside a narrow channel of height H and width B. As shown in Fig. 6.1, the boundaries of the computational domain Ω consist of inflow (Γ_i) and outflow (Γ_o) boundaries, respectively located at $y = 0$ and $y = H$, and solid boundaries (Γ_{ns}) located at $x = 0$ and $x = B$.

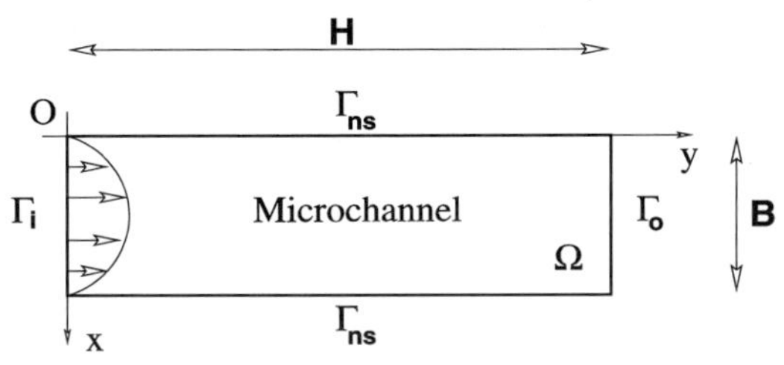

Fig. 6.1 Schematic representation of the computational domain. The inflow and outflow boundaries, respectively Γ_i and Γ_o, are located at $y = 0$ and $y = H$. Γ_{ns} refers to the channel walls, which are located at $x = 0$ and $x = B$. The computational domain consists of the region $\Omega \equiv [0, B] \times [0, H]$. Adapted from [123]

The evolution of the flow within the channel is governed by the Navier-Stokes equations:

$$\begin{cases} \dfrac{\partial \mathbf{u}}{\partial t} + (\mathbf{u} \cdot \nabla)\mathbf{u} = -\nabla p + \nabla \cdot \overline{\boldsymbol{\sigma}}, & \text{(a)} \\ \nabla \cdot \mathbf{u} = 0 & \text{(b)} \end{cases} \tag{6.1}$$

where $\mathbf{u}$ is the velocity field, $\tilde{p}$ is the pressure, ρ is the density, $p \equiv \tilde{p}/\rho$, $\boldsymbol{\tau}$ is the viscous stress, and $\overline{\boldsymbol{\sigma}} \equiv \boldsymbol{\tau}/\rho$. When the dynamic viscosity is uniform, the viscous force can be expressed as:

$$\nabla \cdot \overline{\boldsymbol{\sigma}} \equiv \nabla \cdot \left[\nu \overline{\mathbf{S}}\right] = \nu \nabla^2 \mathbf{u} \tag{6.2}$$

where ν is the kinematic viscosity and $\overline{\mathbf{S}}$ is the symmetric part of the velocity gradient tensor. The governing equations are supplemented with velocity boundary conditions, which consist of imposed velocity at the inflow, outflow conditions on Γ_o, and no-slip conditions on the solid walls.

In addition to the Navier-Stokes equations (6.1), we also consider the temperature distribution within the channel. The evolution of the temperature field, T, is governed by the energy equation:

$$\frac{\partial T}{\partial t} + \nabla \cdot (\mathbf{u}T) = -\nabla \cdot \mathbf{q} \tag{6.3}$$

where $\mathbf{q} \equiv -\kappa \nabla T$ is the heat flux, and κ is the thermal diffusivity. We use adiabatic conditions at the solid walls, outflow condition at Γ_o, and Dirichlet conditions at inflow. Note that when the viscosity is independent of temperature the evolution of the flow field can be determined independently of (6.3). On the other hand, when ν depends on T, a non-trivial coupling exists between the Navier-Stokes and energy equations.

6.1.2 Intrusive Formulation and Solution Scheme

For a general setting, it is useful to introduce the PC representations of all field variables, namely the velocity, pressure, temperature, kinematic viscosity, and thermal diffusivity,

$$T(\mathbf{x}) = \sum_{i=0}^{\mathrm{P}} T_i(\mathbf{x})\Psi_i, \tag{6.4}$$

$$\mathbf{u}(\mathbf{x}) = \sum_{i=0}^{\mathrm{P}} \mathbf{u}_i(\mathbf{x})\Psi_i, \tag{6.5}$$

$$p(\mathbf{x}) = \sum_{i=0}^{\mathrm{P}} p_i(\mathbf{x})\Psi_i, \tag{6.6}$$

$$\nu(\mathbf{x}) = \sum_{i=0}^{\mathrm{P}} \nu_i(\mathbf{x})\Psi_i, \tag{6.7}$$

$$\kappa(\mathbf{x}) = \sum_{i=0}^{\mathrm{P}} \kappa_i(\mathbf{x})\Psi_i \tag{6.8}$$

keeping in mind that some of these representations may not always be necessary, or may simply collapse to a deterministic constant. Furthermore, we assume that the boundary conditions are also appropriately decomposed into similar expansions.

The intrusive approach is based on inserting these PC expansions into the governing equations, and then using a Galerkin formalism in order to determine the unknown coefficients. This results in the following system of governing equations:

$$\frac{\partial \mathbf{u}_k}{\partial t} + \sum_{i=0}^{\mathrm{P}}\sum_{j=0}^{\mathrm{P}} C_{ijk}(\mathbf{u}_i \cdot \nabla)\mathbf{u}_j = -\nabla p_k + \nabla \cdot \overline{\sigma}_k, \tag{6.9}$$

$$\nabla \cdot \mathbf{u}_k = 0, \tag{6.10}$$

$$\frac{\partial T_k}{\partial t} + \sum_{i=0}^{\mathrm{P}}\sum_{j=0}^{\mathrm{P}} C_{ijk}\nabla \cdot \left(\mathbf{u}_i T_j\right) = -\nabla \cdot \mathbf{q}_k \tag{6.11}$$

where

$$C_{ijk} \equiv \frac{\langle \Psi_i \Psi_j \Psi_k \rangle}{\langle \Psi_k \Psi_k \rangle} \tag{6.12}$$

is the multiplication tensor.

It can be observed that the equation system above has a similar form to the original Navier-Stokes equation. Due to the appearance of coupling terms, however, the system size is by a factor of P larger than the corresponding deterministic system. If not addressed properly, the enlargement of the system size in the stochastic formulation can constitute a major drawback, especially when the implementation of a fast $O(N)$ solver is not possible. Another consideration is the desire to base the proposed development on existing deterministic solvers and computer codes.

The present approach to the formulation of the stochastic solver is based on the observation that the velocity divergence constraints are *decoupled*, and this suggests the implementation of a projection scheme [29] in which the advection and diffusion terms are integrated in a first fractional step, and the divergence constraints are then enforced in a second fractional step. Because the divergence constraints are decoupled, this approach results in a set of $\mathrm{P}+1$ decoupled pressure projection steps. Since these steps typically account for the bulk of the computational effort in incompressible flow simulations, the solution of the stochastic system can be at essentially a cost of $\mathrm{P}+1$ deterministic solutions. Coupled with the spectral nature of the stochastic representation, this leads to a highly efficient stochastic solver, as illustrated in the examples of the following section. Note that in the case of multiple random variables, the stochastic solution can be determined at a cost essentially of L

deterministic solutions, where L is the number of stochastic dimensions. Generally, P and L are generally much smaller than the number of independent MC realizations that are needed for adequate representation of the uncertainty. This feature, together with the decoupled structure of the pressure projection steps, is behind the efficiency of SPM.

In the applications below, the stochastic scheme adapts elements of previously developed low-Mach-number solvers in [116, 164]. We rely on discretization of all fields variables using a uniform Cartesian mesh with cell size Δx and Δy in the x and y directions, respectively. The velocity modes $\mathbf{u}_k$ are defined on cell edges, while the scalar fields p_k, T_k, and v_k are defined on cell centers. Spatial derivatives are approximated using second-order centered differences.

As mentioned earlier, the governing equations are integrated using a fractional step projection scheme. In the first fractional step, we integrate the coupled advection-diffusion equations:

$$\frac{\partial \mathbf{u}_k}{\partial t} + \sum_{i=0}^{\mathrm{P}} \sum_{j=0}^{\mathrm{P}} C_{ijk} (\mathbf{u}_i \cdot \nabla) \mathbf{u}_j = \nabla \cdot \overline{\boldsymbol{\sigma}}_k \tag{6.13}$$

for $k = 0, \ldots, \mathrm{P}$. The explicit, second-order Adams-Bashforth scheme is used for this purpose; we thus have:

$$\frac{\mathbf{u}_k^* - \mathbf{u}_k^n}{\Delta t} = \frac{3}{2}\mathbf{H}_k^n - \frac{1}{2}\mathbf{H}_k^{n-1}, \quad k = 0, \ldots, \mathrm{P} \tag{6.14}$$

where $\mathbf{u}_k^*$ are the predicted velocity modes, Δt is the time step,

$$\mathbf{H}_k \equiv \nabla \cdot \overline{\boldsymbol{\sigma}}_k - \sum_{i=0}^{\mathrm{P}} \sum_{j=0}^{\mathrm{P}} C_{ijk} (\mathbf{u}_i \cdot \nabla) \mathbf{u}_j \tag{6.15}$$

and the superscripts refer to the time level. A similar treatment is used for the energy equation, which is integrated using:

$$\frac{T_k^{n+1} - T_k^n}{\Delta t} = \frac{3}{2} J^n - \frac{1}{2} J^{n-1}, \quad k = 0, \ldots, \mathrm{P} \tag{6.16}$$

where

$$J_k \equiv -\nabla \cdot \mathbf{q}_k - \sum_{i=0}^{\mathrm{P}} \sum_{j=0}^{\mathrm{P}} C_{ijk} \mathbf{u}_i \cdot \nabla T_j. \tag{6.17}$$

In the second fractional step, we perform a pressure correction to the predicted velocity in order to satisfy the divergence constraints. Specifically, we have:

$$\frac{\mathbf{u}_k^{n+1} - \mathbf{u}_k^*}{\Delta t} = -\nabla p_k, \quad k = 0, \ldots, \mathrm{P} \tag{6.18}$$

where the pressure fields p_k are determined so that the fields $\mathbf{u}_k^{n+1}$ satisfy the divergence constraints in (6.10), i.e.

$$\nabla \cdot \mathbf{u}_k^{n+1} = 0. \tag{6.19}$$

Combining (6.18) and (6.19) results in the following system of *decoupled* Poisson equations:

$$\nabla^2 p_k = -\frac{1}{\Delta t} \nabla \cdot \mathbf{u}_k^*, \quad k = 0, \ldots, \mathrm{P}. \tag{6.20}$$

Similar to the original projection method, the above Poisson equations are solved, independently, subject to Neumann conditions that are obtained by projecting equation (6.18) in the direction normal to the domain boundary [29, 112]. The weak formulation approach outlined above is used for this purpose. Fast, Fourier-based solvers are employed for the inversion of the discrete operators.

6.1.3 Numerical Examples

In this section, application of the SPM outlined above is illustrated based on simplified examples involving uncertain transport properties or boundary conditions, which in all cases are treated as stochastic quantity generated by a single Gaussian variable. Within this restricted scope, three different problems are considered:

1. In the first example, we consider the case of an uncertain viscosity, which is treated as a Gaussian random variable. The viscosity is assumed to be spatially uniform and solution of the energy equation is not required.
2. In the second example, we also consider uncertainty in viscosity, which is taken to be generated by a Gaussian stochastic temperature. The temperature is assumed to be uncertain, but spatially uniform, and the uncertainty in viscosity is reflected through a nonlinear viscosity law.
3. In the third example, we consider a coupled problem which involves a temperature-dependent viscosity and an evolving temperature field. In this case, the uncertainty is generated by a random boundary condition on the inlet temperature.

6.1.3.1 Example 1

In this example, we consider the case of an uncertain viscosity coefficient that is spatially uniform; it is modeled as a Gaussian random variable:

$$\nu = \nu_0 + \xi_1 \nu_1 \tag{6.21}$$

where ν_0 is the mean viscosity, ξ_1 is a normalized Gaussian variable (having with zero mean and unit variance), while ν_1 is a deterministic coefficient which corresponds to the standard deviation of the viscosity. Note that with the use of Gaussian noise in the viscosity, negative values are, in principle, possible. Below, potential complications arising from this situation are avoided by ensuring that the standard deviation in viscosity is substantially smaller than the mean. In more general situations, a non-Gaussian distribution may be needed to ensure that negative viscosities have negligible likelihood.

Having a single source of uncertainty, the evolution of the velocity field is totally decoupled from that of temperature. Thus, the formulation consists of the momentum conservation equations, (6.10) and (6.9), with the viscous stress term given by:

$$\nabla \cdot \overline{\boldsymbol{\sigma}}_k = \nu_0 \nabla^2 \mathbf{u}_k + \nu_1 \sum_{i=0}^{\mathrm{P}} C_{1ij} \nabla^2 \mathbf{u}_i . \tag{6.22}$$

The advection-diffusion equation (6.11) is simply dropped from the system of governing equations.

Parabolic inflow: We start by examining a simplified case where a deterministic parabolic velocity profile is imposed at the channel inlet. We use:

$$u_{in} = 0, \qquad v_{in} = V_{ref} \left[1 - 4 \left(\frac{x}{B} - 0.5 \right)^2 \right], \tag{6.23}$$

where (u_{in}, v_{in}) are the x and y components of the inlet velocity, V_{ref} is the reference velocity, and B is the channel width. The flow is characterized by the Reynolds number based on B, ν_0 and V_{ref}:

$$\mathrm{Re} \equiv \frac{B V_{ref}}{\nu_0}. \tag{6.24}$$

As mentioned earlier, we are interested in applications with moderate to low Reynolds numbers, so that the flow is stable and laminar.

The (deterministic) solution for steady flow in the channel is the well-known Poiseuille solution [14]:

$$u(x,y) = 0, \qquad v(x,y) = v_{in}(x), \qquad \frac{\partial p}{\partial x} = 0, \qquad \frac{\partial p}{\partial y} = -\frac{4 \mu V_{ref}}{B^2} \tag{6.25}$$

where $\mu = \rho \nu$ is the dynamic viscosity. Thus, when the inlet profile is parabolic, the velocity field is independent of the axial coordinate, y, and, so long as this solution is stable, the velocity is independent of the viscosity as well.

It follows from the above remarks that for the present inlet velocity conditions an uncertainty in the viscosity would only affect the rate of pressure drop. We have:

$$\frac{\partial p}{\partial y} = \frac{-4 \rho V_{ref}}{B^2} \nu = \frac{-4 \rho V_{ref}}{B^2} (\nu_0 + \nu_1 \xi_1) \equiv \frac{\partial p_0}{\partial y} \Psi_0 + \frac{\partial p_1}{\partial y} \Psi_1 . \tag{6.26}$$

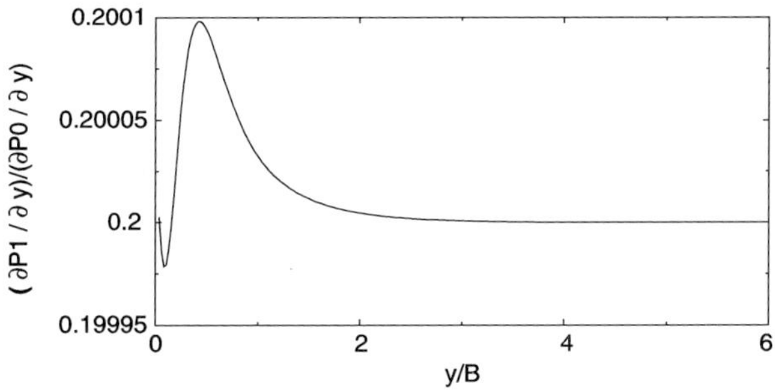

Fig. 6.2 Pressure gradient ratio $(\partial p_1/\partial y)/(\partial p_0/\partial y)$ along the channel centerline versus the normalized streamwise coordinate y/B. Results are obtained for a channel with $H/B = 6$, Re $= 40.62$, and $\nu_1/\nu_0 = 0.2$. The simulation is performed using a grid with 64×256 cells, $\Delta t V_{ref}/B = 10^{-3}$ and a PC expansion with P $= 2$. Adapted from [123]

Using the definitions $\Psi_0 = 1$ and $\Psi_1 = \xi_1$, and exploiting the orthogonality of the Hermite polynomials we get:

$$\frac{\partial p_0}{\partial y} = \frac{-4\rho V_{ref}\nu_0}{B^2}, \tag{6.27}$$

and

$$\frac{\partial p_1}{\partial y} = \frac{-4\rho V_{ref}\nu_1}{B^2} = \frac{\nu_1}{\nu_0}\frac{\partial p_0}{\partial y}. \tag{6.28}$$

Furthermore, the variance $\sigma^2_{\partial p}$ of the pressure gradient is given by:

$$\sigma^2_{\partial p} \equiv \left\langle \left(\frac{\partial p}{\partial y} - \frac{\partial p_0}{\partial y} \right)^2 \right\rangle = \left\langle \left(\frac{\partial p_1}{\partial y}\Psi_1 \right)^2 \right\rangle = \left(\frac{\partial p_1}{\partial y} \right)^2. \tag{6.29}$$

Thus, for Poiseuille flow the effect of uncertainty in viscosity can be characterized analytically.

The analytical expressions derived above are used to verify the predictions of the stochastic projection scheme. Results are obtained for a channel flow with Re $=$ 40.62 and $\nu_1/\nu_0 = 0.2$. The simulations are performed in a domain with aspect ratio $H/B = 6$, using a 64×256 computational grid, a time step $\Delta t = 10^{-3} B/V_{ref}$, and a polynomial chaos expansion with P $= 2$. Figure 6.2 shows the ratio of the computed pressure gradients $\partial p_1/\partial y$ and $\partial p_0/\partial y$, at steady state, along the centerline of the channel. The results are in excellent agreement with the theoretical prediction in (6.28). For $y/B > 2$ the analytical and computed results are practically identical, but tiny differences occur near the domain inlet. The maximum relative error between the exact and computed pressure gradient ratios is, however, quite small and falls below 0.05%.

Uniform inflow: We now consider the case of a uniform inlet velocity profile:

$$u_{in} = 0, \qquad v_{in} = V_{ref}. \tag{6.30}$$

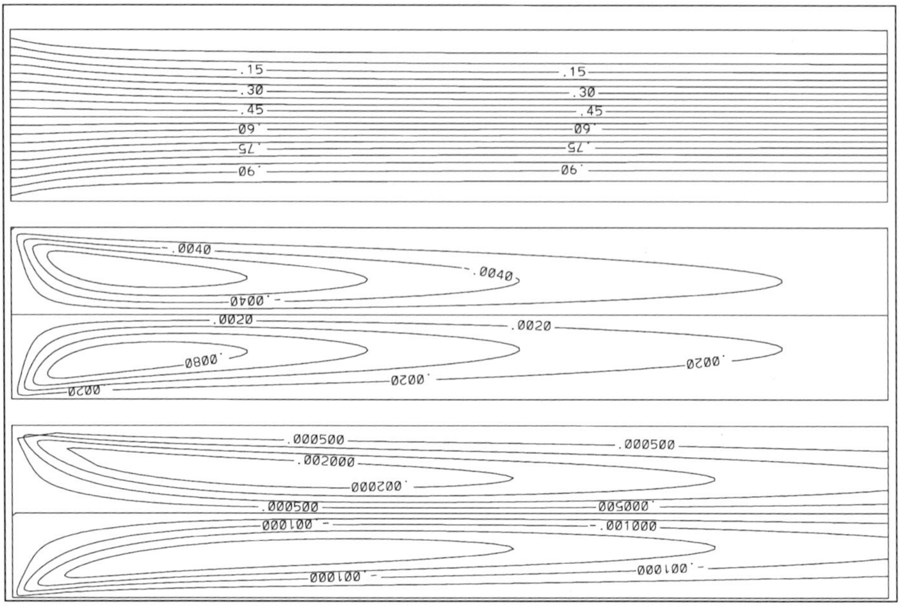

Fig. 6.3 Streamfunction contours corresponding to $\mathbf{u}_0$ (*top*), $\mathbf{u}_1$ (*middle*) and $\mathbf{u}_2$ (*bottom*). Results are obtained for a channel with $H/B = 6$, $\mathrm{Re} = 81.24$, and $\nu_1/\nu_0 = 0.2$. The flow is from left to right along the $+y$ direction; the entire domain ($0 \leq y/B \leq 6$, $0 \leq x/B \leq 1$) is shown. The simulation is performed using a grid with 64×256 cells, $\Delta t\, V_{ref}/B = 2 \times 10^{-3}$ and a PC expansion with $\mathrm{P} = 3$. Adapted from [123]

With this inflow condition, the steady (deterministic) flow gradually evolves towards a parabolic Poiseuille profile. The transition reflects the growth of a laminar boundary layer which eventually fills the channel; this delimits the entrance length, whose value depends on the Reynolds number [180]. Within the transition region, the flow field is no longer uniform so that, in the stochastic case, all of the velocity and pressure modes exhibit a non-trivial behavior.

In order to illustrate this behavior, a simulation is performed for a channel with $\mathrm{Re} = 81.24$ and $\nu_1/\nu_0 = 0.3$. The simulation is performed using a uniform grid with 64×256, a time step $\delta t\, V_{ref}/B = 2 \times 10^{-3}$, and a polynomial chaos expansion with $\mathrm{P} = 3$. The unsteady equations are integrated in time until steady conditions are reached. Results are shown in Figs. 6.3–6.5 which depict contours of the streamfunction, the streamwise velocity and the cross-stream velocity, respectively. The streamfunction is reconstructed from the steady velocity field. In each figure, plots are generated for the mean ($k = 0$) as well as modes 1 and 2.

The results illustrate the growth of the boundary layer in the entrance region. For the mean flow modes ($k = 0$), entrance effects extend up to $y/B \simeq 4$ and the corresponding distribution becomes uniform at larger streamwise locations. The streamwise velocity modes v_1 and v_2 exhibit appreciable variation up to 4–5 chan-

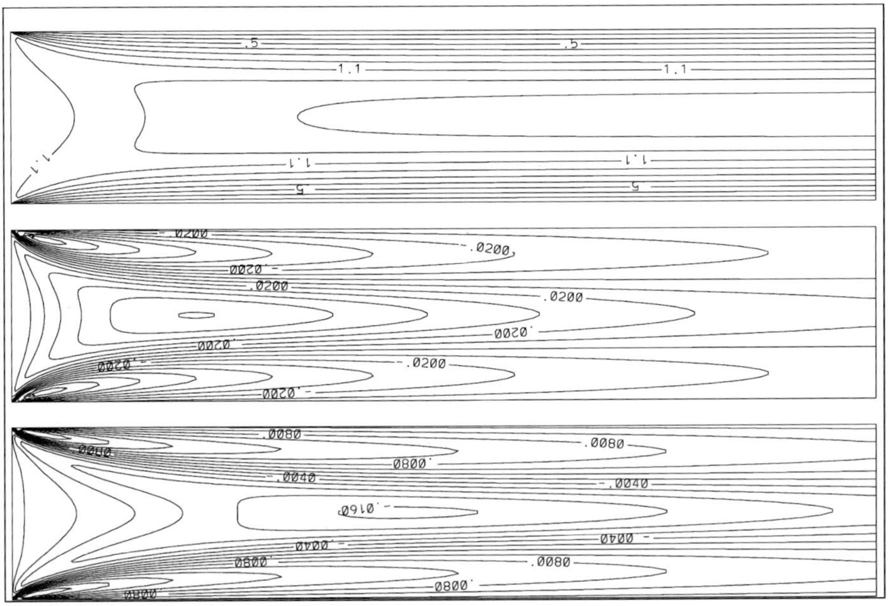

Fig. 6.4 Contour plots of the streamwise velocity components v_0 (*top*), v_1 (*middle*) and v_2 (*bottom*). Same parameters as in Fig. 6.3. Adapted from [123]

nel widths, but the cross-stream velocity modes u_0, u_1 and u_2 have negligible values outside the region $0 \le y/B \le 3$.

Note that the magnitudes of the fields decrease as k increases, which reflects the fast convergence of the spectral representation. The stochastic velocity field resulting from the uncertainty in viscosity is dominated by the first mode, which exhibits recirculation regions near the channel entrance that are symmetric with respect to the centerline. Below, we contrast the present solution with results obtained using a nonlinear viscosity law.

6.1.3.2 Example 2

In this example, we focus once more on a straight channel with uniform inflow but consider that the viscosity is temperature dependent. The temperature field is assumed to be spatially uniform, but uncertain with Gaussian statistics:

$$T = T_0 + T_1 \xi_1, \tag{6.31}$$

where T_0 is the mean (reference) temperature and T_1 the standard deviation. We assume a polynomial representation of the viscosity in the neighborhood of T_0 and

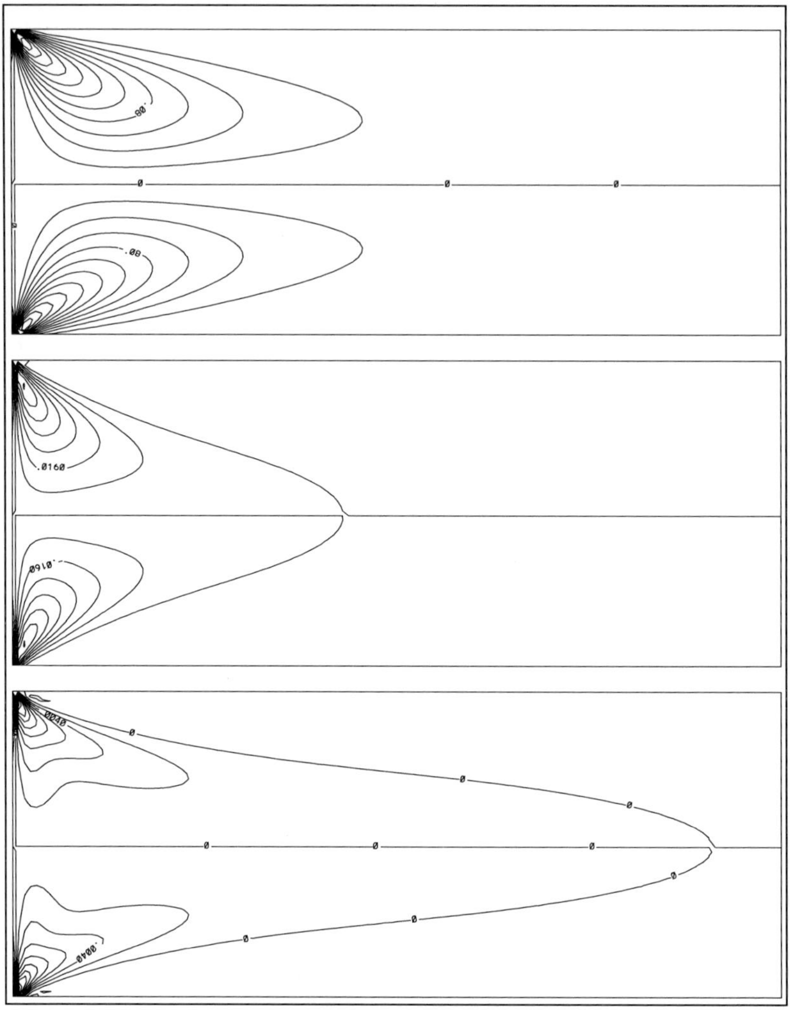

Fig. 6.5 Contour plots of the cross-stream velocity components u_0 (*top*), u_1 (*middle*) and u_2 (*bottom*). Only the first half of the domain ($0 \le y/B \le 3$) is shown. Same parameters as in Fig. 6.3. Adapted from [123]

restrict our attention to the second-order case:

$$\frac{\nu(T)}{\nu_0} = 1 + a_1(T - T_0) + a_2(T - T_0)^2, \tag{6.32}$$

where a_1 and a_2 are given constants and $\nu_0 \equiv \nu(T_0)$.

Substituting (6.31) into (6.32) we obtain the following stochastic representation for ν:

$$\frac{\nu(T_0 + \xi_1 T_1)}{\nu_0} = 1 + \beta_1 \xi_1 + \beta_2 \xi_1^2, \tag{6.33}$$

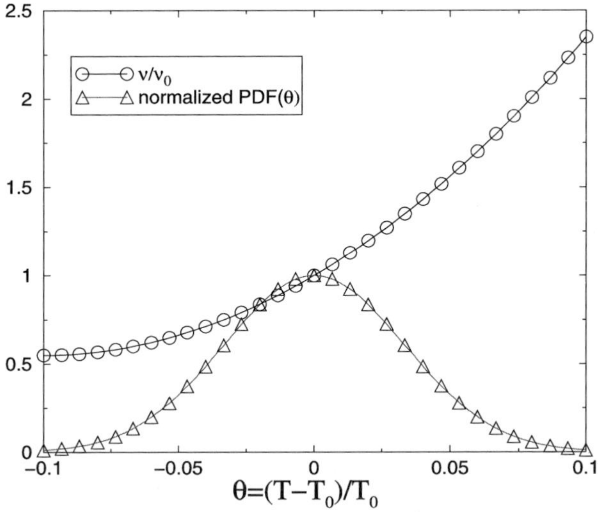

Fig. 6.6 Dependence of viscosity on the temperature given in (6.32) and the scaled probability density function of T. Adapted from [123]

where $\beta_1 \equiv a_1 T_1$ and $\beta_2 \equiv a_2 T_1^2$. It follows that the viscous force in the momentum equation (6.9) can be written as:

$$\nabla \cdot \overline{\sigma}_k = \nu_0(1+\beta_2)\nabla^2 \mathbf{u}_k + \nu_0 \beta_1 \sum_{i=0}^{\mathrm{P}} \nabla^2 \mathbf{u}_i C_{1ik} + \nu_0 \beta_2 \sum_{i=0}^{\mathrm{P}} \nabla^2 \mathbf{u}_i C_{2ik}. \qquad (6.34)$$

Note that if $a_2 = 0$, then the viscosity evolves linearly with T and the first problem is recovered with the choice $\nu_1 = a_1 T_1$. On the other hand, where $a_2 \neq 0$ it is clear that ν is no longer Gaussian. In the computations, we analyze the effect of non-Gaussian statistics by contrasting linear and nonlinear viscosity laws.

In the computations below, the coefficients in the nonlinear viscosity law are $a_1 T_0 = 9$ and $a_2 T_0^2 = 45$. Meanwhile, the standard deviation in the temperature is fixed as $T_1/T_0 = 1/30$. The nonlinear viscosity law is plotted in Fig. 6.6 together with the scale probability density function of the temperature. Note for this choice of parameters, linearization of the viscosity law (i.e. setting $a_2 = 0$) would result in the same problem last considered. Thus the effect of the nonlinearity in the viscosity law can be examined by comparing the results with those given above.

A simulation is performed for a channel at $\mathrm{Re} = V_{ref} B/\mathbf{u}_0 = 81.24$, where $\nu_0 \equiv \nu(T_0)$. The same computational grid as in the linear problem is used, together with a time step $\Delta t V_{ref}/B = 10^{-3}$, and a polynomial chaos expansion with $\mathrm{P} = 3$. Results are given in Figs. 6.7–6.9, which respectively show the distribution of streamfunction, streamwise velocity and cross-stream velocity at steady state. As done earlier, the mean distributions are plotted together with the first two modes. The computed results show that the mean flow behavior in the present case is quite

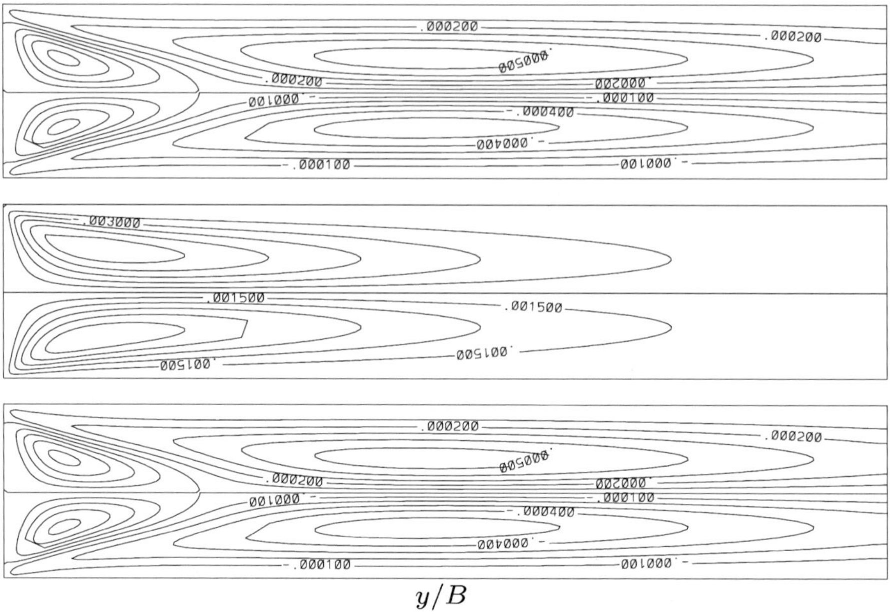

Fig. 6.7 Contour plots of the streamfunction distribution corresponding to $\mathbf{u}_0$ (*top*), $\mathbf{u}_1$ (*middle*) and $\mathbf{u}_2$ (*bottom*). Results are obtained for a channel with $H/B = 6$, Re $= 81.24$, the nonlinear viscosity law and stochastic temperature shown in Fig. 6.6. The simulation is performed using a grid with 64×256 cells, $\Delta t V_{ref}/B = 10^{-3}$ and a polynomial chaos expansion with P $= 3$. Adapted from [123]

similar to that depicted in Figs. 6.3–6.5. In particular, the development of the laminar boundary layer is clearly reflected in the mean streamfunction contours, which are deflected away from solid boundaries as one moves downstream (Fig. 6.7). As y increases, the mean cross-stream velocity component decays rapidly, and a parabolic streamwise velocity profile eventually prevails.

On the other hand, the "stochastic modes" $\mathbf{u}_1$ and $\mathbf{u}_2$ exhibit noticeable differences from the corresponding fields obtained with a linear viscosity law. The distributions reveal a more complex structure in the nonlinear case, especially for the second mode where one can notice the presence of multiples lobes that are symmetrically distributed on both sides of the channel centerline. Thus, the nonlinear term in the viscosity law can have a dramatic impact on the variance fields.

In order to further examine the effects of the viscosity law on the predictions, we linearize the governing equations and thus consider the unsteady Stokes problem. In this formulation, the nonlinear inertial terms are omitted and the evolution of the flow field follows a gradual decay towards the steady at a rate that is governed by the viscous timescale. This simple flow evolution enables us to perform straightforward comparison of different solutions during the transient. In addition, by contrasting the results of the Stokes and Navier-Stokes computations, one can gain additional insight into the role of inertial effects on the structure of the variance fields. Unsteady

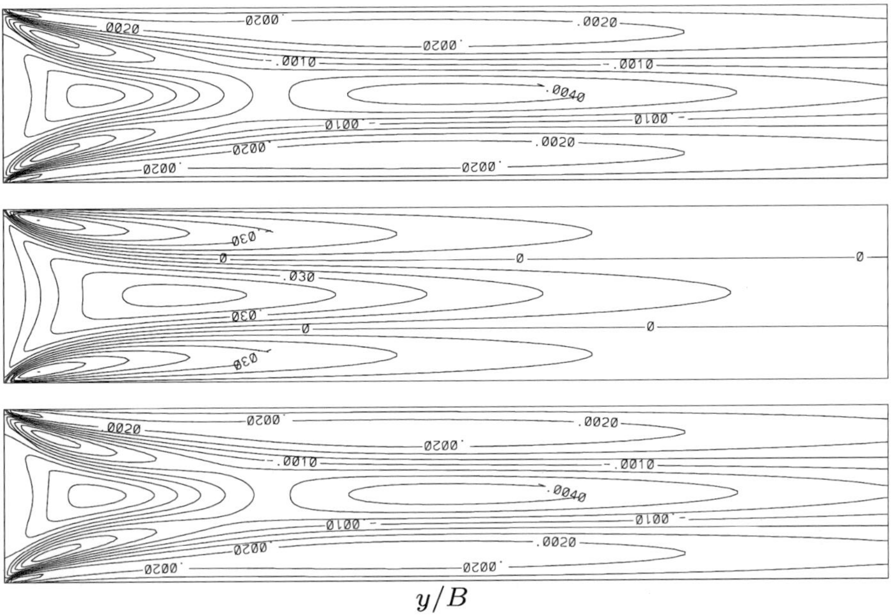

Fig. 6.8 Contour plots of the streamwise velocity components v_0 (*top*), v_1 (*middle*) and v_2 (*bottom*). Same parameters as in Fig. 6.7. Adapted from [123]

Stokes solutions are performed for a viscosity law with $a_2 T_0^2 = 45$ (the nonlinear case) and the predictions are contrasted with results obtained with $a_2 = 0$ (the linear case). For both cases, instantaneous distributions of the standard deviation in the u and v velocity components and in the streamfunction are shown in Fig. 6.10. The simulations are initialized with the fluid at rest and the fields are generated at a fixed time instant $t_a = V_{ref} t / B = 1$, before the decay of the flow transient. The variance σ_f^2 of a generic field variable $f(\mathbf{x}, \xi_1)$ is obtained from the corresponding polynomial chaos expansion using:

$$\sigma_f^2(\mathbf{x}) \equiv \left\langle (f(\mathbf{x}) - f_0(\mathbf{x}))^2 \right\rangle = \left\langle \left(\sum_{i=1}^{P} f_i(\mathbf{x}) \Psi_i \right)^2 \right\rangle = \sum_{i=1}^{P} f_i^2(\mathbf{x}) \left\langle \Psi_i^2 \right\rangle. \tag{6.35}$$

The results in Fig. 6.10 show that significant magnitude differences exist between the predictions obtained using linear and nonlinear viscosity laws. In particular, the difference between the two standard deviation fields exhibits peak values that are comparable to those of the corresponding fields. However, unlike our experiences above using the Navier-Stokes solver, the distributions for the two viscosity laws in the Stokes case have a very similar structure. This indicates that nonlinear advective effects can have a substantial impact on the structure of the variance field.

We also use the Stokes problem to investigate the sensitivity of the computed solution with respect to refinement of the computational grid (h refinement) and the

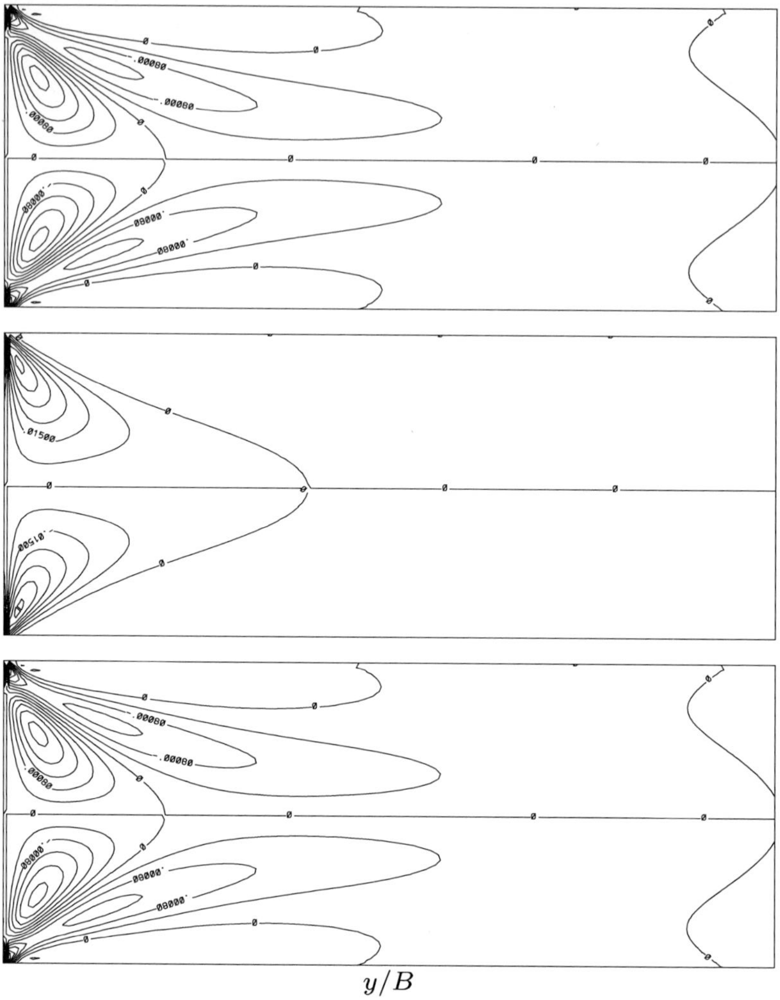

Fig. 6.9 Contour plots of the cross-stream velocity components u_0 (*top*), u_1 (*middle*) and u_2 (*bottom*). Only the first half of the domain ($y/B \leq 3$) is shown. Same parameters as in Fig. 6.7. Adapted from [123]

order of the polynomial chaos expansion (P refinement). Results of this study are given in Fig. 6.11, which depicts distributions of standard deviation in the cross-stream velocity u. The P-refinement tests are based on results obtained with P = 3, 5, and 7, using a 64×256 grid and a time step $\Delta t V_{ref}/B = 10^{-3}$. The h-refinement tests are based on three grids having 32×128, 64×256 and 96×384 cells in the x and y directions, respectively; in these tests, P = 3 and the time steps are $\Delta t V_{ref}/B = 10^{-3}$ for the two coarsest grids and 5×10^{-4} for the finest. The results show that, for the Stokes problem, the standard deviation distribution and peak values are essentially unaffected by the value of P, which demonstrates the fast con-

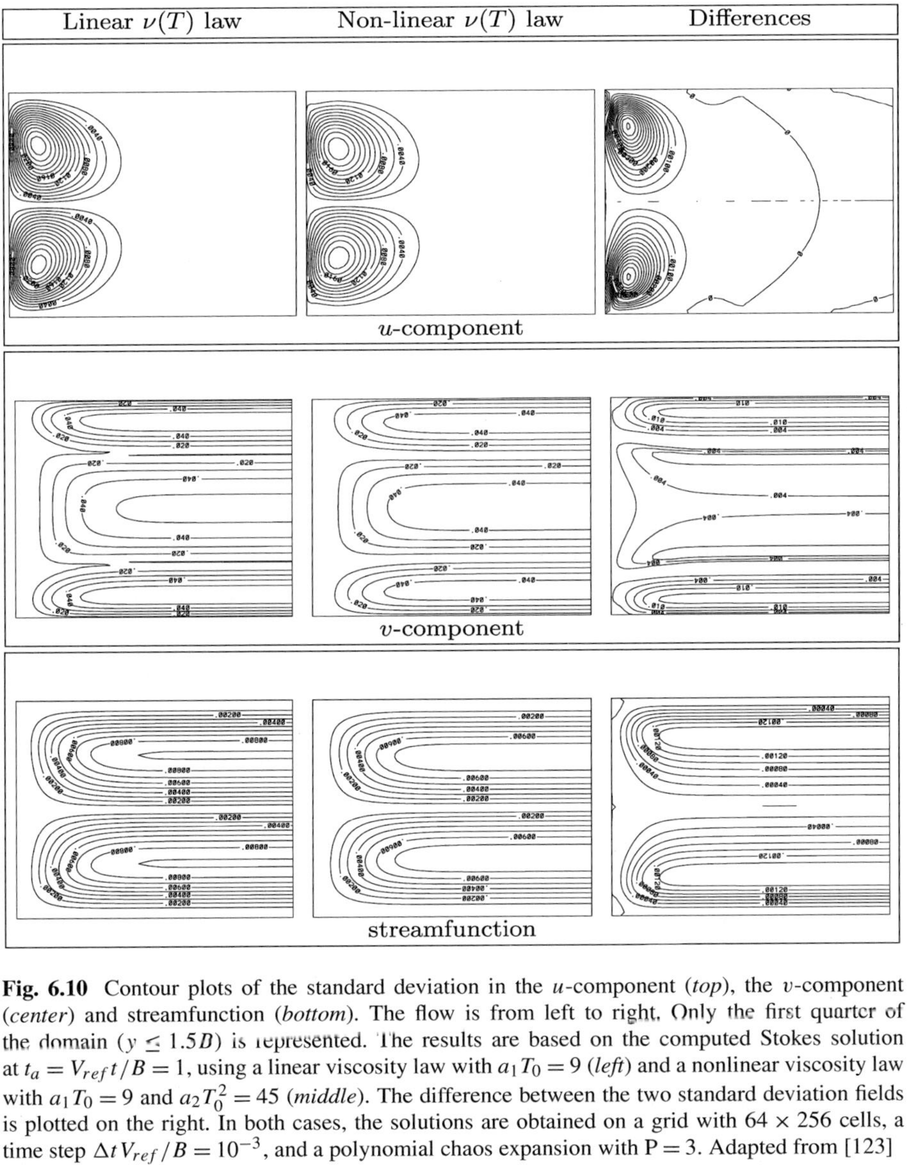

Fig. 6.10 Contour plots of the standard deviation in the u-component (*top*), the v-component (*center*) and streamfunction (*bottom*). The flow is from left to right. Only the first quarter of the domain ($y \leq 1.5B$) is represented. The results are based on the computed Stokes solution at $t_a = V_{ref}t/B = 1$, using a linear viscosity law with $a_1 T_0 = 9$ (*left*) and a nonlinear viscosity law with $a_1 T_0 = 9$ and $a_2 T_0^2 = 45$ (*middle*). The difference between the two standard deviation fields is plotted on the right. In both cases, the solutions are obtained on a grid with 64×256 cells, a time step $\Delta t V_{ref}/B = 10^{-3}$, and a polynomial chaos expansion with $\mathrm{P} = 3$. Adapted from [123]

vergence of the spectral expansion. Figure 6.11 also shows that the predictions at two finest grid levels are nearly identical, while a slightly higher peak in the standard deviation can be observed for the coarsest grid level. Further examination of the results (not shown) reveals that differences between the coarse level predictions and the more refined computations are restricted to a small region near the channel inlet, and that at larger downstream distances the coarse grid provides accurate prediction of the solution.

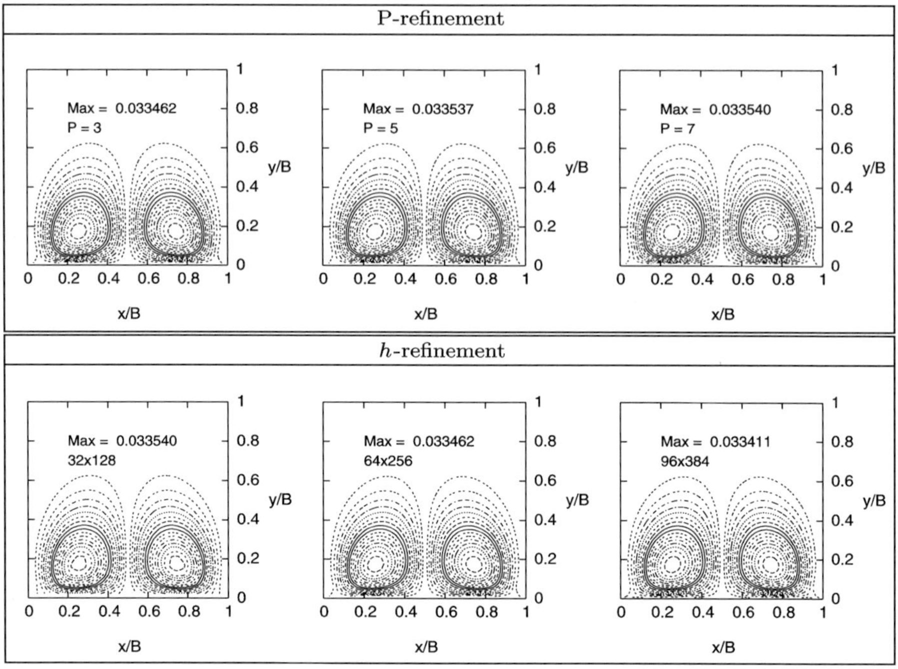

Fig. 6.11 Effect of the order P and cell size h on the standard deviation of the u-velocity. Adapted from [123]

6.1.3.3 Example 3

In the third problem, the temperature is no longer assumed spatially uniform and its evolution is governed by the energy equation (6.3). The kinematic viscosity is assumed to vary linearly with temperature, according to:

$$\frac{\nu(\mathbf{x})}{\nu_0} = 1 + K'\left(T(\mathbf{x}) - T_{ref}\right), \tag{6.36}$$

where T_{ref} is a reference temperature, $\nu_0 \equiv \nu(T_{ref})$, and K' is a constant. Note that the temperature dependence of the viscosity provides a non-trivial coupling between the energy and Navier-Stokes equations. The uncertainty in the process is considered to arise due to a stochastic temperature profile, T_{in}, at the inlet of the microchannel.

For the stochastic data above, the diffusion flux terms in the energy and momentum equations are expressed as:

$$-\boldsymbol{\nabla}\cdot\mathbf{q}_k = \kappa\nabla^2 T_k, \tag{6.37}$$

$$\boldsymbol{\nabla}\cdot\overline{\boldsymbol{\sigma}}_k = \nu_0(1 - K'T_{ref})\nabla^2\mathbf{u}_k + \nu_0 K'\sum_{i=0}^{\mathrm{P}}\sum_{j=0}^{\mathrm{P}}\boldsymbol{\nabla}\cdot\left(T_j\overline{\mathbf{S}}(\mathbf{u}_i)\right)C_{ijk}. \tag{6.38}$$

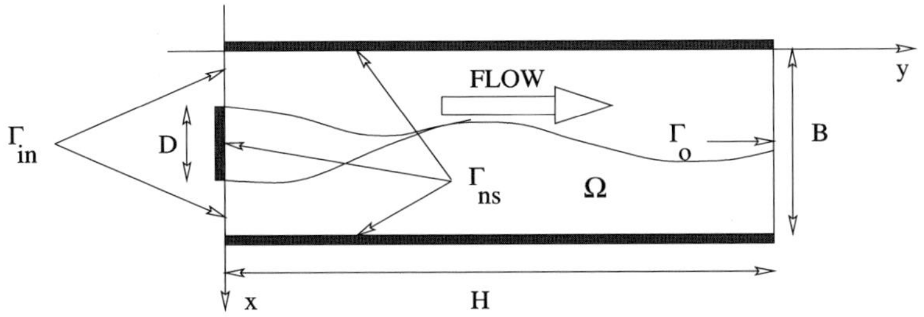

Fig. 6.12 Schematic illustration of the double-inlet microchannel. Adapted from [123]

These representations are incorporated into the coupled system (6.9)–(6.11).

Implementation of the SPM is now illustrated based on simulations of the flow and temperature fields in the double-inlet microchannel schematically shown in Fig. 6.12. The channel inlet consists of two streams having identical parabolic velocity profiles with peak velocity V_{ref}. The two inlet streams are separated by a plate of thickness D. Thus, the problem can be treated as the wake of a slender bluff body of width D that is located at the center of the channel. The flow is characterized by the Reynolds number $\text{Re} \equiv V_{ref} B/\nu_0$, the blockage ratio D/B, and the Prandtl number λ/ν_0. As indicated in Sect. 6.1.3.3, $\nu_0 \equiv \nu(T_{ref})$ is the reference viscosity. Note that the blockage ratio and Re can be combined to define a Reynolds number based on the plate thickness, $\text{Re}_D \equiv V_{ref} B/\nu_0 = \text{Re} D/B$. If Re_D is large enough the wake of the plate is unstable and periodic vortex shedding is observed, at least for small downstream distances. This situation is considered in the example below. Specifically, we consider a doublet inlet microchannel with blockage ratio $D/B = 0.2$ and Reynolds number $\text{Re} = 826$. The Reynolds number based on the plate thickness is $\text{Re}_D = 165.2$.

As mentioned earlier, the uncertainty in this problem is taken to arise due to a stochastic temperature boundary condition. Specifically, the temperature of the first inlet is taken to be deterministic and equal to T_{ref}. Meanwhile, the temperature of the second inlet is treated as a random Gaussian variable, with a mean value of T_{ref} and standard deviation of $0.1T_{ref}$. The fluid viscosity is assumed to depend on the temperature according to (6.36). This provides a strong coupling between the momentum and energy equations, which is examined in the computations by varying the coupling parameter K'. Specifically, results are obtained using $K = 0.1$, 0.2 and 0.4, where $K \equiv K' T_{ref}$. In all cases, the Prandtl number $\lambda/\nu_0 = 6$. The computations are performed in a domain with $H/B = 5$, using a 100×352 grid, a time step $\Delta t V_{ref}/B = 2 \times 10^{-3}$ and a polynomial chaos expansion with $\text{P} = 3$.

Figures 6.13–6.14 depict instantaneous contours of streamwise and cross-stream velocity, respectively, at a dimensionless time $t V_{ref}/B = 100$. Plotted in each figure are distributions of the mean instantaneous prediction together with those of modes 1 and 2; results obtained using $K = 0.4$ are used. The distributions of the mean field exhibit the presence of well defined patches that are arranged in a wavy pattern, which reflects the development of an unstable wake. The results also reveal that the

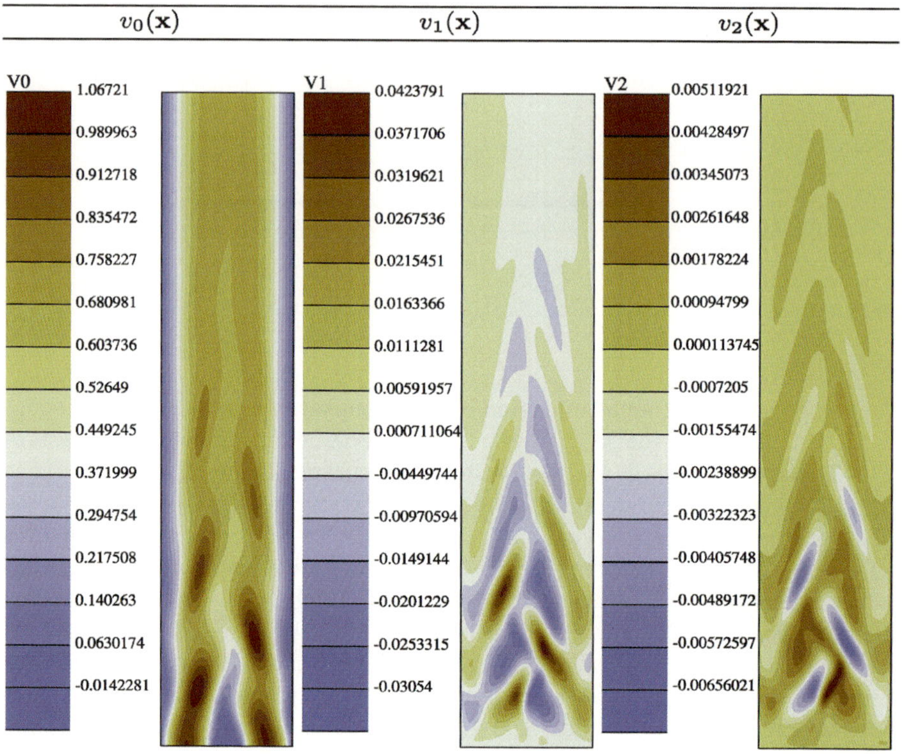

Fig. 6.13 Instantaneous distribution of v_0, v_1 and v_2 at $tV_{ref}/B = 100$. Velocity is normalized using V_{ref}, and results obtained using $K = 0.4$ are used. The flow is from bottom to top, along the $+y$ direction. Adapted from [123]

strengths of the vortices shed into the wake gradually decrease with downstream distance. This effect can be clearly observed in Fig. 6.14 which shows that the magnitude of the cross-stream velocity component decreases with increasing distance from the channel entrance. Thus, the strengths of the vortices decay with y and, for the selected value of the Reynolds number, one would in fact expect a steady parabolic profile at large downstream distances.

Near the channel entrance, the distributions of $\mathbf{u}_1$ and $\mathbf{u}_2$ (Figs. 6.13 and 6.14) also reveal the presence of well-defined structures that are spatially well correlated with those of the mean field. The velocity magnitudes of the first mode are roughly an order of magnitude higher than those of the second mode. With increasing downstream distance, the magnitudes of $\mathbf{u}_1$ and $\mathbf{u}_2$ gradually decrease. This trend is also expected since at large downstream distances one would recover a parabolic velocity profile whose strength is solely determined by the volume flux in the channel. In this problem, the volume flux is deterministic which indicates that all velocity modes with $k \geq 1$ vanish as y increases.

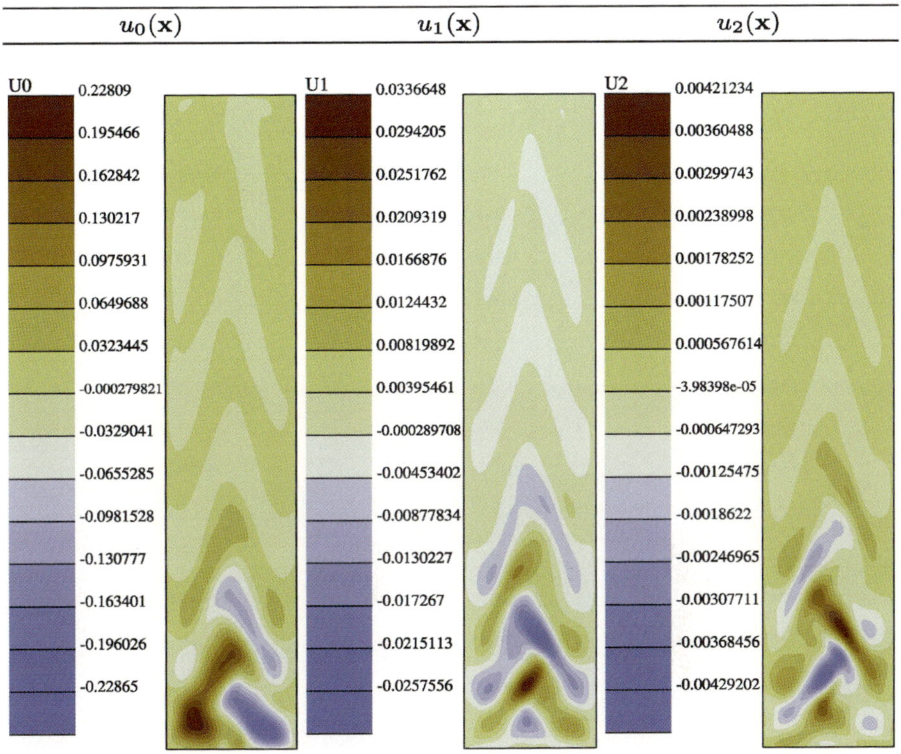

Fig. 6.14 Instantaneous distribution of u_0, u_1 and u_2 at $tV_{ref}/B = 100$. Velocity is normalized using V_{ref}, and results obtained using $K = 0.4$ are used. Adapted from [123]

Figure 6.15 shows instantaneous temperature contours for $K = 0.4$, generated at the same time as in Figs. 6.13 and 6.14. The distributions of T_0 and T_2 show the presence of well-defined patches of alternating sign which are consequently arranged at the center of the domain. The strength of the temperature fluctuations within these patches first increases with y, reaches a maximum value around $y/B \sim 2$ and then decreases as we move further downstream. Near the channel entrance, the distribution of T_1 reflects the inlet temperature conditions, which are deterministic for the first inlet and stochastic in the second; thus, T_1 vanishes near the first inlet, peaks near the second, with a gradual transition region at the face of the solid plate. As one moves downstream, the width of this transition region increases leading to the formation of an asymmetric wavy pattern around the wake centerline, with small positive values near the left wall and high values near the right wall. Note that the peak value decreases as one moves downstream while the minimum increases, which illustrates how the uncertainty in the boundary condition diffuses as it advected by the flow.

The close correspondence between the temperature fluctuations in the distributions of T_0 and T_2 in Fig. 6.15 is remarkable and it is instructive to use the uncertainty representation scheme to interpret the results. The polynomial chaos expan-

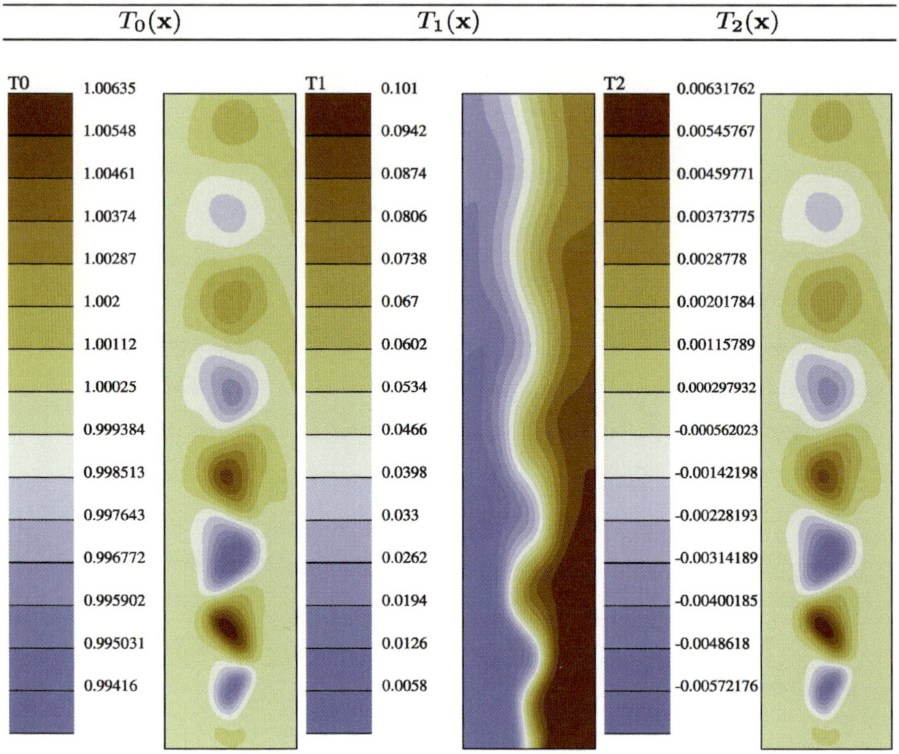

Fig. 6.15 Instantaneous distribution of T_0, T_1 and T_2 at $tV_{ref}/B = 100$. Temperature is normalized using T_{ref}, and results obtained using $K = 0.4$ are used. Adapted from [123]

sion of the temperature field can be written as:

$$\begin{aligned} T(\mathbf{x}, \xi_1) &= T_0(\mathbf{x})\Psi_0(\xi_1) + T_1(\mathbf{x})\Psi_1(\xi_1) + T_2(\mathbf{x})\Psi_2(\xi_1) + \cdots \\ &= T_0(\mathbf{x}) + T_1(\mathbf{x})\xi_1 + T_2(\mathbf{x})(\xi_1^2 - 1) + \cdots . \end{aligned} \tag{6.39}$$

For $\xi_1 = 0$ the two inlet streams have identical temperature, T_{ref}. This implies that in this case the temperature field is uniform and everywhere equal to T_{ref}. Using $\xi_1 = 0$ and truncation at P = 2, (6.39) gives:

$$T(\mathbf{x}, \xi_1 = 0) \simeq T_0(\mathbf{x}) - T_2(\mathbf{x}), \tag{6.40}$$

i.e. the temperature prediction corresponding to $\xi_1 = 0$ is the difference between the mean and second mode. Since, as indicated above, $T(\mathbf{x}, \xi_1 = 0) = T_{ref}$ the fluctuations in T_0 and T_2 should cancel out, so long as the spectral truncation used is valid. This constraint is in fact reflected in the distributions shown in Fig. 6.15, which also indicates that the (truncated) higher modes have little impact on the present predictions. Also note that

$$\langle T(\mathbf{x}) \rangle = T_0(\mathbf{x}) \neq T(\mathbf{x}, \xi_1 = 0), \tag{6.41}$$

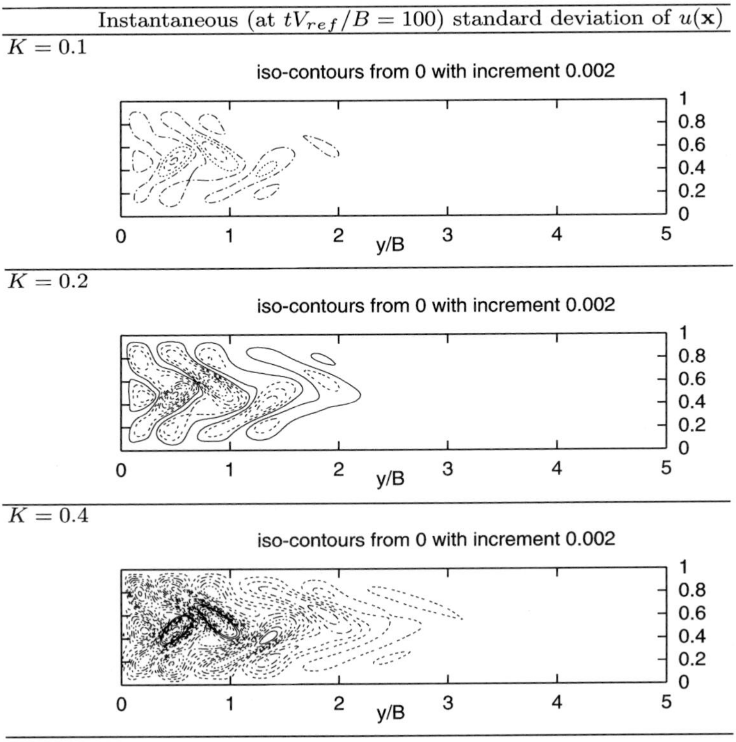

Fig. 6.16 Instantaneous distribution of the standard deviation of the normalized v-velocity field at time $tV_{ref}/B = 100$; *top*: $K = 0.1$, *middle*: $K = 0.2$, *bottom*: $K = 0.4$. Adapted from [123]

which indicates that the expected temperature field does not coincide with the "deterministic" prediction for $\xi_1 = 0$.

Instantaneous distributions of the standard deviation of u, v and T are shown in Figs. 6.16–6.18, respectively. Plotted are results obtained at $tV_{ref}/B = 100$ using $K = 0.1$, 0.2 and 0.4. The results indicate that the normalized standard deviation for the streamwise (v) and cross-stream (u) velocity components increase with increasing with K. As expected, the largest standard deviation values occur in the near wake, where strong vortical structures are present. On the other hand, the contours of the temperature standard deviation exhibit a wavy, asymmetric spreading band near the center of the domain. Unlike the standard deviation of the velocity field, the standard deviation of the temperature is essentially insensitive to the coupling parameter K. Thus, for the present conditions, the propagation of the uncertainty in the temperature field appears to be dominated by the deterministic thermal diffusion coefficient and advection with the (stochastic) mean velocity field.

Profiles of time averaged values of the streamwise velocity, cross-stream velocity and temperature are given in Fig. 6.19. The figure depicts profiles of the first three modes, generated at the streamwise plane $y/B = 1.25$ using simulations with $K = 0.1$, 0.2 and 0.4. The time-averaged profiles reveal similar trends to those ob-

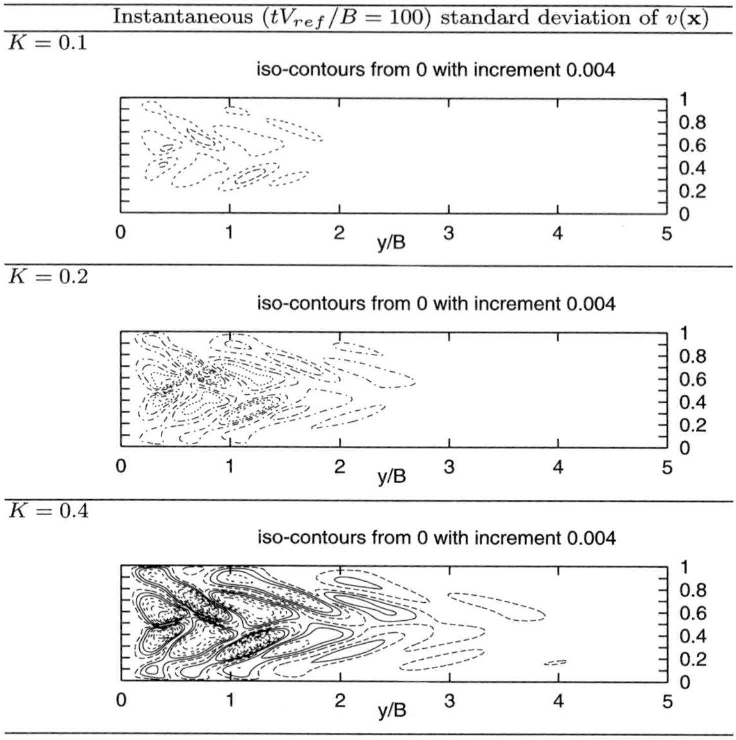

Fig. 6.17 Instantaneous distribution of the standard deviation of the normalized v-velocity field at time $tV_{ref}/B = 100$; *top*: $K = 0.1$, *middle*: $K = 0.2$, *bottom*: $K = 0.4$. Adapted from [123]

served in the instantaneous distributions. The mean velocity profiles clearly reflect the development of the unstable wake. Meanwhile, the uncertainty in the velocity field is dominated by the contribution of the first mode, whose peak values are significantly larger than those of the second mode. The results also indicate that as K increases the uncertainty in the velocity field also increases. This behavior is in sharp contrast with that observed for the temperature profile. The mean temperature prediction exhibits a pronounced dependence on K while the first mode appears to be insensitive to K. As discussed earlier, the fluctuations in the profiles of T_0 and T_2 are quite similar, but are dominated by T_1 which is forced at the inlet boundary. The above trends are also reflected in Fig. 6.20, which depicts profiles of the normalized standard deviation of the mean velocity components and of temperature. Combined with the results in Fig. 6.19 it is evident that the contribution of the first mode to the standard deviation is dominant. One can also observe the insensitivity of the temperature standard deviation to the selected value of K, and the strong dependence of the velocity uncertainty on the coupling parameter.

We conclude the discussion with a brief remark on the possible use of the stochastic predictions. For instance, in the case of the streamwise profile, the standard deviation is vanishingly small at cross-stream locations ($x/B \sim 0.3$ and $x/B \sim 0.7$)

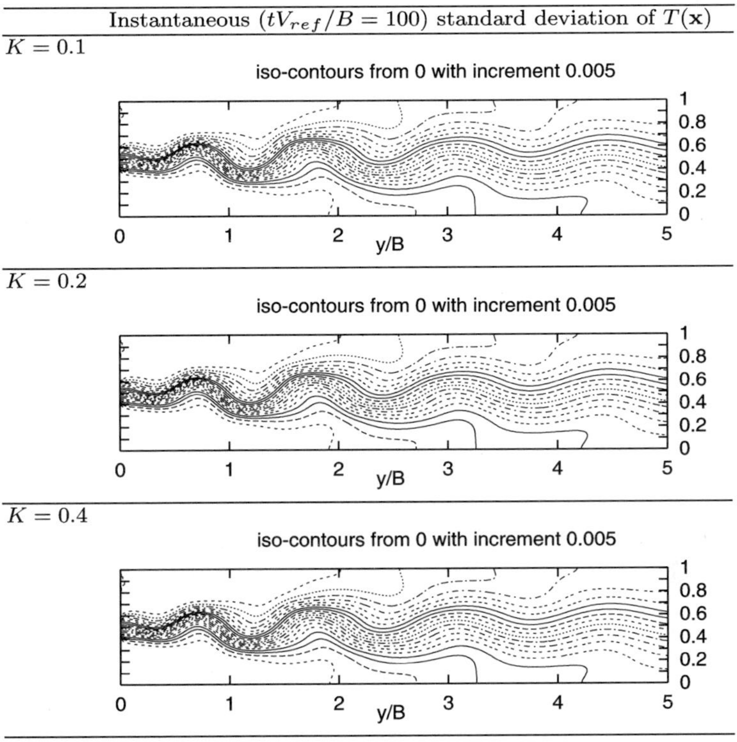

Fig. 6.18 Instantaneous distribution of the standard deviation of the normalized temperature field at time $tV_{ref}/B = 100$; *top*: $K = 0.1$, *middle*: $K = 0.2$, *bottom*: $K = 0.4$. Adapted from [123]

where the mean signal approaches its peak value (compare Figs. 6.20 and 6.19). The ratio of the standard deviation to the mean value is clearly minimized at the corresponding locations. Consequently, these positions provide ideal sites for probing the streamwise velocity, in a fashion that minimizes the effect of the uncertainty in stochastic inlet temperature. This illustrates how the stochastic simulation results may be applied to experiment design.

6.2 Boussinesq Extension

This section aims at generalizing the formulation of the incompressible SPM to account for weak compressibility, and to demonstrate application of the resulting scheme to situations involving multi-dimensional random data.

We outline these extensions by considering the idealized case of natural convection in a square cavity under stochastic boundary conditions. As outlined in Sect. 6.2.1, we restrict the study to natural convection in the limit of small temperature and density gradients. This topic has received considerable attention and various approaches have been proposed, including models based on the well-known

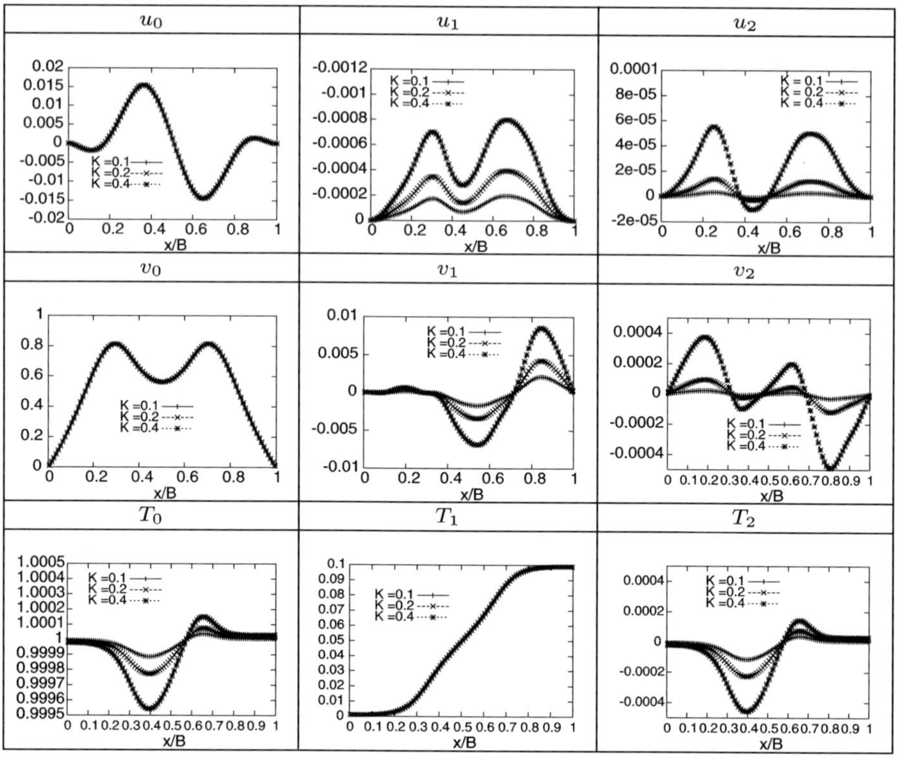

Fig. 6.19 Time-averaged profiles of U (*top*), v (*center*) and T (*bottom*), at the plane $y/B = 1.25$. The modes correspond to $k = 0$ (*left*), $k = 1$ (*center*) and $k = 2$ (*right*). The *curves* depict results obtained for $K = 0.1$, 0.2 and 0.4. Adapted from [123]

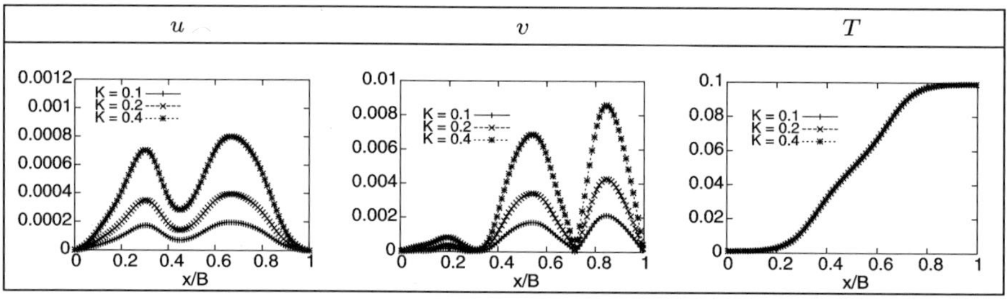

Fig. 6.20 Time-averaged standard deviation profiles at the plane $y = 1.25B$ for the normalized u-velocity (*left*), the normalized v-velocity (*center*) and the normalized temperature (*right*). The *curves* depict results obtained for $K = 0.1$, 0.2 and 0.4. Adapted from [123]

Boussinesq approximation (e.g. [48, 130, 131]) as well as low-Mach-number models (e.g. [27, 132]). In addition, simulations of internal natural convection have been used as benchmark tests for different flow regimes [33, 34, 37, 48, 93, 95, 105, 113, 176, 179, 201, 223, 240, 243].

A stochastic formulation of a driven cavity problem is then introduced in Sect. 6.2.2 which consists in treating the cold wall as having a uniform temperature and imposing a stochastic temperature distribution on the hot vertical boundary. The latter is treated as a Gaussian process which is characterized by its variance and correlation length. The Karhunen-Loève expansion [137] is applied to construct a spectral representation of this process.

Based on this generalized representation, an extended formulation of the incompressible SPM is presented in Sect. 6.2.3. The structure of the resulting stochastic solution scheme shares many similarities with that of its incompressible counterpart, similar to what one observes for deterministic projection-based solvers.

A brief validation study of the deterministic prediction is first performed in Sect. 6.2.4.1, and is used to select an appropriate grid resolution level. The convergence properties of the spectral stochastic scheme are then analyzed in Sect. 6.2.4.2, and the properties of the computed velocity and temperature modes are examined in Sect. 6.2.5. In order to verify the spectral computations, a non-intrusive spectral projection (NISP) approach is applied in Sect. 6.2.6. Two variants are considered which are based on high-order Gauss-Hermite (GH) quadrature [1, 114] or on Latin Hypercube Sampling (LHS) strategy [153]. The predictions of both sampling schemes are contrasted with the spectral computations and used to further examine its properties. In Sect. 6.2.7, a quantitative analysis of the effects of the imposed stochastic temperature profile is finally provided.

6.2.1 Deterministic Problem

We consider a square 2D cavity, of side $\tilde{L}$, filled with a Newtonian fluid of density $\tilde{\rho}$, molecular viscosity $\tilde{\mu}$ and thermal conductivity $\tilde{\kappa}$ (tildes denote here dimensional quantities). The coordinate system is chosen so that y is the vertical direction, pointing upwards, and the x axis is horizontal. The two horizontal walls are assumed adiabatic. The left vertical wall is maintained at uniform temperature $\tilde{T}_h$, and the right vertical wall is maintained at $\tilde{T}_c$. We assume that $\tilde{T}_h > \tilde{T}_c$, so that the left vertical wall (located at $x = 0$) is referred as hot wall, while the right vertical wall is the cold wall. We define the reference temperature $\tilde{T}_{\text{ref}} \equiv (\tilde{T}_h + \tilde{T}_c)/2$ and the reference temperature difference $\Delta\tilde{T}_{\text{ref}} \equiv \tilde{T}_h - \tilde{T}_c$. Variables are normalized with respect to the appropriate combination of reference length, $\tilde{L}$, velocity, $\tilde{V}$, time, $\tau \equiv \tilde{L}/\tilde{V}$, and pressure $\tilde{P} = \tilde{\rho}\tilde{V}^2$.

In the Boussinesq limit, $2(\tilde{T}_h - \tilde{T}_c)/(\tilde{T}_h + \tilde{T}_c) \ll 1$, the normalized governing equations are expressed as [130]:

$$\frac{\partial \mathbf{u}}{\partial t} + \mathbf{u} \cdot \nabla \mathbf{u} = -\nabla p + \frac{\mathrm{Pr}}{\sqrt{\mathrm{Ra}}} \nabla^2 \mathbf{u} + \mathrm{Pr} T \mathbf{y}, \tag{6.42}$$

$$\nabla \cdot \mathbf{u} = 0, \tag{6.43}$$

$$\frac{\partial T}{\partial t} + \nabla \cdot (\mathbf{u} T) = \frac{1}{\sqrt{\mathrm{Ra}}} \nabla^2 T \tag{6.44}$$

where $\mathbf{u}$ is the velocity, t is time, p is pressure, and $T \equiv (\tilde{T} - \tilde{T}_{\text{ref}})/\Delta\tilde{T}_{\text{ref}}$ is the normalized temperature. The normalization leads to the usual definitions of Prandtl and Rayleigh numbers, respectively $\text{Pr} = \tilde{\mu}c_p/\tilde{\kappa}$ and $\text{Ra} = \tilde{\rho}\tilde{g}\beta\Delta\tilde{T}_{\text{ref}}\tilde{L}^3/(\tilde{\mu}\tilde{\kappa})$ where β is the coefficient of thermal expansion and $\tilde{g}$ is gravitational acceleration. Fluid properties are assumed constant, and are evaluated at the reference temperature. In all cases, the deterministic system is integrated from an initial state of rest using $\text{Pr} = 0.71$ and $\text{Ra} = 10^6$. For this choice of physical parameters, a steady laminar recirculating flow regime occurs [130].

6.2.2 Stochastic Formulation

We consider the effect of random fluctuations on the cold wall. The normalized mean wall temperature at $x = 1$ is expressed as:

$$T(x=1,y) = T_c + T'(y) = -\frac{1}{2} + T'(y). \tag{6.45}$$

Using brackets to denote expectations, we set $\langle T(x=1,y)\rangle = T_c$, i.e. T' has vanishing expectation and the mean temperature along the cold wall is independent of y and corresponds to T_c.

The random component is assumed to be given by a stationary Gaussian process which is characterized by its variance σ_T^2, and a correlation function, $\mathcal{K}$, given by:

$$\mathcal{K}(y_1,y_2) \equiv \mathcal{K}(|y_1-y_2|) \equiv \langle T'(y_1)T'(y_2)\rangle = \sigma_T^2 \exp\left[-|y_1-y_2|/L_c\right], \tag{6.46}$$

where L_c is the normalized correlation length. As discussed in Sect. 2.1, this kernel can be expressed in terms of its KL expansion [137]:

$$T'(y) = \sum_{i=1}^{\infty} \sqrt{\lambda_i} u_i(y)\xi_i, \tag{6.47}$$

where the ξ_i's are uncorrelated Gaussian variables having vanishing expectation and unit variance, the eigenfunctions u_i are given by (2.22),

$$\lambda_n = \sigma_\theta^2 \frac{2L_c}{1+(\omega_n L_c)^2}, \tag{6.48}$$

and ω_n are the positive (ordered) roots of the characteristic equation (2.24). Figure 2.1 depicts the first 10 eigenvalues and eigenfunctions for a process with $L_c = 1$.

In numerical implementations, the KL expansion (6.47) is truncated, and the temperature "fluctuation" is approximated as:

$$T' = \sum_{i=1}^{\text{N}} \xi_i \sqrt{\lambda_i} u_i(y), \tag{6.49}$$

where N is the number of modes retained in the computations. The error in this truncation has also been analyzed in detail in Sect. 2.1.

6.2.3 Stochastic Expansion and Solution Scheme

As in the previous section, the dependence of the solution on the uncertain model data is represented in terms of the PC system. We illustrate this representation for a generic field variable, $\Phi(\mathbf{x}, t, \boldsymbol{\xi})$, where $\boldsymbol{\xi} = \xi_1, \ldots, \xi_N$. Φ is decomposed according to:

$$\Phi(\mathbf{x}, t, \boldsymbol{\xi}) = \sum_{i=0}^{P} \Phi_i(\mathbf{x}, t) \Psi_i(\boldsymbol{\xi}), \tag{6.50}$$

where Φ_i are (yet to be determined) deterministic "coefficients", Ψ_i denote the Polynomial Chaos [25, 137, 241], while $P + 1$ is the total number of modes used in the spectral expansion. General expressions for Ψ_i, including higher-order terms, can be found in Sect. 2.2.

We rely on (6.50) to form representations of the stochastic velocity, pressure and temperature distributions. Governing equations for the unknown expansion coefficients are obtained using a Galerkin approach that takes advantage of the orthogonality of the Polynomial Chaoses. This results in the following coupled system:

$$\frac{\partial \mathbf{u}_k}{\partial t} + (\mathbf{u} \cdot \nabla \mathbf{u})_k = -\nabla p_k + \frac{\mathrm{Pr}}{\sqrt{\mathrm{Ra}}} \nabla^2 \mathbf{u}_k + \mathrm{Pr} T_k \mathbf{y}, \tag{6.51}$$

$$\nabla \cdot \mathbf{u}_k = 0, \tag{6.52}$$

$$\frac{\partial T_k}{\partial t} + \nabla \cdot (\mathbf{u}T)_k = \frac{1}{\sqrt{\mathrm{Ra}}} \nabla^2 T_k, \tag{6.53}$$

for $k = 0, \ldots, P$. Here, $\mathbf{u}_k(\mathbf{x}, t)$, $p_k(\mathbf{x}, t)$ and $T_k(\mathbf{x}, t)$ are the coefficients in the PC expansion of the normalized velocity, pressure and temperature fields, respectively. The quadratic velocity-velocity and velocity-temperature products are given by:

$$(\mathbf{u} \cdot \nabla \mathbf{u})_k = \sum_{i=0}^{P} \sum_{j=0}^{P} C_{ijk} \mathbf{u}_i \nabla \mathbf{u}_j \tag{6.54}$$

$$(\mathbf{u}T)_k = \sum_{i=0}^{P} \sum_{j=0}^{P} C_{ijk} \mathbf{u}_i T_j, \tag{6.55}$$

with C_{ijk} defined as in (6.12). Thus, the Galerkin procedure results in a coupled system for the velocity and temperature modes. Note, however, that the velocity divergence constraints are decoupled, which enables us to immediately adapt the incompressible SPM developed in the previous section. This approach is outlined below.

6.2.3.1 Boundary Conditions

Boundary conditions are also treated in a weak sense. Specifically, the PC decomposition is also introduced into the corresponding expressions, and orthogonal projections are used to derive boundary conditions for the velocity and temperature modes. For the setup outlined in Sect. 6.2.2, we obtain:

$$\mathbf{u}_k = 0, \quad k = 0, \dots, \mathrm{P}\ \forall \mathbf{x} \in \partial\Omega\,, \tag{6.56}$$

$$\frac{\partial T_k}{\partial y} = 0, \quad k = 0, \dots, \mathrm{P} \text{ for } y = 0, \text{ and } y = 1, \tag{6.57}$$

$$T_0(x = 0, y) = \frac{1}{2}, \qquad T_0(x = 1, y) = -\frac{1}{2}, \tag{6.58}$$

$$T_k(x = 0, y) = 0, \qquad T_k(x = 1, y) = \sqrt{\lambda_k} u_k(y) \quad \text{for } k = 1, \dots, \mathrm{N}, \tag{6.59}$$

$$T_k(x = 0, y) = T_k(x = 1, y) = 0 \quad \text{for } k > \mathrm{N}. \tag{6.60}$$

Here $\Omega = [0, 1] \times [0, 1]$ denotes the computational domain, and $\partial\Omega$ is its boundary.

6.2.3.2 Solution Method

As mentioned earlier, numerical integration of the governing equations of the stochastic modes follows an explicit fractional step procedure that is based on first advancing the velocity and temperature modes using:

$$\begin{aligned}\tilde{\mathbf{u}}_k = & \frac{4\mathbf{u}_k^n - \mathbf{u}_k^{n-1}}{3} + \frac{2\Delta t}{3}\Bigg[-2(\mathbf{u}\cdot\nabla\mathbf{u})_k^n + (\mathbf{u}\cdot\nabla\mathbf{u})_k^{n-1} \\ & + \frac{\mathrm{Pr}}{\sqrt{\mathrm{Ra}}}\nabla^2\mathbf{u}_k^n + \mathrm{Pr}T_k^n\mathbf{y}\Bigg],\end{aligned} \tag{6.61}$$

$$T_k^{n+1} = \frac{4T_k^n - T_k^{n-1}}{3} + \frac{2\Delta t}{3}\left[\nabla\cdot(-2(\mathbf{u}T)_k^n + (\mathbf{u}T)_k^{n-1}) + \frac{1}{\sqrt{\mathrm{Ra}}}\nabla^2 T_k^n\right], \tag{6.62}$$

where superscripts refer to the time level and Δt is the time step. Note that since we are primarily interested in the steady-state solution of the above system, we have combined explicit second-order time discretization of the convective terms and with first-order discretization of the buoyancy and viscous terms. Spatial derivatives are approximated using second-order centered differences. In the second fractional step, the "intermediate" velocity modes $\tilde{\mathbf{u}}_k$ are updated so as to satisfy the divergence constraints [29, 112]; we use:

$$\mathbf{u}_k^{n+1} = \tilde{\mathbf{u}}_k - \frac{2\Delta t}{3}\nabla p_k^{n+1}, \tag{6.63}$$

where p_k are solutions to the Poisson equations,

$$\nabla^2 p_k^{n+1} = \frac{3}{2\Delta t} \nabla \cdot \tilde{\mathbf{u}}_k, \tag{6.64}$$

with homogeneous Neumann conditions [29, 112]. Note that the above elliptic systems for the various modes are **decoupled**, a key feature in the efficiency of the SPM.

In the implementations below, we rely on a conservative second-order finite-difference discretization on a uniform Cartesian mesh with $(\mathrm{N_x}, \mathrm{N_y})$ cells in the x and y directions respectively. A direct, Fourier-based, fast Poisson solver is used to invert (6.64). Since these inversions account for the bulk of the CPU time, and since systems for individual modes are decoupled, the computational cost scales essentially as $\mathrm{O}(\mathrm{N_x} \times \mathrm{N_y} \times \mathrm{P})$. This estimate is in fact reflected in the tests below.

6.2.4 Validation

6.2.4.1 Deterministic Prediction

We start with a brief discussion of deterministic predictions, obtained by setting the order (No) of the PC expansion to zero. In this case, the stochastic boundary conditions reduce to those of the classical problem with uniform hot and cold wall temperatures, respectively $T_h = 1/2$ and $T_c = -1/2$. The resulting predictions are used to validate the computations and to select a suitable grid size. To this end, the results are compared with the spectral computations of Le Quéré [130]. For $\mathrm{Ra} = 10^6$, Le Quéré found a steady Nusselt number $\mathrm{Nu} = 8.8252$, with the Nusselt number defined by:

$$\mathrm{Nu} \equiv - \int_0^1 \frac{\partial T}{\partial x} \, \mathrm{d}y. \tag{6.65}$$

Following a systematic grid refinement study, we find that a computational grid with $\mathrm{N_x} = 140$ and $\mathrm{N_y} = 100$ is sufficient for accurate predictions. Starting from an initial state of rest, the computations are carried out until steady conditions are reached. Specifically, the computations are stopped when the maximum change in any field quantity falls below a tolerance $\epsilon = 10^{-10}$. For the current grid resolution, the steady Nusselt number is found to be $\mathrm{Nu} = 8.8810$, which is within 0.63% of the prediction of Le Quéré. The structure of the steady field, depicted in Fig. 6.21, reveals thermal boundary layers on the hot wall and cold wall and a clockwise circulation of the fluid; these predictions are also in good agreement with the results reported in [130].

6.2.4.2 Convergence Analysis

An analysis of the convergence of the spectral representation scheme is performed in this section. Following the discussion above, we are presently dealing with a two-

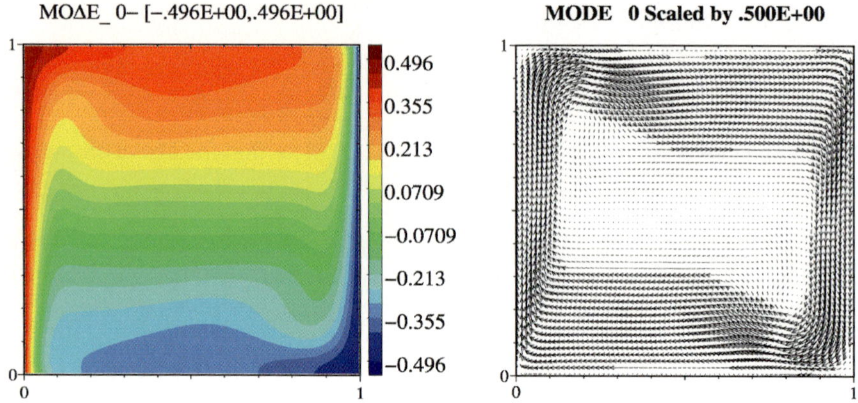

Fig. 6.21 Scaled temperature field (*left*) and velocity vectors (*right*) for the deterministic temperature boundary conditions ($T_h = -T_c = 1/2$) computed using zero-order spectral expansion (No = 0). Adapted from [128]

parameter discretization that involves the number, N, of Karhunen-Loève modes, as well as the order, No, of the PC expansion. As discussed in [90], the total number, P, of orthogonal polynomials increases monotonically with N and No.

Convergence with N**:** In Sect. 2.1, we observed that the KL expansion converged rapidly. Consequently, one would expect that truncation of this expansion would have little effect on the predictions. We now examine this expected trend by computing the mean Nusselt number,

$$\overline{\mathrm{Nu}} = -\int_0^1 \frac{\partial T_0}{\partial x}\,\mathrm{d}y \tag{6.66}$$

and its standard deviation,

$$\sigma(\mathrm{Nu}) = \left(\sum_{i=1}^{\mathrm{P}} \left\{\left[-\int_0^1 \frac{\partial T_i}{\partial x}\,\mathrm{d}y\right]^2 \langle \Psi_i \Psi_i \rangle\right\}\right)^{1/2} \tag{6.67}$$

for N ranging from 2 to 10. For brevity, we restrict our attention to a first-order PC expansion, and results are obtained with fixed $L_c = 1$, and $\sigma_T = 0.25$.

The average of the local heat flux variance along the wall is given by:

$$\overline{\sigma^2}(\partial T/\partial x) = \int_0^1 \sum_{i=1}^{\mathrm{P}} \left[\frac{\partial T_i}{\partial x}\right]^2 \langle \Psi_i \Psi_i \rangle\,\mathrm{d}y, \tag{6.68}$$

and should be carefully distinguished from $\sigma^2(\mathrm{Nu})$. At steady state, the net heat flux on the hot wall equals that on the cold wall; since this relationship holds for arbitrary realization, $\sigma^2(\mathrm{Nu})$ has the same value on the hot wall as on the cold wall. On the

Table 6.1 Effect of N on $\overline{\text{Nu}}$ and $\sigma(\text{Nu})$. Results are obtained for No = 1, $L_c = 1$, and $\sigma_T = 0.25$. The number of modes in the PC expansion is also shown. Adapted from [128]

N	$\overline{\text{Nu}}$	$\sigma(\text{Nu})$	P
2	8.96344	2.47009	2
4	8.97114	2.46979	4
6	8.97179	2.46980	6
8	8.97190	2.46980	8
10	8.97192	2.46980	10

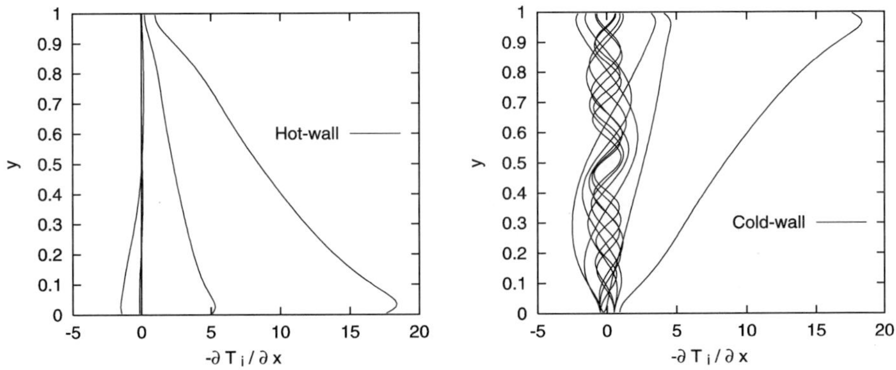

Fig. 6.22 Local heat-fluxes versus y on the hot (*left*) and cold (*right*) walls, for modes 0–10. A first-order expansion is used with N = 10, $L_c = 1$ and $\sigma_T = 0.25$. Adapted from [128]

other hand, $\overline{\sigma^2}(\partial T/\partial x)$ is expected to assume a higher value on the cold wall, where random fluctuations are imposed, than on the hot wall, since these fluctuations are expected to be smoothed out by diffusion.

Computed values of $\overline{\text{Nu}}$ and $\sigma(\text{Nu})$ are reported in Table 6.1. As expected, the results show that for the present conditions $\overline{\text{Nu}}$ and $\sigma(\text{Nu})$ converge rapidly with N. In order to further examine the predictions, we plot in Fig. 6.22 the distribution of the normalized heat flux $-\partial T/\partial x$ along the hot and cold walls as a function of P. (Note that for No = 1, P = N.) Clearly, on the hot wall, only modes 0 and 1 contribute significantly to the local heat flux; for the higher modes, $\partial T_i/\partial x$ is close to zero for all y. This situation is contrasted with the distribution of the heat fluxes on the cold wall, where significant heat flux fluctuations are observed for all the PC modes. However, as noted earlier, the net heat flux across the hot and cold walls are equal at steady state. Thus, when integrated along the boundary, the significant fluctuations of the higher modes on the cold wall tend to cancel out. This explains the rapid convergence of integral quantities in Table 6.1.

The standard deviation of the local heat flux, shown in Fig. 6.23, closely reflects the above trends. In particular, by comparing points symmetrically across the mid-plane, the results clearly show that the values on the cold wall are generally larger than on the hot wall. Also note that the curve for the cold wall exhibits a notice-

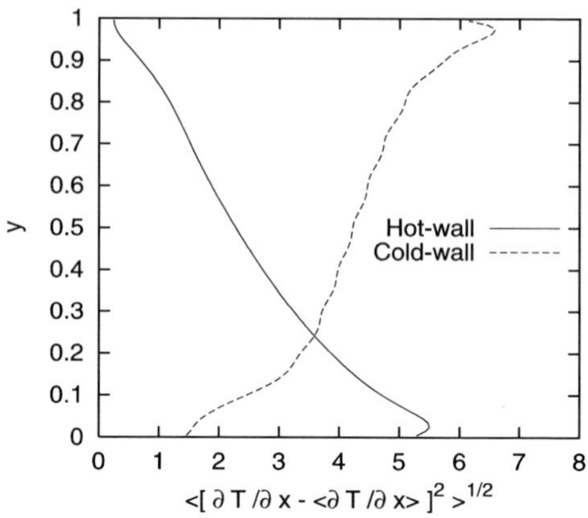

Fig. 6.23 Standard deviation of the local heat fluxes versus y on the hot and cold walls. A first-order expansion is used with N = 10, $L_c = 1$ and $\sigma_T = 0.25$. Adapted from [128]

Table 6.2 Mean Nusselt number and its standard deviation for first, second, and third-order PC expansion with N = 4 and 6. In all cases, $L_c = 1$ and $\sigma_T = 0.25$. The value of P is indicated. Adapted from [128]

	N = 4	N = 6	N = 4	N = 6	N = 4	N = 6
	$\overline{\text{Nu}}$	$\overline{\text{Nu}}$	σ(Nu)	σ(Nu)	P	P
No = 1	8.97114	8.97179	2.46979	2.46980	5	7
No = 2	8.97289	8.97352	2.46323	2.46327	15	27
No = 3	8.97337	8.97340	2.46239	2.46245	34	83

able waviness that corresponds to the imposed conditions, whereas the curve for the hot-wall distribution is smooth.

Convergence with No**:** We now analyze the convergence of the PC expansion by contrasting results obtained with No = 1, 2 and 3. Results are obtaining with $L_c = 1$, and $\sigma_T = 0.25$, using both 4 and 6 KL modes.

The computed values of $\overline{\text{Nu}}$ and σ(Nu) are reported in Table 6.2, together with the number P of polynomials used. The results exhibit a fast convergence as the order of the PC expansion, No, increases. The differences in $\overline{\text{Nu}}$ and σ(Nu) between second- and third-order solutions are less than 0.01% and 0.05%, respectively. The close quantitative agreement between the results for No = 2 and 3 indicates that, at least as far as integral quantities are concerned, a second-order expansion is sufficiently accurate. This fast convergence rate is also indicative of the smooth dependence of the solution with respect to the imposed random temperature fluctuations.

Heat fluxes: Plotted in Fig. 6.24 are the heat flux distributions along the cold (top row) and hot (bottom row) walls for No = 1, 2 and 3. Results are obtained with N = 4 and curves are plotted for every mode in the PC expansion. The local heat flux profiles for the "first-order modes" (index $i \leq 4$) have similar shapes as those reported in Fig. 6.22: these modes have significant amplitude on the cold wall, whereas modes higher than 2 are much less pronounced on the hot wall. On both walls, the first-order modes are slightly influenced by the order of the PC expansion. While

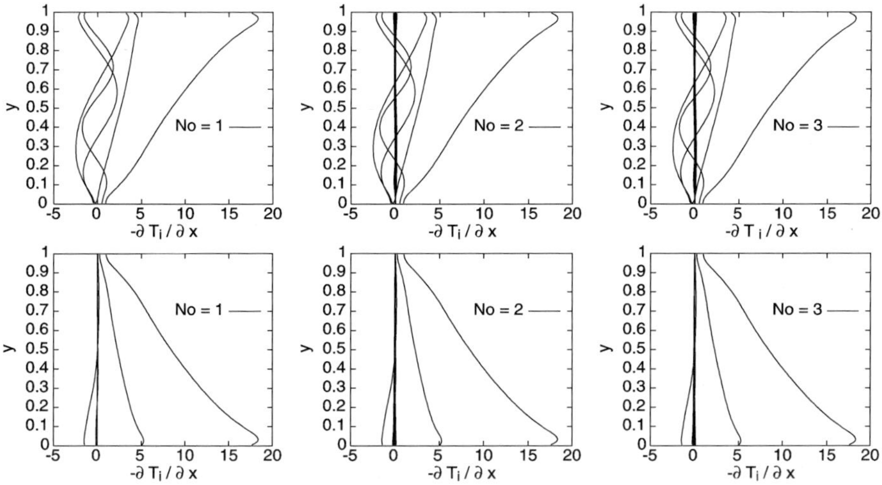

Fig. 6.24 Local heat flux versus y on the cold (*top row*) and hot (*bottom row*) walls. Results are obtained with N = 4, $L_c = 1$, and $\sigma_T = 0.25$. *Curves* are plotted for every mode in the PC expansion. P = 4 for No = 1, P = 14 for No = 2, and P = 34 for No = 3. Adapted from [128]

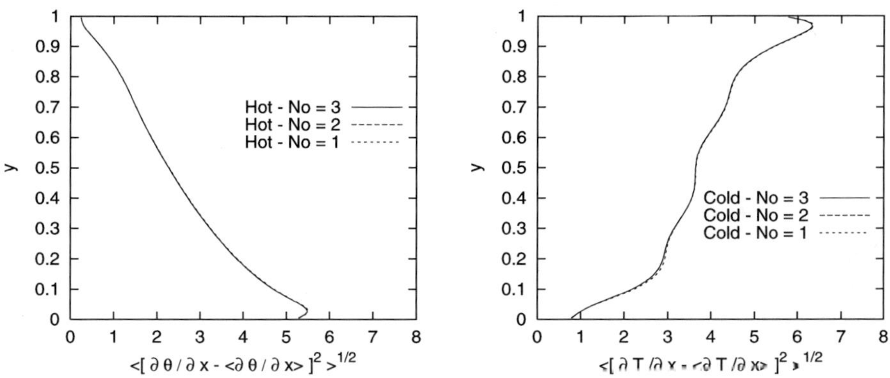

Fig. 6.25 Local standard deviation of the heat-fluxes on the hot (*left*) and cold (*right*) walls, for No = 1, 2 and 3. Results are obtained with N = 4, $L_c = 1$, and $\sigma_T = 0.25$. Adapted from [128]

increasing No introduces more modes in the expansion (P = 14 for No = 2 and P = 34 for No = 3), the heat fluxes associated with these higher-order modes are very low. Consequently the "correction" of the local heat-fluxes, arising when No is increased, is weak whenever No > 1. This fact is also shown in Fig. 6.25 where the local heat-flux standard deviations are plotted for No = 1, 2 and 3.

The present analysis of wall heat fluxes only shows how the solution converges, globally or locally, on the vertical boundaries. To further investigate the behavior of the spectral representation, we analyze the temperature and velocity fields within the cavity. We focus our attention on the distributions of mean quantities and their

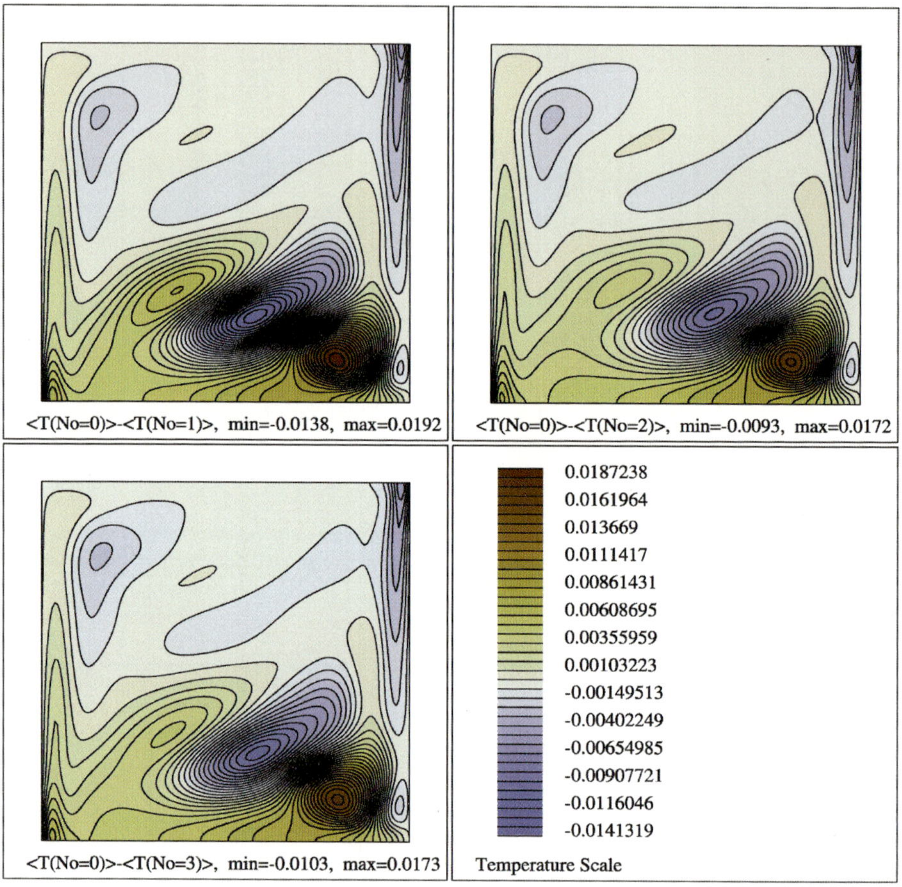

Fig. 6.26 Contours of $T^0 - \langle T^{\mathrm{No}} \rangle$. Plots are generated for No = 1, 2 and 3. Results are obtained with N = 4, $L_c = 1$, and $\sigma_T = 0.25$. Adapted from [128]

standard deviations and postpone to Sect. 6.2.5 the examination of individual mode structure.

Temperature field: We start by noting that, since the natural convection in the cavity is not a linear process, the mean temperature distribution differs from the deterministic prediction corresponding to the mean temperature boundary condition, $T_c = -1/2$. This deterministic prediction, corresponding to $\xi_i = 0$, $i = 1, \ldots, \mathrm{N}$, shall be denoted by T^0. Meanwhile, we shall denote by $\langle T^{\mathrm{No}=1,2,3} \rangle \equiv T_0(\mathrm{No})$ the mean predictions obtained using first, second, and third-order PC expansions, respectively.

Examination of the mean temperature fields obtained with No = 1, 2 and 3 (not shown) reveals that these fields have features similar to those of T^0 (shown earlier in Fig. 6.21). Thus, we have found it more convenient to analyze the difference fields $T^0 - \langle T^{\mathrm{No} \geq 1} \rangle$. These difference fields are plotted in Fig. 6.26 for N = 4. A close

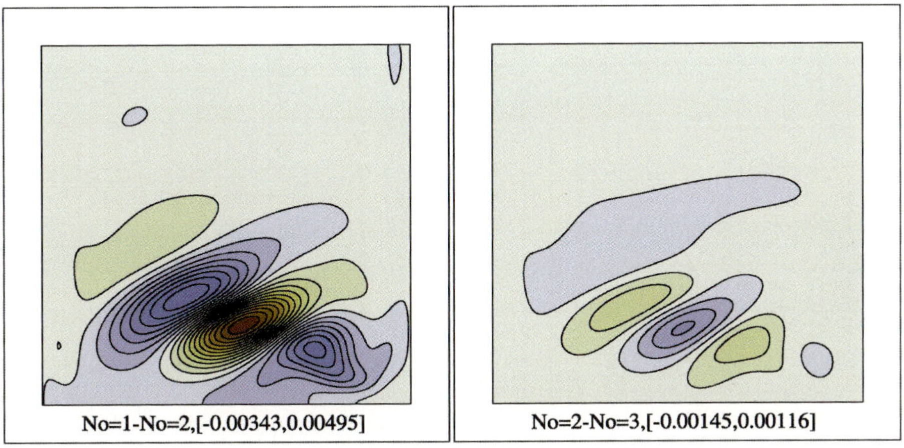

Fig. 6.27 Contour plots of $\langle T^1\rangle - \langle T^2\rangle$ (*left*) and $\langle T^2\rangle - \langle T^3\rangle$ (*right*). Results are obtained with N = 4, $L_c = 1$, and $\sigma_T = 0.25$. Adapted from [128]

agreement is observed between the plots corresponding to the different expansion orders. Only a very weak dependence of the local magnitudes on No can be detected. Thus, increasing No has only a weak effect on the expected temperature field. To further demonstrate the convergence of the spectral representation, the differences $\langle T^{\mathrm{No}=1}\rangle - \langle T^{\mathrm{No}=2}\rangle$ and $\langle T^{\mathrm{No}=2}\rangle - \langle T^{\mathrm{No}=3}\rangle$ are displayed in Fig. 6.27. The results show that, at least as far as the mean field is concerned, the first-order expansion captures most of the effects of uncertainty. The difference $\langle T^{\mathrm{No}=2}\rangle - \langle T^{\mathrm{No}=3}\rangle$ is very small, indicating that the truncated terms have a weak impact on the mean temperature.

Figure 6.26 also shows that the mean temperature along the cold wall is higher than that of T^0. The opposite situation is reported along the hot wall, where the mean temperature is lowered by the uncertainty. These changes are responsible for the improvement of the global heat-transfer coefficient $\overline{\mathrm{Nu}}$ (see further discussion below). In addition, the mean temperature on the bottom of the cavity is significantly lower than that of T^0; in the upper part of the cavity, the mean and deterministic predictions are nearly equal. To explain these trends, one notes that the mean clockwise flow circulation is not altered by the stochastic boundary conditions (as will be shown later). So, on average, the fluid is traveling downwards along the cold wall, where it is affected by random temperature conditions. The random fluctuations are transported across the cavity to the hot wall. As the fluid travels upwards along the hot wall, uncertainty is reduced due to diffusion, so that when reaching the upper part of the cavity, the fluid temperature has lost most of its uncertainty, and its mean value is close to that of T^0. We also observe that the deviation of the mean temperature field from T^0 exhibits a complex structure, with alternating signs, in the lower-right quadrant, where the deviation from T^0 peaks. This pattern is closely "correlated" with the uncertainties in the velocity fields, as further discussed below.

Additional insight into the role of stochastic boundary conditions can be gained from Fig. 6.28, which depicts the temperature standard deviation fields for No = 1,

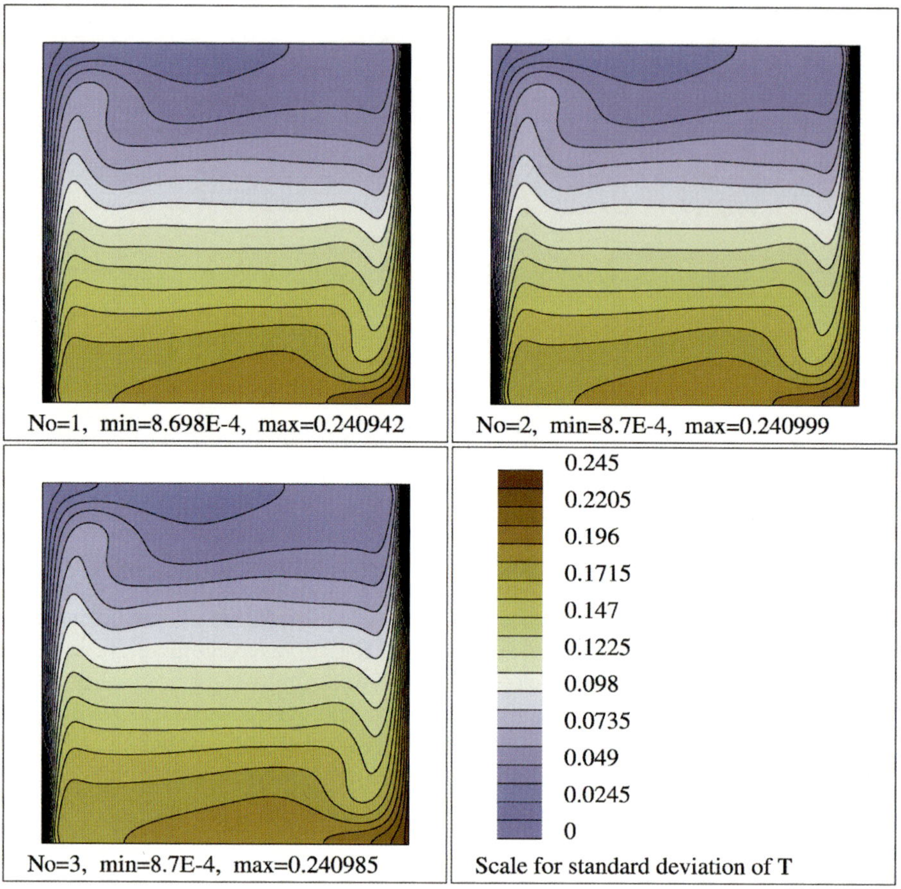

Fig. 6.28 Contours of standard deviation in temperature for No $= 1, 2$ and 3. Results are obtained with N $= 4$, $L_c = 1$, and $\sigma_T = 0.25$. Adapted from [128]

2 and 3. The results show that the standard deviation distribution has a structure similar to that of the mean, with two layers parallel to the vertical walls and a horizontal stratified arrangement from the bottom to the top of the cavity. The standard deviation vanishes on the hot wall, where deterministic conditions are imposed, and reaches its maximum on the cold wall, with values close to σ_T. This spatial distribution is consistent with the above arguments regarding the role of circulation in driving the uncertainty.

Finally, in Fig. 6.29, it is shown that the expansions for No $= 1, 2$ and 3 provide essentially the same estimate of the temperature standard deviation, with differences in the fourth significant digit. These results also demonstrate the fast convergence rate of the spectral expansion, and the fact that in the present case a first-order expansion captures most of the standard deviation.

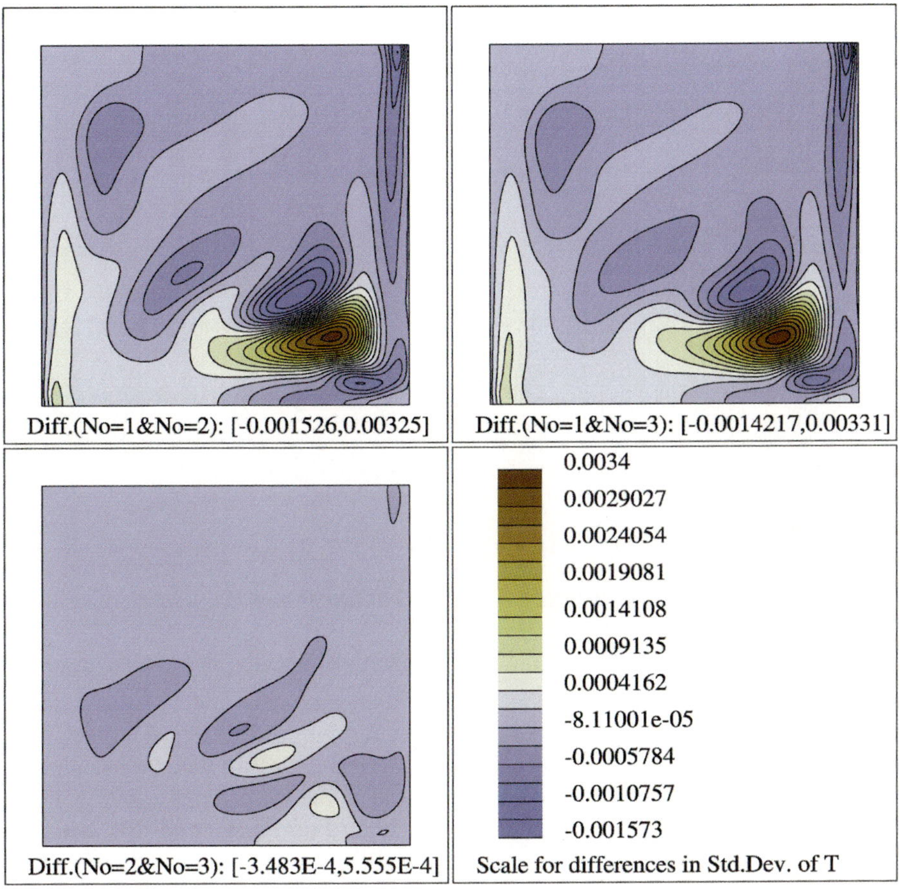

Fig. 6.29 Differences in the temperature standard deviation computed using No = 1, 2 and 3. In all cases, N = 6, $L_c = 1$, and $\sigma_T = 0.25$. Adapted from [128]

Velocity field: As was done for the temperature distribution, we start by examining the deviation of the mean velocity field from $\mathbf{u}^0$, which denotes the deterministic solution corresponding to the mean temperature condition ($T_c = -1/2$). The mean velocity fields corresponding to first, second, and third-order PC expansions will be denoted by $\langle \mathbf{u}^{\mathrm{No}=1,2,3} \rangle$ respectively. For each case, we find that the deviation of the mean solution from $\mathbf{u}^0$ is small, and we consequently focus on the differences $\langle \mathbf{u}^{\mathrm{No}\geq 1} \rangle - \mathbf{u}^0$.

Figure 6.30a shows the distribution of $\langle \mathbf{u}^{\mathrm{No}=3} \rangle - \mathbf{u}^0$ for a simulation with N = 6, $L_c = 1$, and $\sigma_T = 0.25$. The difference field exhibits three complex structures that lie in the lower part of the cavity. While these structures resemble the recirculating eddies of the mean flow, it should be emphasized that the velocity magnitudes have been scaled by a factor of 10 compared to those in Fig. 6.21. Thus, with respect to $\mathbf{u}^0$, the mean field is significantly perturbed in the regions occupied by these

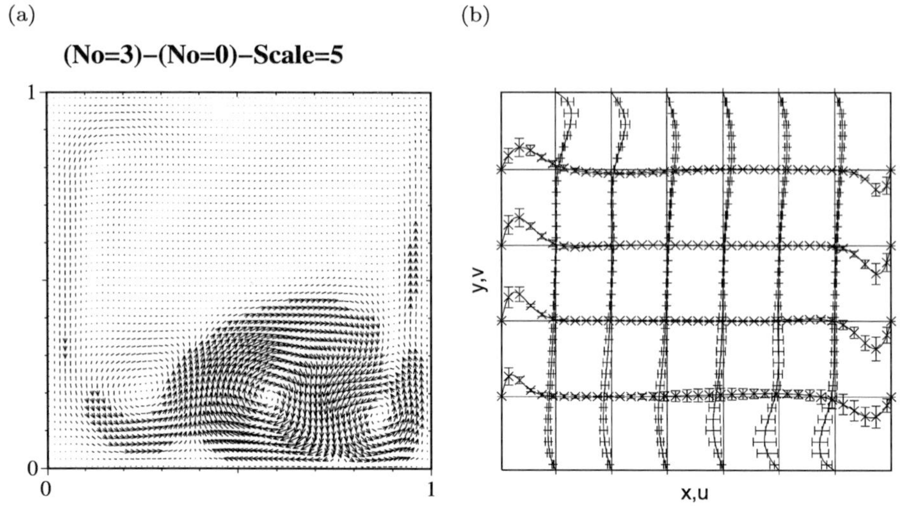

Fig. 6.30 (**a**) Velocity map of the difference $\langle \mathbf{u}^{\mathrm{No}=3} \rangle - \mathbf{u}^0$. (**b**) Profiles of mean horizontal velocity ($\langle u^{\mathrm{No}=3} \rangle$) and mean vertical velocity ($\langle v^{\mathrm{No}=3} \rangle$). The profiles are independently scaled for clarity. The scaled length of the bars corresponds to 6 times the local standard deviation. Results are obtained with $\mathrm{N}=6$, $L_c=1$, and $\sigma_T=0.25$. Adapted from [128]

structures, but it is not actually recirculating. This can be verified by inspecting the mean solution itself, depicted in Fig. 6.30b using the profiles of mean horizontal and mean vertical velocity. The profiles show that the mean flow is not recirculating, but that flow "reversal," hence recirculation, in the lower-right corner is likely to occur. In this region, one observes large standard deviations and low mean velocities, especially outside the boundary layers; this is indicative of large sensitivity to the stochastic boundary conditions. This trend is consistent with earlier observations regarding the deviations $T^0 - \langle T^{\mathrm{No}\geq 1} \rangle$, which exhibited maxima at these same locations.

In order to verify that the behavior of the stochastic solution is well represented, and consequently that the trends above are not an artifact of the method, we inspect in Fig. 6.31 the distribution of $\langle \mathbf{u}^{\mathrm{No}=3} \rangle - \langle \mathbf{u}^{\mathrm{No}=2} \rangle$. The velocity map is generated with a scaling factor that is ten times larger than that used in Fig. 6.30a. The results clearly demonstrate that there are very small differences between the second and third-order solutions, and that both provide accurate representations of the stochastic process.

Remarks: We close this section with two remarks regarding the ability of the spectral representation to accurately reproduce individual events, and regarding the CPU costs of the spectral solution scheme.

Recall that the spectral representation relies on a weighted residual procedure to determine the mode coefficients. This representation is the closest polynomial to the exact response "surface" in the corresponding L_2 norm. Although optimal in this sense, the PC representation does not guarantee that individual "realizations"

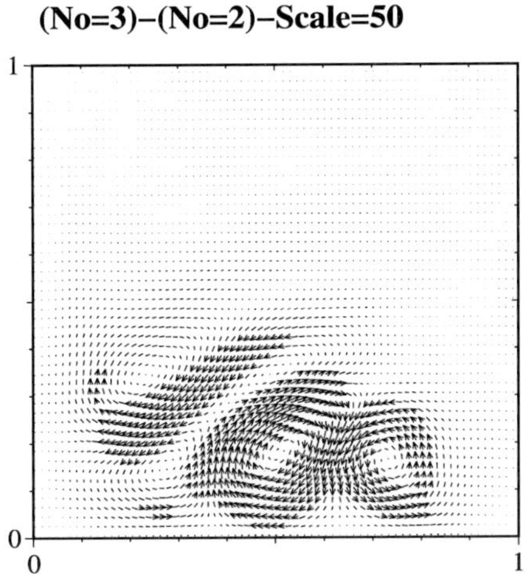

Fig. 6.31 Velocity map of the difference $\langle \mathbf{u}^{\text{No}=3} \rangle - \langle \mathbf{u}^{\text{No}=2} \rangle$. Results are obtained with $\text{N} = 6$, $L_c = 1$, and $\sigma_T = 0.25$. Adapted from [128]

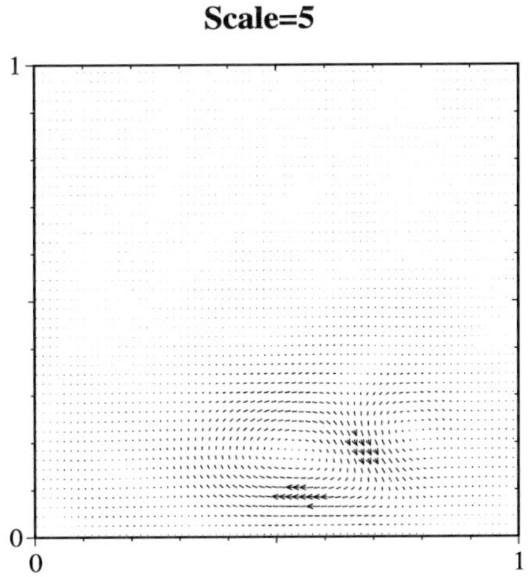

Fig. 6.32 Velocity map of the difference $\mathbf{u}^{\text{No}=2}(\boldsymbol{\xi} = 0) - \mathbf{u}^0$. Results are obtained with $\text{N} = 6$, $L_c = 1$, and $\sigma_T = 0.25$. Adapted from [128]

are exactly interpolated. However, our experiences indicate that when the PC representation is of sufficiently high order, it can also be used to obtain highly accurate estimates of individual realizations. This quality is illustrated in Fig. 6.32, where we plot the difference between $\mathbf{u}^0$ and the second-order solution evaluated at $\boldsymbol{\xi} = 0$, i.e. $\mathbf{u}^{\text{No}=2}(\boldsymbol{\xi} = 0)$. The figure is generated with a scaling factor 10 times larger than that used for the deterministic solution of Fig. 6.21, demonstrating that the agreement between $\mathbf{u}^0$ and $\mathbf{u}^{\text{No}=2}(\boldsymbol{\xi} = 0)$ is indeed very good.

Regarding the performance of the spectral computations, we had anticipated earlier that the CPU cost would scale linearly with P, with near unity coefficient. As shown in Fig. 6.33, this behavior is in fact observed, and, together the spectral behavior of the errors in the spectral approximation, can be used to guide the selection of a suitable stochastic discretization level that properly balances accuracy and CPU cost.

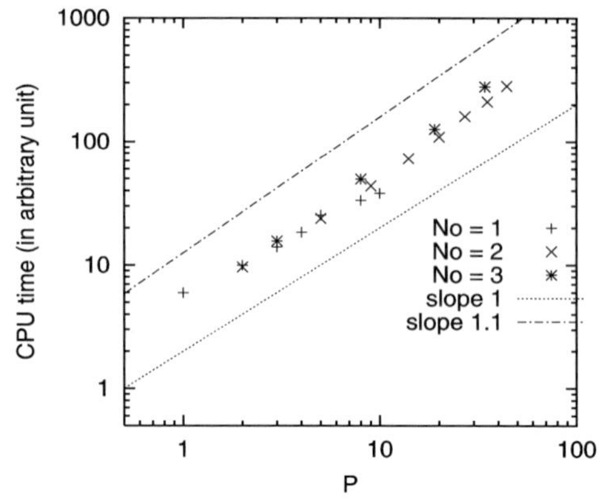

Fig. 6.33 CPU time needed to perform 100 time steps for PC expansions with different No and N. A fixed mesh size of 140 × 100 cells is used. Scaled CPU times are reported as a function of the largest polynomial index P. Adapted from [128]

6.2.5 Analysis of Stochastic Modes

In this section, we examine individual velocity and temperature modes in PC expansion. For brevity, we restrict our attention to spectral predictions obtained with N = 4, No = 2, $L_c = 1$ and $\sigma_T = 0.25$. For this spectral resolution P = 14, giving a total of 15 modes. Thus, we end up with a moderate number of velocity and temperature distributions, which are analyzed below.

6.2.5.1 Velocity Modes

Figure 6.34 provides vector maps for all the modes in the computations. Different scaling factors are used to represent the various fields, as indicated in the labels. Note that the zeroth mode corresponds to the mean velocity field, which has been already studied in Sect. 6.2.4.2. Thus, we shall focus on the higher modes.

The first-order velocity modes, $v_k, k = 1, \dots, 4$ follow mode 0 in the first column of Fig. 6.34. They correspond to the polynomial $\Psi_k = \xi_k$ for $k = 1, \dots, \mathrm{N} = 4$. Thus, these modes reflect the linear response of the stochastic velocity field to the corresponding KL eigenfunctions appearing in (6.49) and plotted in Fig. 2.1. Note that the first KL mode has a nearly uniform, positive value, and that it exhibits a positive velocity along the cold wall, and a negative velocity along the hot wall. This is not surprising since, for $\xi_1 > 0$, the first KL mode tends to decrease the temperature difference between the two walls. However, the structure of $\mathbf{u}_1$ is not similar to that of $\mathbf{u}_0$. The two fields are governed by different dynamics, as can be appreciated from the governing equations of the corresponding modes. This is conveniently illustrated for a first-order expansion, for which the mean modes are governed by:

$$\frac{\partial \mathbf{u}_0}{\partial t} + \sum_{i=0}^{\mathrm{N}} \mathbf{u}_i \cdot \nabla \mathbf{u}_i = -\nabla p_0 + \frac{\mathrm{Pr}}{\sqrt{\mathrm{Ra}}} \nabla^2 \mathbf{u}_0, \tag{6.69}$$

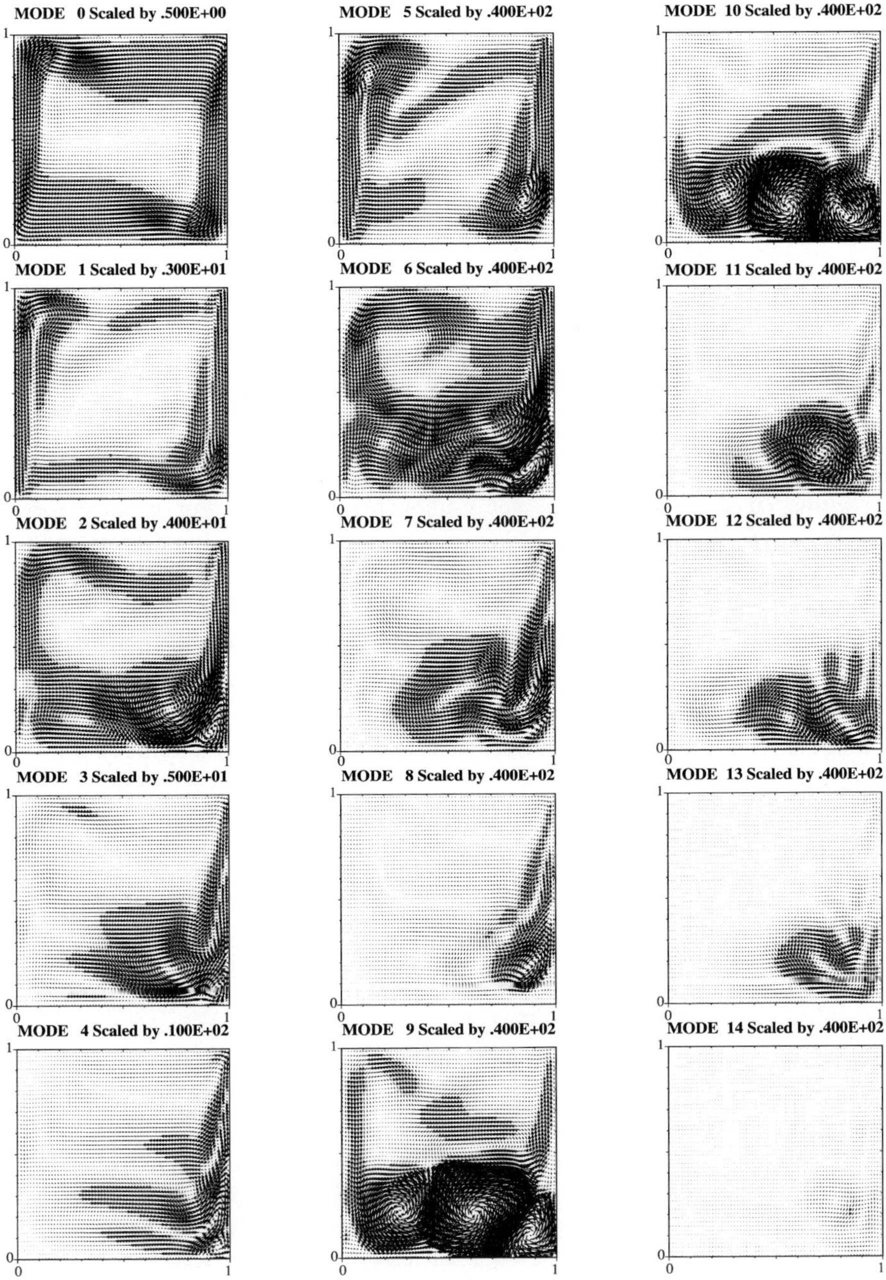

Fig. 6.34 Velocity fields $\mathbf{u}_k$ for $k = 0, \ldots, 14$. Results are obtained with $N = 4$, $No = 2$, $L_c = 1$ and $\sigma_T = 0.25$. Note that different velocity scales are used, as indicated in the labels. Adapted from [128]

$$\frac{\partial T_0}{\partial t} + \sum_{i=0}^{N} \nabla \cdot (\mathbf{u}_i T_i) = \frac{1}{\sqrt{\text{Ra}}} \nabla^2 T_0, \tag{6.70}$$

whereas

$$\frac{\partial \mathbf{u}_k}{\partial t} + \mathbf{u}_0 \nabla \mathbf{u}_k + \mathbf{u}_k \nabla \mathbf{u}_0 = -\nabla p_k + \frac{\text{Pr}}{\sqrt{\text{Ra}}} \nabla^2 \mathbf{u}_k, \tag{6.71}$$

$$\frac{\partial T_k}{\partial t} + \nabla \cdot (\mathbf{u}_0 T_k + \mathbf{u}_k T_0) = \frac{1}{\sqrt{\text{Ra}}} \nabla^2 T_k \tag{6.72}$$

for $k = 1, \ldots, \text{N}$. One observes that the first mode is advected (and stretched) by the mean velocity ($\mathbf{u}_0$) and not by $\mathbf{u}_1$. This observation, which also applies for $\mathbf{u}_2$, $\mathbf{u}_3$ and $\mathbf{u}_4$, remains true for a second-order expansion.

In the neighborhood of the cold wall, all the first-order velocity modes clearly reflect the shape of the corresponding Karhunen-Loève mode. For instance, for $\mathbf{u}_2$, the velocity points upward on the highest part of the cold wall and downward in its lowest part, as the associated temperature perturbation is respectively positive and negative (Fig. 2.1). For $\mathbf{u}_1$ and $\mathbf{u}_2$, the velocity magnitudes are significant near all solid boundaries; on the other hand, for $\mathbf{u}_3$ and $\mathbf{u}_4$, the velocity magnitudes are negligible on the hot wall. For the first order modes, the velocity magnitudes decrease with increasing mode index; note in particular that the scale factor for $\mathbf{u}_4$ is twice that of $\mathbf{u}_3$. If a larger value for N is used, the additional first-order velocity fields are weaker than those retained, and are localized near the cold boundary (not shown). This trend is consistent with our earlier discussion of the weakening effects of the higher-frequency, random fluctuations.

The second-order velocity fields, $\mathbf{u}_k$, $k = 5, \ldots, 14$, are plotted in the center and right columns of Fig. 6.34; the same scaling factor is used for these modes, allowing straightforward comparison. Note that this scaling factor is 80 times larger than that of $\mathbf{u}_0$, and 10 times larger than that of $\mathbf{u}_4$. Thus, the magnitudes of the second-order velocity fields are much smaller than those of zeroth mode and first-order modes. This rapid decay also reflects the rapid convergence of the PC expansion.

The second-order velocity modes have very different patterns, some being only significant along the cold wall, others affecting the entire cavity. Some of these structures can be easily interpreted. For example, $\mathbf{u}_5$, which corresponds to $\Psi_5 = \xi_1^2 - 1$, has a structure similar to that of $\mathbf{u}_1$. For other modes, the structure of the corresponding velocity fields are quite complex and difficult to interpret. It is interesting to note, however, that the velocity fields involving the second KL mode ($\mathbf{u}_6$, $\mathbf{u}_9$, $\mathbf{u}_{10}$ and $\mathbf{u}_{11}$) seem to have the most significant magnitudes, suggesting that this mode has greater impact on the stochastic process than the others. On the other hand, second-order polynomials associated with the fourth KL mode appear to be very weak.

6.2.5.2 Temperature Modes

Figure 6.35 shows contour plots of the temperature modes T_k, $k = 0, \dots, 14$. Since the mean temperature field, T_0, has been analyzed earlier, we will focus on first-order and second-order modes.

The contours of the first mode, T_1, are similar to those of T_0, even though the corresponding values differ. This is not surprising since these two modes obey similar boundary conditions, with T_0 being subjected to a uniform Dirichlet condition on the cold wall, while T_1 is nearly uniform there. However, some differences between the distributions of T_1 and T_0 can be observed at the lower right corner of the cavity. These differences appear to be governed by the circulation of the mean flow in the cavity. To appreciate this effect, we note that it is the mean field, $\mathbf{u}_0$, which contributes to the transport of T_1; the heat flux associated with $\mathbf{u}_1$, which points upwards near the cold wall, is dependent on the mean temperature field T_0. The role of the mean field in the transport of T_2, T_3 and T_4 can also be appreciated from the corresponding contour plots. Note that T_2, T_3 and T_4 are very small in the upper half of the cavity, but have significant values in the lower part of the cavity and/or in the vicinity of the cold wall. In particular, for T_3 and T_4 one observes fluctuations of alternating sign that are localized near the cold boundary, and that coincide with the shape of the corresponding KL mode.

As for velocity, the second-order temperature modes are more difficult to interpret than the first-order modes. The only structures that can be easily identified are the imposed cold-wall distributions. The results indicate that significant mode coupling occurs, which can be detected by inspecting the modes involving mixed products of the ξ_i's. For instance, for T_7 a second-order coupling between ξ_1 and ξ_3 is involved; this mode exhibits three distinct zones along the cold wall, which reflect the shape of the third mode in the KL expansion. Apart from such identifiable features, the second-order modes can have complex distributions, some of which are localized in the lower part of the cavity, while others extend throughout the domain.

Regarding the amplitude of the second-order modes, we note those modes involving ξ_2 and ξ_3, *i.e.* the second and third KL eigenfunctions, are dominant. Thus, not all second-order modes contribute equally to the stochastic process. In general, however, the second-order temperature modes are at least one order of magnitude lower than the first-order modes. This is consistent with earlier observations regarding the convergence of the expansion.

6.2.6 Comparison with NISP

In order to verify the spectral computations of the previous section, a non-intrusive spectral projection (NISP) approach is employed. The modes $\mathbf{u}_i$ and T_i are obtained by projecting deterministic computations onto the PC basis. If $\mathbf{u}^d(\boldsymbol{\xi})$ and $T^d(\boldsymbol{\xi})$ denote the deterministic solution corresponding to a particular realization

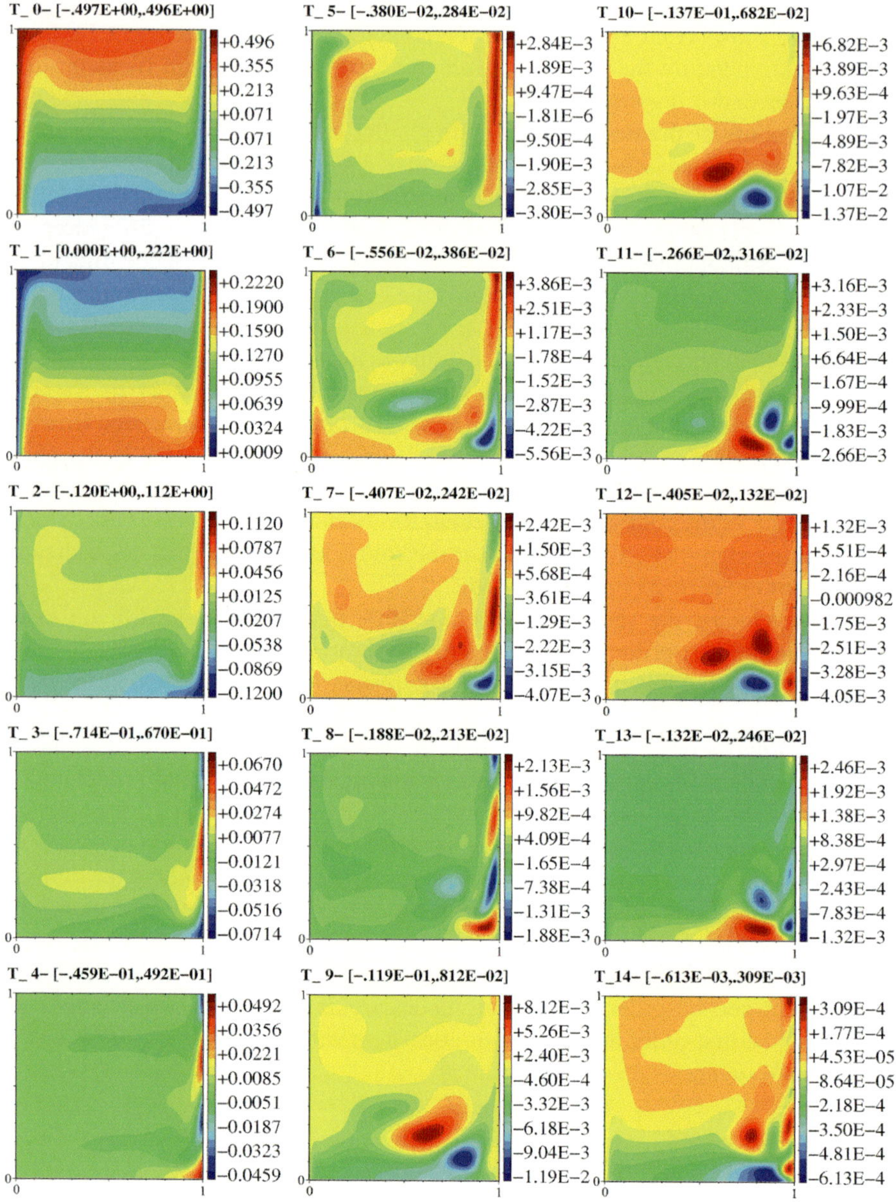

Fig. 6.35 Scaled temperature fields T_k for $k = 0, \dots, 14$. Results are obtained with N = 4, No = 2, $L_c = 1$ and $\sigma_T = 0.25$. Adapted from [128]

$\boldsymbol{\xi} = (\xi_1, \ldots, \xi_N)$, then the polynomial coefficients are, by definition, given by:

$$(\mathbf{u}_i, T_i) = \frac{\langle (\mathbf{u}, T)^d \Psi_i \rangle}{\langle \Psi_i \Psi_i \rangle}$$
$$= \int_{-\infty}^{\infty} d\xi_1 \cdots \int_{-\infty}^{\infty} d\xi_N \left[(\mathbf{u}, T)^d (\boldsymbol{\xi}) \frac{\Psi_i(\boldsymbol{\xi})}{\langle \Psi_i^2 \rangle} \prod_{k=1}^{N} \frac{\exp(-\xi_k^2/2)}{\sqrt{2\pi}} \right]. \quad (6.73)$$

6.2.6.1 Gauss-Hermite Quadrature

For moderate values of N, the above multi-dimensional integration can be efficiently performed using Gauss-Hermite (GH) quadrature [1, 114]. Using n collocation points along each stochastic direction, (6.73) can be approximated as:

$$(\mathbf{u}_i, T_i) = \sum_{n_1=1}^{n} \cdots \sum_{n_N=1}^{n} (\mathbf{u}, T)^d (x_{n_1}, \ldots, x_{n_N}) \frac{\Psi_i(x_{n_1}, \ldots, x_{n_N})}{\langle \Psi_i \Psi_i \rangle} \prod_{k=1}^{N} w_{n_k} \quad (6.74)$$

where (x_k, w_k), $k = 1, \ldots, n$ denote the one-dimensional GH integration points and weights. The quadrature in (6.74) is exact when the integrand is a polynomial of degree of $2n-1$ or less. Thus, the coefficients can be exactly estimated if the process is spanned by polynomials of degree less than or equal to $(2n-1)/2$. In this situation, the number, N_d, of deterministic realizations that is required in the NISP approach for given N and No is $N_d = (2\text{No} - 1)^{\text{N}}$. It should be emphasized that for arbitrary N and No, N_d is always greater than P, the number of polynomials in the spectral approach above. Since the CPU time in the spectral approach is approximately P times that of a deterministic solution, NISP is not as efficient as the spectral approach. Its main advantage, however, is that it makes use of a deterministic solver without the need for any modifications.

NISP/GH computations are performed for a case with $\text{N} = 4$ and $\text{No} = 2$. We use $n = 3$ and so obtain $N_d = 81$ deterministic realizations for the corresponding GH quadrature points. (In contrast, the intrusive spectral approach above has $\text{P} = 14$, for a total of 15 modes.) Velocity and temperature modes obtained using NISP are plotted in Figs. 6.36 and 6.37 respectively. The corresponding results obtained using the intrusive spectral approach were given in Figs. 6.34 and 6.35, and have been extensively discussed in the previous section.

For the velocity fields, we find an excellent agreement between the intrusive spectral results (Fig. 6.34) and the NISP predictions (Fig. 6.36) for the zeroth and first-order modes. For the second-order modes ($\mathbf{u}_k$, $k = 5, \ldots, 14$), small deviations are observed between the two sets, but the primary structure of the modes is quite similar. These small deviations are pronounced for coupled modes involving ξ_2 and ξ_3; the deviations are substantially smaller for the non-mixed quadratic modes. Despite these small deviations, the agreement between the intrusive and NISP/GH predictions is very satisfactory.

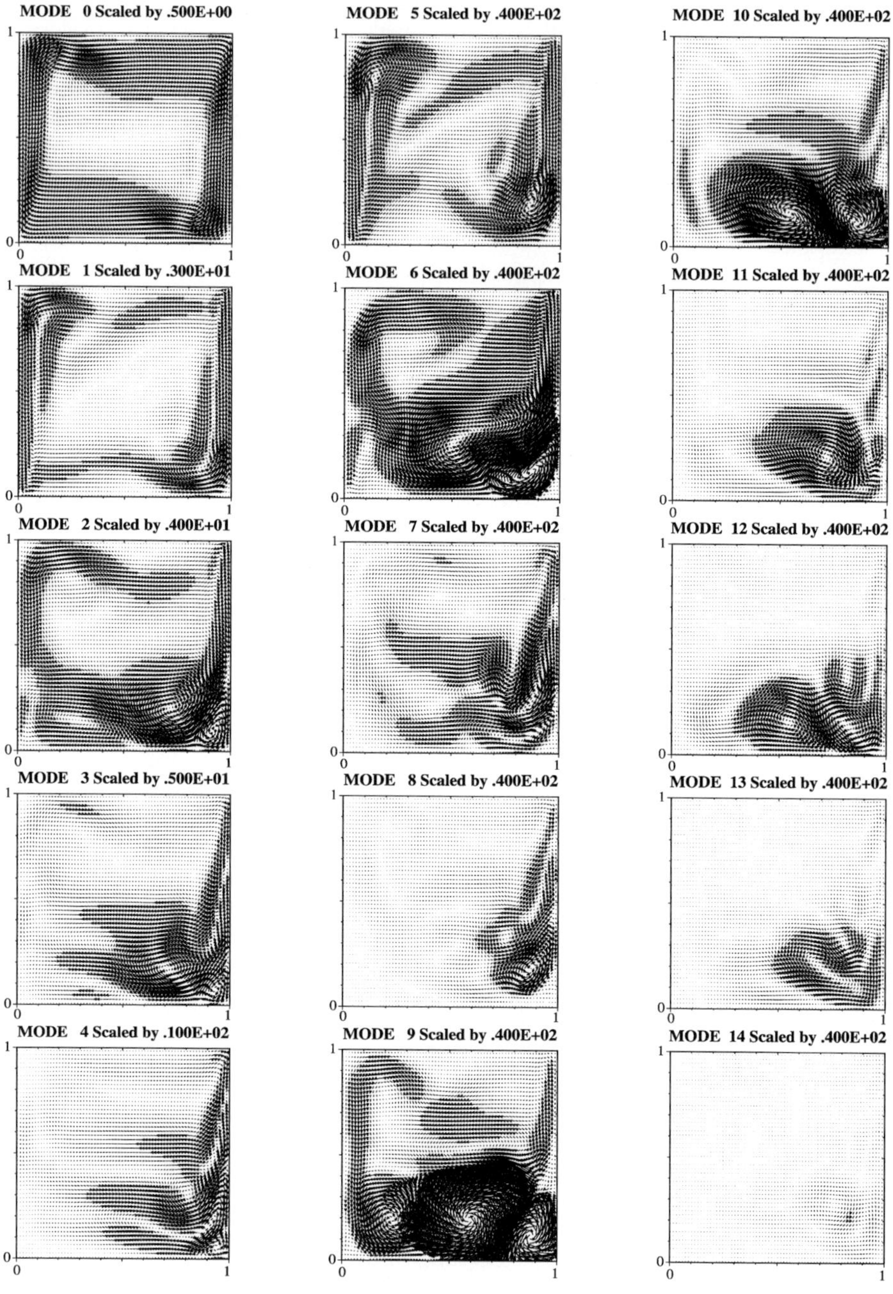

Fig. 6.36 Velocity fields $\mathbf{u}_k$ for N = 4, obtained using NISP/GH predictions with $N_d = 81$. Note that different scale factors apply on vectors magnitude. $L_c = 1$ and $\sigma_T = 0.25$. Adapted from [128]

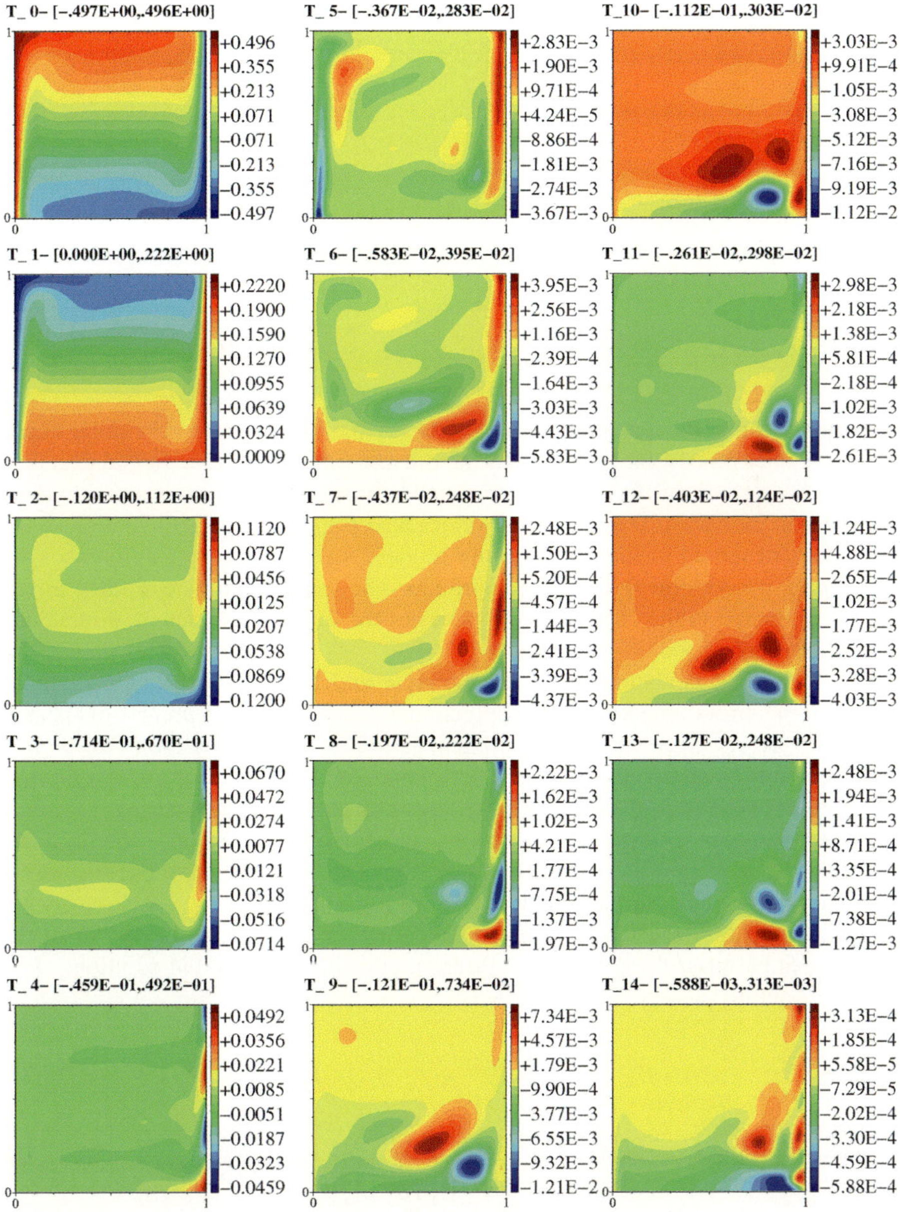

Fig. 6.37 Scaled temperature fields T_k for N = 4, obtained using NISP/GH predictions with $N_d = 81$. $L_c = 1$ and $\sigma_T = 0.25$. Adapted from [128]

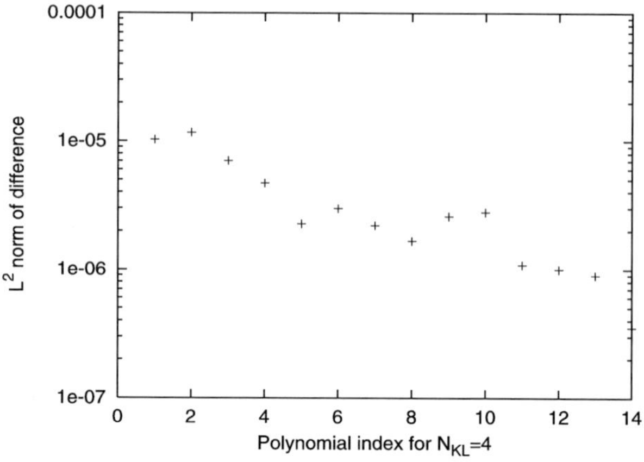

Fig. 6.38 L2 norm of the difference in the common temperature modes obtained with intrusive spectral calculations using N = 4 and N = 6. In both cases, a second-order PC expansion is used. Adapted from [128]

Comparison of the temperature modes in Figs. 6.37 and 6.35 reveal similar trends as the velocity modes. In particular, the zeroth and first-order modes are in excellent quantitative agreement, as can be appreciated by inspecting the maxima and minima reported on individual frames. These values also provide a good illustration of the deviations observed in the second-order modes. Again the largest differences are observed for modes involving mixed products. The small magnitude of these differences, compared to the characteristic values of the first order terms, is evident and should be emphasized.

The origin of deviations between intrusive and NISP/GH predictions can be traced to the errors inherent in both approaches. These primarily consist of spectral truncation errors in the intrusive approach, and aliasing errors in the NISP predictions. Obviously, complete agreement between NISP and spectral computations can only be achieved in the case of a finite spectrum. Since we are presently dealing with second-order spectral representations, agreement would occur if the third and higher-order modes vanish identically, which is clearly not the case: the third-order terms are very small, but not identically vanishing.

In order to further examine these differences, we rely on the L2 norms of the differences between the same temperature modes in two different solutions, $T^{(1)}$ and $T^{(2)}$, defined according to:

$$\mathcal{E}_{ik}^2 \equiv \left[\int\int (T_i^{(1)} - T_k^{(2)})^2 \,\mathrm{d}x\,\mathrm{d}y \right]^{1/2}. \tag{6.75}$$

The indices i and k are selected so that Ψ_i in the PC expansion of $T^{(1)}$ refers to the same polynomial Ψ_k in the polynomial expansion for $T^{(2)}$. Obviously, $i = k$ when $T^{(1)}$ and $T^{(2)}$ have the same number of KL modes, N.

We have first compared modal solutions obtained with intrusive spectral computations using the same order PC expansion but different number of KL modes. In this case, the error measure is only relevant for the modes that are shared in both representations, namely those belonging to the expansion having lower N value. A sample of this exercise is shown in Fig. 6.38, which shows the L2 norm between

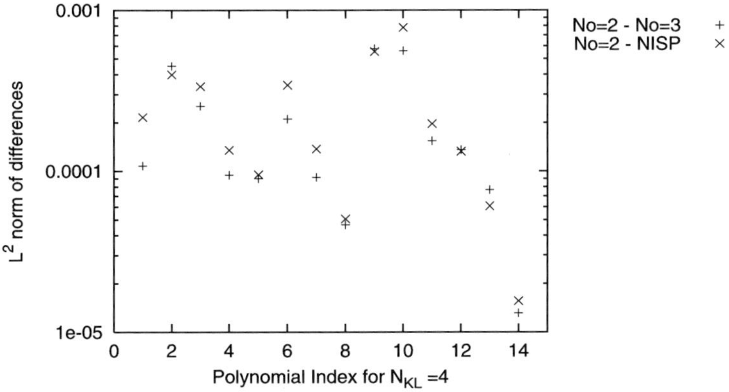

Fig. 6.39 L2 norms of differences in temperature modes obtained with intrusive spectral predictions using second- and third-order PC expansions (+), and between predictions obtained the second-order spectral and second-order NISP/GH predictions (×). In all cases, N = 4. Adapted from [128]

temperature modes obtained using second-order expansions with N = 4 and 6. As is evident in the figure, the L2 errors between the modal solutions are very small, indicating a very good agreement between the predictions. The same analysis was repeated with third-order PC expansions (not shown) and revealed similar trends. This further supports earlier claims that for the present conditions N = 4 is sufficient for adequate representation of the stochastic boundary conditions.

Figure 6.39 shows the L2 norm of the differences between the second-order and third-order intrusive predictions, and between second-order intrusive and second-order NISP/GH results. In all cases, we use N = 4 and L2 norms are shown for all 14 modes in the second-order PC expansion. The results indicate that for all modes the L2 norms are small, with magnitudes following below 10^{-3}. In addition, the differences between second-order NISP and intrusive predictions are comparable to corresponding deviations obtained using intrusive spectral computations with No = 2, and No = 3. Thus, the deviations between the NISP/GH and intrusive spectral predictions are of the same order as the spectral truncation errors in the latter approach.

6.2.6.2 Latin Hypercube Sampling

As mentioned earlier, a Latin hypercube sampling (LHS) approach is also applied in order to determine PC mode distributions. LHS is a stratified sampling technique where the random variable distributions are divided into equal probability intervals, and events are formed by randomly selecting variables within each of these intervals [153]. LHS typically requires fewer samples than simple pseudo-random sampling to reach the same degree of convergence, and a uniform sampling of phase space is assured within the limits of the sample size. In the computations,

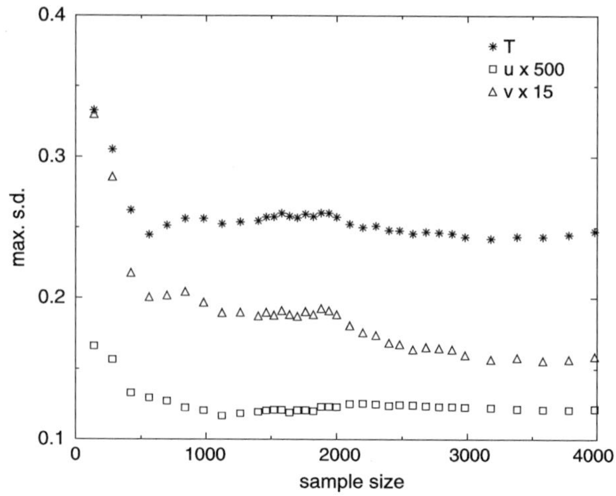

Fig. 6.40 Maximum standard deviation of temperature, u-velocity, and v-velocity over the computational domain plotted versus the sample size. Note the velocity standard deviations are scaled, as indicated in the legend. Adapted from [128]

the DAKOTA toolkit [62, 63, 242] is used to generate the necessary samples of the uncorrelated Gaussian variables appearing in the KL expansion. Individual realizations are then projected onto the PC basis in order to determine the mode distributions.

NISP/LHS computations are performed for a case with N $= 6$, $L_c = 1$, and $\sigma_T = 0.25$. The sampling tools in DAKOTA were used to generate a six-dimensional array of uncorrelated normalized Gaussians. The convergence of the mode amplitudes and the mean Nusselt number were monitored as the number of realizations increased. An example of the convergence diagnostics is given in Fig. 6.40, which shows the maximum standard deviation for temperature and velocity in the entire domain as a function of the sample size. For the present set of conditions, a sample of size of 4000 was deemed sufficient for the analysis, even though statistics are evidently not fully converged as can be appreciated from the figure.

In the following, we discuss results obtained from NISP/LHS computations in light of the above NISP/GH results and the earlier "intrusive" spectral results. The spatial distributions of PC modes of order ≤ 2 obtained using NISP/LHS (not shown) were first compared with corresponding predictions obtained with second- and third-order intrusive computations. The comparison reveals an excellent agreement for the mean and first-order modes but noticeable quantitative and qualitative differences do occur in the second-order modes. We briefly illustrate these differences by plotting in Fig. 6.41 the L2-norm of the differences between (i) the NISP/LHS results and the second-order intrusive predictions, and (ii) the NISP/LHS results and third-order intrusive predictions; the L2-norm of differences between second- and third-order spectral predictions are also shown for comparison. As observed earlier, the second- and third-order predictions are in excellent agreement with each other, with L2-norms falling below 10^{-3}. The differences between the NISP/LHS and spectral predictions are also small, but the corresponding L2-norms are about an order of magnitude larger than those of differences between spectral predictions. It can also be observed that the L2 norms of differences be-

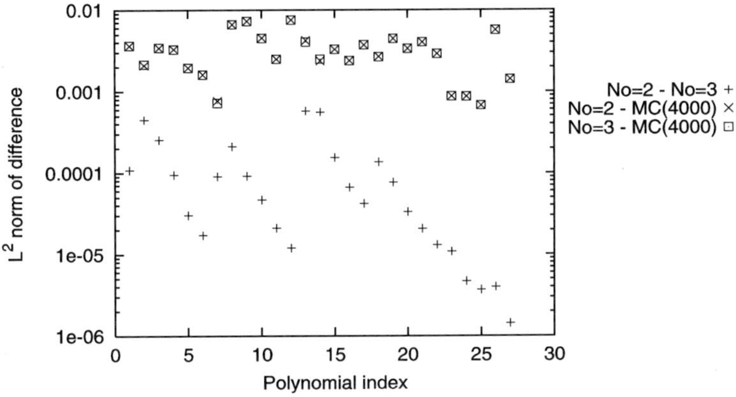

Fig. 6.41 L2 norm of differences in temperature modes obtained with intrusive spectral predictions using second- and third-order PC expansions (+), intrusive second-order and NISP/LHS with 4000 realizations (×), and intrusive third-order and NISP/LHS with 4000 realizations (□). In all cases, N = 6, and the comparison is restricted to second-order modes. Adapted from [128]

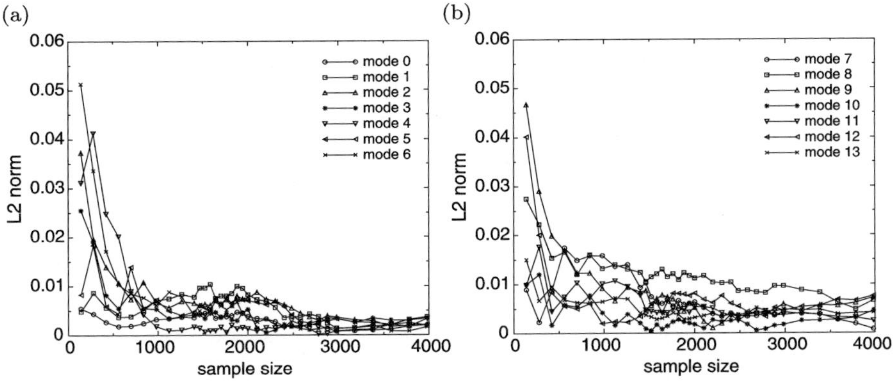

Fig. 6.42 L2 norm of differences in temperature modes obtained with second order intrusive and NISP/LHS predictions for different sample size: (**a**) modes 0–6, (**b**) modes 7–13. In both approaches, N = 6, $L_c = 1$, and $\sigma_T = 0.25$. Adapted from [128]

tween the NISP/LHS and intrusive predictions are nearly the same for both second-order and third-order spectral expansions. This indicates that the differences between NISP/LHS and spectral results are strongly affected by the sampling errors in the NISP/LHS approach and that, although still small, these errors are substantially larger than spectral truncation errors.

Additional insight into the convergence of the NISP/LHS computations can be gained from Fig. 6.42, which shows the L2-norm of the differences in mode distributions between the NISP/LHS and second-order intrusive results, as a function of the sample size. Plotted in Fig. 6.42a are L2-norms for the mean and first order modes; results for modes 7–13 are shown in Fig. 6.42b. Generally, the difference between

NISP/LHS and spectral predictions diminishes quickly, but a residual difference remains for all modes as the sample size increases. The difference decays quicker for the mean and the first-order modes (Fig. 6.42a), than for modes 7–13 (Fig. 6.42b). As can be observed in Fig. 6.41, the differences between NISP/LHS and intrusive spectral predictions are such that L2 norms corresponding to the mean and first-order modes are comparable to or smaller than those corresponding to some of the second-order modes. Since the latter are significantly weaker than the former, this indicates that the NISP/LHS predictions of the higher-order modes have large relative errors and are not well converged. This also shows that the sampling errors in NISP/LHS are behind the observed differences in the distributions of second-order modes.

6.2.7 Uncertainty Analysis

We conclude this study with a quantitative analysis of the effects of the stochastic boundary conditions on heat transfer statistics within the cavity. We rely on spectral computations using N = 6, No = 2 and a 140 × 100 computational grid. Results are obtained for three different correlations lengths and standard deviations, namely $L_c = 0.5$, 1 and 2, and $\sigma_T = 0.125$, 0.25 and 0.5.

Computed values of $\overline{\mathrm{Nu}}$ and $\sigma(\mathrm{Nu})$ are reported in Tables 6.3 and 6.4, respectively. Table 6.3 provides the mean Nusselt number along with the difference $\overline{\mathrm{Nu}} - \mathrm{Nu}^0$, where Nu^0 denotes the Nusselt number corresponding to the deterministic prediction with $T_c = -1/2$. The results show that $\overline{\mathrm{Nu}}$ is larger than Nu^0. For fixed correlation length, $\overline{\mathrm{Nu}} - \mathrm{Nu}^0$ increases approximately as σ_T^2. In contrast, $\overline{\mathrm{Nu}}$ exhibits a weaker dependence on L_c. This is not surprising since, in the range considered, the eigenvalues λ_i of KL modes vary slowly with the correlation length.

Unlike $\overline{\mathrm{Nu}}$, for fixed L_c the standard deviation $\sigma(\mathrm{Nu})$ exhibits an approximately linear dependence on σ_T, as shown in Table 6.4. Furthermore, compared with the mean, $\sigma(\mathrm{Nu})$ exhibits a more pronounced dependence on L_c. This trend is consistent with variations of the KL mode amplitudes with the correlation length. As L_c

Table 6.3 Mean Nusselt number for different values of L_c and σ_T. Spectral results with N = 6 and No = 2 are used. Adapted from [128]

$\overline{\mathrm{Nu}}$	$\sigma_T = 0.125$	$\sigma_T = 0.25$	$\sigma_T = 0.5$
$L_c = 0.5$	8.902	8.967	9.228
$L_c = 1$	8.904	8.974	9.268
$L_c = 2$	8.905	8.977	9.293
$\overline{\mathrm{Nu}} - \mathrm{Nu}^0$	$\sigma_T = 0.125$	$\sigma_T = 0.25$	$\sigma_T = 0.5$
$L_c = 0.5$	0.021	0.086	0.347
$L_c = 1$	0.023	0.093	0.387
$L_c = 2$	0.024	0.096	0.412

Table 6.4 Standard deviation of the Nusselt number for different values of L_c and σ_T. Spectral results with N = 6 and No = 2 are used. Adapted from [128]

$\sigma(\mathrm{Nu})$	$\sigma_T = 0.125$	$\sigma_T = 0.25$	$\sigma_T = 0.5$
$L_c = 0.5$	1.097	2.186	4.334
$L_c = 1$	1.236	2.463	4.859
$L_c = 2$	1.322	2.634	5.178

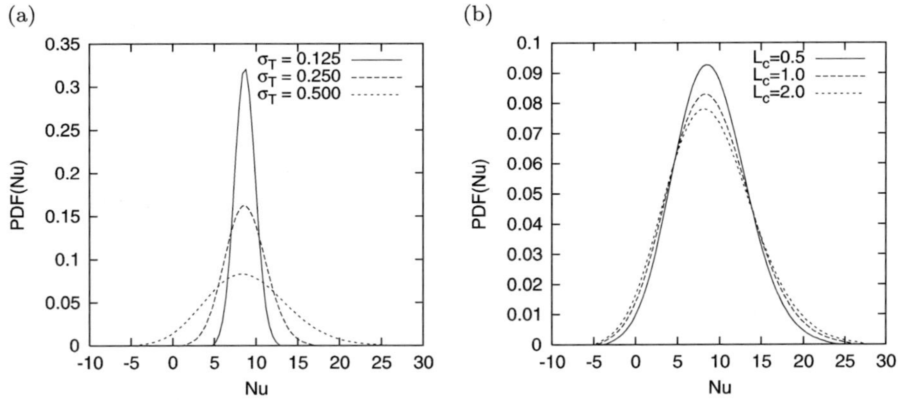

Fig. 6.43 Pdfs of the Nusselt number computed from the spectral simulations using N = 6 and No = 2: (**a**) $L_c = 1$ and $\sigma_T = 0.125$, 0.25 and 0.5; (**b**) $\sigma_T = 0.5$ and $L_c = 0.5$, 1 and 2. Adapted from [128]

increases, the magnitude of the first KL modes increases, and since these modes have a dominant impact on the uncertainty, so does $\sigma(\mathrm{Nu})$.

Figure 6.43 depicts pdf's of the Nusselt number computed from the spectral solution. Figure 6.43a shows that the most likely value of Nu is not significantly affected by σ_T, showing a slight decrease as σ_T increases. On the other hand, the skewness of the pdf increases with σ_T. In particular, for $\sigma_T = 0.5$, one observes a flatter tail at high Nu values than for the lower values. These trends are consistent with earlier results in Table 6.3, which show that $\overline{\mathrm{Nu}} - \mathrm{Nu}_0$ increases substantially as σ_T increases.

The effect of L_c on the pdf of the Nusselt number is depicted in Fig. 6.43b for fixed $\sigma_T = 0.5$. Consistent with the results of Table 6.4 the pdf becomes wider as L_c increases. Besides this trend, L_c appears to have a weak direct influence on the shape of the pdf.

Finally, we note that at $\sigma_T = 0.5$, the pdf can extend into the negative Nu range. This indicates that in extreme situations, the "mean" temperature on the right vertical wall may exceed the constant value on the left vertical wall, leading to a reversal of the circulation within the cavity and in the wall heat transfer. While such extremes have low probability and consequently a small contribution to low-order statistics, they demonstrate the capability of the present method of treating situations with large uncertainty. To illustrate these large changes, we plot in Fig. 6.44 the velocity profiles across the cavity corresponding for fixed $L_c = 1$ and three different stan-

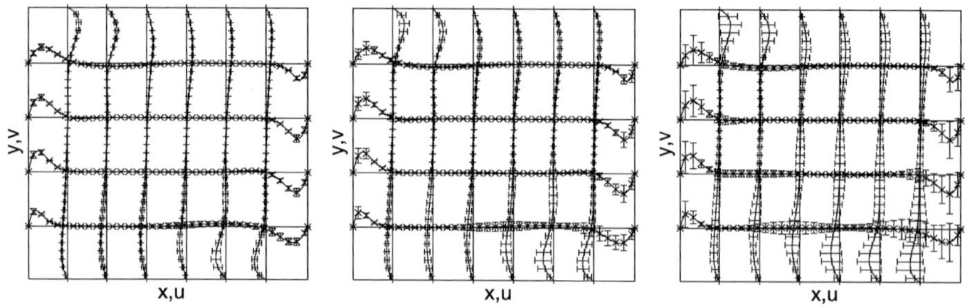

Fig. 6.44 Mean velocity profiles across the cavity for $\sigma_T = 0.125$ (*left*), 0.25 (*center*) and 0.5 (*right*). The error-bars correspond to 6 times the local standard deviation. The same scaling is used for all three *plots*. Spectral results with $L_c = 1$, N = 6, and No = 2 are used. Adapted from [128]

dard deviations, $\sigma_T = 0.125$, 0.25 and 0.5. The length of the "uncertainty" bars is proportional to six times the local standard deviation. Clearly, the uncertainty bars increase as σ_T increases. In particular, for $\sigma_T = 0.5$ the uncertainty bars suggest that events with upward velocity near the cold wall become probable. In contrast, one observes that the mean flow field is not strongly affected by σ_T.

6.3 Low-Mach Number Solver

In this section, we generalize the Boussinesq scheme above to stochastic zero-Mach-number solver. To illustrate the development, we focus on the same physical setup introduced in the previous section, but the assumption of weak temperature differences is no longer invoked. A zero-Mach-number model [116, 132, 140, 164] is introduced in Sect. 6.3.1, and later used in Sect. 6.3.2 to construct a stochastic projection solver. In Sect. 6.3.3, a brief validation study of the stochastic solver is performed, based on comparing the resulting predictions with available results from the literature. The scheme is then applied in Sect. 6.3.4 to analyze the behavior of steady-state heat transfer and of the velocity and temperature fields within the cavity under stochastic, non-Boussinesq conditions.

6.3.1 Zero-Mach-Number Model

In the zero-Mach-number limit [116, 132, 140, 164], the action of acoustic waves is ignored and the pressure is decomposed into a hydrodynamic component $\Pi(\mathbf{x}, t)$ and a spatially uniform thermodynamic component $\overline{P}(t)$. Following these assumptions, we will thus be concerned with the numerical solutions of the following set of normalized governing equations:

$$\frac{\partial \rho}{\partial t} = -\nabla \cdot \rho \mathbf{u}, \tag{6.76}$$

$$\frac{\partial(\rho u)}{\partial t} = -\frac{\partial(\rho u^2)}{\partial x} - \frac{\partial(\rho uv)}{\partial y} - \frac{\partial \Pi}{\partial x} + \frac{1}{\sqrt{\mathrm{Ra}}}\Phi_x, \tag{6.77}$$

$$\frac{\partial(\rho v)}{\partial t} = -\frac{\partial(\rho uv)}{\partial x} - \frac{\partial(\rho v^2)}{\partial y} - \frac{\partial \Pi}{\partial y} + \frac{1}{\sqrt{\mathrm{Ra}}}\Phi_y - \frac{1}{\mathrm{Pr}}\frac{\rho - 1}{2\epsilon}, \tag{6.78}$$

$$\frac{\partial T}{\partial t} = -\mathbf{u}\cdot\nabla T + \frac{1}{\rho \mathrm{Pr}\sqrt{\mathrm{Ra}}}\nabla\cdot(\kappa\nabla T) + \frac{\gamma - 1}{\rho\gamma}\frac{\mathrm{d}\overline{P}}{\mathrm{d}t}, \tag{6.79}$$

$$\overline{P} = \rho T \tag{6.80}$$

where ρ is the density, $\mathbf{u} = (u, v)$ is the velocity field, Π is the hydrodynamic pressure, $\mathrm{Ra} = \tilde{g}\beta\Delta\tilde{T}\tilde{L}^3/\tilde{\nu}_0\tilde{\alpha}_0$ is the Rayleigh number, β is thermal expansion coefficient, $\tilde{g}$ is the gravitational acceleration, Φ_x and Φ_y are the viscous stress terms in the x and y directions, respectively, $\mathrm{Pr} = \tilde{\nu}_0/\tilde{\alpha}_0$ is the Prandtl number (here equal to 0.7), $\kappa \equiv \tilde{\kappa}/\tilde{\kappa}_0$ is the normalized thermal conductivity, γ is the specific heat ratio, and $\overline{P}(t)$ is the thermodynamic pressure [132]. Tildes are used to denote dimensional quantities and the subscript 0 is used to denote reference quantities. Variables are normalized with respect to the appropriate combination of the reference temperature $\tilde{T}_0$, density $\tilde{\rho}_0$, length $\tilde{L}$, and velocity $\tilde{V}_0 \equiv \tilde{\nu}_0\sqrt{\mathrm{Ra}}/\tilde{L}$. The reference temperature $\tilde{T}_0$ is defined as $\tilde{T}_0 \equiv (\tilde{T}_h + \tilde{T}_c)/2$, where $\tilde{T}_h$ is the hot wall temperature, and $\tilde{T}_c$ is the (average) cold wall temperature. We also define the ratio $\epsilon \equiv \Delta\tilde{T}/2\tilde{T}_0$, where $\Delta\tilde{T} \equiv \tilde{T}_h - \tilde{T}_c$, and note that the Boussinesq limit is recovered as $\epsilon \to 0$.

The viscous stress terms are given by the divergence of the viscous stress tensor,

$$\tau = \frac{\mu}{2}\left[(\nabla\times\mathbf{u}) + (\nabla\times\mathbf{u})^T\right] \tag{6.81}$$

where $\mu \equiv \tilde{\mu}/\tilde{\mu}_0$ is the normalized viscosity. In the computations, the viscosity and thermal conductivity are assumed to be constant, or to depend on temperature according to the Sutherland law [132]:

$$\kappa(T) = T^{3/2}\frac{1 + S_\kappa}{T + S_\kappa}, \qquad \mu(T) = T^{3/2}\frac{1 + S_\mu}{T + S_\mu} \tag{6.82}$$

where $S_\kappa = 0.648$ and $S_\mu = 0.368$. For brevity, most of the results presented below correspond to the constant property case.

The above system of equations is supplemented with no-slip boundary conditions on velocity, while adiabatic conditions ($\partial T/\partial y = 0$) are used on horizontal walls. For the left and right vertical boundaries, Dirichlet conditions on temperature are used; in the deterministic case, we set $T = 1 + \epsilon$ and $T = 1 - \epsilon$ at the left and right vertical walls, respectively.

As in the previous section, we also consider the effect of "random" fluctuations on the cold wall, and assume that the random component is a Gaussian process with an auto-correlation function, $\mathcal{K}(y_1, y_2) \equiv \langle T'(y_1)T'(y_2)\rangle = \epsilon^2\sigma_T^2\exp[-|y_1 - y_2|/L_c]$, where L_c is the normalized correlation length and σ_T^2 is the normalized variance. Similar to Sect. 6.2, the normalized wall temperature at $x = 1$ is expressed in terms of its truncated KL expansion:

$$T(x=1,y) = T_c + T'(y) = (1-\epsilon) + \epsilon \sum_{i=1}^{\mathrm{N}} \xi_i \sqrt{\lambda_i} u_i(y), \tag{6.83}$$

where the eigenfunctions $u_i(y)$ are given by (2.22), $\lambda_n = 2L_c\sigma_T^2/[1+(\omega_n L_c)^2]$ and ω_n are the (ordered) positive roots of (2.24).

6.3.2 Solution Method

One fundamental difference between the present situation and that in Sect. 6.2 arises due to the fact that higher-order nonlinearities arise in the present set of governing equations whereas only second-order nonlinearities were present earlier. Additional means are thus required to estimate the moments corresponding to these higher-order nonlinearities. An additional, more delicate, complication is that these higher-order nonlinearities appear in stochastic divergence constraints. As shown below, this necessitates the introduction of a specially tailored procedure in order to ensure that the corresponding solvability constraints are exactly satisfied.

6.3.2.1 Stochastic System

For the purpose of computational convenience, the governing equations are first recast as [116, 164]:

$$\frac{\partial \rho}{\partial t} = \frac{1}{\gamma T}\frac{\partial \overline{P}}{\partial t} + \frac{1}{T}\left(\rho \mathbf{u}\cdot\nabla T - \frac{1}{\mathrm{Pr}\sqrt{\mathrm{Ra}}}\nabla\cdot(\kappa\nabla T)\right), \tag{6.84}$$

$$\frac{\partial \overline{P}}{\partial t} = -\gamma \frac{\int_\Omega \frac{1}{T}(\rho\mathbf{u}\cdot\nabla T - \frac{1}{\mathrm{Pr}\sqrt{\mathrm{Ra}}}\nabla\cdot(\kappa\nabla T))\,\mathrm{d}\Omega}{\int_\Omega \frac{1}{T}\,\mathrm{d}\Omega}, \tag{6.85}$$

$$\frac{\partial(\rho u)}{\partial t} = -\frac{\partial(\rho u^2)}{\partial x} - \frac{\partial(\rho u v)}{\partial y} - \frac{\partial \Pi}{\partial x} + \frac{1}{\sqrt{\mathrm{Ra}}}\Phi_x, \tag{6.86}$$

$$\frac{\partial(\rho v)}{\partial t} = -\frac{\partial(\rho u v)}{\partial x} - \frac{\partial(\rho v^2)}{\partial y} - \frac{\partial \Pi}{\partial y} + \frac{1}{\sqrt{\mathrm{Ra}}}\Phi_y - \frac{1}{\mathrm{Pr}}\frac{\rho-1}{2\epsilon}, \tag{6.87}$$

$$T = \frac{\overline{P}}{\rho}. \tag{6.88}$$

As in [162], the evolution equation for the thermodynamic pressure has been derived by enforcing global mass conservation over the whole domain.

We rely on PC expansions of the thermodynamic pressure and all field variables, and governing equations for the unknown expansion coefficients are obtained using a Galerkin approach. Taking advantage of the orthogonality of the PC basis, we formally obtain:

$$\frac{\partial \rho_k}{\partial t} = \mathcal{H}_k \tag{6.89}$$

$$\frac{\partial \overline{P}_k}{\partial t} = \mathcal{G}_k, \tag{6.90}$$

$$\frac{\partial (\rho u)_k}{\partial t} = \mathcal{X}_k - \frac{\partial \Pi_k}{\partial x}, \tag{6.91}$$

$$\frac{\partial (\rho v)_k}{\partial t} = \mathcal{Y}_k - \frac{\partial \Pi_k}{\partial y}, \tag{6.92}$$

$$T_k = \left(\frac{\overline{P}}{\rho}\right)_k, \tag{6.93}$$

where

$$\mathcal{H}_k \equiv \frac{1}{\gamma}\left(\frac{1}{T}\frac{\partial \overline{P}}{\partial t}\right)_k + \left(\frac{1}{T}\left[\rho \mathbf{u}\cdot\nabla T - \frac{1}{\mathrm{Pr}\sqrt{\mathrm{Ra}}}\nabla\cdot(\kappa\nabla T)\right]\right)_k, \tag{6.94}$$

$$\mathcal{G}_k \equiv -\gamma\left[\frac{\int_\Omega \frac{1}{T}(\rho\mathbf{u}\cdot\nabla T - \frac{1}{\mathrm{Pr}\sqrt{\mathrm{Ra}}}\nabla\cdot(\kappa\nabla T))\,\mathrm{d}\Omega}{\int_\Omega \frac{1}{T}\,\mathrm{d}\Omega}\right]_k, \tag{6.95}$$

$$\mathcal{X}_k \equiv -\frac{\partial (\rho u^2)_k}{\partial x} - \frac{\partial (\rho u v)_k}{\partial y} + \frac{1}{\sqrt{\mathrm{Ra}}}(\Phi_x)_k, \tag{6.96}$$

$$\mathcal{Y}_k \equiv -\frac{\partial (\rho u v)_k}{\partial x} - \frac{\partial (\rho v^2)_k}{\partial y} + \frac{1}{\sqrt{\mathrm{Ra}}}(\Phi_y)_k - \frac{1}{\mathrm{Pr}}\frac{\rho_k - 1}{2\epsilon}, \tag{6.97}$$

and the subscript k refers to the mode index. The subscript notation denoting mode indices is used both for variables and expressions; in the latter case, we have

$$(\mathcal{E})_k \equiv \frac{\langle(\mathcal{E})\Psi_k\rangle}{\langle\Psi_k^2\rangle}$$

where $(\mathcal{E})$ denotes a generic expression.

6.3.2.2 Boundary Conditions

When random forcing is considered, the boundary conditions on the velocity and temperature modes, derived based on a Galerkin formalism, are expressed as:

$$\mathbf{u}_k = 0, \quad k = 0, \dots, \mathrm{P}\ \forall \mathbf{x} \in \partial\Omega, \tag{6.98}$$

$$\frac{\partial T_k}{\partial y} = 0, \quad k = 0, \ldots, \mathrm{P} \text{ for } y = 0, \text{ and } y = 1, \tag{6.99}$$

$$T_0(x = 0, y) = 1 + \epsilon, \qquad T_0(x = 1, y) = 1 - \epsilon, \tag{6.100}$$

$$T_k(x = 0, y) = 0, \qquad T_k(x = 1, y) = \epsilon\sqrt{\lambda_k} f_k(y) \quad \text{for } k = 1, \ldots, \mathrm{N}, \tag{6.101}$$

$$T_k(x = 0, y) = T_k(x = 1, y) = 0 \quad \text{for } k > \mathrm{N}. \tag{6.102}$$

Here $\Omega = [0, 1] \times [0, 1]$ denotes the computational domain, and $\partial\Omega$ is its boundary.

6.3.2.3 Solution Method

The solution scheme is adapted from the variable density projection method developed in [116, 164]. For spatial discretization, we rely on a uniform, Cartesian, staggered grid with N_x and N_y cells in the x and y directions respectively. Velocity components are specified at cell edges, while scalar variables are defined at cell centers. Second-order conservative centered differences are used to approximate spatial derivatives.

For the present setup, an explicit time integration scheme proves suitable. We use the second-order Adams-Bashforth scheme to update the density field and thermodynamic pressure, according to:

$$\rho_k^{n+1} = \rho_k^n + \Delta t \left(\frac{3}{2}\mathcal{H}_k^n - \frac{1}{2}\mathcal{H}_k^{n-1} \right), \quad k = 0, \ldots, \mathrm{P}, \tag{6.103}$$

$$\overline{P}_k^{n+1} = \overline{P}_k^n + \Delta t \left(\frac{3}{2}\mathcal{G}_k^n - \frac{1}{2}\mathcal{G}_k^{n-1} \right), \quad k = 0, \ldots, \mathrm{P} \tag{6.104}$$

where Δt denotes the time step and superscripts refer to the time level. Using the updated density field and thermodynamic pressure, the temperature at the new time level is obtained from the equation of state:

$$T_k^{n+1} = \left(\frac{\overline{P}}{\rho} \right)_k^{n+1}, \quad k = 0, \ldots, \mathrm{P}. \tag{6.105}$$

Next, we integrate the pressure-split momentum equation using:

$$(\rho u)_k^* = (\rho u)_k^n + \Delta t \left(\frac{3}{2}\mathcal{X}_k^n - \frac{1}{2}\mathcal{X}_k^{n-1} \right), \quad k = 0, \ldots, \mathrm{P}, \tag{6.106}$$

$$(\rho v)_k^* = (\rho v)_k^n + \Delta t \left(\frac{3}{2}\mathcal{Y}_k^n - \frac{1}{2}\mathcal{Y}_k^{n-1} \right), \quad k = 0, \ldots, \mathrm{P}. \tag{6.107}$$

As in the SPM outlined in Sects. 6.1 and 6.2 [123, 128], the pressure field is then obtained by inverting the following **decoupled** elliptic systems for the pressure modes:

$$\nabla^2 \Pi_k = \frac{1}{\Delta t} \left[\nabla \cdot (\rho \mathbf{u})_k^* + \left. \frac{\partial \rho_k}{\partial t} \right|^{n+1} \right], \quad k = 0, \ldots, \mathrm{P}, \tag{6.108}$$

with homogeneous boundary Neumann conditions on all the modes. In (6.108), the time derivative is obtained from the second-order difference:

$$\left.\frac{\partial \rho_k}{\partial t}\right|^{n+1} = \frac{3\rho_k^{n+1} - 4\rho_k^n + \rho_k^{n-1}}{\Delta t}, \quad k = 0, \ldots, \mathrm{P}. \tag{6.109}$$

A pressure correction step is then implemented in order to enforce local continuity constraints:

$$(\rho u)_k^{n+1} = (\rho u)_k^* - \Delta t \frac{\partial \Pi_k}{\partial x}, \quad k = 0, \ldots, \mathrm{P}, \tag{6.110}$$

$$(\rho v)_k^{n+1} = (\rho v)_k^* - \Delta t \frac{\partial \Pi_k}{\partial y}, \quad k = 0, \ldots, \mathrm{P}, \tag{6.111}$$

and the updated velocity field is finally obtained from:

$$u_k^{n+1} = \left(\frac{(\rho u)^{n+1}}{\rho^{n+1}}\right)_k, \quad k = 0, \ldots, \mathrm{P}, \tag{6.112}$$

$$v_k^{n+1} = \left(\frac{(\rho v)^{n+1}}{\rho^{n+1}}\right)_k, \quad k = 0, \ldots, \mathrm{P}. \tag{6.113}$$

6.3.2.4 Galerkin and Pseudo-spectral Evaluation of Nonlinear Terms

The above system involves various nonlinear combinations of stochastic quantities. The latter include quadratic and higher-order products, as well as inverse functions involving temperature or density.

For quadratic products, the Galerkin approach can be easily implemented in its true Galerkin form. For cubic and higher-order products, the Galerkin procedure becomes computationally cumbersome and inefficient. Instead, a pseudo-spectral approach is used for nonlinear expressions. For product expressions involving more than two stochastic quantities, we resort to repeated applications of "binary" Galerkin evaluations. For example, for the triple product $d = abc$, a two step approach is utilized, where we first compute $d' = ab$ using the Galerkin formula, and then apply the same formula to obtain $d = d'c$. This approach is immediately generalized to products involving an arbitrary number of stochastic quantities.

The only remaining nonlinear expressions involve inverse operations (of temperature or density). Unless otherwise noted, these expressions are approximated using a truncated Taylor series expansion around the mean of the stochastic quantity. For instance, the inverse $1/a$ of a stochastic quantity $a = \sum_{i=0}^{\mathrm{P}} a_i \Psi_i$ is approximated as:

$$\frac{1}{a} \simeq \frac{1}{a_0} - \frac{1}{a_0^2}(a - a_0) + \frac{1}{a_0^3}(a - a_0)^2 - \frac{1}{a_0^3}(a - a_0)^3 + \cdots, \tag{6.114}$$

where the higher-order exponentiations are evaluated using the pseudo-spectral approach just introduced.

Note that, as expected, the pseudo-spectral evaluation of stochastic quantities introduces aliasing errors. When the spectral representation is sufficiently resolved, however, these errors do not degrade the spectral convergence of the scheme. This claim may be verified by systematic refinement of the representation, as performed for instance in Sect. 6.2.

6.3.2.5 Pressure Solvability Constraints

In the development of the above stochastic solver, one unanticipated difficulty arose during the enforcement of stochastic divergence constraints associated with the pressure Poisson equation. As reflected in (6.108), the pressure modes are obtained as the solution of Poisson equations with homogeneous Neumann conditions. These equations are thus subject to the following solvability constraints:

$$\int_\Omega \frac{1}{\Delta t}\left[\nabla\cdot(\rho\mathbf{u})_k^* + \left.\frac{\partial\rho_k}{\partial t}\right|^{n+1}\right] d\Omega = 0. \tag{6.115}$$

For a closed domain with rigid boundaries, the divergence term appearing in the integral vanishes identically, both in the continuous limit and for the present staggered discretization. Thus, the only remaining concern is the annihilation of the second term, in other words exact enforcement of global mass conservation over the entire domain.

Unfortunately, the right hand side density evolution equation (6.89) involves complex combinations of stochastic quantities which, as mentioned earlier, are only approximately estimated. Consequently, without special care, the solvability constraints could only be approximately satisfied. Even though errors associated with solvability constraints were generally minute, these always led to the blow-up of the computations.

To overcome these difficulties, special care in the evaluation of the thermodynamic pressure source term was implemented. The procedure is based on rewriting (6.95) as:

$$\mathcal{T}\mathcal{G} = \mathcal{S} \tag{6.116}$$

where $\mathcal{G}$ refers to the stochastic quantity with coefficients $(\mathcal{G}_1, \dots, \mathcal{G}_k)$,

$$\mathcal{T}_k \equiv \int_\Omega \left[\frac{1}{T}\right]_k d\Omega, \tag{6.117}$$

and

$$\mathcal{S}_k = \int_\Omega \left[\frac{1}{T}\left(\rho\mathbf{u}\cdot\nabla T - \frac{1}{\mathrm{Pr}\sqrt{\mathrm{Ra}}}\nabla\cdot(\kappa\nabla T)\right)\right]_k d\Omega. \tag{6.118}$$

Next, instead of using the approximate Taylor series approach (6.114) to determine $\mathcal{G}$, the latter is obtained by inverting the linear system:

$$\sum_{i=0}^{\mathrm{P}}\sum_{j=0}^{\mathrm{P}} C_{ijk}\mathcal{T}_i\mathcal{G}_j = \mathcal{S}_k, \quad k = 0, \dots, \mathrm{P}, \tag{6.119}$$

which may be alternatively written as:

$$\mathbf{A}\mathcal{G} = \mathcal{S}, \tag{6.120}$$

where $\mathbf{A}$ is the matrix given by:

$$A_{ij} = \sum_{k=0}^{\mathrm{P}} C_{ijk}\mathcal{T}_k, \quad 0 \le i, j \le \mathrm{P}. \tag{6.121}$$

It is readily verified that when the $\mathcal{G}_k$'s are obtained as the solution of the above equation the integral constraints in (6.115) are exactly satisfied. (To this end, it is sufficient to note that the above inverse procedure is an exact discrete de-convolution of the Galerkin product.) Furthermore, with the resulting scheme, stable numerical solutions are obtained. Thus, the present approach provides a simple and effective means for obtaining a pressure solution that ensures that local divergence constraints are satisfied.

6.3.3 Validation

6.3.3.1 Boussinesq Limit

As a first validation, the results obtained with the spectral solver are compared with previous stochastic spectral computations reported in [128] for the Boussinesq limit. In all cases, the computations are carried out starting from a fluid at rest, with uniform temperature ($T = 1$) and pressure ($P = 1$). The unsteady system of equations is integrated in time till steady conditions are reached. For clarity of the presentations, the quantities given in this subsection are re-scaled consistently with the Boussinesq normalization.

Zero-order spectral expansion: We set $\epsilon = 0.001$, and use the temperature independent properties ($\kappa = \mu = 1$). Then the grid convergence of the solution is analyzed for increasingly refined spatial discretization, and zero-order spectral expansion (i.e. the deterministic problem). Results are reported in Table 6.5, for $\mathrm{Ra} = 10^6$, in terms of wall-averaged Nusselt number together with its minimal and maximal

Table 6.5 Mean, minimum and maximum Nusselt numbers for deterministic computations with $\epsilon = 0.001$ and Ra $= 10^6$. Adapted from [127]

$N_x \times N_y$	40×40	80×80	120×120	160×160	Spectral [130]
Nu_{av}	9.426	8.982	8.895	8.865	8.825
Nu_{min}	0.939	0.971	0.976	0.978	–
Nu_{max}	20.50	18.74	18.09	17.85	–

Table 6.6 Deterministic predictions of mean, minimum and maximum Nusselt numbers for Ra $= 10^5$. Results are compared with reported Boussinesq predictions. Adapted from [127]

	Model	Nu_{av}	Nu_{min}	Nu_{max}
De Vahl Davis [49]	Boussinesq	4.519	0.729	7.717
Le Quéré *et al.* [131]	Boussinesq	4.523	0.728	7.720
Chenoweth *et al.* [27]	Boussinesq	4.520	–	–
Hortmann *et al.* [103]	Boussinesq	4.522	–	7.720
Paillere *et al.* [177] (80×80)	zero-Mach ($\epsilon = 0.01$)	4.523	0.738	7.68
Present (80×80)	zero-Mach ($\epsilon = 0.001$)	4.547	0.726	7.840

values along the vertical walls, respectively:

$$\begin{aligned} \mathrm{Nu}_{av} &= \frac{1}{2\epsilon} \int_0^1 -\kappa \frac{\partial T(0, y)}{\partial x} \, \mathrm{d}y, \\ \mathrm{Nu}_{min,max} &= \frac{1}{2\epsilon} \min, \max \left(-\kappa \frac{\partial T(0, y)}{\partial x} \right). \end{aligned} \tag{6.122}$$

Note that these quantities are computed along the hot wall; for the present small value of ϵ, the solution is nearly symmetrical with respect to the central vertical plane. In fact, the agreement between hot and cold walls values is within 0.1% for the minima and maxima of the Nusselt, while the wall-averaged Nusselt numbers on both walls are equal (for any ϵ) at steady state. Table 6.5 shows that as the grid is refined, the present finite difference results rapidly approach the spectral results of Le Quéré [130]. Moreover, a detailed analysis of the solution (not shown) shows that the temperature and velocity fields are also in very good agreement with the results reported in [130]. In Table 6.6, we compare the computed values of the wall-averaged Nusselt number at Ra $= 10^5$, $\epsilon = 0.001$, obtained using a 80×80 grid, with some reference results based on the Boussinesq approximation. Again, a good agreement with reported results is observed.

Stochastic Boussinesq computations: The predictions of the stochastic zero-Mach code for small ϵ are now validated against the spectral Boussinesq computations presented in Sect. 6.2 [128]. Results are obtained with $L_c = 1$ and $\sigma_T = 0.5$. The random temperature fluctuations are represented using a KL expansion with N $= 4$, and a second-order PC expansion is used. Accordingly, P $= 14$ *i.e.* the PC expansion has 15 polynomials.

Table 6.7 Comparison of stochastic zero-Mach predictions for $\epsilon = 0.001$ with Boussinesq results from Sect. 6.2. Adapted from [127]

	N.B. 80×80	N.B. 140×100	Boussinesq 140×100
$\langle \mathrm{Nu}_{av} \rangle$	9.0794	8.9716	8.9729
$\sigma(\mathrm{Nu}_{av})$	2.4993	2.4602	2.4632

In order to compare the zero-Mach stochastic predictions with Boussinesq results, we choose a small value of the Boussinesq parameter, $\epsilon = 0.001$. The constant-property formulation is used, and the governing equations are time-integrated up to steady state, using 80×80 and 140×100 grids. As shown in Table 6.7, when the same spatial grid resolution is used, there is excellent agreement between the Boussinesq and zero-Mach predictions of the mean Nusselt number, $\langle \mathrm{Nu}_{av} \rangle$ and its standard deviation, $\sigma(\mathrm{Nu}_{av})$. Specifically, for a 140×100 grid, the differences between the zero-Mach and Boussinesq predictions are less than 0.15%, while differences between the zero-Mach predictions at different resolution levels are less than 1.6%. The higher value of $\langle \mathrm{Nu}_{av} \rangle$ for the lower grid resolution is not surprising, as the convergence analysis provided above (Table 6.5) has shown an increasing over-estimation of Nu_{av} with decreasing grid resolution for the deterministic case. In addition to the comparison in Table 6.7, detailed analysis (not shown) of the individual modes in the spectral expansion also reveals excellent agreement between zero-Mach and Boussinesq predictions. In particular, when the same grid resolution is used, the relative difference between corresponding second-order modes in the Boussinesq and zero-Mach solutions is everywhere less than 1%. Thus, at small ϵ, close agreement between the zero-Mach and Boussinesq predictions is observed both for integral quantities and local field values.

6.3.3.2 Non-Boussinesq Regime

In this section we examine the behavior of the zero-Mach code in the non-Boussinesq regime, i.e. for moderate values of ϵ. Since, to our knowledge, no results are available for the stochastic problem, the analysis is limited to the deterministic case. The effects of the Boussinesq parameter ϵ on the flow statistics in the stochastic case are examined below.

We start with a brief examination of the effects of spatial resolution on steady-state predictions. To this end, we set $\epsilon = 0.6$, $\mathrm{Ra} = 10^6$ and rely on the variable-property model. In Table 6.8, the steady-state Nusselt number is computed with spatial resolutions, corresponding to grids with 40×40, 80×80, 120×120, and 160×160 cells. Also shown are the maximum and minimum values of Nu on the hot and cold walls. Note that for this large ϵ case, the local heat flux distributions on the hot and cold walls differ, which leads to different extrema. However, at steady state the wall-averaged Nusselt number is identical for both the hot and cold walls. The results of Table 6.8 also show that as the grid is refined, the predictions tend towards a fixed value. In particular, when the number of cells along each direction is larger or equal to 80, the predicted values of Nu_{av} vary by less than 1%.

Table 6.8 Mean, minimum and maximum Nusselt numbers on the hot and cold walls for deterministic predictions with $\epsilon = 0.6$, $\mathrm{Ra} = 10^6$ and a variable-property model. Adapted from [127]

$N_x \times N_y$	40×40	80×80	120×120	160×160
Nu_{av}	8.600	8.744	8.688	8.651
Nu_{min}-(hot/cold)	(0.987-2.037)	(1.057-0.663)	(1.064-0.677)	(1.064-0.691)
Nu_{max}-(hot/cold)	(23.86-12.48)	(21.81-14.77)	(21.00-15.38)	(20.70-15.48)

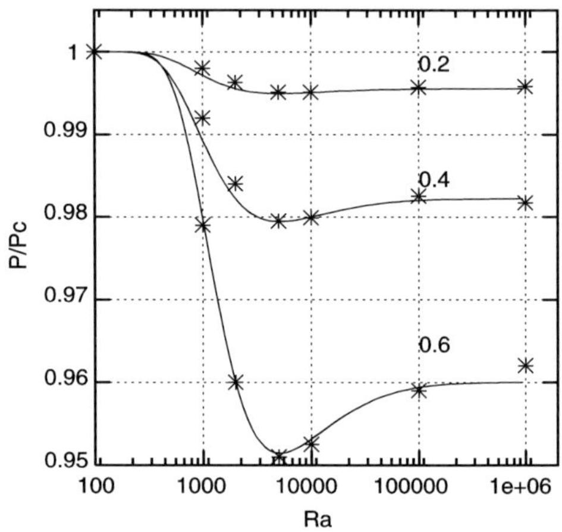

Fig. 6.45 $\overline{P}/\overline{P}_c$ (*symbols*) versus Ra for $\epsilon = 0.2,\ 0.4$ and 0.6. *Solid lines* reflect the analytical results of [27]. A computational grid with 80×80 cells is used. Adapted from [127]

In order to gain additional confidence in the computations, we contrast the predicted value of the steady-state thermodynamic pressure, $\overline{P}$, with the analytical predictions of Chenoweth & Paolucci [27]. This is a stringent test because the steady-state pressure is obtained by time integration of the unsteady pressure field (see (6.85)), and is thus potentially affected by the accumulation of time integration errors. Several investigations (e.g. [27, 132, 177]) have in fact pointed out to severe difficulties in computing the static pressure by direct integration, due to inaccurate time integration schemes and/or inconsistency of the overall scheme. For the present scheme, however, the steady $\overline{P}$ is accurately predicted, as illustrated in Fig. 6.45. The latter depicts curves of $\overline{P}(\epsilon)/\overline{P}_c(\epsilon)$, where $\overline{P}$ is the steady pressure and $\overline{P}_c$ is the pressure corresponding to a purely conductive solution [27]. Results are generated for $\epsilon = 0.2$, 0.4, 0.6 and Rayleigh numbers in the range $10^2 \le \mathrm{Ra} \le 10^6$. The simulations were performed on a 80×80 grid using the variable-property model. In each case, the numerical time-steps was selected so as to satisfy the stability constraints of the explicit time integration scheme. In the figure, the static pressures computed at $\mathrm{Ra} = 10^2$ were used to estimate $P_c(\epsilon)$, since they were found to be in excellent agreement with the analytical expressions of [27]. As shown in Fig. 6.45, an excellent agreement with the analytical results [27] is obtained, but small deviations are observed at $\mathrm{Ra} = 10^6$ and $\epsilon = 0.6$ where the grid may not be sufficiently refined.

6.3.4 Uncertainty Analysis

Following the brief validation study above, the stochastic zero-Mach-number code is applied in this section to analyze the effect of the Boussinesq parameter on the statistics of the stochastic cavity flow. We set $\mathrm{Ra} = 10^6$, $L_c = 1$, $\sigma_T = 0.5$, and consider four different values of ϵ, $\epsilon = 0.01$, 0.1, 0.2, and 0.3. We rely on previous experiences in Sect. 6.2 and restrict the computations to a second-order PC expansion, and perform computations on a grid with $N_x = 120$ and $N_y = 100$. For brevity, only the constant-property model is used.

6.3.4.1 Heat Transfer Characteristics

In order to compare the solutions for different values of the Boussinesq parameter, including the Boussinesq regime at very small ϵ, it is convenient to rescale the temperature modes according to:

$$T_0 \longrightarrow T_0^\epsilon \equiv 1 + \frac{T_0 - 1}{2\epsilon},$$

$$T_k \longrightarrow T_k^\epsilon \equiv \frac{T_k}{2\epsilon}, \quad k = 1, \ldots, \mathrm{P}.$$

Based on this normalization, the wall-averaged and local Nusselt numbers are given by:

$$\mathrm{Nu}_{av}(\boldsymbol{\xi}) = -\sum_{k=0}^{\mathrm{P}} \int_0^1 \frac{\partial T_k^\epsilon}{\partial x}(x, y) \Psi_k(\boldsymbol{\xi})\, \mathrm{d}y,$$

$$\mathrm{Nu}(y, \boldsymbol{\xi}) = -\sum_{k=0}^{\mathrm{P}} \frac{\partial T_k^\epsilon}{\partial x}(x, y) \Psi_k(\boldsymbol{\xi})$$

with $x = 0$ for the hot wall and $x = 1$ for the cold wall. As mentioned earlier, the local values of $\mathrm{Nu}(y)$ may differ on both walls, but the average values are equal at steady state.

In Table 6.9, the expected values and standard deviations of Nu_{av} and $\overline{P}$ at steady state are given for $\epsilon = 0.01$, 0.1, 0.2 and 0.3, for the spectral solutions obtained using first- and second-order PC expansions, $\mathrm{No} = 1$ and 2 respectively. The results show that $\langle \mathrm{Nu}_{av} \rangle$ and $\sigma(\mathrm{Nu}_{av})$ increase with ϵ. The results also show that the first-order and second-order PC expansions yield estimates of $\langle \mathrm{Nu}_{av} \rangle$ that are in close agreement, but differences in the corresponding standard deviations can be noted. For $\epsilon \le 0.1$, the first-order PC expansion slightly overestimates $\sigma(\mathrm{Nu}_{av})$, while the reverse trend occurs at higher ϵ. Table 6.9 also shows that the mean thermodynamic pressure and its standard deviation exhibit a nonlinear dependence on ϵ. The differences between first and second-order PC predictions are more pronounced than for the Nusselt number. In particular, for $\epsilon \ge 0.2$, a second-order PC expansion is found

Table 6.9 Expectations ($\langle . \rangle$) and standard deviations ($\sigma(.)$) of Nu and $\overline{P}$. Deterministic predictions of the mean Nusselt number are also reported in the right column. Adapted from [127]

	No = 1				No = 0
	$\langle \mathrm{Nu}_{av} \rangle$	$\sigma(\mathrm{Nu}_{av})$	$\langle \overline{P} \rangle$	$\sigma(\overline{P})$	Nu_{av}
$\epsilon = 0.01$	8.990	2.479	0.9999	0.0022	8.871
$\epsilon = 0.10$	9.018	2.531	0.9959	0.0232	8.872
$\epsilon = 0.20$	9.055	2.591	0.9833	0.0501	8.874
$\epsilon = 0.30$	9.103	2.653	0.9612	0.0819	8.880
	No = 2				No = 0
	$\langle \mathrm{Nu}_{av} \rangle$	$\sigma(\mathrm{Nu}_{av})$	$\langle \overline{P} \rangle$	$\sigma(\overline{P})$	Nu_{av}
$\epsilon = 0.01$	8.992	2.472	0.9999	0.0022	8.871
$\epsilon = 0.10$	9.019	2.529	0.9959	0.0232	8.872
$\epsilon = 0.20$	9.058	2.598	0.9832	0.0538	8.874
$\epsilon = 0.30$	9.108	2.676	0.9609	0.0829	8.880

Table 6.10 Local heat-flux statistics along the hot and cold walls. Results were obtained using a second-order PC expansion. Adapted from [127]

	Hot wall		Cold wall	
	$\min\langle \mathrm{Nu} \rangle$	$\max\langle \mathrm{Nu} \rangle$	$\min\langle \mathrm{Nu} \rangle$	$\max\langle \mathrm{Nu} \rangle$
$\epsilon = 0.01$	0.981	18.50	0.917	18.41
$\epsilon = 0.10$	0.999	18.82	0.890	18.30
$\epsilon = 0.20$	1.019	19.24	0.853	18.21
$\epsilon = 0.20$	1.041	19.76	0.803	18.17
	$\min[\sigma(\mathrm{Nu})]$	$\max[\sigma(\mathrm{Nu})]$	$\min[\sigma(\mathrm{Nu})]$	$\max[\sigma(\mathrm{Nu})]$
$\epsilon = 0.01$	0.244	5.559	0.805	6.358
$\epsilon = 0.10$	0.255	5.871	0.816	6.540
$\epsilon = 0.20$	0.268	6.302	0.834	6.814
$\epsilon = 0.20$	0.282	6.831	0.866	7.228

necessary for accurate prediction of the standard deviation of the thermodynamic pressure.

The results presented in Table 6.9 provide information on the effect of ϵ of the statistics of the overall heat transfer. We now turn our attention to the statistics of the local heat flux distributions on the hot and cold walls. Provided in Table 6.10 are the minima and maxima of the mean heat flux and its standard deviation for both the hot and cold walls. The results indicate that there are opposite trends with ϵ concerning the maxima and minima of $\langle \mathrm{Nu} \rangle$ on the two walls. Specifically, the minimum and maximum of values of $\langle \mathrm{Nu} \rangle$ increase with increasing ϵ on the hot wall, while they decrease with increasing ϵ on the cold wall. In addition, the effect

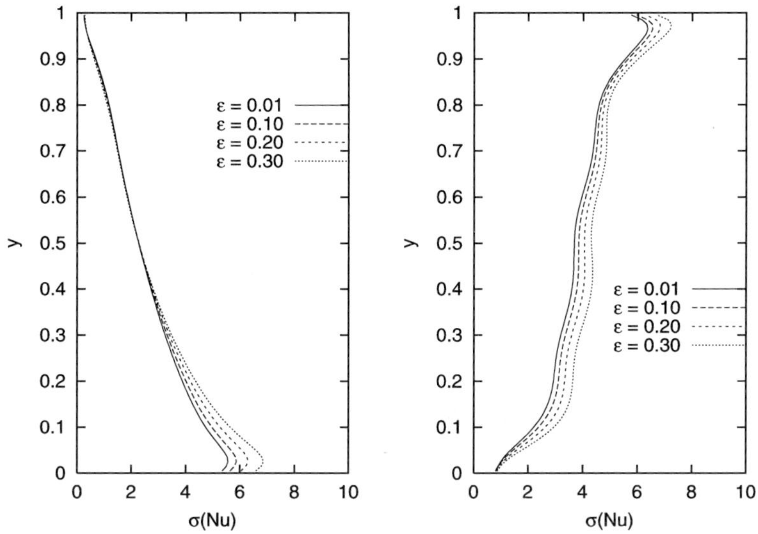

Fig. 6.46 Standard deviation of the local heat flux across the hot (*left plot*) and cold (*right plot*) walls. Results were obtained using a second-order PC with N = 4 and a constant-property model. Adapted from [127]

of ϵ on the maximum expected heat flux is more pronounced on the hot wall than it is on the cold wall. Meanwhile, the maximum and minimum values of $\sigma(\text{Nu})$ increase with ϵ on both the hot and cold walls. This trend is consistent with our previous observation that the standard deviation increases with ϵ. It is also interesting to note that, as shown in Fig. 6.46, the values of $\sigma(\text{Nu})$ are generally larger on the cold wall, where the uncertainty is applied, than on the hot wall. In addition, the figure shows that on the cold wall $\sigma(\text{Nu})$ is everywhere affected by ϵ, while on the hot wall the variations of $\sigma(\text{Nu})$ with ϵ are limited to the lower part of the wall.

6.3.4.2 Mean Fields

In this section we examine the dependence of the mean temperature and velocity fields within the cavity on ϵ. In Fig 6.47, the mean rescaled temperature fields T_0^ϵ are plotted for $\epsilon = 0.01, 0.1, 0.2$ and 0.3. Thanks to the scaling, in all cases $T_0^\epsilon = 1.5$ on the left(hot) wall, $T_0^\epsilon = 0.5$ on the right (cold) wall. Thus, direct comparison for different ϵ is possible.

Figure 6.47 shows that the impact of ϵ on the scaled temperature field is noticeable at the lower part of the and upper parts of the cavity. The effects are most pronounced around the upper left corner of the cavity near the hot wall and the lower right corner of the cavity near the cold wall. With increasing ϵ the fluid traveling at the bottom of the cavity has, on the mean, a lower scaled temperature; the same trend applies in the upper part of the cavity. In the core of the flow, the scaled temperature distributions exhibit similar shape for all ϵ, reflecting the classical patterns of

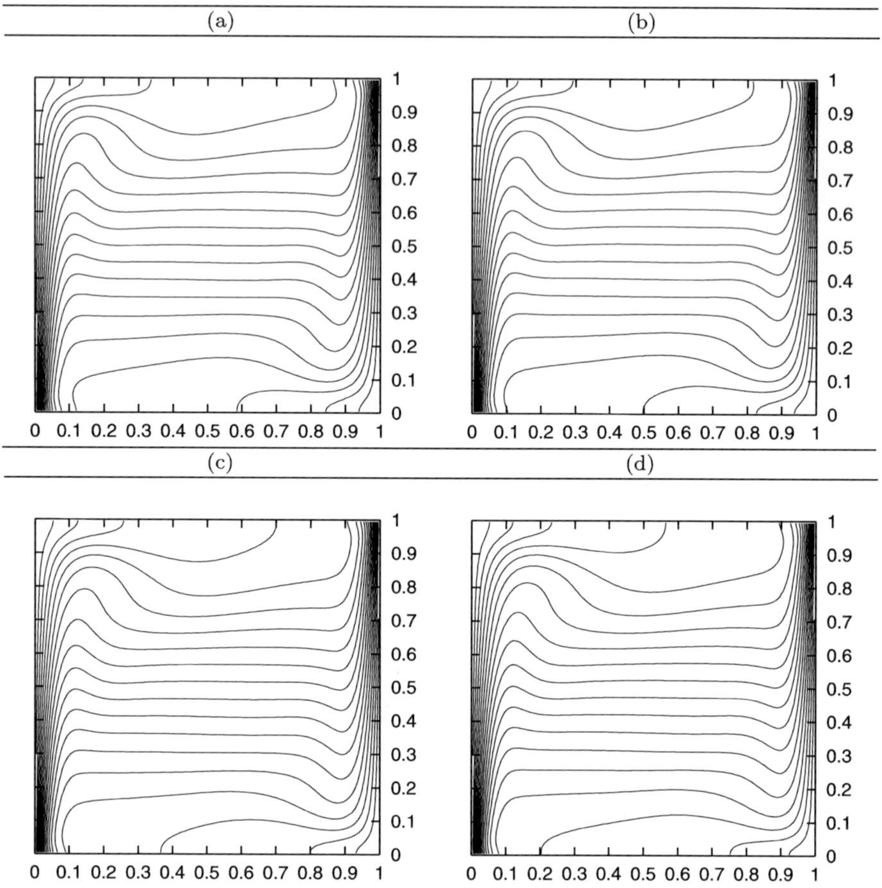

Fig. 6.47 Contours of the mean scaled temperature T_0^ϵ for: (**a**) $\epsilon = 0.01$, (**b**) $\epsilon = 0.1$, (**c**) $\epsilon = 0.2$, and (**d**) $\epsilon = 0.3$. Contours range from the maximum value ($T_0^\epsilon = 1.5$) for the hot wall ($x = 0$) to the minimum ($T_0^\epsilon = 0.5$) on the cold wall ($x = 1$) with increments 0.05. Adapted from [127]

natural convection in a square cavity. In this region, the mean thermal stratification ($\partial T_0^\epsilon / \partial y$) is weakly dependent on ϵ.

To further analyze the influence of ϵ on T_0^ϵ, in Fig. 6.48a we plot the difference between the two averaged fields computed for $\epsilon = 0.01$ and 0.3. From this plot, one can observe that on average the scaled temperature in the cavity is globally lower for $\epsilon = 0.3$ than for $\epsilon = 0.01$, except along the two vertical boundary layers where it is higher. The figure also reveals that large temperature differences, with amplitude as high as 5% of ΔT, occur in the neighborhood of the top left corner and bottom right corner. The difference in mean solutions should not be fully attributed to a different response to the temperature boundary condition uncertainty. To establish this claim, Fig. 6.48b shows the differences in scaled temperature fields, for the same values ϵ, between deterministic computations. The similarity between the two frames in

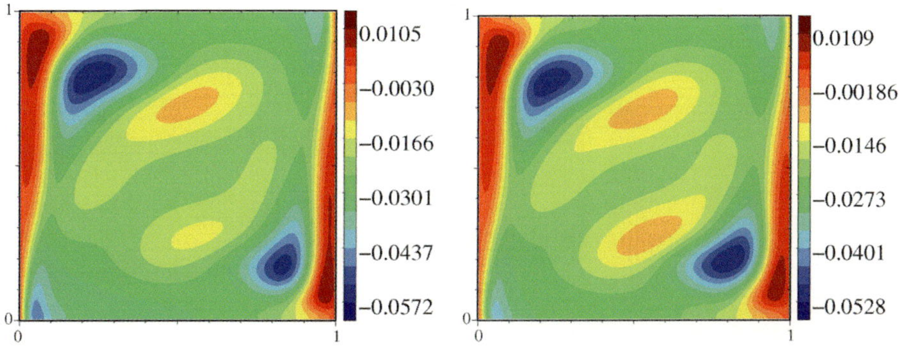

Fig. 6.48 Contours of differences in scaled stochastic temperature fields. *Left*: $\langle T^{\epsilon}(\epsilon = 0.3)\rangle - \langle T^{\epsilon}(\epsilon = 0.01)\rangle$; *Right*: $T^{\epsilon}(\epsilon = 0.3) - T^{\epsilon}(\epsilon = 0.01)$ for the deterministic problem. Adapted from [127]

Fig. 6.48 indicates that imposition of stochastic conditions result, on average, in only weak amplification of the differences that occur in the deterministic solution. Similar observations can be drawn by inspection of the mean and deterministic velocity fields plotted in Fig. 6.49. Note the strong spatial correlation of the differences between the solution for $\epsilon = 0.01$ and 0.3 for the scaled temperature and velocity fields.

It follows from the discussion above that the influence of ϵ on the mean solutions in the stochastic case is better understood when the corresponding deterministic, or zero order, solutions are first subtracted from the averaged fields. Results of such exercise are given in Fig. 6.50, based on second-order PC computations for $\epsilon = 0.3$ and $\epsilon = 0.01$. There are striking similarities between the distributions for both values of ϵ. Specifically, for both cases the differences in the velocity fields exhibit three recirculation zones, with alternated sign of circulation, along the lower wall. The strengths of these structures, which have been observed in the Boussinesq analysis, are amplified with increasing ϵ. A similar effect is also visible for the scaled temperature fields plotted in Fig. 6.50. With respect to the deterministic solution, the application of stochastic conditions results in lower mean temperature at the bottom of the cavity, with a maximum amplitudes of less than 1.8% for $\epsilon = 0.01$ and greater than 3% for $\epsilon = 0.3$. Thus, the impact of stochastic temperature fluctuations exhibits a nonlinear dependence on ε.

6.3.4.3 Standard Deviations

The dependence of the flow statistics on ϵ is briefly illustrated by analyzing the standard deviations of the scaled temperature fields. The latter are plotted in Fig. 6.51 for $\epsilon = 0.01$, 0.1, 0.2 and 0.3. As a result of this scaling, for all cases the (theoretical) value of $\sigma(T^{\epsilon})$, the standard deviation of T^{ϵ}, is 0 on the hot wall and 0.25 on the cold wall. One can observe in Fig. 6.51 that the standard deviations are not exactly equal to 0.25 on the cold wall but slightly lower. As discussed in Sect. 6.2, this small

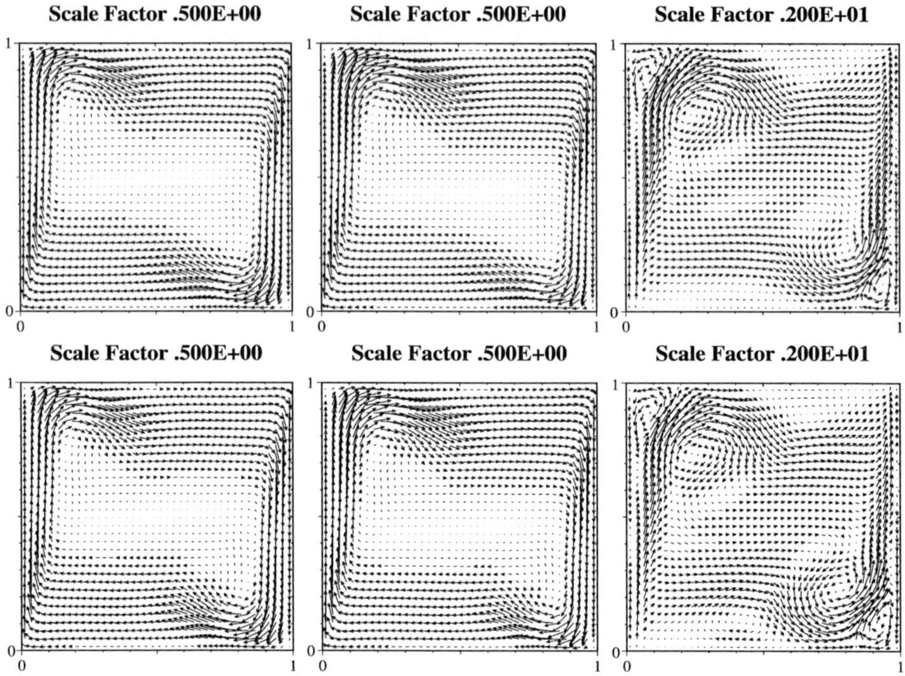

Fig. 6.49 *Top row*: Mean velocity field for $\epsilon = 0.01$ (*left*), $\epsilon = 0.3$ (*center*) and the corresponding difference field (*right*). *Bottom row*: Deterministic velocity field for $\epsilon = 0.01$ (*left*), $\epsilon = 0.3$ (*center*) and the corresponding difference field (*right*). Adapted from [127]

discrepancy is due to the truncation (N = 4) of the KL expansion. As observed earlier, the distribution of $\sigma(T^{\epsilon})$ exhibits a recirculation pattern that is similar to that of the mean temperature field.

To obtain a better appreciation of the dependence of $\sigma(T^{\epsilon})$ on ϵ, the difference between the standard deviation fields computed for $\epsilon = 0.01$ and $\epsilon = 0.3$ is plotted in Fig 6.52. The figure indicates that in the interior of the cavity, larger values of $\sigma(T^{\epsilon})$ occur for $\epsilon = 0.3$ than for $\epsilon = 0.01$. This trend does not hold along the vertical walls, where the impact of ϵ is much weaker. Also note that the differences in $\sigma(T^{\epsilon})$ between the solutions for $\epsilon = 0.3$ and $\epsilon = 0.01$ can be substantial, reaching approximately 14% of the imposed value along the cold wall.

6.3.5 Remarks

In this section, the SPM was extended to zero-Mach-number flows, based on a mass-conservative formulation which is exactly implemented using a specially-tailored stochastic inverse procedure. The approach results in *decoupled* mass divergence constraints, which are inverted using a fast Poisson solver. Thus, a stable and efficient stochastic SPM for zero-Mach-number flows is constructed.

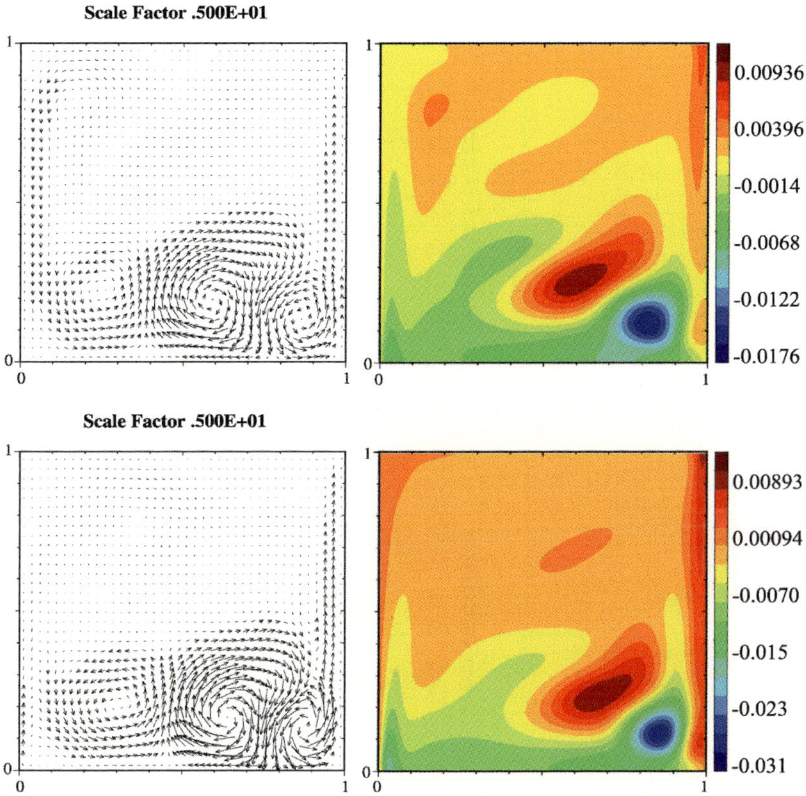

Fig. 6.50 Differences between mean and deterministic velocity (*left*) and temperature (*right*). *Top row*: $\epsilon = 0.01$; *bottom row*: $\epsilon = 0.3$. Adapted from [127]

One of the key aspects of the present construction is that the mass conservation constraints associated with the decoupled Neumann problems for the individual pressure modes are satisfied to machine precision. Numerical tests indicate that, when this is the case, a stable and accurate solution scheme is obtained. On the other hand, when mass conservation constraints are only approximately satisfied, the computations are unstable. Thus, suitable treatment of the mass conservation constraints is an essential item in the extension of Boussinesq solvers to variable-density zero-Mach-number flows.

6.4 Stochastic Galerkin Projection for Particle Methods

The principal objective of this section is to explore the application of PC methods in conjunction with Lagrangian particle approximations of the Navier-Stokes equations. As discussed in the extensive exposition of Cottet and Koumoutsakos [41], particle methods have a long history [30, 133, 198], and are theoretically well-

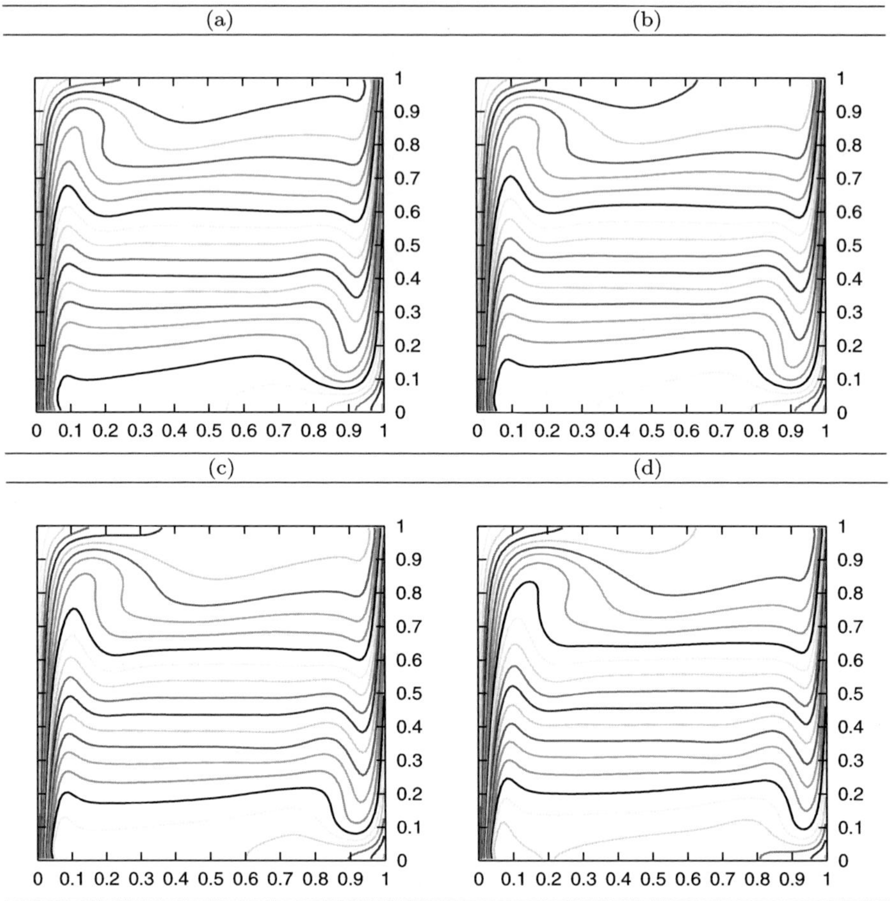

Fig. 6.51 Standard deviation in scaled temperature T^{ϵ} for (**a**) $\epsilon = 0.01$, (**b**) $\epsilon = 0.1$, (**c**) $\epsilon = 0.2$, and (**d**) $\epsilon = 0.3$. Note that $\sigma(T^{\epsilon}) = 0$ on the hot-wall ($x = 0$), where a deterministic temperature is imposed, and that it peaks on the cold wall ($x = 1$) where the uncertain fluctuations are imposed. The same contour increment is used in all frames. Adapted from [127]

grounded [42, 194] with available convergence results [15, 28, 40, 53]. Their advantages include the flexibility to treat complex and moving boundary problems, the ability to tackle in a natural fashion problems in infinite domains, and the ability to deal with problem with low or even vanishing diffusivity. This last feature is particularly attractive as stabilized Eulerian deterministic convection schemes typically rely on upwinding, which are difficult to extend to a stochastic setting, especially when the velocity has large uncertainty. In contrast, Lagrangian particle methods handle convection in a stable and non-diffusive way. This aspect will consequently be a focus of this section.

In Sect. 6.4.1 we provide a summary of the formulation of a deterministic particle method, whose extension to stochastic problems is later carried out in Sect. 6.4.2.

The extension strategy is based on the use of a single set of particles to transport all the uncertain modes of the solution. In Sect. 6.4.3, a validation of the proposed method is presented by considering two simple problems: the purely diffusive evolution of an uncertain Gaussian vortex, and the non-diffusive convection of a passive scalar by an uncertain velocity field. The validation is performed by comparison with exact solutions. To demonstrate the effectiveness of the proposed technique, the simulation of the natural convection of a localized patch of heated fluid in an infinite domain is considered in Sect. 6.4.4.

6.4.1 Particle Method

In this section, we discuss the formulation of the particle method for the deterministic problem considered, together with acceleration techniques and implementation details. The deterministic governing equations are first presented in Sect. 6.4.1.1. In Sect. 6.4.1.2, the particle approximation is outlined, using integral representations of the diffusion and buoyancy terms. The derivation of these integral kernels is detailed in Sect. 6.4.1.3. Diffusion is accounted for using the Particle Strength Exchange (PSE) method of Degond and Mas-Gallic [53], which provides a suitable approach for extending the scheme to the stochastic setting. Buoyancy terms are similarly discretized in a conservative way following the recent work of Eldredge *et al.* [65]. Section 6.4.1.4 discusses the implementation of a particle-mesh method devised to reduce the computational cost of the particle velocity evaluations. Finally, a remeshing procedure for the deterministic problem is briefly discussed in Sect. 6.4.1.5.

6.4.1.1 Boussinesq Equations in Rotation Form

In the Boussinesq limit, the equations of motion in primitive $(\mathbf{u}, p, T)$ variables were given in Sect. 6.2.1. Here, we consider an infinite 2D domain with no internal boundary and so supplement the governing equations (6.42)–(6.44) with the far-field boundary conditions:

$$\mathbf{u}(\mathbf{x}, t) = 0, \qquad T(\mathbf{x}, t) = 0 \quad \text{as } |\mathbf{x}| \to \infty. \tag{6.123}$$

Defining the vorticity $\boldsymbol{\omega} = \boldsymbol{\nabla} \wedge \mathbf{u} = \omega \mathbf{e}_z$, and taking the curl of the momentum equation we obtain:

$$\frac{\partial \omega}{\partial t} + \boldsymbol{\nabla} \cdot (\mathbf{u}\omega) = \frac{\mathrm{Pr}}{\sqrt{\mathrm{Ra}}} \Delta\omega + \mathrm{Pr}\frac{\partial T}{\partial x}, \tag{6.124}$$

with initial conditions $\omega(\mathbf{x}, 0) = (\boldsymbol{\nabla} \wedge \mathbf{u}(\mathbf{x}, 0)) \cdot \mathbf{e}_z$ and boundary condition $\omega(\mathbf{x}, t) = 0$ as $|\mathbf{x}| \to \infty$. Next, we introduce the streamfunction $\psi(\mathbf{x}, t)$, defined by:

$$\mathbf{u} = \boldsymbol{\nabla} \wedge (\psi \mathbf{e}_z), \qquad \Delta\psi = -\omega. \tag{6.125}$$

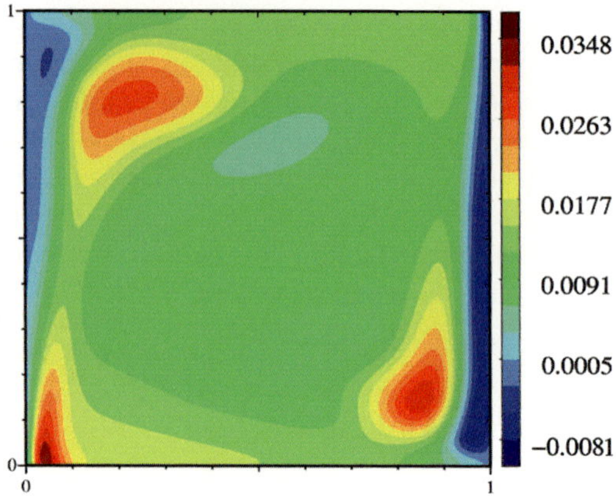

Fig. 6.52 Difference field $\sigma(T^{\epsilon})(\epsilon = 0.3) - \sigma(T^{\epsilon})(\epsilon = 0.01)$. Adapted from [127]

The normalized governing equations to be solved are then given by:

$$\frac{\partial \omega}{\partial t} + \boldsymbol{\nabla} \cdot (\mathbf{u}\omega) = \frac{\mathrm{Pr}}{\sqrt{\mathrm{Ra}}} \Delta\omega + \mathrm{Pr} \frac{\partial T}{\partial x}, \tag{6.126}$$

$$\frac{\partial T}{\partial t} + \boldsymbol{\nabla} \cdot (\mathbf{u}T) = \frac{1}{\sqrt{\mathrm{Ra}}} \Delta T, \tag{6.127}$$

$$\Delta \psi = -\omega, \tag{6.128}$$

$$\mathbf{u} = \boldsymbol{\nabla} \wedge (\psi \mathbf{e}_z), \tag{6.129}$$

$$\omega(\mathbf{x}, 0) = (\boldsymbol{\nabla} \wedge \mathbf{u}(\mathbf{x}, 0)) \cdot \mathbf{e}_z \tag{6.130}$$

$$\mathbf{u}, \omega, T \to 0 \quad \text{as } |\mathbf{x}| \to \infty. \tag{6.131}$$

6.4.1.2 Particle Formulation

The principle of particle methods is to discretize the fluid domain into Lagrangian elements (or particles), which in the present case carry vorticity and temperature. The position $\mathbf{x}_p$ of a particle obeys:

$$\frac{\mathrm{d}\mathbf{x}_p}{\mathrm{d}t} = \mathbf{u}(\mathbf{x}_p, t). \tag{6.132}$$

The velocity and vorticity fields are related through the Biot-Savart integral:

$$\mathbf{u} = \frac{-1}{2\pi} \mathcal{K} \star \boldsymbol{\omega} = \frac{-1}{2\pi} \int_{\mathbb{R}^2} \mathcal{K}(\mathbf{x}, \mathbf{y}) \wedge \boldsymbol{\omega} \, \mathrm{d}\mathbf{y}, \tag{6.133}$$

with the 2D kernel given by:

$$\mathcal{K}(\mathbf{x}, \mathbf{y}) = (\mathbf{x} - \mathbf{y}) / |\mathbf{x} - \mathbf{y}|^2 .$$

As further discussed below, the Laplacian and gradient operators can be approximated by integral conservative operators of the form:

$$\Delta q \approx \int_{\mathbb{R}^2} \mathcal{L}(\mathbf{x}-\mathbf{y})\left[q(\mathbf{y})-q(\mathbf{x})\right]\mathrm{d}\mathbf{y},$$

$$\frac{\partial q}{\partial x} \approx \int_{\mathbb{R}^2} \mathcal{G}^x(\mathbf{x}-\mathbf{y})\left[q(\mathbf{y})+q(\mathbf{x})\right]\mathrm{d}\mathbf{y}. \tag{6.134}$$

With these representations, the transport equations for the vorticity and heat can be recast as:

$$\frac{\mathrm{d}\mathbf{x}_p}{\mathrm{d}t} = \frac{-1}{2\pi}\int_{\mathbb{R}^2} \mathcal{K}(\mathbf{x}_p,\mathbf{y}) \wedge \boldsymbol{\omega}(\mathbf{y})\,\mathrm{d}\mathbf{y}, \tag{6.135}$$

$$\frac{\mathrm{d}\omega}{\mathrm{d}t} = \frac{\mathrm{Pr}}{\sqrt{\mathrm{Ra}}}\int_{\mathbb{R}^2} \mathcal{L}(\mathbf{x}_p-\mathbf{y})\left[\omega(\mathbf{y})-\omega(\mathbf{x}_p)\right]\mathrm{d}\mathbf{y} + \mathrm{Pr}\int_{\mathbb{R}^2} \mathcal{G}^x(\mathbf{x}_p-\mathbf{y})\left[T(\mathbf{y})+T(\mathbf{x}_p)\right]\mathrm{d}\mathbf{y}, \tag{6.136}$$

$$\frac{\mathrm{d}T}{\mathrm{d}t} = \frac{1}{\sqrt{\mathrm{Ra}}}\int_{\mathbb{R}^2} \mathcal{L}(\mathbf{x}_p-\mathbf{y})\left[T(\mathbf{y})-T(\mathbf{x}_p)\right]\mathrm{d}\mathbf{y}, \tag{6.137}$$

where $\mathrm{d}/\mathrm{d}t$ is the Lagrangian (material) derivative.

The particle discretization relies on a set of Np Lagrangian elements, with associated position $\mathbf{X}_i(t)$, circulation $\Gamma_i(t)$ and heat $H_i(t)$. We denote by ϵ the core radius of the particles, and introduce the smoothing function $\zeta_\epsilon(\mathbf{x})$ such that $\zeta_\epsilon(\mathbf{x}) \to \delta(\mathbf{x})$ as $\epsilon \to 0$, where δ is the Dirac delta function. The smoothed approximation of the vorticity and temperature fields are

$$\omega(\mathbf{x},t) = \sum_{i=1}^{\mathrm{Np}} \Gamma_i(t)\zeta_\epsilon(\mathbf{x}-\mathbf{X}_i(t)), \qquad T(\mathbf{x},t) = \sum_{i=1}^{\mathrm{Np}} H_i(t)\zeta_\epsilon(\mathbf{x}-\mathbf{X}_i(t)). \tag{6.138}$$

Denoting $\mathcal{K}_\epsilon$ the regularized version of $\mathcal{K}$, the evolution equations for the particle positions and strengths are:

$$\frac{\mathrm{d}\mathbf{X}_i}{\mathrm{d}t} = \frac{-1}{2\pi}\sum_{j=1}^{\mathrm{Np}} \Gamma_j \mathcal{K}_\epsilon(\mathbf{X}_i,\mathbf{X}_j), \tag{6.139}$$

$$\frac{\mathrm{d}\Gamma_i}{\mathrm{d}t} = \frac{\mathrm{Pr}}{\sqrt{\mathrm{Ra}}}\sum_{j=1}^{\mathrm{Np}} \mathcal{L}(\mathbf{X}_i-\mathbf{X}_j)S[\Gamma_j-\Gamma_i] + \mathrm{Pr}\sum_{j=1}^{\mathrm{Np}} \mathcal{G}^x(\mathbf{X}_i-\mathbf{X}_j)S[H_j+H_i], \tag{6.140}$$

$$\frac{\mathrm{d}H_i}{\mathrm{d}t} = \frac{1}{\sqrt{\mathrm{Ra}}} \sum_{j=1}^{\mathrm{Np}} \mathcal{L}(\mathbf{X}_i - \mathbf{X}_j) S \left[H_j - H_i \right]. \tag{6.141}$$

In the equations above, we used S to denote the volume of the particles, assumed to be the same for all particles. It is seen that the particles move with the local fluid velocity, while their vorticity and temperature evolve due to diffusion and buoyancy. The initial conditions for the discrete system equations are:

$$\Gamma_i(0) \approx \omega(\mathbf{X}_i(0), 0) S, \qquad H_i(0) \approx T(\mathbf{X}_i(0), 0) S. \tag{6.142}$$

These initial conditions, together with (6.139)–(6.141), constitute a system of ODEs that approximates the continuous system in (6.126)–(6.131).

6.4.1.3 Approximation of Diffusion and Buoyancy Terms

As mentioned above, integral representations of the diffusion operator and buoyancy terms are used in the computations. Before outlining these representations, a brief discussion is provided of the rationale behind the present approach. For brevity, the discussion focuses on the diffusion term only. One first notes that the representation announced in (6.140), (6.141) is not the only means of dealing with diffusion in the context of particle methods. Alternative treatments include random walk and diffusion velocity methods.

In the random walk method [30], diffusion is simulated using random displacement of the particles having a zero mean, and variance proportional to the diffusivity and time step. The strength of the particles, on the other hand, remains unchanged. This approach offers several attractive features, including ease of implementation, the obvious conservative character of the algorithm, and the independence of the random displacements so that particle-particle interactions do not intervene. Unfortunately, extension of the random walk algorithm to stochastic problems tackled using PC representations is generally difficult. An example where severe difficulties arise concerns the case of an uncertain diffusivity, because in this case the variance of the random displacements would be uncertain.

In the diffusion velocity method [16, 54, 196], the diffusion term is recast as a transport term, and diffusion is consequently accounted for by moving the particles using the sum of local convection and diffusion velocities. The particle strength is not affected, and so the method is conservative. Unfortunately, the diffusion velocity method does not appear to be generally well-suited for extension to the uncertain case. As for random walk, uncertainty in the diffusion velocity makes it difficult to devise a particle transport scheme. This difficulty is further compounded by the nonlinearities appearing in the definition of the diffusion velocity. Being stochastic, the diffusion velocity must be represented in terms of PC expansions, which adds complexity and may require significant computational overhead.

In order to avoid the difficulties outlined above, the PSE method is adopted for the purpose of modeling diffusion. As further discussed below, PSE amounts to an update of the particle strengths but not of their positions. As further discussed in Sect. 6.4.2 below, this feature provides a distinct advantage in a stochastic setting, as it makes possible to define representations involving a single set of particles. The PSE method relies on an integral approximation of the Laplacian, which rely on the definition of a radially-symmetric smoothing function, η_ϵ, defined according to:

$$\eta_\epsilon(\mathbf{x}) \equiv \frac{1}{\epsilon^2}\eta\left(|\mathbf{x}|/\epsilon\right),$$

where η is a radial function and ϵ is the core parameter. Following [53], η is assumed to satisfy the moment conditions:

$$\int_{\mathbb{R}^2} x^2\eta(\mathbf{x})\,\mathrm{d}\mathbf{x} = \int_{\mathbb{R}^2} y^2\eta(\mathbf{x})\,\mathrm{d}\mathbf{x} = 2,$$

$$\int_{\mathbb{R}^2} x^{\alpha_1}y^{\alpha_2}\eta(\mathbf{x})\,\mathrm{d}\mathbf{x} = 0, \quad 1 \le \alpha_1+\alpha_2 \le m+1,\ \alpha_1,\alpha_2 \ne 2.$$

Based on these definitions, it can be shown [53] that the Laplacian of a generic scalar field c can be approximated by:

$$\Delta c = \frac{1}{\epsilon^2}\left(\eta_\epsilon \star c - c\right) + \mathcal{O}(\epsilon^m), \tag{6.143}$$

where $\star$ denotes the convolution operator.

Applying the approximation above to the vorticity and temperature fields associated with the corresponding smoothed particle representations leads to the following definition of the diffusion kernel,

$$\mathcal{L}(\mathbf{x}-\mathbf{y}) = \frac{1}{\epsilon^2}\eta_\epsilon(\mathbf{x}-\mathbf{y}), \tag{6.144}$$

which is used in the evolution equations for the particle strengths (6.140), (6.141). Note that due to the symmetry of the kernel, the diffusion treatment in (6.140), (6.141) is conservative.

The generalization of the integral approximation of the diffusion operator to derivatives of arbitrary order was recently conducted by Eldredge *et al.* [65], who provide the following integral approximation:

$$D^\beta f \equiv \frac{\partial^{|\beta|}}{\partial x_1^{\beta_1}\cdots\partial x_d^{\beta_d}} f(\mathbf{x}) \approx \frac{1}{\epsilon^{|\beta|}}\int [f(\mathbf{y}) + (-1)^{|\beta|+1} f(\mathbf{x})]\eta_\epsilon^{(\beta)}(\mathbf{x}-\mathbf{y})\,\mathrm{d}\mathbf{y}, \tag{6.145}$$

for positive d-dimensional multi-index β, $|\beta| = \sum_{i=1}^{d}\beta_i$. They derived constraints on the moments of $\eta^{(\beta)}$ to achieve approximation of arbitrary order, and proposed a set of 2D kernels, for both first derivatives and the Laplacian, with accuracy up to eighth order. In this work, we use the second-order kernel [65]:

$$\eta(\mathbf{x}) = \frac{4}{\pi}\exp[-|\mathbf{x}|^2]. \tag{6.146}$$

Applying the integral approximation of $\partial/\partial x$ to the particle representation leads to:

$$\mathcal{G}^x(\mathbf{X}_i, \mathbf{X}_j) = \frac{1}{\epsilon}\eta_\epsilon^{(x)}(\mathbf{X}_i - \mathbf{X}_j), \tag{6.147}$$

where

$$\eta^{(x)}(\mathbf{x}) \equiv -\frac{2x}{\pi}\exp[-|\mathbf{x}|^2].$$

The summations in (6.140), (6.141) suggest that the complexity of the evaluation of the integral diffusion and buoyancy terms are $\mathcal{O}(\mathrm{Np}^2)$. However, the kernels have fast decay with $\mathbf{x}-\mathbf{y}$, and so the quadrature can in fact be truncated to particles in a finite neighborhood of the local evaluation point. Specifically, interactions between particles separated by a distance greater than 4ϵ are neglected in the computations. This results in significant reduction of the CPU cost by keeping track of the list of interacting particles, which allows discretizations using a large number of particles.

6.4.1.4 Acceleration of Velocity Computation

The computation of the velocity for all the particles using (6.139) requires $\mathcal{O}(\mathrm{Np}^2)$ operations. For a large number of particles, the resulting CPU cost is prohibitive. Consequently, fast methods have been proposed to speed-up the evaluation of particles velocity. These include hybrid, or particle-mesh, methods and fast multipole methods.

The multipole methods are based on the idea that one can globally approximate the velocity induced by a cluster of particles, provided that the observation point is located far enough from the cluster. This is usually achieved through multipole expansions, using a hierarchical data structure of the particles distribution. Depending on their construction, the multipole techniques exhibit theoretical computational costs scaling as $\mathcal{O}(\mathrm{Np})$ or $\mathcal{O}(\mathrm{Np}\log\mathrm{Np})$.

The hybrid methods rely on a mesh to compute the velocity field by solving a Poisson's equation for the streamfunction [32]. The procedure involves three steps: the projection step, the resolution of the Poisson's equation and the interpolation step. In the projection step, a mesh covering the particles is constructed and the values of the vorticity field are estimated at the mesh points using a projection operator, which distributes the particles' vorticity onto fixed grid points, $\mathbf{x}_g = (x_g, y_g)$, as follows:

$$\omega(\mathbf{x}_g) = \sum_{i=1}^{M} \frac{\Gamma_i}{h_g^2}\Lambda\left(|x_g - X_i|/h_g\right)\Lambda\left(|y_g - Y_i|/h_g\right),$$

where h_g is the (uniform) mesh size, and

$$\Lambda(u) = \begin{cases} (1-u^2)(2-u)/2, & \text{if } 0 \le u < 1; \\ (1-u)(2-u)(3-u)/6, & \text{if } 1 \le u \le 2; \\ 0 & \text{otherwise.} \end{cases}$$

The strength of a particle is thus distributed on the 16 nearest mesh points. The Poisson equation $\Delta\psi = -\omega$ is then solved on the mesh. We rely on Dirichlet boundary conditions that are obtained from the integral representation of the streamfunction:

$$\psi(\mathbf{x}) = \int_{\mathbb{R}^2} \mathcal{H}(\mathbf{x},\mathbf{y})\omega(\mathbf{y})\,\mathrm{d}\mathbf{y} \approx \sum_{j=1}^{\mathrm{Np}} \Gamma_j \mathcal{H}_\epsilon(\mathbf{x},\mathbf{X}_j). \tag{6.148}$$

Once the streamfunction is obtained, numerical differentiation is used to determine the velocity at the mesh points. Second-order centered differences are used for this purpose. The final step consists in interpolating the mesh velocity on the particle positions. Once again, this is achieved using the interpolation formula:

$$\mathbf{u}(\mathbf{X}_i) = \sum_{g=1}^{\mathrm{M}} \mathbf{u}(\mathbf{x}_g)\Lambda\left(|x_g - X_i|/h_g\right)\Lambda\left(|y_g - Y_i|/h_g\right),$$

where M is the number of mesh points. For a uniform mesh, with constant spacing h_g, a fast FFT-based solver is employed to solve the Poisson streamfunction equation. The overall CPU cost of the velocity computation thus scales as $\mathcal{O}(\mathrm{Np} + \mathrm{M}\log\mathrm{M})$.

The stochastic spectral expansion of the problem, introduced below, makes use of a set of particles having deterministic location and random strength. As a result, both multipole expansions and particle-mesh methods can be used for the stochastic velocity computation, by applying the respective scheme to each particle circulation mode. This feature allows for a straightforward parallelization of the computation of velocity modes. Note that in particle-mesh schemes the projection coefficients depend on the deterministic particle locations only; their computation is thus factorized over the stochastic modes. Multipole coefficients combine both the (deterministic) locations and (stochastic) strengths such that only the construction of the data structure of the particles distribution can be factorized. For this reason, the use of a particle-mesh method appeared preferable to us over the multipole expansion approach, and the former is used exclusively in the computations below.

6.4.1.5 Remeshing

A difficulty inherent to the PSE method is the need to introduce new particles to properly account for the diffusion outside the initial support of ω and T. Furthermore, as the particles are convected, their distribution can be so highly strained and distorted that they cease to overlap in some region of the flow. This calls for a remeshing procedure where the fields are periodically rediscretized (see [41] and references therein). This is achieved by considering periodically a new set of particles uniformly distributed on a grid with uniform spacing $\sqrt{S}$ covering the current set of particles. The new particle strengths are computed from the interpolation scheme used for the projection of the vorticity on the velocity mesh. This remeshing introduces a numerical diffusion, inherent to the interpolation procedure, which must be kept as small as possible. In the computations below, this is achieved by remeshing infrequently. Specifically, numerical diffusion due to remeshing was found to be negligibly small, and consequently no special means were needed to reduce it further.

6.4.2 Stochastic Formulation

In this section, we extend the particle method described in the previous section to uncertain flow conditions using PC expansions. Essential ingredients for the construction of the stochastic basis and the PC expansion are given in Sect. 6.4.2.1. We then proceed in Sect. 6.4.2.2 to a straightforward stochastic extension of the deterministic particle equations obtained in Sect. 6.4.2.1 and show that it results in a cumbersome, impractical problem. This first attempt however clearly indicates which characteristics of the particle discretization are desirable to be conserved in the process of PC expansion. With these characteristics in mind, we then propose in Sect. 6.4.2.3 an alternative expansion of the particle problem that exhibits the desired properties. Finally, we draw a few conclusions concerning the resulting particle formulation of the uncertain flow problem and its connection with the solution method of the deterministic case.

6.4.2.1 Stochastic Basis and PC Expansion

Let $\boldsymbol{\xi}$ denote the second-order random vector $\boldsymbol{\xi} = (\xi_1 \dots \xi_{\mathrm{N}})$, with independent components, defined on a probability space (Θ, Σ, P), with value in Ξ, and prescribed probability. Without loss of generality, we shall assume that the ξ_i's are identically distributed, with a density $p(\xi_i)$. Therefore, the density of $\boldsymbol{\xi}$ is

$$p_{\Xi}(\boldsymbol{\xi}) = \prod_{i=1}^{\mathrm{N}} p(\xi_i). \tag{6.149}$$

Any well-behaved random quantity a, (*e.g.* second-order one), can be expanded in a convergent Fourier-like series of the form

$$a = \sum_{k=0}^{\infty} [a]_k \Psi_k(\boldsymbol{\xi}), \tag{6.150}$$

where the Ψ_k's are functionals of the random vector $\boldsymbol{\xi}$ and $[a]_k$ are the expansion coefficients. In this section, we shall consider $\boldsymbol{\xi}$ uniformly distributed in $[-1, 1]^{\mathrm{N}}$ with a multidimensional Legendre polynomial expansion [246], $p(\xi_i) = 1/2$ for

$\xi_i \in [-1, 1]$ and 0 otherwise. Adopting the convention consisting in ordering the polynomial indexes with increasing polynomial order, *i.e.*

$$\Psi_0 = 1, \qquad \Psi_{i\in[1,\mathrm{N}]} = \xi_i, \qquad \Psi_{\mathrm{N}+1} = \xi_1^2 - 1/2, \qquad \ldots, \tag{6.151}$$

the polynomial expansion of a truncated to order No is

$$a \approx \sum_{k=0}^{\mathrm{P}} [a]_k \Psi_k(\boldsymbol{\xi}), \quad \mathrm{P}+1 = \frac{(\mathrm{N}+\mathrm{No})!}{\mathrm{N}!\mathrm{No}!}. \tag{6.152}$$

Thus, the mean and variance of a can be respectively expressed as:

$$\langle a \rangle = [a]_0, \qquad \left\langle (a - \langle a \rangle)^2 \right\rangle \approx \sum_{k=1}^{\mathrm{P}} [a]_k^2 \left\langle \Psi_k^2 \right\rangle. \tag{6.153}$$

If some data of the problem is uncertain (for instance the initial conditions or some transport properties), the data will be considered as random and consequently represented by suitable PC expansions. As a result, the flow variables will also be functionals of $\boldsymbol{\xi}$, and the system (6.126)–(6.131) becomes:

$$\frac{\partial \omega(\mathbf{x}, t, \boldsymbol{\xi})}{\partial t} + \nabla \cdot (\mathbf{u}(\mathbf{x}, t, \boldsymbol{\xi}) \omega(\mathbf{x}, t, \boldsymbol{\xi})) = \frac{\mathrm{Pr}(\boldsymbol{\xi})}{\sqrt{\mathrm{Ra}(\boldsymbol{\xi})}} \Delta \omega(\mathbf{x}, t, \boldsymbol{\xi}) + \mathrm{Pr}(\boldsymbol{\xi}) \frac{\partial T(\mathbf{x}, t, \boldsymbol{\xi})}{\partial x}, \tag{6.154}$$

$$\frac{\partial T(\mathbf{x}, t, \boldsymbol{\xi})}{\partial t} + \nabla \cdot (\mathbf{u}(\mathbf{x}, t, \boldsymbol{\xi}) T(\mathbf{x}, t, \boldsymbol{\xi})) = \frac{1}{\sqrt{\mathrm{Ra}(\boldsymbol{\xi})}} \Delta T(\mathbf{x}, t, \boldsymbol{\xi}), \tag{6.155}$$

$$\Delta \psi(\mathbf{x}, t, \boldsymbol{\xi}) = -\omega(\mathbf{x}, t, \boldsymbol{\xi}), \tag{6.156}$$

$$\mathbf{u}(\mathbf{x}, t, \boldsymbol{\xi}) = \nabla \wedge (\psi(\mathbf{x}, t, \boldsymbol{\xi}) \mathbf{e}_z), \tag{6.157}$$

$$\omega(\mathbf{x}, 0, \boldsymbol{\xi}) = (\nabla \wedge \mathbf{u}(\mathbf{x}, 0, \boldsymbol{\xi})) \cdot \mathbf{e}_z, \tag{6.158}$$

$$\mathbf{u}(\mathbf{x}, t, \boldsymbol{\xi}), \omega(\mathbf{x}, t, \boldsymbol{\xi}), T(\mathbf{x}, t, \boldsymbol{\xi}) \to 0 \quad \text{a.s. as } |\mathbf{x}| \to \infty. \tag{6.159}$$

The above system has to be solved in $\mathbb{R}^2 \times [0, T] \times [-1, 1]^{\mathrm{N}}$. It is assumed that the probabilistic model of the data are such that the second-order moment of $\mathrm{Ra}^{-1/2}(\boldsymbol{\xi})$ is finite. It is remarked that (6.154)–(6.159) involves no differential operator in $\boldsymbol{\xi}$, a characteristic expressing the independence of deterministic realizations of the flow for different realizations of the data. It is also clear that a numerical approximation of (6.154)–(6.159) will require an efficient discretization to account for the dependence of the solution on $\boldsymbol{\xi}$. By expanding the flow variables using the PC basis,

$$\left[\omega, T, \mathbf{u}, \psi\right](\mathbf{x}, t, \boldsymbol{\xi}) = \sum_{k=0}^{\infty} \left[\omega, T, \mathbf{u}, \psi\right]_k (\mathbf{x}, t) \Psi_k(\boldsymbol{\xi}), \tag{6.160}$$

it is expected that a low polynomial order will be sufficient to properly represent the solution. In other words, it is expected that the series (6.160) is quickly convergent such that only a small number P of stochastic modes can be actually retained in the computation. We now focus on the derivation of a particle approximation of the stochastic flow.

6.4.2.2 Straightforward Particle Formulation

A straightforward approach to derive a particle formulation of (6.154)–(6.159) would consist in considering the particle formulation of the deterministic problem, given by (6.139)–(6.142), and then to allow for a dependence of the variables on $\boldsymbol{\xi}$. This appears feasible as (6.154)–(6.159) has no differential operator in the stochastic dimensions. It would result in the following system of stochastic ODEs:

$$\frac{\mathrm{d}\mathbf{X}_i(t,\boldsymbol{\xi})}{\mathrm{d}t} = \frac{-1}{2\pi}\sum_{j=1}^{\mathrm{Np}} \Gamma_j(t,\boldsymbol{\xi})\mathcal{K}_\epsilon(\mathbf{X}_i(t,\boldsymbol{\xi}),\mathbf{X}_j(t,\boldsymbol{\xi})), \tag{6.161}$$

$$\begin{aligned}\frac{\mathrm{d}\Gamma_i(t,\boldsymbol{\xi})}{\mathrm{d}t} &= \frac{\mathrm{Pr}(\boldsymbol{\xi})}{\sqrt{\mathrm{Ra}(\boldsymbol{\xi})}}\sum_{j=1}^{\mathrm{Np}} S\mathcal{L}(\mathbf{X}_i(t,\boldsymbol{\xi})-\mathbf{X}_j(t,\boldsymbol{\xi}))\left[\Gamma_j(t,\boldsymbol{\xi})-\Gamma_i(t,\boldsymbol{\xi})\right]\\ &\quad + \mathrm{Pr}(\boldsymbol{\xi})\sum_{j=1}^{\mathrm{Np}} S\mathcal{G}^x(\mathbf{X}_i(t,\boldsymbol{\xi})-\mathbf{X}_j(t,\boldsymbol{\xi}))\left[H_j(t,\boldsymbol{\xi})+H_i(t,\boldsymbol{\xi})\right],\end{aligned} \tag{6.162}$$

$$\frac{\mathrm{d}H_i(t,\boldsymbol{\xi})}{\mathrm{d}t} = \frac{1}{\sqrt{\mathrm{Ra}(\boldsymbol{\xi})}}\sum_{j=1}^{\mathrm{Np}} S\mathcal{L}(\mathbf{X}_i(t,\boldsymbol{\xi})-\mathbf{X}_j(t,\boldsymbol{\xi}))\left[H_i(t,\boldsymbol{\xi})-H_j(t,\boldsymbol{\xi})\right], \tag{6.163}$$

$$\Gamma_i(0,\boldsymbol{\xi}) = \omega(\mathbf{X}_i(0,\boldsymbol{\xi}),0,\boldsymbol{\xi})S, \qquad H_i(0,\boldsymbol{\xi}) = T(\mathbf{X}_i(0,\boldsymbol{\xi}),0,\boldsymbol{\xi})S. \tag{6.164}$$

A particle representation is thus obtained in which the particles have both random position and strengths. To derive a weak formulation of this problem, the truncated PC expansions of the particles position and strengths,

$$\mathbf{X}_i(\boldsymbol{\xi}) = \sum_{k=0}^{\mathrm{P}} [\mathbf{X}_i]_k\Psi_k(\boldsymbol{\xi}), \qquad \Gamma_i(\boldsymbol{\xi}) = \sum_{k=0}^{\mathrm{P}} [\Gamma_i]_k\Psi_k(\boldsymbol{\xi}), \qquad H_i(\boldsymbol{\xi})\sum_{k=0}^{\mathrm{P}} [H_i]_k\Psi_k(\boldsymbol{\xi}), \tag{6.165}$$

are introduced into (6.161)–(6.164), which are then projected onto the PC basis. This yields a system of P + 1 coupled problems for the stochastic modes. For instance, the equations for the k-th mode of the particles position are:

$$\left\langle\Psi_k^2\right\rangle\frac{\mathrm{d}[\mathbf{X}_i]_k}{\mathrm{d}t} = \frac{-1}{2\pi}\sum_{j=1}^{\mathrm{Np}}\left\langle\Psi_k\sum_{l=0}^{\mathrm{P}}\left[\Gamma_j\right]_l\Psi_l\mathcal{K}_\epsilon\left(\sum_{m=0}^{\mathrm{P}}[\mathbf{X}_i]_m\Psi_m,\sum_{n=0}^{\mathrm{P}}\left[\mathbf{X}_j\right]_n\Psi_n\right)\right\rangle. \tag{6.166}$$

We already see from this equation that the approach is cumbersome as the nonlinearities of the smoothing kernel $\mathcal{K}_\epsilon$ make it difficult to obtain a true Galerkin representation. A pseudo-spectral technique was proposed in [120] to find the stochastic modes of $\mathcal{K}_\epsilon(\mathbf{X}_i(\boldsymbol{\xi}), \mathbf{X}_j(\boldsymbol{\xi}))$, but this results in a prohibitively expensive strategy when the number of particles is large. It can also be remarked that in the computational strategy for the deterministic problem discussed above, the velocity kernel is never evaluated (if one excepts the boundary conditions for ψ) as the velocity is computed by solving the streamfunction Poisson equation. However, the particle-mesh strategy cannot be immediately applied to the present formulation as the stochastic positions of the particles make it difficult to project the vorticity on the mesh; similar difficulties would also arise in the interpolation of the velocity modes at random particle positions. Additional hurdles also include the evaluation of the diffusion and buoyancy terms.

Based on the discussion above, it was concluded that an efficient formulation would involve a set of particles with **deterministic** positions but stochastic strengths. This would avoid the difficulties above, and further enable the use of fast particle-mesh techniques.

6.4.2.3 Particle Discretization of the Stochastic Flow

Let us go back to the Eulerian stochastic transport equations for the vorticity and heat. Dropping from the notation the spatial and time dependence of the variables, the conservative form is written as:

$$\left\{ \begin{aligned} \frac{\partial \omega(\boldsymbol{\xi})}{\partial t} + \nabla \cdot (\mathbf{u}(\boldsymbol{\xi})\omega(\boldsymbol{\xi})) &= \frac{\mathrm{Pr}(\boldsymbol{\xi})}{\sqrt{\mathrm{Ra}(\boldsymbol{\xi})}} \Delta\omega(\boldsymbol{\xi}) + \mathrm{Pr}(\boldsymbol{\xi}) \frac{\partial T(\boldsymbol{\xi})}{\partial x}, \\ \frac{\partial T(\boldsymbol{\xi})}{\partial t} + \nabla \cdot (\mathbf{u}(\boldsymbol{\xi}) T(\boldsymbol{\xi})) &= \frac{1}{\sqrt{\mathrm{Ra}(\boldsymbol{\xi})}} \Delta T(\boldsymbol{\xi}). \end{aligned} \right. \tag{6.167}$$

For expansions truncated at order No, the weak formulation of the system above is given by the following evolution equations for the stochastic modes:

$$\begin{aligned} &\frac{\partial [\omega]_k}{\partial t} + \sum_{l=0}^{\mathrm{P}} \sum_{m=0}^{\mathrm{P}} C_{klm} \nabla \cdot ([\mathbf{u}]_l [\omega]_m) \\ &\quad = \sum_{l=0}^{\mathrm{P}} \sum_{m=0}^{\mathrm{P}} C_{klm} \left(\left[\frac{\mathrm{Pr}}{\sqrt{\mathrm{Ra}}} \right]_l \nabla^2 [\omega]_m + [\mathrm{Pr}]_l \frac{\partial [T]_m}{\partial x} \right), \\ &\frac{\partial [T]_k}{\partial t} + \sum_{l=0}^{\mathrm{P}} \sum_{m=0}^{\mathrm{P}} C_{klm} \nabla \cdot ([\mathbf{u}]_l [T]_m) = \sum_{l=0}^{\mathrm{P}} \sum_{m=0}^{\mathrm{P}} C_{klm} \left[\frac{1}{\sqrt{\mathrm{Ra}}} \right]_l \nabla^2 [T]_m, \end{aligned}$$

where $C_{klm} \equiv \langle \Psi_k \Psi_l \Psi_m \rangle / \langle \Psi_k^2 \rangle$ is the multiplication tensor. Since by convention $\Psi_0 = 1$, we have $C_{k0k} = 1$ and $C_{k0m} = 0$ for $k \neq m$; Thus the previous equations can be rewritten as:

$$\frac{\partial[\omega]_k}{\partial t} + \nabla\cdot([\mathbf{u}]_0[\omega]_k) = -\sum_{l=1}^{\mathrm{P}}\sum_{m=0}^{\mathrm{P}} C_{klm}\nabla\cdot([\mathbf{u}]_l[\omega]_m) + \sum_{l=0}^{\mathrm{P}}\sum_{m=0}^{\mathrm{P}} C_{klm}\left(\left[\frac{\mathrm{Pr}}{\sqrt{\mathrm{Ra}}}\right]_l \nabla^2[\omega]_m + [\mathrm{Pr}]_l\frac{\partial[T]_m}{\partial x}\right),$$

$$\frac{\partial[T]_k}{\partial t} + \nabla\cdot([\mathbf{u}]_0[T]_k) = -\sum_{l=1}^{\mathrm{P}}\sum_{m=0}^{\mathrm{P}} C_{klm}\nabla\cdot([\mathbf{u}]_l[T]_m) + \sum_{l=0}^{\mathrm{P}}\sum_{m=0}^{\mathrm{P}} C_{klm}\left[\frac{1}{\sqrt{\mathrm{Ra}}}\right]_l \nabla^2[T]_m.$$

Consequently, we can write:

$$\frac{D\mathbf{x}_p}{Dt} = [\mathbf{u}]_0, \tag{6.168}$$

$$\begin{aligned}\frac{D[\omega]_k}{Dt} =& -\sum_{l=1}^{\mathrm{P}}\sum_{m=0}^{\mathrm{P}} C_{klm}\int_{\mathbb{R}^2}\mathcal{G}^x(\mathbf{x}_p-\mathbf{y})\left([u]_l[\omega]_m(\mathbf{x}_p)+[u]_l[\omega]_m(\mathbf{y})\right)\mathrm{d}\mathbf{y}\\ &-\sum_{l=1}^{\mathrm{P}}\sum_{m=0}^{\mathrm{P}} C_{klm}\int_{\mathbb{R}^2}\mathcal{G}^y(\mathbf{x}_p-\mathbf{y})\left([v]_l[\omega]_m(\mathbf{x}_p)+[v]_l[\omega]_m(\mathbf{y})\right)\mathrm{d}\mathbf{y}\\ &+\sum_{l=0}^{\mathrm{P}}\sum_{m=0}^{\mathrm{P}} C_{klm}\left[\frac{\mathrm{Pr}}{\sqrt{\mathrm{Ra}}}\right]_l\int_{\mathbb{R}^2}\mathcal{L}(\mathbf{x}_p-\mathbf{y})\left([\omega]_m(\mathbf{y})-[\omega]_m(\mathbf{x}_p)\right)\mathrm{d}\mathbf{y}\\ &+\sum_{l=0}^{\mathrm{P}}\sum_{m=0}^{\mathrm{P}} C_{klm}[\mathrm{Pr}]_l\int_{\mathbb{R}^2}\mathcal{G}^x(\mathbf{x}_p-\mathbf{y})\left([T]_m(\mathbf{x}_p)+[T]_m(\mathbf{y})\right)\mathrm{d}\mathbf{y},\end{aligned} \tag{6.169}$$

$$\begin{aligned}\frac{D[T]_k}{Dt} =& -\sum_{l=1}^{\mathrm{P}}\sum_{m=0}^{\mathrm{P}} C_{klm}\int_{\mathbb{R}^2}\mathcal{G}^x(\mathbf{x}_p-\mathbf{y})\left([u]_l[T]_m(\mathbf{x}_p)+[u]_l[T]_m(\mathbf{y})\right)\mathrm{d}\mathbf{y}\\ &-\sum_{l=1}^{\mathrm{P}}\sum_{m=0}^{\mathrm{P}} C_{klm}\int_{\mathbb{R}^2}\mathcal{G}^y(\mathbf{x}_p-\mathbf{y})\left([v]_l[T]_m(\mathbf{x}_p)+[v]_l[T]_m(\mathbf{y})\right)\mathrm{d}\mathbf{y}\\ &+\sum_{l=0}^{\mathrm{P}}\sum_{m=0}^{\mathrm{P}} C_{klm}\left[\frac{1}{\sqrt{\mathrm{Ra}}}\right]_l\int_{\mathbb{R}^2}\mathcal{L}(\mathbf{x}_p-\mathbf{y})\left([T]_m(\mathbf{y})-[T]_m(\mathbf{x}_p)\right)\mathrm{d}\mathbf{y}.\end{aligned} \tag{6.170}$$

In the previous equation, we have denoted by

$$\frac{Df}{Dt} \equiv \frac{\partial f}{\partial t} + [\mathbf{u}]_0\cdot\nabla f, \tag{6.171}$$

the material derivative defined with regard to the mean velocity. The stochastic velocity modes have the integral representation

$$[\mathbf{u}]_k(\mathbf{x}) = \frac{-1}{2\pi}\int_{\mathbb{R}^2} \mathcal{K}(\mathbf{x},\mathbf{y})[\omega]_k(\mathbf{y})\,d\mathbf{y}. \tag{6.172}$$

Introducing the smoothed particle approximations of the stochastic vorticity and temperature fields,

$$\omega(\mathbf{x},\boldsymbol{\xi}) = \sum_{i=1}^{\mathrm{Np}}\sum_{k=0}^{\mathrm{P}}[\Gamma_i]_k\Psi_k(\boldsymbol{\xi})\zeta_\epsilon(\mathbf{x}-\mathbf{X}_i), \tag{6.173}$$

$$T(\mathbf{x},\boldsymbol{\xi}) = \sum_{i=1}^{\mathrm{Np}}\sum_{k=0}^{\mathrm{P}}[H_i]_k\Psi_k(\boldsymbol{\xi})\zeta_\epsilon(\mathbf{x}-\mathbf{X}_i), \tag{6.174}$$

and defining the stochastic velocity modes on particle $\mathbf{X}_i$ as

$$\mathbf{U}_i(\boldsymbol{\xi}) \equiv \sum_{k=0}^{\mathrm{P}}[\mathbf{U}_i]_k\Psi_k(\boldsymbol{\xi}), \quad [\mathbf{U}_i]_k = \frac{-1}{2\pi}\sum_{j=1}^{\mathrm{Np}}\left[\Gamma_j\right]_k\mathcal{K}_\epsilon(\mathbf{X}_i,\mathbf{X}_j), \tag{6.175}$$

the particle approximation of (6.168)–(6.170) is given by:

$$\frac{D\mathbf{X}_i}{Dt} = [\mathbf{U}_i]_0, \tag{6.176}$$

$$\begin{aligned}\frac{D[\Gamma_i]_k}{Dt} =& -\sum_{j=1}^{\mathrm{Np}}\sum_{l=1}^{\mathrm{P}}\sum_{m=0}^{\mathrm{P}} C_{klm}S\left\{\mathcal{G}^x(\mathbf{X}_i-\mathbf{X}_j)\left([U_i]_l[\Gamma_i]_m+\left[U_j\right]_l\left[\Gamma_j\right]_m\right)\right.\\ &\left.+\,\mathcal{G}^y(\mathbf{X}_i-\mathbf{X}_j)\left([V_i]_l[\Gamma_i]_m+\left[V_j\right]_l\left[\Gamma_j\right]_m\right)\right\}\\ &+\sum_{j=1}^{\mathrm{Np}}\sum_{l=0}^{\mathrm{P}}\sum_{m=0}^{\mathrm{P}} C_{klm}S\left[\frac{\mathrm{Pr}}{\sqrt{\mathrm{Ra}}}\right]_l\mathcal{L}(\mathbf{X}_i-\mathbf{X}_j)\left(\left[\Gamma_j\right]_m-[\Gamma_i]_m\right)\\ &+\sum_{j=1}^{\mathrm{Np}}\sum_{l=0}^{\mathrm{P}}\sum_{m=0}^{\mathrm{P}} C_{klm}S[\mathrm{Pr}]_l\mathcal{G}^x(\mathbf{X}_i-\mathbf{X}_j)\left([H_i]_m+\left[H_j\right]_m\right),\end{aligned} \tag{6.177}$$

$$\begin{aligned}\frac{D[H_i]_k}{Dt} =& -\sum_{j=1}^{\mathrm{Np}}\sum_{l=1}^{\mathrm{P}}\sum_{m=0}^{\mathrm{P}} C_{klm}S\left\{\mathcal{G}^x(\mathbf{X}_i-\mathbf{X}_j)\left([U_i]_l[H_i]_m+\left[U_j\right]_l\left[H_j\right]_m\right)\right.\\ &\left.+\,\mathcal{G}^y(\mathbf{X}_i-\mathbf{X}_j)\left([V_i]_l[H_i]_m+\left[V_j\right]_l\left[H_j\right]_m\right)\right\}\\ &+\sum_{j=1}^{\mathrm{Np}}\sum_{l=0}^{\mathrm{P}}\sum_{m=0}^{\mathrm{P}} C_{klm}S\left[\frac{1}{\sqrt{\mathrm{Ra}}}\right]_l\mathcal{L}(\mathbf{X}_i-\mathbf{X}_j)\left(\left[H_j\right]_m-[H_i]_m\right),\end{aligned} \tag{6.178}$$

for $i = 1, \dots, \mathrm{Np}$ and $k = 0, \dots, \mathrm{P}$. The initial conditions for the above system of coupled ODEs are:

$$[\Gamma_i]_k(0) = S\frac{\langle \Psi_k \omega(\mathbf{X}_i(0), 0, \boldsymbol{\xi})\rangle}{\langle \Psi_k^2\rangle}, \qquad [H_i]_k(0) = S\frac{\langle \Psi_k T(\mathbf{X}_i(0), 0, \boldsymbol{\xi})\rangle}{\langle \Psi_k^2\rangle}, \quad (6.179)$$

for $k = 0, \dots, \mathrm{P}$ and $i = 1, \dots, \mathrm{Np}$.

Comments: As seen from (6.176), the particles are displaced with the mean velocity field. The stochastic modes of the particle strengths now evolve according to two distinct mechanisms. The first is the diffusion operator which couples all of the stochastic modes, unless the Rayleigh and Prandtl numbers are both certain. In this case, the stochastic modes independently diffuse with all the same diffusion coefficient $\mathrm{Pr}/\sqrt{\mathrm{Ra}}$ or $1/\sqrt{\mathrm{Ra}}$, for the vorticity and temperature respectively. The second corresponds to variation in the mode strengths, due to the convection by the stochastic velocity field. If the velocity field is certain, $[\mathbf{U}_i]_{k>0} = 0$ for all particles, and only the diffusion and buoyancy terms remain in the right-hand-side of (6.177)–(6.178). Note that $[\mathbf{U}_i]_{k>0} = 0$ implies that $[\Gamma_i]_{k>0} = 0$ as well.

An important remark concerns the cost of the evaluation of the stochastic modes of the particles' velocity. Because the particle positions are deterministic, the hybrid mesh-particle method is still practical. The determination of the projection mesh-points and coefficients for each particle is in fact identical to that of the deterministic method. As a consequence, the CPU cost of the computation of velocity modes is roughly $(\mathrm{P}+1)$ times larger than the cost of the deterministic evaluation for the same number of particles (in fact slightly less as the computation of the projection coefficients is factored on the $\mathrm{P}+1$ modes). This CPU cost is essentially dominated by the resolution of the Poisson streamfunction equations, which are decoupled. Thus, straightforward parallelization strategies can be envisioned. Similarly, the velocity interpolation at the particle centers involves a unique evaluation of the interpolants which are the same for all the stochastic modes.

Another remark concerns the size of the system of coupled ODEs. For Np particles and $\mathrm{P}+1$ stochastic modes, we have to advance in time 2Np positions and $2(\mathrm{P}+1)\mathrm{Np}$ strengths, giving a total of $2(\mathrm{P}+2)\mathrm{Np}$ variables. The efficiency of the method will thus depend on the ability of the selected basis to represent the uncertainty with a minimal number of stochastic modes. To integrate these ODEs, the deterministic time-stepping scheme can still be used, thus preserving the whole structure of existing deterministic particles code. Only the evaluation of the right-hand-side of the ODEs is altered. The cost of the ODEs' right-hand-side evaluation is obviously critical for the performance of the method. We have already mentioned that the velocity calculations likely scales with the number of modes for given number of particles, and can be parallelized, at least in the particle-mesh approach. At first glance, it appears that the CPU cost of the evaluation of the right-hand-side of the strength ODEs is $(\mathrm{P}+1)^3$ times greater than for its deterministic counterpart. This is an overly conservative estimate, because (1) the multiplication tensor is sparse, and (2) since the particle positions are deterministic many of the integral kernels are identical. In the computations below, we take advantage of this feature by computing and storing the kernels when the near-neighbors lists are constructed. To further improve CPU performance, the stochastic products $\mathbf{U}_i\Gamma_i$, $\mathbf{U}_i H_i$, $(\mathrm{Pr}/\sqrt{\mathrm{Ra}})\Gamma_i$, $\mathrm{Pr}\Gamma_i$ and $(1/\sqrt{\mathrm{Ra}})H_i$ are first computed, in an efficient vectorized way (with inner loop on particle index) before considering the right-hand-side assembly. Doing so, applications with a large number of particles and high-order expansions are possible even on small platforms, as illustrated in Sect. 6.4.4.

6.4.3 Validation

In this section, we present two computational examples of particle simulations with polynomial chaos expansion. Our objective is to validate the proposed extension of the deterministic particle scheme to stochastic situations. To allow for a detailed analysis of the treatment of the stochastic diffusion and convection terms, we consider two examples consisting of the purely diffusive evolution of a circular vortex, and of the purely convective transport of a passive scalar.

6.4.3.1 Diffusion of a Circular Vortex

We consider the problem of the diffusion of a circular exponential vortex without thermal effect. The governing equation for the vorticity field is

$$\frac{\partial \omega}{\partial t} + \mathbf{u} \cdot \nabla \omega = \nu \nabla^2 \omega, \tag{6.180}$$

with an initial condition of the form

$$\omega(r, t=0) = \frac{\exp[-r^2/d]}{\pi d}, \quad r = |\mathbf{x}|. \tag{6.181}$$

For this setting, the vortex induces a circular velocity field, with azimuthal component $v(r,t)$ and no radial component. The vortex shape is preserved, but due to diffusion the vortex core spreads with time, and so the velocity field is time dependent. The convective term vanishes, and the diffusion coefficient ν can be lumped with time. Consequently, for uncertain ν, the problem corresponds to uncertainty in the core spreading time scale. We shall consider an uncertain diffusion coefficient of the form:

$$\nu(\xi) = \nu_0 + \nu_1 \xi, \quad \nu_0 = 0.005, \, \nu_1 = \nu_0/2, \tag{6.182}$$

where ξ is uniformly distributed on $[-1, 1]$. Thus, the diffusion coefficient has a uniform distribution in the range $[0.0025, 0.0075]$. The total circulation of the vortex $\int \omega d\mathbf{x} = 1$ is an invariant of the flow, while the total circulations of the stochastic vorticity modes ($k > 0$) are all zero, *i.e.*

$$\int_{\mathbb{R}^2} [\omega]_k(\mathbf{x}, t)\, d\mathbf{x} = 0, \quad \forall t \text{ and } k \geq 1. \tag{6.183}$$

These identities can also be verified based on the exact expression of the vorticity:

$$\omega_e(r,t,\xi) = \frac{1}{\pi(d+4\nu(\xi)t)} \exp\left[\frac{-r^2}{d+4\nu(\xi)t}\right], \tag{6.184}$$

or the azimuthal velocity:

$$v_e(r,t,\xi) = \frac{1}{2\pi r}\left(1-\exp\left[\frac{-r^2}{d+4\nu(\xi)t}\right]\right).$$

For the particle simulation, the complete set of particle equations is solved, including the convection of the particles with the mean velocity field and stochastic coupling terms $\nabla \cdot ([\mathbf{u}]_l[\omega]_m)$, even though the latter are not expected to contribute because all stochastic modes of the velocity should be orthogonal to the stochastic modes of the vorticity gradient. The problem is solved for an initial condition corresponding to $d = 10\pi\nu_0$. The evolution equations are integrated using a third-order Runge-Kutta scheme with time step $\Delta t = 0.02$. The polynomial order $\mathrm{No} = 5$, so that the number of terms in the stochastic polynomial expansion is $\mathrm{P}+1 = 6$. The particle approximation uses a smoothing parameter $\epsilon = 0.05$. The mesh for the Poisson solver has a spacing $h_g = \epsilon$. Remeshing is performed every 10 time steps; the distance between neighboring particles $\sqrt{S} = \epsilon/2$ after remeshing.

In Fig. 6.53, we compare the computed and exact values of the mean and standard deviation of the vorticity as a function of the distance to the vortex center, r, and at different times, $1 \le t \le 30$. The computed values reported correspond to the particle approximation on the semi-line $y = 0$, $x \ge 0$, while the exact values are obtained by means of accurate Gauss-Lobatto integrations of the analytic solution (6.184):

$$\langle \omega_e(r,t,\xi)\rangle = \frac{1}{2}\int_{-1}^{1} \omega_e(r,t,\xi)\,\mathrm{d}\xi, \tag{6.185}$$

$$\sigma^2(\omega_e(r,t,\xi)) = \frac{1}{2}\int_{-1}^{1} \omega_e^2(r,t,\xi)\,\mathrm{d}\xi - \langle \omega_e(r,t,\xi)\rangle^2. \tag{6.186}$$

An excellent agreement is observed for all cases shown in Fig. 6.53. The plots show that as the mean vorticity field spreads, its variance increases (up to $t \approx 10$), followed by a slower decay. The presence of a node point where the standard deviation of $\omega(r)$ exhibits a local minimum is clearly visible. This node slowly moves to larger distance from the vortex center as time progresses.

Figure 6.54 compares computed and exact radial profiles of the mean and standard deviation of the azimuthal velocity. Again, an excellent agreement between computed and exact solutions is observed. The plots show that at any fixed r, the mean velocity decays monotonically with time. On the contrary, for given $r > 0$ there is a first period of time where the velocity standard deviation increases, followed by a second stage where it decays as one may have expected.

Figure 6.55 shows the evolution of the first five stochastic modes $[\omega]_k(r,t)$ for $0 \le r \le 2$. Also shown for comparison are profile of the mean mode $[\omega]_0$. Mode 1

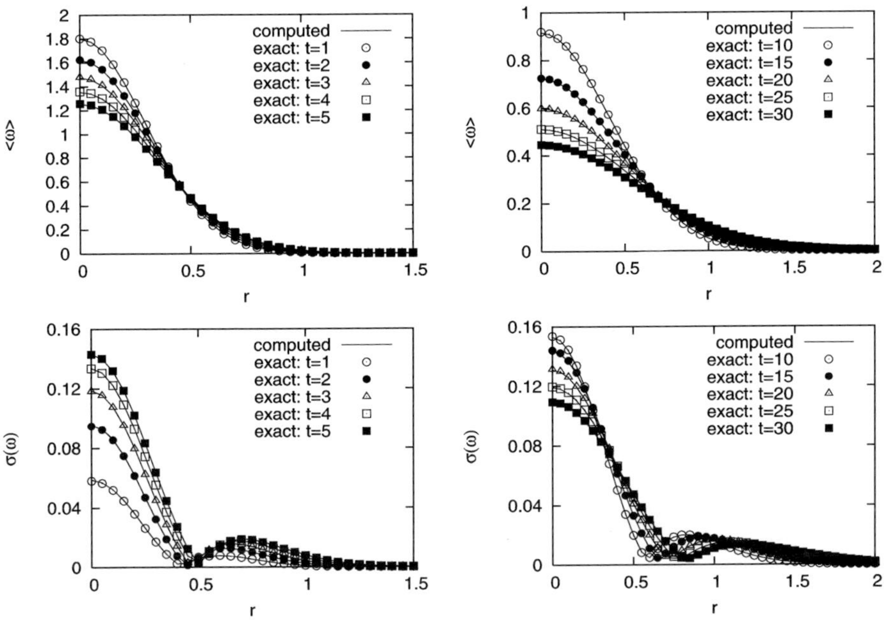

Fig. 6.53 Comparison of numerical (*lines*) and exact (*symbols*) values of the mean value of the vorticity $\langle\omega(r,t)\rangle$ (*top row*) and corresponding standard deviations $\sigma(\omega(r,t))$ (*bottom row*). Results are presented at time $t = 1, 2, 3, 4, 5$ (*left*) and $t = 10, 15, 20, 25$ and 30 (*right*). Adapted from [121]

expresses the linear departure from the mean solution as the first Legendre polynomial $\Psi_1 = \xi$. The negative values of $[\omega]_1$ at early times in the neighborhood of the vortex center express the fact that when ν (*i.e.* ξ) increases the diffusion becomes more active and so the vorticity in this region experiences lower values. At larger distance from the vortex center, an increase in the diffusion coefficient yields on the contrary larger vorticity values, and so $[\omega]_1 > 0$ for larger r. It is interesting to note that as time increases, first $[\omega]_1$ quickly increases in magnitude for $r \simeq 0$ before leveling-off for $10 \leq t \leq 30$, and then undergoes a slow decay. It is noted that as time increases, the support of $[\omega]_1$ becomes broader. Higher-order stochastic modes also exhibit similar three-stages dynamics (initial increase, leveling-off and slow decay) and broader support as time increases. For higher modes, however, the initial increase is delayed. Furthermore, one observes that $[\omega]_k$ at $r = 0$ is negative for k odd and positive for k even. Finally, one can appreciate the convergence of the stochastic polynomial expansion by noticing the decay in the magnitudes of $[\omega]_k$ as k increases. In fact, a computation with lower-order $\text{No} = 3$ revealed no significant differences on the resulting mean and standard deviation fields as reported in Figs. 6.53 and 6.54. In this simulation, the number of particles increases (due to vorticity spreading) from initially $\text{Np} = 8,192$, to $\text{Np} = 38,720$ when the computations were stopped. The CPU time was roughly 3 hours on a desktop PC.

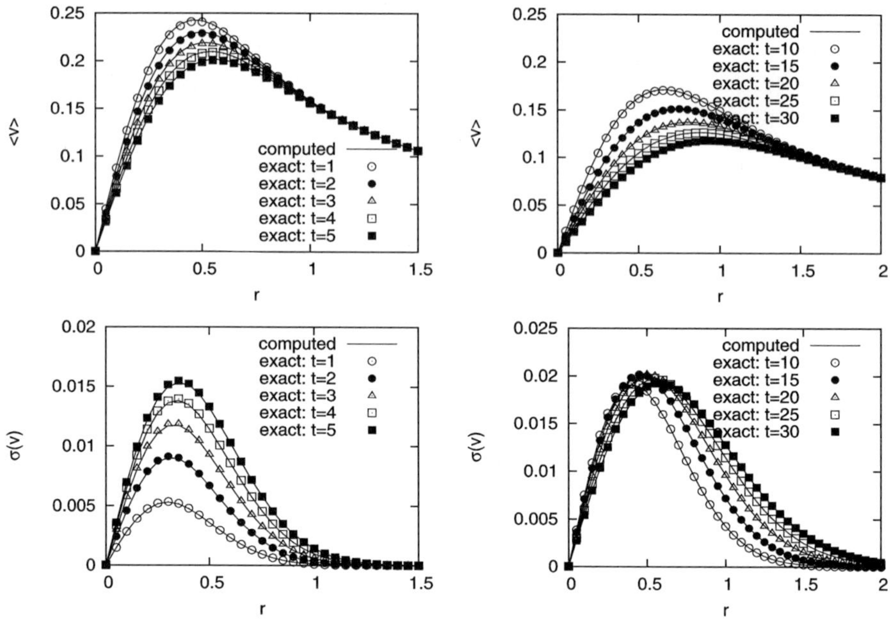

Fig. 6.54 Radial profiles of numerical (*lines*) and exact (*symbols*) values of the mean azimuthal velocity $\langle v(r,t)\rangle$ (*top row*) and its standard deviation $\sigma(v(r,t))$ (*bottom row*) at corresponding times. Profiles are plotted at time $t = 1, 2, 3, 4, 5$ (*left*) and $t = 10, 15, 20, 25$ and 30 (*right*). Adapted from [121]

6.4.3.2 Convection of a Passive Scalar

In this section, a second test problem is considered which consists of the convection of a passive scalar. The stochastic problem is specified in terms of the transport equation:

$$\frac{\partial c}{\partial t} + \mathbf{u} \cdot \nabla c = 0, \tag{6.187}$$

with given, uncertain, divergence-free velocity field $\mathbf{u}$ and the deterministic initial condition:

$$c(\mathbf{x}, 0) = \frac{\exp[-r^2/d]}{\pi d}, \quad r = |\mathbf{x} - \mathbf{c}_0|. \tag{6.188}$$

We set $\mathbf{c}_0 = \mathbf{e}_y$, $d = 0.05$ and

$$\mathbf{u}(\mathbf{x}) = -\mathbf{x} \wedge \mathbf{e}_z(1 + 0.075\xi). \tag{6.189}$$

Again, the random variable ξ is assumed to be uniformly distributed over $[-1, 1]$, and so the convective field corresponds to solid rotation about the origin with an uncertain rotational speed of 1 ± 0.075 round(s) per 2π units of time. The center of

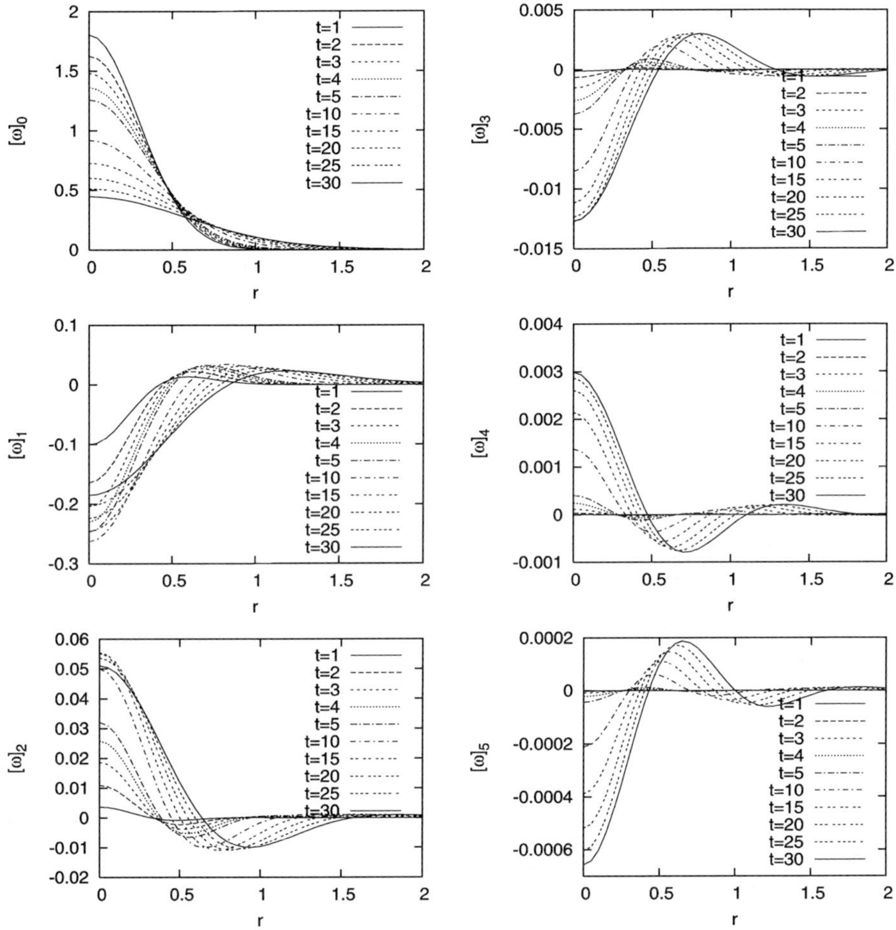

Fig. 6.55 Radial profiles of the vorticity modes $[\omega]_k(r,t)$ plotted at time $t = 1, 2, 3, 4, 5, 10, 15, 20, 25, 30$. Adapted from [121]

the concentration field, **c**, evolves according to:

$$\mathbf{c}(t,\xi) = \cos\left[\pi/2 + (1+0.075\xi)t\right]\mathbf{e}_x + \sin\left[\pi/2 + (1+0.075\xi)t\right]\mathbf{e}_y, \quad (6.190)$$

and the exact scalar distribution may be expressed as:

$$c_e(\mathbf{x},t,\xi) = \frac{1}{2d}\exp\left[\frac{-|\mathbf{x}-\mathbf{c}(t,\xi)|^2}{d}\right]. \quad (6.191)$$

For the inputs above, at $t = k(2\pi)$ the center of the scalar distribution is uniformly distributed over an arc with length $0.015(2\pi)k \sim 1k$, while the radius of the scalar distribution is estimated as $\sim 2\sqrt{d} \sim 0.5$. These estimates show that the uncertain velocity field induces in just one revolution an uncertainty in the scalar

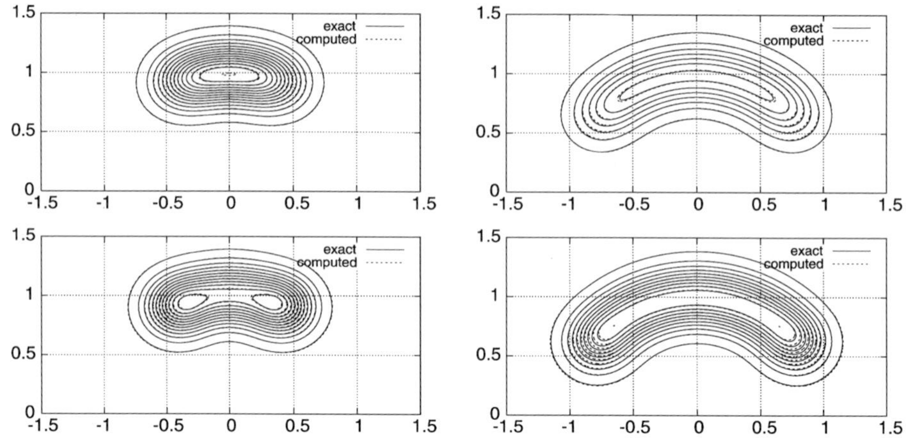

Fig. 6.56 Mean (*top row*) and standard deviation (*bottom row*) of the scalar field after 1 revolution (*left*) and 2 revolutions (*right*). Plotted are the particle solution (No $= 20$) and the exact solution. Contour levels start at $\Delta/2$ with fixed increments Δ, where $\Delta = 0.2$. Adapted from [121]

field location which is of the same order as the diameter of the distribution of c. Thus, the present problem constitutes a challenging test, as high-order expansions are needed in order to represent solution at even moderate times. Note that these challenges are inherent to the stochastic nature of the problem, and are not associated with the selected Lagrangian discretization scheme. Specifically, high-order PC expansions would also be needed if an Eulerian discretization scheme is used. Note, however, that in the latter case the numerical solution would face additional difficulties associated with the transport of a non-diffusing scalar, and that these difficulties would also arise in a deterministic setting. To address these difficulties, Eulerian approaches rely on elaborate discretizations (using for instance flux limiters), but extension of these discretizations to situations involving random velocity fields is not immediate. Particle methods, on the other hand, are well-suited for convection-dominated problems. One of the objectives of the present tests is to verify that this remains the case when the convective field is uncertain.

The problem is first solved for a particle discretization with $\epsilon = 0.025$, a third-order Runge-Kutta scheme with $\Delta t = 2\pi/400$, and large polynomial order No $= 20$. Remeshing is applied every 10 iterations, with spacing $\sqrt{S} = \epsilon/2$. The computed means and standard deviations of the scalar fields after 1 and 2 revolutions, together with the corresponding exact solutions, are plotted in Fig. 6.56. The agreement between particle and exact solutions is again excellent.

We present in Fig. 6.57 the time-evolution of the spectrum $E_2(k)$, defined according to:

$$E_2(k) = \sum_{i=1}^{\mathrm{Np}} [C_i]_k^2, \tag{6.192}$$

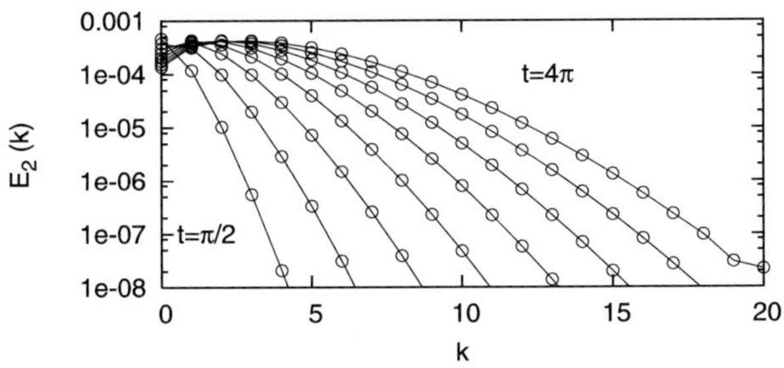

Fig. 6.57 Spectrum $E_2(k)$, as defined in (6.192), of the particle solution for No = 20. Curves are generated at every quarter revolution. Adapted from [121]

where C_i is the scalar strength of the i-th particle. E_2 essentially measures the energy in the individual scalar modes. One observes that the energy of mode zero decays, which reflects increasing uncertainty in the location where the scalar is concentrated. The energies of the higher modes $k > 0$ on the contrary steadily increase. It is however observed that the spectra monotonically decrease as k increases, denoting the convergence of the polynomial expansion. However, as time progresses, the decay of the spectrum with k becomes slower, indicating that the number of stochastic modes needed to suitably represent the solution increases with time.

We now focus on the conservation of first invariants in the numerical solution. Since the initial condition is deterministic, we have for all time

$$I(0) = \sum_{i=1}^{\mathrm{Np}} [C_i]_0 = 1, \qquad I(k) = \sum_{i=1}^{\mathrm{Np}} [C_i]_k = 0, \quad k \geq 1. \tag{6.193}$$

The proposed method being conservative at the discrete level, it is expected that these invariants are also conserved in the computations. However, due to remeshing, this is not necessarily the case. Specifically, in order to avoid an unreasonable increase in the number of particles, only particles with strengths exceeding a predetermined threshold are retained after remeshing, while others are discarded. As a result, the invariants are not exactly conserved. The error that is incurred depends on the threshold value used to decide whether particles are kept or omitted. In the present computation, this criterion consists in discarding particles whose strength satisfy $|[C_i]_k| < 10^{-8}$ for all k. Figure 6.58 shows the evolution $|I(k)|$ as a function of the number of revolutions for even modes, $k \geq 2$. It is observed that the invariants $I(k > 0)$ do in fact vary with time. Similar observation is made for the odd modes, which however have smaller magnitudes. However, the conservation "errors" associated with the first invariants are small, particularly compared to the energies $E_2(k)$, and consequently deemed acceptable. Thus, further refinement of the remeshing procedure was not attempted. Note that due to the uncertainty in the convective field, the domain covered by the particles must extend beyond the support of a single realization (or in other words the support of the initial distribution). This is achieved during remeshing, specifically through the introduction of additional particles around the boundaries of the prevailing particle distribution. As is the case for

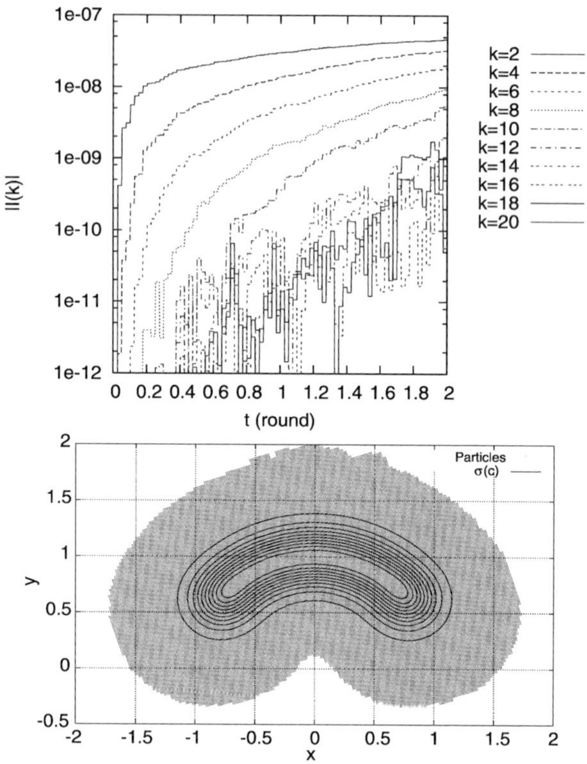

Fig. 6.58 *Top*: evolution of the first invariants of the even modes $I(k)$. *Bottom*: distribution of the particles after 2 revolutions. Also plotted are the isolines of the standard deviation of c at the same time; contour levels start at 0.15 with constant increment 0.3. The number of particles $\mathrm{Np} \sim 39500$, whereas only 15000 particles were present at the start of the computations. Adapted from [121]

particle removal, the introduction of new particles depend on the selected tolerance. Numerical tests have shown that conservation of the invariants considered above improves by lowering the tolerance level. However, this improvement comes at the cost of a significant increase in the number of particles and with only small improvement of the statistics, as more resources are added in regions where the scalar modes are very small. This is illustrated in Fig. 6.58, which simultaneously depicts the particle distribution and contours of the variance in the scalar field. In particular, the figure indicates that the region where significant variance occurs is contained well within the region covered by the particles, and thus the addition of new particles is beyond the point of diminishing returns. These observations also justify the selection of the tolerance value.

In another series of numerical tests, the order of the PC expansion was progressively decreased while keeping constant the parameters of the particle discretization. The aim was to assess the robustness of the predictions to under-resolved polynomial expansions. Spectra for $\mathrm{No} = 10$ and $\mathrm{No} = 5$ are plotted in Fig. 6.59, and are contrasted with previous predictions obtained with $\mathrm{No} = 20$. The comparison shows that with $\mathrm{No} = 10$ the predictions can be considered to be suitably resolved up to 1.5 revolutions, while for $\mathrm{No} = 5$ the resolution becomes insufficient after just 3/4 of revolution. Truncation to lower-order essentially affects the highest mode, while for $k < \mathrm{No}$ the energy is nearly unaffected by the selected value of No. Additional

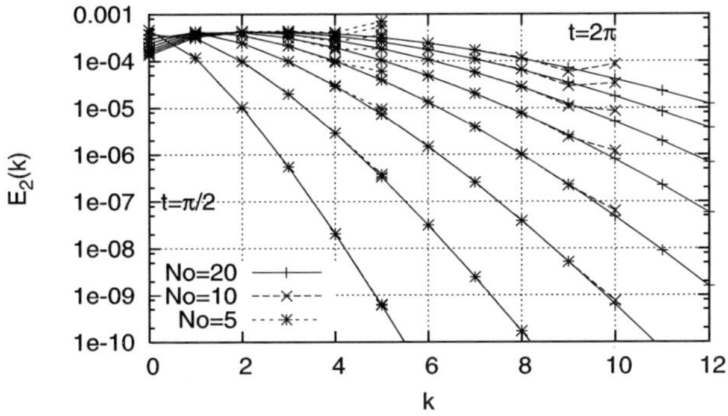

Fig. 6.59 Comparison of the spectra $E_2(k)$ at different times and computed for stochastic expansions truncated at different order No = 20, 10 and 5. Adapted from [121]

insight into the impact of expansion order can be gained from Fig. 6.60, which depicts contours at $t = 4\pi$ of mode $k = 4$ computed using No = 5 and 10, and of mode $k = 9$ computed with No = 10 and No = 20. It is seen that with No = 5 the prediction of mode 4 is clearly corrupted. For mode 9, the predictions obtained with No = 10 and 20 are close, though noticeable differences are still observed. The impact of resolution effects can also be appreciated from Fig. 6.61, which illustrates the standard deviation fields for different orders at the end of the computations.

6.4.4 Application to Natural Convection Flow

In this section, we discuss the implementation of the stochastic mesh-particle scheme to simulate the evolution of a localized hot patch of fluid in an infinite domain. The initial conditions are:

$$\omega(\mathbf{x}, t = 0) = 0, \qquad T(\mathbf{x}, t = 0) = \exp\left(-10r^8\right), \quad r = |\mathbf{x}|. \tag{6.194}$$

For the range of Rayleigh numbers considered here, the evolution of the flow that can be summarized as follows. At early stages, the hot patch starts to rise due to buoyancy, and a pair of counter-rotating vortices is created on its sides. The flow induced by the vortices distorts the patch, and the latter experiences higher velocities at its centerline. As time increases, the distortion becomes more pronounced and the patch is strained in a filament trapped in the two rolling vortices. At the centerline, a smaller secondary patch of hot fluid subsequently detaches, and a second pair of counter-rotating vortices is formed; see Figs. 6.63 and 6.64 below.

One can easily draw a qualitative picture of the dependence with the Rayleigh number of these processes summarized above. For instance, at lower Rayleigh numbers one would expect less energetic vortices, as higher mixing rates would occur during the rollup process. The eventual interruption of successive detachments

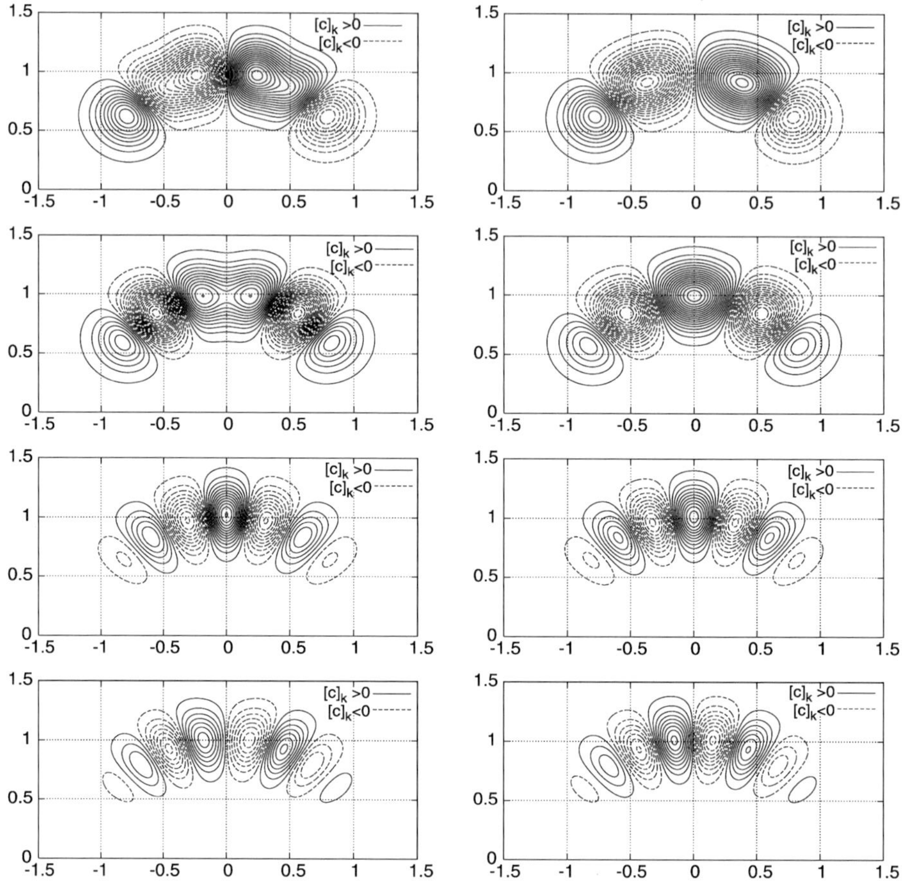

Fig. 6.60 *Top row*: isolines of $[c]_4$, computed with expansion orders No $= 5$ (*left*) and No $= 10$ (*right*). *Second row*: isolines of $[c]_5$, computed with expansion orders No $= 5$ (*left*) and No $= 10$ (*right*). *Third row*: isolines of $[c]_9$, computed with expansion orders No $= 10$ (*left*) and No $= 20$ (*right*). *Fourth row*: isolines of $[c]_{10}$, computed with expansion orders No $= 10$ (*left*) and No $= 20$ (*right*). In all cases, the computed solutions at $t = 4\pi$ are used; isolines start at ± 0.1 with constant increment ± 0.2. Adapted from [121]

and rollups due to viscous stabilization would also be anticipated. However, from a quantitative perspective, prediction of the dependence of even integral quantities on variability in the Rayleigh number is more challenging. Indeed, variability in the Rayleigh number not only affects diffusion rates and the convective field (Sects. 6.4.3.1 and 6.4.3.2), but also the complex coupling between the thermal and convective fields.

To investigate these effects, the full set of Boussinesq equations is solved. The Prandtl number $\mathrm{Pr} = 0.71$. An uncertain Rayleigh number is considered, $\mathrm{Ra} = 2.5 \times 10^5 \pm 5 \times 10^4$, and once again a uniform pdf is assumed, leading to the following

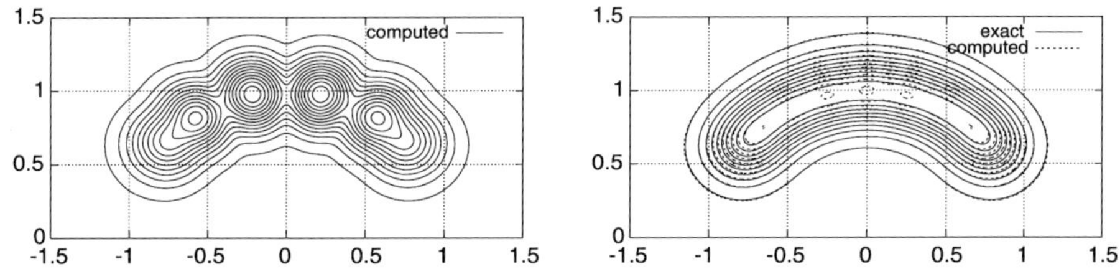

Fig. 6.61 Isolines of standard deviation in c at $t = 4\pi$, computed with expansions truncated at No = 5 (*left*) and No = 10 (*right*). Also plotted on the right is the standard deviation of the exact solution. Isolines start at 0.1 with constant increment 0.2. Adapted from [121]

one-dimensional Legendre expansion:

$$\mathrm{Ra}(\xi) = [\mathrm{Ra}]_0 + [\mathrm{Ra}]_1\xi = 2.5 \times 10^5 + 5 \times 10^4\xi,$$

$$p(\xi) = \begin{cases} 1/2, & \xi \in [-1, 1] \\ 0, & \text{otherwise} \end{cases}.$$

Evaluation of the PC expansion of the factor $1/\sqrt{\mathrm{Ra}}$, which features in the governing equations, is performed in two steps. First, the expansion of $1/\mathrm{Ra}$ is computed by inverting the exact Galerkin product [52, 127]:

$$\mathrm{Ra} \times \frac{1}{\mathrm{Ra}} = \sum_k \left(\sum_l \sum_m C_{klm} [\mathrm{Ra}]_l \left[\frac{1}{\mathrm{Ra}}\right]_m \right) \Psi_k \equiv 1\Psi_0. \tag{6.195}$$

Second, the expansion of the square root of $1/\mathrm{Ra}$ is extracted by solving a nonlinear system of equations expressing again the Galerkin product [52]:

$$\sum_l \sum_m C_{klm} \left[\frac{1}{\sqrt{\mathrm{Ra}}}\right]_l \left[\frac{1}{\sqrt{\mathrm{Ra}}}\right]_m = \left[\frac{1}{\mathrm{Ra}}\right]_k, \quad k = 0, \ldots, \mathrm{No}. \tag{6.196}$$

To extract the "positive" square root, an iterative Newton method is employed with starting values $[1/\sqrt{\mathrm{Ra}}]_0 = 1/\sqrt{[\mathrm{Ra}]_0}$ and $[1/\sqrt{\mathrm{Ra}}]_{k>0} = 0$. The resulting probability density function of $1/\sqrt{\mathrm{Ra}}$ is plotted in Fig. 6.62; a PC expansion with No = 12 is used. The PC approximation of $1/\sqrt{\mathrm{Ra}}$ is also reported in Fig. 6.62, which indicates that the flow will be less diffusive as ξ increases. Also note that $\langle 1/\sqrt{\mathrm{Ra}} \rangle \sim 2.01 \times 10^{-3}$ slightly differs from $1/\sqrt{\langle \mathrm{Ra} \rangle} = 2 \times 10^{-3}$.

In the computations, we set $\epsilon = 1/30$, and remesh spacing $\sqrt{S} = \epsilon/2$. The mesh has spacing $h_g = \epsilon$, giving an average of 4 particles per cell. Time integration is performed by applying the second-order Adams-Bashforth scheme, with time step $\Delta t = 0.2$, except for iterations that immediately follow remeshing, where a second-order Runge-Kutta scheme is employed. Remeshing is performed every 4 iterations.

Figure 6.63 compares the mean temperature field $[T]_0$ with the deterministic temperature field corresponding to the mean Rayleigh number $\langle \mathrm{Ra} \rangle$, at times $t = 10$ and $t = 20$. At $t = 10$, the two fields are nearly identical, the stochastic mean field

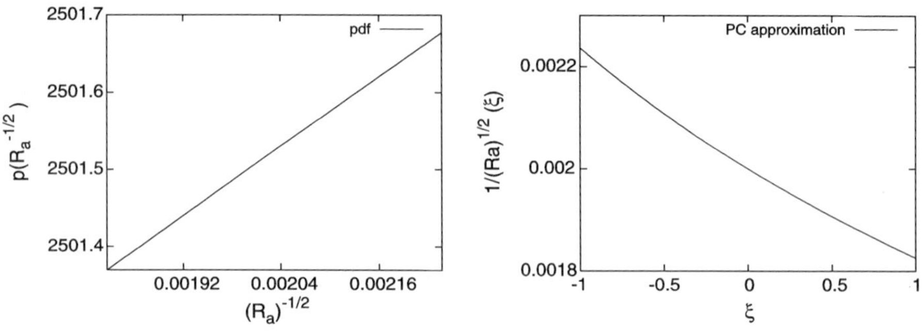

Fig. 6.62 *Left*: probability density function of $1/\sqrt{\mathrm{Ra}}$. *Right*: PC approximation of $1/\sqrt{\mathrm{Ra}}(\xi)$. Adapted from [121]

just being slightly smoother than the deterministic prediction. On the contrary, at $t = 20$ noticeable differences appear, particularly in the zone of high shear around the primary vortex, in the core of the secondary vortices, and at the centerline of the leading hot spot. As discussed previously, these differences arise due to the nonlinearity of the flow, as well as the small differences in the mean diffusion coefficient and the square root of the inverse mean Rayleigh number. Similar conclusions can be drawn from the inspection of the corresponding mean and deterministic vorticity fields compared in Fig. 6.64. Note how the size and location of the mean secondary vortices are affected by the variability in Ra.

Additional insight into the variability of the flow is sought through analysis of standard deviation fields. Figure 6.65 shows the mean and standard deviation of the temperature field, and the magnitude and standard deviation of the vorticity field at $t = 20$. Only half of the domain is shown, as the temperature field is symmetric with respect to the axis $x = 0$, whereas the vorticity field is antisymmetric. For the temperature field, it is observed that the areas of greatest variability are located on the centerline of the leading patch of hot fluid and on the external sides of the fluid entrained by the primary vortex. Note that in some areas the temperature standard deviation is as high as one third of the local mean. In other areas, however, such as the core of the primary vortex, only small variability of the temperature occurs. Similar trends can be observed for the vorticity field, though higher variation is noted around the edges of the primary vortex where vorticity filaments of opposite sign are rolling up. In these regions, variability in Ra has a significant influence on the vortex roll-up, and the location of the filament structures has a high level of uncertainty.

To assess the performance of the particle-mesh scheme, we plot in Fig. 6.66 the evolution of Np as well as the particle distribution at $t = 20$. The number of particles increases steadily with time, which is expected since the support of the stochastic solution is also increasing. At $t = 20$, the number of particles $\mathrm{Np} \sim 15000$ whereas $\mathrm{Np} = 13000$ at the start of the computations. Also note that the particle distribution properly covers the support of the mean and standard deviation fields depicted earlier, and even extends into regions where the solution modes are essentially vanishing. Thus, a less conservative criterion for particle removal could have safely been

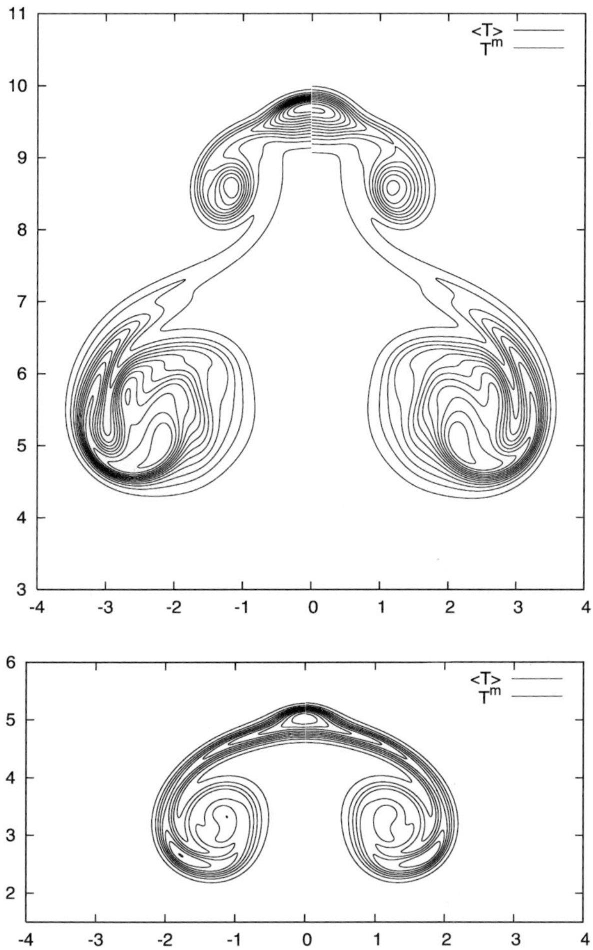

Fig. 6.63 Contours of the mean temperature field $\langle T \rangle$, for $x > 0$ (*right half*), and of the deterministic temperature field T^m, for $x < 0$ (*left half*) corresponding to the mean Rayleigh number $\langle \text{Ra} \rangle$. *Top*: solutions at $t = 20$ with levels starting at 0.01 with constant increment 0.02. *Bottom*: solutions at $t = 10$, with levels starting at 0.03 with constant increment 0.06. Adapted from [121]

used. Also note that the computation simulation was actually carried up to $t = 27.5$, time at which the number of particles exceeded than 200000. The CPU time used on a desktop PC was about 10 hours. Thus, the present experience shows that the overhead due to stochastic polynomial expansion can be manageable, even when a large number of particles is used.

As mentioned previously, a PC expansion with $\text{No} = 12$ was used in the present simulation. In order to verify the suitability of the expansion, we plot in Fig. 6.67 the spectra of the vorticity and temperature fields. From these spectra, it can be concluded that the expansion is clearly over-resolved at early times ($t \leq 7$), where a third- or fourth-order expansion would have been sufficient. The energy of the low-order stochastic modes slowly increases up to $t \sim 7$, when a first transition occurs, after which the energy increases at a higher pace. This time actually corresponds to the formation of the primary vortices. The primary vortices then experience a lower rise velocity than the bulk of the hot fluid, and filamentation around its core occurs

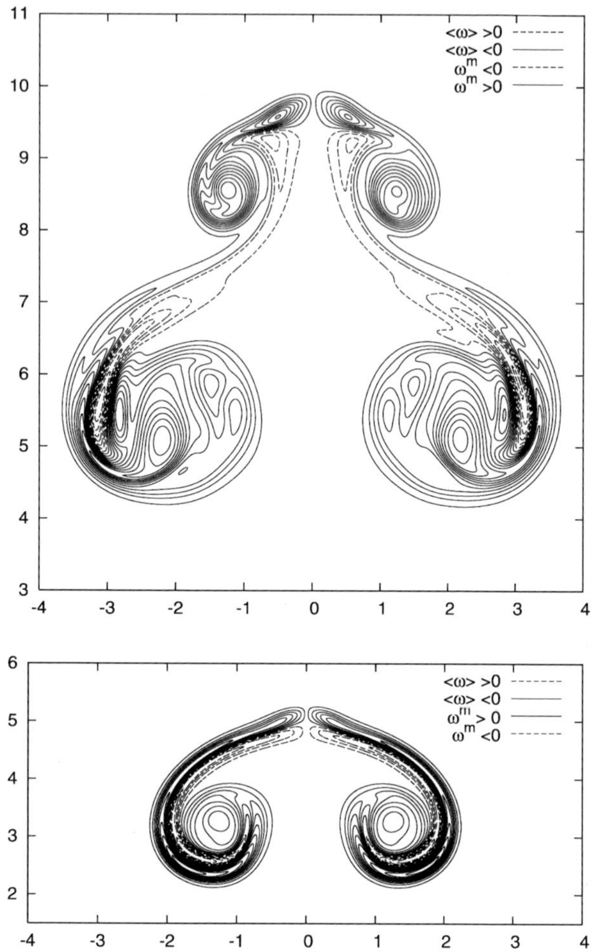

Fig. 6.64 Contours of the mean stochastic vorticity field $\langle\omega\rangle$, for $x > 0$ (*right half*), with the deterministic vorticity field ω^m, for $x < 0$ (*left half*), corresponding to the simulation using $\langle\text{Ra}\rangle$. *Top*: solutions at $t = 20$ with levels starting at ± 0.1 with constant increment ± 0.2. *Bottom*: solutions at $t = 10$ with levels starting at 0.25 with constant increment 0.5. Adapted from [121]

(see Fig. 6.64). This is accompanied by the sharp increase of the mode energies during the period $t \in [8, 13]$. At later stages, after detachment of the primary vortices, the energy of the stochastic modes levels off. Around $t = 17$–18, the formation of the secondary vortices begins, and the higher-order modes $k > 8$ again experience a rapid growth up to the time at which the secondary vortices detach. Note that during this process the lower-order modes are less affected than the higher-order modes. Consequently, the stochastic dynamics lead to flatter and broader spectra as time advances, denoting the need for additional modes for suitable resolution. For instance, between $t = 10$ and $t = 20$ the energy ratio between modes 1 and 12 has reduced from more than 10 orders of magnitude to about 6. Overall, the expansion order was deemed sufficient, and this was verified through inspection of the spatial structure of the stochastic modes (not shown), and their rapid decay with increasing mode number.

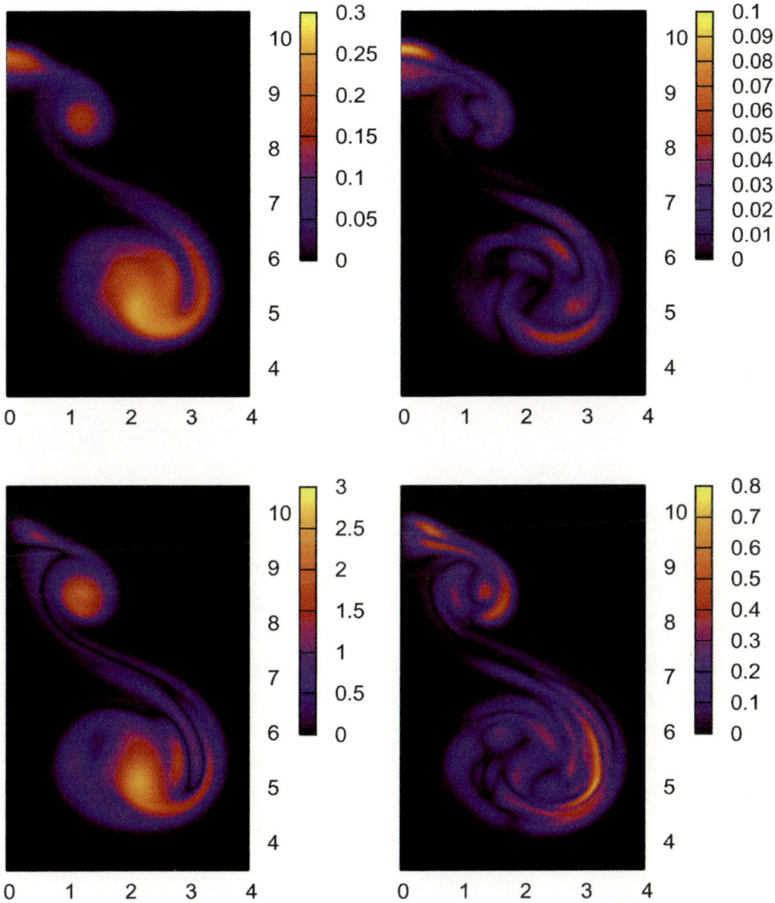

Fig. 6.65 *Top*: mean temperature field $[T]_0$ (*left*) and standard deviation (*right*) at $t = 20$. *Bottom*: magnitude of the mean vorticity field $|[\omega]_0|$ (*left*) and standard deviation of ω (*right*) at $t = 20$. Adapted from [121]

As a final check of the simulation accuracy, the errors on the first invariant of the temperature modes $I_T(k)$ are plotted in Fig. 6.68. The evolution of the invariant errors is reported only for the temperature modes, as the vorticity modes are anti-symmetric and thus the first-order errors tend to naturally balance out. Figure 6.68 shows that the errors increase steadily, with small modulation during vortex formation. As discussed in the previous section, these errors are entirely due to the remeshing scheme, which in particular removes particles with low strengths (a discussion on the effect of particle removal in the deterministic context can be found in [64]). However, it is seen that the error on the invariants are low and are smaller than 10^{-7} for all modes. Compared to the mean thermal energy of the system, $I_T(k = 0) = 1.6$, the first-order errors are evidently negligible.

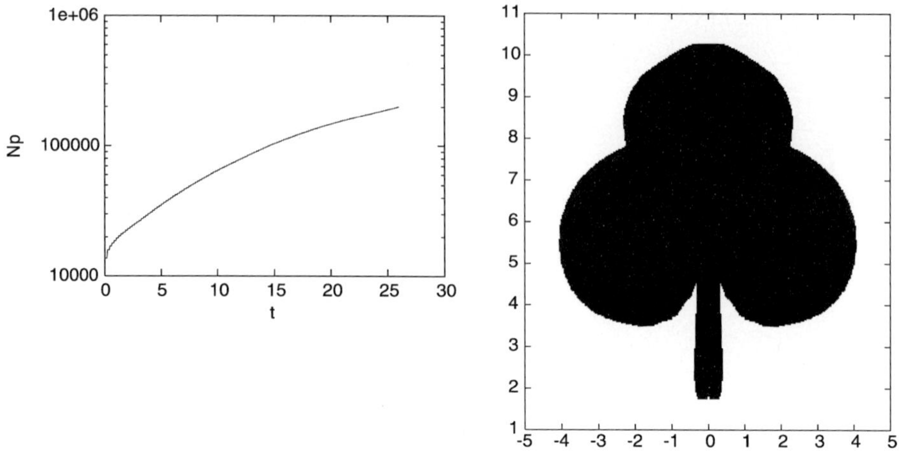

Fig. 6.66 Evolution of the number of particles in the simulation, and the particle distribution at $t = 20$. Adapted from [121]

In the analysis above, we have focused on the first and second statistical moments of the solution. One should note, however, that the PC expansion provides the local probability law of any observable (functional) of the flow variables. In particular, since in the present problem the input uncertainty is bounded, the stochastic representation readily yields estimates of temperature maxima at any point of the domain. As an example, we show in Fig. 6.69 the local sensitivities S_T and S_ω of the temperature and vorticity fields at $t = 20$. The sensitivity of T is defined by:

$$S_T = \left.\frac{\partial T}{\partial \mathrm{Ra}}\right|_{\mathrm{Ra}=\langle \mathrm{Ra}\rangle} = \left.\frac{\partial T}{\partial \xi}\left(\frac{\partial \mathrm{Ra}}{\partial \xi}\right)^{-1}\right|_{\xi=0} = \left.\frac{1}{[\mathrm{Ra}]_1}\frac{\partial T}{\partial \xi}\right|_{\xi=0}.$$

Using the PC expansion of T and the particle approximation, the local sensitivity at point $\mathbf{x}$ around $\langle \mathrm{Ra}\rangle$ is

$$S_T(\mathbf{x}) = \sum_{k=0}^{\mathrm{No}}\sum_{i=1}^{\mathrm{Np}} \eta_\epsilon(\mathbf{x}-\mathbf{X}_i)[H_i]_k \frac{\partial \Psi_k}{\partial \xi}(\xi = 0).$$

Note that this sensitivities could be estimated at any Ra value in the uncertainty range.

6.4.5 Remarks

In this section, a stochastic Lagrangian method was outlined for UQ in fluid flow. The method combines the advantages and flexibility of particle discretizations

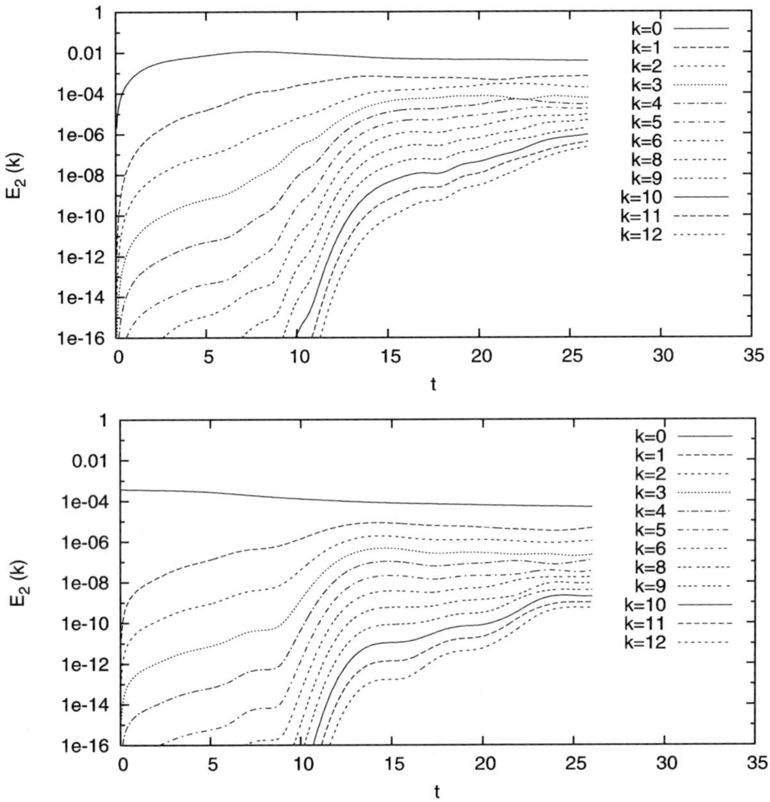

Fig. 6.67 Evolution of energies of the stochastic modes of vorticity (*top*) and temperature (*bottom*). Adapted from [121]

(mesh-less, robustness, stability, low diffusivity,...) with efficient uncertainty representations involving PC expansions. Its essential feature is the use of a single set of particles to transport the stochastic modes of the solution. This approach overcomes the difficulties that would arise if different sets of particles are defined for different modes. It further allows for the re-use of fast methods, including particle-mesh techniques. Another key aspect of the method is the conservative character at the discrete level of the treatment of the interaction between stochastic modes. The discretization was found to be stable and essentially diffusion-free, thus allowing uncertainty quantification in slightly viscous flow, and transport at high, or even infinite Péclet numbers by uncertain velocity fields.

The discussion above has focused on a basic construction that combines a particle-mesh scheme with uncertainty quantification method based on the PC formalism, and applications of the resulting stochastic scheme were restricted to unbounded flows. Additional work is needed to extend the formulation to more general conditions and gain better understanding of the range of possible applications. For instance, generalization of the formulation to accommodate solid boundaries

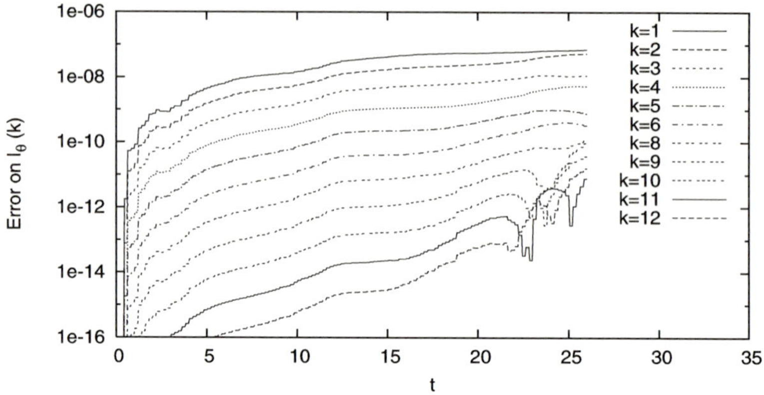

Fig. 6.68 Evolution with time of the absolute error in the first invariant of the temperature modes. Adapted from [121]

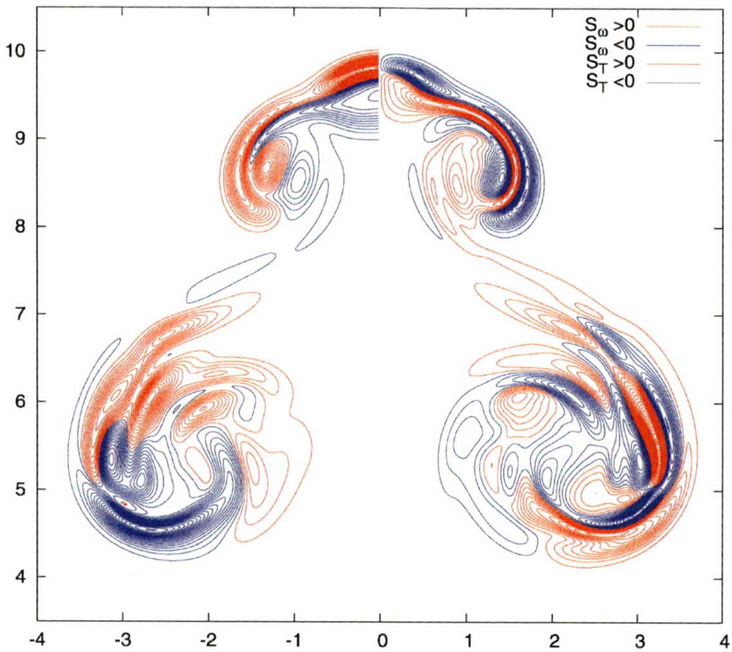

Fig. 6.69 Iso-contours of local sensitivity in temperature (*left half of the plot*, $x < 0$) and vorticity (*right half*, $x > 0$) with regard to Ra about $\langle \mathrm{Ra} \rangle$. Contours starts at $\pm\Delta$ with successive increments Δ. $\Delta = 1.5 \times 10^{-7}$ for temperature and 2×10^{-6} for vorticity. Adapted from [121]

would be desirable. This could be achieved, for instance, by adapting the scheme of Koumoutsakos *et al.* [118], in which a linear problem for the vorticity flux across the boundary is solved at each time-step. Application of the PC formalism to this scheme would result in a set of decoupled problems for the stochastic vorticity modes, thus preserving the features of the deterministic formulation. Another possible refinement of the present construction concerns the remeshing algorithm. In particular, more elaborate constructions should be sought, which would improve accuracy and allow conservation of flow invariants. In particular, accurate approaches developed for deterministic flows [41] appear to be readily amenable to PC expansions using the presently developed framework.

6.5 Mulitphysics Example

In this section, we provide an illustration of PC-based UQ in a complex setting, namely in the context of protein labeling reactions in microchannel flows. These flows generally involve electroosmotic flow of charged components in an electrolyte buffer, and are generally characterized by strong coupling between multiple physical and chemical processes [225]. Numerical simulations for detailed studies of phenomena such as analyte dispersion therefore require accurate models for the fluid flow, species transport, chemical reactions, buffer equilibrium, protein ampholytic behavior, electrostatic field strength, wall layer, and many other processes [197]. Most of these processes are well understood and adequate models are generally available. Many simulations of microchannel flow can be found in the literature, with varying detail in the resolution of the ongoing physical processes [5, 22, 66–68, 142, 152, 178, 182, 205, 226]. However, simulations that take into account the full range of coupled processes in microchannel flow are less common.

Below, we summarize recent experiences at coupling spectral UQ techniques with a detailed model of electroosmotic and pressure-driven flow in a microchannel filled with an electrolyte buffer and model protein analyte samples. First we outline the formulation of the governing equations that constitute the deterministic system model. As further discussed below, the formulation considers the fully coupled momentum, species transport, and electrostatic field equations, including a model for the dependence of the zeta potential on pH and buffer molarity. A mixed finite-rate, partial-equilibrium formulation is applied for the chemical reactions. In particular, "fast" electrolyte reactions are described by associated equilibrium constraints, while the remaining "slow" protein labeling reactions are modeled with finite-rate kinetics. Next, we implement the stochastic uncertainty quantification method to reformulate these equations into evolution equations for the spectral mode strengths. We then proceed to the description of the numerical construction used to integrate the resulting set of equations, highlighting particular developments necessary for handling the coupled evolution of momentum, species, and the electrostatic field. The methodology is then applied to model protein labeling reactions in homogeneous systems as well as two-dimensional microchannel flows. The results illustrate the convergence of the construction as well as the propagation/growth of uncertainty in the simulations. The detailed physical model gives insight in important microfluidic sample dispersion mechanisms.

6.5.1 Physical Models

6.5.1.1 Momentum

The continuity and momentum equations for a two-dimensional flow field in the (x, y) plane, with uniform density and viscosity are given by [192]

$$\nabla \cdot \mathbf{u} = 0, \tag{6.197}$$

$$\frac{\partial \mathbf{u}}{\partial t} + \mathbf{u} \cdot \nabla \mathbf{u} = -\nabla p + \nu \nabla^2 \mathbf{u} \tag{6.198}$$

where $\mathbf{u}$ is the velocity, p is the pressure normalized by density, and ν is the kinematic viscosity.

The microchannel flows in this section are electroosmotically driven with an applied electrostatic field in the x-direction. Assuming a double layer that is thin with respect to the channel size, the effect of wall electrostatic forces can be represented in terms of a wall slip velocity u_w, using the Helmholtz-Smoluchowski relationship [192]

$$u_w = \frac{\epsilon \zeta}{\mu} \nabla_t \phi_w \tag{6.199}$$

where ϵ is the permittivity of the fluid, ζ is the zeta potential, ϕ_w is the electrostatic field potential at the wall, and μ is the dynamic viscosity. Since both the electrostatic field and the ζ potential depend on the fluid composition, (6.199) represents a major coupling between the flow velocity and the species transport.

The ζ potential is a function of the wall material and fluid characteristics [152, 237]. In this work, a relationship for ζ as a function of the local pH and buffer molarity was obtained from empirical data for the zeta potential of a fused silica capillary in an aqueous solution of KCl, as shown in Fig. 6.70 [202]. This data was

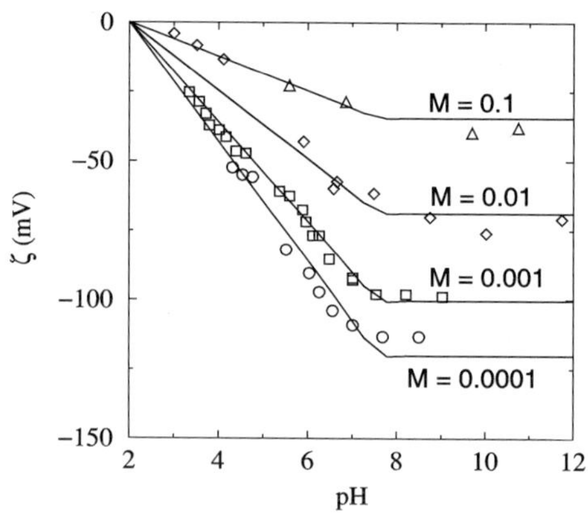

Fig. 6.70 Empirical data and curve fit for the ζ potential of a fused silica capillary versus pH in an aqueous solution of KCl at various molarities. Adapted with permission from [202], Copyright 1992 American Chemical Society

curve-fitted into the following relationship:

$$\zeta(\text{pH}, \mathcal{M}) = \left\{ -(\text{pH}-2) + \left(\frac{1}{2} + \frac{1}{2} \tanh(5(\text{pH}-7.5)) \right) (\text{pH}-7.6) \right\} \times \left(-2.7 \ln(\mathcal{M} + 2.3 \times 10^{-4}) \right) \tag{6.200}$$

where $\mathcal{M}$ is the molarity of the KCl solution. The quantitative accuracy of this curve-fit is obviously limited to systems similar to the one considered in [202]. However, (6.200) qualitatively gives the correct behavior of $\zeta(\text{pH}, \mathcal{M})$ for various other systems [152, 237].

6.5.1.2 Species Concentrations

A variety of species are considered in this work, ranging from model proteins and dyes in samples, to the ions of aqueous buffer solutions. The transport of these species is governed by [192]:

$$\frac{\partial c_i}{\partial t} + \nabla \cdot \left[c_i (\mathbf{u} + \mathbf{u}_i^e) \right] = \nabla \cdot (D_i \nabla c_i) + \hat{w}_i \tag{6.201}$$

where c_i is the concentration of species i, and D_i is the corresponding diffusivity. The electromigration velocity $\mathbf{u}_i^e$ accounts for the electrophoretic movement of electrically charged species relative to the bulk flow. This velocity is given by [192]

$$\mathbf{u}_i^e = -\beta_i z_i F \nabla \phi \tag{6.202}$$

where β_i is the electrophoretic mobility for species i, z_i is the charge number, F is the Faraday constant (9.648×10^4 C/mol), and ϕ is the electrostatic field potential. The term $\hat{w}_i$ is a source term from the chemical and electrochemical reactions in which species i is involved. Note that for each species, the diffusivity D_i and the mobility β_i are coupled through the Nernst-Einstein [192]

$$D_i = RT\beta_i \tag{6.203}$$

where R is the universal gas constant and T the temperature.

The integration of (6.201) is performed differently depending on the chemical time scales involved. In general, electrolyte association and dissociation reaction rates are several orders of magnitude faster than electrophoretic phenomena [225] and typical sample-processing reactions. Thus, direct integration of fast reactions would impose severe time step restrictions. In order to avoid these difficulties, an equilibrium approach for the electrolyte reactions is implemented. For example, consider a weak acid HA, which dissociates according to

$$\text{HA} \overset{K_A}{\longleftrightarrow} \text{H}^+ + \text{A}^- \tag{6.204}$$

where

$$K_A \equiv \frac{[\mathrm{H}^+][\mathrm{A}^-]}{[\mathrm{HA}]} \tag{6.205}$$

is the corresponding dissociation constant. Instead of integrating (6.201) for the concentrations of species HA and A^- individually, consider the combined concentration of both of these quantities $\rho_a = [\mathrm{HA}] + [\mathrm{A}^-]$. The source terms for [HA] and $[\mathrm{A}^-]$ from the electrolyte reaction (6.204) cancel out in the ρ_a transport equation, which is the sum of the transport equations for the two individual quantities.

$$\begin{aligned}\frac{\partial \rho_a}{\partial t} &+ \nabla \cdot \left[c_{\mathrm{HA}}(\mathbf{u} + \mathbf{u}^e_{\mathrm{HA}}) + c_{\mathrm{A}^-}(\mathbf{u} + \mathbf{u}^e_{\mathrm{A}^-})\right] \\ &= \nabla \cdot \left[D_{\mathrm{HA}} \nabla c_{\mathrm{HA}} + D_{\mathrm{A}^-} \nabla c_{\mathrm{A}^-}\right].\end{aligned} \tag{6.206}$$

Therefore, barring any other chemical reactions involving these species, ρ_a is a conserved quantity and can be integrated with (6.206) without a chemical source term [66, 178, 197]. Note that if the chemical source terms for HA or A^- in (6.201) do include participation by reactions other than the HA buffer chemistry, *e.g.* by (typically slow) sample chemistry, then the utilization of ρ_a is still advantageous in that it eliminates the fast electrolyte reactions, but in this case ρ_a is no longer a conserved scalar. In either case, one arrives at a governing equation for ρ_a, which does not include the fast reaction terms. Once ρ_a is known, the concentrations of the individual components of the weak acid are obtained from:

$$[\mathrm{HA}] = \frac{[\mathrm{H}^+]}{[\mathrm{H}^+] + K_A} \rho_a \equiv \alpha_{\mathrm{HA}} \times \rho_a, \tag{6.207}$$

$$[\mathrm{A}^-] = \frac{K_A}{[\mathrm{H}^+] + K_A} \rho_a \equiv \alpha_{\mathrm{A}^-} \times \rho_a. \tag{6.208}$$

Note that this construction is equally useful for buffers with multiple dissociation states, where ρ_a is the sum of concentrations of the weak acid and all of its dissociated states. Since the mobilities and diffusivities are generally different for the species that make up ρ_a, the convection and diffusion terms in the transport equation for ρ_a are calculated as the sum of the convection and diffusion for each species in ρ_a. A similar approach holds for weak bases.

For strong acids and bases, which are fully dissociated in the solution, or for other species that do not take part in electrolyte dissociation and association reactions, (6.201) can be integrated directly. The model proteins and fluorescent dyes in this work are assumed to have a fixed charge, so their concentrations are integrated using (6.201), with an appropriate finite-rate chemical source term. However, a complete ampholyte description for proteins can readily be formulated with a similar framework as is used for the weak acids and bases [69, 159, 160, 197]. In the simulations below, proteins are assumed to take part in a finite rate, irreversible labeling reaction of the form

$$\mathrm{U} + \mathrm{D} \xrightarrow{k_L} \mathrm{L} \tag{6.209}$$

with a pH-dependent reaction rate $k_L = k_L(\mathrm{pH})$. In (6.209), U is the unlabeled protein, D the fluorescent dye, and L the labeled protein.

Since a thin double layer is assumed, the system is also assumed to satisfy the electroneutrality condition

$$\sum_i z_i c_i = 0 \tag{6.210}$$

everywhere in the domain. The concentrations of H^+ and OH^- are obtained from this electroneutrality condition and the water dissociation constant

$$[\mathrm{H}^+][\mathrm{OH}^-] = K_w. \tag{6.211}$$

Note that the composition, and therefore the total charge, of weak acids and bases in the system depends on the H^+ concentration; see (6.207)–(6.208). The substitution of (6.208) and (6.211) into the electroneutrality condition (6.210) to account for the dependence of $[A^-]$ and $[OH^-]$ on $[H^+]$, introduces nonlinear terms in this equation. For buffers with multiple dissociation states, even more nonlinear terms are introduced. Therefore, an iterative solution of the electroneutrality condition for $[H^+]$ is usually required.

6.5.1.3 Electrostatic Field Strength

Allowing for concentration field gradients, the electrostatic field potential, ϕ is obtained from the current continuity constraint [192],

$$\nabla \cdot (\sigma \nabla \phi) = -F \sum_i z_i \nabla \cdot (D_i \nabla c_i). \tag{6.212}$$

This equation is coupled to the species concentrations through the right hand side (diffusion of charge) and the electrical conductivity σ of the solution

$$\sigma = F^2 \sum_i z_i^2 \beta_i c_i. \tag{6.213}$$

The electrostatic field strength is then obtained as $\mathbf{E} = -\nabla \phi$.

6.5.2 Stochastic Formulation

To propagate uncertainty from the input parameters to the stochastic model results, we apply a Galerkin formalism to obtain governing equations for the unknown PC coefficients. This results in:

$$\frac{\partial \mathbf{u}_k}{\partial t} + \sum_{i=0}^{\mathrm{P}} \sum_{j=0}^{\mathrm{P}} C_{ijk} (\mathbf{u}_i \cdot \nabla) \mathbf{u}_j = -\nabla p_k + \sum_{i=0}^{\mathrm{P}} \sum_{j=0}^{\mathrm{P}} C_{ijk} \nu_i \nabla^2 \mathbf{u}_j, \tag{6.214}$$

$$\nabla \cdot \mathbf{u}_k = 0, \tag{6.215}$$

$$\frac{\partial c_{m,k}}{\partial t} + \sum_{i=0}^{\mathrm{P}} \sum_{j=0}^{\mathrm{P}} C_{ijk} \nabla \cdot \left(c_{m,i} (\mathbf{u}_j + \mathbf{u}^e_{m,j}) \right)$$
$$= \sum_{i=0}^{\mathrm{P}} \sum_{j=0}^{\mathrm{P}} C_{ijk} \nabla \cdot (D_{m,i} \nabla c_{m,j}) + \hat{w}_{m,k} \tag{6.216}$$

where

$$\mathbf{u}^e_{m,j} = \frac{\langle \Psi_j \mathbf{u}^e \rangle}{\langle \Psi_j^2 \rangle} = \sum_{k=0}^{\mathrm{P}} \sum_{i=0}^{\mathrm{P}} C_{kij} \beta_k z F \nabla \phi_i, \tag{6.217}$$

$$\hat{w}_{m,k} = \frac{\langle \Psi_k \hat{w} \rangle}{\langle \Psi_k^2 \rangle}. \tag{6.218}$$

Equations (6.217) and (6.218) represent the (pseudo-spectral) projection of the electrophoretic velocities and the chemical source terms onto the PC basis. Finally, the electrostatic field equation (6.212) becomes

$$\sum_{i=0}^{\mathrm{P}} \sum_{j=0}^{\mathrm{P}} C_{ijk} \nabla \cdot (\sigma_i \nabla \phi_j) = -F \sum_m z_m \sum_{i=0}^{\mathrm{P}} \sum_{j=0}^{\mathrm{P}} C_{ijk} \nabla \cdot (D_{m,i} \nabla c_{m,j}). \tag{6.219}$$

The modes σ_i of the electrical conductivity are obtained from

$$\sigma_i = F^2 \sum_m z_m^2 \sum_{j=0}^{\mathrm{P}} \sum_{k=0}^{\mathrm{P}} C_{jki} \beta_{m,j} c_{m,k}. \tag{6.220}$$

Equations (6.214), (6.215), (6.216), and (6.219) each represent a set of P+1 coupled equations to be solved for the mode strengths u_k, p_k, $c_{m,k}$, and ϕ_k, $k = 0, \dots, \mathrm{P}$.

6.5.3 Implementation

6.5.3.1 Data Structure

As described in Sect. 6.5.1.2, species concentrations are integrated differently, based on whether or not they take part in equilibrium reactions. For instance, for components of weak acids or bases, which typically serve as buffers, only the combined concentration of all components is integrated directly. The total charge associated with the buffer components is required for the enforcement of the electroneutrality equation (6.210). For a given buffer, this total charge can be obtained from the total buffer concentration ρ and [H^+] through buffer-specific equations such as (6.207), (6.208).

To make the treatment of weak acids or bases as general as possible, separate objects are used in the current code to represent these components. Each object contains all the species properties for the weak acid or base it represents, as well as the dissociation constants for the electrolyte reactions between its species. Specific functions are also associated with each object to return the total charge or other information about the weak acid or base, given its total concentration and [H^+]. This way, different buffers can be included in the simulations by simply including different objects, without the need for specific code modifications.

6.5.3.2 Spatial Discretization

The computational domain is discretized using a Cartesian mesh with uniform cell size Δx and Δy in the x and y direction respectively. Vector fields, such as the velocity and the electrostatic field strength, are defined on the cell faces. Scalar fields, such as pressure and species concentrations are defined at the cell centers. Spatial derivatives are discretized with second-order central differences.

6.5.3.3 Electroneutrality

As explained in Sect. 6.5.1.2, the individual concentrations of the buffer ions and [H^+] are obtained from the electroneutrality condition (6.210). This results in a set of nonlinear algebraic relations between $P+1$ stochastic modes. This coupled nonlinear system of equations is iteratively solved at each point in the domain, using a Newton solver from the NITSOL package [185]. The solver uses an inexact Newton method with backtracking. Using the solution from the previous time step as initial guess, the convergence is generally very fast.

6.5.3.4 Electrostatic Field Strength

To obtain the electrostatic field potential ϕ, the set of $P+1$ equations (6.219) needs to be solved over the domain. Since these equations are coupled, an iterative solution method was developed, consisting of Gauss-Seidel iterations over the spatial dimensions in combination with SOR iterations over the stochastic dimensions. To accelerate the convergence speed, spatial coarsening with a multigrid approach is applied. A detailed description of this algorithm can be found in Chap. 7; see also [122]. The electrostatic field strength is computed in turn as the gradient of the electrostatic potential.

6.5.3.5 Time Integration

The time integration algorithm in this work is based on a previously developed stochastic projection method for the momentum equations in incompressible flow [123]. This momentum solver uses a time splitting approach in which the convection and diffusion terms are integrated in a first fractional step, and the continuity constraints are then enforced in a pressure projection step [29]. Since the continuity constraints (6.197) are decoupled in the stochastic dimension, this leads to a set of P + 1 decoupled Poisson problems.

In the current work, this method is expanded to the integration of the coupled momentum and species transport equations, in combination with the electrostatic field solution. For brevity, the equations for the stochastic mode k of the species concentrations and the velocity can be written as:

$$\frac{\partial c_k}{\partial t} = -\mathcal{C}sp_k + \mathcal{D}sp_k + \mathcal{S}sp_k, \tag{6.221}$$

$$\frac{\partial \mathbf{u}_k}{\partial t} = -\mathcal{C}m_k + \mathcal{D}m_k - \nabla p_k \tag{6.222}$$

where $\mathcal{C}sp_k$, $\mathcal{D}sp_k$, and $\mathcal{S}sp_k$ represent the convection, diffusion, and chemical source terms in the species equation (6.216). Similarly, $\mathcal{C}m_k$ and $\mathcal{D}m_k$ represent the convection and diffusion terms in the momentum equation (6.214). Using the projection scheme for momentum, in combination with a Runge Kutta (RK) time integration scheme, (6.221) and (6.222) are discretized between t^n and the RK stage time level $t^{(s)} = t^n + \Delta t^{(s)}$ as

$$\frac{c_k^{(s)} - c_k^n}{\Delta t^{(s)}} = -\mathcal{C}sp_k^{(s-1)} + \mathcal{D}sp_k^{(s-1)} + \mathcal{S}sp_k^{(s-1)} \equiv \mathcal{F}sp_k^{(s-1)}, \tag{6.223}$$

$$\frac{\mathbf{u}_k^{(s),*} - \mathbf{u}_k^n}{\Delta t^{(s)}} = -\mathcal{C}m_k^{(s-1)} + \mathcal{D}m_k^{(s-1)} \equiv \mathcal{F}m_k^{(s-1)}, \tag{6.224}$$

$$\frac{\mathbf{u}_k^{(s)} - \mathbf{u}_k^{(s),*}}{\Delta t^{(s)}} = -\nabla p_k^{(s)} \tag{6.225}$$

where $\mathcal{F}sp_k$ and $\mathcal{F}m_k$ represent the full right hand sides in the corresponding time integration steps. Equation (6.225) is the pressure correction step, which requires the pressure to be solved for first. The equation for pressure is obtained by substituting (6.225) into the stochastic form of the continuity equation for $\mathbf{u}^{(s)}$

$$\nabla \cdot \mathbf{u}_k^{(s)} = 0, \tag{6.226}$$

resulting in the following set of Poisson equations

$$\nabla^2 p_k^{(s)} = -\frac{1}{\Delta t^{(s)}} \nabla \cdot \mathbf{u}_k^{(s),*}, \quad k = 0, \ldots, \mathrm{P}. \tag{6.227}$$

As discussed in earlier, these P + 1 Poisson equations are decoupled; therefore, each can be solved individually using existing Poisson solvers for deterministic flow problems. In the current work, the same Fast Fourier Transform based flow solver is used as in [123].

The time integration of (6.223) and (6.224) is performed using the 4-stage, 4th-order Runge-Kutta scheme (RK4) [97], which was selected because of its good stability for convection dominated problems. Keeping in mind the coupling between the equations, the computations during the subsequent stages of the RK4 integration over a time step Δt from time t^n to $t^{n+1} = t^n + \Delta t$ can be represented with the following pseudo-code. The superscripts $^{(s)}$ denote the Runge-Kutta stage number. For clarity, the subscripts for the mode strength k have been dropped.

- Stage $s = 1$; $t = t^n$
  Calculate the right-hand sides in (6.223) and (6.224) using the species concentrations, velocities and electrostatic field strength at time $t = t^n$:
  - $\mathcal{F}sp^{(1)} = \mathcal{F}sp\,(c(t^n), \mathbf{u}(t^n), \mathbf{E}(t^n))$
  - $\mathcal{F}m^{(1)} = \mathcal{F}m\,(\mathbf{u}(t^n), u_w(t^n))$ where u_w is the electroosmotic wall velocity.
- Stage $s = 2, 3, 4$; $t = t^n + \Delta t^{(s)}$
  Update species concentrations to the current time level:
  - $c^{(s)} = c(t^n) + \Delta t^{(s)} \mathcal{F}sp^{(s-1)}$ for all directly integrated species
  - Solve electroneutrality constraint to obtain $[H^+]^{(s)}$
  - Update concentrations of weak acids and/or bases

  Update electrostatic field strength and velocity boundary conditions using the updated concentrations:
  - $\mathbf{E}^{(s)} = \mathbf{E}(c^{(s)})$
  - $u_w^{(s)} = u_w(c^{(s)}, \mathbf{E}^{(s)})$

  Update velocities to the current time level:
  - Update the velocities to their intermediate (*) values at the current time level: $\mathbf{u}^{(s),*} = \mathbf{u}(t^n) + \Delta t^{(s)} \mathcal{F}m^{(s-1)}$
  - Apply the boundary conditions $u_w^{(s)}$ to the $\mathbf{u}^{(s),*}$ velocity field
  - Solve for pressure at this time level using (6.227): $p^{(s)} = p(\mathbf{u}^{(s),*})$
  - Apply the pressure correction to $\mathbf{u}^{(s),*}$ to obtain $\mathbf{u}^{(s)}$: $\mathbf{u}^{(s)} = \mathbf{u}^{(s),*} - \Delta t^{(s)} \nabla p^{(s)}$

  Calculate the new right hand sides in equations (6.223) and (6.224) using the updated species concentrations, velocities and electrostatic field strength:
  - $\mathcal{F}sp^{(s)} = \mathcal{F}sp(c^{(s)}, \mathbf{u}^{(s)}, \mathbf{E}^{(s)})$
  - $\mathcal{F}m^{(s)} = \mathcal{F}m(\mathbf{u}^{(s)}, u_w^{(s)})$
- Final update to time $t^{n+1} = t^n + \Delta t$
  Update species concentrations to t^{n+1}
  - $c(t^{n+1}) = c(t^n) + \Delta t(\frac{1}{6}\mathcal{F}sp^{(1)} + \frac{2}{6}\mathcal{F}sp^{(2)} + \frac{2}{6}\mathcal{F}sp^{(3)} + \frac{1}{6}\mathcal{F}sp^{(4)})$ for all directly integrated species
  - Solve electroneutrality constraint to obtain $[H^+]$ at t^{n+1}
  - Update concentrations of weak acids and/or bases

  Update electrostatic field strength and velocity boundary conditions using the updated concentrations:
  - $\mathbf{E}(t^{n+1}) = \mathbf{E}(c(t^{n+1}))$

- $u_w(t^{n+1}) = u_w(c(t^{n+1}), \mathbf{E}(t^{n+1}))$

Update velocities to t^{n+1}

- Update the velocities to the intermediate (*) values at t^{n+1}:
  $\mathbf{u}^*(t^{n+1}) = \mathbf{u}(t^n) + \Delta t(\frac{1}{6}\mathcal{F}m^{(1)} + \frac{2}{6}\mathcal{F}m^{(2)} + \frac{2}{6}\mathcal{F}m^{(3)} + \frac{1}{6}\mathcal{F}m^{(4)})$
- Apply the boundary conditions $u_w(t^{n+1})$ to the $\mathbf{u}^*(t^{n+1})$ velocity field
- Solve for pressure at t^{n+1} using (6.227): $p(t^{n+1}) = p(\mathbf{u}^*(t^{n+1}))$
- Apply the pressure correction to $\mathbf{u}^*(t^{n+1})$ to obtain $\mathbf{u}(t^{n+1})$:
  $\mathbf{u}(t^{n+1}) = \mathbf{u}^*(t^{n+1}) - \Delta t \nabla p(t^{n+1})$

In the above integration scheme, the respective time steps $\Delta t^{(s)}$ of the Runge-Kutta stages s = 2, 3, and 4 are given by $\frac{1}{2}\Delta t$, $\frac{1}{2}\Delta t$, and Δt.

6.5.3.6 Estimates of Nonlinear Transformations

The governing equations for the spectral mode strengths include various nonlinear terms, including products, exponentials, and inverse transformations. These are accounted for in a pseudo-spectral fashion.

In order to calculate products of three or more stochastic variables, we rely on repeated applications of the Galerkin formula for a quadratic product, as outlined in Sect. 6.3.2.4. Another frequent operation is the calculation of the inverse of a stochastic quantity. This is implemented using the matrix approach outlined in Sect. 6.3.2.5.

More challenging are estimates of non-polynomial functions of stochastic variables such as the exponential, which show up in the calculation of the protein labeling reaction rate with (6.229), or the logarithm in the calculation of pH. These transformations are estimated using the Taylor series approach.

In the applications below, all of the operations outlined above, as well as fundamental PC definitions defined in Appendix C have been implemented using the UQ toolkit library.[1]

6.5.4 Validation

This subsection presents some results of test problems illustrating the spatial and temporal convergence properties of the developed code. Figure 6.71 shows the geometry considered for these test problems, consisting of a rectangular microchannel in which a protein U and dye D react to form a labeled protein L. An external electrostatic potential is applied across the system to generate an electroosmotic flow in the x-direction. The unlabeled protein U has a charge of $+1$ versus a charge of -1 for the dye D, so electrophoresis will move U forward and D backward, relative

[1] B.J. Debusschere *et al.*, http://www.sandia.gov/UQToolkit/.

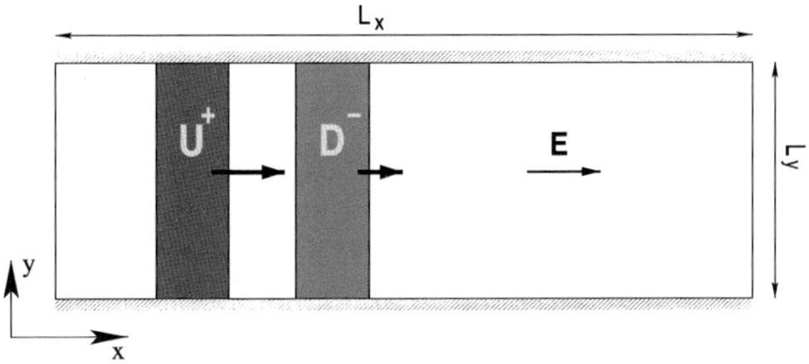

Fig. 6.71 Geometry for the numerical test problems: a plug of protein U and dye D are introduced in a rectangular microchannel and react to form a labeled protein L. Adapted from [51]

to the bulk flow. For all cases simulated in this work, an aqueous potassium phosphate (KH_2PO_4) buffer solution is considered. Therefore, the species in the solution are the proteins U and L, the dye D, the electrolytes H^+, OH^-, K^+ as well as the components of phosphoric acid H_3PO_4, $H_2PO_4^-$, HPO_4^{--}, and PO_4^{---}.

As mentioned in Sect. 6.5.1.2, the proteins in this solution are assumed to have a fixed charge and can therefore be integrated using (6.201) with a chemical reaction source term $\hat{\omega}_i$ according to a model irreversible labeling reaction

$$\mathrm{U} + \mathrm{D} \xrightarrow{k_L} \mathrm{L}. \tag{6.228}$$

The rate constant k_L in this reaction is pH dependent, given by the following equation:

$$k_L = k_L^0 + d_L e^{-(\mathrm{pH}-\mathrm{pH}_0)^2/\delta_{\mathrm{pH}}^2}. \tag{6.229}$$

The Gaussian dependence of this relationship on pH is based on the shape of the measured pH-dependence of the rate of production of the high-fluorescence-efficiency species from the reaction of Naphthalene-2,3-dicarboxaldehyde (NDA) with amino acids in the presence of CN^- [156]. Unless stated otherwise, the values for the reaction rate parameters are chosen in this section as $k_L^0 = 0.25 \times 10^6\ \mathrm{mol}^{-1}\,\mathrm{s}^{-1}$, $d_L = 2.15 \times 10^6\ \mathrm{mol}^{-1}\,\mathrm{s}^{-1}$, $\mathrm{pH}_0 = 7.40$, and $\delta_{\mathrm{pH}} = 0.85$. The chemical source terms used in (6.201) are correspondingly

$$\hat{\omega}_U = \hat{\omega}_D = -\hat{\omega}_L = -k_L[\mathrm{U}][\mathrm{D}]. \tag{6.230}$$

The concentration of the K^+ ion, which is fully dissociated and is a conserved quantity can also be integrated by (6.201) directly (without a source term). Phosphoric acid, however, is a weak acid and will dissociate according to the following electrolyte reactions

$$H_3PO_4 \overset{K_1}{\longleftrightarrow} H^+ + H_2PO_4^-, \tag{6.231}$$

$$H_2PO_4^- \overset{K_2}{\longleftrightarrow} H^+ + HPO_4^{--}, \tag{6.232}$$

$$HPO_4^{--} \overset{K_3}{\longleftrightarrow} H^+ + PO_4^{3-} \tag{6.233}$$

where the K_i are the corresponding dissociation constants. As discussed in Sect. 6.5.1.2, an equilibrium formulation is used for these fast electrolyte reactions. Therefore, we consider the total concentration of this weak acid

$$\rho_a = [H_3PO_4] + [H_2PO_4^-] + [HPO_4^{--}] + [PO_4^{---}] \tag{6.234}$$

whose transport equation is obtained similarly to (6.206) by adding up the transport equations for all the components in ρ_a so the dissociation reaction source terms disappear. The concentrations of the individual components of ρ_a are then calculated as $c_i = \alpha_i \rho_a$, where the α_i are calculated from the equilibrium expressions for the dissociation reactions (6.231)–(6.233) and can be written as a function of $[H^+]$ and the dissociation constants only:

$$\alpha_{H_3PO_4} = \frac{[H^+]^3}{[H^+]^3 + K_1[H^+]^2 + K_1K_2[H^+] + K_1K_2K_3}, \tag{6.235}$$

$$\alpha_{H_2PO_4^-} = \frac{K_1[H^+]^2}{[H^+]^3 + K_1[H^+]^2 + K_1K_2[H^+] + K_1K_2K_3}, \tag{6.236}$$

$$\alpha_{HPO_4^{--}} = \frac{K_1K_2[H^+]}{[H^+]^3 + K_1[H^+]^2 + K_1K_2[H^+] + K_1K_2K_3}, \tag{6.237}$$

$$\alpha_{PO_4^{---}} = \frac{K_1K_2K_3}{[H^+]^3 + K_1[H^+]^2 + K_1K_2[H^+] + K_1K_2K_3}. \tag{6.238}$$

As discussed in Sect. 6.5.1.1, (6.200) is used to model the dependence of the zeta potential on pH and buffer molarity. The concentration of the fully dissociated potassium ion, $[K^+]$, is used for the local buffer molarity $\mathcal{M}$ along the walls. The temperature is assumed constant in this work, with all species properties and reaction rate constants evaluated at 298 K.

For the computations in this paper, all parameters and field variables, were represented with third-order polynomial chaos expansions. The highest-order stochastic modes in the expansions of the predicted field variables were significantly lower than the lower-order modes, indicating that the third-order expansions were sufficiently accurate.

Convergence with grid spacing: To test the spatial convergence rate of the code, simulations of the test case described above were run on a domain with $L_x = 1$ cm and $L_y = 0.25$ cm. The potassium phosphate buffer solution was initialized with a uniform concentration of 10^{-3} mol/l and a pH of 7.25. The unlabeled protein U and the dye D were initialized with a profile, Gaussian in x and uniform in y, both with a maximum concentration of 10^{-5} mol/l at $x = 4$ mm and a width of 1 mm. The labeled protein concentration was initialized to zero. The electrostatic potential difference ΔV between the inlet and exit of the domain was set to 10 V, creating an average field strength of 0.01 kV/cm. An uncertainty of 1% was assumed in the mobilities of both U and D, in the labeling rate parameter pH_0 of (6.229), and in the potential difference ΔV. Using third-order polynomial chaos expansions, these

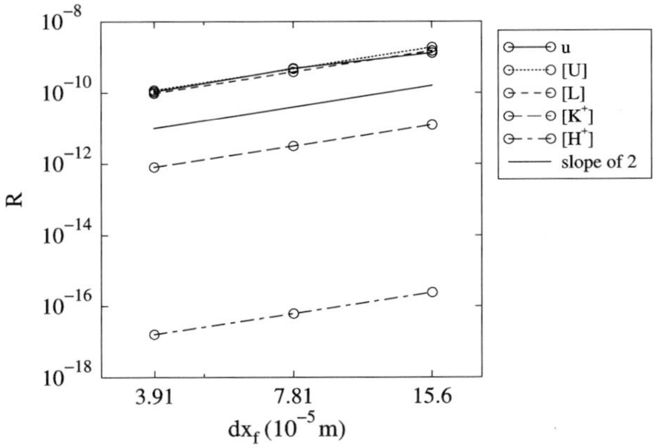

Fig. 6.72 L^2 norm of the difference between solutions on successive grids as a function of the fine grid spacing dx_f. The slope of the lines shows a second order spatial convergence rate for various species concentrations as well as the streamwise velocity. Adapted from [51]

4 uncertain parameters led to 4 stochastic dimensions with a total of $P+1=35$ stochastic modes.

Four runs were performed, with uniform grid spacings in x and y doubling between each run, from 3.91×10^{-5} m in the finest grid to 3.13×10^{-4} m in the coarsest grid (corresponding respectively to 256×64, 128×32, 64×16, and 32×8 cells in $x \times y$). Each run used the same time step of 10^{-4} seconds for a total of 200 time steps. Figure 6.72 shows the L^2 norm of the difference between the solutions for the streamwise velocity u as well as several species concentrations at successive grid spacings. To monitor the spatial convergence of the full stochastic solution, the L^2 norm was calculated over all points in space and all $P+1$ stochastic modes. Clearly, the slope of the curves in Fig. 6.72 shows an overall second order convergence rate with grid spacing, consistent with the spatial differencing scheme used.

Convergence with time step: The temporal convergence behavior of the code was studied with a similar test case as in the previous section. Referring to Fig. 6.71, the domain sizes were chosen as $L_x = 2$ cm and $L_y = 0.25$ cm. The buffer initialization was the same as in the previous case. For the unlabeled protein U and the dye D, however, the peak concentrations were raised to 10^{-4} mol/l, located at $x = 4$ mm and $x = 6$ mm respectively. The electrostatic potential difference ΔV across the domain was set to 2000 V, giving an average field strength of 1 kV/cm. A slightly higher uncertainty of 2% was assumed in the mobilities of both U and D, the parameters pH_0 and ΔV, as well as the bulk kinematic viscosity. These 5 stochastic dimensions with third order polynomial chaos expansions led to a total of $P+1=56$ stochastic modes.

This test case was run for a total time of 0.5 s, with 5 different time steps, ranging in factors of 2 from 6.25×10^{-4} s up to 1.00×10^{-2} s. In each case, the number of cells was 128×16 in $x \times y$. Figure 6.73 shows the L^2 norm of the difference between the solutions for the streamwise velocity u as well as several species concentrations

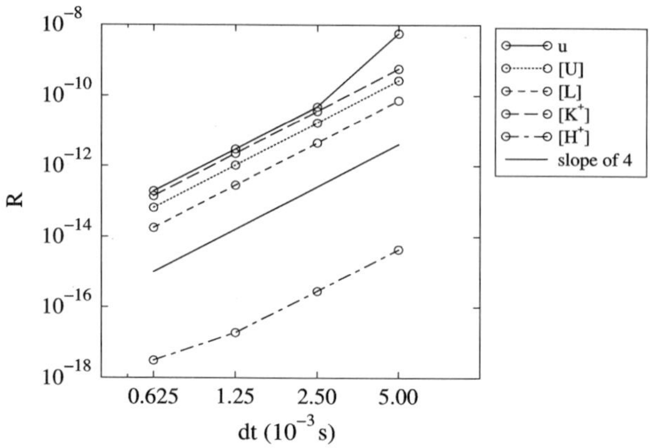

Fig. 6.73 L^2 norm of the difference between solutions at successive time steps as a function of the shorter time step dt. The slope of the lines shows a fourth order temporal convergence rate for various species concentrations as well as the streamwise velocity. Adapted from [51]

at successive time steps.The fourth order temporal convergence rate observed in Fig. 6.73 is consistent with the Runge Kutta scheme used in the time integration.

Protein labeling in a homogeneous buffer: To illustrate the stochastic uncertainty quantification methodology, this section describes protein labeling in a simple homogeneous system. Figure 6.74 shows the time evolution of the concentrations of the unlabeled and labeled protein in a homogeneous potassium phosphate buffer at a pH of 8.25. In this problem, the dye D was assumed to be present in abundance so that the source term for the labeled protein in 6.201 can be written as

$$\hat{w}_L = k_L[\mathrm{U}]. \tag{6.239}$$

The same expression as before, (6.229), was used for the reaction rate, but with the following parameters: $k_L^0 = 0.25 \times 10^{-3}\ \mathrm{s}^{-1}$, $d_L = 2.15\ \mathrm{s}^{-1}$, $\mathrm{pH}_0 = 9.25$, and $\delta_{\mathrm{pH}} = 0.85$. Both proteins U and L, as well as the dye D were assumed to have no charge, and therefore the buffer equilibrium and pH did not change with time. For this simulation, a standard deviation of 1% was assumed for all parameters in the rate expression (6.229), as well as for the electrolyte dissociation constants. Third-order PC expansions were used.

The resulting uncertainty in the protein concentrations is indicated in Fig. 6.74 with "error bars" that span the $\pm 3\sigma$ range, where σ indicates the standard deviation. Clearly, uncertainty in the input parameters causes large uncertainties in the simulated concentrations. At the point where $[\mathrm{U}] = 0.5$, a standard deviation of 1% in the parameter pH_0 is magnified about 16 times in the standard deviation of [U].

Note that after about 3 seconds, the range of the $\pm 3\sigma$ "error bars" becomes so large that it seems to include concentrations for U that are negative, which is clearly not physically possible. However, the interval $\pm 3\sigma$ around the mean value properly

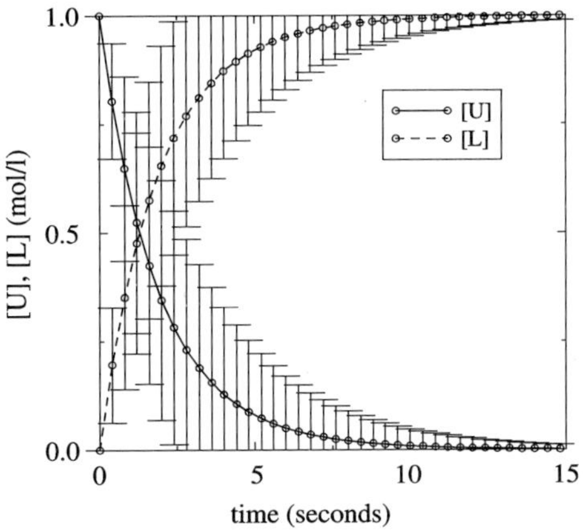

Fig. 6.74 Time evolution of U and L concentrations in a homogeneous protein labeling reaction. The uncertainty in these concentrations, due to a 1% uncertainty in the labeling reaction rate parameters, is indicated by $\pm 3\sigma$ "error bars". Adapted from [51]

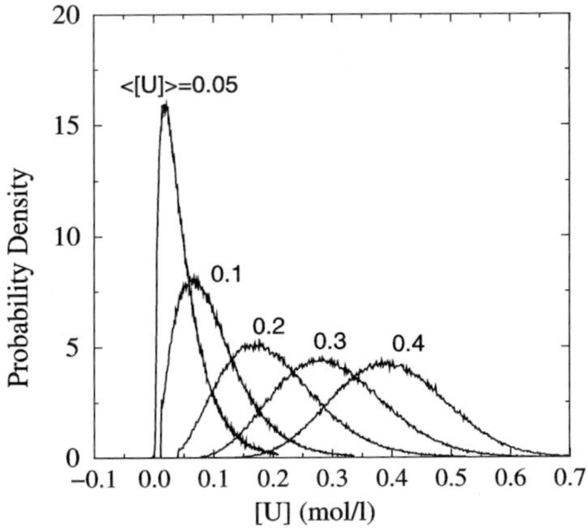

Fig. 6.75 PDF of the unlabeled protein concentration at different mean values. As the unlabeled protein reacts away, its PDF becomes narrower and more skewed. Adapted from [51]

represents the full range of possibilities for a certain variable only when its probability density function is Gaussian, and therefore symmetric. Figure 6.75 shows the probability density function of [U], generated from its PC expansion at various points in time. When the mean value of [U] is sufficiently far away from zero, this PDF has a Gaussian shape. However, for mean values of [U] closer to zero, the PDF becomes narrower and more skewed. This predicted uncertainty properly reflects the physical system behavior where all unlabeled protein reacts away, but its concentration cannot be negative.

6.5.5 Protein Labeling in a 2D Microchannel

In this section, the simulation and uncertainty quantification code is used to tackle a more physically challenging problem of protein labeling in a two-dimensional microchannel. The problem set-up is similar to the numerical test problems described

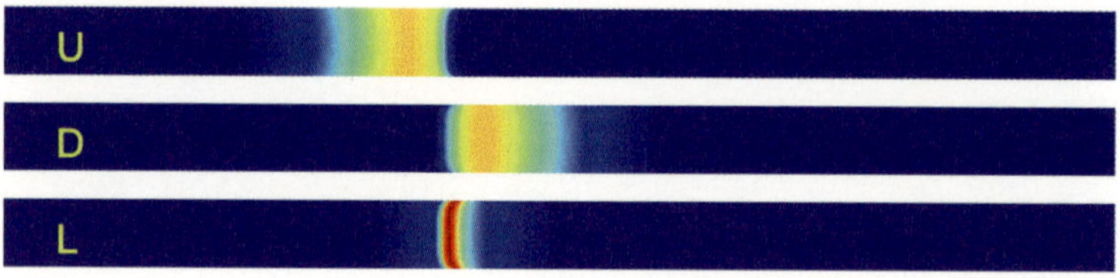

Fig. 6.76 Mean concentrations of proteins U, L, and dye D at $t = 0.12$ s. U and D just met and L is produced at their interface. The values of the contour levels go linearly from 0 (*blue*) to 1.3×10^{-4} mol/l (*red*). In this figure, as well as in all subsequent contour plots, the full physical domain is shown, from 0 to 1 cm in x and from 0 to 1 mm in y. Adapted from [51]

in Sect. 6.5.4. The labeling reaction is the same as equation (6.228) with the reaction rate k_L and the corresponding source terms as in (6.229) and (6.230). Again, a charge of $+1$ is assumed for the unlabeled protein U and a charge of -1 for the dye D, resulting in a neutral labeled protein L.

Referring to Fig. 6.71, a microchannel was considered with a length $L_x = 1$ cm and a height $L_y = 1$ mm. The potassium phosphate buffer solution was initialized with a uniform concentration of 10^{-3} mol/l and a pH of 7.25. The Gaussian profiles for the initial U and D concentrations had peak concentrations of 10^{-4} mol/l, located at $x = 2.5$ mm and $x = 4$ mm respectively, and a width in x of 0.75 mm. The electrostatic potential difference ΔV across the domain was set to 1000 V, giving an average field strength of 1 kV/cm. An uncertainty of 1% was assumed in the mobility of U, in the labeling rate parameter pH_0, the dissociation constant K_2, and the potential difference ΔV. Third-order polynomial chaos expansions were used in the computations with a total of 35 stochastic modes. The time step was set to 2×10^{-4} s and the domain was discretized with 512×32 cells in x and y.

Figure 6.76 shows a contour plot of the mean concentrations of the proteins and dye at $t = 0.12$ s. At this point in time, the plugs of U and D have just met at $x \approx 4$ mm, and labeled protein is generated at the interface. Note that the labeling reaction is fast compared to the electroosmotic and electrophoretic transport. Consequently, U and D react as soon as they meet, resulting in almost no overlap between the U and D profiles, and a sharp profile for L. Since L is neutral, it travels with the bulk convective velocity, which is the average of the total convective velocities of U and D. Therefore the peak value of L is always located at the interface of U and D, and since L is generated in that same area, its peak concentration will keep increasing. At $t = 0.12$ s, the peak concentration for L is 1.3×10^{-4} mol/l, which is already higher than the peak concentrations of 9.4×10^{-5} mol/l for U and D.

The standard deviations in the concentrations of Fig. 6.76 are given in Fig. 6.77. The highest uncertainties appear in the reaction zone at the interface between U and D, with a maximum coefficient of variation of about 20% in the L concentration. Even though Fig. 6.77 only shows the overall uncertainty in the concentrations, a strong feature of the PC formalism is that the contributions of the uncertainty in individual parameters to this overall uncertainty can easily be retrieved, as explained in Sect. 6.5.2. Figure 6.78, for example, shows the contributions from each of the 4 uncertain input parameters to the standard deviation of the L concentration, in the area around the reaction zone, at $y = 0.5$ mm.

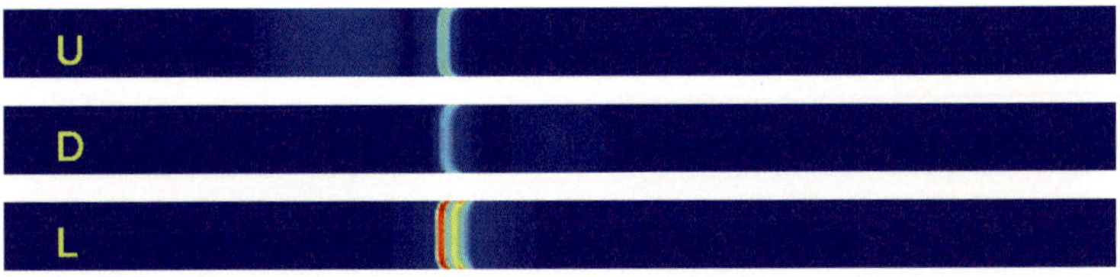

Fig. 6.77 Standard deviation of the protein and dye concentrations at $t = 0.12$ s. The values of the contour levels go linearly from 0 (*blue*) to 1.1×10^{-5} mol/l (*red*). The largest uncertainties are found in the reaction zone. Adapted from [51]

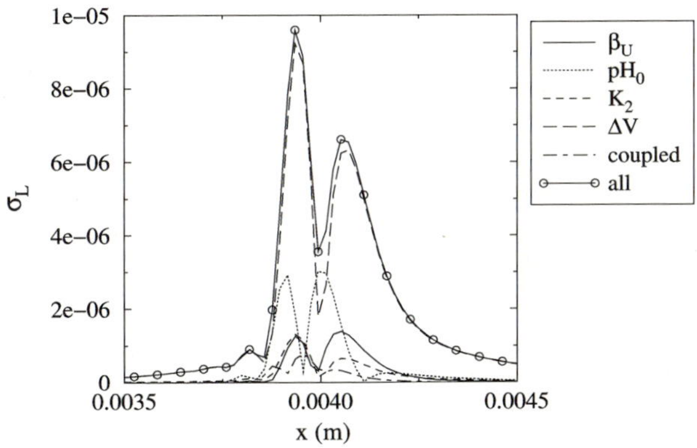

Fig. 6.78 Major contributions of individual input parameters to the overall standard deviation in [L] in the area around the reaction zone at $t = 0.12$ s, $y = 0.5$ mm. The uncertainty in the applied voltage potential "ΔV" has the most dominant contribution to the overall standard deviation in [L]. Adapted from [51]

The total standard deviation of [L] is given by the curve labeled "all" in this figure. This overall standard deviation has a profile with a double peak, which for a single peak mean species profile, is characteristic of uncertainty caused by the convection velocity. When a single peak species profile is transported by an uncertain convection velocity, the uncertainty in the position of the peak at a given point in time will cause the most variability at the sides of the peak, where the profile has a steep slope in the x-direction. At the top of the profile, there is no concentration gradient and uncertainties in peak position cause little uncertainty in the observed concentrations at that location.

As indicated by the curve labeled "ΔV", the uncertainty in the applied electrostatic field potential has the most dominant contribution to the overall standard deviation. Since both the electroosmotic and electrophoretic velocities are directly proportional to ΔV, the uncertainty caused by this parameter naturally shows a double peak, characteristic of convection velocity uncertainty. Similarly, the parameter β_U affects the electrophoretic transport of the reactant U and its resulting contribution to the standard deviation of [L] also has a double peak, albeit smaller than the ΔV contribution.

The contribution of parameter pH_0 also shows a double peak, but with its center located on the left side of the [L] profile, where the gradient of [L] in x is very steep. The steepness of the [L] profile in that area is largely determined by the speed of the labeling reaction compared to the convection speed, with a faster reaction rate

Fig. 6.79 Mean (*top*) and standard deviation (*bottom*) of the labeled protein concentration L at $t = 0.50$ s. The initially flat profiles are now severely distorted. The values of the contour levels go linearly from 0 (*blue*) to 3.2×10^{-4} mol/l (*red*) in the *top plot* and from 0 (*blue*) to 10^{-4} mol/l (*red*) in the *bottom plot*. Adapted from [51]

leading to a sharper increase in [L]. With the pH in this area between 7.0 and 7.1 (not shown), (6.229) predicts significant variability in k_L for changes in pH_0. So the uncertainty in pH_0 mainly affects the slope of the [L] profile on the left side, consistent with the observed contribution of parameter pH_0 in Fig. 6.78.

Figure 6.78 further shows more minor contributions, from the dissociation parameter K_2 and from the coupled terms. Even though their contribution is small in this case, those coupled terms are interesting from a theoretical point of view, as they represent coupled effects of independent parameters. In the current figure, those terms represent the sum of three different coupled effects: the coupled effect of ΔV and β_U, of ΔV and pH_0, and of ΔV and K_2.

As time goes on and the U and D plugs cross each other, nearly all U and D are consumed in the labeling reaction. At $t = 0.50$ s, only labeled protein L remains, with its mean concentration and standard deviation as shown in Fig. 6.79. The maximum mean concentration of L at this point in time is 2.4×10^{-4} mol/l in the center of the channel, and about 3.2×10^{-4} mol/l near the walls. So the L concentration is up to three times as large as the initial U and D concentrations. The standard deviation in L, as shown in the bottom plot of Fig. 6.79, is very large near the wall, with maximum values up to 10^{-4} mol/l and coefficients of variation up to 100%. Again, the standard deviation in [L] exhibits the double peak near the centerline, which is characteristic of uncertainty caused by the convection velocity.

What is particularly significant though, is the major distortion of the L plug, as opposed to the straight profile observed at early times. This distortion is caused by the disturbance of the buffer electrolyte, in response to the movement and annihilation of the charged protein U and the dye D. To explain why this is physically happening, consider Fig. 6.80, which shows the mean and standard deviation of the electrical conductivity σ of the electrolyte solution at $t = 0.50$ s. Because two charged molecules are used up for every new labeled protein, the area around the L plug has a reduced concentration of ions, with a mean electrical conductivity of almost a third lower than in the undisturbed buffer. Upstream of the L plug, the electrical conductivity shows some smaller fluctuations, which stem from shifts in the buffer equilibrium. Since the buffer ions are primarily negatively charged, those disturbances travel slower than the labeled protein plug. The bottom plot of Fig. 6.80 shows that the highest uncertainties in the electrical conductivity are found around the L plug, near the center and especially at the walls.

The large spatial variations in the electrical conductivity in turn cause nonuniformities in the electrical field strength, as shown in Figs. 6.81 and 6.82. Near

Fig. 6.80 Mean (*top*) and standard deviation (*bottom*) of the electrical conductivity of the electrolyte solution at $t = 0.50$ s. Annihilation of ions in the labeling reaction results in a significantly lower mean electrical conductivity near the L plug. The values of the contour levels go linearly from 7.1×10^{-3} S/m (*blue*) to 1.3×10^{-2} S/m (*red*) in the *top plot* and from 0 (*blue*) to 1.5×10^{-3} S/m (*red*) in the *bottom plot*. Adapted from [51]

Fig. 6.81 Mean (*top*) and standard deviation (*bottom*) of the electrical field strength in the x-direction at $t = 0.50$ s. Near the L plug, the mean streamwise electrical field strength is about 40% higher than in the undisturbed flow. The values of the contour levels go linearly from 91.4 kV/m (*blue*) to 146 kV/m (*red*) in the *top plot* and from 0.20 kV/m (*blue*) to 13 kV/m (*red*) in the *bottom plot*. Adapted from [51]

Fig. 6.82 Mean (*top*) and standard deviation (*bottom*) of the electrical field strength in the y-direction at $t = 0.50$ s. The magnitude of the mean of this field strength is up to 15% of the initial field strength in the x-direction. The values of the contour levels go linearly from −16.3 kV/m (*blue*) to 16.3 kV/m (*red*) in the *top plot* and from 0 (*blue*) to 5.8 kV/m (*red*) in the *bottom plot*. Adapted from [51]

the L plug, the mean electrostatic field strength in the x-direction reaches a value up to 40% higher than in the undisturbed flow. This increase strongly affects the local electroosmotic and electrophoretic velocities, causing an increased wall velocity, leading to the observed distortion of the L plug. The largest uncertainties are again found near the L plug, with maxima up to 10%. Even though the initial field strength in the y-direction was zero, Fig. 6.82 shows that this y-component is quite significant at $t = 0.50$ s. The magnitude of this field strength is up to 15% of the initial, streamwise electrostatic field strength for the mean value. Even though this y-component does not affect the electroosmotic flow velocity directly, it does provide electrophoretic ion transport in the wall-normal direction, which can further distort sample profiles.

As indicated by (6.199), the electroosmotic wall velocity depends on both the local electrostatic field strength and ζ potential, which in turn depends on the pH and the buffer molarity, as modeled by (6.200). Since all these variables are disturbed by the charged protein movement and annihilation, the electroosmotic wall velocity

Fig. 6.83 Mean (*top*) and standard deviation (*bottom*) of the streamwise velocity at $t = 0.50$ s. The local increase in the electroosmotic wall velocity leads to recirculation zones near the L plug. The largest uncertainties are found near the wall. The values of the contour levels go linearly from 6.8 mm/s (*blue*) to 9.1 mm/s (*red*) in the *top plot* and from 5.6×10^{-3} mm/s (*blue*) to 0.59 mm/s (*red*) in the *bottom plot*. Adapted from [51]

Fig. 6.84 Mean (*top*) and standard deviation (*bottom*) of the wall-normal velocity at $t = 0.50$ s. The mean of this velocity has a magnitude of up to 6% of the initial streamwise velocity. The values of the contour levels go linearly from −0.56 mm/s (*blue*) to 0.56 mm/s (*red*) in the *top plot* and from 0 (*blue*) to 0.26 mm/s (*red*) in the *bottom plot*. Adapted from [51]

varies in the streamwise direction. These wall velocity changes in turn cause pressure gradients and local recirculation zones, as indicated by the velocity fields in Figs. 6.83 and 6.84. Figure 6.83 shows the streamwise velocity field, which has a mean wall velocity that is up to 20% higher near the L plug. The wall-normal velocity field shows positive and negative velocities near the L plug, with magnitudes up to 6% of the initial streamwise velocity.

Clearly, the recirculation zones in the flow field will distort initially flat sample profiles. This increases the hydrodynamic dispersion, on top of the electrokinetic dispersion caused by non-uniformities in the electrophoretic transport.

6.6 Concluding Remarks

In this chapter we have outlined the development and implementation of stochastic solvers for UQ in incompressible and weakly compressible flows. Attention has been focused on intrusive formulations. Within this framework, we have outlined the development of finite-difference solvers that are based on a pressure projection formalism. For the class of methods considered here, an essential element of practical implementations concerns the decoupled nature of mass conservation constraints, which preserves the efficiency of the deterministic solver and thus allows us to fully exploit the rapid convergence of the spectral representations.

We have also explored the incorporation of the intrusive formalism into Lagrangian particle methods. A key feature of the constructions outlined and tested above concerns the definition of a single set of particles, having stochastic strengths, but moving with the mean wind velocity. Once again, this methodology enabled us to preserve the numerical features of deterministic particle methods and thus establish robust and efficient computational schemes.

By illustrating implementation of the stochastic schemes in different settings, and analyzing their performance, the discussion aimed at providing a well established foundation for extension of the present formulations to a wider class of applications. While some of these may be immediate, hurdles can also be anticipated. We postpone discussion of such opportunities and anticipated challenges to Chap. 10.

Part II
Advanced Topics

Chapter 7
Solvers for Stochastic Galerkin Problems

The first chapters of this book were about the stochastic discretization of models involving some uncertain parameters. The focus was on the derivation of numerical or mathematical approaches for the definition and determination of the spectral coefficients for the stochastic solutions. In this chapter, we specifically focus on some numerical techniques and algorithms dedicated to the resolution of the mathematical problems arising from the stochastic Galerkin projection. These problems are characterized by their size, which is typically $(\mathrm{P}+1)$ times larger than their deterministic counterpart, where $(\mathrm{P}+1)$ is the dimension of the stochastic basis used. Since the deterministic problems of interest are often large and P can be large as well, the resolution of the stochastic Galerkin problems is generally costly and requires efficient numerical strategies and appropriate numerical algorithms. In fact, even for simple linear problems, the assembly of the matrix representing the discrete system of equations to be solved is often impossible and its direct inversion is generally not an option.

In Sect. 7.1, we quickly review some Krylov methods for the resolution of large sparse linear problems. The methods discussed here are quite general and lead to iterative algorithms for the resolution of large systems of linear equations. Their application to systems arising from the Galerkin projection is discussed and a particular attention is given to the definition of appropriate preconditioners.

Section 7.2 describes a multigrid technique for the resolution of stochastic diffusion equations with random diffusivity. The multigrid algorithms rely on a coarsening of the spatial discretization to accelerate the convergence rate of the iterative solvers.

In Sect. 7.3 we detail Newton iterations for the resolution of the steady stochastic incompressible Navier-Stokes equations. This nonlinear solver combines Newton iterations with Krylov methods for non-symmetric problems discussed in Sect. 7.1, preconditioned by the linearized stochastic Navier-Stokes equations. This large nonlinear problem also illustrates the implementation of Krylov methods where the actual assembly of the system is never actually performed.

O.P. Le Maître, O.M. Knio, *Spectral Methods for Uncertainty Quantification*,
Scientific Computation,
DOI 10.1007/978-90-481-3520-2_7,

7.1 Krylov Methods for Linear Models

Linear models are of considerable importance as they represent a large variety of physical systems. In addition, the resolution of nonlinear models are generally based on series of corrections to the approximate solution, each consisting in the resolution of a linear problem such as in Newton-Raphson iterations. Therefore, solution techniques for linear systems of equations are a key ingredient of numerical algorithms and their efficiency, both in terms of computational time and memory requirement, are essential in high performance computing. This section discusses few iterative strategies for the resolution of large linear systems arising from the Galerkin projection of linear models. To recall notations and the structure of linear system of equations, we first briefly summarize the stochastic Galerkin projection procedure in the case of a linear model, which was detailed in Chap. 4 and fully exemplified in Chap. 5.

We assume that the stochastic problem has been already discretized at the deterministic level (using finite element, finite difference or any other technique), yielding a linear stochastic problem on the probability space $(\Xi, \mathcal{B}_\Xi, P_\Xi)$. This semi-discrete problem has the form

$$A(\boldsymbol{\xi})u(\boldsymbol{\xi}) = b(\boldsymbol{\xi}), \quad \text{a.s.} \tag{7.1}$$

where $A(\boldsymbol{\xi}) \in \mathbb{R}^{m\times m} \otimes L_2(\Xi, P_\Xi)$ is a stochastic matrix that will be assumed sparse, $b(\boldsymbol{\xi}) \in \mathbb{R}^m \otimes L_2(\Xi, P_\Xi)$ is the stochastic right-hand-side, and $u(\boldsymbol{\xi}) \in \mathbb{R}^m \otimes L_2(\Xi, P_\Xi)$ is the model solution. The stochastic discretization is performed by introducing a basis of orthogonal stochastic functionals spanning a finite dimensional stochastic subspace $\mathcal{S}_\mathrm{P}$ of $L_2(\Xi, P_\Xi)$:

$$\mathcal{S}_\mathrm{P} = \operatorname{span}\{\Psi_0, \dots, \Psi_\mathrm{P}\} \subset L_2(\Xi, P_\Xi), \qquad \dim \mathcal{S}_\mathrm{P} = (\mathrm{P}+1). \tag{7.2}$$

The developments in this section are independent of the stochastic basis, and we simply assume that by convention $\Psi_0 = 1$, *i.e.* mode 0 corresponds to the mean mode. To lighten the notation we consider orthonormal bases:

$$\langle \Psi_i, \Psi_j \rangle = \int_\Xi \Psi_i(\mathbf{y})\Psi_j(\mathbf{y}) p_\Xi(\mathbf{y})\,d\mathbf{y} = \delta_{i,j}, \quad \forall i, j = 0, \dots, \mathrm{P}. \tag{7.3}$$

The stochastic expansion of the solution on $\mathcal{S}_\mathrm{P}$ is expressed as:

$$u(\boldsymbol{\xi}) \approx \sum_{k=0}^{\mathrm{P}} u_k \Psi_k(\boldsymbol{\xi}), \quad u_k \in \mathbb{R}^m,\ k = 0, \dots, \mathrm{P}. \tag{7.4}$$

Proceeding with the stochastic Galerkin projection, (7.1) becomes:

$$\sum_{i=0}^{\mathrm{P}} \langle A(\boldsymbol{\xi})\Psi_i, \Psi_k \rangle\, u_i = \langle b(\boldsymbol{\xi}), \Psi_k \rangle = b_k, \quad k = 0, \dots, \mathrm{P}. \tag{7.5}$$

This set of coupled systems of linear equations can be recast as a single deterministic system, aggregating the stochastic modes of the solution and of the right-hand-side in vectors **u** and **b** respectively, called hereafter the Galerkin system,

$$\mathbf{A}\mathbf{u} = \mathbf{b}. \tag{7.6}$$

In (7.6), the deterministic matrix **A** has a $(\mathrm{P}+1) \times (\mathrm{P}+1)$ block structure, the blocks being expressed as:

$$\mathbf{A}_{i,j} = \left\langle A(\boldsymbol{\xi})\Psi_j, \Psi_i \right\rangle \in \mathbb{R}^{m \times m}, \quad i, j = 0, \dots, \mathrm{P}. \tag{7.7}$$

The total dimension of the Galerkin system is denoted $n = m \times (\mathrm{P}+1)$, where both m and P can be large.

7.1.1 Krylov Methods for Large Linear Systems

In this section, we introduce the Krylov methods for the resolution of large linear systems. Our objective is limited to the exposition of the essential ideas underlining Krylov and subspace techniques for the iterative resolution of linear systems. We are not providing actual algorithms which can be found in classical books dealing with this matter. The reader interested in more details, specific algorithms and extensive discussions of implementation issues may consult the references [92, 199].

As discussed previously, the size n of system in (7.6) will be large, due either to the dimensions of the deterministic discretization space (m), of stochastic space $(\mathrm{P}+1)$, or both. At any rate, the direct resolution of the system through

$$\mathbf{u} = \mathbf{A}^{-1}\mathbf{b}, \tag{7.8}$$

by means of the inversion of the matrix **A** (direct solver) will be assumed unfeasible from the numerical point of view. This is first due to the number of operations in the computation of $\mathbf{A}^{-1}$, which scales in $\mathcal{O}(n^3)$ and so becomes quickly prohibitive, and also because although **A** can be sparse its inverse is generally much denser: storage of $\mathbf{A}^{-1}$ (and intermediate matrices during the inversion process) is prohibitively demanding.

A key idea to overcome the actual inversion of **A** is to approach $\mathbf{A}^{-1}$ by a polynomial $q_l(\mathbf{A})$, with degree $(l-1)$, and to define the corresponding approximation $\mathbf{u}^{(l)}$ of **u** as:

$$\mathbf{u} = \mathbf{A}^{-1}\mathbf{b} \approx q_l(\mathbf{A})\mathbf{b} = \mathbf{u}^{(l)}. \tag{7.9}$$

Denoting $\mathbf{u}_0$ an arbitrary vector of $\mathbb{R}^m$, we can rewrite $\mathbf{u}^{(l)}$ as

$$\mathbf{u}^{(l)} - \mathbf{u}^{(0)} = q_l(\mathbf{A})\mathbf{r}^{(0)}, \quad \mathbf{r}^{(0)} = \mathbf{b} - \mathbf{A}\mathbf{u}^{(0)}. \tag{7.10}$$

This last expression shows that $\mathbf{u}^{(l)} - \mathbf{u}^{(0)}$ is in the vector space spanned by the monomial $(\mathbf{A}^i \mathbf{r}^{(0)})$. This space is called the Krylov space, denoted

$$\mathcal{K}_l(\mathbf{A}, \mathbf{r}^{(0)}) = \text{span}\{\mathbf{r}^{(0)}, \mathbf{A}\mathbf{r}^{(0)}, \mathbf{A}^2\mathbf{r}^{(0)}, \dots, \mathbf{A}^{l-1}\mathbf{r}^{(0)}\}. \tag{7.11}$$

The solution space being defined, $\mathbf{u}^{(0)}$ can be determined by means of Petrov-Galerkin orthogonality condition on the equation residual. Specifically, the equation residual $\mathbf{A}\mathbf{u}^{(l)} - \mathbf{b}$ is required to be orthogonal to a secondary space $\mathcal{L}_l$ having the same dimension as $\mathcal{K}_l$. Denoting $\mathbf{V}_l$ and $\mathbf{W}_l$ two bases of $\mathcal{K}_l$ and $\mathcal{L}_l$ respectively, the approximate solution is

$$\mathbf{u}^{(l)} - \mathbf{u}^{(0)} = \mathbf{V}_l \left(\mathbf{W}_l^T \mathbf{A} \mathbf{V}_l\right)^{-1} \mathbf{W}_l^T \mathbf{r}^{(0)}. \tag{7.12}$$

This expression shows that $\mathbf{u}^{(l)}$ depends on the operator $\mathbf{A}$, on the right-hand-side $\mathbf{b}$ and initial vector $\mathbf{u}^{(0)}$ which together define $\mathcal{K}_l$ and $\mathbf{r}^{(0)}$, and finally on the selected projection space $\mathcal{L}_l$. Since $\mathbf{u}^{(0)}$ is chosen arbitrarily, and since $\mathbf{b}$ and $\mathbf{A}$ are given, the only degree of freedom we have is in the definition of the projection space $\mathcal{L}_l$ which results in different approaches.

An important point to note is that if the selection $\mathcal{L}_l$ theoretically completely defines $\mathbf{u}^{(l)}$ through (7.12), the solution actually computed usually also depends on the numerical algorithms used: theoretically equivalent methods result in different numerical solutions depending on the algorithm and implementation followed. Krylov algorithms aim at constructing the bases and solving the projected system in the most efficient way. Usually, Arnoldi or Lanczos methods are implemented for the sequential construction of the bases, while the projection is performed progressively as the size l of the Krylov space is increased. The resolution of the projected system is usually performed at the end of the bases construction. For some particular operators and algorithms, it is possible to construct the bases such that the projected system is diagonal; it thus allows for the sequential updates of the solution and residual, as the size of Krylov space increases, and so to monitor the convergence of the solution with the index l which can be considered as an iteration index.

7.1.1.1 GMRes Method

The Generalized Minimal Residual [200] method applies to any system $\mathbf{A}$. It is based on the selection of the projection space $\mathcal{L}_l = \mathbf{A}\mathcal{K}_l$ and the basis $\mathbf{V}_l$ uses an Arnoldi procedure. The corresponding solution solves the least square residual problem:

$$\mathbf{u}^{(0)} = \text{Arg} \min_{\mathbf{v} \in \mathcal{K}_l} \|\mathbf{A}\mathbf{v} - \mathbf{b}\|_2. \tag{7.13}$$

The main limitation of GMRes comes from the need to store the Krylov basis $\mathbf{V}_l$ in order to reconstruct the solution after the resolution of the projected system. This may be an issue if the dimension l of the Krylov space needed to obtain a sufficiently small residual becomes large: l vectors of $\mathbb{R}^n$ have to be stored. To overcome this limitation, restart strategies have been proposed in the literature.

7.1.1.2 Conjugate Gradient Method

The Conjugate Gradient method applies to symmetric positive definite (SPD) systems and corresponds to a projection space $\mathscr{L}_l = \mathscr{K}_l$. For SPD systems, one can construct sequentially the vectors of $\mathbf{V}_l$ which are $\mathbf{A}$-orthogonal, leading to a diagonal projected system. This diagonal structure allows for successive updates of the solution $\mathbf{u}^{(l)}$ and residual $\mathbf{r}^{(l)}$ as the dimension of the Krylov space increases, making the Conjugate Gradient really an iterative method and limiting the storage requirement to only few vectors of $\mathbb{R}^n$.

7.1.1.3 Bi-Conjugate Gradient Method

The Conjugate Gradient (CG) method possesses very attractive characteristics, in particular in terms of memory requirements and convergence rate. However, it is limited to SPD matrices. Efforts have then been deployed to derive algorithms that retain the advantages of the CG method while making it applicable to general non-symmetric systems. An example of such algorithms is the Bi-Conjugate Gradient method, which roughly consists in constructing a projection space $\mathscr{L}$ based on $\mathbf{A}^T$ instead of $\mathbf{A}$. Again, efficient implementations can be derived to obtain a sequence of updates for the approximate solution and residual vectors without having to store the full sequence of basis vectors, and limited storage requirement as a result.

All the methods discussed above have algorithmic variants and can incorporate subtle features to improve their computational efficiency. However, one essential feature of all these algorithms is that they do not need the actual assembly of the matrix, but require only to evaluate the effect of the operator $\mathbf{A}$ (and eventually its transpose) on given vectors. This is due to the fact that contrary to direct methods, Krylov methods do not rely on transformations or factorization of $\mathbf{A}$.

7.1.2 Preconditioning

The convergence of Krylov methods primary depends on the condition number of the matrix $\mathbf{A}$ defined as the ratio of its largest to smallest singular values,

$$\mathrm{Cond}(\mathbf{A}) = \sigma_{\max}/\sigma_{\min}, \tag{7.14}$$

and to a lesser extend on the distribution of the spectrum (cluster of eigen values). The lower the condition number, the faster the convergence rate of the Krylov method. A key idea is then to transform system (7.6) to a equivalent system associated to a matrix having a lower condition number. This can be achieved by considering for instance the (left-)preconditioned system

$$\mathbf{P}^{-1}\mathbf{A}\mathbf{u} = \mathbf{P}^{-1}\mathbf{b} \tag{7.15}$$

in place of the original one. In (7.15), $\mathbf{P}$ is a non-singular matrix (so $\mathbf{P}^{-1}$ exists) such that

$$\text{Cond}(\mathbf{P}^{-1}\mathbf{A}) < \text{Cond}(\mathbf{A}). \tag{7.16}$$

The optimal preconditioner is $\mathbf{P}^{-1} = \mathbf{A}^{-1}$, for which the condition number is 1. However, $\mathbf{A}^{-1}$ being unknown the direct definition of $\mathbf{P}^{-1}$ is not possible. In fact, it is more convenient to construct the inverse of the preconditioner $\mathbf{P}$, based on our knowledge of $\mathbf{A}$, *i.e.* to set $\mathbf{P} = \mathbf{P}(\mathbf{A})$.

Examples of preconditioners are discussed below. Before doing so, we detail the expected properties of preconditioners. An important aspect in the construction of $\mathbf{P}$ is that the extension of Krylov algorithms to the preconditioned system essentially amounts to the redefinition of the original residuals $\mathbf{r}_k$ to the residual of the precondition system, $\mathbf{z}_k$, through the relation

$$\mathbf{P}\mathbf{z}^{(l)} = \mathbf{r}^{(l)}. \tag{7.17}$$

The sequence of systems (7.17) have to be solved for $\mathbf{z}^{(l)}$, once or twice per Krylov step depending on the actual algorithm used, so $\mathbf{P}$ should be easily inverted in order to limit the computational overhead arising from the preconditioning. Otherwise, any improvement in the convergence rate would be degraded by the increase in the computational times of the Krylov basis vectors. Another point to consider when constructing preconditioners is the memory requirement for their storage. To summarize, the preconditioners $\mathbf{P}(\mathbf{A})$ should be such that:

1. $\text{Cond}(\mathbf{P}^{-1}\mathbf{A})$ is as close as possible to 1,
2. the resolution of $\mathbf{P}\mathbf{z} = \mathbf{r}$ should involve minimal overhead,
3. storage of $\mathbf{P}$ should incur limited additional memory allocations.

7.1.2.1 Jacobi Preconditioner

Diagonal preconditioners are the simplest one to invert. A straightforward and $\mathbf{A}$-based diagonal preconditioner construction consists in setting

$$\mathbf{P} = \mathbf{D}(\mathbf{A}), \tag{7.18}$$

where $\mathbf{D}$ is a diagonal matrix containing the diagonal entries $(\mathbf{A})_{i,i}$ of the initial system. The Jacobi preconditioner is obviously sparse and does not require additional memory allocation, since non-zero entries of $\mathbf{A}$ are readily available. Beside this advantage, the Jacobi preconditioner will bring significant improvement of the convergence rate if and only if the matrix $\mathbf{A}$ is diagonally dominant; this is the case for instance of stochastic linear diffusion equations with appropriate deterministic discretization [250].

7.1.2.2 ILU Preconditioners

The LU decomposition of an invertible matrix $\mathbf{A}$ is expressed as:

$$\mathbf{A} = \mathbf{L}\mathbf{U}, \tag{7.19}$$

where $\mathbf{L}$ and $\mathbf{U}$ are lower and upper triangular matrices, respectively. Given a matrix $\mathbf{A}$, a systematic algorithm can be applied to compute its LU decomposition in a finite number of operations. From the LU decomposition of $\mathbf{A}$, the solution to (7.6) can be obtained in two steps:

$$\mathbf{L}\mathbf{v} = \mathbf{b}, \quad \text{and then} \quad \mathbf{U}\mathbf{u} = \mathbf{v}. \tag{7.20}$$

We see that $\mathbf{u}$ is obtained by solving two systems of equations with triangular matrix structure. This amounts a forward elimination and backward substitution steps, with a maximum operation count that is $\mathcal{O}(n^2)$ for dense systems. Taking $\mathbf{P} = \mathbf{L}\mathbf{U}$ would meet the first two requirements for an efficient preconditioner $\mathbf{P}$. Indeed, $\mathbf{P} = \mathbf{L}\mathbf{U}$ is optimal and the exact solution is obtained in just one iteration (the LU decomposition is in fact a direct method). However, factorization of $\mathbf{A}$ and storage of $\mathbf{L}$ and $\mathbf{U}$, requires a significant computational resources and memory allocation because the sparse structure of A is destroyed during the factorization process. The latter point naturally leads to the design of construction techniques for $\mathbf{P}$ which yields acceptable sparse lower and upper triangular factors. These techniques are known as incomplete LU decomposition of $\mathbf{A}$; we write

$$\mathbf{P} = \mathbf{L}^{inc}\mathbf{U}^{inc}(\mathbf{A}), \tag{7.21}$$

where $\mathbf{L}^{inc}$ and $\mathbf{U}^{inc}$ are sparse triangular approximations of $\mathbf{L}$ and $\mathbf{U}$. Classically, these approximates are constructed so as to introduce no additional non-zero entries compared to $\mathbf{A}$. As a result, the resolution of $\mathbf{L}^{inc}\mathbf{U}^{inc}\mathbf{z} = \mathbf{r}$ remains easy thanks to the triangular structure of the factors, gains computational efficiency from the sparse structure of L^{inc} and U^{inc} and finally requires a memory allocation which is exactly equal to that of the original system $\mathbf{A}$. We also remark that for SPD systems, the LU decomposition reduces to the Cholesky decomposition $\mathbf{A} = \mathbf{L}\mathbf{L}^T$ with corresponding incomplete Cholesky decomposition factor $\mathbf{L}^{inc}$ and subsequent savings in memory allocation.

Because the ILU decomposition $(\mathbf{L}^{inc}\mathbf{U}^{inc})^{-1} \neq \mathbf{A}^{-1}$, the exact solution is not obtained in just one iteration, but the convergence rate of the corresponding Krylov method generally improves significantly compared to the unpreconditioned version of the algorithms. Note that in addition to the memory and resolution overheads, one also has to perform the ILU decomposition of $\mathbf{A}$. This decomposition is computed in a preprocessing stage so the corresponding computational time can eventually be factored if one has to solve (7.6) for multiple right-hand-sides. This is typically the case for evolution problems where the discrete operator $\mathbf{A}$ to be inverted at each time iteration is time independent.

7.1.3 Preconditioners for Galerkin Systems

In the previous paragraphs, we have described some Krylov methods and preconditioners suitable for the resolution of generic large, sparse systems of equations. These methods and associated algorithms are generic in the sense that they apply to large classes of linear systems.

In practice, the selection of the preconditioner is crucial for the overall efficiency of the solver. In fact, in the context of deterministic models numerous highly optimized preconditioners have been proposed, allowing for significant computational savings compared to the generic ones exposed above. These specialized preconditioners are tailored for specific types of model and discretization methods, for which *a priori* information regarding the spectrum of the discretized linear operator and/or some characteristics of the solution are available and exploited. This specialization of the preconditioners highlights their critical importance for large scale computations where the numerical efficiency (combining computational time, complexity and memory requirements) is essential. From this perspective, the construction of efficient preconditioners is even more important in the context of stochastic simulations, which are known to be even more computationally intensive: in the stochastic Galerkin approach for linear problems, where the size of the resulting problem is $\mathrm{P}+1$ times larger than its deterministic counterpart, any gain in efficiency brought by a preconditioner will be magnified. Therefore, it is not surprising that researchers have focused on the derivation of *ad hoc* preconditioners optimized for these specific types of problems.

Intuitively, the design of specialized preconditioners for Galerkin systems should incorporate both the specificities of the underlining deterministic problems and of the particular structure resulting from the stochastic Galerkin projection. The latter depends on the stochastic discretization basis. In fact it appears that these two aspects can be separated to some extent.

7.1.3.1 Block-Jacobi Preconditioners

We have seen that the Galerkin system has a (symmetric) block structure,

$$\mathbf{A} = \begin{pmatrix} \mathbf{A}_{00} & \cdots & \mathbf{A}_{0\mathrm{P}} \\ \vdots & \ddots & \vdots \\ \mathbf{A}_{\mathrm{P}0} & \cdots & \mathbf{A}_{\mathrm{PP}} \end{pmatrix}, \quad \mathbf{A}_{ij} = \mathbf{A}_{ji} = \left\langle A(\xi)\Psi_i\Psi_j \right\rangle. \tag{7.22}$$

This structure can be exploited to construct a block-Jacobi preconditioner, where we define

$$\mathbf{P} = \begin{pmatrix} \mathbf{A}_{00} & [0] & \cdots & \cdots & [0] \\ [0] & \ddots & \ddots & & \vdots \\ \vdots & \ddots & \mathbf{A}_{ii} & \ddots & \vdots \\ \vdots & & \ddots & \ddots & [0] \\ [0] & \cdots & \cdots & [0] & \mathbf{A}_{\mathrm{PP}} \end{pmatrix}. \tag{7.23}$$

Clearly, this block diagonal preconditioner incorporates more information than the (diagonal) Jacobi preconditioner and so yields better convergence properties. This is obtained at the price of the allocation of (P + 1) sparse matrices of $\mathbb{R}^{m \times m}$ and an increased computational complexity. Indeed, the resolution of the preconditioning problems (7.17) is more demanding for the block-diagonal preconditioner than for its purely diagonal (Jacobi) form. Still, this resolution is uncoupled and can be performed in parallel for the individual stochastic modes:

$$\mathbf{P}\mathbf{z}^{(l)} = \mathbf{r}^{(l)} \Rightarrow z_i^{(l)} = \mathbf{P}_{ii}^{-1} r_i^{(l)}, \quad i = 0, \ldots, \mathrm{P}. \tag{7.24}$$

The decoupling arising from the block diagonal structure of the preconditioner can be even more exploited by substituting the matrices $\mathbf{P}_{ii} = \mathbf{A}_{ii}$ by their respective incomplete LU decomposition, minimizing further the overhead of the preconditioner inversion. Alternatively, one can rely on an approximate resolution of the preconditioning problems through any deterministic linear solver, ending with composite iterations: iterations on the Krylov space dimension in an outer-loop and inner (uncoupled) iterations for the preconditioning problems treated mode by mode.

7.1.3.2 Operator Expectation Preconditioning

In some specific cases, the block diagonal preconditioning approach just described yields an even simpler preconditioner structure. In particular, if the stochastic operator $A(\boldsymbol{\xi})$ is affine in $\boldsymbol{\xi}$, it can be expressed as

$$A(\boldsymbol{\xi}) = \overline{A} + \sum_{i=1}^{\mathrm{N}} \overline{A}_i \xi_i. \tag{7.25}$$

Such operators have first-order PC expansions:

$$A(\boldsymbol{\xi}) = A_0 \Psi_0(\boldsymbol{\xi}) + \sum_{i=1}^{\mathrm{N}} A_i \Psi_i(\boldsymbol{\xi}), \tag{7.26}$$

where we have assumed that Ψ_i is a polynomial of degree one in ξ_i. By virtue of orthonormality of the basis, and the convention $\Psi_0 = 1$, we have

$$\langle A(\boldsymbol{\xi})\Psi_i\Psi_i\rangle = A_0 \langle \Psi_i\Psi_i\rangle + \sum_{j=1}^{\mathrm{N}} A_j \left\langle \Psi_j\Psi_i\Psi_i\right\rangle = A_0 = \langle A\rangle, \tag{7.27}$$

such that the blocks $\mathbf{P}_{ii}$ are all identical and equal to the mean operator $\langle A\rangle$. This particular situation limits the allocation memory for the preconditioner to a unique sparse matrix of $\mathbb{R}^{m\times m}$. Also, the resolution of (7.24) can be optimized by considering solution methods tailored for multiple right-hand-sides (vector version of linear sparse solvers). Even though in general the stochastic operator $A(\boldsymbol{\xi})$ is not an affine function of $\boldsymbol{\xi}$, a common situation is when the stochastic operator present small deviations from its mean $\langle A\rangle$. Then, one can consider the following preconditioner:

$$\mathbf{P} = \begin{pmatrix} \langle A\rangle & [0] & \cdots & \cdots & [0] \\ [0] & \langle A\rangle & \ddots & & \vdots \\ \vdots & \ddots & \ddots & \ddots & \vdots \\ \vdots & & \ddots & \langle A\rangle & [0] \\ [0] & \cdots & \cdots & [0] & \langle A\rangle \end{pmatrix}. \tag{7.28}$$

For this definition of $\mathbf{P}$, each preconditioning step of the Krylov method requires the resolution of $(\mathrm{P}+1)$ uncoupled deterministic systems of equation, with size m, which in turn can rely on nested efficient linear solvers, with eventually a vectorization with regards to the stochastic modes of the residuals or alternatively a parallel implementation.

This block diagonal preconditioning, based on the operator expectation, was first proposed in [88], in the context of linear symmetric positive definite equations with CG iteration, and later extended to nonsymmetric problems in [183]. Although this idea appears natural, the notion of small deviations of the discrete operator from its expected value is a loose concept that may be misleading, especially in case ξ_i is defined on an unbounded domain. This is also an issue when the linear system to be solved corresponds to a linearization of a nonlinear system. In [190, 191] for instance, the authors show that for a fixed SPD operator $A(\boldsymbol{\xi})$, representing a stochastic elliptic operator discretized by linear finite-elements in space, the efficiency of the block diagonal preconditioner in (7.28) decreases for Hermite bases when the expansion order increases. This is due to the continuous introduction of new eigenvalues in the system when the stochastic basis dimension increases. Clearly, since the preconditioner's block is independent of the expansion order, the additional eigenvalues are not properly accounted for.

7.1.3.3 Specialized Block Diagonal Preconditioners

As mentioned previously, a great deal of effort has been dedicated to the construction of specialized preconditioners for deterministic problems. It is obvious that these efforts should be capitalized on when considering the stochastic version of the same problems. In fact, the structure of the stochastic Galerkin problems allows for the extension of these specialized preconditioners. In [71, 76], the authors derived a general framework for the construction of a preconditioner for the Galerkin system, on the basis of a preconditioner tailored for the underlined deterministic model.

The construction relies first on an expression of the Galerkin system using a Kronecker product; this expression highlights, through the product form, the contribution of the stochastic basis on the structure of the system. This separation facilitates the analysis of the Galerkin system properties (*e.g.* providing simple bounds for the system eigenvalues), and leads naturally to an expression of the preconditioner following a similar Kronecker product form. The end result of the construction is again a repeated block-diagonal structure. Numerical tests presented in [71] have shown the robustness against the expansion order of this type of preconditioner.

To close this discussion on preconditioners, we would like to remark that their mathematical analysis is today limited to operators having a linear dependence with regards to the uncertain random variables ξ_i (*i.e.* they have a first order expansion on the corresponding stochastic polynomial bases). Extension of the theory to more general forms of stochastic linear operators will be necessary, for instance to improve the convergence rate of Newton-type iterations in nonlinear solvers. This will likely be the object of significant developments in a near future.

7.2 Multigrid Solvers for Diffusion Problems

This section addresses the challenge of solving large systems arising from the discretization of stochastic parabolic and elliptic equations. Specifically, we describe the adaptation of a (deterministic) multigrid (MG) technique [228] for the solution of the system of equations arising from the finite difference discretization of the spectral representation of the stochastic diffusion equation.

We focus on a "generic" diffusion problem for a quantity u, with a random, spatially-varying isotropic diffusion coefficient λ, inside a two-dimensional domain $\mathcal{D}$. The general form of the governing equation for this problem can be expressed as:

$$\alpha \frac{\partial u(\mathbf{x},t,\theta)}{\partial t} = \nabla \cdot [\lambda(\mathbf{x},\theta)\nabla u(\mathbf{x},t,\theta)] + s(\mathbf{x},t,\theta), \tag{7.29}$$

where $\alpha = 0, 1$ in the steady and unsteady cases, respectively, s is a given stochastic source term, and θ denotes a random event defined on an abstract probability space. The formulation is completed by specifying Neumann and/or Dirichlet boundary conditions for u, as well as an initial condition in the unsteady case. The elliptic (steady) limit of (7.29) has been thoroughly analyzed from the mathematical point of view; see for instance [10, 50, 102, 224].

Equation (7.29) becomes stochastic whenever λ, s or the initial/boundary conditions on u are uncertain. In addition to this specific example, (7.29) appears, by itself or as part of a larger system, in the formulation of many problems involving gradient diffusion processes [51, 101, 102, 107, 123, 135, 165, 210], as well as a variety of problems such as 1D-linear elasticity problems [57] and electromagnetism [211].

For brevity, but without loss of generality, the diffusivity field will be represented in terms of the KL expansion, which is assumed to be known. We focus our attention on constructing a solver for u, which is sought in terms of its Polynomial Chaos (PC) representation [90, 102, 204]. While the source term s may also be given in terms of a KL representation or more generally by a PC expansion, we will restrict our attention in the numerical tests to the case where $s \equiv 0$, *i.e.* to the homogeneous form of (7.29). For the purpose of the present construction, this enables us to avoid unnecessary details associated with setting up a stochastic source field. Both of these restrictions, however, can be easily relaxed within the framework of the construction.

In Sect. 7.2.1, we recall the basic concepts and properties of the PC expansion of a stochastic process. Using these concepts, the stochastic spectral formulation of (7.29) is derived in Sect. 7.2.2 and the difficulties inherent to the solution of the spectral equations are discussed. Next, the finite difference discretization of the stochastic system is introduced (Sect. 7.2.3), and an iterative technique is proposed to solve the resulting set of equations. In Sect. 7.2.6, a multigrid technique, based on spatial coarsening, is developed to improve the convergence rate of the previous iterative method. The MG algorithm is applied to selected test problems in Sect. 7.2.7, and the tests are used to examine its efficiency and scalability properties.

7.2.1 Spectral Representation

We briefly recall the essential ingredients of the spectral expansion, limiting ourself for simplicity to Wiener-Hermite expansions. We assume a parametrization of the problem using a set $\boldsymbol{\xi} = \{\xi_1, \ldots, \xi_{\mathrm{N}}\}$ of N independent normalized Gaussian variables. The solution of (7.29) is approximated as

$$u(\mathbf{x}, t, \theta) \approx u(\mathbf{x}, t, \boldsymbol{\xi}) \equiv \sum_{k=0}^{\mathrm{P}} u_k(\mathbf{x}, t)\Psi_k(\boldsymbol{\xi}), \tag{7.30}$$

where u_k are deterministic coefficients and $\{\Psi_0, \ldots, \Psi_{\mathrm{P}}\}$ is a (truncated) orthogonal basis consisting of multidimensional Hermite polynomials in ξ_i. The truncation is such that the degree of the polynomials is at most equal to No, the order of the expansion. We recall that the total number of modes, $\mathrm{P}+1$, depends on N and No, according to:

$$\mathrm{P}+1 = \frac{(\mathrm{N}+\mathrm{No})!}{\mathrm{N}!\mathrm{No}!}. \tag{7.31}$$

In general, all field variables exhibit a stochastic character and are therefore represented with PC expansions. In particular, the diffusivity and source fields are expressed as:

$$\lambda(\mathbf{x}, t, \theta) = \sum_{k=0}^{\mathrm{P}} \lambda_k(\mathbf{x}, t)\Psi_k(\theta), \qquad s(\mathbf{x}, t, \theta) = \sum_{k=0}^{\mathrm{P}} s_k(\mathbf{x}, t)\Psi_k(\theta), \tag{7.32}$$

respectively.

The formulation above is quite general, and enables us to accommodate situations where the initial and boundary conditions on u, the diffusivity field, λ, and source field, s, are all uncertain. While the general case may be of interest, its treatment would require a detailed analysis of the source of uncertainty, which would distract from the present objective. Thus, in order to limit the scope of the simulations, while at the same time provide a meaningful test to the solver developed below, we restrict our attention to the case of a random diffusivity, deterministic boundary conditions, and vanishing source field.

The diffusivity is assumed to be log-normal with stationary distribution in space:

$$\log(\lambda(\mathbf{x}, \cdot)) \equiv \tilde{\lambda}(\mathbf{x}, \cdot) \sim \mathrm{N}(\mu, \sigma_\lambda), \quad \forall \mathbf{x} \in \mathcal{D}, \tag{7.33}$$

where μ is related to the median value of $\lambda(\cdot, \theta)$, hereafter fixed to 1 (so $\mu = 0$), and σ_λ depends on the coefficient of variation (COV) of λ. An exponentially decaying covariance function is assumed for $\tilde{\lambda}(\mathbf{x}, \cdot)$:

$$C(\mathbf{x}, \mathbf{x}') = \sigma_\lambda^2 \exp - \frac{\|\mathbf{x} - \mathbf{x}'\|}{L_c}, \tag{7.34}$$

where L_c is the correlation length and σ_λ is the standard deviation of $\tilde{\lambda}$. The truncated KL expansion [89, 137] of $\tilde{\lambda}(\mathbf{x}, \theta)$ is:

$$\tilde{\lambda}(\mathbf{x}, \theta) = \mu + \sum_{k=1}^{\mathrm{N}} \sqrt{\beta_k}\tilde{\lambda}_k(\mathbf{x})\xi_k(\theta), \tag{7.35}$$

where β_k and $\tilde{\lambda}_k$ are, respectively, the leading eigenvalues and eigenfunctions appearing in the spectral representation of C:

$$C(\mathbf{x}, \mathbf{x}') = \sum_{k=1}^{\infty} \beta_k \tilde{\lambda}_k(\mathbf{x})\tilde{\lambda}_k(\mathbf{x}'). \tag{7.36}$$

In the computations below, the leading eigenvalues and eigenfunctions are obtained with the Galerkin procedure described in [85, 87, 89]. Also note that the first N Polynomial Chaoses coincide with the normalized Gaussian variables ξ_k, *i.e.* $\Psi_k(\theta) = \xi_k$ for $k = 1, \ldots, \mathrm{N}$. Thus, the KL representation of $\tilde{\lambda}$ can be formally viewed as a special case of a PC representation in which polynomials of degree larger than one have vanishing coefficient. Different cases are considered by varying the COV and L_c, and analyzing their effect on the performance of the solver.

When the expansion of $\tilde{\lambda}(\mathbf{x}, \theta)$ is known, the expansion of its exponential is determined for all $\mathbf{x}$ using one of the techniques proposed in Sect. 4.5; this yields:

$$\lambda(\mathbf{x}, \boldsymbol{\xi}) \approx \exp\left(\mu + \sum_{k=1}^{\mathrm{N}} \sqrt{\beta_k} \tilde{\lambda}_k(\mathbf{x}) \Psi_k(\boldsymbol{\xi})\right) \approx \sum_{k=0}^{\mathrm{P}} \lambda_k(\boldsymbol{\xi}) \Psi_k(\boldsymbol{\xi}). \tag{7.37}$$

7.2.2 *Continuous Formulation and Time Discretization*

7.2.2.1 Stochastic Galerkin Projection

The Galerkin projection of (7.29) results in (see Chap. 5):

$$\alpha \frac{\partial u_i(\mathbf{x}, t)}{\partial t} = \sum_{l=0}^{\mathrm{P}} \sum_{m=0}^{\mathrm{P}} C_{ilm} \nabla \cdot [\lambda_l(\mathbf{x}) \nabla u_m(\mathbf{x}, t)] + s_i(\mathbf{x}, t)$$
$$\text{for } i = 0, \dots, \mathrm{P}. \tag{7.38}$$

The multiplication tensor C_{ilm} is given by:

$$C_{ilm} \equiv \frac{\langle \Psi_i \Psi_l \Psi_m \rangle}{\langle \Psi_i \Psi_i \rangle}. \tag{7.39}$$

7.2.2.2 Boundary and Initial Conditions

Boundary conditions, and, when relevant, initial conditions, are needed to solve (7.38). These are also implemented in a "weak sense", *i.e.* the boundary conditions are also projected onto the PC basis, leading to explicit conditions for each of the u_i's. As noted earlier, the boundary conditions in the present study can be either of the Neumann or Dirichlet type. For brevity, we assume here that the boundary conditions are deterministic. Denoting by $\partial\mathcal{D}_D$ and $\partial\mathcal{D}_N$ the part of the boundaries of $\mathcal{D}$ where Dirichlet and Neumann conditions apply, respectively, the boundary conditions for all modes are given by:

$$u_0(\mathbf{x}, t) = u^D(\mathbf{x}, t), \qquad u_{i \in [1,\mathrm{P}]}(\mathbf{x}, t) = 0 \quad \forall \mathbf{x} \in \partial\mathcal{D}_D, \tag{7.40}$$

$$\frac{\partial u_0}{\partial n}(\mathbf{x}, t) = g_0^{\mathrm{N}}(\mathbf{x}, t), \qquad \frac{\partial u_k}{\partial n}(\mathbf{x}, t) = 0, \quad k = 1, \dots, \mathrm{P} \; \forall \mathbf{x} \in \partial\mathcal{D}_N \tag{7.41}$$

where n denotes the direction normal to the boundary.

Note that for steady problems involving only Neumann conditions, the modes of the source field must satisfy the integral constraints

$$\int_{\mathcal{D}} s_k(\mathbf{x})\, \mathrm{d}\mathbf{x} = \int_{\partial\mathcal{D}} \sum_{l=0}^{\mathrm{P}} C_{kl0} \lambda_l g_0^{\mathrm{N}}\, \mathrm{d}s \quad \text{for } k = 1, \dots, \mathrm{P}. \tag{7.42}$$

Here, ds is the surface element along $\partial\mathcal{D}$. For unsteady problems, an initial condition for u is required. This initial condition may be deterministic or uncertain. In the former case, we have

$$u_0(\mathbf{x}, t=0) = u_0^0(\mathbf{x}), \qquad u_k(\mathbf{x}, t=0) = 0 \quad \text{for } k = 1, \ldots, \mathrm{P}. \tag{7.43}$$

On the other hand, when the initial condition is uncertain, initial conditions for all the modes u_k need to be specified.

7.2.2.3 Implicit Time Discretization

For unsteady problems, the use of an explicit time integration scheme for (7.38), as proposed in [123] in the context of the Navier-Stokes equations, leads to a simple algorithm that requires direct evaluation of the coupling terms. The use of explicit time-schemes has shown its efficiency for transient computations, but explicit stability restrictions on the time step can be prohibitive on fine grids. For steady-state problems, a pseudo-transient approach may also be conceived, but in this case as well stability restrictions may lead to poor computational efficiency. Consequently, the development of an efficient numerical solver for the coupled system of equations is needed. This approach is adopted in the development below.

A simple, generic, example of an implicit time integration method is the Euler backward scheme, whose application to (7.38) results in the following semi-discrete form:

$$\frac{\alpha}{\Delta t} u_i^{n+1} - \sum_{l=0}^{\mathrm{P}} \sum_{m=0}^{\mathrm{P}} C_{ilm} \nabla \cdot \left[\lambda_l(\mathbf{x}) \nabla u_m^{n+1}\right] = s_i^{n+1}(\mathbf{x}) + \frac{\alpha}{\Delta t} u_i^n \tag{7.44}$$

where Δt is the time step and the superscripts refer to the time level.

7.2.3 Finite Difference Discretization

7.2.3.1 Spatial Discretization

Let $\mathcal{D} \equiv [0, L] \times [0, H]$ be a rectangular domain discretized in a set of $\mathrm{N_x} \times \mathrm{N_y}$ non-overlapping cells with uniform size $\Delta x = L/\mathrm{N_x}$ and $\Delta y = H/\mathrm{N_y}$ in the x and y directions. We denote by $(\Phi)_{i,j}$, for $i = 1, \ldots, \mathrm{N_x}$ and $j = 1, \ldots, \mathrm{N_y}$ the cell-averaged value of Φ,

$$(\Phi)_{i,j} \equiv \frac{1}{\Delta x \Delta y} \int_{(i-1)\Delta x}^{i\Delta x} \int_{(j-1)\Delta y}^{j\Delta y} \Phi(\mathbf{x})\, dx\, dy, \tag{7.45}$$

where Φ stands for any of the field variables u_k, λ_k and s_k. Using this convention, we rely on the following centered, second-order spatial discretization of (7.44):

$$\begin{aligned}
&\frac{\alpha}{\Delta t}(u_k^{n+1})_{i,j} - \sum_{l=0}^{\mathrm{P}}\sum_{m=0}^{\mathrm{P}} C_{klm}\left[\frac{(\lambda_l)_{i+1,j}+(\lambda_l)_{i,j}}{2}\frac{(u_m^{n+1})_{i+1,j}-(u_m^{n+1})_{i,j}}{\Delta x^2}\right.\\
&\quad - \frac{(\lambda_l)_{i,j}+(\lambda_l)_{i-1,j}}{2}\frac{(u_m^{n+1})_{i,j}-(u_m^{n+1})_{i-1,j}}{\Delta x^2}\\
&\quad + \frac{(\lambda_l)_{i,j+1}+(\lambda_l)_{i,j}}{2}\frac{(u_m^{n+1})_{i,j+1}-(u_m^{n+1})_{i,j}}{\Delta y^2}\\
&\quad \left. - \frac{(\lambda_l)_{i,j}+(\lambda_l)_{i,j-1}}{2}\frac{(u_m^{n+1})_{i,j}-(u_m^{n+1})_{i,j-1}}{\Delta y^2}\right]\\
&= (s_k^{n+1})_{i,j} + \frac{\alpha}{\Delta t}(u_k^n)_{i,j}, \quad k=0,\ldots,\mathrm{P}. \qquad (7.46)
\end{aligned}$$

The above equation can be recast in the following generic form:

$$\begin{aligned}
&\sum_{l=0}^{\mathrm{P}}\sum_{m=0}^{\mathrm{P}} C_{klm}\Big[(W_l)_{i,j}(u_m^{n+1})_{i+1,j} + (E_l)_{i,j}(u_m^{n+1})_{i-1,j}\\
&\quad + (N_l)_{i,j}(u_m^{n+1})_{i,j+1} + (S_l)_{i,j}(u_m^{n+1})_{i,j-1} + (C_l^k)_{i,j}(u_m^{n+1})_{i,j}\Big]\\
&= (f_k^{n+1})_{i,j}, \quad k=0,\ldots,\mathrm{P} \qquad (7.47)
\end{aligned}$$

which shows that a linear system of $\mathrm{N_x} \times \mathrm{N_y} \times (\mathrm{P}+1)$ equations must be solved in order to advance the solution by one time step. Of course, in the steady case ($\alpha = 0$) this system is solved only once, and the superscripts indicating the time level are no longer needed.

7.2.3.2 Treatment of Boundary Conditions

Both Dirichlet and Neumann conditions are implemented using ghost cell techniques. For the case of a Dirichlet condition, a ghost cell is introduced at the mirror image with respect to the boundary point of the neighboring interior cell. The value of the solution at the ghost cell is then determined by linearly extrapolating the solution from the interior, leading to a linear combination of the known value at the boundary and the neighboring interior node. Using this relationship, the ghost variables are then eliminated from the equation system. A similar approach is used in the case of a Neumann condition, based on expressing the known value of the normal derivative in terms of a second-order centered difference formula involving the solution at the neighboring internal node and the corresponding ghost node. The resulting relationship is then substituted into the equation system in order to eliminate the ghost variable. This approach results in a modified system of the form,

$$\sum_{l=0}^{\mathrm{P}}\sum_{m=0}^{\mathrm{P}} C_{klm}\Big[(\tilde{W}_l)_{i,j}(u_m^{n+1})_{i+1,j} + (\tilde{E}_l)_{i,j}(u_m^{n+1})_{i-1,j} + (\tilde{N}_l)_{i,j}(u_m^{n+1})_{i,j+1} + (\tilde{S}_l)_{i,j}(u_m^{n+1})_{i,j-1} + (\tilde{C}_l^k)_{i,j}(u_m^{n+1})_{i,j}\Big] = (\tilde{f}_k^{n+1})_{i,j}, \tag{7.48}$$

where the tildes are used to indicate the modified values after implementation of the boundary conditions.

7.2.4 Iterative Method

Since the size of system (7.48) is large for most applications, iterative solution methods are preferred over direct schemes. In this work, Gauss-Seidel iterations are used [220].

7.2.4.1 Outer Iterations

Let us denote by $(\tilde{u}_m)_{i,j}^{ou}$ the estimate of $(u_m^{n+1})_{i,j}$ after the ou-th Gauss-Seidel iteration. This estimate can be computed by applying the following algorithm, called outer iterations, in contrast with the inner iterations described later:

- Loop on ou (Gauss-Seidel index)
  - For $i = 1$ to $\mathrm{N_x}$, do
    - For $j = 1$ to $\mathrm{N_y}$, do
      - Find $(\tilde{u}_k)_{i,j}^{ou+1}$ such that:

$$\sum_{l=0}^{\mathrm{P}}\sum_{m=0}^{\mathrm{P}} C_{klm}(\tilde{C}_l^k)_{i,j}(\tilde{u}_m)_{i,j}^{ou+1} = (\tilde{f}_k^n)_{i,j} - \sum_{l=0}^{\mathrm{P}}\sum_{m=0}^{\mathrm{P}} C_{klm}\left[(\tilde{W}_l)_{i,j}(\tilde{u}_m)_{i+1,j}^{ou} + (\tilde{E}_l)_{i,j}(\tilde{u}_m)_{i-1,j}^{ou+1} + (\tilde{N}_l)_{i,j}(\tilde{u}_m)_{i,j+1}^{ou} + (\tilde{S}_l)_{i,j}(\tilde{u}_m)_{i,j-1}^{ou+1}\right] \equiv (Q_k)_{i,j}^{ou}, \quad k = 0,\ldots,\mathrm{P} \tag{7.49}$$

    - End of loop on j
  - End of loop on i
- End of loop on ou

Thus,

$$(R_k)^{ou}_{i,j} = (Q_k)^{ou}_{i,j} - \sum_{l=0}^{\mathrm{P}} \sum_{m=0}^{\mathrm{P}} C_{klm} (\tilde{C}_l)_{i,j} (\tilde{u}_m)^{ou}_{i,j},$$

is the local residual of (7.49), for the kth mode, at the ou-th Gauss-Seidel iteration.

7.2.4.2 Inner Iterations

For each point in space, (7.49) can be rewritten in vector form as:

$$\begin{bmatrix} \sum_{l=0}^{\mathrm{P}} C_{00l}(\tilde{C}_l) & \cdots & \sum_{l=0}^{\mathrm{P}} C_{0\mathrm{P}l}(\tilde{C}_l) \\ \vdots & \ddots & \vdots \\ \sum_{l=0}^{\mathrm{P}} C_{\mathrm{P}0l}(\tilde{C}_l) & \cdots & \sum_{l=0}^{\mathrm{P}} C_{\mathrm{PP}l}(\tilde{C}_l) \end{bmatrix} \cdot \begin{pmatrix} (\tilde{u}_0)^{ou+1} \\ \vdots \\ (\tilde{u}_{\mathrm{P}})^{ou+1} \end{pmatrix} = \begin{pmatrix} (Q_0)^{ou} \\ \vdots \\ (Q_{\mathrm{P}})^{ou} \end{pmatrix} \quad (7.50)$$

where the grid-point indices have been dropped for clarity. Thus, at this stage, one has to solve a system of $\mathrm{P}+1$ equations to compute $(\tilde{u}_{k=0,\dots,\mathrm{P}})^{ou+1}_{i,j}$ from (7.49). A standard relaxation method (SOR) [220] is employed for this purpose. Denoting by ω the over-relaxation parameter, and by $[A_{km}]$ the system matrix corresponding to (7.50), the iterations are performed according to:

- Loop over in (SOR index)
  - Do $k = 0, \dots, \mathrm{P}$

    Compute a new estimate of $(\tilde{u}_k)$ solution of (7.50) using

$$(\tilde{u}_k)_{in+1} = (1-\omega)(\tilde{u}_k)_{in} + \frac{\omega}{A_{kk}} \left((Q_k)_{in} - \sum_{m=0}^{k-1} A_{km} (\tilde{u}_m)_{in+1} - \sum_{m=k+1}^{\mathrm{P}} A_{km} (\tilde{u}_m)_{in} \right) \quad (7.51)$$

  - End of loop over k.
- End of loop over in.

Note that for convenience, the Gauss-Seidel index ou has been dropped in (7.51).

Remark The above decomposition of the iterative scheme into outer and inner loops may appear artificial, since a global iteration on the three-dimensional system for $(u_k^{n+1})_{i,j}$ could be constructed. However, in view of the implementation of the multigrid scheme, which is based on spatial coarsening, it is found more convenient to clearly distinguish the inner iterations—which locally update the spectral coefficients of the solution, from the outer iterations—which account for the spatial coupling. In addition, computational tests (not shown) indicate that the convergence of the outer GS iteration is greatly improved when a more accurate estimate of the exact solution of (7.50) is used.

7.2.5 Convergence of the Iterative Scheme

The efficiency of the overall iterative method proposed above is estimated through the convergence rate of $(\tilde{u})^{ou}$ towards $(u)^{n+1}$, as the number of iterations ou increases. This convergence rate depends on the spectral radius of the system (7.48). Since C_{klm} does not depend on the solution variables or parameters, the spectral radius is only a function of the stochastic diffusivity field $\lambda(\mathbf{x}, \boldsymbol{\xi})$, of the time-step Δt (if relevant), and of Δx and Δy. For the deterministic problem ($\mathrm{P} = 0$), it is known that $(\lambda)_{i,j} \geq 0$ for all i, j is necessary to ensure convergence, and that the convergence rate deteriorates when $\alpha/\Delta t$ decreases. In the stochastic case, the situation is more complex. In particular, the computations below indicate that as COV increases, the convergence rate of the present iterative scheme deteriorates. These experiences appear to be consistent with the theoretical results in [9], where uniformly-distributed random variables were used to ensure positivity.

The measure of convergence is obtained through the L^2-norm of the residual for a given mode l which is expressed as:

$$\mathcal{N}_l \equiv \left\{ \sum_{i=1}^{\mathrm{N_x}} \sum_{j=1}^{\mathrm{N_y}} \left[(R_l)_{i,j} \right]^2 \Delta x \Delta y \right\}^{1/2} . \tag{7.52}$$

The convergence of the iterative method will be further analyzed in Sect. 7.2.7 by monitoring the evolution of the maximum (over all modes) normalized residual:

$$\mathcal{R}_p \equiv \frac{\max_l [\mathcal{N}_l(p)]}{\max_l [\mathcal{N}_l(p=0)]} \tag{7.53}$$

where the index p refers to the MG cycle index (defined below).

7.2.6 Multigrid Acceleration

It is known from the analysis of deterministic diffusion equations that the convergence rate is a function of spatial frequencies. Specifically, the longest wavelengths exhibit the lowest convergence rate, while short scales converge faster. To improve convergence, acceleration techniques based on spatial coarsening have been proposed in the literature [39, 74, 228]. Below, we outline a multigrid technique for the stochastic case.

The basic idea of the multigrid technique is to treat the modes with low spatial frequencies on coarser grids, since fine spatial resolution is not required for these modes. The gain of the method is due to the faster convergence of the long-wave modes on the coarser grids, as well as the lower CPU cost of the corresponding iterations. Since multigrid methods are widely used, we will just recall the main ingredients of the approach, namely (i) the definition of the grid levels, (ii) the projection step and (iii) the prolongation procedure.

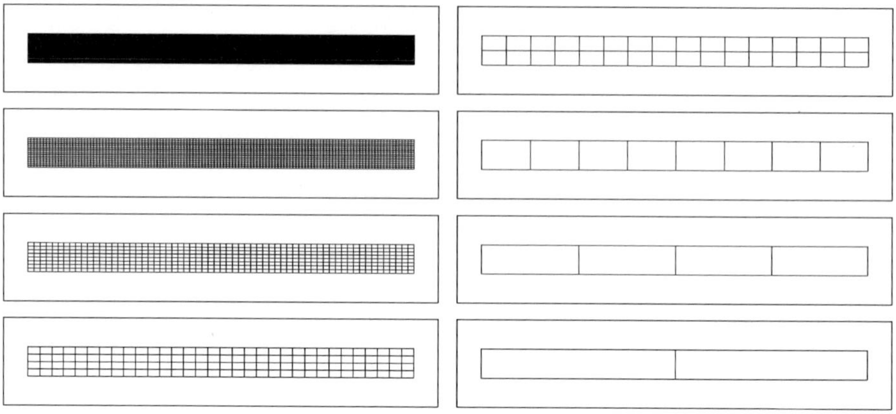

Fig. 7.1 Example of grid coarsening used for the multigrid method. The base grid consists of 32×256 cells (*top left*). The mesh is first coarsened by merging 4 (two in each direction) cells to form a coarser child cell. When the number of cells in one direction is odd, the coarsening process switches to 1D merging as in the last 3 grid levels. Adapted from [122]

7.2.6.1 Definition of Grid Levels

Thanks to the regular structure of the computational grid, the coarsening is made by merging a set of neighboring grid cells to give a single cell on the next (coarser) grid level. This leads to a hierarchical set of grids. In the current implementation, a coarsening step consists of merging 4 cells (2 in each direction) with surface areas $\Delta x^k \times \Delta y^k$ each, to obtain a child cell with surface area $\Delta x^{k+1} \times \Delta y^{k+1} = 4\Delta x^k \times \Delta y^k$, the superscripts denoting the respective grid levels. Thus, starting from a grid level k, made of $\mathrm{N_x}^k \times \mathrm{N_y}^k$ cells, the next grid level contains $\mathrm{N_x}^{k+1} \times \mathrm{N_y}^{k+1} = (\mathrm{N_x}^k \times \mathrm{N_y}^k)/4$ cells. Clearly, this process can be repeated as long as $\mathrm{N_x}^k$ and $\mathrm{N_y}^k$ are even numbers. Whenever one of the number of cells in a direction is odd, the coarsening automatically switches to a one-dimensional coarsening procedure in which only two cells are merged to make a child cell. This procedure is illustrated in Fig. 7.1, where the successive grid levels are plotted. Clearly the procedure is optimal when $\mathrm{N_x}$ and $\mathrm{N_y}$ are powers of 2.

7.2.6.2 Projection and Prolongation Procedures

On the finest grid level, a small number N_{ou} of outer iterations is first performed. This provides approximate solutions $(\tilde{u}_m)_{i,j}^{N_{ou}}$ with residuals $(R_m)_{i,j}^{N_{ou}}$. These residuals are then projected onto the next coarser grid, where problem (7.48) is considered, with $(R_m)_{i,j}^{N_{ou}}$ as the right-hand-side (in lieu of $(f_k)_{i,j}$), and with the same but homogeneous boundary conditions. (In other words, on the coarser grids, the residual equation is solved.) To do so, one has to provide an estimate of (λ) and (R_m) on consecutive grid levels. This is achieved by averaging their respective values over the parent cells as illustrated in the top scheme of Fig. 7.2.

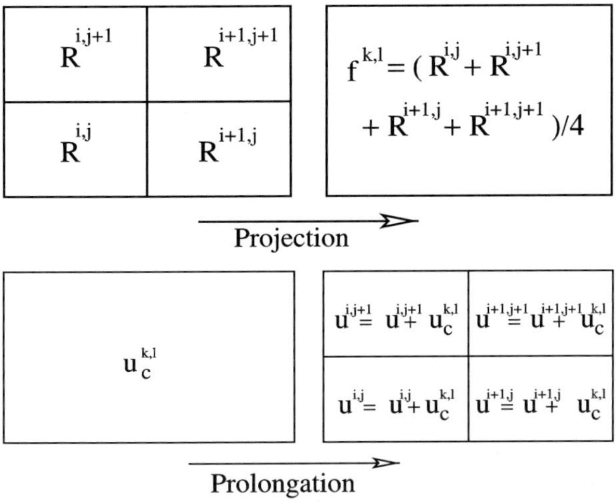

Fig. 7.2 Illustration of the projection (*top*) and the prolongation (*bottom*) procedures to transfer data between two successive grid levels. In the projection step, the residual on a given grid is transferred to the next (*coarser*) grid level by spatial averaging. The same methodology is used to transfer the diffusivity data. For the prolongation of the solution from one grid level to the next (*finer*) one, simple addition is used. Adapted from [122]

On the new grid level, a few outer iterations are performed, following the same methodology, to obtain an approximate solution and a residual. The projection process is repeated until the last grid level is reached. Then, from the coarser grid level, where an estimate solution for the residual equation has been obtained, it is first transferred to the previous grid level through a prolongation procedure and then used to correct the solution on that finer grid level. In the current implementation, this is achieved by summing the cell averaged solution at level k with the solution of its parent cell, as shown in Fig. 7.2. When the solution has been prolongated onto level $k-1$, a few outer iterations are performed (smoothing step) the process is repeated until the initial, fine, grid is reached.

7.2.6.3 Multigrid Cycles

Starting from the original grid, the application of successive projections up to the coarsest grid level, followed by successive prolongations up to the starting grid, is referred to as a cycle. Different kinds of cycles may be used [228], according to the excursion path along the grid levels. For instance, the so-called W-cycles have been designed to improve the convergence rate of the multigrid method, and many other examples can be found in the literature. Since our objective here is to outline a multigrid methodology for stochastic diffusion equations, we limit ourselves to the simplest case of the V-cycle as described above. Moreover, we use a constant number, N_{ou}, of Gauss-Seidel iterations on every grid level, after every projection or prolongation step.

7.2.6.4 Implementation of the Multigrid Scheme

Implementation of the resulting MG scheme can be summarized as follows:

1. **Initialization:**
   - Determine the spectral basis, compute and store the multiplication tensor C.
   - Compute the KL decomposition of the λ. (Alternatively the PC expansion of λ is imported, as in [51], or otherwise set by the user.)
   - Determine the system coefficients for all grid levels:
     `For ig = 1,...,Ng`
     – Determine the grid properties: $\Delta x^{ig} = 2^{ig-1}\Delta x,\ \Delta y^{ig} = 2^{ig-1}\Delta y, \mathrm{N_x}^{ig} = \mathrm{N_x}/2^{ig-1},\ \mathrm{N_y} = \mathrm{N_y}/2^{ig-1}$.
     – Compute the cell averaged diffusion field $(\lambda_l)_{i,j}^{ig}$, for $i = 1,\dots,\mathrm{N_x}^{ig}$, $j = 1,\dots,\mathrm{N_y}^{ig}$.
     – Using (7.46) and (7.47), determine the system coefficients $(C_l^k)_{i,j}^{ig}$, $(E_l)_{i,j}^{ig}$, $(W_l)_{i,j}^{ig}$, $(N_l)_{i,j}^{ig}$, and $(S_l)_{i,j}^{ig}$.
     – Compute and store the modified system coefficients accounting for the boundary conditions: $(\tilde{C}_l^k)_{i,j}^{ig}$, $(\tilde{E}_l)_{i,j}^{ig}$, $(\tilde{W}_l)_{i,j}^{ig}$, $(\tilde{N}_l)_{i,j}^{ig}$, $(\tilde{S}_l)_{i,j}^{ig}$.
     `End of loop over ig`
   - Initialize solution.

2. **Loop over time index n:**
   a. Compute the right-hand side of system (7.48) on the first grid level: $(\tilde{f}_k^n)_{i,j}^{ig=1}$.
   b. Initialize solution on the first grid level: $(\tilde{u}_k)_{i,j}^{ig=1} = (\tilde{u}_k^n)_{i,j}$.
   c. **Beginning of V-cycle**
      `For ig = 1 ,..., Ng` (coarsening)

      - If $ig > 1$ then initialize the solution $(\tilde{u})^{ig}$ to zero.
      - **Outer loop :**
        `For ou = 1,...,Nou`
        – **Loop over spatial indices**
          `For i = 1,...,`$\mathrm{N_x}^{ig}$
          `For j = 1,...,`$\mathrm{N_y}^{ig}$
          · Using (7.49), compute the right-hand side of (7.50).
          · **Inner loop**
            `For in = 1,...,Nin`
            · **Loop over mode index**
              `For k = 0,...,P`
              Apply (7.51) to $(\tilde{u}_k)_{i,j}^{ig}$
              `End of loop over k`
            `End of loop over in`
          `End of loop over i,j`
        `End of loop over ou`
      - If $ig < Ng$, then
        – Compute the local residual $(R_k)_{i,j}^{ig}$ of (7.48) on the current grid level.
        – Project the local residuals to compute the right-hand side of (7.48) at the next grid level $ig+1$, *i.e.* determine $(\tilde{f}_k)_{i,j}^{ig+1}$.

```
End of loop over ig
For ig = Ng - 1 ,..., 1 (refinement)
```

- Update solution $(\tilde{u}_k)^{ig}$ through the prolongation of $(\tilde{u}_k)^{ig+1}$.
- **Outer loop:**
  `For ou = 1,...,Nou`
  - **Loop over spatial indices**
    `For i = 1,...,` $\mathrm{N_x}^{ig}$
    `For j = 1,...,` $\mathrm{N_y}^{ig}$
    · Using (7.49), compute the right-hand side of (7.50).
    · **Inner loop**
    `For in = 1,...,Nin`
    · **Loop over mode index**
    `For k = 0,...,P`
    Apply (7.51) to $(\tilde{u}_k)^{ig}_{i,j}$
    `End of loop over k`
    `End of loop over in`
    `End of loop over i,j`
  `End of loop over ou`

`End of loop over ig`

Compute the local residual $(R_k)_{i,j}$ of (7.48) on the first grid level. If one of the norms $\mathcal{N}_l$ from (7.52) is greater than the prescribed threshold, then a new V-cycle is performed starting from (c).

d. Determine the solution: $(\tilde{u}_k^{n+1})_{i,j} = (\tilde{u}_k)^{ig=1}_{i,j}$, for $i = 1,\dots,\mathrm{N_x}$, $j = 1,\dots,\mathrm{N_y}$ and $k = 1,\dots,\mathrm{P}$.

3. **End of time loop**

7.2.7 Results

We now present test results that show the behavior and convergence properties of the multigrid method. For the test cases below, we set $\alpha = 0$ and study the stochastic diffusion in a square domain, with unit edge-length and with no source term ($s \equiv 0$). Deterministic boundary conditions are used with Dirichlet conditions on $x = 0$ (where $u = 1$) and $x = 1$ (where $u = 0$), and homogeneous Neumann boundary conditions for the $y = 0$ and $y = 1$ edges. To analyze the performance of the scheme, we monitor the evolution of the maximum (over all the modes) of the L^2-norms of the normalized residuals, more specifically the decay of the peak residual as the number of multigrid cycles increases.

7.2.7.1 Multigrid Acceleration

Dependence on grid size: We start by examining the dependence of the convergence rate on the number of points involved in the spatial discretization. The computations use a diffusivity field with $COV = 2.0$, $L_c = 5$, a KL expansion with

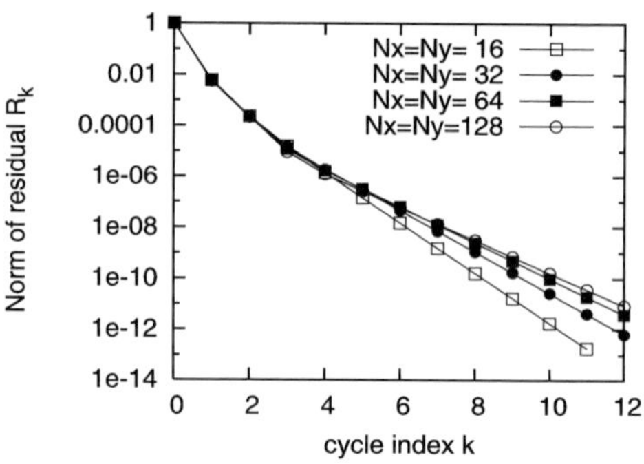

Fig. 7.3 Normalized residual versus cycle index for grids with $N_x = N_y = 16, 32, 64$ and 128. The corresponding number of grid levels are $Ng = 4, 5, 6$ and 7. $COV = 2.0$, $L_c = 5$, $N = 5$, $No = 2$ ($P = 20$), $\omega = 1.5$, $N_{ou} = N_{in} = 3$

$N = 5$ modes, together with a second-order PC expansion ($No = 2$). With these parameters, $P = 20$ and so the total number of modes equals 21. The multigrid parameters are $N_{ou} = N_{in} = 3$ and $\omega = 1.5$. Computations are performed for grids with $N_x = N_y = 16, 32, 64$ and 128; the corresponding number of grid levels are $Ng = 4, 5, 6$ and 7.

For each case, the maximum normalized residual is plotted against cycle number in Fig. 7.3. The results clearly show the quasi-independence of the convergence rate with respect to the spatial discretization. There is a weak improvement in the convergence rate at the lower values of N_x and N_y, which may be attributed to the lack of proper resolution of the KL modes on the coarser meshes. This claim is supported by the observation that the convergence rate tends to be grid-size independent when N_x and N_y increase. The weak dependence of the convergence rate on the grid size also highlights the excellent scalability of the method concerning the spatial discretization, as the CPU time scales roughly as $N_x \times N_y$. Note that the relaxation parameter, ω, and number of inner and outer iterations, N_{ou} and N_{in}, have been selected based on systematic tests (not shown) to determine their optimal value. While further refinement of these parameters may be possible, these values will be kept the same for the remaining cases below, unless explicitly stated.

Effect of grid levels on MG acceleration: Figure 7.4 shows the evolution of the peak normalized residual with the number of cycles for a fixed grid with $N_x = N_y = 32$. Results obtained for different numbers of grid levels in the V-cycles are shown, namely $N_g = 1, 2, 3, 4$ and 5.

The results illustrate the effect of MG acceleration with increasing number of grid levels. The setting $N_g = 1$ corresponds to the Gauss-Seidel iteration, applied to the initial system of equations with no coarsening. Thus, after the first V-cycle (that is $2N_{ou} = 6$ GS iterations) the short-scales in the residual (mostly related to the Dirichlet boundary conditions) have been reduced and the convergence rate falls dramatically. This clearly illustrates the lower convergence rate of the larger length scales. When the number of grid levels is increased to $N_g = 2$ the convergence rate is slightly improved, but the iterative method is still inefficient. In fact, the first significant improvement is reported for $N_g = 3$, where one observes a residual reduction factor per V-cycle of approximately 0.78. With $N_g = 4$, the convergence

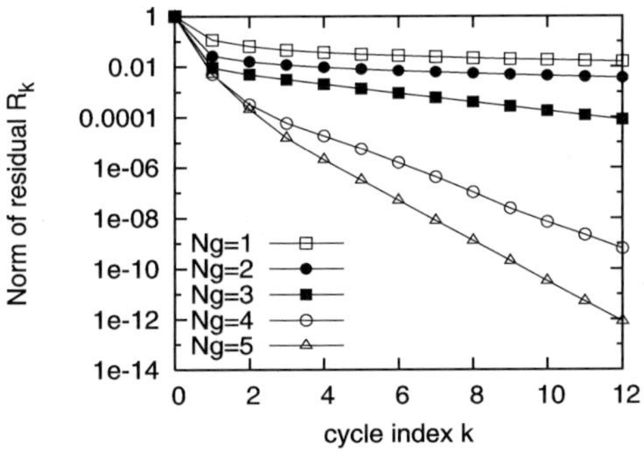

Fig. 7.4 Normalized residual versus cycle index for $N_x = N_y = 32$, and $Ng = 1$, 2, 3, 4 and 5. COV = 2.0, $L_c = 5$, N = 10, No = 2 (P = 65), $\omega = 1.5$, $N_{ou} = N_{in} = 3$

rate is much larger, as the residual reduction factor per V-cycle is approximately 0.2. As expected, the largest convergence rate is observed for $N_g = 5$, with a residual reduction factor close to 0.1. These tests show that the discretization parameters N_x and N_y should be selected, to the extent possible, so that the coarsest grid level has a minimum number of cells in each direction. Note, in particular, that the large improvement in convergence rate between $N_g = 3$ and $N_g = 5$ is achieved at a very low additional CPU cost, since the fourth and last grid levels only involve 16 and 4 cells, respectively.

7.2.7.2 Influence of Stochastic Representation Parameters

We now investigate the effect of the stochastic representation parameters, namely the number, N, of KL modes used in the representation of the stochastic diffusivity field, and the order, No, of the PC expansion. In the tests below, the spatial discretization parameters are held fixed, as are the over-relaxation parameter, $\omega = 1.5$, the number of grid levels, $Ng = 5$, and the number of iterations performed on each grid level, $N_{in} = N_{ou} = 3$. The coefficient of variation, COV = 2, and correlation length, $L_c = 1$, are also held fixed.

Effect of N**:** Plotted in Fig. 7.5 is the peak normalized residual against cycle number for (a) a first-order PC expansion with N ranging from 10 to 80, and (b) a second-order PC expansion with N = 10, 15 and 20 (P = 65, 135 and 230, respectively). For both first and second-order expansions, the evolution of the residual is independent of N, again showing the efficiency of the MG scheme. Note that in the present case one cannot infer from this behavior a linear relationship between N and the CPU time. The latter is in fact a strong function of the number of non-zero terms in C, which depends on both N and No. This contrasts with previous observation regarding scalability of the scheme with respect to the number of grid points.

Effect of No**:** The results of the previous section show a dependence of the convergence rate of the multigrid method on the order of the PC expansion. This dependence is further investigated by setting N = 10 and varying No from 1 to 3; with

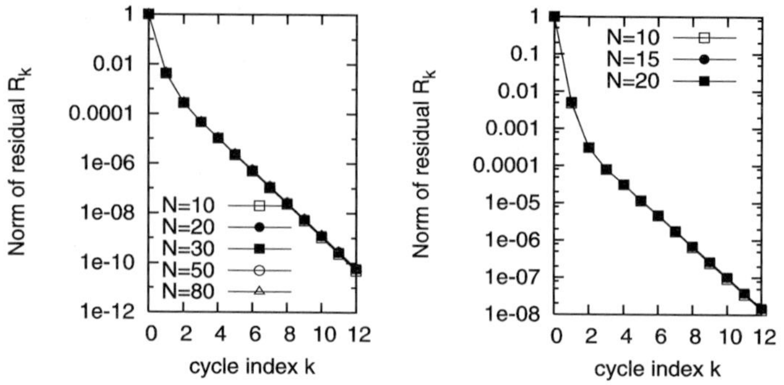

Fig. 7.5 Peak normalized residual versus cycle number. *Left*: first-order PC expansion with $N = 10, 20, 30$, 50 and 80; *right*: second-order PC expansion with $N = 10$ ($P = 65$), 15 ($P = 135$) and 20 ($P = 230$). In both cases, $N_x = N_y = 32$, $L_c = 1$, $N_{in} = N_{ou} = 3$, and $\omega = 1.5$

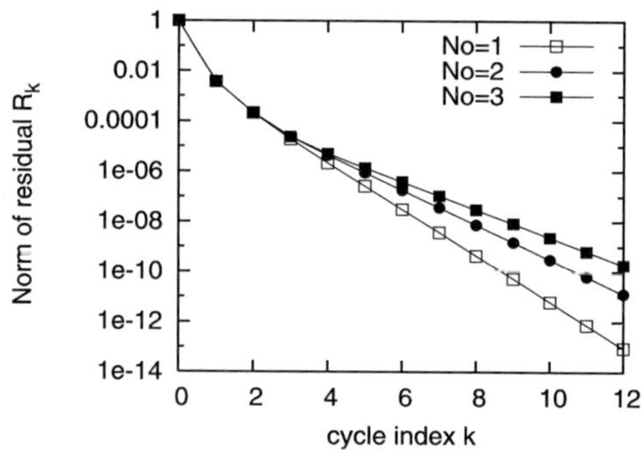

Fig. 7.6 Peak normalized residual versus cycle number for $No = 1$ ($P = 10$), $No = 2$ ($P = 65$), and $No = 3$ ($P = 285$). In all cases, $N = 10$, $N_x = N_y = 32$, $L_c = 1$, $N_{in} = N_{ou} = 3$, and $\omega = 1.5$

$P = 10$, 65 and 285, respectively. The convergence of the iterations for these cases is illustrated in Fig. 7.6, which depicts the behavior of the peak residual as the number of cycles increases.

The results indicate that the convergence rate decreases slightly as the order of the PC expansion increases. The residual reduction factor per cycle is about 0.09 for $No = 1$ and approximately 0.2 for $No = 3$. In light of the experiences above, it is evident that the present reduction in convergence rate is not due to the increase in number of modes P, but rather to the need for additional cycles in order to propagate the residual for coupled terms of different order. For the present examples, the convergence rate is still satisfactory for $No = 3$. In situations requiring higher-order expansions, however, further improvement may be required. This could be achieved for instance by blending the (spatial) MG concepts with a spectral (mode) coarsening procedure.

7.2.7.3 Effects of Diffusivity Field Statistics

We now analyze the effects of the diffusivity field characteristics on the convergence rate, by varying its statistical parameters. We recall that the diffusivity is parametrized using its COV, which represents the normalized local statistical spread of the log-normal distribution about the median value of λ, and the correlation length, which accounts for the spatial variability of the process. The effects of these two parameters are analyzed separately below.

Effect of diffusivity variability: Tests on the effect of the variability of the diffusivity field are performed using $L_c = 1$, a KL expansion with $N = 25$, and PC expansion of first and second-order ($P = 25$ and 350, respectively). A 32×32 computational grid is used and the MG parameters are as follows: $\omega = 1.5$, $N_{ou} = N_{in} = 3$ and $N_g = 5$. Results with different values of COV are reported in Fig. 7.7.

The results show that, as expected, the convergence rate is strongly dependent on the variance of the diffusivity field. Actually, the highest convergence rate is achieved for the lowest values of COV (COV = 1.15). Consistent with previous findings, for the same values of COV the second-order PC expansion exhibits slower convergence rate than the first-order scheme. Moreover, when COV increases, the convergence rate decreases for both the first- and second-order expansions, but the reduction is more substantial for the latter. It should be emphasized, however, that the deterioration of convergence rate with increasing COV is due to the extreme behavior of the corresponding problem, and therefore is not inherent to the present multigrid scheme.

Influence of the correlation length: The effect of the correlation length is analyzed by performing computations with a fixed variability (COV = 1.5) but varying L_c. As illustrated in Fig. 7.8, as L_c decreases, the spectrum of λ broadens with higher amplitudes in the small scales. Since the variance is fixed, however, the "energy" content of the spectrum remains constant.

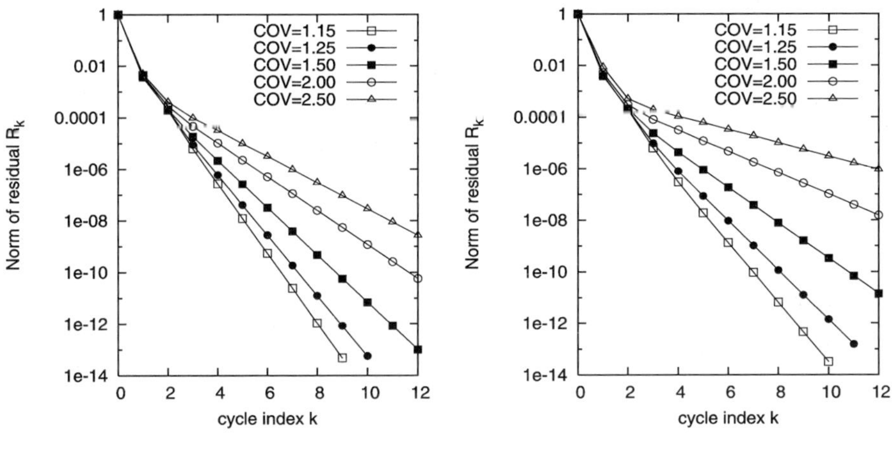

Fig. 7.7 Peak residual versus number of cycles for different values of λ's COV; *left*: first-order PC expansion, *right*: second-order PC expansion. In all cases, $L_c = 1$, $N = 25$, $\omega = 1.5$, $N_{ou} = N_{in} = 3$, $Ng = 5$, and $N_x = N_y = 32$

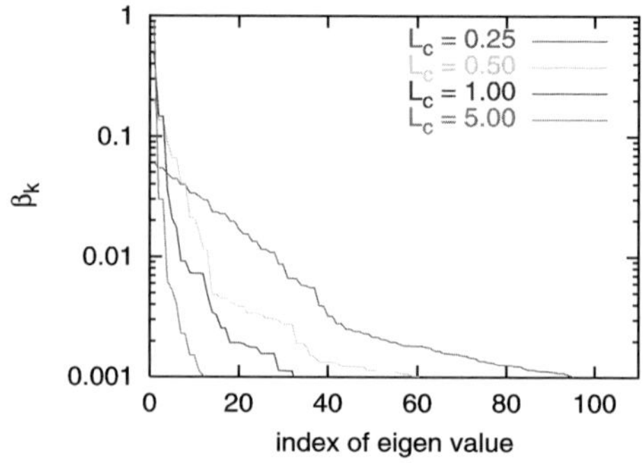

Fig. 7.8 Spectra of the eigenvalues, β_k, of the KL expansion for different correlation lengths $L_c = 0.25$, 0.5, 1, and 5. Adapted from [122]

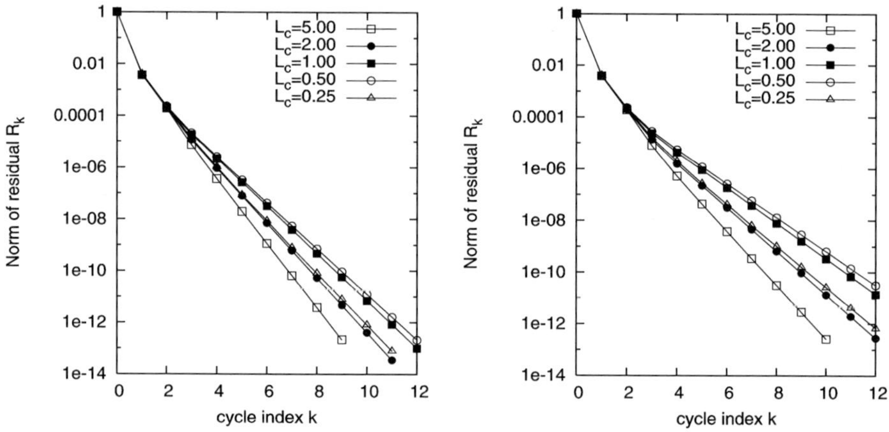

Fig. 7.9 Peak residual versus number of cycles for N = 20 KL modes and different values of L_c; *left*: first-order PC expansion (P = 20), *right*: second-order PC expansion (P = 350). In all cases, COV = 1.5, $N_{in} = N_{out} = 3$, $\omega = 1.5$, and $N_x = N_y = 32$

Figure 7.9 shows the convergence rate of the MG iterations for different correlation lengths, $L_c = 0.25$, 0.5, 1, 2 and 5. Plotted are results obtained using both first- and second-order PC expansions. The results show a weak dependence of the convergence rate on L_c, indicating that the MG method effectively maintains its good convergence properties even as small-scale fluctuations in λ increase. The weak dependence of the convergence rate on spatial lengthscales of λ has also been observed in deterministic simulations (not shown). This shows that the extension of the MG scheme to stochastic problems does not adversely affect its effectiveness in dealing with spatially varying diffusivity. Closer analysis of the results in Fig. 7.9 also supports our previous observation that the convergence rate for first-order PC expansions is larger than for the second-order case, but differences are now less significant due to a lower COV.

As shown in Fig. 7.8, decreasing L_c results in a broader eigenvalue spectrum, which raises the question whether N = 20 KL modes is sufficient to capture all the relevant scales of λ. This question arises because truncation of the KL expansion removes the highest spatial frequencies, and leads to under-estimation of the variance. To verify that the near collapse of the curves in Fig. 7.9 with decreasing L_c is not due to such truncation, simulations were repeated using a first-order PC expansion and a higher number of KL modes, N = 105. The results (not shown) exhibit essentially the same convergence rate as with N = 20. This indicates that in this case the truncation of the KL expansion does not affect the convergence rate of the multigrid solver.

7.2.7.4 Selection of Multigrid Parameters

The computational tests above were performed with fixed MG parameters, which enabled direct comparison between various cases and thus simplified the analysis. It is evident, however, that tuning these parameters can improve the efficiency of the method. For the test cases in the previous sections, selecting $\omega \in [1.2, 1.7]$ results in convergent iterations, but varying ω within this range affected the convergence rate. Specifically, for fixed tolerance on the peak residual (10^{-10}), the number of cycles needed to achieve this level varied between 3–5 cycles. Consequently, tests should generally be conducted in order to select the optimal ω value for the problem at hand. A similar optimization process should also be conducted for proper selection of the number of inner and outer iterations. Clearly, using a large number of outer iterations results in an inefficient method, since one does not want to perform a large number of outer iterations on the initial grid level. At the same time, a minimal number of outer iterations is required during prolongation in order to smooth the solution sufficiently before switching to the next grid level. Thus, N_{ou} should be carefully optimized. Meanwhile, N_{in} should be set to the minimum value above which the convergence rate starts to decrease.

Lastly, the efficiency of the multigrid procedure can also be drastically improved by designing cycles with a more complex structure than the simple V-cycle used in the present work. To illustrate the improvement that can be achieved by adaptation of cycle structure, a line-coarsening strategy, designed for highly stretched grids and/or domains with high aspect ratios, is briefly outlined below. Assuming that $\Delta x \gg \Delta y$, the strategy consists of (i) performing a 1D-coarsening along the y-direction only, which eventually leads to a quasi-1D problem in x; (ii) applying a 1D MG approach in the x-direction, which is iterated until the overall tolerance level is reached; and (iii) performing a prolongation in the y direction only. This cycle, whose structure is schematically illustrated in Fig. 7.10, is repeated until the residual on the original fine grid drops below the desired tolerance level.

In Fig. 7.11 we contrast the convergence rates of the MG scheme using the V-cycles and of the adapted MG scheme outlined above. In both cases, the number of grid cells is fixed, $N_x = 128$ and $N_y = 32$, but the aspect ratio, L/H, of the domain is varied. Note that the case $L/H = 4$ corresponds to a grid with square

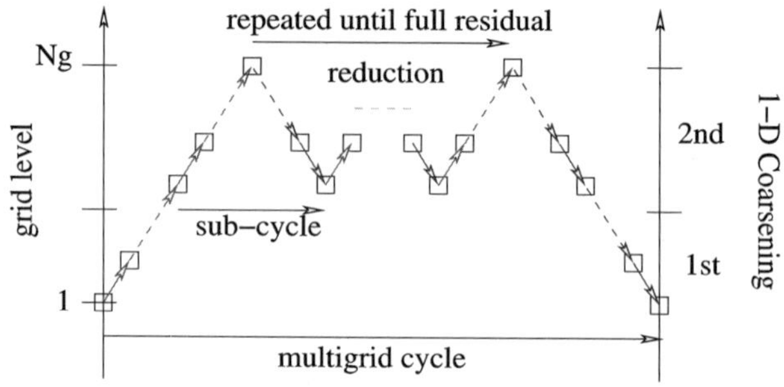

Fig. 7.10 Example of line-coarsening strategy for highly stretched grids. Instead of the usual V-cycle, the coarsening is first performed along the well resolved direction, and next in the second direction. Treatment (sub-cycles) of the second direction is repeated until the residual is reduced to the selected tolerance level. Adapted from [122]

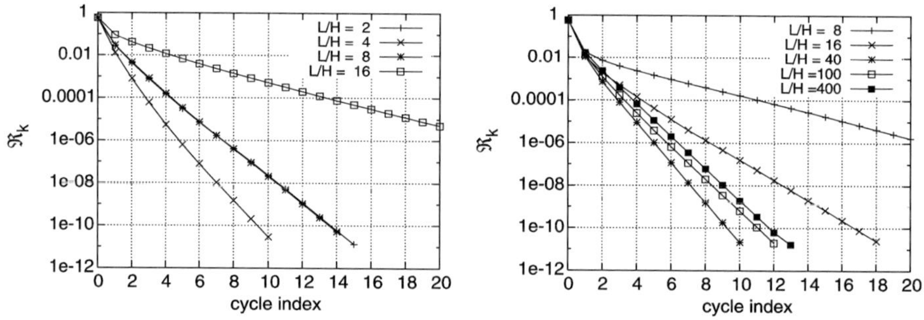

Fig. 7.11 Convergence rate of MG iterations: V-cycle approach (*left*) and line coarsening strategy (*right*). Curves are generated for solutions obtained in domains with different aspect ratio, L/H. In both simulations, $N_x = 128$, $N_y = 32$, $N_{ou} = 3$, $N_{in} = 2$, $\omega = 1.5$, $No = 2$, and $N = 3$. Adapted from [122]

cells, *i.e.* $\Delta x = \Delta y$. For the V-cycle iterations, the results indicate that the convergence rate deteriorates as the cell aspect ratio increases. Meanwhile, with the line-coarsening strategy, the convergence rate improves as L/H is increased from 8 to 40; for higher aspect ratios, up to $L/H = 400$, the convergence rate decreases slightly, but remains at a satisfactory level. In contrast, for such high aspect ratios, the regular V-cycle iterations are quite inefficient.

7.3 Stochastic Steady Flow Solver

In this section, we outline the development and implementation of stochastic solver for the steady Navier-Stokes equation with random data. Specifically, we describe a Newton scheme that enables us to solve efficiently the spectral problem arising from the Galerkin projection of the steady stochastic Navier-Stokes equations onto a PC basis. The scheme is a direct extension of the techniques originally designed for deterministic problems [60, 230, 231]. The Newton method uses the unsteady equations to derive a linear equation for the stochastic Newton increments. This linear equation is subsequently solved using a matrix-free strategy, where the iterations consist in performing integrations of the linearized unsteady Navier-Stokes equations, with an appropriate time stepping to allow for a decoupled integration of the stochastic modes. Thus, the method is particularly appealing in the stochastic spectral context as it requires little modifications of the unsteady stochastic solver, while preserving the key feature that stochastic modes are resolved in a decoupled fashion. In addition to its efficiency, examples are provided that demonstrate that the method is able to provide the stochastic steady solution even for flow regimes where the steady solution is likely to be unstable. Thus, a key advantage is achieved over pseudo-transient approaches, whose application in these situations can be highly problematic.

We outline the construction of the steady stochastic solver in the context of the incompressible Navier-Stokes equations. In Sect. 7.3.1 we recall the incompressible Navier-Stokes equations and introduce the class of time integration schemes to be used in the Newton method. Accommodation of random inputs is then considered in Sect. 7.3.2, which outlines a numerical method for the integration of unsteady modes appearing in the PC expansion of the solution. Special emphasis is placed on time discretizations that decouple the integration of the spectral modes. Section 7.3.3 then outlines the construction of a Newton method for the resolution of the stochastic steady Navier-Stokes equations. The Newton iterations, together with the equations satisfied by the Newton increments are first described in detailed. A matrix-free method is then outlined which enables their computation. The Newton method is then tested in Sects. 7.3.4 and 7.3.5 which provide examples of increasing stochastic dimension and complexity. These examples are used to assess the efficiency and robustness of the Newton iterations with regard to numerical and discretization parameters, flow variability and stability of the stochastic steady solution.

7.3.1 Governing Equations and Integration Schemes

We consider the flow of an incompressible, Newtonian fluid, with uniform density and kinematic viscosity in a bounded domain Ω_x. The flow is governed by the dimensionless unsteady Navier-Stokes equations:

$$\begin{cases} \dfrac{\partial \mathbf{u}}{\partial t} + \mathbf{u} \cdot (\boldsymbol{\nabla} \mathbf{u}) = -\boldsymbol{\nabla} p + \dfrac{1}{\mathrm{Re}} \nabla^2 \mathbf{u} + \mathbf{f}, \\ \boldsymbol{\nabla} \cdot \mathbf{u} = 0, \end{cases} \tag{7.54}$$

where $\mathbf{f}$ is a normalized force field and Re is the Reynolds number. Without loss of generality, we shall consider Dirichlet boundary conditions, and so boundary and initial velocity conditions are expressed as:

$$\begin{cases} \mathbf{u}(\mathbf{x} \in \Gamma, t) = \mathbf{u}_\Gamma(\mathbf{x}), \\ \mathbf{u}(\mathbf{x}, t = 0) = \mathbf{u}_0(\mathbf{x}). \end{cases} \tag{7.55}$$

Note that we restrict ourselves to time independent boundary conditions, as we are interested in steady solutions. For the resolution of (7.54), we focus on time discretizations involving an explicit treatment of the nonlinear terms and an implicit (or semi-implicit) treatment of the linear ones. For simplicity, we consider a simple first-order Euler scheme:

$$\begin{cases} \left(\mathbf{I} - \frac{\Delta t}{\mathrm{Re}} \boldsymbol{\nabla}^2\right) \mathbf{u}^{n+1} + \Delta t \boldsymbol{\nabla} p^{n+1} = \mathbf{u}^n - \Delta t \mathbf{u}^n \cdot \boldsymbol{\nabla} \mathbf{u}^n + \Delta t \mathbf{f}^{n+1}, \\ \boldsymbol{\nabla} \cdot \mathbf{u}^{n+1} = 0, \end{cases} \tag{7.56}$$

where the superscripts refer to the time level and Δt is the time-step size. It is seen that this time-discretization, results in a linear problem (namely the Stokes problem) for the unknown solution at time level $n + 1$.

To remain as general as possible, we recast the semi-discrete system (7.56) as:

$$\mathcal{I}\left(\mathcal{U}^{n+1}\right) = \mathcal{L}(\mathcal{U}^n, \mathcal{U}^{n-1}, \ldots) + \mathcal{N}\left(\mathcal{U}^n, \mathcal{U}^{n-1}, \ldots\right) + \mathcal{S}(\mathbf{f}^{n+1}), \tag{7.57}$$

where $\mathcal{U}^{n+1}$ is the solution at $t = (n+1)\Delta t$. (7.57) has to be solved for $\mathcal{U}^{n+1} \in \mathcal{V}_{\mathbf{u}_\Gamma}$, where $\mathcal{V}_{\mathbf{u}_\Gamma}$ is a suitable functional space for $\mathcal{U}$ satisfying the boundary conditions. Note that $\mathcal{I}$ and $\mathcal{L}$ are linear operators, whereas $\mathcal{N}$ is a nonlinear operator. The last term $\mathcal{S}$ accounts for the force field. We assume the availability of a deterministic solver for the resolution of (7.57). In addition, we shall consider (7.57) as a generic form of the semi-discrete incompressible Navier-Stokes equations, where the formal notation $\mathcal{U}$ stands for the relevant set of variables involved in the actual formulation, so that $\mathcal{U} = (\mathbf{u}, p)$ in primitive variables formulations, $\mathcal{U} = (\omega)$ in vorticity formulations, $\mathcal{U} = (\omega, T)$ when the Navier-Stokes equations are complemented with an energy equation (see the Boussinesq example below).

7.3.2 Stochastic Spectral Problem

When random data is involved, the solution will also be a random quantity and so it is formally expanded as:

$$\mathcal{U}(\theta) = \mathcal{U}(\boldsymbol{\xi}(\theta)) = \sum_{i=0}^{\infty} (\mathcal{U})_i \, \Psi_i(\boldsymbol{\xi}(\theta)) \in \mathcal{V}_{\mathbf{u}_\Gamma} \otimes L_2(\Xi, P_\Xi), \tag{7.58}$$

where $\boldsymbol{\xi}(\theta) \in \Xi$ is a stochastic vector describing the random inputs and Ψ refers to the PC basis. Introducing the above expansion into (7.57), we get:

$$\mathcal{I}\left(\sum_i \left(\mathcal{U}^{n+1}\right)_i \Psi_i\right) = \mathcal{L}\left(\sum_i (\mathcal{U}^n)_i \Psi_i, \sum_i \left(\mathcal{U}^{n-1}\right)_i \Psi_i, \dots\right) + \mathcal{N}\left(\sum_i (\mathcal{U}^n)_i \Psi_i, \sum_i \left(\mathcal{U}^{n-1}\right)_i \Psi_i, \dots\right) + \mathcal{S}\left(\sum_i \left(\mathbf{f}^{n+1}\right)_i \Psi_i\right). \tag{7.59}$$

To solve (7.59), the stochastic expansion has to be truncated. To remain general with regard to the stochastic discretization used, we denote

$$\mathcal{S}^{\mathrm{P}} \equiv \mathrm{span}\{\Psi_0, \dots, \Psi_{\mathrm{P}}\} \subset L_2(\Xi, P_\Xi) \tag{7.60}$$

the finite dimensional stochastic approximation space with $\dim(\mathcal{S}^{\mathrm{P}}) = \mathrm{P}+1$. Substituting $\mathcal{U}$ with its truncated expansion, (7.59) is not satisfied in general, but yields a stochastic residual. A weak solution of the problem is then sought by requiring the residual to be orthogonal to $\mathcal{S}^{\mathrm{P}}$. With this constraint, we have for $i = 0, \dots, \mathrm{P}$

$$\left\langle \Psi_i, \mathcal{I}\left(\sum_{j=0}^{\mathrm{P}} (\mathcal{U})_j^{n+1} \Psi_j\right)\right\rangle = \left\langle \Psi_i, \mathcal{L}\left(\sum_{j=0}^{\mathrm{P}} (\mathcal{U})_j^n \Psi_j\right)\right\rangle + \left\langle \Psi_i, \mathcal{N}\left(\sum_{j=0}^{\mathrm{P}} (\mathcal{U})_j^n \Psi_j\right)\right\rangle + \left\langle \Psi_i, \mathcal{S}\left(\sum_{j=0}^{\mathrm{P}} \left(\mathbf{f}^{n+1}\right)_j \Psi_j,\right)\right\rangle, \tag{7.61}$$

which must be solved for $\mathcal{U}^{n+1} \in \mathcal{V}_{\mathbf{u}_\Gamma} \otimes \mathcal{S}^{\mathrm{P}}$. For the sake of simplicity we have considered a single time discretization level.

Since we are considering situations where the coefficients (*e.g.* the fluid properties such as density and viscosity) of the Navier-Stokes equations are certain, the operator $\mathcal{I}$ is deterministic and we have immediately

$$\left\langle \Psi_i, \mathcal{I}\left(\sum_{j=0}^{\mathrm{P}} (\mathcal{U})_j^{n+1} \Psi_j\right)\right\rangle = \langle \Psi_i \Psi_i \rangle \, \mathcal{I}\, (\mathcal{U})_i^{n+1}. \tag{7.62}$$

In more general situations, the linear operator $\mathcal{I}$ depends on the random event, as for instance when the Reynolds number of the flow is random. In this case, one can formally expand the random operator $\mathcal{I}(\boldsymbol{\xi})$ on the stochastic basis:

$$\mathcal{I}(\boldsymbol{\xi}) = \sum_{i=0}^{\mathrm{P}} \mathcal{I}_i \Psi_i(\boldsymbol{\xi}). \tag{7.63}$$

As a result, application of $\mathcal{I}$ to $\mathcal{U}$ couples all the stochastic modes and (7.62) does not hold anymore. Instead, we have

$$\left\langle \Psi_i, \mathcal{I}\left(\sum_{j=0}^{P} (\mathcal{U})_j^{n+1} \Psi_j\right)\right\rangle = \sum_{j=0}^{P}\sum_{l=0}^{P} \langle \Psi_i \Psi_j \Psi_l \rangle \mathcal{I}_l \, (\mathcal{U})_j^{n+1}. \tag{7.64}$$

This coupling of the stochastic modes is not desirable as it makes it significantly more difficult to invert $\mathcal{I}$. This difficulty can be easily overcome by using a semi-implicit treatment of the linear terms, leading to a deterministic operator $\mathcal{I} \equiv \mathcal{I}_0$ and a modified operator $\mathcal{L}$ accounting for modes $\mathcal{I}_{i>0}$. An example of such a procedure to enforce a decoupled stochastic modes integration is provided in Sect. 7.3.5.

Assuming that (7.62) holds, and making use of the orthogonality of the stochastic basis functions, (7.61) becomes

$$(\mathcal{U})_i^{n+1} = \mathcal{I}^{-1}\left\{\mathcal{L}_i(\mathcal{U}^n) + \mathcal{N}_i(\mathcal{U}^n) + \mathcal{S}_i^{n+1}\right\}, \quad i = 0, 1, \ldots, P. \tag{7.65}$$

This equation highlights the decoupling since the determination of $(\mathcal{U})_i^{n+1}$ is independent of $(\mathcal{U})_{j\neq i}^{n+1}$: the spectral problem has been factored in a series of $(P+1)$ problems of smaller size. Comparison of (7.65) with (7.57) also reveals that the overall computational cost of the solution procedure will be $(P+1)$ times greater than for the deterministic problem, with some additional overheads arising from the projection of the explicit terms $\mathcal{L}_i(\cdot)$, $\mathcal{S}_i(\cdot)$ and $\mathcal{N}_i(\cdot)$.

7.3.3 Resolution of Steady Stochastic Equations

The decoupled resolution of the stochastic modes is an attractive feature of the solution method described above. However, one is often interested in finding solutions of the steady Navier-Stokes equation. Although the time integration of the unsteady equation may provide the steady solution as $t \to \infty$, there are many situations where computation of steady solution via unsteady integration is not desirable or practical, because of the slow decay of the unsteady solution toward a steady state (requiring integration over long periods of time) or because the steady solution sought is unstable to perturbations. In these cases, the resolution of the steady equations has to be considered.

Direct determination of the solution to the steady equations is a difficult task, because of the size of the nonlinear problem and of the non-trivial coupling between the stochastic modes. The question is therefore: how to take advantage of the decoupled time-marching schemes to compute stochastic solutions of the steady equations, even unstable ones? This question has long been addressed in the deterministic context [60, 230], and we propose in the following an extension of these techniques to stochastic flows.

7.3.3.1 Newton Iterations

It is clear that the determination of a (weak) solution of the steady stochastic Navier-Stokes equations consists of finding stationary points of (7.65) or, more specifically, determining $\mathcal{U} \in \mathcal{V}_{\mathbf{u}_\Gamma} \otimes \mathcal{S}^{\mathrm{P}}$ such that

$$\frac{\partial \mathcal{U}}{\partial t} \approx \mathcal{F}(\mathcal{U}) \equiv \frac{\mathcal{I}^{-1}\{\mathcal{L}(\mathcal{U}) + \mathcal{S}(\mathbf{f}) + \mathcal{N}(\mathcal{U})\} - \mathcal{U}}{\Delta t_n} = 0. \tag{7.66}$$

It is remarked that the "time-step" size is here denoted Δt_n, because it is a numerical parameter of the steady solver and not an actual time-step used to advance the solution in time. To solve (7.66), we rely on Newton iterations. Let $\mathcal{U}^k \in \mathcal{V}_{\mathbf{u}_\Gamma} \otimes \mathcal{S}^{\mathrm{P}}$ be the approximate solution of (7.66) after the k-the Newton iteration. The next Newton iterate, $\mathcal{U}^{k+1} \in \mathcal{V}_{\mathbf{u}_\Gamma} \otimes \mathcal{S}^{\mathrm{P}}$, is

$$\mathcal{U}^{k+1}(\boldsymbol{\xi}) = \mathcal{U}^k(\boldsymbol{\xi}) + \delta\mathcal{U}^k(\boldsymbol{\xi}), \tag{7.67}$$

where the stochastic Newton increment $\delta\mathcal{U}^k$ satisfies

$$\mathcal{J}(\mathcal{U}^k)\delta\mathcal{U}^k = -\mathcal{F}(\mathcal{U}^k). \tag{7.68}$$

Here, $\mathcal{J}(\mathcal{U})$ is the Jacobian of $\mathcal{F}$ at $\mathcal{U}$. Clearly, as $\mathcal{U}^k$ satisfies the boundary conditions, the Newton increment satisfies homogeneous Dirichlet velocity boundary conditions:

$$\delta\mathcal{U}^k \in \mathcal{V}_0 \otimes \mathcal{S}^{\mathrm{P}}. \tag{7.69}$$

The computation of $\mathcal{J}(\mathcal{U})$ is not an option because of its size, which is $(\mathrm{P}+1)$ times larger than its deterministic counterpart. However, it appears that $\mathcal{J}(\mathcal{U})\delta\mathcal{U}$ can be computed without making the Jacobian explicit as soon as one recognizes that

$$\mathcal{J}(\mathcal{U})\delta\mathcal{U} = \frac{\mathcal{I}^{-1}\{\mathcal{L}(\delta\mathcal{U}) + \overline{\mathcal{N}}(\mathcal{U})\delta\mathcal{U}\} - \delta\mathcal{U}}{\Delta t_n}, \tag{7.70}$$

where $\overline{\mathcal{N}}(\mathcal{U})\delta\mathcal{U}$ are the nonlinear terms linearized at $\mathcal{U}$. For the Navier-Stokes equations, the nonlinearity comes from the convective terms. Although the actual forms of the convective terms depend on the selected formulation (*e.g.* on the variables in $\mathcal{U}$), we shall use the abusive notation $\mathcal{U}\nabla\mathcal{U}$ for the convective terms. However, consistent with this notation, the linearized nonlinear terms is expressed as:

$$\overline{\mathcal{N}}(\mathcal{U})\,\delta\mathcal{U} = -\mathcal{U}\nabla\delta\mathcal{U} - \delta\mathcal{U}\nabla\mathcal{U}. \tag{7.71}$$

In other words, $\mathcal{I}^{-1}\{\mathcal{L}(\delta\mathcal{U}) + \overline{\mathcal{N}}(\mathcal{U})\delta\mathcal{U}\}$ is the result of the time-integration (over a single time step) of the linearized stochastic Navier-Stokes equations, for the initial condition $\delta\mathcal{U} \in \mathcal{V}_0 \otimes \mathcal{S}^{\mathrm{P}}$ and homogeneous source term and velocity boundary conditions.

7.3.3.2 Stochastic Increment Problem

At this stage, we have derived an equation,

$$\frac{\mathcal{I}^{-1}\{\mathcal{L}(\delta\mathcal{U}^k) + \overline{\mathcal{N}}(\mathcal{U}^k)\delta\mathcal{U}^k\} - \delta\mathcal{U}^k}{\Delta t_n} + \mathcal{F}(\mathcal{U}^k) = 0, \tag{7.72}$$

for the stochastic Newton increment $\delta\mathcal{U}^k$. It remains to solve efficiently (7.72). To this end, consider the truncated expansions of the Newton iterates and increments:

$$\mathcal{U}^k(\boldsymbol{\xi}) = \sum_{i=0}^{\mathrm{P}} (\mathcal{U}^k)_i \Psi_i(\boldsymbol{\xi}), \qquad \delta\mathcal{U}^k(\boldsymbol{\xi}) = \sum_{i=0}^{\mathrm{P}} (\delta\mathcal{U}^k)_i \Psi_i(\boldsymbol{\xi}). \tag{7.73}$$

The projection of (7.66) on the stochastic basis gives for $i = 0, \dots, \mathrm{P}$:

$$\Delta t_n \mathcal{F}_i(\mathcal{U}^k) = \mathcal{I}^{-1}\left\{\mathcal{L}_i(\mathcal{U}^k) + (\mathcal{S})_i + \mathcal{N}_i(\mathcal{U}^k)\right\} - (\mathcal{U}^k)_i. \tag{7.74}$$

The Galerkin projection of (7.70) gives in turn for $i = 0, \dots, \mathrm{P}$:

$$\Delta t_n \left(\mathcal{J}(\mathcal{U}^k)\delta\mathcal{U}^k\right)_i = \mathcal{I}^{-1}\left\{\mathcal{L}_i(\delta\mathcal{U}^k) + \left(\overline{\mathcal{N}}(\mathcal{U}^k)\delta\mathcal{U}^k\right)_i\right\} - (\delta\mathcal{U}^k)_i. \tag{7.75}$$

Finally, the equations to be solved for the Newton increment modes $(\delta\mathcal{U})_l$ are for $i = 0, \dots, \mathrm{P}$:

$$\mathcal{I}^{-1}\left\{\mathcal{L}_i(\delta\mathcal{U}^k) + \left(\overline{\mathcal{N}}(\mathcal{U}^k)\delta\mathcal{U}^k\right)_i\right\} - (\delta\mathcal{U}^k)_i = -\Delta t_n \mathcal{F}_i(\mathcal{U}^k). \tag{7.76}$$

It is seen that, although this equation is linear in the Newton increment, the determination of the stochastic modes $(\delta\mathcal{U})_i$ is coupled through the linearized nonlinear term (except if $\mathcal{U}^k$ is actually deterministic). To gain further insight about this coupling, we make explicit the i-th mode of the linearized nonlinear term. From (7.71), we have

$$\left(\overline{\mathcal{N}}(\mathcal{U})\delta\mathcal{U}\right)_i = -\frac{\langle(\mathcal{U}\nabla\delta\mathcal{U} + \delta\mathcal{U}\nabla\mathcal{U})\Psi_i\rangle}{\langle\Psi_i^2\rangle}, \tag{7.77}$$

and introducing the stochastic expansions of $\mathcal{U}$ and $\delta\mathcal{U}$, one obtains

$$\begin{aligned}\left(\overline{\mathcal{N}}(\mathcal{U})\delta\mathcal{U}\right)_i &= -\sum_{j=0}^{\mathrm{P}}\sum_{l=0}^{\mathrm{P}} C_{ijl}\left((\mathcal{U})_j\nabla(\delta\mathcal{U})_l + (\delta\mathcal{U})_j\nabla(\mathcal{U})_l\right)\\ &= \sum_{j=0}^{\mathrm{P}}\sum_{l=0}^{\mathrm{P}} C_{ijl}\mathfrak{C}\left[(\mathcal{U})_j, (\delta\mathcal{U})_l\right],\end{aligned} \tag{7.78}$$

where the bilinear convection operator $\mathfrak{C}[.,.]$ is given by:

$$\mathfrak{C}[(\mathcal{U})_j, (\delta\mathcal{U})_l] \equiv (\mathcal{U})_j\nabla(\delta\mathcal{U})_l + (\delta\mathcal{U})_l\nabla(\mathcal{U})_j. \tag{7.79}$$

With these notations, (7.76) can be rearranged to yield: for $i = 0, \dots, \mathrm{P}$

$$\mathcal{I}^{-1}\left\{\mathcal{L}_i(\delta\mathcal{U}^k) + \sum_{j,l=0}^{\mathrm{P}} C_{ijl}\mathcal{C}\left[(\mathcal{U}^k)_j, (\delta\mathcal{U}^k)_l\right]\right\} - (\delta\mathcal{U}^k)_i = -\Delta t_n \mathcal{F}_i(\mathcal{U}^k). \tag{7.80}$$

This shows that it is not possible to decouple the problem for each mode $(\delta\mathcal{U})_i$.

7.3.3.3 Matrix Free Solver

We define the linear operator $\mathcal{G} : (\mathcal{V}, \mathcal{U}) \in (\mathcal{V}_0 \otimes \mathcal{S}^{\mathrm{P}}) \times (\mathcal{V}_{\mathbf{u}_\Gamma} \otimes \mathcal{S}^{\mathrm{P}}) \mapsto \mathcal{G}(\mathcal{V}, \mathcal{U}) \in \mathcal{V}_0 \otimes \mathcal{S}^{\mathrm{P}}$ according to:

$$\mathcal{G}(\mathcal{V}, \mathcal{U}) \equiv \mathcal{I}^{-1}\left\{\mathcal{L}(\mathcal{V}) + \overline{\mathcal{N}}(\mathcal{U})\mathcal{V}\right\} - \mathcal{V}. \tag{7.81}$$

Thus, one has to solve $\mathcal{G}(\mathcal{V}, \mathcal{U}) = -\Delta t_n \mathcal{F}(\mathcal{U})$ for $\mathcal{V}$, given $\mathcal{U}$. As discussed previously, the first term in the expression of $\mathcal{G}$ in (7.81) is the result of a time integration of the linearized Navier-Stokes equations from the initial condition $\mathcal{V}$: a **decoupled** time integration can be employed for its evaluation. This observation suggests to use a matrix free method to solve (7.80) at the discrete level, *i.e.* after spatial discretization. By matrix-free method, we mean that the large system of algebraic equations corresponding to the discrete version of (7.80) is not constructed and inverted; this is essential due to the size of the discrete system, which we recall is $(\mathrm{P}+1)$ times larger than in the deterministic context. Instead iterative techniques are considered for the determination of stochastic increment $\delta\mathcal{U}$, where for a given discrete iterate $\mathcal{V}$ the effect of the linear operator $\mathcal{G}(., \mathcal{U})$ on $\mathcal{V}$ is obtained by means of a time-integration of the linearized Navier-Stokes equations. As a result, the computational cost to perform the pseudo matrix-vector product amounts to the resolution of one time step of the discrete unsteady linearized Navier-Stokes equation which, we emphasize, can be performed in a decoupled fashion. When selecting such matrix free iterative solver, it is important to recognize that the operator $\mathcal{G}$ is not self-adjoint. Two subspace methods for non-symmetric linear systems have been tested in this work for the resolution of the discrete version of (7.80): BiCGStab algorithm [207] and GMRes algorithm [200].

The efficiency of these algorithms is strongly related to the spectrum of $\mathcal{G}$, and the computational cost of the resolution mainly scales with the number of pseudo matrix-vector products (*i.e.* integrations of the linearized discrete Navier-Stokes equations) needed to obtain the discrete increment within a given error tolerance. In fact, the overall computational cost to obtain the steady stochastic solution will depend on (a) the convergence rate of the Newton iterations (which is independent of the iterative algorithm) and (b) the number of integrations of the linearized Navier-Stokes equations to estimate the Newton increments (which depends on the iterative algorithm). It is expected that the convergence of the Newton iterations improves when Δt_n increases. In fact, Δt_n can be selected arbitrarily large, since no stability constraint holds. However, the choice of the time step Δt_n also affects the spectrum of $\mathcal{G}$: using larger Δt_n will usually require more integrations of the linearized Navier-Stokes equations to obtain the increment. Thus, we expect a trade-off on Δt_n, to balance the convergence of the Newton iterations with the numerical cost of computing the increments. This point is essential for the efficiency of the stochastic Newton solver and will be numerically illustrated in the next section.

Also related to the iterative algorithm used to compute the increments are memory requirements. The memory requirements are a key aspect, as the size of the discrete solutions are $(\mathrm{P}+1)$ times larger than for the deterministic problem. The GMRes algorithm requires the storage of successive solution vectors to span the Krylov subspace. In contrast, BiCGStab algorithm needs to store of a fixed number (4) of solution vectors. Krylov subspaces of significant dimension are expected in order to approach the solution of (7.80) using GMRes. This makes the BiCGStab algorithm *a priori* more suited from memory requirements point of view. However, a comparison of the computational cost and memory requirements of the two methods is needed to decide which of the two algorithms should be preferred. This aspect is also investigated in the next section.

7.3.4 Test Problem

7.3.4.1 Problem Definition

The Newton method is tested on the normalized Boussinesq equations (Beqs) describing the natural convection inside a 2-dimensional square cavity: $(x, y) \in \Omega_x = [0, 1]^2$. The unsteady flow is governed by:

$$\begin{cases} \dfrac{\partial \mathbf{u}}{\partial t} + \mathbf{u} \cdot \nabla \mathbf{u} = -\nabla p + \dfrac{\mathrm{Pr}}{\sqrt{\mathrm{Ra}}} \nabla^2 \mathbf{u} + \mathrm{Pr} T \mathbf{y}, \\ \dfrac{\partial T}{\partial t} + \mathbf{u} \cdot \nabla T = \dfrac{1}{\sqrt{\mathrm{Ra}}} \nabla^2 T, \\ \nabla \cdot \mathbf{u} = 0, \end{cases} \tag{7.82}$$

where T is the normalized temperature, Ra is the Rayleigh number, $\mathrm{Pr} = 0.71$ is the Prandtl number of the fluid (air) and $\mathbf{y}$ is the gravity direction. Boundary conditions for the velocity are $\mathbf{u} = 0$ on Γ. For the temperature boundary conditions we assume adiabatic walls at $y = 0$ and 1 (top and bottom boundaries of the cavity) and stochastic temperatures at $x = 0$ and 1 (vertical walls):

$$\begin{cases} \nabla T \cdot \mathbf{y} = 0, & y = 0 \text{ and } 1, \\ T = T_\Gamma(\boldsymbol{\xi}), & x = 0 \text{ and } 1. \end{cases} \tag{7.83}$$

The linearized Boussinesq equations (LBeqs), around $(\mathbf{u}, T)$ satisfying the boundary conditions are

$$\begin{cases} \dfrac{\partial(\delta\mathbf{u})}{\partial t} + (\delta\mathbf{u})\cdot\nabla\mathbf{u} + \mathbf{u}\cdot\nabla(\delta\mathbf{u}) = -\nabla(\delta p) + \dfrac{\mathrm{Pr}}{\sqrt{\mathrm{Ra}}}\nabla^2(\delta\mathbf{u}) + \mathrm{Pr}(\delta T)\,\mathbf{y}, \\ \dfrac{\partial(\delta T)}{\partial t} + (\delta\mathbf{u})\cdot\nabla T + \mathbf{u}\cdot\nabla(\delta T) = \dfrac{1}{\sqrt{\mathrm{Ra}}}\nabla^2(\delta T), \\ \nabla\cdot(\delta\mathbf{u}) = 0, \end{cases} \tag{7.84}$$

with the homogeneous boundary conditions:

$$\begin{cases} \delta\mathbf{u} = 0, & \mathbf{x}\in\Gamma, \\ \nabla(\delta T)\cdot\mathbf{y} = 0, & y = 0 \text{ and } 1, \\ \delta T = 0, & x = 0 \text{ and } 1. \end{cases} \tag{7.85}$$

These equations are solved in vorticity streamfunction formulation (see [188]), on a uniform, staggered grid, using second-order centered differences for the convective and viscous terms. Fast FFT-based solvers are used for the inversion of the heat and Poisson operators. Boundary conditions on the vorticity $\omega = (\nabla\wedge\mathbf{u})\cdot\mathbf{z}$ are determined through an influence matrix technique [46]. Using the formal notations, the solution of the Boussinesq equations is $\mathcal{U} = (\omega, T)$. For the purpose of the analysis, we define the stochastic norms $\|\cdot\|_{\Omega_x}^{\mathcal{S}^{\mathrm{P}}}$ by:

$$\left(\|f\|_{\Omega_x}^{\mathcal{S}^{\mathrm{P}}}\right)^2 \equiv \int_{\Omega_x}\left\langle f^2(\mathbf{x})\right\rangle d\mathbf{x} = \sum_{i=0}^{\mathrm{P}}\left\langle\Psi_i^2\right\rangle\int_{\Omega_x}(f(\mathbf{x}))_i^2\, d\mathbf{x}. \tag{7.86}$$

To monitor the convergence of the steady solution with the iterations, we use the norms of the steady vorticity and temperature equation residuals, denoted respectively $\|R_\omega\|_{\Omega_x}^{\mathcal{S}^{\mathrm{P}}}$ and $\|R_T\|_{\Omega_x}^{\mathcal{S}^{\mathrm{P}}}$, where using formal notations $R_\omega = \mathcal{F}_\omega(\mathcal{U})$ and $R_T = \mathcal{F}_T(\mathcal{U})$.

7.3.4.2 Unsteady Simulations

In a preliminary test, the uncertain temperature boundary conditions are parameterized as follows:

$$\begin{cases} T_\Gamma(\xi) = 1/2 + 1/20\xi_1, & x = 0, \\ T_\Gamma(\xi) = -1/2 + 1/20\xi_2, & x = 1, \end{cases} \tag{7.87}$$

where ξ_1 and ξ_2 are independent and uniformly distributed on $[-1, 1]$. For these settings, the vertical walls support uniform independent random temperatures, with respective expectations equal to $\pm 1/2$, uncertainty levels $\pm 10\%$ and uniform probability densities. The stochastic dimension of the problem is then $\mathrm{N} = 2$, and the orthogonal basis of $\mathcal{S}^{\mathrm{P}}$ is the set of 2-dimensional Legendre polynomials with degree less or equal to a prescribed expansion order No. The spectral problem is time-integrated from the initial conditions $\mathbf{u} = 0$, $T = 0$, using the decoupled strategy

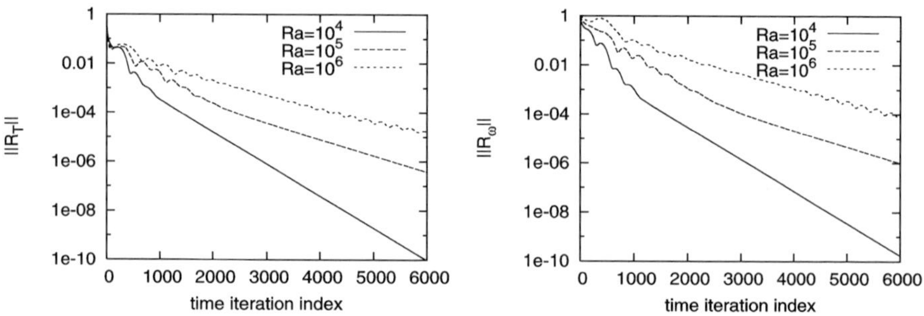

Fig. 7.12 Decay with time-iterations of the steady residual norms $\|R_T\|^{\mathcal{S}^{\mathrm{P}}}_{\Omega_x}$ (*left*) and $\|R_\omega\|^{\mathcal{S}^{\mathrm{P}}}_{\Omega_x}$ (*right*) in unsteady simulations of the Boussinesq equations for $\mathrm{Ra} = 10^4$, 10^5 and 10^6 as indicated. The time-step is $\Delta t = 0.02$. Adapted from [119]

described in Sect. 7.3.2, with a first-order backward Euler scheme with $\Delta t = 0.02$ on a 128×128 spatial grid.

In Fig. 7.12, the convergence of the flow toward the steady state is monitored by plotting the steady residual norms $\|R_T\|^{\mathcal{S}^{\mathrm{P}}}_{\Omega_x}$ and $\|R_\omega\|^{\mathcal{S}^{\mathrm{P}}}_{\Omega_x}$ as functions of the time iteration index. Different Rayleigh numbers are tested: $\mathrm{Ra} = 10^4$, 10^5 and 10^6. The expansion order is set to $\mathrm{No} = 3$ in all the simulations, so $\dim(\mathcal{S}^{\mathrm{P}}) = 10$. For all values of Ra, the norms decay monotonically to 0. However, the decay rate becomes slower as Ra increases, due to a weaker viscous damping of inertial waves. This behavior illustrates how the determination of a steady solution *via* time-integration, although possible in the range of Ra and temperature differences considered, becomes inefficient when Ra increases.

7.3.4.3 Newton Iterations

The Newton iterations are now applied to the problem defined above. The spatial and stochastic discretizations are kept the same, except that $\Delta t_n = 5$ is now used. Newton iterations at a given Ra are initialized with the stochastic steady solution at $\mathrm{Ra}/2$.

Convergence of Newton iterations: Figure 7.13 depicts the convergence with the Newton iterations of the steady residual norms $\|R_T\|^{\mathcal{S}^{\mathrm{P}}}_{\Omega_x}$ and $\|R_\omega\|^{\mathcal{S}^{\mathrm{P}}}_{\Omega_x}$. The computation of the Newton increments uses BiCGStab with a stopping criterion $\epsilon = 10^{-3}$ (see definition below). In these plots, the symbols correspond to the Newton iterates. For the three Ra tested, it is seen that 5 to 6 Newton iterations are needed to achieve a reduction of the residuals below 10^{-12}. The asymptotic convergence rate with the Newton iterations is found weakly dependent on Ra. However, the CPU costs are dependent on Ra, as seen from the curves where the steady residual norms are reported as functions of the total number of LBeqs integrations performed. This trend denotes the degradation of the conditioning of the problem for the Newton

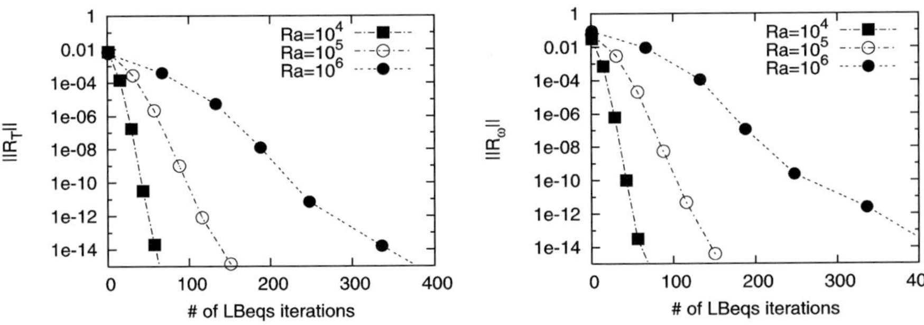

Fig. 7.13 Convergence of the steady residual norms with the number of LBeqs iterations and for Ra $= 10^4$, 10^5 and 10^6 as indicated. Newton increments are computed using BiCGStab with $\Delta t_n = 5$ and stopping criterion $\epsilon = 10^{-3}$. Adapted from [119]

increments when Ra increases. Specifically, it is seen that for Ra $= 10^4$ about 10 iterations on the LBeqs are needed by BiCGStab to solve (7.68) within the requested tolerance, while about 70 iterations are needed for the same tolerance criterion when Ra $= 10^6$. However, keeping in mind that one iteration on the linearized equations (LBeqs) amounts essentially to the computational cost of one iteration of the unsteady equations (Beqs), the efficiency of the Newton method can be directly appreciated from a comparison of the decay rates of the residuals in Figs. 7.13 and 7.12. For instance, when Ra $= 10^6$ about 5000 unsteady iterations yield a reduction of the residual by a factor of roughly 10^3, to be compared with the reduction by a factor of roughly 10^{10} obtained in only 400 LBeqs iterations for the Newton method.

Stopping criterion: The Newton method being iterative, the increments $\delta\mathcal{U}^k(\boldsymbol{\xi})$ need not be exactly computed; approximate increments can be used instead, with the objective of reducing the CPU cost of their determination. In fact, iterations on the BiCGStab algorithm are performed as long as the probabilistic norms of the discrete equations $\mathcal{J}(\mathcal{U}^k)\delta\mathcal{U} = -\mathcal{F}(\mathcal{U}^k)$ (on T and ω), normalized by the norms of the respective right-hand sides, are such that

$$\frac{\|\mathcal{J}(\mathcal{U}^k)\delta\mathcal{U} + \mathcal{F}(\mathcal{U}^k)\|_{\Omega_x}^{\mathcal{S}^{\mathrm{P}}}}{\|\mathcal{F}(\mathcal{U}^k)\|_{\Omega_x}^{\mathcal{S}^{\mathrm{P}}}} > \epsilon, \tag{7.88}$$

for some small positive ϵ. In Fig. 7.14, we compare the convergence of the steady residual norms at Ra $= 10^6$ obtained using different ϵ in BiCGStab. It is seen that the drop in residual between two successive Newton iterates (symbols) improves when ϵ is lowered. However, this improvement becomes negligible when ϵ goes to zero and comes with a significant increase in the number of BiCGStab iterations. Therefore, there is a trade-off between the accuracy in the computed increment and the numerical cost of its resolution. For the present problem, the optimal trade-off in terms of number of LBeqs integrations is for $\epsilon \sim 10^{-1}, 10^{-2}$: for a similar reduction of the residuals, twice as many Newton iterations are needed than for $\epsilon = 10^{-4}$, but increments are obtained within 3 to 4 times less iterations.

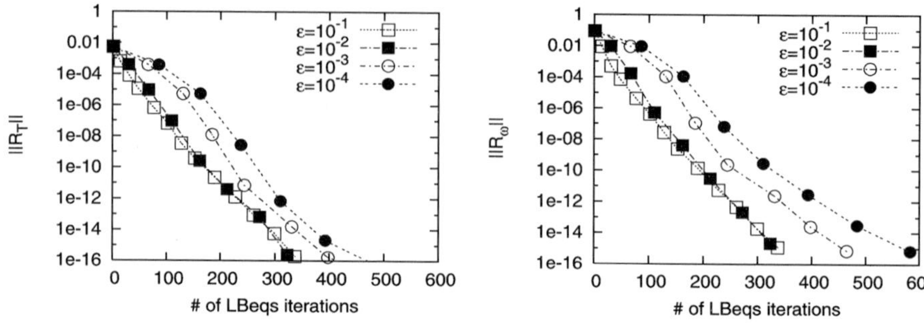

Fig. 7.14 Convergence of the steady residual norms with the number of LBeqs iterations, for different stopping criteria ϵ as indicated and BiCGStab. $\mathrm{Ra} = 10^6$, $\Delta t_N = 5$. Adapted from [119]

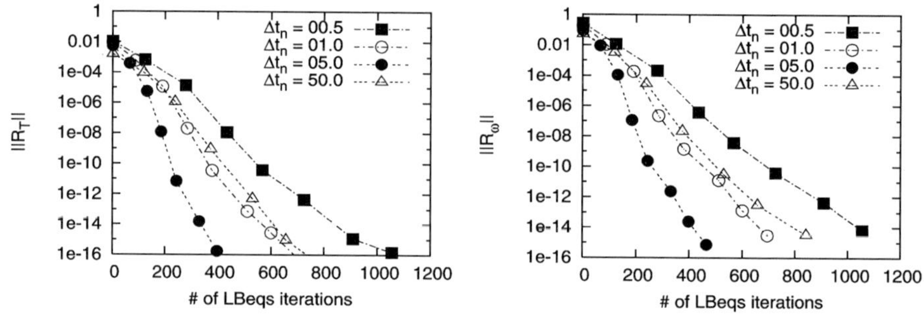

Fig. 7.15 Convergence of steady residual norms with the number of LBeqs iterations using different Newton time-steps Δt_n as indicated. $\mathrm{Ra} = 10^6$, BiCGStab algorithm with $\epsilon = 10^{-3}$. Adapted from [119]

Newton time step Δt_n: We recall that the selection of Δt_n is not subjected to stability restrictions, and that the convergence of the Newton iterations is expected to improve when Δt_n increases. However, the conditioning of the problem on the increments is expected to degrade for increasing Δt_n, with a larger number of time integrations of the linearized problem as a result. These expectations are verified in the following tests for the solution at $\mathrm{Ra} = 10^6$ using BiCGStab with different time steps and $\epsilon = 10^{-3}$. The convergence of the steady residual norms, reported in Fig. 7.15, clearly demonstrates the expected trends: there is an optimal Δt_n balancing the steady residuals reduction, from a Newton iterate to the next, with the number of iterations needed to compute the increments. For the present test, $\Delta t_n \sim 5$ seems optimal in terms of number of iterations on the LBeqs. It is noted that *a priori* determination of the optimal Newton time step, as well as the optimal stopping criterion, remains an open question. Furthermore, the respective CPU costs for the different values of Δt_n are here proportional to the number to LBeqs iterations, thanks to the use of FFT-based direct solvers for the diffusion and Poisson equations. In general, when using iterative solvers for the integration of the linearized

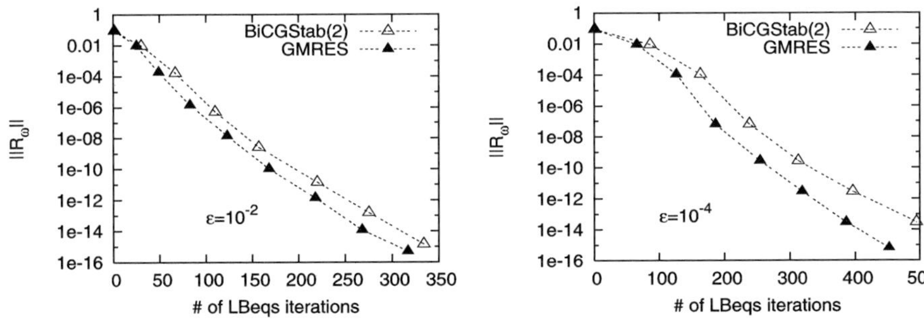

Fig. 7.16 Comparison of convergences of the vorticity steady residual norm with the number of LBeqs iterations for BiCGStab (*open symbols*) and GMRes (*filled symbols*) algorithms and using different stopping criteria $\epsilon = 10^{-2}$ (*left*) and $\epsilon = 10^{-4}$ (*right*). Parameters are $\mathrm{Ra} = 10^6$, $\Delta t_n = 5$. Adapted from [119]

equations such equivalence will not be true, and the comparison will have to be based on the actual CPU times.

GMRes *vs* BiCGStab: To complete this first series of tests, we provide a comparison of the respective efficiencies of BiCGStab and GMRes for the problem at $\mathrm{Ra} = 10^6$. Recall that in addition to the CPU times, memory requirement is an important concern when designing solvers for stochastic spectral problems. Here, we take advantage of the relatively low dimensionality of the stochastic approximation space, $\dim(\mathcal{S}^{\mathrm{P}}) = 10$, to compare the respective efficiencies of the GMRes and BiCGStab algorithms. Indeed, for this problem we were able to construct Krylov subspaces sufficiently large to avoid the need of relying on restart procedures in GMRes.

Figure 7.16 compares the convergence of the steady residual norms using GMRes and BiCGStab, for two stopping criteria $\epsilon = 10^{-2}$ and $\epsilon = 10^{-4}$ and using $\Delta t_n = 5$. It is seen that for the two stopping criteria, GMRes requires less iterations than BiCGStab to approximate the Newton increments. Although significantly more efficient with $\epsilon = 10^{-4}$, one can observe that GMRes constructs Krylov subspaces with dimension up to 80, such that 80 solution vectors have to be stored, to be compared with only 4 in BiCGStab. Furthermore, since as previously noted the increments need not to be accurately computed, BiCGStab is preferred and will be used systematically in the computations below.

7.3.4.4 Influence of the Stochastic Discretization

So far, we have verified that the behavior of the Newton method is consistent with our expectations based on theoretical analysis and our experience with deterministic problems. This was investigated by varying the parameters of the Newton method. This section aims at assessing the efficiency and robustness of the Newton method with regard to the stochastic discretization, *i.e.* when the stochastic approximation space $\mathcal{S}^{\mathrm{P}}$ changes.

To this end, we consider a more complex parameterization of the temperature boundary conditions. The temperature on the cold wall is now certain and equal to $T(x = 1) = -1/2$, while the hot wall temperature is modeled as a stationary Gaussian stochastic process. This uncertainty setting corresponds to the problem treated in Sect. 6.2 which is now briefly summarized. The mean of the Gaussian stochastic process is $1/2$ with a standard deviation $\sigma_T = 0.1$: $T(x = 0, y, \theta) \sim \mathrm{N}(1/2, \sigma_T^2)$. The two-point correlation function of the random temperature along the wall is assumed to decay exponentially with a characteristic length-scale $L = 1$. The KL expansion of the process is

$$T(x = 0, y, \theta) = 1/2 + \sum_{k=1}^{\infty} \sqrt{\lambda_k} T_k(x = 0, y) \xi_k(\theta), \tag{7.89}$$

where the normalized functions $T_k(x = 0, y)$ are the deterministic KL modes and the ξ_k are uncorrelated (so independent) normalized centered Gaussian random variables:

$$\xi_k(\theta) \sim \mathrm{N}(0, 1), \qquad \langle \xi_k \xi_l \rangle = \delta_{kl}. \tag{7.90}$$

Expressions for the deterministic KL modes can be obtained from Sect. 6.2.2. Ordering the KL modes of the temperature boundary condition such that $\lambda_1 \geq \lambda_2 \geq \cdots$, the KL expansion is truncated after the N-th term. We have:

$$T(x = 0, y, \theta) \approx T(x = 0, y, \boldsymbol{\xi}(\theta)) = 1/2 + \sum_{k=1}^{\mathrm{N}} \sqrt{\lambda_k} T_k(x = 0, y) \xi_k(\theta) \tag{7.91}$$

where $\boldsymbol{\xi} = \{\xi_1, \ldots, \xi_{\mathrm{N}}\}$. For the stochastic discretization, we rely on a Wiener-Hermite basis of $L^2(\Xi, P_\Xi)$. Truncating the basis to order No yields the stochastic approximation space $\mathcal{S}^{\mathrm{P}}$, where P is given by (2.53). The expansions of the temperature and vorticity fields on the Wiener-Hermite basis are:

$$T(x, y, \theta) = \sum_{k=0}^{\mathrm{P}} T_k(x, y) \Psi_k(\boldsymbol{\xi}), \qquad \omega(x, y, \theta) = \sum_{k=0}^{\mathrm{P}} \omega_k(x, y) \Psi_k(\boldsymbol{\xi}). \tag{7.92}$$

The stochastic approximation space $\mathcal{S}^{\mathrm{P}}$ can be refined by increasing N and/or No. Our objective when selecting this problem was not to determine the minimal stochastic discretization to achieve a given accuracy (this aspect was investigated in Sect. 6.2 where an unsteady integration was used), but rather to analyze the impact of the stochastic discretization on the efficiency of the Newton method.

Influence of the stochastic order No**:** In a series of simulations, we set $\mathrm{N} = 4$, $\sigma_T = 0.1$ and compute the steady solution for $\mathrm{Ra} = 10^6$ and different stochastic orders $\mathrm{No} = 1, \ldots, 5$. Each computation uses the corresponding steady solution for

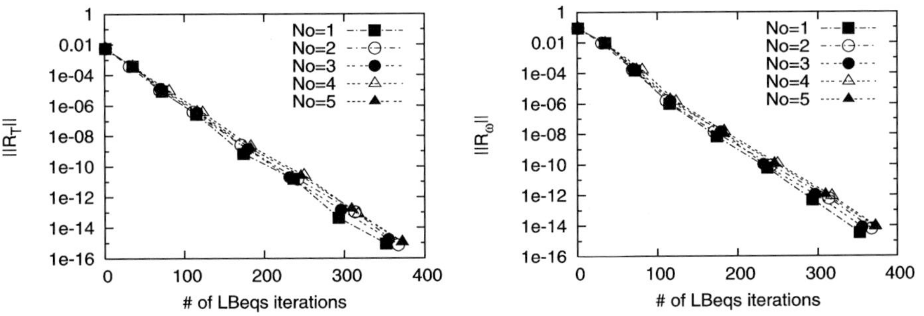

Fig. 7.17 Convergence of the steady residual norms with the number of LBeqs iterations for different stochastic orders No $= 1, \ldots, 5$. Computations use N $= 4$, BiCGStab with $\epsilon = 0.01$ and $\Delta t_n = 5$. Adapted from [119]

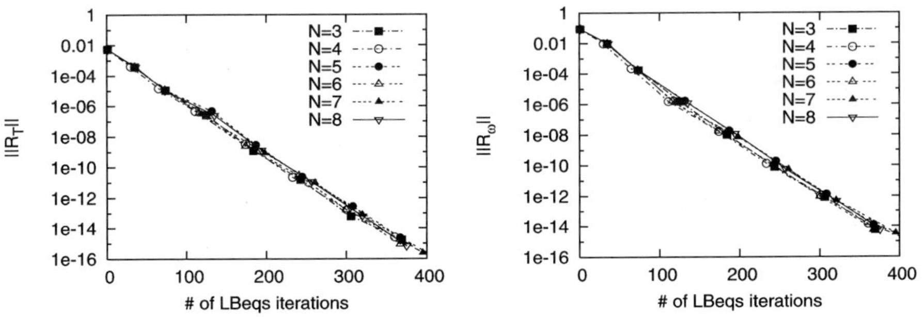

Fig. 7.18 Convergence of the steady residual norms with the number of LBeqs iterations for different KL expansions of the temperature BCs with N $= 3, \ldots, 8$. Computations use No $= 3$, BiCGStab with $\epsilon = 0.01$ and $\Delta t_n = 5$. Adapted from [119]

Ra $= 5 \times 10^5$ to initialize the Newton iterations. The increments are computed using BiCGStab with $\epsilon = 10^{-2}$ and $\Delta t_n = 5$. Figure 7.17 shows the convergence of the steady residuals for the different No. It shows that the convergence rate of the Newton iterations is essentially independent of No, and so is the number of BiCGStab iterations needed to compute the increments.

Influence of the stochastic dimension N**:** In a second series of tests, we fix the stochastic order to No $= 3$ and increase the number of stochastic dimensions, *i.e.* the number of KL modes used to model the stochastic boundary conditions, from N $= 3$ to 8. Again, the Newton iterations are initialized with the respective steady solutions for Ra $= 5 \times 10^5$, and BiCGStab is used with $\epsilon = 10^{-2}$ and $\Delta t_n = 5$. Inspection of the results reported in Fig. 7.18 shows that the convergence of the Newton iterations, as well as the number of BiCGStab iterations for the increments computation, is essentially insensitive to the number N of stochastic dimensions.

Comments: The two previous numerical experiments have shown a convergence of the Newton iterations essentially independent of the stochastic discretization. This

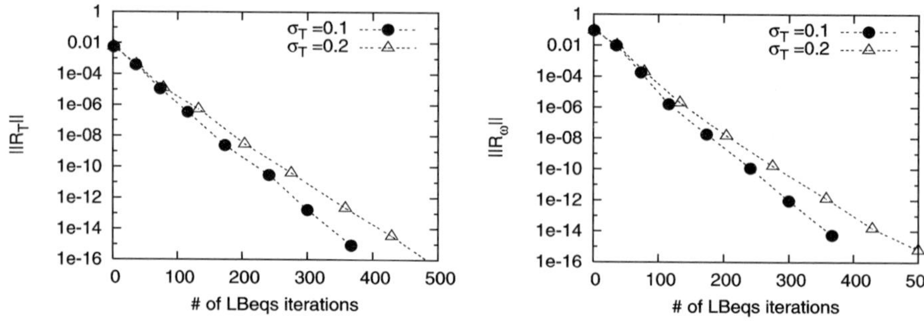

Fig. 7.19 Convergence of the steady residual norms with the number of LBeqs iterations, for temperature BCs with standard deviation $\sigma_T = 0.1$ and 0.2. The stochastic discretization uses No = 3 and N = 6. Other parameters are given in the text. Adapted from [119]

is an interesting finding that was not necessarily expected. In fact, in view of (7.76) which exhibits coupling between the spectral modes of the Newton increments, one may have anticipated an impact of the stochastic discretization on the number of BiCGStab iterations needed for the resolution of (7.68) as N and No increase. Such a trend is not observed in our simulations, denoting robustness with regard to the stochastic discretization. In fact, the coupling between the stochastic modes in (7.68) is more related to the uncertainty level (or variability of the stochastic flow) than to the stochastic discretization. To support this claim, we consider the same problem and numerical parameters, but the standard deviation of the temperature boundary condition is doubled, $\sigma_T = 0.2$. Doubling the variability of the boundary condition increases the variability of the solution and so the magnitude of the stochastic modes and their nonlinear interaction. In Fig. 7.19, we compare the convergence of the steady residuals in the simulations for $\sigma_T = 0.1$ and $\sigma_T = 0.2$ using No = 3 and N = 6. It is seen that the decay of the steady residuals is roughly unaffected when doubling σ_T, but that more BiCGStab iterations are needed to compute the increments.

7.3.4.5 Computational Time

The independence with regard to the stochastic discretization of the residuals decay should not hide the fact that the computational cost significantly increases when the stochastic discretization is refined. We report in Table 7.1 the measured CPU times of the simulations for $\sigma_T = 0.1$. All simulations were sequentially performed on a 64-bit bi-processor workstation (AMD Opteron 250, 2.4 GHz with 4 Gb RAM). The variability in reported CPU-times is estimated to be ±10% (due to other running processes and time measurement errors). In addition to the CPU time, Table 7.1 also provides the corresponding dimensions of the stochastic approximation spaces, *i.e* the number of the stochastic modes in the solution. It is first remarked that for the largest stochastic space there are 165 stochastic modes, so the steady solution in-

Table 7.1 CPU times for the computation of the steady solutions at $\mathrm{Ra} = 10^6$ using different expansion orders (No) and truncation (N) of the KL expansion for the temperature boundary condition. Adapted from [119]

	N = 4					
	No = 1	No = 2	No = 3	No = 4	No = 5	
$\dim(\mathcal{S}^{\mathrm{P}})$	5	15	35	70	126	
CPU times (s)	42	190	901	3482	12891	
	No = 3					
	N = 3	N = 4	N = 5	N = 6	N = 7	N = 8
$\dim(\mathcal{S}^{\mathrm{P}})$	20	35	56	84	120	165
CPU times (s)	373	901	1740	2832	4861	7394

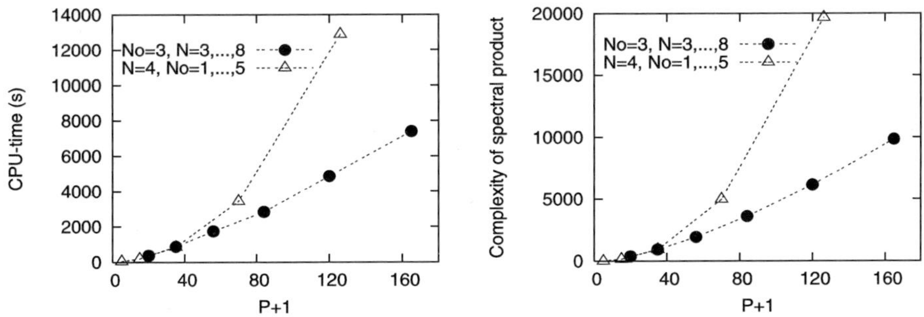

Fig. 7.20 CPU-times for the resolution of the steady problem (*left*) and complexity $\mathcal{C}$ of the Galerkin product (*right*) for different stochastic discretizations. Adapted from [119]

volves roughly 2.7 millions degrees of freedom and the steady solution is computed in slightly more than 2 hours of CPU.

To gain a better appreciation of the evolution of the CPU time with the stochastic discretization, we have plotted in the left part of Fig. 7.20 the measured CPU times as a function of the dimension $(\mathrm{P}+1)$ of the stochastic approximation space. It is seen that for No = 3, the CPU times scale asymptotically as $\dim(\mathcal{S}^{\mathrm{P}})$ when N increases. On the contrary, when N = 4 and No increases, a polynomial scaling of the CPU time with $\dim(\mathcal{S}^{\mathrm{P}})$ is reported. To explain these observations, it is first remarked that the total CPU time is mostly spent in two distinct tasks. First, a large fraction of the CPU time is spent solving the decoupled Poisson and diffusion operators for the stochastic solution modes: a scaling in $\mathcal{O}\dim(\mathcal{S}^{\mathrm{P}})$ for this contribution to the CPU time is expected. The second time-consuming part of the computation comes with the projection of the explicit terms. For the LBeqs, the cost for the projection of the explicit terms is dominated by the stochastic products, which scale with the complexity $\mathcal{C}$ of the Galerkin product. This complexity can be measured [125] as the number of non-zero entries in the multiplication tensor

$C_{klm} = \langle \Psi_k \Psi_l \Psi_m \rangle / \langle \Psi_k^2 \rangle$. The complexity $\mathcal{C}$ is a function of both No and N. With these notations, the total CPU time scales as $\mathcal{O}\dim(\mathcal{S}^{\mathrm{P}}) + \mathcal{O}(\mathcal{C}(\mathrm{N}, \mathrm{No}))$. The right plot in Fig. 7.20 shows $\mathcal{C}$ for the stochastic discretizations tested. It is seen that for fixed expansion order (No = 3) the scaling of $\mathcal{C}$ with $\dim(\mathcal{S}^{\mathrm{P}})$ is asymptotically linear, while for a fixed number of stochastic dimensions (N = 4) the scaling is polynomial in $\dim(\mathcal{S}^{P})$ as No increases. These two trends explain the reported evolutions of the total CPU-time with the stochastic discretization: for fixed expansion order the CPU-time is essentially proportional to the stochastic basis dimension, while for fixed N the contribution of the explicit terms projection to the CPU-time becomes more and more important as No increases.

7.3.5 *Unstable Steady Flow*

In this section, we provide a final example of the application of the Newton method. The objective of this example is two fold. First, it aims at demonstrating the effectiveness of the proposed method when dealing with steady flows that are likely to be unstable to perturbations. Second, the example is used illustrate the decoupling strategy mentioned in Sect. 7.3.2.

7.3.5.1 Uncertainty Settings

We consider the normalized two-dimensional flow around a circular cylinder. The flow is characterized by the Reynolds number $\mathrm{Re} = U_\infty D/\nu$, where U_∞ is the free-stream velocity, D the cylinder diameter and ν the fluid viscosity. It is well known that this flow is unstable for Re > 48. We assume deterministic free-stream velocity and cylinder diameter, and consider uncertainty in the fluid viscosity ν, modeled as a log-normal random variable with median value $\overline{\nu}$ and coefficient of variation $a > 1$; *i.e.* ν is expected to be in the range $[\overline{\nu}/a, \overline{\nu}a]$ with 99.5% probability. The random viscosity can be parameterized using a single random variable ξ as follows:

$$\nu(\xi) = \exp(-\mu + \sigma\xi), \quad \mu = \log\overline{\nu}, \sigma = \log(a)/2.85. \tag{7.93}$$

The stochastic basis consists of one-dimensional Hermite polynomials. On this basis, ν has for stochastic expansion:

$$\nu(\xi) = \sum_i \nu_i \Psi_i(\xi) = \exp(\mu + \sigma^2/2) \sum_i \frac{\sigma^i}{\langle \Psi_i^2 \rangle} \Psi_i(\xi), \tag{7.94}$$

where Ψ_i is the Hermite polynomial with degree i. The median value of the viscosity is set such that the median Reynolds number $\overline{\mathrm{Re}} = U_\infty D/\overline{\nu} = 60$ and the coefficient of variation is fixed to $a = 3/2$. Therefore, most realizations of the stochastic flow are above the critical Reynolds number as one can appreciate from Fig. 7.21 which depicts the probability density function of Re.

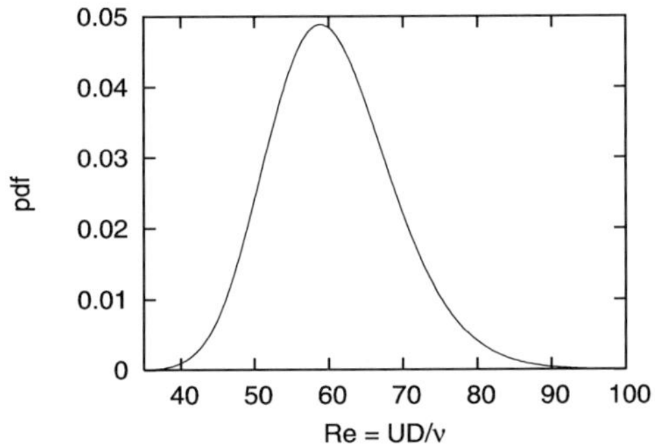

Fig. 7.21 Probability density function of the stochastic Reynolds number of the flow. Adapted from [119]

7.3.5.2 Flow Equations and Stochastic Decoupling

The vorticity streamfunction formulation of the flow is considered in an annular domain extending from the cylinder boundary Γ_c to an external circular boundary Γ_∞ located at $25D$ from the cylinder center. The governing equations are

$$\begin{cases} \dfrac{\partial \omega}{\partial t} = -\mathbf{u} \cdot \nabla \omega + \dfrac{1}{\mathrm{Re}} \nabla^2 \omega, \\ \nabla^2 \psi = -\omega, \\ \mathbf{u} = \nabla \wedge (\psi \mathbf{k}), \end{cases} \tag{7.95}$$

where ω is the vorticity and ψ the streamfunction. Natural boundary conditions are $\mathbf{u} = 0$ on Γ_c and $\mathbf{u} = U_\infty \mathbf{i}$ on Γ_∞. The flow variables are expanded on the Hermite basis and introduced in (7.95), which are in turn projected on the spectral basis. For a truncation of the stochastic basis to stochastic order No, we have: for $i = 0, \ldots, \mathrm{P} = \mathrm{No}$,

$$\begin{cases} \dfrac{\partial (\omega)_i}{\partial t} = -\sum_j \sum_l C_{ijl} \left[(\mathbf{u})_j \nabla (\omega)_l + \left(\dfrac{1}{\mathrm{Re}} \right)_j \nabla^2 (\omega)_l \right], \\ \nabla^2 (\psi)_i = -(\omega)_i, \\ (\mathbf{u})_i = \nabla \wedge ((\psi)_i \mathbf{k}). \end{cases} \tag{7.96}$$

The next step is to select a time-discretization that decouples the integration of the stochastic modes. This is achieved by splitting the stochastic diffusion term as follows:

$$\begin{aligned} &\sum_j \sum_l C_{ijl} \left(\frac{1}{\mathrm{Re}} \right)_j \nabla^2 (\omega)_l \\ &\quad = \sum_l C_{i0l} \left(\frac{1}{\mathrm{Re}} \right)_0 \nabla^2 (\omega)_l + \sum_{j>0} \sum_l C_{ijl} \left(\frac{1}{\mathrm{Re}} \right)_j \nabla^2 (\omega)_l, \\ &\quad = \left(\frac{1}{\mathrm{Re}} \right)_0 \nabla^2 (\omega)_i + \sum_{j>0} \sum_l C_{ijl} \left(\frac{1}{\mathrm{Re}} \right)_j \nabla^2 (\omega)_l. \end{aligned} \tag{7.97}$$

Introducing a first-order time discretization, the semi-discrete equation for the vorticity mode $(\omega)_i$ is

$$\left(\frac{1}{\Delta t}-\left(\frac{1}{\mathrm{Re}}\right)_0\nabla^2\right)(\omega^{n+1})_i=\frac{(\omega^n)_i}{\Delta t}+\sum_{j>0}\sum_l C_{ijl}\left(\frac{1}{\mathrm{Re}}\right)_j\nabla^2(\omega^n)_l$$
$$-\sum_j\sum_l C_{ijl}\left(\mathbf{u}^n\right)_j\nabla\left(\omega^n\right)_l, \qquad (7.98)$$

or using the formal notations

$$\mathcal{I}\left(\omega\right)_i^{n+1}=\mathcal{L}_i\left(\omega^n\right)+\mathcal{N}_i(\omega^n). \qquad (7.99)$$

The annular domain is conformally mapped to a rectangular mathematical domain where the equations are solved. On the opposite sides of the mathematical domain corresponding to Γ_c and Γ_∞, boundary conditions for the vorticity modes $(\omega)_i$ are determined by means of an influence matrix technique [46, 129, 206, 232]. The spatial discretization uses second-order centered differences on a uniform grid with 256×512 points, allowing for fast FFT-based solvers for the inversion of the Poisson and diffusion operators. The Hermite expansion is truncated at $\mathrm{No}=4$ so there are $\mathrm{P}+1=5$ modes in the solution.

7.3.5.3 Results

We apply the Newton method to solve the steady stochastic flow, with $\Delta t_n=10$ and $\epsilon=0.01$ in BiCGStab algorithm. Newton iterations are initialized with $\omega^0(\xi)=0$. For this initialization a limiter on the Newton increments is needed. This limiter rescales the increment by a small positive constant when the norm $\|\delta\mathcal{U}\|_{\Omega_x}^{\mathcal{S}^\mathrm{P}}$ is deemed too large. In the computation presented, the limiter acts only during the first three Newton iterations. Figure 7.22 presents the convergence of the steady residual with the Newton iterations. It is interesting to note that the number of BiCGStab iterations needed to satisfy the stopping criterion increases with the Newton index (not shown). This can be explained by the degradation of the conditioning of the

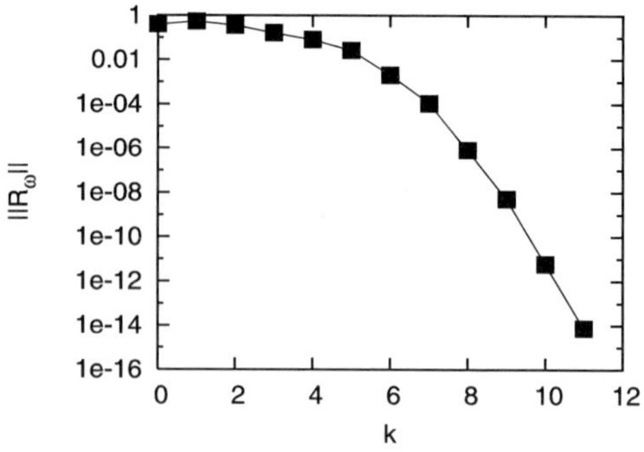

Fig. 7.22 Convergence of the steady equation residual as a function of the Newton iteration index k. Adapted from [119]

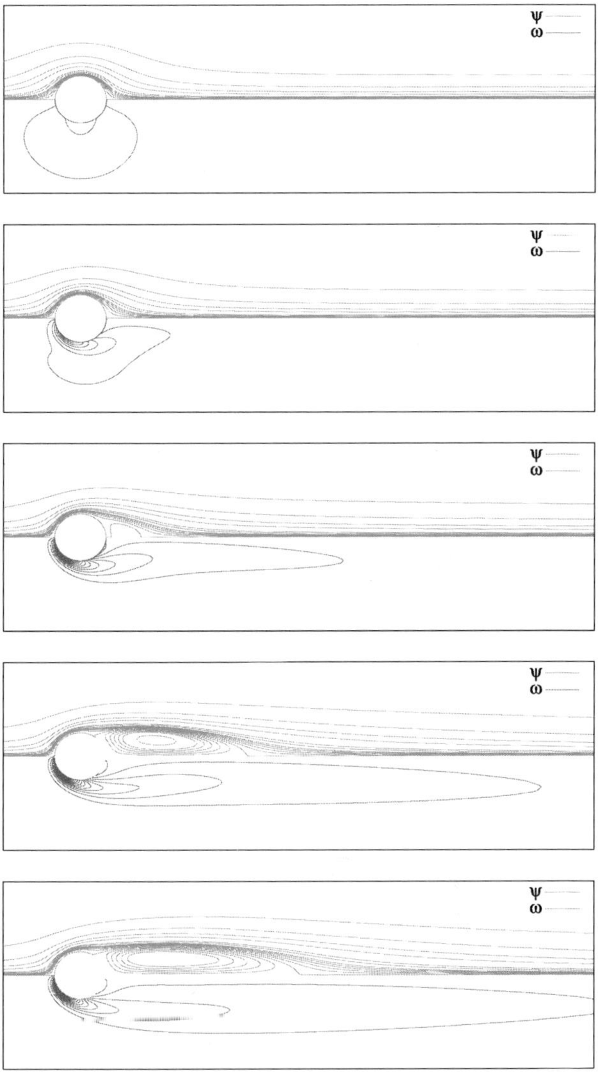

Fig. 7.23 Evolution with the first Newton iterates ($k = 1, \ldots, 5$ *from top to bottom*) of the averaged flow streamlines (*top part of the plots*) and vorticity contours (*bottom part of the plots*). The flow is from left to right. Adapted from [119]

increment problem, as the stochastic modes develop. A total of 1,175 linearized time-integrations was necessary to obtain a residual below 10^{-14}.

Figure 7.23 presents the evolution during the first five Newton iterations of the averaged flow, *i.e.* the mode $i = 0$ of the solution. Plotted are the streamlines (top part of the plots) and vorticity contours (bottom part of the plots). The figure shows the development of the two symmetric recirculation zones and the convection in the downstream direction of the vorticity.

Figure 7.24 presents the stochastic modes of the streamfunction $(\psi)_i$ (top part of the plots) and vorticity field $(\omega)_i$ (bottom part of the plots) of the steady solution, for modes $i = 0, \ldots, 3$. The contour levels have been adapted in each plot to

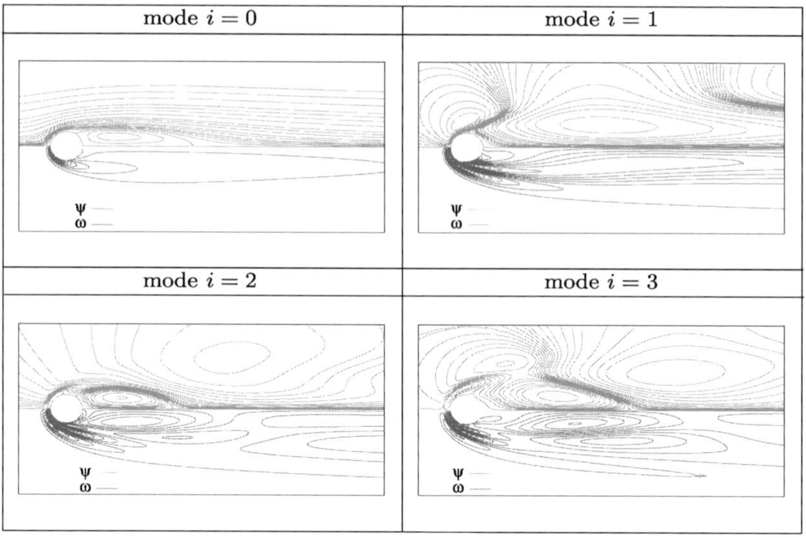

Fig. 7.24 Stochastic modes $i = 0, \ldots, 3$ of the steady streamfunction $(\psi)_i$ (*top part of the plots*) and vorticity field $(\omega)_i$ (*bottom part of the plots*). The flow goes from left to right. Adapted from [119]

highlight the flow structure, but it is emphasized that the magnitude of the fields decays by a factor of 50 to 100 between two successive modes; this highlights the convergence of the stochastic expansion. The plots show that the uncertainty in the fluid viscosity essentially affects the vorticity field in the boundary layers and in the cylinder wake. On the contrary, the streamfunction modes denote an impact of the uncertainty which extends far from the cylinder, even though it primarily affects the magnitude and spatial extension of the recirculation zones behind the cylinder.

To gain a better appreciation of the steady flow variability with regard to the uncertain viscosity, we present in Fig. 7.25 the standard deviation fields of the stochastic vorticity, streamfunction and velocity components. The standard deviation of the vorticity field confirms that the uncertainty essentially affects the vorticity in the boundary layers and cylinder wake. The standard deviation of the streamfunction highlights the variability in the intensity and spatial extension of the symmetric recirculation downstream of the cylinder. The standard deviation of the velocity component u, parallel to the inflow velocity, shows that it is mostly affected by the uncertainty along the flow symmetry axis and in the recirculation zones. Contrary to u, the transverse fluid velocity v exhibits a maximum variability in the immediate neighborhood of the upstream boundary layers; it vanishes on the axis of symmetry, since all realizations of the flow are symmetric.

The top plot in Fig. 7.26 shows the viscous stress distribution over the cylinder boundary, using the classical representation with mean value and uncertainty bounds extending to ± 3 standard deviations. The bottom plot of Fig. 7.26 provides an analysis of the flow recirculation statistics. Specifically, it shows as a function of the downstream distance x/D from the cylinder center, the longitudinal velocity u

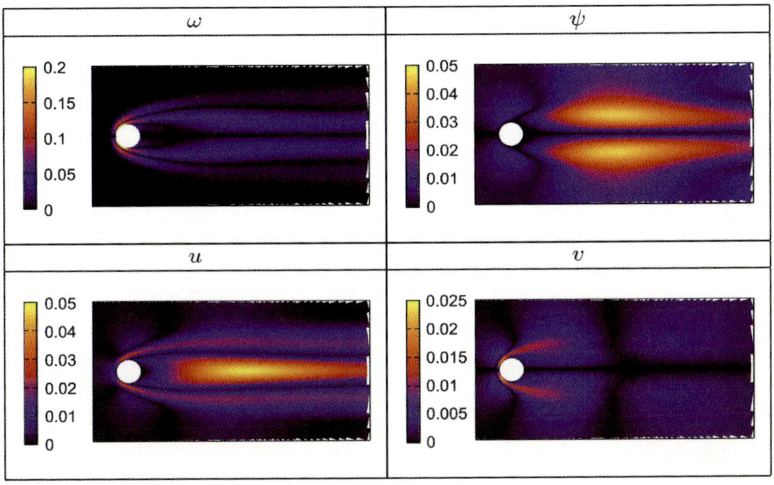

Fig. 7.25 Standard deviation of the vorticity, streamfunction and velocity components u (parallel to the inflow) and v (in the transverse direction). Adapted from [119]

(mean value and ± 3 standard deviations bounds), the probability of u to be negative (*i.e.* the probability of point x/D to be in the recirculation zone) and the probability density (rescaled by its maximum) of having $u(x/D) = 0$ (*i.e.* the probability density that the recirculation extends up to x/D).

7.4 Closing Remarks

In this chapter, we have briefly illustrated the application of matrix-free methods for the solution of large equation systems. Two classes of problems were specifically considered. The first concerned parabolic or elliptic equations arising from transient or steady heat conduction problems. The second considered the steady Navier-Stokes equations, and consequently focused on large linear systems arising in the iterative solutions of the associated discrete nonlinear system.

A geometric multigrid approach was selected for the simulation of heat conduction problems. Specifically, the MG methodology was applied to the Galerkin system associated with truncated spectral representation of the stochastic solution, and the performance of the resulting scheme was demonstrated through practical applications. Several extensions and generalizing building on the present developments can be identified. An attractive approach that is currently being explored is based on exploiting the structure and sparseness of the stochastic system. This structure suggests a hierarchical iterative strategy [183], which has been successfully exploited in the context of stochastic finite elements. In particular, it appears that incorporation of such an approach into the present MG framework could lead to a substantial performance enhancement. Another avenue that is potentially worthwhile to explore concerns the applications of existing software libraries for MG computations. The

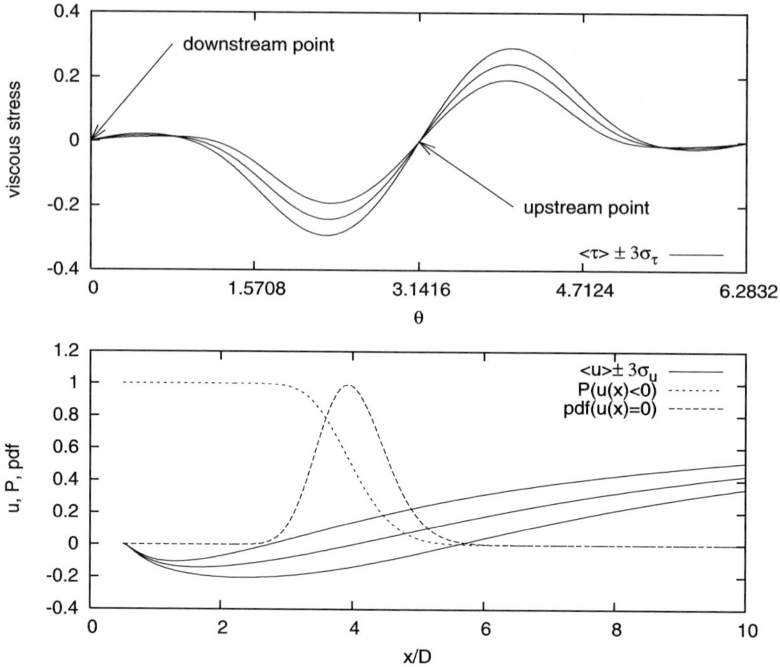

Fig. 7.26 *Top plot*: representation of the viscous stress distribution along the cylinder boundary using mean value ± 3 standard deviations bounds representation. *Bottom plot*: mean and ± 3 standard deviations bounds of the longitudinal velocity u, probability of $u < 0$ and probability density of $u = 0$ as a function of the downstream distance x/D from the cylinder center. Adapted from [119]

latter include both commercial and open source software implementing geometric and algebraic MG methods, as well as various preconditioners for optimizing their performance. In other words, the potential for capitalizing on the investment made in deterministic MG methods appears to be an attractive area in the development of stochastic Galerkin methods. Due to its fundamental relevance, this topic is further highlighted in Chap. 10.

A Newton method has been proposed for the resolution of the stochastic incompressible steady Navier-Stokes equations. It relies on an appropriate time discretization of the unsteady equations to derive a convenient stochastic spectral problem for the Newton increments of the steady solution. The method leads to matrix-free strategies, where Newton increments are successively computed by solving a series of spectral problems consisting in (pseudo-) time integration of the linearized unsteady stochastic flow equations. Similar to experiences with steady and transient heat problems, further improvements of the present construction can also be conceived, namely by optimizing the computation of Newton increments. Specifically, since these increments are the solution of a linear problem, the generalized spectral decomposition methodology [168] outlined in Chap. 9 may provide a particularly attractive means to drastically reduce both CPU times and memory requirements for their approximation.

We close by pointing out two interesting alternative approaches that substantially differ from the class of Krylov methods considered in this chapter, even though they still belong to the broad family of subspace methods. One approach concerns the subspace methods proposed by Nair and Keane [161] for linear models, in which the solution is sought in a subspace of *random vectors*. This subspace construction has to be contrasted with the Krylov space in (7.11), which is spanned by *deterministic vectors*. This stochastic Krylov space also results from successive transformations of an initial random vector through the application of the (preconditioned) random operator $A(\boldsymbol{\xi})$ and not with the stochastically discretized operator $\mathbf{A}$. The sequence of random vectors is then used to project the stochastic system and determine the coordinates of the solution in the so-called stochastic reduced basis. Two definitions were considered in [161] for the subspace solution, corresponding to the condition of a residual orthogonal to the stochastic Krylov subspace in the mean or with probability one. For the first definition, the coordinates of the solution are deterministic coefficients, whereas for the second one they are random variables. Although interesting, this reduced basis approximation appears to be difficult to extend to general linear stochastic operators.

A second approach concerns the generalized spectral decomposition methodology proposed by Nouy [167, 168] for the solution of linear stochastic models. In this approach, the solution is sought in the dominant subspace of the stochastic operator but, contrary to Nair and Keane approximation, the reduced basis consists of deterministic vectors. The coordinates of the approximate solution in the subspace, which are random variables, are once again determined by solving a projected problem. Additional details on this decomposition methodology are provided in Sect. 9.4, where it is extended to nonlinear models.

Chapter 8
Wavelet and Multiresolution Analysis Schemes

An essential aspect of spectral representations analyzed so far is the projection of the process or solution on a polynomial basis, namely on a vector space spanned by infinitely differentiable functions. It is known following the works of Wiener [241] and Cameron and Martin [25] that such spectral representation converges in a mean square sense as $N, No \to \infty$. In the case of numerical simulation, however, one must necessarily rely on truncation, and in this case the convergence may be seriously compromised due to truncation errors and aliasing, and in extreme cases these phenomena may lead to breakdown of the computations.

To illustrate potential difficulties, consider the case of a chemical system whose state at time t is specified using the vector, $c(t)$, of the concentrations of individual species. Suppose also that the system evolves according to a mechanism in which one elementary reaction has an uncertain rate, modeled as a random variable $q(\theta)$ in an abstract probability space (Θ, Σ, P). In this case, we seek the stochastic solution $c(t, q(\theta))$, which involves a development along a single stochastic dimension, *i.e.* $\mathrm{N} = 1$, $\boldsymbol{\xi} = \{\xi\}$. Truncated to order No, the spectral expansion is expressed as:

$$c(t, q(\theta)) \approx u(t, \xi(\theta)) = \sum_{k=0}^{\mathrm{P=No}} u_k(t)\Psi_k(\xi(\theta)).$$

One immediately notes that $u(t, \xi)$ is infinitely differentiable in ξ. Suppose now that the reaction mechanism admits a critical value of q, denoted by q_c, such that

$$\lim_{q \to q_c^-} c(t, q) \neq \lim_{q \to q_c^+} c(t, q),$$

i.e. the state c exhibits a parametric bifurcation for $q = q_c$. One runs into such situations when ignition (or extinction) of one or more reactions occurs, due for instance to threshold effects. If q_c is realizable, one effectively seeks to approximate, using infinitely differentiable functions, a functional admitting a bifurcation or a discontinuity. As for Fourier expansions where Gibbs phenomena are frequently illustrated, the lack of suitability of the spectral basis (too high of a regularity) leads to a very slow convergence in the case of infinite expansions, and to parasitic oscillations in

O.P. Le Maître, O.M. Knio, *Spectral Methods for Uncertainty Quantification*, Scientific Computation,
DOI 10.1007/978-90-481-3520-2_8, © Springer Science+Business Media B.V. 2010

the finite case. Even for simple chemical systems [195], these effects can have catastrophic impact on the computations. Of course, these phenomena can also be present in other systems where bifurcations or steep variations with random inputs can occur. A well-known example concerns complex systems involving shock formation or an energy cascade [31], where PC expansions based on smooth polynomials can lose their advantages or cease to be useful.

In this chapter, we address difficulties arising when random data assume values in the neighborhood of a critical point (or more generally a critical surface), or exhibit steep variations. Attention is focused on particular situations where the solutions remain smooth and well behaved, but which can change dramatically, or even discontinuously, according to specific values of the uncertain data. Such sensitivity of the solution with regard to the random data can be viewed as a parameter shock or bifurcation. For reasons alluded to above, one may expect that Wiener-Hermite or other spectral expansions based on smooth functions would fail to adequately describe the steep (or discontinuous) dependence of the solution on the random data. This chapter explores the possibility of overcoming these difficulties by using a PC expansions based on Haar wavelets [47, 174, 234, 235] or multiwavelets [3]. In doing so, we combine concepts of generalized PC expansions (see Sect. 2.3), and of using of piecewise functions in stochastic Galerkin methods [50]. In contrast to global basis functions, wavelet and MW representations naturally lead to localized decompositions which suggest the possibility of a more robust behavior albeit at the expense of slower rate of convergence.

In Sect. 8.1, we focus our attention on the simplest family of wavelets, namely the Haar basis [47, 174, 234, 235]. We introduce the Wiener-Haar (WHa) expansion of solutions dependent on random data, and briefly outline the salient features of the resulting PC representation. Section 8.2 then illustrates applications of WHa expansions. We first address a simple model problem consisting of a dynamical system having two isolated, stable fixed points. Depending on specific realizations of the random initial conditions, the solution converges to one stable fixed point or the other. Numerical simulations indicate that the WHa scheme can effectively resolve the random process, but that a Wiener-Legendre (WLe) expansion is unsuitable in this case. A more complex situation is then considered, based on a stochastic version of the Rayleigh-Bénard problem. Specifically, we consider two cases involving a random Rayleigh number, with the uncertainty range either containing the critical point or lying entirely above the critical value.

In Sect. 8.3, we generalize the methodology of Sect. 8.1 to arbitrary polynomial order expansions according to the framework proposed by Alpert in [3], the WHa expansion corresponding to the zero order case. The goal of the present generalization is to explore the possibility of combining the advantages of a higher order convergence rate resulting from higher-order polynomial expansion (p-convergence), and of the robustness of local decompositions. A multiresolution analysis (MRA) scheme is specifically constructed, allowing for refinement of the expansion by increasing the number of resolution levels (h-refinement) and/or the polynomial order (p-refinement). In Sect. 8.4, the resulting scheme is tested based on application to the Lorentz system having a single random parameter. In particular, the convergence of the solution is analyzed with respect to the number of refinement levels and the expansion order, and the convergence rate of the MRA scheme is also contrasted with that of two Monte-Carlo (MC) sampling strategies.

Results WHa or MRA computations are finally used in Sect. 8.5 to draw conclusions concerning the balance between accuracy, computational cost, and complexity, and to motivate further developments.

8.1 The Wiener-Haar expansion

In this section, we construct a WHa representation of a random variable. We start in Sect. 8.1.1 with the decomposition of a one-dimensional probability space using Haar's wavelets, and construct in Sect. 8.1.2 the associated orthonormal decomposition of a random process. The construction is then generalized to the multidimensional case in Sect. 8.1.3. Solution methods are given in Sect. 8.1.4, together with a brief discussion of similarities and differences with global PC constructions [123, 128].

8.1.1 Preliminaries

Let ξ be a $\mathbb{R}$-valued random variable with given statistics. We denote by $F_\xi(x)$ the cumulative distribution function of ξ, and assume that $F_\xi(x)$ is a continuous strictly increasing function of x defined on a real interval (a,b), such that $-\infty \le a < b \le \infty$, $F_\xi(a) \equiv 0$ and $F_\xi(b) \equiv 1$. Although extension to the infinite case is possible, we shall exclusively deal with the situation where a and b are finite. Using $F_\xi(x)$, the probability density function of x on (a,b) is given by:

$$p_\xi(x) \equiv \frac{\mathrm{d}F_\xi(x)}{\mathrm{d}x} > 0, \quad \forall x \in (a,b);$$
$$p_\xi(x) \equiv 0, \quad \forall x \notin (a,b).$$

Based on the assumed properties of $F_\xi(x)$, it follows that for all $y \in [0,1]$ there is a unique $x \in [a,b]$ such that $F_\xi(x) = y$. Consequently, we define the one to one mapping

$$y \in [0,1] \longrightarrow x \equiv F_\xi^{-1}(y) \in [a,b]. \tag{8.1}$$

8.1.1.1 Haar Scaling Functions

The scaling function of the Haar system, denoted by $\phi^{\mathrm{w}}(y)$, is given by [47, 174, 234, 235]:

$$\phi^{\mathrm{w}}(y) = \mathbf{I}_{[0,1)}(y) = \begin{cases} 1, & 0 \le y < 1, \\ 0, & \text{otherwise.} \end{cases} \tag{8.2}$$

Introducing the scaling factor j and the sliding factor k, we denote by

$$\phi^{\mathrm{w}}_{j,k}(y) = 2^{j/2}\phi^{\mathrm{w}}(2^j y - k), \tag{8.3}$$

the scaled Haar functions. Now, let $\{\mathbf{V}_j\}_{j=0}^{\infty}$ be the sequence of function spaces defined by $\mathbf{V}_j = \mathrm{span}\{\phi^{\mathrm{w}}_{j,k}, k \in [0, 2^j - 1]\}$, and denote by $\mathbb{P}_j f$ the projection of f onto the space $\mathbf{V}_j$; we thus have:

$$\mathbb{P}_j f = \sum_{k=0}^{2^j-1} f_{j,k}\phi^{\mathrm{w}}_{j,k}(y), \tag{8.4}$$

where the coefficients are given by:

$$f_{j,k} = \int_0^1 f(y)\phi^{\mathrm{w}}_{j,k}(y)\,\mathrm{d}y. \tag{8.5}$$

8.1.1.2 Haar Wavelets

The detail function $g^{j-1} \in \mathbf{V}_j$ is defined as the difference between two successive resolution levels, namely:

$$g^{j-1} = \mathbb{P}_j f - \mathbb{P}_{j-1} f. \tag{8.6}$$

To obtain an expression of the detail function, we introduce the Haar function:

$$\psi^{\mathrm{w}}(y) \equiv \frac{1}{\sqrt{2}}\phi^{\mathrm{w}}_{1,0}(y) - \frac{1}{\sqrt{2}}\phi^{\mathrm{w}}_{1,1}(y) = \begin{cases} 1, & 0 \le y < \frac{1}{2}, \\ -1, & \frac{1}{2} \le y < 1, \\ 0, & \text{otherwise.} \end{cases} \tag{8.7}$$

The Haar function is the *mother wavelet* that generates the wavelet family:

$$\psi^{\mathrm{w}}_{j,k}(y) = 2^{j/2}\psi^{\mathrm{w}}(2^j y - k), \quad j = 0, 1, \ldots, \text{ and } k = 0, \ldots, 2^j - 1. \tag{8.8}$$

From this definition we have

$$\int_0^1 \psi^{\mathrm{w}}_{j,k}(y)\,\mathrm{d}y = 0, \quad \text{and} \quad \int_0^1 \psi^{\mathrm{w}}_{j,k}\psi^{\mathrm{w}}_{l,m}\,\mathrm{d}y = \delta_{jl}\delta_{km}. \tag{8.9}$$

Consequently, the set $\{\psi^{\mathrm{w}}_{j,k};\ j = 0, 1, \ldots, \infty;\ k = 1, \ldots, 2^j - 1\}$ is an orthonormal system, and any function $f \in L_2([0, 1])$ can be arbitrarily well approximated by the sum of its mean and a finite linear combination of the $\psi^{\mathrm{w}}_{j,k}(y)$.

In terms of wavelets, the detail function can be expressed as:

$$g^{j-1}(y) = \sum_{k=0}^{2^j-1} d_{j,k}\psi^{\mathrm{w}}_{j-1,k}(y), \tag{8.10}$$

while $\mathbb{P}_j f$ is given by:

$$\mathbb{P}_j f = \mathbb{P}_0 f + \sum_{l=0}^{j-1} \sum_{k=0}^{2^l-1} d_{l,k} \psi_{l,k}^{\mathrm{w}}(y). \tag{8.11}$$

8.1.2 Wavelet Approximation of a Random Variable

We now seek a wavelet representation of the second-order $\mathbb{R}$-valued functional $X(\xi)$, where $\xi \in (\Theta, \Sigma, P)$ satisfies the assumptions of the previous section. Specifically, we consider an expansion of the form:

$$X(\xi(\theta)) = X_0 + \sum_{j=0}^{\infty} \sum_{k=0}^{2^j-1} X_{j,k}^{\mathrm{w}} \mathrm{W}_{j,k}(\xi(\theta)), \tag{8.12}$$

where $X_{j,k}^{\mathrm{w}}$ are the coefficients of the wavelet approximation of $X(\xi)$,

$$\mathrm{W}_{j,k}(\xi \in [a,b]) \equiv \psi_{j,k}^{\mathrm{w}}(F_\xi(\xi)), \tag{8.13}$$

and the equality is interpreted in mean-square sense. Equation (8.12) can be rewritten as:

$$X(\xi(\theta)) = X_0 + \sum_{j=0}^{\infty} \sum_{k=0}^{2^j-1} X_{j,k}^{\mathrm{w}} \psi_{j,k}^{\mathrm{w}}(F_\xi(\xi(\theta))), \tag{8.14}$$

where $X_0 \equiv \mathbb{P}_0 X(\xi)$. Moreover, the orthonormality of the Haar wavelets ensures that

$$\int_{[a,b]} \mathrm{W}_{j,k}(y)\mathrm{W}_{l,m}(y) p_\xi(y)\,\mathrm{d}y = \int_0^1 \psi_{j,k}^{\mathrm{w}}(x)\psi_{l,m}^{\mathrm{w}}(x)\,\mathrm{d}x \equiv \delta_{j,l}\delta_{k,m}. \tag{8.15}$$

This shows that the set of wavelets $\{\mathrm{W}_{j,k}, j = 0, \ldots, \infty; k = 0, \ldots, 2^j - 1\}$ forms an orthonormal system with respect to the inner product:

$$\langle f, g \rangle \equiv \int_{[a,b]} f(y)g(y)p_\xi(y)\,\mathrm{d}y,$$

and that $\langle f \rangle = \langle f, 1 \rangle$ coincides with the mean or expectation. The wavelet set $\{\mathrm{W}_{j,k}, j = 0, \ldots, \infty; k = 0, \ldots, 2^j - 1\}$ in fact form a basis of the space of second-order random variables $\{X : \langle X, X \rangle < +\infty\}$ [47, 174].

Let us denote by ∇ the set of index integers λ concatenating the scale index j and the space index k: $\nabla \equiv \{\lambda : \lambda = 2^j + k; j = 0, \ldots, \infty; k = 0, \ldots, 2^j - 1\}$. The resolution level will be denoted by $|\lambda|$. Using these conventions, the one-dimensional wavelet expansion of X can be expressed as:

$$X(\xi(\theta) \in [a,b]) = X_0 + \sum_{\lambda \in \nabla} X_\lambda \mathrm{W}_\lambda(\xi(\theta)). \tag{8.16}$$

Moreover,

$$X_0 \equiv \int_0^1 X(F_\xi^{-1}(x)) \phi_{0,0}^{\mathrm{w}}(x)\,\mathrm{d}x = \int_{[a,b]} X(y) p_\xi(y)\,\mathrm{d}y = \langle X \rangle \tag{8.17}$$

is the expected value of X. Consequently, setting $\mathrm{W}_0 \equiv 1$, and denoting ∇_0 the extension of ∇ to include the index 0, the 1-D wavelet expansion becomes:

$$X(\xi(\theta) \in [a,b]) = \sum_{\lambda \in \nabla_0} X_\lambda \mathrm{W}_\lambda(\xi(\theta)), \tag{8.18}$$

where

$$\begin{aligned} X_\lambda &\equiv \int_0^1 X(F_\xi^{-1}(x)) \psi_\lambda^{\mathrm{w}}(x)\,\mathrm{d}x \\ &= \int_{[a,b]} X(y) \mathrm{W}_\lambda(y) p_\xi(y)\,\mathrm{d}y = \langle X, \mathrm{W}_\lambda \rangle . \end{aligned} \tag{8.19}$$

The expansion in (8.18) is the wavelet analogue of the generalized PC expansion used in stochastic spectral methods.

8.1.3 Multidimensional Case

In this section, we extend the WHa expansion to the multidimensional case, and focus for simplicity on a vector $\boldsymbol{\xi}$ of independent random components $\{\xi_1(\theta), \dots, \xi_{\mathrm{N}}(\theta)\}$ obeying:

$$\langle \xi_i \rangle = 0, \quad i = 1, \dots, N, \qquad \text{and} \qquad \left\langle \xi_i \xi_j \right\rangle = \left\langle \xi_i^2 \right\rangle \delta_{ij}, \quad 1 \le i, j \le N.$$

We now consider the multidimensional index $\boldsymbol{\lambda} = (\lambda_1, \lambda_2, \dots, \lambda_{\mathrm{N}})$, and define the sequence

$$\mathcal{W}_n \equiv \left\{ \prod_{k=1}^{\mathrm{N}} \mathrm{W}_{\lambda_k}(\xi_k) : \sum_{k=1}^{\mathrm{N}} |\lambda_k| = n \right\}$$

to be the set of multidimensional wavelets having resolution n. The multidimensional wavelet expansion of $X(\xi)$ can now be formally written as [90, 241]:

$$
\begin{aligned}
X(\xi_1(\theta), \ldots, \xi_N(\theta)) \\
&= X_0\Gamma_0 + \sum_{i_1=1}^{N} c_{i_1}\Gamma_1(\xi_{i_1}(\theta)) \\
&+ \sum_{i_1=1}^{N}\sum_{i_2=1}^{i_1} c_{i_1 i_2}\Gamma_2(\xi_{i_1}(\theta), \xi_{i_2}(\theta)) \\
&+ \sum_{i_1=1}^{N}\sum_{i_2=1}^{i_1}\sum_{i_3=1}^{i_2} c_{i_1 i_2 i_3}\Gamma_3(\xi_{i_1}(\theta), \xi_{i_2}(\theta)\xi_{i_3}(\theta)) + \cdots
\end{aligned}
\tag{8.20}
$$

where $\Gamma_0 \equiv 1$, and $\Gamma_k \in \mathcal{W}_k$ denotes a multidimensional wavelet of resolution k. In practice, the wavelet expansion must be truncated, and different strategies may be used for this purpose. The most intuitive approach is to retain wavelets of resolution n, *i.e.* we retain vectors $\boldsymbol{\lambda}$ such that $\sum_{k=1}^{N} |\lambda_k| \le n$. In this case, the multidimensional resolution level n plays a similar role as the order in Wiener-Hermite expansions [90, 241]. Another possibility is to use the "spherical truncation", e.g. by retaining vectors $\boldsymbol{\lambda}$ satisfying $(\sum_{k=1}^{N} |\lambda_k|^2)^{1/2} \le n$.

Regardless of the truncation strategy, the truncated expansion may be conveniently rewritten as a single-index summation, according to:

$$
X(\boldsymbol{\xi}) \approx \sum_{k=0}^{\mathrm{Nw}} X_k^{\mathrm{w}} \mathcal{H}a_k(\boldsymbol{\xi}),
\tag{8.21}
$$

where $\mathrm{Nw}+1$ is the dimension of the truncated basis, $\{\mathcal{H}a_k, k = 0, \ldots, \mathrm{Nw}\}$. The wavelet expansion can be immediately extended to the representation of second-order random vectors and stochastic processes, similar to the PC expansion discussed in Sect. 2.4.

8.1.4 Comparison with Spectral Expansions

Similar to the WHe [90, 241] and other spectral representations [245], the wavelet expansion (8.21) is an orthonormal approximation of the random variable X. This property may be immediately exploited to extract the expectation of X,

$$
\langle X \rangle = X_0^{\mathrm{w}},
$$

and its variance

$$
\sigma^2(X) = \sum_{k=1}^{\mathrm{Nw}} (X_k^{\mathrm{w}})^2 .
$$

Despite the formal similarities of the corresponding expansions, fundamental differences between wavelet and spectral representations should be noted. In the latter case, global orthogonal polynomials are specifically selected so that, when appropriate smoothness conditions are satisfied, an "infinite-order" convergence rate results. Such convergence rate is not expected for the WHa expansion, in which the basis functions are localized. Specifically, the WHa expansion derived above can be viewed as an "orthogonal sampling" or a local decomposition of the solution into piecewise constant processes. One may expect that in situations where the response of the system shows a localized sharp variation or a discontinuous change, the wavelet decomposition may be more efficient than a spectral expansion, whose convergence could dramatically deteriorate due to Gibbs-type phenomena. Another distinctive feature of the WHa expansion concerns products of piecewise constant processes. For instance, the product xy of two elements x and y of $\mathbf{V}_j$ also belong to the same space. In contrast, the product of two polynomials of degree less than or equal to n does not necessarily belong to the space of polynomials having degree less than or equal to n. Thus, one may also expect that for problems exhibiting steep dependence on random data the WHa scheme is less susceptible to aliasing errors than a spectral scheme. Below, we address these questions by considering situations involving both smooth and discontinuous dependence on the random data, and contrast the behavior of WHa and WLe expansions.

8.2 Applications of WHa Expansion

We first apply the WHa decomposition to a simple dynamical system that involves a discontinuous dependence of the process on the random data, and later consider the case of Rayleigh-Bénard convection with random data in the neighborhood of a critical Rayleigh number.

8.2.1 Dynamical System

Consider the following deterministic differential equation,

$$\frac{\mathrm{d}^2 x}{\mathrm{d}t^2} + f\frac{\mathrm{d}x}{\mathrm{d}t} = -\frac{\mathrm{d}h}{\mathrm{d}x}, \tag{8.22}$$

with parameters $f > 0$ and $\mathrm{d}h/\mathrm{d}x$. The problem requires two initial conditions: $x(t=0) = x_0$ and $v(t=0) \equiv \mathrm{d}x/\mathrm{d}t\,(t=0) = v_0$. The system can be interpreted as the governing equation for a particle moving under the influence of a potential field and of a friction force. In the computations below, we set $h(x) = \frac{35}{8}x^4 - \frac{15}{4}x^2$ so that the differential equation has two stable fixed points ($x = \pm\sqrt{15/35}$), and an unstable fixed point at $x = 0$. The potential field $h(x)$ and the function $\mathrm{d}h/\mathrm{d}x$ are plotted in Fig. 8.1.

A stochastic variant of the above system is constructed by considering an uncertain initial position x_0. On the other hand, the particle is always released with a

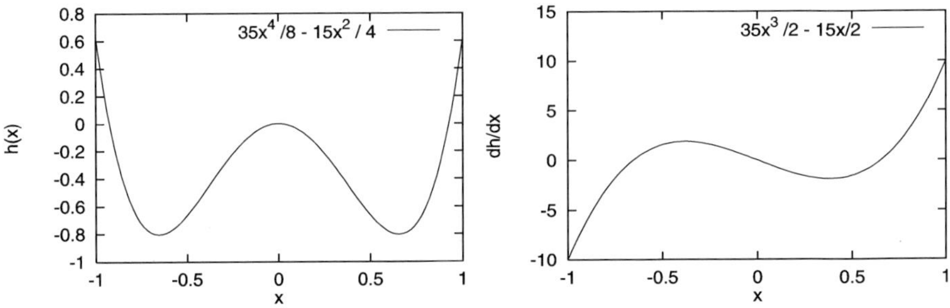

Fig. 8.1 Profiles of $h(x)$ and of $\mathrm{d}h/\mathrm{d}x$ for the model problem of Sect. 8.2.1. Adapted from [124]

vanishing velocity. In the computations, we assume that the initial position is uniformly distributed over the interval $[x_1, x_2]$. The stochastic initial conditions can be expressed as:

$$X(t=0,\xi) = X_0 + \Delta X \xi, \qquad \left.\frac{\mathrm{d}X}{\mathrm{d}t}\right|_{t=0} = 0,$$

where $X(t,\xi)$ denotes the response of the stochastic system, $X_0 \equiv (x_1 + x_2)/2$, $\Delta X = |x_1 - x_2|/2$, and ξ is uniformly distributed over its support $[-1, 1]$ with density $p_\xi = 1/2$. Thus, the stochastic system can be formulated as:

$$\frac{\mathrm{d}^2 X}{\mathrm{d}t^2} + f\frac{\mathrm{d}X}{\mathrm{d}t} = -\frac{35}{2}X^3 + \frac{15}{2}X, \tag{8.23}$$

$$X(t=0,\xi) = X_0 + \Delta X \xi, \tag{8.24}$$

$$\left.\frac{\mathrm{d}X}{\mathrm{d}t}\right|_{t=0} = 0. \tag{8.25}$$

8.2.1.1 Solution Method

In this section, we outline the method used to integrate the stochastic formulation based on the WHa representation. The solution method for the WLe representation is similar and is consequently omitted.

The truncated WHa expansion of the solution process is expressed as:

$$X(t,\xi(\theta)) \approx \tilde{X}(t,\xi) = \sum_{k=0}^{\mathrm{Nw}} X_k^{\mathrm{w}}(t)\mathcal{H}a_k(\xi).$$

Governing equations for the wavelet coefficients X_k^{w} are derived in two steps. The truncated wavelet expansion is first inserted into (8.23), and projections onto the wavelet basis are then performed. The latter step is implemented by multiplying the expanded system by $\mathcal{H}a_l$ and then forming the expectation. This leads to a system of Nw + 1 coupled ODE's for the coefficients:

$$\frac{d^2X_l^w}{dt^2} + f\frac{dX_l^w}{dt} = -\frac{35}{2}\left\langle \tilde{X}^3, \mathcal{H}a_l \right\rangle + \frac{15}{2}X_l^w, \tag{8.26}$$

for $l = 0, \ldots, \mathrm{Nw}$. A similar Galerkin approach is used to derive initial conditions for the individual modes; we get:

$$X_0^w(t=0) = X_0; \qquad X_l^w(t=0) = \Delta X \langle \xi, \mathcal{H}a_l \rangle \quad \text{for } l = 1, \ldots, \mathrm{Nw}, \tag{8.27}$$

$$\left.\frac{dX_l^w}{dt}\right|_{t=0} = 0 \quad \text{for } l = 0, \ldots, \mathrm{Nw}. \tag{8.28}$$

Equation (8.26) can be easily integrated once the cubic term, $\langle \tilde{X}^3, \mathcal{H}a_l \rangle$ is determined. Two approaches are considered here, based on (i) a Galerkin approach, and (ii) a pseudo-spectral approximation. In order to outline these approaches, we first recall the "multiplication tensor",

$$C_{ijk} \equiv \left\langle \mathcal{H}a_i \mathcal{H}a_j \mathcal{H}a_k \right\rangle,$$

and "the triple-product tensor",

$$T_{ijkl} \equiv \left\langle \mathcal{H}a_i \mathcal{H}a_j \mathcal{H}a_k \mathcal{H}a_l \right\rangle.$$

One can readily show (see [90, 123, 128] and Chap. 4) that the Galerkin approximation of quadratic term X^2 is given by:

$$(X^2)_k^w \equiv \left\langle \tilde{X}^2, \mathcal{H}a_k \right\rangle = \sum_{i=0}^{\mathrm{Nw}} \sum_{j=0}^{\mathrm{Nw}} C_{ijk} X_i^w X_j^w.$$

In the Galerkin approach, the cubic term is obtained through convolution involving the triple tensor, specifically using the following triple sum:

$$(X^2)_l^w \equiv \left\langle \tilde{X}^3, \mathcal{H}a_l \right\rangle = \sum_{i=0}^{\mathrm{Nw}} \sum_{j=0}^{\mathrm{Nw}} \sum_{k=0}^{\mathrm{Nw}} T_{ijkl} X_i^w X_j^w X_k^w.$$

On the other hand, for the pseudo-spectral approach, the cubic term is approximated through repeated application of the binary multiplication operator, according to:

$$(X^3)_k^w \equiv \left\langle (X^3), \mathcal{H}a_k \right\rangle = \left\langle (X)(X^2), \mathcal{H}a_k \right\rangle \simeq \sum_{i=0}^{\mathrm{Nw}} \sum_{j=0}^{\mathrm{Nw}} C_{ijk} X_i^w (X^2)_j^w.$$

The tensor C_{ijk} (and when present T_{ijkl}) is evaluated in a pre-processing step and then stored for later use in the simulations (Appendix C). Note that, similar to the Hermite and Legendre systems, the multiplication tensor in the Haar system is sparse and can be evaluated exactly. Also note that in the case of the WHa expansion, the pseudo-spectral approach coincides with the Galerkin estimate. This is the case because, as previously noted, the product of elements of $\mathbf{V}_j$ also belongs to $\mathbf{V}_j$. For the WLe and other polynomial representations, on the other hand, repeated binary products introduce aliasing errors due to truncation at the intermediate stages (see discussion in Sect. 4.5).

The time integration of (8.26) is performed using a fourth-order Runge-Kutta scheme. In all computations described in this section, a small value of the time step was used, $\Delta t = 0.001$. This value was selected following a straightforward analysis in which the time step was systematically reduced until it had negligible impact on the predictions.

Below, we compare solutions obtained using the WHa expansion, truncated to a given resolution level Nr, with results obtained using both Galerkin and pseudo-spectral WLe expansions, truncated to order No. Since we are dealing with a single stochastic dimension, the total number of basis functions in the WHa expansion is equal to 2^{Nr}; in the case of WLe, it is equal to $\mathrm{No}+1$.

8.2.1.2 Results

WLe scheme: The WLe scheme is applied to the model problem above, with stochastic initial conditions specified by $X_0 = 0.05$ and $\Delta X = 0.2$. A relatively large value of the friction coefficient is selected, $f = 2$, so that a steady solution is achieved in a short time. For the present conditions, the analytical prediction of the steady state is given by:

$$\begin{cases} X(t \to \infty, \xi) = -\sqrt{15/35} & \text{for } \xi < -0.25, \\ X(t \to \infty, \xi) = \sqrt{15/35} & \text{for } \xi > -0.25, \end{cases}$$

which results in the following statistical moments, $\langle X(t \to \infty)\rangle = 0.163663$ and $\sigma(t \to \infty) = 0.633865691$.

Figures 8.2 and 8.3 respectively depict the pseudo-spectral and Galerkin WLe solutions for different expansion orders. Plotted are the short-time evolution of $X(t,\xi)$, the solution at $t = 10$, and the corresponding "steady-state" probability density functions (pdfs) of X. The results indicate that, regardless of the order of the expansion, the WLe scheme does not provide an adequate representation of the behavior of the system. Specifically, unphysical oscillations in $X(\xi)$ are observed, which are more pronounced in the pseudo-spectral computations than in the Galerkin results. These oscillations are still present in the "large time" steady solution, as shown in Fig. 8.4. Around the equilibrium points $X = \pm\sqrt{15/35}$, the amplitude of the wiggles decreases slightly as No increases, but their frequency increases. The manifestation of these wiggles is reminiscent of the Gibbs phenomenon which occurs in spectral decompositions of discontinuous signals. The impact of this phenomenon can also be appreciated in the predicted steady-state pdf of X, shown in the right column of Figs. 8.2 and 8.3. For the present stochastic problem, the analytical pdf consists of two Dirac masses of unequal strength, located at the stable

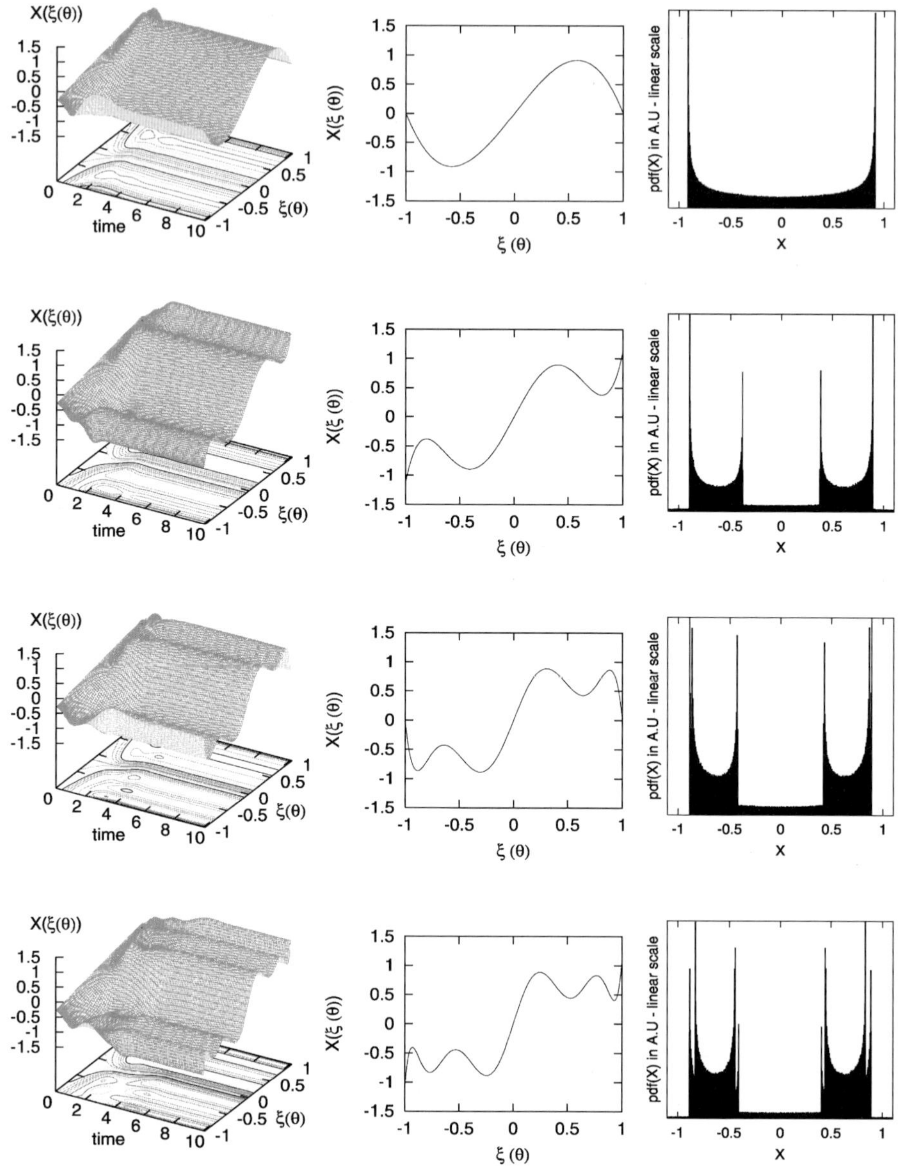

Fig. 8.2 Pseudo-spectral WLe solution for the model problem of Sect. 8.2.1. The *left column* shows the evolution of $X(t, \xi)$ for $0 \leq t \leq 10$. The solution $X(t = 10)$ is plotted in the *middle column* and the steady-state pdf is shown in the *right column*. Results are obtained for expansion orders No = 3, 5, 7, and 9, arranged *from top to bottom*. Adapted from [124]

equilibrium points $X = \pm\sqrt{15/35}$. The WLe predictions differ significantly from the analytical prediction; around the equilibrium points, they exhibit a broad spec-

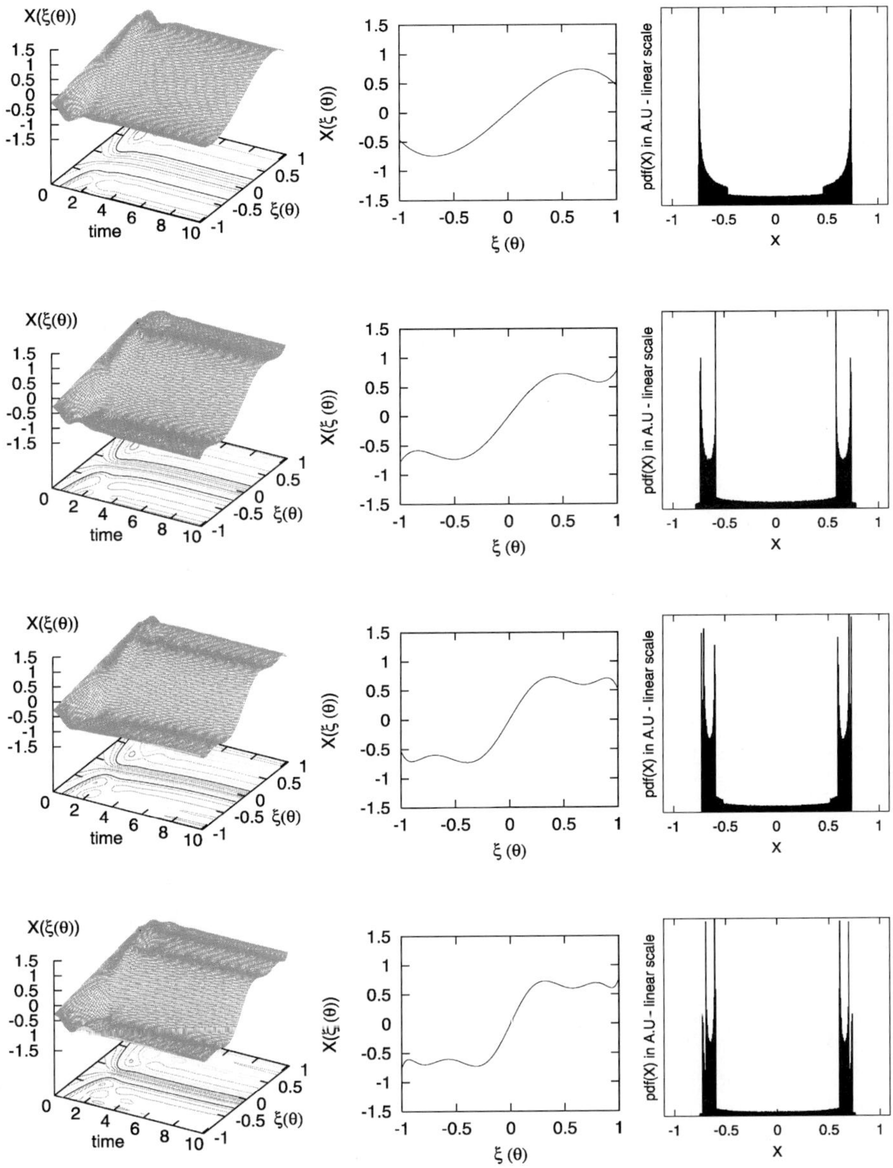

Fig. 8.3 Galerkin WLe solution for the model problem of Sect. 8.2.1. The *left column* shows the evolution of $X(t, \xi)$ for $0 \leq t \leq 10$. The solution $X(t = 10)$ is plotted in the *middle column* and the steady-state pdf is shown in the *right column*. Results are obtained for expansion orders No = 3, 5, 7, and 9, arranged *from top to bottom*. Adapted from [124]

trum with multiple peaks. This is also characteristic of the application of a spectral representation to a discontinuous problem.

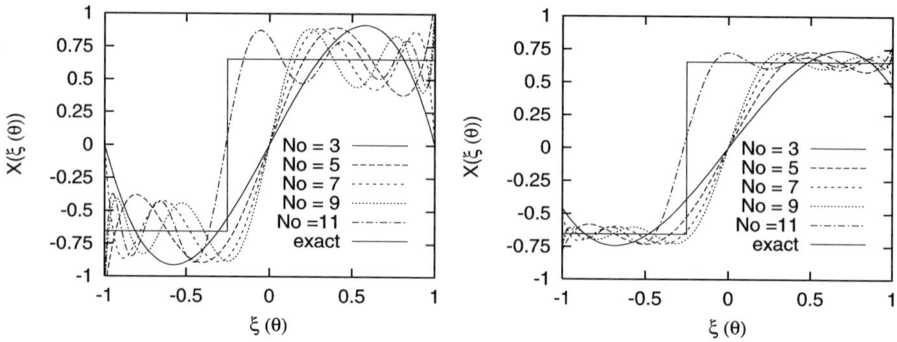

Fig. 8.4 Steady solution $X(t \to \infty)$ of the model problem of Sect. 8.2.1, obtained using WLe expansions with No = 3, 5, 7, 9 and 11. Plotted are curves obtained using a pseudo-spectral (*left*) and Galerkin (*right*) approximation. Adapted from [124]

In addition to poor representation of the asymptotic pdf, in the present case the WLe scheme also fails to provide accurate predictions of some of the low-order moments. To illustrate this claim, we provide in Fig. 8.5 the WLe predictions of the mean and standard deviation of X at steady state, for $\text{No} = 3, \ldots, 31$. The results demonstrate that the mean response is poorly estimated for both the Galerkin and pseudo-spectral approximations, and that it fluctuates substantially with No. Better, though still inadequate, predictions of the standard deviation are obtained. Similar to the mean, these predictions also fluctuate with No.

WHa scheme: The WHa scheme is now applied to the same problem of Sect. 8.2.1. Results are obtained for an increasing number Nr of resolution levels. Figure 8.6 shows the evolution of $X(t, \xi)$ for $0 \le t \le 10$; results obtained with Nr = 2, 3, 4 and 5 are depicted. The results indicate that, so long as $\text{Nr} > 2$, the WHa scheme correctly captures the bifurcation dividing the trajectories converging to the two stable equilibrium points. The transition is first captured at Nr = 3 and further increase of the value of Nr only affects the smoothness of the solution during the initial transient. For Nr = 2, the WHa scheme yields an incorrect result for $-0.5 \le \xi \le 0$, predicting in this range that the position rapidly equilibrates at $X = 0$. This corresponds to a physical but unstable equilibrium point. This erroneous prediction is obtained because the mean initial position over the corresponding uncertainty range is zero. It is interesting to note that, although the prediction for Nr = 2 is incorrect for $-0.5 \le \xi \le 0$, it is correct for the remaining parts of the uncertainty range. Thus, errors incurred at specific values of the random data do not appear to pollute the entire prediction; this (desirable) property of WHa schemes has been observed in a large number of (under-resolved) computations.

The mean and standard deviation of X at steady state are reported in Table 8.1 for all considered values of Nr. We observe that for $\text{Nr} \ge 3$ the analytical prediction is exactly recovered. This complete agreement with the analytical solution is due to the fact that the discontinuity, located at $\xi = -0.25$, is "naturally" captured for

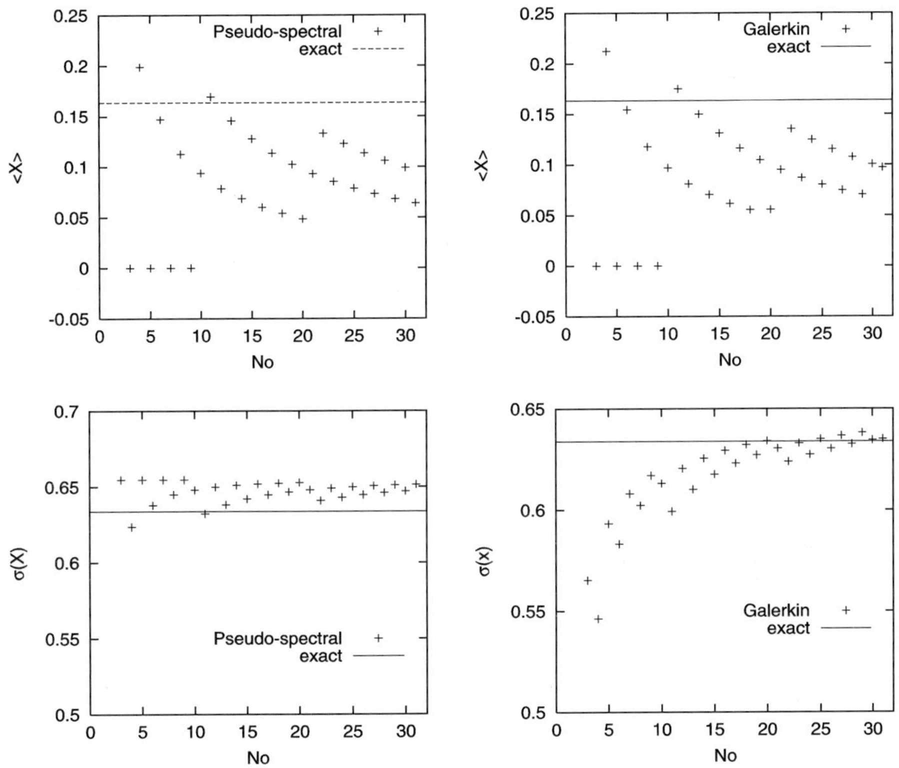

Fig. 8.5 Convergence with No of the computed mean (*top*) and standard deviation (*bottom*) of $X(t \to \infty)$ using WLe expansion for the model problem of Sect. 8.2.1. Plotted are curves obtained using a pseudo-spectral (*left*) and Galerkin (*right*) approximation. The analytical predictions are given by $\langle X \rangle = 0.16366$ and $\sigma(X) = 0.63387$. Adapted from [124]

Table 8.1 Mean and standard deviation of X at steady state. Results are obtained using the WHa scheme with different Nr. The analytical predictions are give by $\langle X(\theta) \rangle = 0.1636634$ and $\sigma(X(\theta)) = 0.633865691$. Adapted from [124]

Nr	$\langle X(\theta) \rangle$	$\sigma(X(\theta))$
2	0.3273268	0.566946718
3	0.1636634	0.633865691
4	0.1636634	0.633865691

$\text{Nr} \geq 3$; it is specifically located at the edge of neighboring wavelets at the level $j = 3$. Thus, the steady-state solution has vanishing details for scales (or resolution indices) $j > 3$.

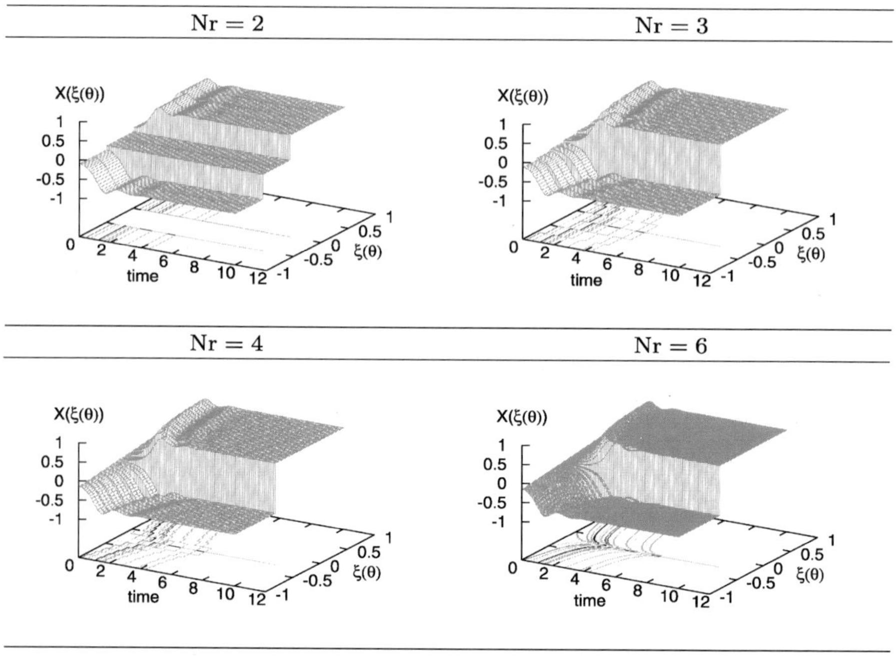

Fig. 8.6 WHa solution for the model problem of Sect. 8.2.1. The *plots* show the evolution of $X(t,\xi)$ for different resolution levels, $\mathrm{Nr} = 2$, 3, 4 and 6. Adapted from [124]

Highly discontinuous solution: The example above showed that in the case of a single point of discontinuity, the WHa decomposition can provide an accurate representation of the stochastic process, but that the WLe scheme proved inadequate. To have a finer appreciation of the properties of the WHa scheme, we now consider a more difficult problem obtained simply by reducing the friction coefficient, f. Specifically, we set $f = 0.05$ and focus on stochastic initial conditions given by $X_0 = 1$ and $\Delta X = 0.1$. As in the previous case, the particle is released from a state of rest, i.e. the initial velocity is deterministic and equal to 0.

The reduction of the friction coefficient, together with the higher initial energy of the system, result in a complex response. Specifically, the particle oscillates for several cycles between the two potential wells, before reaching a final equilibrium position. Furthermore, the inverse map between each of two stable equilibrium points and the corresponding initial positions (which as previously mentioned is assumed to be uniformly distributed between 0.9 and 1.1), results in a union of several disjoint intervals. Thus, the situation is more complex than that of the previous case, where two intervals were obtained. Consequently, one anticipates that a significantly higher resolution level would be needed to correctly characterize the behavior of the stochastic system.

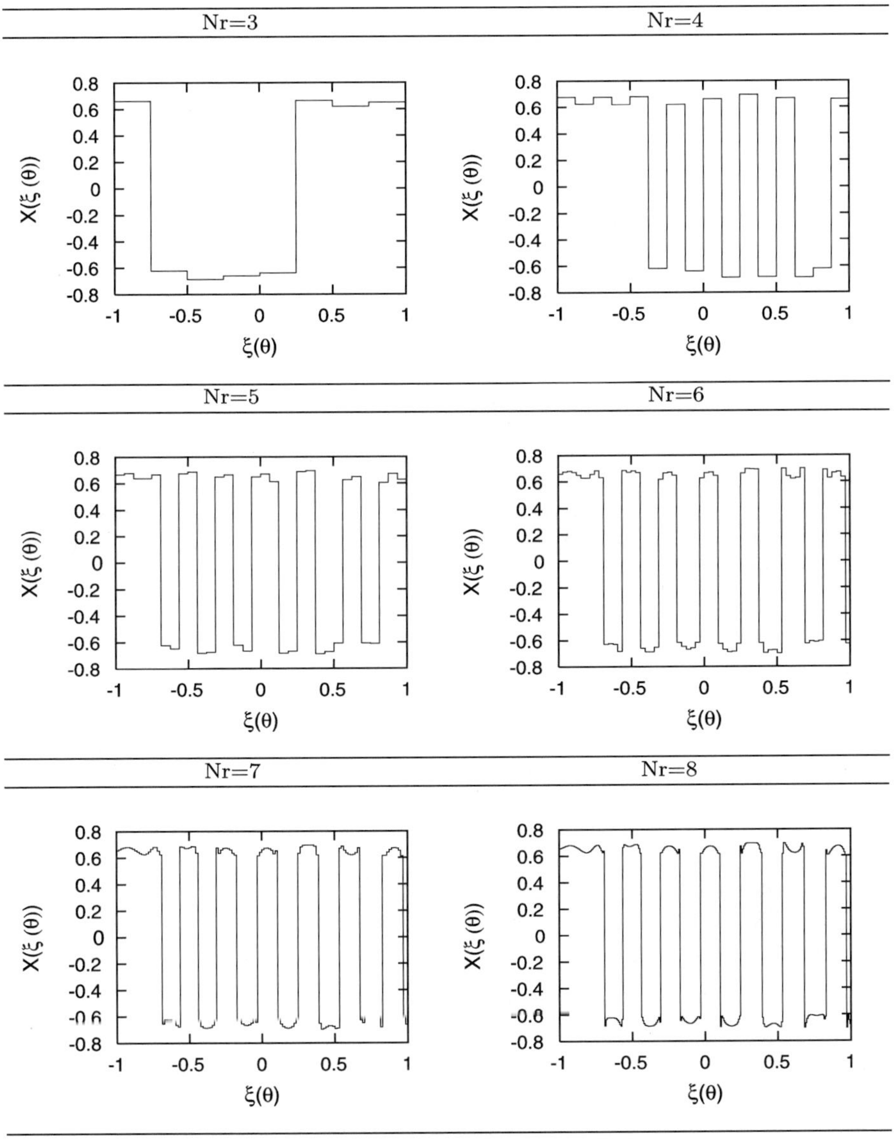

Fig. 8.7 WHa solution at $t = 100$ for the model problem of Sect. 8.2.1 with $f = 0.05$. The plots show curves of $X(t = 100, \xi)$ versus ξ, computed using different resolution levels, Nr = 3, 4, 5, 6, 7 and 8 as indicated. Adapted from [124]

To illustrate the convergence of the WHa scheme in the present case, results were obtained with a wider range of resolution levels, $3 \leq \text{Nr} \leq 8$. Results are plotted in Fig. 8.7, which depicts $X(\xi)$ at $t = 100$ for all considered values of Nr. The results indicate that for the present conditions, 6 resolution levels are needed to capture the

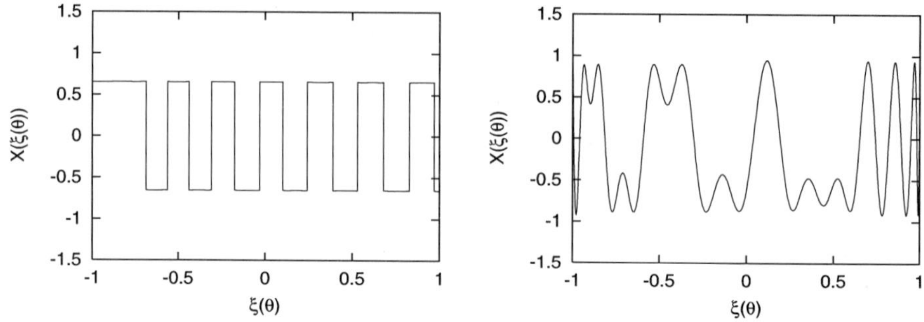

Fig. 8.8 Large-time (steady) solutions for the problem of Sect. 8.2.1 with $f = 0.05$. Plotted are curves for $X(t \to \infty, \xi)$ versus ξ. *Left*: WHa expansion with Nr = 6; *right*: pseudo-spectral WLe expansion with No = 32. Adapted from [124]

response of the system, and particularly all the corresponding discontinuities. By increasing the resolution level beyond Nr = 6, one obtains additional details on the response of the system within the regions of continuity, as well as a slight refinement of the locations of the discontinuities.

One also notes that, even when the resolution level is too low to correctly capture all of the discontinuities, the WHa expansion still provides a meaningful prediction, in the sense that steady-state realizations do in fact correspond to a stable equilibrium point. In other words, resolution errors do not lead to an unphysical prediction. This robustness of the WHa expansion is further illustrated in Fig. 8.8 which compares the WHa solution using Nr = 7 with the pseudo-spectral WLe prediction with No = 32. The results are generated at $t = 250$, where a stationary state is nearly reached. The figure shows that the pseudo-spectral WLe prediction is everywhere polluted by wiggles and yields predictions that are far from true equilibrium; meanwhile, the Galerkin WLe scheme predicts a constant solution $X(\xi) = \sqrt{15/35}$ (not shown). With the WHa scheme, on the other hand, the correct result is obtained. Naturally, the robustness of the WHa predictions should be carefully exploited, so as to ensure that the process is adequately resolved and that statistical moments are accurately computed. This can be achieved by systematic refinement of resolution level, as performed in the example above.

8.2.2 Rayleigh-Bénard Instability

In this section, we address a more complex problem that consists of a stochastic Rayleigh-Bénard flow in the Boussinesq limit, with random data in the neighborhood of the critical point. Specifically, we consider a closed rectangular 2D cavity of height $\tilde{H}$ and length $\tilde{L}$. The cavity is filled with air (Pr = 0.7), and has aspect ratio $A \equiv \tilde{L}/\tilde{H} = 2$ (tildes indicate dimensional quantities). Gravity points downward, and the bottom wall of the cavity is maintained at a hot temperature $\tilde{T}_h$ while the top wall is maintained at a cold temperature $\tilde{T}_c$. The vertical thermal gradient can also be characterized in terms of the reference temperature $\tilde{T}_{\text{ref}} \equiv (\tilde{T}_h + \tilde{T}_c)/2$, and the temperature difference $\Delta\tilde{T}_{\text{ref}} \equiv \tilde{T}_h - \tilde{T}_c$. The vertical walls are assumed adiabatic.

Deterministic system: In the deterministic case, the governing equations for the flow are identical to those given in Sect. 6.2.1, namely (6.42)–(6.44). We thus limit the present discussion to specify the boundary conditions in terms of non-dimensional quantities.

Let Ω denote the computational domain, $\partial\Omega$ its boundary. We denote by $\partial\Omega_c$ the cold wall, $\partial\Omega_h$ the hot wall, and $\partial\Omega_v$ the vertical walls; we have $\partial\Omega = \partial\Omega_h \cup \partial\Omega_c \cup \partial\Omega_v$. Using this notation, the boundary conditions are expressed as:

$$\mathbf{u}(\mathbf{x}, t) = 0 \quad \forall \mathbf{x} \in \partial\Omega, \qquad T(\mathbf{x}, t) \equiv T_h = \frac{1}{2} \quad \forall \mathbf{x} \in \partial\Omega_h, \tag{8.29}$$

$$T(\mathbf{x}, t) \equiv T_c = -\frac{1}{2} \quad \forall \mathbf{x} \in \partial\Omega_c, \qquad \frac{\partial T(\mathbf{x}, t)}{\partial x} = 0 \quad \forall \mathbf{x} \in \partial\Omega_v. \tag{8.30}$$

The stability of the Rayleigh-Bénard problem summarized above has been extensively analyzed. In particular, results [59, 80] reveal the existence of a critical value, Ra_c, of the Rayleigh number, below which the flow is stable. In this regime, the fluid velocity vanishes identically, and the temperature distribution exhibits a linear variation vertically across the cavity, which is representative of a purely conductive system. Above the critical value, the flow is unstable and the growth of the instability leads to the establishment of recirculation zones, which results in enhanced heat transfer across the cavity.

Stochastic formulation: We consider a stochastic variant of the Rayleigh-Bénard problem, namely under deterministic cold-wall (normalized) temperature T_c but random hot-wall temperature T_h. The statistics of T_h are assumed to be such that both stable and unstable behavior occur with finite probability. We model the uncertainty by decomposing T_h as $T_h(\theta) \equiv T_h(\xi(\theta)) = \frac{1}{2} + T_r \xi(\theta)$. (In the stochastic case, the quantities $\Delta\tilde{T}_{\text{ref}}$ and $\tilde{T}_{\text{ref}}$ appearing in the normalization of the governing equations are defined using the mean hot wall temperature, $\langle \tilde{T}_h \rangle$.) Thus, T_r characterizes the random fluctuations of the normalized temperature around its mean. The random variable ξ is assumed to be uniformly distributed on the interval $[-1, 1]$.

Similar to the approach of the previous subsection, both the WLe and WHa expansions are implemented. Thus, in the former case the solution is represented in terms of Legendre polynomials $\mathcal{L}e_k$, and in the latter in terms of Haar wavelets, $\mathcal{H}a_k$. The governing equations for the velocity, temperature, and pressure modes are identical to those given in Sect. 6.2.3, namely (6.51)–(6.53), the only differences residing in the definition of the multiplication tensor. Thus, to complete the formulation, it is sufficient to specify boundary conditions for velocity and temperature.

Since we are dealing with a closed cavity, the velocity boundary conditions on $\partial\Omega$ are $\mathbf{u}_i = 0$ for $i = 0, \ldots, \mathrm{P}$. For the scaled temperature, we have:

$$T_0 = -\frac{1}{2}, \qquad T_{i=1,\ldots,P} = 0 \quad \forall \mathbf{x} \in \partial\Omega_c, \tag{8.31}$$

$$T_i = \frac{\langle T_h(\xi)\Psi_i(\xi)\rangle}{\langle \Psi_i \Psi_i \rangle} \quad \text{for } i = 0, \ldots, P \ \forall \mathbf{x} \in \partial\Omega_h, \tag{8.32}$$

$$\frac{\partial T_i}{\partial x} = 0 \quad \text{for } i = 0, \ldots, P \ \forall \mathbf{x} \in \partial\Omega_v. \tag{8.33}$$

Numerical method and baseline results: The governing equations (6.51)–(6.53) are integrated using the stochastic projection method (SPM) described in Sect. 6.2. The present generalization primarily concerns the adaptation of the multiplication tensor to the selected basis function expansion.

Prior to performing stochastic simulations, deterministic computations were performed (simply by setting $\mathrm{P} = 0$). We set $\mathrm{Ra} = 2150$, *i.e.* the Rayleigh number is slightly larger than critical. The simulations were then performed by perturbing the hot wall temperature so as to determine the critical conditions for the instability. In these computations, the initial condition consists of the purely conductive solution, which is perturbed using a low-energy white-noise perturbation. As illustrated in Fig. 8.9, after a short time the kinetic energy exhibits an exponential growth or decay, depending on the (perturbed) value of the hot wall temperature. The critical temperature is determined by computing the growth rate for different values of T_h, as shown in Fig. 8.9. The curve is then interpolated in order to locate the value where the growth rate vanishes. The results indicate that, for the present parameters, a critical value $T_{\text{crit}} = 0.4301$ is obtained.

Consequently, for the presently selected conditions, the overall heat transfer corresponds to the purely conductive solution whenever $T_h \le T_{\text{crit}}$. The rate of heat transfer is characterized using the Nusselt number:

$$\mathrm{Nu} \equiv \frac{1}{A(T_h - T_c)} \int_0^A \left. \frac{\partial T}{\partial y} \right|_{y=0} \mathrm{d}x.$$

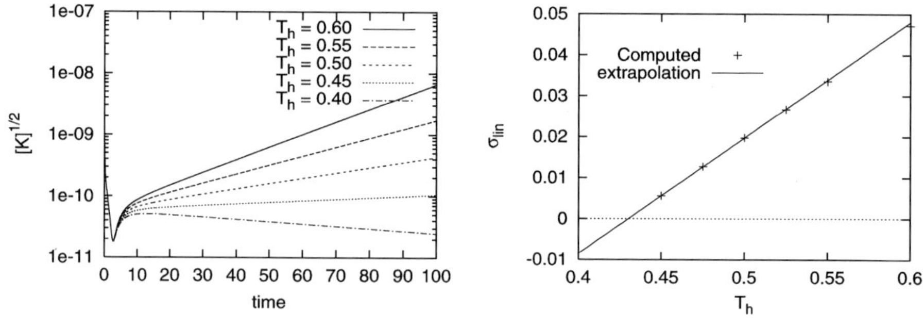

Fig. 8.9 *Left*: Square root of kinetic energy versus time for the linearized Rayleigh-Bénard problem. Plotted are curves for different hot-wall temperatures. *Right*: linear growth rate versus hot-wall temperature. The critical value, determined by linear extrapolation, is $T_{\text{crit}} = 0.4301$. The computations were performed on a uniform mesh having 60 grid points in the x-direction and 30 grid points in the y-direction. Adapted from [124]

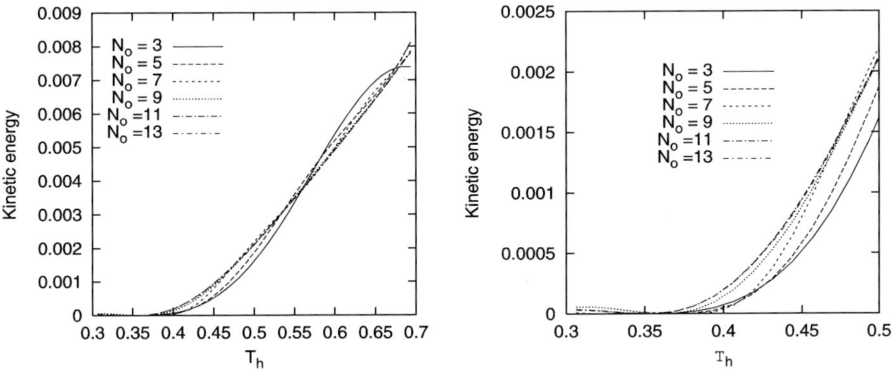

Fig. 8.10 Steady-state kinetic energy versus hot wall temperature using WLe expansions with No = 3, 5, 7, 9, 11 and 13. The *right plot* shows a detailed view in the neighborhood of the critical value. Adapted from [124]

Clearly, in the stable (conductive) regime, Nu = 1. For $T_h > T_{\text{crit}}$, heat transfer enhancement occurs so that $\text{Nu}(T_h) > 1$. Thus, the difference $\delta\text{Nu}(T_h) \equiv \text{Nu}(T_h) - 1$ provides a measure of the heat transfer enhancement.

8.2.2.1 WLe Expansion

We now apply the WLe scheme to simulate the response of the stochastic flow. As mentioned earlier, we focus on the case of uncertain hot wall temperature, which is assumed to be uniformly distributed in the range $[0.3, 0.5]$. Figure 8.10 shows kinetic energy as a function of T_h for different values of No. The curves are reconstructed from the steady-state WLe coefficients. The results indicate that the curves approach each other as No increases, which suggests that the WLe computations are converging. Unfortunately, individual realizations obtained for $T_h < T_{\text{crit}}$ do not exhibit a vanishing kinetic energy, as one would expect based on the stability considerations above. In other words, if the WLe computations are in fact converging, they do not appear to be converging to the exact solution.

In order to gain additional insight into the behavior of WLe predictions, we plot in Fig. 8.11 curves for δNu as function of $T_h(\xi)$. As before, results are generated for different expansion orders, No. As for the kinetic energy, the observed behavior of δNu appears to converge with increasing No, but not to the exact solution. Furthermore, at low values of T_h, an unphysical effect, corresponding to negative values of δNu, can be observed. Specifically, the negative values of δNu indicate that, for the corresponding realizations, an overall heat transfer rate is predicted that is *smaller* than that of the conductive solution! This unphysical response occurs over a substantial band of possible realizations, which extends over about 25% of the entire range of possible realizations. The origin of the unphysical response is further analyzed in Figs. 8.12 and 8.13, which show the steady-state velocity and temperature fields (reconstructed from the WLe expansion) for selected values of T_h. In particular, the

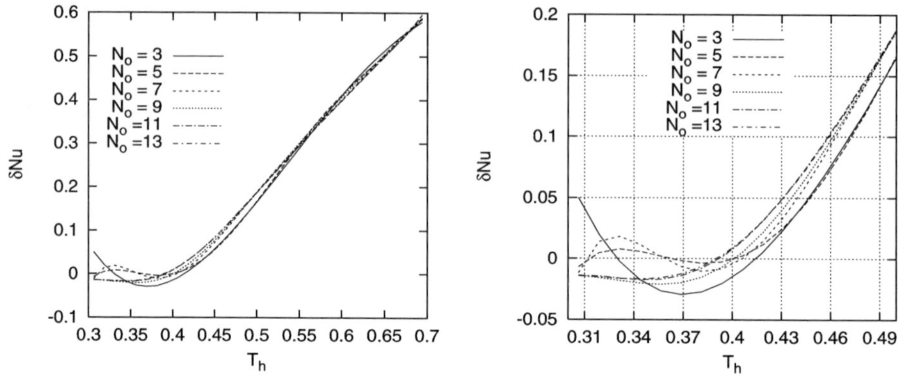

Fig. 8.11 $\delta\text{Nu}(T_h)$ versus T_h using WLe expansions with No = 3, 5, 7, 9, 11 and 13. The *right plot* shows a detailed view in the neighborhood of the critical value. Adapted from [124]

Table 8.2 Mean and standard deviation of the overall heat transfer rate across the cavity. Shown are results obtained using the WLe expansion with different expansion order, No. Adapted from [124]

No	$\langle \int_0^A \partial T/\partial y dx \rangle$	Standard deviation
1	2.24405	0.4750
3	2.23657	0.4667
5	2.23929	0.4647
7	2.24254	0.4619

figure shows that for sub-critical values of T_h, instead of vanishing, the predicted velocity exhibits a recirculating flow pattern with a *reverse* sign. Thus, the "energy leakage" that was earlier observed in Fig. 8.10 for small T_h, is accompanied by a severe breakdown of the WLe prediction.

Figure 8.14 shows the pdf of δNu for WLe expansion with No = 3, 5, 7 and 9. The results illustrate the difficulties of the computations in approaching the exact solution, which should exhibit a singular spike at $\delta\text{Nu} = 0$. Another symptom of the inefficiency of the WLe expansion for the present problem is the loss of spectral convergence. This can be appreciated from Table 8.2, which provides the mean values of overall heat transfer rate, $\langle \int_0^A \partial T/\partial y \, \mathrm{d}x \rangle$, and of the corresponding standard deviation for different No.

The present experiences indicate that for problems involving bifurcations or loss of smoothness with respect to the random data, the WLe expansion may be essentially impractical. Similar limitations are expected for other spectral representations [245] based on global basis functions.

8.2.2.2 WHa Expansion

In this section, we apply the WHa scheme to compute the stochastic, steady-state response of the Rayleigh-Bénard flow. We start by examining global properties of the flow field, and then analyze the stochastic velocity and temperature distributions.

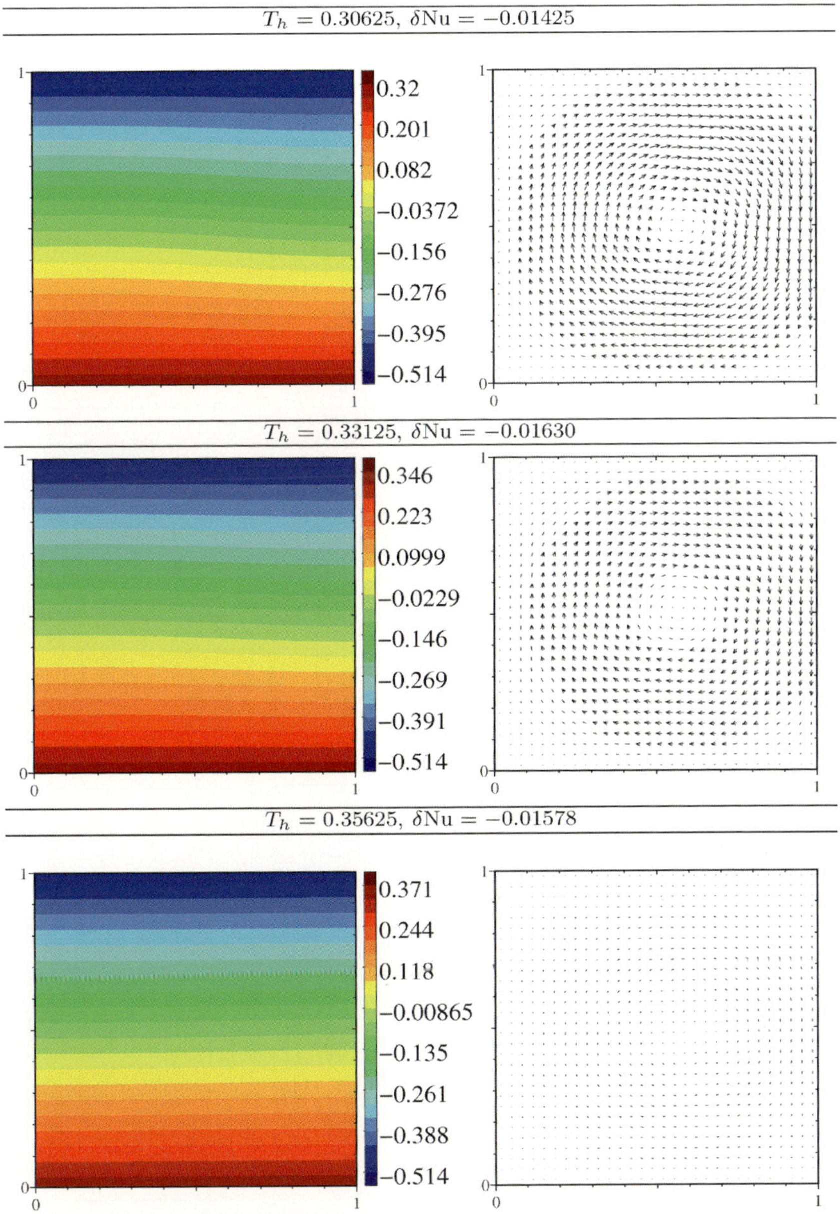

Fig. 8.12 Individual realizations of the steady-state temperature and velocity distributions, as predicted using a WLe expansion with No = 9. The selected values of T_h are indicated. Due to symmetry with respect to the mid vertical plane, only the left half of the cavity is plotted. Adapted from [124]

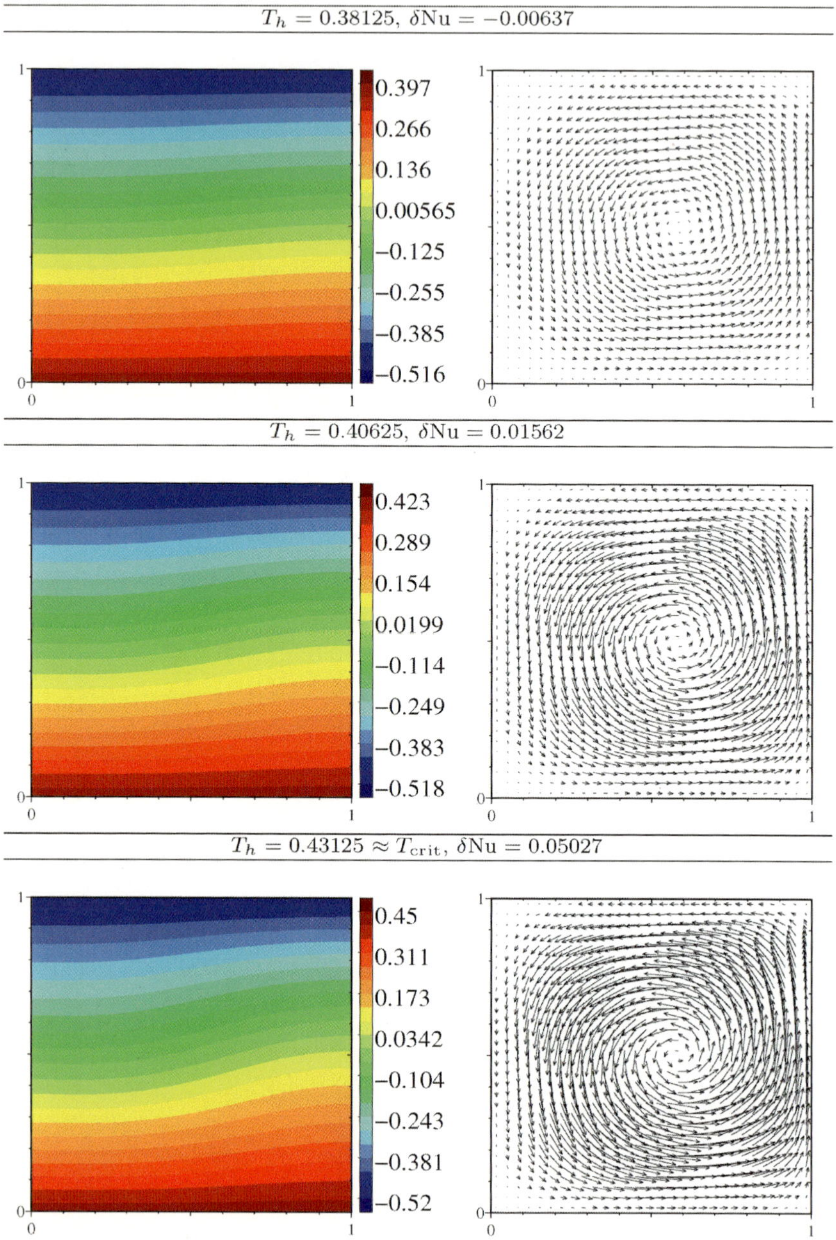

Fig. 8.13 Individual realizations of the steady-state temperature and velocity distributions, as predicted using a WLe expansion with $\text{No} = 9$. The selected values of T_h are indicated. Due to symmetry with respect to the mid vertical plane, only the left half of the cavity is plotted. Adapted from [124]

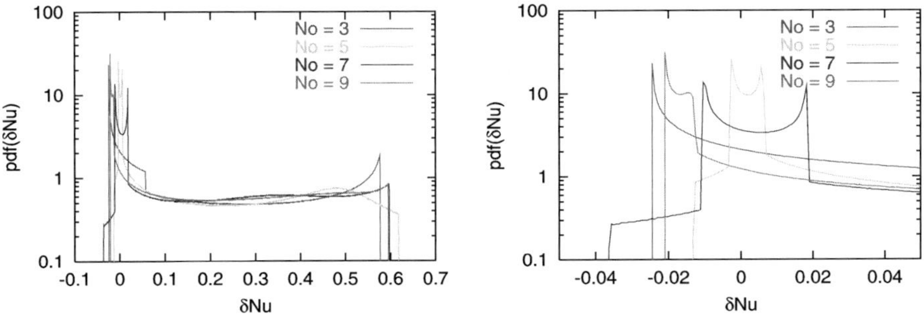

Fig. 8.14 Probability density functions of δNu computed using WLe expansions with No = 3, 5, 7 and 9. The *right plot* shows details of the left tail of the pdf's. Adapted from [124]

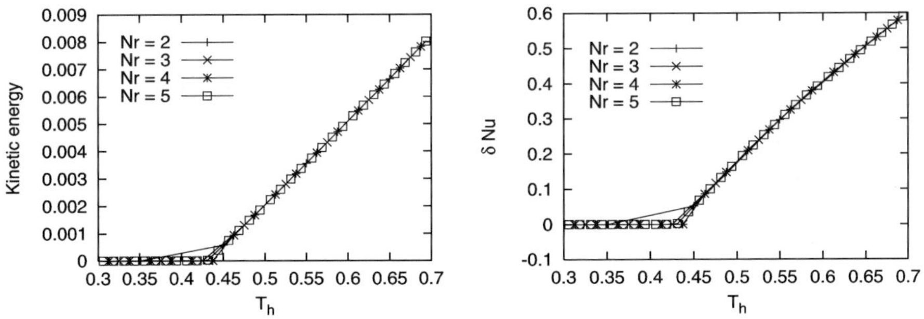

Fig. 8.15 Kinetic energy (*left*) and $\delta\mathrm{Nu}(T_h)$ (*right*) versus hot-wall temperature using WHa expansions with Nr = 2, 3, 4 and 5. Adapted from [124]

Kinetic energy and heat transfer: Figure 8.15 provides curves of the kinetic energy and δNu plotted against T_h. The curves are reconstructed based on the wavelet coefficients, and results are shown for expansion using Nr = 2, 3, 4 and 5. As far as these integral measures are concerned, the results indicate that the WHa scheme is much better adapted than the WLe scheme at capturing the transition between conductive and convective regimes, even when a coarse stochastic discretization with Nr = 2 is considered. As Nr increases, the computations provide an increasingly accurate estimate of the location of the critical point, as illustrated in Fig. 8.16. The latter provides an enlarged view of the local behavior of $\delta\mathrm{Nu}(T_c)$ near the critical point, as well as the dependence of δNu on ξ. In the latter format, the results illustrate the piecewise constant nature of the WHa expansion, as well as the essential concept of the approximation scheme which relies on projecting the solution onto the space $\mathbf{V}^{\mathrm{Nr}}$. The robustness of the WHa expansion in capturing the bifurcation can be appreciated by noting that the reconstructed curves capture the correct behavior on both sides of the critical point. Specifically, as predicted by the theory and confirmed by the perturbation analysis above, vanishing values of the kinetic energy and of δNu are predicted for subcritical values of T_h, while for supercritical values, an essentially linear increase of δNu with T_h is observed.

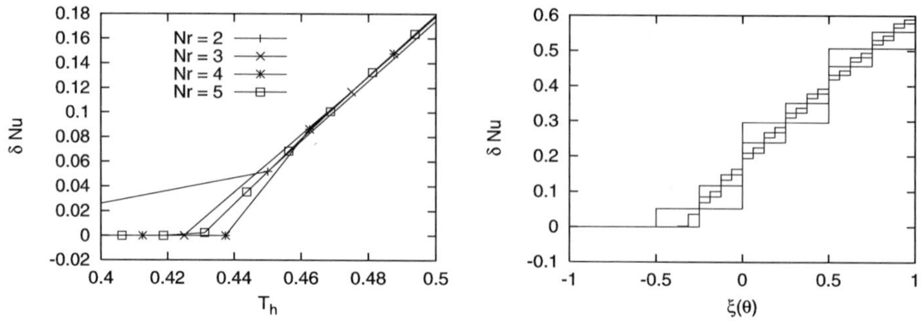

Fig. 8.16 *Left*: detailed view of δNu versus T_h in the neighborhood of the critical value. *Right*: detailed view of δNu versus ξ near the critical value. Plotted are curves obtained using WHa expansions with Nr = 2, 3, 4 and 5. Adapted from [124]

Table 8.3 Mean and standard deviation of the overall heat transfer rate across the cavity. Shown are results obtained using the WHa expansion with different Nr. Adapted from [124]

Nr	$\langle \int_0^A \partial T/\partial y \, dx \rangle$	Standard deviation
1	2.22300	0.4230
2	2.23588	0.4524
3	2.23791	0.4627
4	2.23795	0.4653
5	2.23906	0.4652

To gain further appreciation of the robustness of the WHa scheme, Table 8.3 provides the mean values of the overall heat transfer rate, $\langle \int_0^A \partial T/\partial y \, dx \rangle$, and of the corresponding standards deviation for different Nr. The results indicate that the predictions are close to one another and tend to cluster as Nr increases.

Detail distributions of velocity and temperature: Figure 8.17 shows the computed velocity fields corresponding to the detail coefficients. Recall that for the present 1D WHa expansion, the velocity is expressed in terms of the details $\mathbf{u}_{j,k}$ according to:

$$\mathbf{u}(\mathbf{x},\xi) = \sum_{k=0}^{2^{\mathrm{Nr}}-1} \mathbf{u}_k(\mathbf{x}) \mathcal{H}a_k(\xi) \equiv \mathbf{u}_0 + \sum_{j=1}^{\mathrm{Nr}} \sum_{k=0}^{2^{j-1}-1} \mathbf{u}_{j,k} \psi^{\mathrm{w}}_{j,k}(\mathrm{p}(\xi)).$$

In the left column of Fig. 8.17, we show the mean field $\mathbf{u}_0(\mathbf{x}) \equiv \langle \mathbf{u} \rangle$; in the second column, the detail corresponding to the difference of the mean with $\mathbb{P}_1\mathbf{u}$. As expected, the latter indicates that the circulation inside the cavity increases with ξ, and accordingly with T_h. The second detail fields, which are plotted in the third column, correspond to differences between $\mathbb{P}_2\mathbf{u}$ and $\mathbb{P}_1\mathbf{u}$; they reveal a similar recirculating pattern as the first detail and point to a similar trend. More interesting trends can be observed from the detail fields at the next level, $j = 3$. Specifically, while the details corresponding to the highest values of T_h (appearing on top) exhibit similar patterns and trends as those for $j = 2$ and $j = 1$, those corresponding to the low

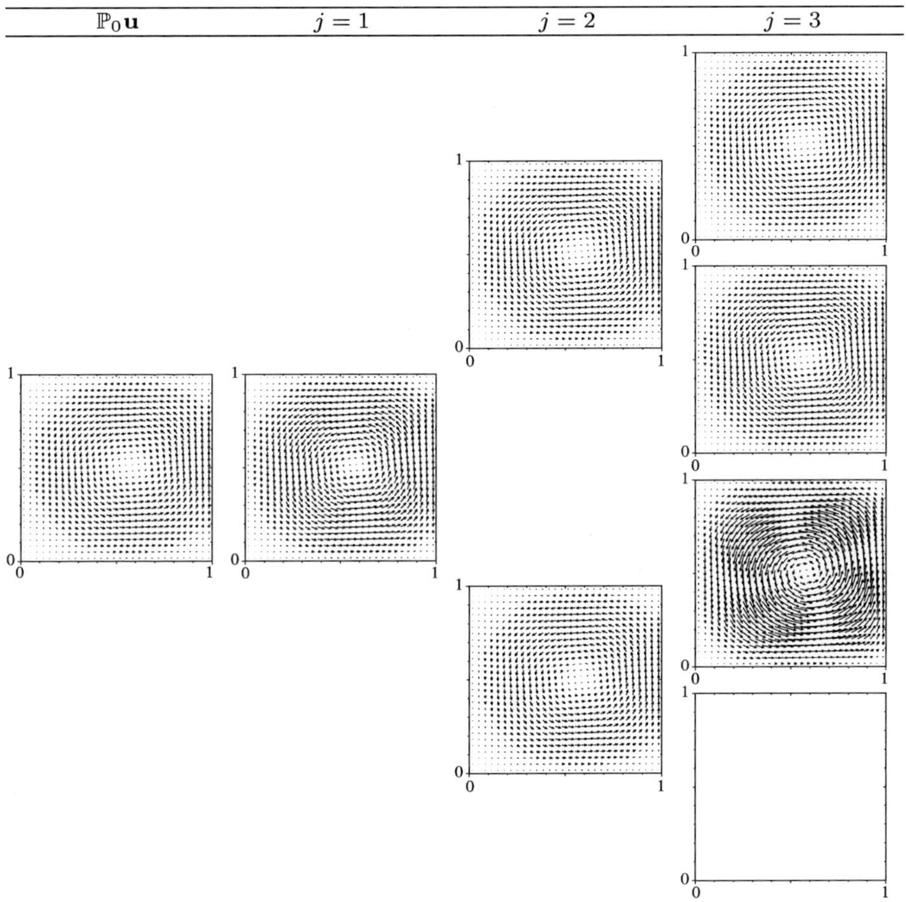

Fig. 8.17 Mean velocity field ($\mathbb{P}_0\mathbf{u}$, *left*) and first wavelet modes, $\mathbf{u}_{j,k}$ using a WHa expansion with $\mathrm{Nr} = 5$. The scale indices, j, are indicated. At each scale index, frames for different space indices $k = 0, \ldots, 2^{j-1} - 1$ are plotted, and are arranged *from bottom to top*. The magnitude of the vector is normalized using a factor equal to 2^{j+1}. Due to symmetry with respect to the mid vertical plane, only the left half of the cavity is plotted. Adapted from [124]

values have different structure. Specifically, for $k = 0$, the field vanishes, indicating that no correction is needed for the corresponding "realizations". Furthermore, the detail for $k = 1$ has larger magnitude than the other details at the same level; this increased "activity" coincides with the location of the transition between conductive and convective regimes. On the next detail level, Fig. 8.18 shows that no corrections are needed for the velocity fields at the lowest two temperature bands, whereas the corrections corresponding to the remaining fields reflect a trend of increasing circulation with higher temperatures. Similar trends can be observed by inspections of the details at the last level, $j = 5$, (not shown).

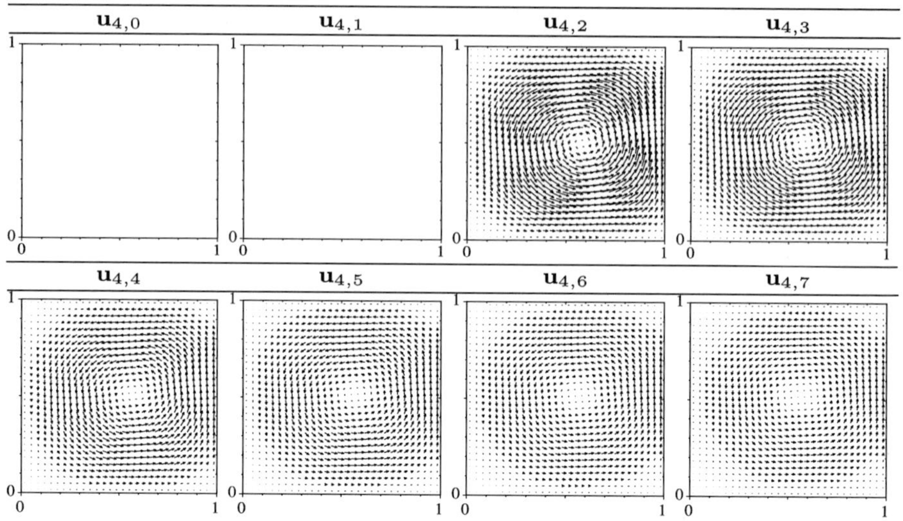

Fig. 8.18 Wavelet modes for the velocity field at a scale index $j = 4$. The magnitude of the vector is normalized using a factor equal to 2^{j+1}. Results are obtained using a WHa expansion with $\mathrm{Nr} = 5$. The space index k is indicated, and only the left half of the cavity is plotted. Adapted from [124]

The stochastic temperature field is analyzed following the same approach used for the velocity. In Fig. 8.19, we plot distributions of the mean velocity as well as the details for levels $j = 1, 2$, and 3; details at the next level and the highest level, $j = 4$ and $j = 5$, are plotted in Figs. 8.20 and 8.21, respectively. Note that, unlike the velocity field, the detail distributions for temperature do not vanish, even at the highest resolution level. This is the case because inhomogeneous boundary conditions prevail at the hot wall, regardless of whether the flow is stable or not. Nonetheless, the transition from a conductive to a convective regime can still be detected. In the former case, the distributions are characterized by parallel horizontal contour lines while severe distortions of this pattern occur as the flow transitions. As for the velocity field, the results indicate that the transition is first captured at $j = 3$; at this level, the detail corresponding to $k = 0$ exhibits flat horizontal contours, while the detail temperature fields for $k > 0$ exhibit a distorted pattern that is characteristic of a recirculating flow field. It is interesting to note that at higher levels (Figs. 8.19–8.21), the detail distributions corresponding to supercritical temperature values exhibit a similar spatial distribution. This suggests that in this situation the higher levels primarily introduce an amplitude correction to the prevailing recirculating flow.

8.2.2.3 Continuous Problem

The results of the previous sections demonstrate that the WHa expansion provides a robust and well-suited approach for analyzing stochastic processes involving bifurcations or discontinuous dependence on random data. On the other hand, when the

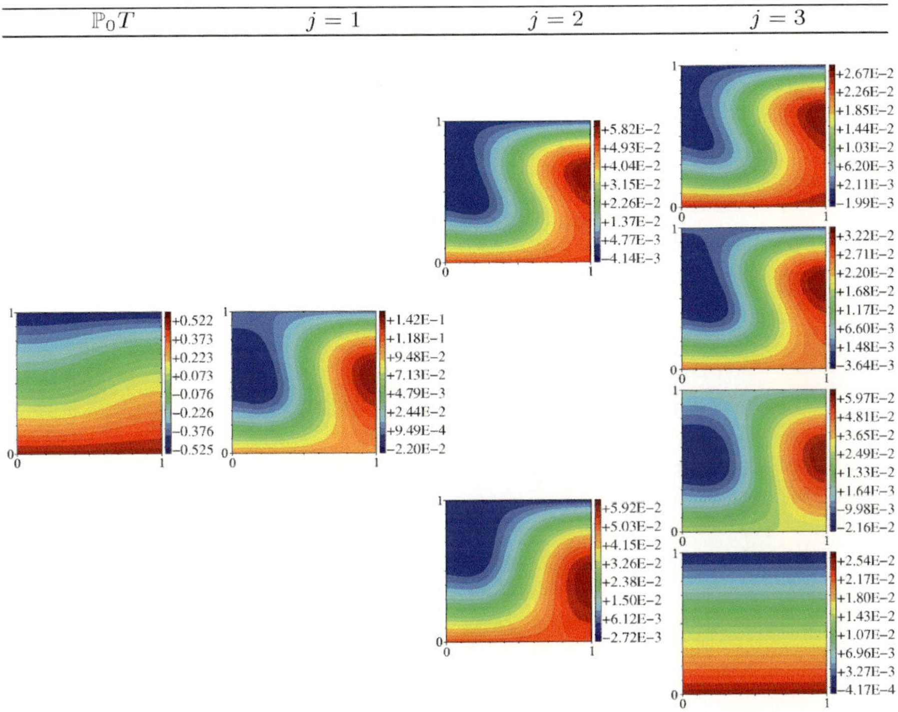

Fig. 8.19 Mean temperature field ($\mathbb{P}_0 T$, *left*) and first wavelet modes, $T_{j,k}$ using a WHa expansion with Nr $= 5$. The scale indices, j, are indicated. At each scale index, frames for different space indices $k = 0, \ldots, 2^{j-1} - 1$ are plotted, and are arranged from bottom to top. Due to symmetry with respect to the mid vertical plane, only the left half of the cavity is plotted. Adapted from [124]

process depends smoothly on the random data, global spectral expansions are expected to be substantially more efficient than wavelet representations [246]. Specifically, for spectral representations a fast, "infinite-order" convergence is expected, while for the Haar representation errors are expected to decay as the inverse of Nr.

We briefly illustrate the convergence of the WHa and WLe schemes for a problem involving a smooth dependence on the random data. To this end, we consider once again the same stochastic Rayleigh-Bénard problem, but increase the Rayleigh number to Ra $= 3000$. For this value of the Rayleigh number, the convective regime always prevails as all possible realizations of T_h are larger than the critical value.

Table 8.4 shows predictions of the mean overall heat transfer and of the corresponding standard deviation, using the WHa scheme with increasing Nr and the WLe scheme with increasing No. The results indicate that the WLe predictions rapidly become independent of No; in particular, identical predictions of the mean heat transfer and its standard deviation are obtained with No $= 4$ and 5. The WHa predictions also appear to be converging as Nr increases, though at an appreciably smaller rate. To gain additional insight into the convergence of both predictions,

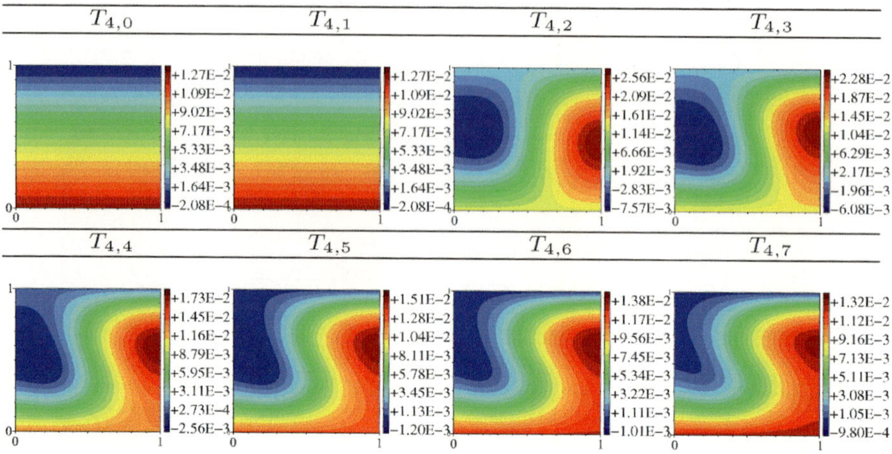

Fig. 8.20 Wavelet modes for the temperature field at a scale index $j = 4$. Results are obtained using a WHa expansion with Nr $= 5$. The space index k is indicated, and only the left half of the cavity is plotted. Adapted from [124]

Table 8.4 Mean and standard deviation of the overall heat transfer rate across the cavity. Shown are results obtained using the WHa expansion with different resolution levels, Nr, and the WLe scheme with different values of No. The Rayleigh number Ra $= 3000$, and unstable conditions prevails for all possible realizations of the hot wall temperature. Adapted from [124]

Nr	$\langle \int_0^A \partial T/\partial y \, dx \rangle$	Standard deviation
1	2.92239879	0.49257723
2	2.92350067	0.55054599
3	2.92377786	0.56409585
4	2.92384732	0.56743188
5	2.92386463	0.56826288
6	2.92386897	0.56847041
No	$\langle \int_0^A \partial T/\partial y \, dx \rangle$	Standard deviation
1	2.92384455	0.56871803
2	2.92387023	0.56854003
3	2.92387042	0.56853954
4	2.92387042	0.56853957
5	2.92387042	0.56853957

we plot in Fig. 8.22 an approximate error estimate, defined as the absolute value of the difference between a given prediction and the WLe result with No $= 5$. Thus, in these estimates, the 5th-order WLe solution is used as surrogate for the exact solution. The results of Fig. 8.22 illustrate the fast decay of the error for the WLe scheme. For the WHa scheme, the error also decays with increasing Nr, though at an appreciably smaller rate. The moderate rate of convergence of the WHa computations can also be appreciated in Fig. 8.23, which depicts the computed distributions of δNu for different values of Nr. The results illustrate the "staircase" Haar approx-

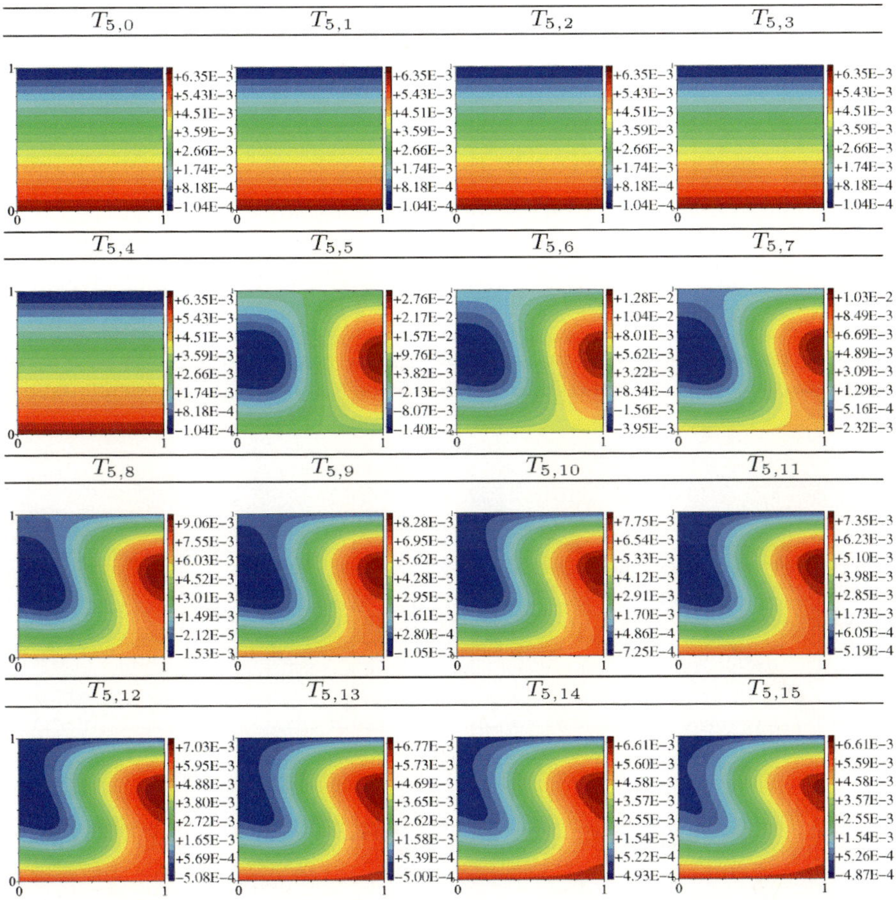

Fig. 8.21 Wavelet modes for the temperature field at a scale index $j = 5$. Results are obtained using a WHa expansion with $\mathrm{Nr} = 5$. The space index k is indicated, and only the left half of the cavity is plotted. Adapted from [124]

imation of the continuous curve expressing the dependence of $\delta\mathrm{Nu}$ on T_h and on ξ. Thus, for the present conditions, the WLe scheme outperforms the WHa scheme.

8.3 Multiresolution Analysis and Multiwavelet Basis

In this section, we shall focus on extending the WHa formulation introduced earlier to multiwavelet (MW) basis function expansion. To this end, we consider a generic stochastic process $U(\boldsymbol{\xi})$, whose evolution is governed by:

$$\mathcal{O}(U(\boldsymbol{\xi}), \boldsymbol{\xi}) = 0, \tag{8.34}$$

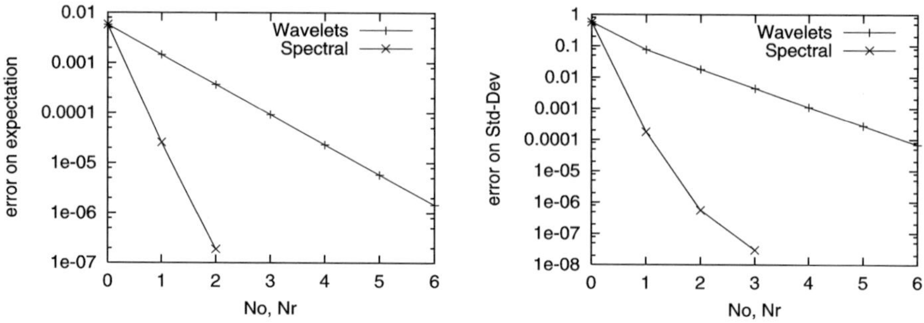

Fig. 8.22 Convergence of the expected (*left*) and standard deviation (*right*) of Nu with increasing No for the WLe expansion and Nr for the WHa expansion. The errors is estimated using the absolute value of the difference between the solution and the WLe prediction with No = 5. Adapted from [124]

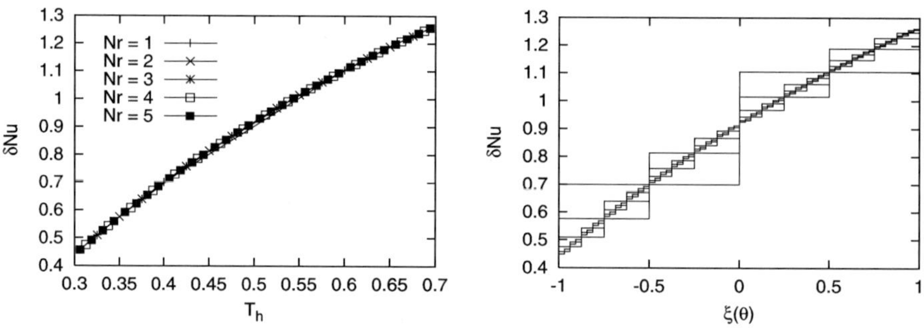

Fig. 8.23 δNu versus T_h (*left*) and versus ξ (*right*) using WHa expansions with Nr = 1, 2, 3, 4 and 5. Adapted from [124]

where $\mathcal{O}$ is a (deterministic) nonlinear operator and $\boldsymbol{\xi}$ is the random coefficient vector with independent components. To simplify the presentation, we shall first focus on the case of a single uncertain parameter, ξ; generalization to the multidimensional case is then addressed in Sect. 8.3.4.

8.3.1 Change of Variable

Let $F_\eta(y)$ denote the distribution function, giving the probability $P(\xi \le y)$. As for the Haar expansion in the previous section, we assume that F_ξ is a continuous, monotonically increasing function of y over the interval (a,b), $-\infty \le a < b \le \infty$, and that $F_\xi(a) = 0$ and $F_\xi(b) = 1$. Based on the assumed properties of $F_\xi(y)$, it follows that for all $x \in [0, 1]$ there is a unique $y \in [a, b]$ such that $F_\xi(y) = x$. In addition, if η is a uniformly distributed random variable on [0, 1], then $F_\xi^{-1}(\eta)$ is a random variable on (a, b) having the same distribution as ξ [94]. Consequently, instead of expanding the random process in terms of ξ, we alternatively develop a representation with respect to η.

8.3.2 *Multiresolution Analysis*

In this section, we recall some properties of the multiwavelet bases introduced by Alpert in [3] (see also [4]). The application to the representation of random processes is considered in Sect. 8.3.3.

8.3.2.1 Vector Spaces

For $\mathrm{No} = 0, 1, \ldots$ and $k = 0, 1, 2, \ldots,$ we define the space $\mathbf{V}_k^{\mathrm{No}}$ of piecewise-continuous polynomials, according to:

$$\begin{aligned}\mathbf{V}_k^{\mathrm{No}} \equiv \{f :\ & \text{the restriction of } f \text{ to the interval } (2^{-k}l, 2^{-k}(l+1)) \\ & \text{is a polynomial of degree} \le \mathrm{No}, \text{ for } l = 0, \ldots, 2^k - 1 \\ & \text{and } f \text{ vanishes outside the interval } [0,1]\}.\end{aligned} \tag{8.35}$$

Thus, $\mathbf{V}_k^{\mathrm{No}}$ has dimension $(\mathrm{No}+1)(2^k)$ and $\mathbf{V}_0^{\mathrm{No}} \subset \mathbf{V}_1^{\mathrm{No}} \subset \cdots \subset \mathbf{V}_k^{\mathrm{No}} \subset \cdots$. Denoting $\mathbf{V}^{\mathrm{No}}$ the union all of spaces $\mathbf{V}_k^{\mathrm{No}}$, $\mathbf{V}^{\mathrm{No}} = \overline{\bigcup_{k\ge 0} \mathbf{V}_k^{\mathrm{No}}}$ [3], we remark that $\mathbf{V}^{\mathrm{No}}$ is dense in $L_2([0,1])$ with respect to the norm $\|f\|^2 = \langle f, f\rangle_{[0,1]}$ where

$$\langle f, g\rangle_{[0,1]} = \int_0^1 f(x)g(x)\mathrm{d}x. \tag{8.36}$$

The multiwavelets (MW) subspace $\mathbf{W}_k^{\mathrm{No}}$, $k = 0, 1, 2, \ldots$ is defined as the orthogonal complement of $\mathbf{V}_k^{\mathrm{No}}$ in $\mathbf{V}_{k+1}^{\mathrm{No}}$; we write:

$$\mathbf{V}_k^{\mathrm{No}} \oplus \mathbf{W}_k^{\mathrm{No}} = \mathbf{V}_{k+1}^{\mathrm{No}}, \quad \mathbf{W}_k^{\mathrm{No}} \perp \mathbf{V}_k^{\mathrm{No}}. \tag{8.37}$$

From this construction, we have:

$$\mathbf{V}_0^{\mathrm{No}} \bigoplus_{k\ge 0} \mathbf{W}_k^{\mathrm{No}} = L_2([0,1]). \tag{8.38}$$

8.3.2.2 Multiwavelet Basis

An orthonormal basis, $\{\psi_0, \psi_1, \ldots, \psi_{\mathrm{No}}\}$, of $\mathbf{W}_0^{\mathrm{No}}$ is introduced. The ψ_i's are piecewise polynomial functions of degree less than or equal to No. From the orthonormality condition, we have:

$$\langle \psi_i, \psi_j\rangle_{[0,1]} = \delta_{ij}. \tag{8.39}$$

Since $\mathbf{W}_0^{\mathrm{No}} \perp \mathbf{V}_0^{\mathrm{No}}$, the first $\mathrm{No}+1$ moments of the ψ_i vanish, *i.e.*

$$\left\langle \psi_j, x^i \right\rangle_{[0,1]} = 0, \quad 0 \le i, j \le \mathrm{No}. \tag{8.40}$$

As further discussed below, (8.39) and (8.40) result in a system of polynomial equations whose solution yields the $(\mathrm{No}+1)$ ψ_i functions.

The space $\mathbf{W}_k^{\mathrm{No}}$, whose dimension is $(\mathrm{No}+1)(2^k)$, is spanned by the multiwavelets, ψ_{jl}^k, which are translated and dilated version of the ψ_i's. The ψ_{jl}^k are given by:

$$\psi_{jl}^k(x) = 2^{k/2}\psi_j(2^k x - l), \quad j = 0, \dots, \mathrm{No}, \text{ and } l = 0, \dots, 2^k - 1 \tag{8.41}$$

and their support is $\mathrm{Supp}(\psi_{jl}^k) = [2^{-k}l, 2^{-k}(l+1)]$. Due to the orthonormality of the ψ_i's, we have:

$$\left\langle \psi_{il}^k, \psi_{jm}^{k'} \right\rangle_{[0,1]} = \delta_{ij}\delta_{lm}\delta_{kk'}. \tag{8.42}$$

A basis $\{\phi_0, \dots, \phi_{k-1}\}$ for $\mathbf{V}_0^k$ is also constructed. Rescaled Legendre polynomials are used for this purpose. Letting $\mathcal{L}e_i$ denote the Legendre polynomial [1] of degree i, defined over $[-1, 1]$, we set:

$$\phi_i(x) = \frac{\mathcal{L}e_i(2x-1)}{L_i}, \quad i = 0, 1, \dots, \mathrm{No}, \tag{8.43}$$

where L_i is a normalization factor selected such that:

$$\langle \phi_i, \phi_j \rangle_{[0,1]} = \delta_{ij} \quad \text{for } i, j = 0, \dots, \mathrm{No}. \tag{8.44}$$

The space $\mathbf{V}_k^{\mathrm{No}}$, whose dimension is $2^k(\mathrm{No}+1)$, is spanned by the polynomials ϕ_{il}^k,

$$\phi_{il}^k(x) = 2^{k/2}\phi_i(2^k x - l), \quad i = 0, \dots, \mathrm{No} \text{ and } l = 0, \dots, 2^k - 1 \tag{8.45}$$

which are translated and dilated versions of the ϕ_i's.

8.3.2.3 Construction of the ψ_j's

The methodology for the constructing the polynomial functions $\psi_i(x)$, $i = 0, \dots, k-1$ satisfying (8.39), (8.40) has been proposed by Alpert [3]. It is briefly summarized as follows.

The starting point is two sets of polynomial functions $p_i(x)$, and $\tilde{q}_j(x)$, defined for $x \in [0, 1]$:

$$\begin{cases} p_i(x) = x^i & \text{for } i = 0, \dots, k-1, \\ \tilde{q}_i(x) = \begin{cases} p_i(x) & \text{if } x \ge 1/2, \\ -p_i(x) & \text{if } x < 1/2, \end{cases} & \text{for } i = 0, \dots, k-1. \end{cases} \tag{8.46}$$

Step 1: In a first step, each function of the set $\{q_i(x), i = 0, \ldots, k-1\}$ is orthogonalized with respect to all the functions $p_i(x)$ for $i = 0, \ldots, k-1$. This is equivalent to the determination of the set of coefficients α_{ij} such that

$$q_j(x) = \tilde{q}_j(x) + \sum_{i=0}^{k-1} \alpha_{ij} p_i(x), \tag{8.47}$$

solves

$$\langle q_j, p_i \rangle_{[0,1]} = 0, \quad \text{for } i, j = 0, 1, \ldots, k-1. \tag{8.48}$$

At the end of this step, the functions $\{q_j, j = 0, \ldots, k-1\}$ have k vanishing moments (so satisfy (8.40)), but are not orthogonal: $\langle q_i, q_j \rangle_{[0,1]} \neq 0$ for $i \neq j$.

Step 2: The orthogonalization of the set of functions $q_i(x)$ is enforced through the following Gram-Schmidt procedure (see for instance [221]):

1. $r_{k-1}(x) = q_{k-1}(x)$.
2. $j = k-1$
3. $j \leftarrow j-1$.
4. Orthogonalization of $q_j(x)$ with respect to the set of functions $\{r_{j+1}, \ldots, r_{k-1}\}$. This is achieved by determining the coefficients β_l so that

$$r_j(x) = q_j(x) + \sum_{l=j+1}^{k-1} \beta_l r_l(x), \tag{8.49}$$

   satisfies $\langle r_j, r_l \rangle_{[0,1]} = 0$ for $l = j+1, \ldots, k-1$.
5. If $j > 0$ continue at 3, else terminate.

Since the polynomial functions r_j, $j = 0, \ldots, k-1$ are simply linear combinations of the $q_l(x)$ functions, they also have k vanishing moments but they are now *mutually orthogonal*.

Step 3: Finally, the desired functions $\psi_i(x)$ are obtained by normalizing the $r_i(x)$,

$$\psi_i(x) = \frac{r_i(x)}{\langle r_i, r_i \rangle_{[0,1]}}. \tag{8.50}$$

The computation of the ψ_i functions can use of Gauss-Legendre quadrature rules (see Appendix B), allowing for exact estimate of the inner products $\langle f, g \rangle_{[0,1]}$ up to round-off errors which are negligible for moderate k. In Fig. 8.24, the functions $\psi_{j=0,\ldots,k-1}(x)$ are plotted for $k = 2, 3, 4$ and 6. Note that for $k = 1$, one obtains the Haar mother wavelet.

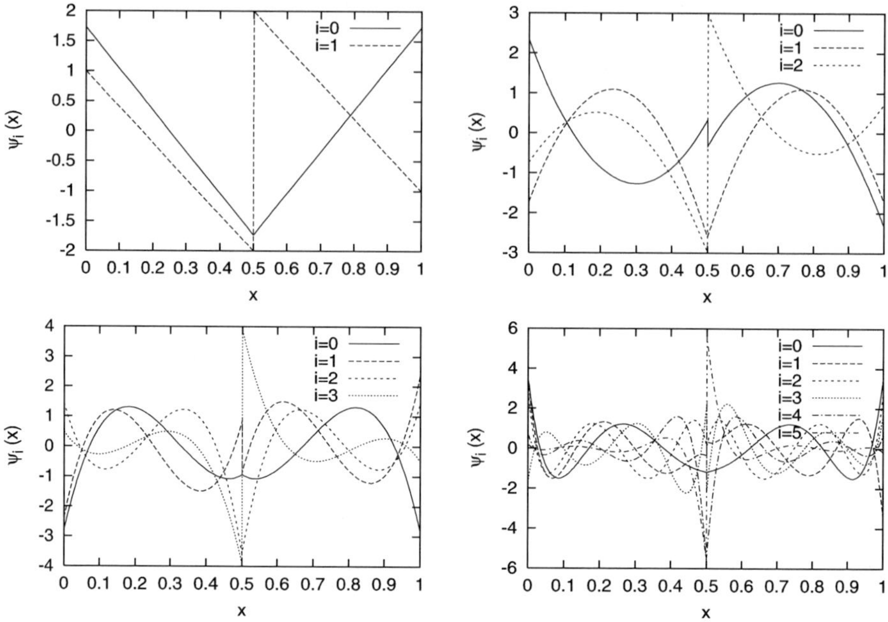

Fig. 8.24 $\psi_i(x)$ for expansion orders No $= 1, 2, 3$ and 5. Adapted from [125]

8.3.2.4 MW Expansion

A function $f(x) \in L_2([0, 1])$ can be arbitrary well approximated using the MRA scheme constructed above. We denote $f^{\mathrm{No,Nr}}$ the projection of f on $\mathbf{V}_{\mathrm{Nr}}^{\mathrm{No}}$; we have:

$$f^{\mathrm{No,Nr}} \equiv \mathbb{P}_{\mathrm{Nr}}^{\mathrm{No}}[f] = \sum_{l=0}^{2^{\mathrm{Nr}}-1} \sum_{i=0}^{\mathrm{No}} \langle \phi_{il}^{\mathrm{Nr}}, f \rangle_{[0,1]} \phi_{il}^{\mathrm{Nr}}(x)$$
$$= \sum_{l=0}^{2^{\mathrm{Nr}}-1} \sum_{i=0}^{\mathrm{No}} \overline{f}_{il}^{\mathrm{Nr}} \phi_{il}^{\mathrm{Nr}}(x). \tag{8.51}$$

An alternative expression for $f^{\mathrm{No,Nr}}$, valid for all $\mathrm{Nr} \geq 1$, in terms of multiwavelets is:

$$f^{\mathrm{No,Nr}}(x) \equiv \mathbb{P}_0^{\mathrm{No}}[f(x)] + \sum_{k=0}^{\mathrm{Nr}-1} \sum_{l=0}^{2^k-1} \left(\sum_{i=0}^{\mathrm{No}} df_{il}^k \psi_{il}^k(x) \right). \tag{8.52}$$

The MW coefficients df_{il}^k appearing in (8.52) are given by:

$$df_{il}^k = \left\langle \left\{ \mathbb{P}_{k+1}^{\mathrm{No}}[f] - \mathbb{P}_k^{\mathrm{No}}[f] \right\}, \psi_{il}^k \right\rangle_{[0,1]}. \tag{8.53}$$

Denoting by $\delta_{\text{No},\text{Nr}}$ the L_2-norm of the approximation error of f on $\mathbf{V}_{\text{Nr}}^{\text{No}}$, $\delta^2_{\text{No},\text{Nr}} \equiv \left\langle f - f^{\text{No},\text{Nr}}, f - f^{\text{No},\text{Nr}} \right\rangle_{[0,1]}$, convergence is characterized by the decay of δ with increasing polynomial order No (p convergence) or with increasing resolution levels Nr (h convergence).

8.3.3 Expansion of the Random Process

Consider a random process $U(\xi)$, where ξ is a random variable satisfying the assumption of Sect. 8.3.1. Further, we assume that $U(\xi)$ is a second-order process, *i.e.*

$$\left\langle U^2 \right\rangle = \int_a^b U(y)^2 p_\xi(y)\,\mathrm{d}y < +\infty. \tag{8.54}$$

Using the change of variables introduced in Sect. 8.3.1, we express $U(\xi)$ in terms of $\eta \sim \mathcal{U}[0,1]$ as:

$$U(\xi) = U(F_\xi^{-1}(\eta)) = \tilde{U}(\eta). \tag{8.55}$$

Introducing this change of variable in (8.54) gives

$$\int_a^b U(y)^2 p_\xi(y)\,\mathrm{d}y = \int_0^1 \tilde{U}(x)^2\,\mathrm{d}x = \left\langle \tilde{U}, \tilde{U} \right\rangle_{[0,1]} < +\infty, \tag{8.56}$$

showing that $\tilde{U} \in L_2([0,1])$. Thus, $\tilde{U}$ can be expanded according to (8.52).

Let us denote by Δ the set of index integers λ concatenating the scale index k, support index l and MW index i:

$$\Delta \equiv \{\lambda : \lambda = (\text{No}+1)(2^k + l - 1) + i; k = 0, \ldots, \infty; \\ l = 0, \ldots, 2^k - 1; i = 0, \text{No}\}. \tag{8.57}$$

The resolution level k for any $\lambda \in \Delta$ will be denoted by $|\lambda|$. Using this convention, the MW expansion of $\tilde{U}(\eta)$ can be expressed as:

$$\tilde{U}(\eta) = \mathbb{P}_0^{\text{No}}[\tilde{U}] + \sum_{\lambda \in \Delta} \tilde{U}_\lambda \psi_\lambda(\eta) = \sum_{i=0}^{\text{No}} \overline{\tilde{U}}_{i0}^{0} \phi_i(\eta) + \sum_{\lambda \in \Delta} \tilde{U}_\lambda \psi_\lambda(\eta). \tag{8.58}$$

Letting $\Delta_0 = \{-\text{No}-1, -\text{No}, \ldots, -1\}$ and $\diamond = \Delta \cup \Delta_0$, we can rewrite (8.58) as:

$$\tilde{U}(\eta) = \sum_{\lambda \in \diamond} \tilde{U}_\lambda W_\lambda(\eta), \tag{8.59}$$

where

$$\begin{cases} W_\lambda(\eta) = \psi_\lambda(\eta) & \text{for } \lambda \in \Delta, \\ W_\lambda(\eta) = \phi_{1-\lambda}(\eta) & \text{for } \lambda \in \Delta_0. \end{cases} \tag{8.60}$$

Thus, the process can be expanded as:

$$U(\xi) = \tilde{U}(\eta) = \sum_{\lambda \in \diamond} \tilde{U}_\lambda W_\lambda(\eta).$$

8.3.4 The Multidimensional Case

Extension of the 1D MW expansion to the N-dimensional case is now considered. For simplicity, we focus on a vector $\boldsymbol{\xi}$ with random components $\{\xi_1(\theta), \ldots, \xi_{\mathrm{N}}(\theta)\}$ with range $\Xi \subseteq \mathbb{R}^{\mathrm{N}}$. We assume that the components of $\boldsymbol{\xi}$ are uncorrelated and independent, so that:

$$p_\Xi(\mathbf{y}) = \prod_{d=0}^{\mathrm{N}} p_d(y_d),$$

where p_d denotes the probability density function of the ξ_d. Consistent with the 1D case, we also assume that:

$$p_d(\xi_d) = \begin{cases} > 0 & \text{if } \xi \in (a_d, b_d), \\ 0 & \text{if } \xi \notin (a_d, b_d), \end{cases} \tag{8.61}$$

so that $\forall x_d \in [0, 1]$ there is a unique $y_d \in (a_d, b_d)$ such that

$$F_d(y_d) \equiv \int_{a_d}^{y_d} p_d(y)\,\mathrm{d}y = x_d.$$

We now consider the multi-index $\boldsymbol{\lambda} = (\lambda_1, \ldots, \lambda_{\mathrm{N}})$, and define the set

$$\Lambda_k = \left\{ \boldsymbol{\lambda} : \sum_{d=1}^{\mathrm{N}} |\lambda_d| = k \right\}. \tag{8.62}$$

Let $M_k = \mathrm{Card}(\Lambda_k)$, and define the set

$$\mathcal{W}_k \equiv \left\{ \prod_{d=1}^{\mathrm{N}} W_{\lambda_d} \; : \boldsymbol{\lambda} \equiv (\lambda_1, \ldots, \lambda_{\mathrm{N}}) \in \Lambda_k \right\} \tag{8.63}$$

of multidimensional multiwavelets having resolution level k. The MW expansion of $\tilde{U}(\boldsymbol{\eta})$ can now be formally written as:

$$\tilde{U}(\boldsymbol{\eta}) = \sum_{i=1}^{M_0} c_i^0 \Gamma_i^0(\eta_1, \ldots, \eta_{\mathrm{N}}) + \sum_{i=1}^{M_1} c_i^1 \Gamma_i^1(\eta_1, \ldots, \eta_{\mathrm{N}}) + \sum_{i=1}^{M_2} c_i^2 \Gamma_i^2(\eta_1, \ldots, \eta_{\mathrm{N}}) + \cdots \tag{8.64}$$

where $\Gamma^k(\boldsymbol{\eta}) \in \mathcal{W}_k$ denotes a multidimensional wavelet of resolution k. In practice, the MW expansion has to be truncated. Here, we choose to retain all multi-indices $\boldsymbol{\lambda}$ such that $|\boldsymbol{\lambda}| \equiv \sum_{d=1}^{\mathrm{N}} |\lambda_d| \leq \mathrm{Nr}$, where Nr is a prescribed resolution level. After truncation, the finite expansion may be rewritten in a single-index form as:

$$\tilde{U}(\boldsymbol{\eta}) \approx \sum_{i=0}^{\mathrm{P}} \tilde{U}_i \mathcal{M}^w{}_i(\eta_1, \ldots, \eta_{\mathrm{N}}), \tag{8.65}$$

where $\mathrm{P}+1 = \sum_{i=0}^{\mathrm{Nr}} M_i$ is the dimension of the truncated basis $\{\mathcal{M}^w{}_i, i = 0, \ldots, \mathrm{P}\}$. Further, we use the convention that the indexing is performed in such a way that the first element of the basis is $\mathcal{M}^w{}_0 = 1$.

8.3.4.1 Mean and variance

The mean of $U(\boldsymbol{\xi})$ is, by definition, given by:

$$\begin{aligned}\langle U\rangle &= \int_{a_1}^{b_1} \mathrm{d}y_1 \cdots \int_{a_{\mathrm{N}}}^{b_{\mathrm{N}}} \mathrm{d}y_{\mathrm{N}} \left[U(\mathbf{y}) \prod_{d=1}^{\mathrm{N}} p_d(y_d) \right] \\ &= \int_0^1 \mathrm{d}x_1 \cdots \int_0^1 \mathrm{d}x_{\mathrm{N}} \left[\tilde{U}(x_1, \ldots, x_{\mathrm{N}}) \right]. \end{aligned} \tag{8.66}$$

Introducing the MW expansion (8.65) in the above one immediately obtains:

$$\langle U\rangle = \tilde{U}_0.$$

Expression for the variance of U may be obtained in an analogous fashion. By definition, we have:

$$\begin{aligned}\sigma^2(U) &= \int_{a_1}^{b_1} \mathrm{d}y_1 \cdots \int_{a_{\mathrm{N}}}^{b_{\mathrm{N}}} \mathrm{d}y_{\mathrm{N}} \left[U(\mathbf{y}) - \langle U\rangle \right]^2 \prod_{i=1}^{\mathrm{N}} p_i(y_i) \\ &= \int_0^1 \mathrm{d}x_1 \cdots \int_0^1 \mathrm{d}x_{\mathrm{N}} \left[\tilde{U}(x_1, \ldots, x_{\mathrm{N}}) - \tilde{U}_0 \right]^2. \end{aligned} \tag{8.67}$$

Introducing the expansion (8.65) in the previous expression, and taking into account the orthonormality of the basis functions, we get:

$$\sigma^2(U) = \sum_{i=1}^{\mathrm{P}} \tilde{U}_i^2.$$

8.4 Application to Lorenz System

The MRA scheme above is first applied to the stochastic Lorenz system:

$$\begin{cases} \dfrac{\partial X}{\partial t} = \rho(Y - X), \\ \dfrac{\partial Y}{\partial t} = \mathrm{Ra}(\xi)X - Y - XZ, \\ \dfrac{\partial Z}{\partial t} = -aZ + XY, \end{cases} \tag{8.68}$$

where $\rho = 10$, $a = 8/3$ and $\mathrm{Ra}(\xi)$ is a random parameter with uniform distribution over $[15, 21]$. Deterministic initial conditions are used, according to $X(t = 0) = Y(t = 0) = Z(t = 0) = 1$. For the present setting, the Lorenz system exhibits damped oscillations for all possible values of Ra, leading to an asymptotically steady solution as $t \to \infty$. However, the damping time-scale exhibits sharp dependence on Ra. Below, we shall focus on the statistics of X at time $t = 25$. At this value of t, the solution for the lower values of Ra has nearly achieved a steady state, while the solution for higher values of Ra still exhibits large-amplitude oscillations. Such sharp variation with Ra requires a refined discretization along the random dimension in order to properly represent the local dynamics. In Sect. 8.4.1 the $h-$ and $p-$ convergence of the computed expectation and standard deviation of X are investigated. Then in Sect. 8.4.2, the same problem is solved using two MC sampling techniques in order to assess the efficiency of the MRA scheme.

8.4.1 h–p Convergence of the MW Expansion

8.4.1.1 Solution Method

For given expansion order and resolution level, the solution is approximated using:

$$(X, Y, Z)(t, \xi) = (\tilde{X}, \tilde{Y}, \tilde{Z})(t, \eta) \approx \sum_{\beta=0}^{P} (\tilde{X}, \tilde{Y}, \tilde{Z})_\beta(t)\mathcal{M}^w{}_\beta(\eta), \tag{8.69}$$

where we have suppressed, for clarity, the explicit dependence on Ra. Introducing these expansions into the Lorenz equations, and projection on the PC basis, one obtains the following coupled system for the MW coefficients:

$$\begin{cases} \dfrac{\partial \tilde{X}_\beta}{\partial t} = \rho(\tilde{Y}_\beta - \tilde{X}_\beta), \\ \dfrac{\partial \tilde{Y}_\beta}{\partial t} = \widetilde{(\mathrm{Ra}X)}_\beta - \tilde{Y}_\beta - \widetilde{(XZ)}_\beta, \\ \dfrac{\partial \tilde{Z}_\beta}{\partial t} = -a\tilde{Z}_\beta + \widetilde{(XY)}_\beta, \end{cases} \tag{8.70}$$

which need to be solved for all $\beta = 0, \ldots, P$. The initial conditions are also obtained by application of the Galerkin approach, resulting in:

$$\begin{cases} \tilde{X}_\beta(t=0) = \tilde{Y}_\beta(t=0) = \tilde{Z}_\beta(t=0) = 1 & \text{for } \beta = 0, \\ \tilde{X}_\beta(t=0) = \tilde{Y}_\beta(t=0) = \tilde{Z}_\beta(t=0) = 0 & \text{for } \beta = 1, \ldots, P. \end{cases} \tag{8.71}$$

The time integration of the $3 \times \mathrm{Card}(\mathrm{Nr}, \mathrm{No}) = 3(\mathrm{No}+1)2^{\mathrm{Nr}}$ unknown MW coefficients is performed using a fourth-order Runge-Kutta scheme, with a time step $\Delta t = 0.005$. This was selected based on successive refinement until the solution became essentially independent of Δt. Note that in order to integrate the above system, one needs to evaluate the MW expansion of the products of two stochastic quantities, as in the quadratic term $\widetilde{XZ}$. To this end, we rely on an "exact" Galerkin procedure, according to:

$$\begin{aligned} (\widetilde{XZ})_\beta = \left\langle \tilde{X}\tilde{Z}, \mathcal{M}^w{}_\beta \right\rangle_{[0,1]} &= \left\langle \sum_{\lambda=0}^{P}\sum_{\gamma=0}^{P} (\tilde{X}_\lambda \tilde{Z}_\gamma)\mathcal{M}^w{}_\lambda \mathcal{M}^w{}_\gamma, \mathcal{M}^w{}_\beta \right\rangle_{[0,1]} \\ &= \sum_{\lambda=0}^{P}\sum_{\gamma=0}^{P} \tilde{X}_\lambda \tilde{Z}_\gamma \left\langle \mathcal{M}^w{}_\lambda \mathcal{M}^w{}_\gamma, \mathcal{M}^w{}_\beta \right\rangle_{[0,1]}. \end{aligned} \tag{8.72}$$

As usual, the multiplication tensor $\langle \mathcal{M}^w{}_\alpha \mathcal{M}^w{}_\beta, \mathcal{M}^w{}_\gamma \rangle_{[0,1]}$ is solution independent and is computed and stored in a pre-processing stage. The sparse structure of the multiwavelet multiplication tensor is further discussed below.

8.4.1.2 Convergence Results

Figure 8.25 shows the computed values of $X(t, \mathrm{Ra})$ for $20 \le t \le 25$; plotted are results obtained with $\mathrm{Nr} = 1, \ldots, 4$ and $\mathrm{No} = 0, 4$. In all the frames, the same contour levels are used. The results illustrate the convergence of the predictions as Nr and No increase. The plots appear to indicate that convergence is faster for the lower values of Ra, where the solution has essentially decayed, than for the higher values, where X still exhibits large oscillations. This reveals an interesting feature of the MW representation (which has also been observed for WHa expansions [124]) namely that errors can be localized to regions where insufficient resolution is provided. This is in contrast to classical spectral representations, where under-resolution typically results in global breakdown of the solution.

Additional quantitative assessment of the h–p convergence of the MRA scheme is obtained by analyzing the computed solution statistics at $t = 25$. Table 8.5 shows the computed values of $\langle X \rangle$ at $t = 25$, for expansion orders $0 \le \mathrm{No} \le 6$ and resolution levels $0 \le \mathrm{Nr} \le 6$. The results indicate that the predictions convergence as Nr and No are increased. For $\mathrm{Nr} = 0$, however, p-convergence is very slow and accuracy up to two digits is not achieved even for $\mathrm{No} = 8$ (not shown). In particular, the results also indicate that in order to achieve accuracy up to the 3rd digit, a fifth-order expansion is needed for $\mathrm{Nr} = 1$, while using $\mathrm{No} = 0$ with $\mathrm{Nr} = 6$ provides similar

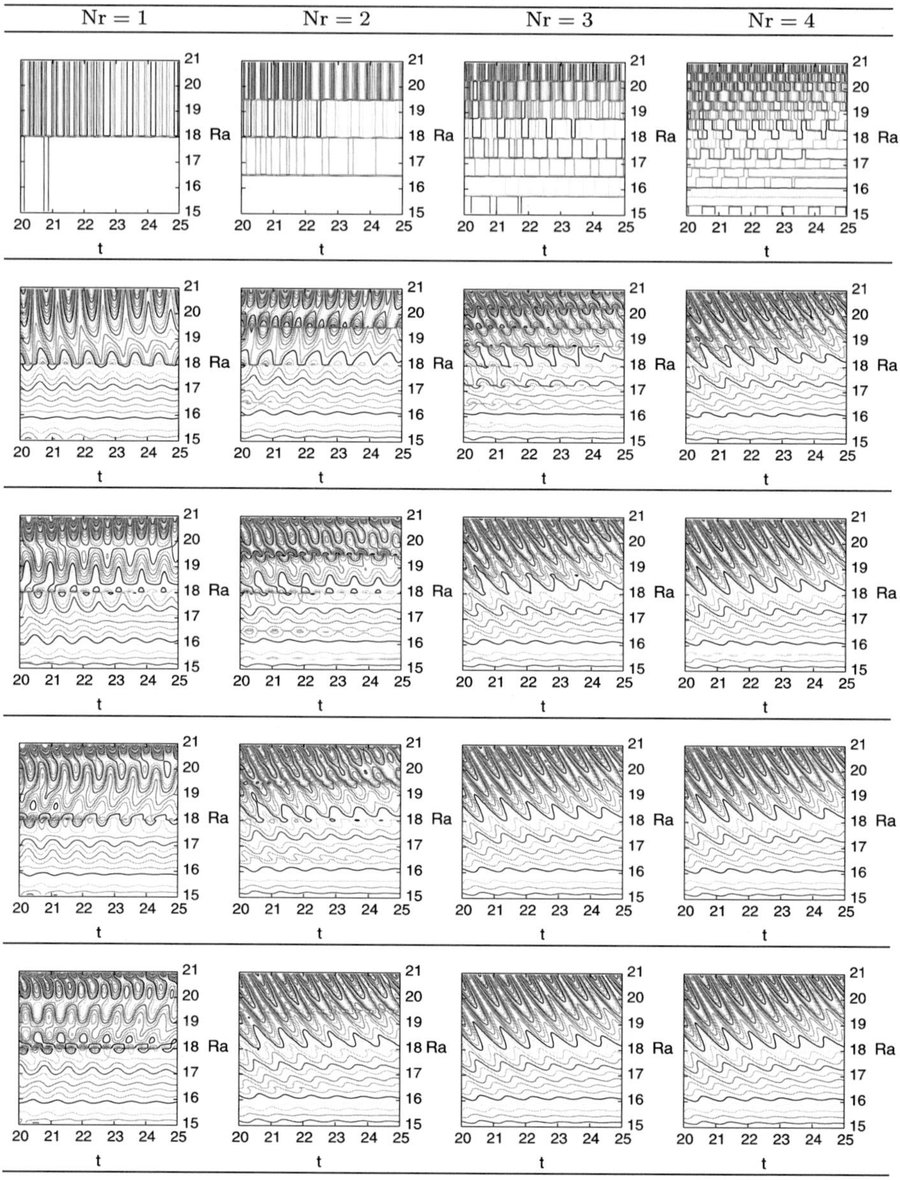

Fig. 8.25 Isolines of $X(t \in [20, 25], \mathrm{Ra})$ of the stochastic Lorenz problem, for $\mathrm{Nr} = 1, \ldots, 4$ (*left to right*) and $\mathrm{No} = 0, \ldots, 4$ (*top to bottom*). Time increases along the horizontal axis, while the vertical axis corresponds to $\mathrm{Ra} \in [15, 21]$. The same contour levels are used for all plots. Adapted from [125]

accuracy. Similar trends are also observed when analyzing the predictions of the standard deviation $\sigma(X)$, which are reported in Table 8.6.

Table 8.5 Computed values of $\langle X \rangle$ at $t = 25$ for the stochastic Lorenz system with different values of Nr and No. Adapted from [125]

$-\langle X \rangle$	No = 0	No = 1	No = 2	No = 3	No = 4	No = 5	No = 6
Nr = 0	6.739074	6.702924	3.657155	5.305756	6.844937	6.575842	6.669102
Nr = 1	6.754146	6.742118	6.702877	6.731783	6.741929	6.730446	6.730405
Nr = 2	6.716436	6.740495	6.733795	6.731213	6.730252	6.730173	6.730114
Nr = 3	6.738296	6.729341	6.730230	6.730139	6.730108	6.730105	–
Nr = 4	6.731381	6.730081	6.730106	6.730105	6.730105	–	–
Nr = 5	6.730400	6.730104	6.730105	6.730105	–	–	–
Nr = 6	6.730177	6.730105	6.730105	–	–	–	–

Table 8.6 Computed values of $\sigma(X)$ at $t = 25$ for the stochastic Lorenz system with different values of Nr and No. Adapted from [125]

$\sigma(X)$	No = 0	No = 1	No = 2	No = 3	No = 4	No = 5	No = 6
Nr = 0	0.000000	0.316249	5.660838	3.312410	0.518508	0.220683	0.287331
Nr = 1	0.327423	0.375252	0.312143	0.381295	0.371085	0.360037	0.360215
Nr = 2	0.329662	0.373340	0.361711	0.360578	0.357961	0.357668	0.357666
Nr = 3	0.371072	0.355050	0.357989	0.357654	0.357668	0.357667	–
Nr = 4	0.358382	0.357673	0.357667	0.357668	0.357667	–	–
Nr = 5	0.357853	0.357666	0.357667	0.357667	–	–	–
Nr = 6	0.357713	0.357667	0.357667	–	–	–	–

The results above indicate that in order to improve the accuracy of the predictions, one may either increase the number of resolution levels (h-refinement) or the expansion order (p-refinement). Since for complex problems it is usually desirable to maintain a moderate expansion order, typically No ≤ 4, increasing Nr provides suitable means to enhance the predictions. This raises the question concerning how Nr and No should be selected so as to achieve target accuracy at the lowest possible CPU cost. For the present simple setting, this question may be addressed by relating CPU cost to discretization parameters. A rough initial estimate is the size of the ODE system, which is proportional to Card(Nr, No) $= (\text{No} + 1)2^{\text{Nr}}$ (Table 8.7). However, since the computational load is in fact dominated by the evaluation of quadratic Galerkin products, a sharper estimate is based on the number of operations actually performed in these products, which is proportional to the number of non-zero entries in the multiplication tensor $\langle \mathcal{M}^w{}_\lambda \mathcal{M}^w{}_\gamma, \mathcal{M}^w{}_\beta \rangle_{[0,1]}$. This number, which is used as surrogate measure for CPU, is denoted by $\mathcal{C}(\text{Nr}, \text{No})$ and reported in Table 8.8 for different values of Nr and No. Comparison of Tables 8.7 and 8.8 shows that with increasing refinement $\mathcal{C}(\text{Nr}, \text{No})$ increases much quickly than Card(Nr, No).

Next, we seek a relationship between CPU and accuracy by plotting in Fig. 8.26 the absolute errors in $\langle X \rangle$ and $\sigma(X)$ at $t = 25$ against $\mathcal{C}(\text{Nr}, \text{No})$. For the purpose

Table 8.7 Values of Card($\overline{\diamond}$) for different order and number of resolution levels. Adapted from [125]

Card(Nr, No)	No = 0	No = 1	No = 2	No = 3	No = 4	No = 5	No = 6
Nr = 0	1	2	3	4	5	6	7
Nr = 1	2	4	6	8	10	12	16
Nr = 2	4	8	12	16	20	24	28
Nr = 3	8	16	24	32	40	48	56
Nr = 4	16	32	48	64	80	96	112
Nr = 5	32	64	96	128	160	–	–
Nr = 6	64	128	192	256	–	–	–

Table 8.8 Values of $\mathcal{C}$(Nr, No) for different order and number of resolution levels. Adapted from [125]

$\mathcal{C}$(Nr, No)	No = 0	No = 1	No = 2	No = 3	No = 4	No = 5	No = 6
Nr = 0	1	4	11	29	42	69	106
Nr = 1	4	22	69	166	308	531	880
Nr = 2	16	154	611	1 538	3 084	5 457	9 034
Nr = 3	52	706	2 883	7 354	14 936	26 541	43 254
Nr = 4	148	2 578	10 625	27 434	56 040	99 561	160 018
Nr = 5	388	8 242	34 305	89 094	181 216	–	–
Nr = 6	964	24 178	101 501	262 826	–	–	–

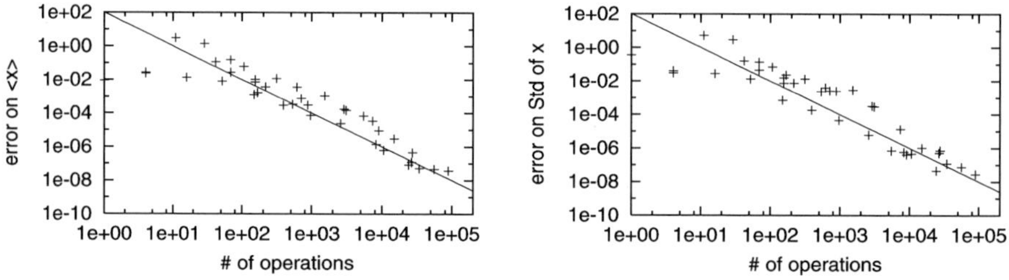

Fig. 8.26 Absolute errors on the expectation and standard deviation of X, at $t = 25$, as a function of the number of operations in the spectral product: Card(Nr, No). The exact solution is taken to be the computed values for Nr = 6 and No = 2, as reported in Tables 8.5 and 8.6. The *solid lines* are $\sim x^{-2}$. Adapted from [125]

of estimating errors, the solution obtained using Nr = 6 and No = 2 has been used as surrogate for the exact solution (MC predictions below support this approximation). The results indicate that for the present parameter range the "errors" on the first two statistical moments of X decay roughly as $\mathcal{C}(\mathrm{Nr}, \mathrm{No})^{-2}$. While the details of this relationship may be dependent on the problem and parameter range, the collapse of the data clearly indicates that in the present case the "complexity estimate" $\mathcal{C}(\mathrm{Nr}, \mathrm{No})$ provides a good measure for both the CPU and the accuracy of the predictions. In addition, the results provide a convenient means for selecting Nr and No, and indicate that multiple alternatives may exist for achieving a target accuracy level.

8.4.2 *Comparison with Monte Carlo Sampling*

In this section, we compare the efficiency of the MRA scheme above with that of MC sampling strategies. The latter rely on computing deterministic solutions for specific values of the random parameters, and then evaluating the desired moments through a collocation approach [94, 134]. Since individual solutions are deterministic, the evaluation of the product XY of X and Y requires a single operation, so that the "complexity" of obtaining a single MC realization is taken to be 1. This provides a consistent approach for comparing the efficiency of MRA and MC. In other words, the CPU load in MRA is gauged using $\mathcal{C}(\mathrm{Nr}, \mathrm{No})$ while in MC it is gauged using the number, m, of realizations.

8.4.2.1 Classical Sampling Strategy

A classical MC sampling strategy is first applied. The approach is based on using a random number generator to generate m independent realizations of the stochastic parameter, Ra_i, $i = 1, \dots, m$, that are uniformly distributed over the interval $[15, 21]$. For each realization, the corresponding deterministic Lorenz system is integrated up to $t = 25$, resulting in particular in a set of predictions, $X_i, i = 1, \dots, m$. The first two statistical moments of X are then estimated using:

$$\langle X \rangle_m = \frac{1}{m} \sum_{l=1}^{m} X_i$$

and

$$\sigma_m^2(X) \approx \left\langle (X - \langle X \rangle_m)^2 \right\rangle^2 = \frac{1}{m-1} \sum_{i=1}^{m} X^2 - \langle X \rangle_m^2 \, .$$

In Fig. 8.27, the absolute errors in $\langle X \rangle$ and $\sigma(X)$ plotted against the number of MC realizations, $1 \le m \le 10^7$. As in the analysis above, the MRA solution with $\mathrm{Nr} = 6$, $\mathrm{No} = 2$ is used as surrogate for the exact solution for the definition of the errors. The results indicate that the errors in $\langle X \rangle$ and $\sigma(X)$ decay as $m^{-1/2}$, as expected for an unbiased sampling strategy. (This behavior in fact justifies the use of the resolved MRA estimate as a substitute of the exact solution.) The low convergence rate of the present sampling scheme is contrasted with that of the MRA scheme, where errors would scale as m^{-2}. Comparison of Figs. 8.26 and 8.27 also shows that the

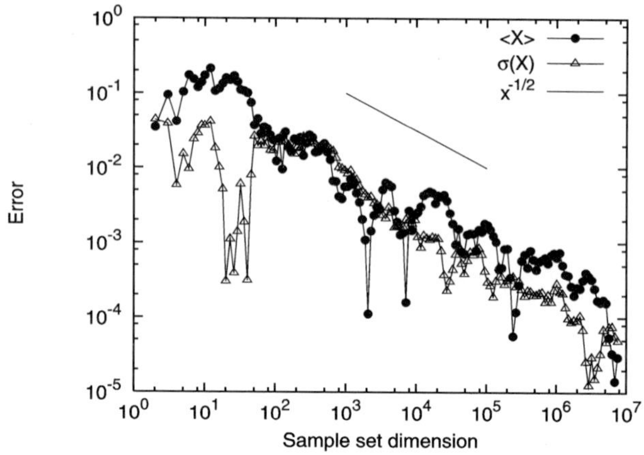

Fig. 8.27 Absolute errors on $\langle X\rangle$ and $\sigma(X)$ at $t = 25$ versus the number of MC samples. Adapted from [125]

number of operations needed to achieve error levels smaller than 10^{-2} is much smaller for MRA than for MC. Thus, for the present stochastic Lorenz problem, the MRA scheme appears to be substantially more efficient than conventional MC.

8.4.2.2 Latin Hypercube Sampling

A Latin Hypercube Sampling (LHS) strategy [153] is considered in this section in order to improve the convergence of the MC approach. The essential feature of LHS is to divide the probability domain into a set, (Nbin), of bins having equal probability, and then to perform classical MC sampling on each of the bins. This strategy enhances the effectiveness of the MC computations, especially at low values of m, by forcing the sampling scheme to visit all the bins, before hitting a given bin a second time.

The convergence of LHS scheme is illustrated in Fig. 8.28, where the errors in $\langle X\rangle$ and $\sigma(X)$ at $t = 25$ are plotted against the total number of samples, m. Shown are results obtained using $\text{Nbin} = 10,\ 100$ and 1000. The results indicate that LHS enhances the convergence of the MC simulations, especially for low values m ($m < 10000$). On the other hand, when m/Nbin becomes large, the asymptotic convergence rate of classical MC method, in $1/\sqrt{m}$ is recovered. Despite the improvement over the standard MC approach, the performance of the LHS is significantly lower than that of the MRA.

8.5 Closing Remarks

In this chapter, we have explored the use of wavelet and MW representations in multiresolution PC representations. These developments have resulted in a number of attractive features, of which we highlight following two. First, and perhaps most important, is the capability of addressing problems involving steep or discontinuous

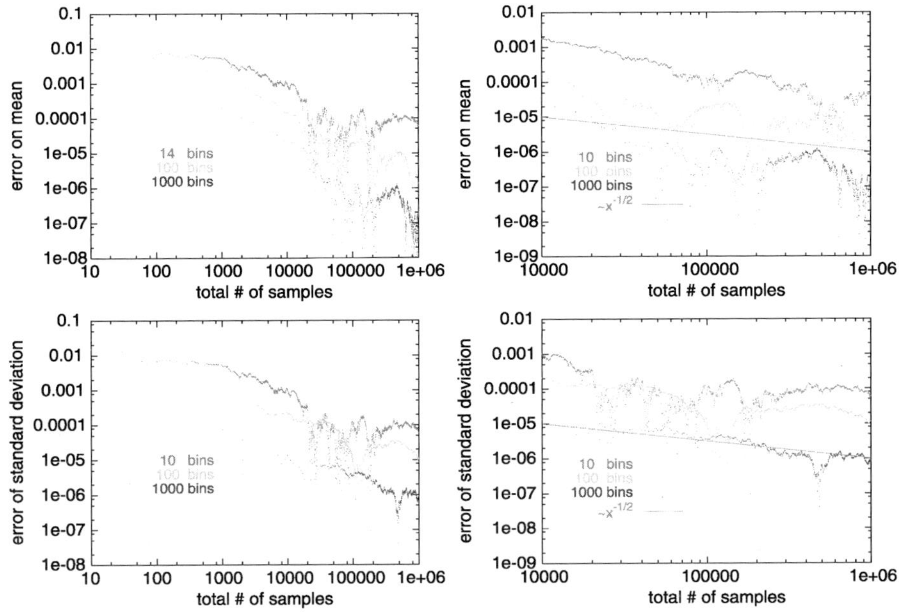

Fig. 8.28 Absolute errors on $\langle X \rangle$ (*top*) and $\sigma(X)$ (*bottom*) for LHS at $t = 25$. The *right plots* show enlarged views for $10^4 \leq m \leq 10^6$. Adapted from [125]

dependence of the stochastic solution on the random input data. In particular, this capability was demonstrated based on model computations of an idealized model of a particle moving under the action of an imposed potential and friction, and of near-critical Rayleigh-Bénard flow. For both problems, the simulations indicate that implementation of a WHa representation effectively captures steep transitions and discontinuities that develop in the stochastic solution, and result in accurate predictions of the corresponding statistics. In these complex situations, representations based on global basis functions suffered from several limitations, including poor prediction of transition, of the state of the system, and of associated statistics.

The second feature we wish to highlight concerns the MRA scheme, which was tested through application to a simplified Lorenz system. The numerical tests illustrated the convergence of the MW expansion with increasing order No and resolution levels Nr, and also revealed that for the present setting, the MRA scheme offers substantial improvement in efficiency over MC approaches. On the other hand, the tests also indicated that the computational overheads of MRA rapidly increase as the MW parameters are refined. This points to a potentially severe limitation in the extension of MRA to stochastic problems with multiple random parameters. Means to overcome such limitations are addressed in the next chapter, where adaptive strategies are considered in single- and multi-dimensional settings.

Chapter 9
Adaptive Methods

Adaptive schemes generally aim at reducing CPU cost by adjusting the quality of the representation to capture essential features of the solution. In this chapter, we explore four different strategies for performing such refinement. The following developments are motivated in part by the experiences with MW representations outlined in the previous chapter. In particular, these indicated that while "refinement" can be naturally implemented in conjunction with wavelet representation, for instance by increasing the levels of details, a uniform or brute-force refinement is likely to require excessive CPU resources, even for simple examples. Adaptive schemes consequently strive to perform such refinements locally where needed, in an effort to minimize necessary computational resources.

The first two strategies considered in this chapter specifically concern MRA schemes. In Sect. 9.1, we explore the possibility of a local refinement of the MW expansion to adequately capture the solution. We restrict ourselves to adaptation of the resolution level only, though combined adaptation of both resolution level and expansion order is also conceivable within the same framework. The strategy in Sect. 9.1 essentially consists in disregarding the basis components corresponding to low magnitude MW coefficients. To test the resulting algorithm, application to the Rayleigh-Bénard problem initially tackled in Sect. 8.2.2 is once again considered. The applications are used to demonstrate the capability of the adaptive scheme in automatically locating the bifurcation, and the savings resulting from the concentration of computational resources to capture the associated transition.

Section 9.2 also considers MW expansions, but alternatively explores a second strategy based on the adaptive partitioning of the random parameter space. The strategy is based on analysis of contributions of 1D details to the variance within individual sub-domains in random parameter space. A directional subdivision algorithm is then implemented whenever the contribution from the 1D details exceeds a user-specified threshold. A numerical study of the resulting adaptive scheme is then conducted, based on applications to a chemical kinetics problem. Computations are in particular used to demonstrate the ability of the scheme to localize refinements in regions of steep variation, and to quantify the computational gains achieved using these local refinements.

O.P. Le Maître, O.M. Knio, *Spectral Methods for Uncertainty Quantification*,
Scientific Computation,
DOI 10.1007/978-90-481-3520-2_9,

In Sect. 9.3, an adaptive method is presented based *a posteriori* error estimation for the numerical solution of a stochastic variational problem. The discretization of the stochastic variational problem uses standard finite elements in space and piecewise continuous orthogonal polynomials in the stochastic domain. The *a posteriori* methodology consists in measuring the error as the functional difference between the continuous and discrete solutions. This functional difference is approximated using the discrete solution of the primal stochastic problem and two discrete adjoint solutions of the associated dual stochastic problem. The dual problem being linear, the error estimation results in a limited computational overhead. Based on this error estimate, an adaptive refinement of the approximation space is performed. The refinement can concern the spatial or stochastic approximations, and can consist in increasing the approximation order or in using finer elements. The efficiency of the resulting refinement strategy is verified based on computational tests of the stochastic Burgers equation.

Section 9.4 discusses the Generalized Spectral Decomposition (GSD) method for the resolution of nonlinear stochastic problems. The GSD method consists in the construction of a reduced basis approximation of the Galerkin solution. Two power-type algorithms are considered for the sequential construction of the successive "generalized spectral modes," which involves decoupled resolution of a series of deterministic and low dimensional stochastic problems. The resulting schemes are tested on two model problems, namely the 1D steady viscous Burgers equation and a 2D nonlinear diffusion problem. These computations highlight the effectiveness of the GSD algorithms, which exhibit convergence rates that naturally depend on the spectrum of the stochastic solution, but are independent of the dimension of the stochastic approximation space.

Experiences gained through the applications in Sects. 9.1–9.4 are finally used in Sect. 9.5 to briefly motivate further developments.

9.1 Adaptive MW Expansion

As outlined above, the present approach aims at reducing the CPU cost by adaptively refining the MW expansion exposed in Chap. 8. For simplicity, discussion is restricted to the one-dimensional case. The essential concept, which is similar to compression in image processing, is outlined as follows. Given the highest resolution level allowed, Nr, we denote $\overline{\diamondsuit}$ the restriction of the index set $\diamondsuit$ to MW (see Sect. 8.3.3) corresponding to resolution levels $\leq$ Nr. Then, the solution $\tilde{U}(\eta)$ is expanded using a reduced basis:

$$U(\xi) = \tilde{U}(\eta) \approx \sum_{\lambda \in \overline{\diamondsuit}_a \subseteq \overline{\diamondsuit}} \tilde{U}_\lambda W_\lambda(\eta), \tag{9.1}$$

where $\overline{\diamondsuit}_a \subseteq \overline{\diamondsuit}$ is a reduced set of MW indices such that

$$\overline{\diamondsuit}_a \equiv \{\lambda \in \overline{\diamondsuit} : |\tilde{U}_\lambda| > \varepsilon_a\}. \tag{9.2}$$

Recall that the random variables ξ and η are related through $\xi = F_\xi^{-1}(\eta)$, where F_ξ is the distribution function of ξ and $\eta \sim \mathcal{U}[0, 1]$. In the definition of the index set (9.2), ε_a is a prescribed threshold parameter that may be a function of the resolution level, i.e. $\varepsilon_a = \varepsilon_a(|\lambda|)$. Clearly, for this scheme, the expansion retains only details with significant "energy". Obviously, since $\tilde{U}(\eta)$ is yet to be determined, it is not possible to reduce $\overline{\diamondsuit}$ *a priori*. To overcome this difficulty, an iterative scheme is constructed, where the representation is successively refined, by locally increasing the level of resolution only where needed.

9.1.1 Algorithm for Iterative Adaptation

The algorithm below is used for this purpose:

1. **Initialization.** We start by computing an initial coarse approximation of $\tilde{U}(\eta)$, denoted $\tilde{U}^{(l=0)}(\eta)$, on $\overline{\diamondsuit}_a^{(l=0)} \equiv \Delta_0$.
2. **Analysis.** Let us denote $\delta\overline{\diamondsuit}^{(l)}$ the set of integer indices defined as

$$\delta\overline{\diamondsuit}^{(l)} \equiv \overline{\diamondsuit}_a^{(l)} \cap \{\beta : |\beta| = l, |\tilde{U}_\beta^{(l)}| > \varepsilon_a(l)\}. \tag{9.3}$$

   We consider on the following two situations:

   - If $\delta\overline{\diamondsuit}^{(l)} = \emptyset$ the adaptive scheme has converged, $\tilde{U}^{(l)}(\eta)$ is the final solution and the iterations are stopped.
   - Otherwise, we have $\delta\overline{\diamondsuit}^{(l)} = \{\beta_0, \ldots, \beta_q\}$ then the solution $\tilde{U}^{(l)}$ needs refinement over the union of the supports of the elements of $\delta\overline{\diamondsuit}^{(l)}$,

$$\mathrm{Supp}[\delta\overline{\diamondsuit}^{(l)}] \equiv \bigcup_{\beta \in \delta\overline{\diamondsuit}^{(l)}} \mathrm{Supp}(\mathrm{W}_\beta).$$

3. **Refinement.** We set

$$\overline{\diamondsuit}_a^{(l+1)} = \overline{\diamondsuit}_a^{(l)} \cup \delta\overline{\diamondsuit}_a^{(l)},$$

   where

$$\delta\overline{\diamondsuit}_a^{(l)} \equiv \left\{\lambda \in \Delta : |\lambda| = l + 1, \text{ and } \mathrm{Supp}(\mathrm{W}_\lambda) \subseteq \mathrm{Supp}[\delta\overline{\diamondsuit}^{(l)}]\right\}.$$

4. **Computation.** Compute $\tilde{U}^{(l+1)}$ the approximation of s on the subspace spanned by $\{W_\lambda, \lambda \in \overline{\diamondsuit}_a^{(l+1)}\}$. If $l < \mathrm{Nr}$, where Nr is a prescribed maximal resolution level, set $l = l + 1$ and go back to step 2.

Note that the iterative scheme above generates a continuous cascade of details at successive resolution levels. Specifically, for any index $\lambda \in \overline{\diamondsuit}_a^{(l>0)}$ with $|\lambda| \geq 1$, $\exists \beta \in \overline{\diamondsuit}_a^{(l)}$ such that $|\beta| = |\lambda| - 1$ and $\mathrm{Supp}(\mathrm{W}_\lambda) \subset \mathrm{Supp}(\mathrm{W}_\beta)$. In other words, any

MW of resolution level $k > 1$ of the adaptive basis has parents. Thus, one may not be able to guarantee that the iterations give $\lim_{l\to\infty} \overline{\Diamond}_a^{(l)} = \Diamond_a$, for every function ε_a and process $U(\xi)$. Moreover, while more elaborate versions of the scheme above can be conceived, the simple approach above may still be well-suited for most applications. Below, it is applied to a stochastic Rayleigh-Bénard problem.

9.1.2 Application to Rayleigh-Bénard Flow

To test the adaptive refinement scheme, we consider the problem initially treated in Sect. 8.2.2 [124]. The setup consists of a 2D rectangular cavity with insulated vertical walls. The upper boundary is maintained at a (normalized) deterministic cold temperature, $T_c = -1/2$, while the bottom wall has a spatially-uniform random (normalized) temperature $T_h(\eta) = 1/2 + T_r\eta$, $\eta \sim \mathcal{U}[0,1]$. The system is characterized by three non-dimensional parameters: the Rayleigh number $\mathrm{Ra} = 2,150$, the Prandtl number $\mathrm{Pr} = 0.71$, and the cavity aspect ratio $A = 2$. These non-dimensional parameters are based on the appropriate combination of reference quantities, selected as in Sects. 8.2.2 and 6.2. We assume that $T_h(\xi)$, is uniformly distributed on the interval $[0.3, 0.7]$. The difficulty in the present setup arises from the presence of a critical point, corresponding to $T_h = 0.4301$ (see Sect. 8.2.2 and [124]). For $T_h < 0.4301$ the system has a stable solution corresponding to a pure conductive regime with vanishing fluid velocity, while for $T_h > 0.4301$ two steady recirculation zones exist, enhancing the heat transfer across the cavity compared to the conductive regime.

The flow inside the cavity is modeled using the Navier-Stokes equations in the Boussinesq limit [128]. The governing equations are solved on 60×30 computational grid using a second-order scheme. The stochastic projection method (SPM) is used for this purpose (see Sect. 6.1 and [123, 128]). It was shown in Sect. 8.2.2 that a WLe expansion (corresponding to $\mathrm{Nr} = 0$ in the present MRA construction) fails to correctly represent the bifurcation in the uncertainty range, while a WHa expansion (corresponding to $\mathrm{No} = 0$ in the MRA scheme) provides robust estimates. The adaptive scheme is now employed using $\mathrm{No} = 1, 2$ and 3, and a maximal resolution level $\mathrm{Nr} = 6$. For the purpose of the analysis step of the adaptive scheme, a stochastic observable has to be selected. In the computations below, we rely on the overall heat transfer rate, i.e. we set

$$\tilde{U}^{(l)}(\eta) = \sum_{\lambda \in \overline{\Diamond}_a^{(l)}} W_\lambda(\eta) \frac{1}{A} \int_0^A \frac{\partial \tilde{T}_\lambda^{(l)}}{\partial z} \, dy, \tag{9.4}$$

where z and y are the normalized vertical and horizontal coordinates. For the refinement threshold, we set $\epsilon_a(\lambda) = 0.03 \times 2^{-|\lambda|}$. When the set of indices is increased from $\overline{\Diamond}_a^{(l)}$ to $\overline{\Diamond}_a^{(l+1)}$, the newly created MW coefficients are initialized to zero, while the others are kept at their previously computed values. Then the flow equations are

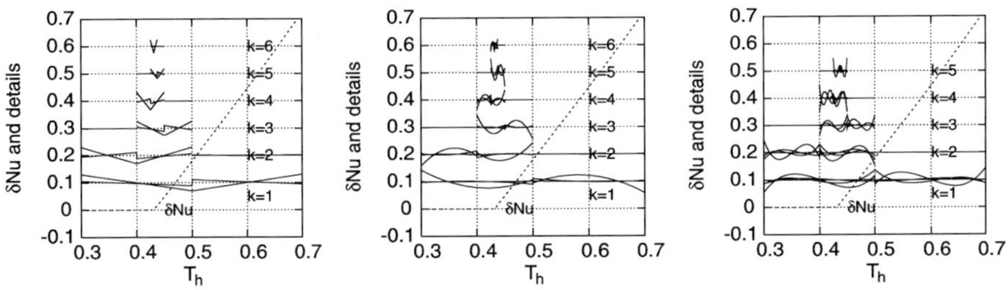

Fig. 9.1 Computed values δNu versus T_h, for No = 1 (*left*), 2 (*middle*) and 3 (*right*). The *plots* also depict the details $\tilde{U}^{(l)}$ at different resolution levels, k, as indicated. The details are shifted in the vertical direction for clarity. Adapted from [125]

time-integrated up to the steady state, before being further analyzed for refinement. Note that the simulation time needed to reach the steady state increases with the iteration index l, since the newly added MW are increasingly localized around the critical point, where the growth rate of the instability vanishes. As in Sect. 8.2.2, the normalized heat transfer enhancement, defined as [124]:

$$\delta\tilde{\text{Nu}}^{(l)}(\eta) \equiv \tilde{U}^{(l)}(\eta)/(\tilde{T}_h(\eta) - T_c) - 1,$$

is monitored for the purpose of analyzing the predictions.

Figure 9.1 shows the values of δNu obtained the end of the iterations are plotted against T_h. As for the WHa solution in [124], the discontinuity at the critical temperature is well captured. Also plotted in Fig. 9.1 are the (rescaled) details of $\tilde{U}^{(l)}$ at different resolution levels ($k > 1$), as obtained at the end of the refinement. The results illustrate how the solution is refined in the neighborhood of the critical point, and only there. In fact, at each iteration l, refinement occurs within the support of the current-level MWs that overlap the critical point. Moreover, for the selected threshold function ϵ_a, it is observed that the iterative scheme proceeds up to the maximal resolution level allowed (Nr = 6) for No = 1 and 2, but stops at level $k = \text{Nr} - 1$ for No = 3. Also note that the predictions for the three expansion orders considered are in excellent agreement with each other, suggesting that in all cases accurate predictions are obtained.

For a better appreciation of the efficiency of the adaptive scheme, we provide in Table 9.1 an analysis of the size of the "adapted" basis. The first three columns provide, respectively, the highest resolution level at the end of the iterations, the dimension of the basis (Card_a) and the corresponding value of $\mathcal{C}_a$ which measures the number of operations in the Galerkin product (see discussion in Sect. 8.4.1.2). The efficiency of the adaptive refinement is estimated by the "compression" ratios, $\text{Card}_a/\text{Card}(\text{Nr}_f, \text{No})$ and $\mathcal{C}_a/\mathcal{C}(\text{Nr}_f, \text{No})$, between the adaptively reduced basis and the full basis. As shown in Table 9.1, these ratios are quite small, and thus reveal a substantial reduction in CPU cost.

The experiences above suggest that in situations requiring a high level of local refinement, lower-order expansions are preferable to higher-order expansions as the former are more likely to have lower compression ratios than the latter. While

Table 9.1 Properties of the reduced basis obtained by adaptive refinement. Provide are the maximal refinement level, Nr_f at the end of the iterations, the dimension of the basis, and the corresponding value of $\mathcal{C}_a$. Also shown are the "compression" ratios, $\mathrm{Card}_a/\mathrm{Card}(\mathrm{Nr}_f, \mathrm{No})$ and $\mathcal{C}_a/\mathcal{C}(\mathrm{Nr}_f, \mathrm{No})$. Adapted from [125]

Order	Nr_f	Card_a	$\mathcal{C}_a$	$\mathrm{Card}_a/\mathrm{Card}(\mathrm{Nr}_f, \mathrm{No})$	$\mathcal{C}_a/\mathcal{C}(\mathrm{Nr}_f, \mathrm{No})$
No = 1	6	24	2,602	0.1875	0.1076
No = 2	6	36	10,855	0.1875	0.1070
No = 3	5	40	17,174	0.3125	0.1928

very low compression ratios may be expected in problems with large number N of stochastic dimensions, extension of the present adaptive refinement scheme to the multidimensional case may prove difficult to implement. In addition, since as noted earlier the dimension of basis and the number of Galerkin operation increase rapidly with N, small compression ratios may not be sufficient to overcome the added complexity in multi-dimensional problems. These observations in part motivate the alternative approach of the following section.

9.2 Adaptive Partitioning of Random Parameter Space

The experiences of the previous adaptive section indicate that it is possible to minimize the CPU load of MRA of stochastic problems by limiting the number of overlapping MW supports. However, one drawback of the scheme above is that the complexity $\mathcal{C}_a$ of the spectral product, which reflects the coupling between MW coefficients, is quickly increasing. This is due to the generation of a **continuous** cascade of details at successive scales. This observation raises the question whether a truly local analysis, that decouples the representation at different scales, can lead to even more efficient computations.

In this section, we develop a local refinement scheme based on the expansion in (8.51). Comparison of the two expansions in (8.51) and (8.52) shows that the former, in terms of ϕ functions, does not involve any summation over the scale indices, contrary to the expansion in terms of details, ψ. This difference stems from the fact that the basis functions ϕ_{il}^k, $i = 0, \dots, \mathrm{No}$ and $l = 0, \dots, 2^k - 1$, couple with only No other components, namely those having the same sliding index l. In contrast, the ψ_{il}^k couple with many other components. This suggests an adaptive strategy based on successive partitions of the random parameter space, through the determination of a **local** resolution level. These considerations lead us to the second adaptive scheme below.

9.2.1 Partition of the Random Parameter Space

Let $\Xi = [a_1, b_1] \times \cdots \times [a_{\mathrm{N}}, b_{\mathrm{N}}]$ be the range of random parameters. Let Ξ^m, $m = 1, \ldots, \mathrm{Nb}$, be a finite partition of Ξ in Nb non-overlapping subdomains:

$$\begin{cases} \Xi^m = [a_1^m, b_1^m] \times \cdots \times [a_{\mathrm{N}}^m, b_{\mathrm{N}}^m], \\ \Xi = \bigcup_{m=1}^{\mathrm{Nb}} \Xi^m, \\ \Xi^m \cap \Xi^{m'} = \varnothing \quad \text{if } m \neq m'. \end{cases} \tag{9.5}$$

On each of the subdomains Ξ^m we define the local probability density function of $\boldsymbol{\xi}$, denoted by p^m. Since the components of $\boldsymbol{\xi}$ are independent, we have:

$$p^m(\mathbf{y}) = \prod_{d=1}^{\mathrm{N}} p_d^m(y_d), \tag{9.6}$$

where $p_d^m(y)$ is defined according to

$$p_d^m(y) \equiv \frac{p_d(y)}{F_{\xi_d}(b_d^m) - F_{\xi_d}(a_d^m)}, \tag{9.7}$$

where F_{ξ_d} is the distribution function of the random variable ξ_d. Clearly, we have

$$p^m(\mathbf{y}) > 0 \quad \text{for } \mathbf{y} \in \Xi^m \text{ and } \int_{\Xi^m} p^m(\mathbf{y})\,\mathrm{d}\mathbf{y} = 1. \tag{9.8}$$

Moreover, defining

$$F_d^m(y_d \in [a_d^m, b_d^m]) \equiv \int_{a_d^m}^{y_d} p_d^m(y')\,\mathrm{d}y' = x_d^m, \tag{9.9}$$

we see that $x_d^m \in [0, 1]$. Now, if η_d^m is uniformly distributed over $[0, 1]$, the random variable $(F_d^m)^{-1}(\eta_d^m) \in [a_b^m, b_d^m]$ and has the same distribution as ξ_d^m. Thus, a second-order stochastic process can be locally expanded on Ξ^m, in terms of the random vector $\boldsymbol{\eta}^m$, having uniformly distributed and independent components: $(\eta_1^m, \ldots, \eta_{\mathrm{N}}^m) \sim \mathcal{U}[0, 1]^{\mathrm{N}}$.

9.2.2 Local Expansion Basis

Let $\boldsymbol{\gamma}$ be the set of multidimensional indices

$$\boldsymbol{\gamma} = \left\{ (\gamma_1, \ldots, \gamma_{\mathrm{N}}) : \sum_{d=1}^{\mathrm{N}} \gamma_d \leq \mathrm{No} \right\}.$$

For $\boldsymbol{\xi} \in \Xi^m$, we build the local projection basis as

$$\mathcal{B}_p(\Xi^m) \equiv \left\{ \Phi^m_{\boldsymbol{\lambda} \in \boldsymbol{\gamma}}(\eta^m_1, \ldots, \eta^m_{\mathrm{N}}) \equiv \prod_{d=1}^{\mathrm{N}} \phi_{\lambda_d}(\eta^m_d) \right\},$$

and the details directional basis $\mathcal{B}^d_a$, $d = 1, \ldots, \mathrm{N}$ as

$$\mathcal{B}^d_a(\Xi^m) \equiv \left\{ \psi_i(\eta^m_d), i = 0, \ldots, \mathrm{No} \right\}.$$

The complete local expansion basis will be the union of $\mathcal{B}_p$ and $\mathcal{B}^d_a$:

$$\mathcal{B}(\Xi^m) = \mathcal{B}_p(\Xi^m) + \bigcup_{d=1}^{\mathrm{N}} \mathcal{B}^d_a(\Xi^m).$$

Finally, the multidimensional process $U(\boldsymbol{\xi})$ will have the following local expansion on Ξ^m:

$$\begin{aligned} U(\boldsymbol{\xi} \in \Xi^m) = \tilde{U}(\boldsymbol{\eta}^m) \approx \sum_{\boldsymbol{\lambda} \in \boldsymbol{\gamma}} \tilde{U}^m_{\lambda_1, \ldots, \lambda_{\mathrm{N}}} \Phi_{\lambda_1, \ldots, \lambda_{\mathrm{N}}}(\eta^m_1, \ldots, \eta^m_{\mathrm{N}}) \\ + \sum_{d=1}^{\mathrm{N}} \sum_{i=0}^{\mathrm{No}} \tilde{U}^m_{d,i} \psi_i(\eta^m_d). \end{aligned} \tag{9.10}$$

Note that the local basis $\mathcal{B}(\Xi^m)$, spanning the local expansion of U according to (9.10), is in fact the rescaled Legendre polynomials basis ($\mathcal{B}_p(\Xi^m)$) augmented with the first-level detail basis, $\mathcal{B}^d_a$, $d = 1, \ldots, \mathrm{N}$. Thus, $\tilde{U}(\boldsymbol{\eta}^m)$, approximated by (9.10), is the local Wiener-Legendre projection of order No, plus one dimensional details. The utility of the one dimensional details will be made clear soon.

The local expectation of U is given by $\langle U \rangle_{\Xi^m} = U^m_{0,\ldots,0}$, and its local variance is

$$\sigma^2_{\Xi^m}(U) \approx \left(\hat{\sigma}_{\Xi^m}\right)^2 + \sum_{d=1}^{\mathrm{N}} \left(\sigma^d_{\Xi^m}\right)^2,$$

where we have denoted

$$\left(\hat{\sigma}_{\Xi^m}\right)^2 \equiv \sum_{\boldsymbol{\lambda} \in \boldsymbol{\gamma}_p} (U^m_{\boldsymbol{\lambda}})^2, \qquad \left(\sigma^d_{\Xi^m}\right)^2 \equiv \sum_{i=0}^{\mathrm{No}} \left(U^m_{d,i}\right)^2,$$

with $\boldsymbol{\gamma}_p = \boldsymbol{\gamma} - \{0, \ldots, 0\}$.

The total expectation of the process is given by the volume-weighted summation of the local expectations:

$$\langle U \rangle = \sum_{m=1}^{\mathrm{Nb}} \langle U \rangle_{\Xi^m} \mathrm{Vol}^m, \tag{9.11}$$

where Vol^m is the volume of the m-th subdomain of the random parameter space:

$$\mathrm{Vol}^m = \prod_{d=1}^{N} (F_{\xi_d}(b_d^m) - F_{\xi_d}(a_d^m)).$$

Finally the total variance of the process is given by

$$\sigma^2(U) = \sum_{m=1}^{\mathrm{Nb}} \left[\sigma^2_{\Xi^m}(U) + \left(U^m_{0,\dots,0} - \langle U \rangle\right)^2\right] \mathrm{Vol}^m. \tag{9.12}$$

9.2.3 Error Indicator and Refinement Strategy

Assume that the current partition of Ξ involves Nb subdomains, *i.e.* $\Xi = \bigcup_{m=1}^{\mathrm{Nb}} \Xi^m$. On each subdomain Ξ^m, the process is expanded on the local basis $\mathcal{B}(\Xi^m)$, the spectral coefficients being computed through Galerkin projection methods.[1] To decide if a given subdomain m needs more refinement, and to determine which stochastic directions need such refinement, we consider the following test:

$$\text{if } \frac{\sigma^d_{\Xi_m}}{\sigma_{\Xi_m}} \geq \varepsilon_2(\mathrm{Vol}^m). \tag{9.13}$$

If the inequality is satisfied, the subdomain is refined along the d-th dimension. Here $\varepsilon_2(\mathrm{Vol}^m) < 1$ is a prescribed threshold function. Note that the test compares the "energy" of the one-dimensional details along the d-th stochastic direction with the local variance of the solution. In other words, the one-dimensional details coefficients are used as indicators of the quality of the representation along their respective stochastic direction. A new partition of Ξ is then constructed, by splitting Ξ^m into smaller subdomains. Specifically, if we assume the inequality (9.13) is satisfied for a single dimension d, then refinement of $\Xi^m = [a_1^m, b_1^m] \times \cdots \times [a_N^m, b_N^m]$ will give birth to two new subdomains $\Xi^{m'}$ and $\Xi^{m''}$, defined by:

$$\begin{cases} \Xi^{m'} = [a_1^{m'}, b_1^{m'}] \times \cdots \times [a_N^{m'}, b_N^{m'}] \\ \qquad = [a_1^m, b_1^m] \times \cdots \times [a_d^m, (a_d^m + b_d^m)/2] \times \cdots \times [a_N^m, b_N^m], \\ \Xi^{m''} = [a_1^{m''}, b_1^{m''}] \times \cdots \times [a_N^{m''}, b_N^{m''}] \\ \qquad = [a_1^m, b_1^m] \times \cdots \times [(a_d^m + b_d^m)/2, b_d^m] \times \cdots \times [a_N^m, b_N^m]. \end{cases} \tag{9.14}$$

Then, local expansions of the process on the newly created subdomains are computed, before being analyzed to determine whether additional refinement is needed. This sequence of analysis and refinement steps is repeated up to convergence. It is

[1]Non-intrusive approaches could alternatively be used.

emphasized that, during refinement, computations are performed in newly created subdomains only, since the local solutions over other subdomains are unaffected, the expansions being local. Note also that this methodology is well suited for parallel implementation, since local computations are independent of each other.

9.2.4 Example

We consider the following test problem:

$$\begin{cases} \dfrac{d\rho}{dt} = \alpha(1-\rho) - \gamma\rho - \beta(\rho-1)\rho^2, \\ \rho(t=0) = \rho_0 \end{cases} \tag{9.15}$$

which models the time-evolution of the surface coverage $\rho \in [0,1]$ for a given species [141]. Coefficient α is the surface absorption rate, γ is the desorption rate, and β is the recombination rate (the exponent 2 is due to the need of two sites for a recombination) [141]. This problem has one or two fixed points according to the value of β and it exhibits smooth dependence on the other parameters. In Sect. 9.2.4.1, the statistics of the solution at $t = 1$ are investigated considering uncertainties in the initial coverage ρ_0 and in the reaction parameter β. Next in Sect. 9.2.4.2, uncertainty in α is also considered, increasing the number of stochastic dimensions to 3. We shall consider that ρ_0 is uniformly distributed in the range $[0,1]$; the distributions of other random parameters are specified later. In order to propagate the uncertainty and determine its impact on the solution, the Galerkin scheme is applied to the governing equation (9.15). The resulting coupled system of ODEs is integrated in order to determine the evolution of the expansion coefficients. A fourth-order Runge-Kutta scheme is used for this purpose, with a time step $\Delta t = 0.01$.

9.2.4.1 Two-Dimensional Problem

In this section, deterministic values of the absorption and desorption rates are used, respectively $\alpha = 1$ and $\gamma = 0.01$. Meanwhile, β is assumed to be uniformly distributed in the interval $[0,20]$. Thus the problem has two stochastic dimensions, $\boldsymbol{\xi} = \{\xi_1, \xi_2\}$ with ξ_1 uniformly distributed in $[0,1]$ and ξ_2 uniformly distributed in $[0,20]$.

Computations are performed with an expansion order No $= 3$, and the results are used to analyze the statistics of the surface coverage, ρ, at time $t = 1$. Figure 9.2 illustrates the computed solution, obtained with different values of the threshold ε_2. The latter is expressed as $\varepsilon_2(\text{Vol}) = C/\sqrt{\text{Vol}^m}$, where C is a prescribed constant. Four values are considered, namely $C = 0.5, 0.1, 0.01$, and 0.001. Shown in Fig. 9.2 are the predicted response surface of ρ plotted against the random input data, the partitions of random parameter space as determined by the adaptive scheme, and

Table 9.2 Computed values of $\langle\rho\rangle$ and $\sigma(\rho)$ for No = 3 and different threshold functions. Also provided are the number of subdomains Nb at the end of the refinement, and estimates of the CPU load and memory requirement. Adapted from [125]

$\varepsilon_2(\text{Vol}^m)$	Nb	$\langle\rho\rangle$	$\sigma(\rho)$	CPU	Memory
$5.10^{-1}/\sqrt{\text{Vol}^m}$	1	0.338,366,540	0.329,585,455	8 301	48
$1.10^{-1}/\sqrt{\text{Vol}^m}$	5	0.350,449,337	0.332,070,259	58 107	240
$1.10^{-2}/\sqrt{\text{Vol}^m}$	16	0.350,410,081	0.332,172,422	224 127	768
$1.10^{-3}/\sqrt{\text{Vol}^m}$	47	0.350,410,331	0.332,171,412	622 575	2 256
$1.10^{-4}/\sqrt{\text{Vol}^m}$	151	0.350,410,331	0.332,171,421	1 884 327	7 248
$1.10^{-5}/\sqrt{\text{Vol}^m}$	439	0.350,410,331	0.332,171,421	5 486 961	21 072

the density of $\rho(t = 1)$. The latter is computed by direct sampling of the spectral expansions using 10^6 samples.

The results show that with $\varepsilon_2 = 0.5/\sqrt{\text{Vol}^m}$ only one domain is generated, *i.e.* no subdivisions are performed. The corresponding response surface and density reflect a poor approximation, particularly since unphysical realizations with $\rho < 0$ and $\rho > 1$ are predicted. When the threshold parameter is decreased to $0.1/\sqrt{\text{Vol}^m}$, one observes that the solution involves Nb = 5 subdomains. Note that refinement is concentrated in areas where the solution exhibits the steepest dependence on random data, as one would expect based on the construction of the adaptive scheme. Also note that no unphysical realizations are predicted for this threshold function, as all predicted value of ρ fall between 0 and 1. As ε_2 is further decreased (last two rows of Fig. 9.2), the partition becomes increasingly more refined, particularly along directions of sharp variation with respect to the random data. At the lower values of C, the insensitivity of the solution density to the selected value of C can be easily appreciated.

For a better appreciation of the convergence of the adaptive scheme, we report in Tables 9.2–9.4 the computed mean and standard deviation of $\rho(t = 1)$ for decreasing values of C, respectively for third, second, and first-order expansions. The tables also provide the number of subdomains generated during adaptive refinement, as well as estimates of the CPU load and the memory requirement. The CPU load is estimated as the product of complexity $\mathcal{C}(1, \text{No})$ (see Sect. 8.4.1.2) with the *total* number of subdomains computed in order for the adaptive refinement to converge. The latter includes the intermediate subdomains that have been subsequently refined. The memory requirement is based on the number of MW coefficients of the respective solution, equal to the product of the final number of subdomains with the dimension of the expansion basis over a single subdomain.

One observes from Tables 9.2–9.4 that for all expansion orders considered the adaptive predictions appear to converge to same values of mean and standard deviation, and that the solution for No = 3 appears most efficient, as the stopping criterion is achieved with the smallest number of subdomains, as well as smaller CPU and memory requirement. This behavior suggests that the solution is essentially smooth everywhere except on localized areas which are well captured and refined by the

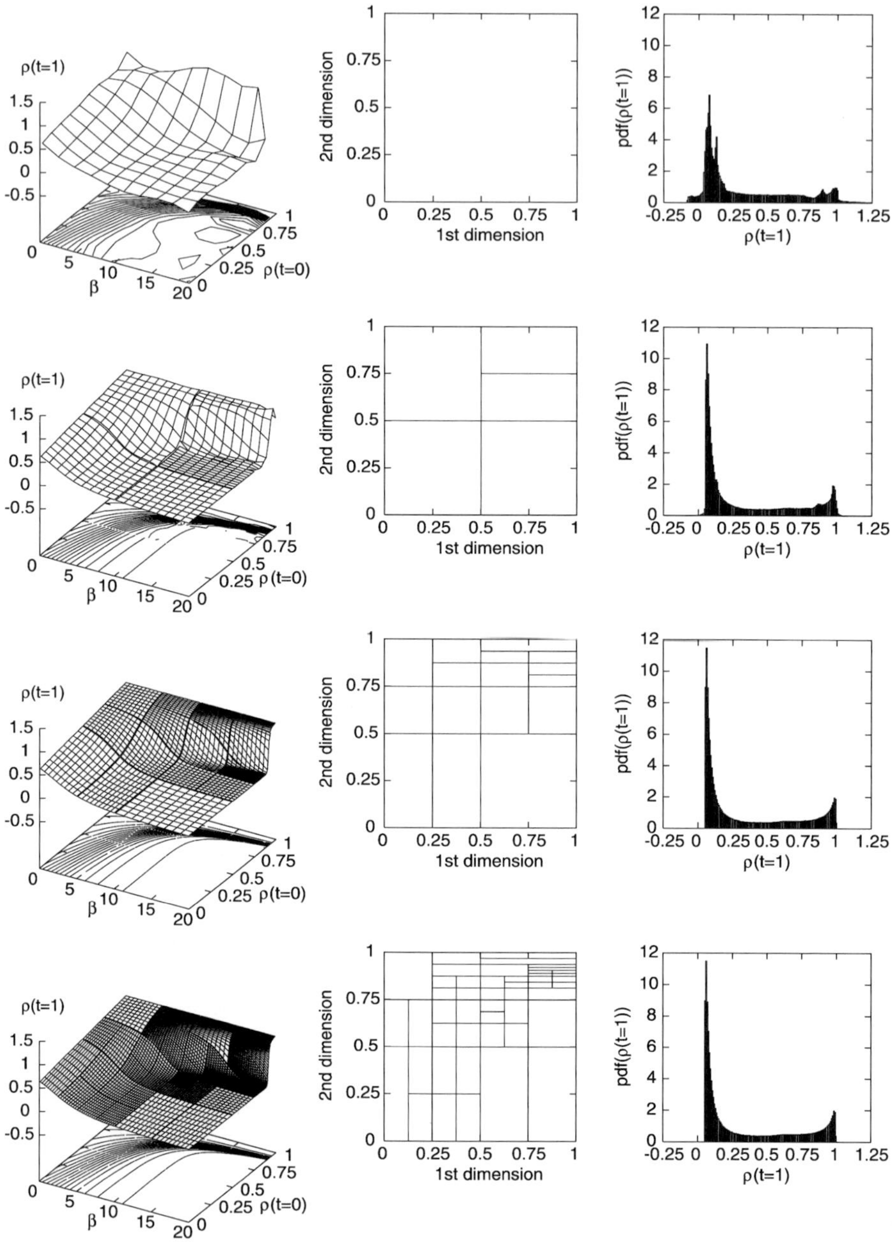

Fig. 9.2 Surface response (*left*), partition of the random parameter space (*middle*) and estimated probability density function (*right*) of the solution at $t = 1$. Results are obtained for $\varepsilon_2(\mathrm{Vol}^m) = 0.5/\sqrt{\mathrm{Vol}^m}, 0.1/\sqrt{\mathrm{Vol}^m},\ 0.01/\sqrt{\mathrm{Vol}^m}$ and $0.001/\sqrt{\mathrm{Vol}^m}$, arranged from top to bottom. The expansion order No = 3. Adapted from [125]

Table 9.3 Computed values of $\langle\rho\rangle$ and $\sigma(\rho)$ for No = 2 and different threshold functions. Also provided are the number of subdomains Nb at the end of the refinement, and estimates of the CPU load and memory requirement. Adapted from [125]

$\varepsilon_2(\mathrm{Vol}^m)$	Nb	$\langle\rho\rangle$	$\sigma(\rho)$	CPU	Memory
$5.10^{-1}/\sqrt{\mathrm{Vol}^m}$	1	0.346,230,683	0.335,764,259	1 667	27
$1.10^{-1}/\sqrt{\mathrm{Vol}^m}$	6	0.350,244,115	0.332,167,461	15 003	162
$1.10^{-2}/\sqrt{\mathrm{Vol}^m}$	34	0.350,410,968	0.332,173,480	85 017	918
$1.10^{-3}/\sqrt{\mathrm{Vol}^m}$	129	0.350,410,326	0.332,171,421	325 065	3 483
$1.10^{-4}/\sqrt{\mathrm{Vol}^m}$	514	0.350,410,330	0.332,171,421	1 291 925	13 878
$1.10^{-5}/\sqrt{\mathrm{Vol}^m}$	1 823	0.350,410,331	0.332,171,421	4 322 531	49 221

Table 9.4 Computed values of $\langle\rho\rangle$ and $\sigma(\rho)$ for No = 1 and different threshold functions. Also provided are the number of subdomains Nb at the end of the refinement, and estimates of the CPU load and memory requirement. *With $\varepsilon_2 = 1.10^{-4}/\sqrt{\mathrm{Vol}^m}$, stopping criterion was not reached; further refinement would have required a larger number of subdomains than allowed (5000). Adapted from [125]

$\varepsilon_2(\mathrm{Vol}^m)$	Nb	$\langle\rho\rangle$	$\sigma(\rho)$	CPU	Memory
$5.10^{-1}/\sqrt{\mathrm{Vol}^m}$	1	0.333,617,359	0.318,197,117	190	12
$1.10^{-1}/\sqrt{\mathrm{Vol}^m}$	8	0.349,221,435	0.330,668,616	2 470	96
$1.10^{-2}/\sqrt{\mathrm{Vol}^m}$	103	0.350,410,367	0.332,170,754	30 970	1 236
$1.10^{-3}/\sqrt{\mathrm{Vol}^m}$	889	0.350,410,309	0.332,171,494	265 050	10 668
$(1.10^{-4}/\sqrt{\mathrm{Vol}^m})^*$	3301	0.350,410,329	0.332,171,426	892 050	39 612

adaptive refinement scheme, as illustrated in Fig. 9.3. Comparison of the results in Fig. 9.3 also suggests that further enhancement of the present scheme could be achieved based on generalizing the refinement strategy so that the local expansion order is also adapted (reduced) during refinement. Excessive refinement in smooth regions observed with No = 1 also indicates that the threshold function may need to be related to the expansion order. Such generalizations will be considered in future work.

In the numerical experiments below, the threshold parameter is decreased before the refinement is stopped. Specifically, starting from $\varepsilon_2(\mathrm{Vol}^m) = C/\sqrt{\mathrm{Vol}^m}$, with $C = 0.5$, the adaptive refinement is carried out until the criterion is satisfied over all subdomains. Then, the constant C is multiplied by a factor of 0.8, and the analysis is repeated. We monitor the convergence of the solution by computing the errors in the expectation and standard deviation of ρ at $t = 1$, using the solution obtained at the end of the refinement with No = 3 as surrogate for the exact solution. In Fig. 9.4, we plot the absolute values of $\langle\rho\rangle - \langle\rho\rangle_{ex}$ and of $\sigma(\rho) - \sigma_{ex}(\rho)$ as a function of the number of subdomains Nb. These plots allow us to estimate how quickly the solutions converge as the partition is refined. The results reveal that the errors decay as $\sim \mathrm{Nb}^{-\mathrm{No}-1}$ for the three tested cases. Note that if the partition were uniform, then the number of subdomains needed to achieve a similar level of accuracy would be

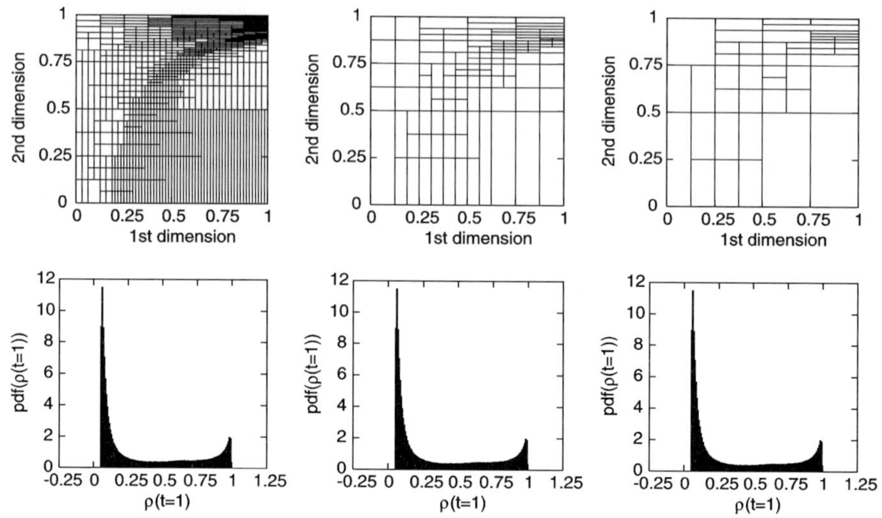

Fig. 9.3 Partition of random parameter space (*top row*) and density of $\rho(t=1)$ (*bottom row*), for No = 1, 2 and 3 (*left to right*). Results were obtained with $\varepsilon_2(\text{Vol}^m) = 10^{-3}/\sqrt{\text{Vol}^m}$. Adapted from [125]

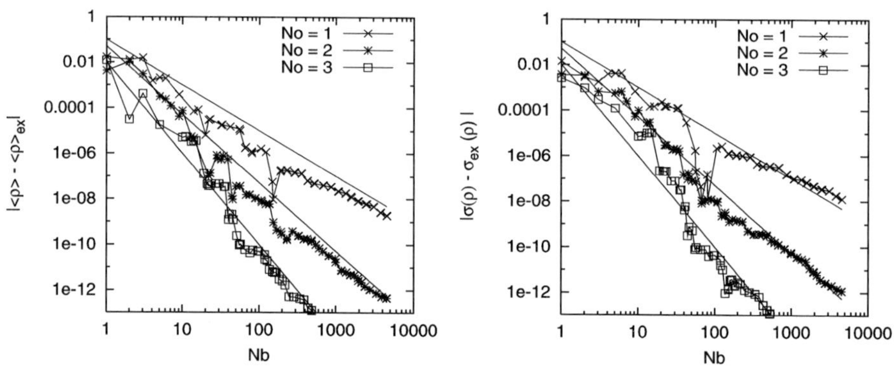

Fig. 9.4 Estimates of the error in $\langle\rho\rangle$ (*left*) and $\sigma(\rho)$ (*right*) at $t = 1$ versus the number Nb of subdomains. Plotted are results obtained with No = 1, 2 and 3. The refinement criterion is progressively made more stringent, as described in the text. *Solid lines* illustrate decay rates proportional to $\sim \text{Nb}^{-(\text{No}+1)}$. Adapted from [125]

much larger than that computed by the adaptive scheme. In particular, if the MRA of Sect. 8.4 were to be applied without refinement at the highest refinement level reached in the adaptive scheme, the computations would require excessive CPU

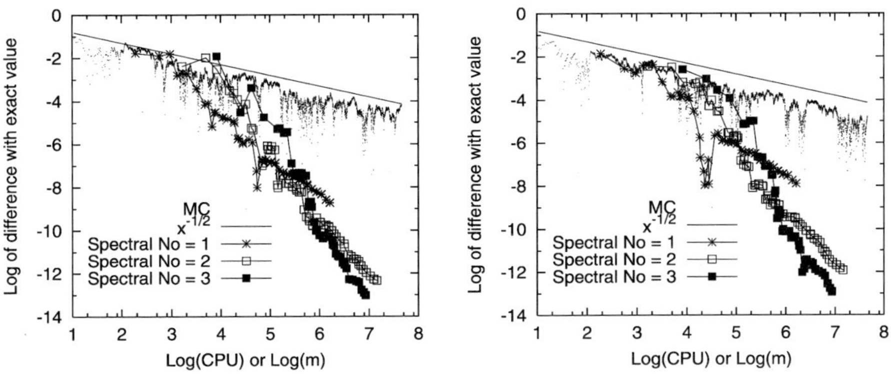

Fig. 9.5 Estimates of the error in $\langle\rho\rangle$ (*left*) and $\sigma(\rho)$ (*right*) at $t = 1$ versus the CPU load in the adaptive computations. Plotted are results obtained with No $= 1$, 2 and 3. The refinement criterion is progressively made more stringent, as described in the text. Also shows are errors in MC simulations, plotted against the number of samples, m. The *solid line* illustrates a decay rate proportional to $m^{-1/2}$. Adapted from [125]

and memory requirements and thus would prove impractical. This highlights the efficiency of the adaptive computations.

Another assessment of the efficiency of the adaptive scheme is performed by estimating the CPU cost needed to achieve a given level of accuracy. Since actual CPU cost is generally dependent on the computing platform and implementation details, the analysis below is based on the CPU load; as in previous sections, the latter is estimated as the product of $\mathcal{C}$ with the total number of subdomains that are computed. This enables us direct comparison with MC approach, for which the CPU load is simply the number of realizations. Figure 9.5 shows the error in the mean and standard deviation of $\rho(t = 1)$ as a function of the CPU-load, denoted CPU, of the spectral adaptive computations using No $= 1,\ 2,$ and 3. Also shown is the error in the MC simulation, plotted against the sample size m. The results indicate that for the present setup the adaptive computations result in very small error estimates. In contrast, the small convergence rate of MC simulations practically limits the accuracy of the corresponding predictions. The figures show that in the intermediate range of CPU, lower errors are achieved with lower expansion order; as the CPU increases, however, accuracy improves with increasing expansion order. This trend supports earlier suggestion that a more efficient approach may be constructed in which the expansion order is also adapted during refinement.

Additional insight into the behavior of the errors is gained by repeating the same experiment, using second, third, and fourth-order expansion, and extending the simulation up to $t = 2$. As shown in [141], as time increases the solution develops sharper fronts, eventually becoming discontinuous at large time. Thus, one would expect that the convergence of higher-order expansions deteriorates as t increases. This is in fact observed in Fig. 9.6, where errors in $\langle\rho\rangle$ and $\sigma(\rho)$ at $t = 2$ are plotted against CPU load. In particular, the results show that in most of the CPU range considered, second-order predictions have smaller errors than third- and fourth-order

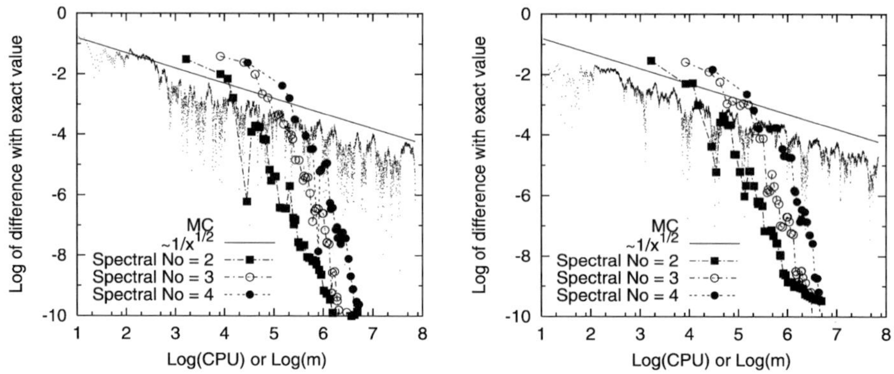

Fig. 9.6 Estimates of the error in $\langle\rho\rangle$ (*left*) and $\sigma(\rho)$ (*right*) at $t = 2$ versus the CPU load in the adaptive computations. Plotted are results obtained with No $= 2$, 3 and 4. The refinement criterion is progressively made more stringent, as described in the text. Also shown are errors in MC simulations, plotted against the number of samples, m. The *solid line* illustrates a decay rate proportional to $m^{-1/2}$. Adapted from [125]

computations. Also note that regardless of the expansion order, the adaptive computations clearly outperform MC simulations.

9.2.4.2 Higher Dimensional Problems

We conclude this section by briefly demonstrating the application of the adaptive scheme to higher dimensional problem. Specifically, results are presented for the case of three random dimensions. The same surface reaction model is used, but uncertainty in the reaction rate is now considered, with α assumed to be uniformly distributed in the range $[0.1, 2]$. In Fig. 9.7, we plot the partition generated during refinement with different threshold functions. The figure shows that the adaptive scheme can accommodate multiple dimensions, and illustrates how refinement is selectively applied along different directions in the space of random data. Also plotted in Fig. 9.7 is the density of the edge-size of the subdomains along the different dimensions; the results are generated using the finest partition ($C = 10^{-3}$). The plot illustrates that more refinement has been applied along the second direction (β) than along the first (initial condition, $\rho(t = 0)$), and that more refinement has been applied on the first direction than the third (α). This last result is not surprising since the dependence of ρ on α, in the range considered, is everywhere smoother than along the other two directions.

9.3 *A posteriori* Error Estimation

As discussed in Sect. 2.3 and Chap. 8, piecewise polynomials [238] and multiwavelets [124, 125] can be used to construct suitable PC bases. A key aspect of these

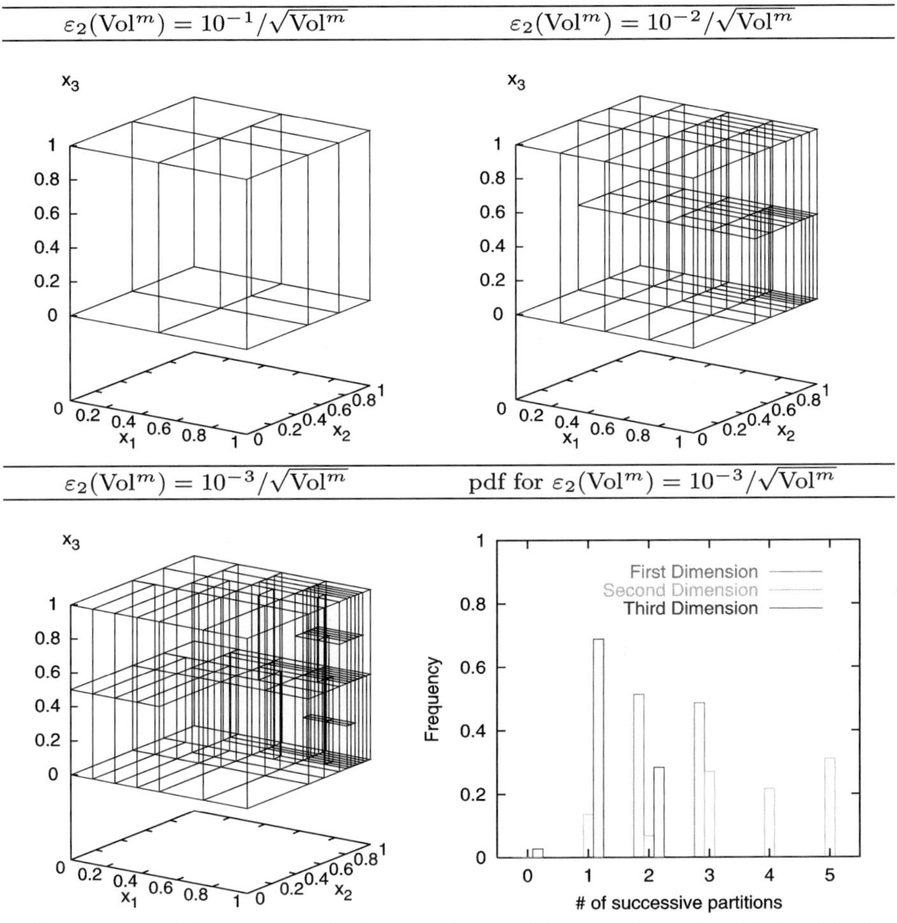

Fig. 9.7 Partition of 3D random parameter space using the adaptive scheme. The refinement threshold is indicated. Also shown in the pdf of the edge-size of the subdomains along the three directions. Adapted from [125]

discontinuous stochastic approximations is that they naturally provide a vehicle for a local adaptation of the representation. This results in improvements in efficiency, namely by maintaining the size of the representation at a reasonable level. The refinement of the stochastic approximation space can consist in of an increase of the local expansion order (p refinement) or in smaller supports (h refinement). We have just illustrated a strategy based on partitioning the domain of the random parameters into subdomains over which independent discontinuous low-order expansions are employed. Heuristic criteria, based on the spectrum of the local expansion, were used to decide whether the local expansion is sufficient or whether it should be improved by means of h refinement, *i.e.* by splitting the subdomain into smaller ones, and along which dimension of the stochastic space. A similar approach was followed [238] but in the context of $h-p$ spectral approximations. The refinement there is also based on heuristic arguments involving the relative contribution of the higher-order terms to the local solution expansion.

Although these approaches provide significant improvements over global PC expansions, in terms of robustness (see e.g. [126]) and computational efficiency, they still lack rigorous criteria for triggering the refinement. Our aim in this section is to outline the construction of a rigorous error estimator, to be used in place of the heuristic error indicators. Specifically, we extend the dual-based *a posteriori* error technique commonly used in the (deterministic) finite element community. This choice is motivated by the firm theoretical foundations of this error estimation technique, and because of its variational framework which makes it suitable for extension to the Galerkin projection of stochastic problems.

In Sect. 9.3.1, the variational formulation of a generic stochastic problem, based on a mathematical model involving parametric (data) uncertainties, is considered. A finite element discretization in space and a piecewise continuous approximation along the stochastic dimensions are then discussed. In Sect. 9.3.2, the dual-based *a posteriori* error estimation is introduced. The methodology makes use of a differentiable functional to measure the difference between the exact (continuous) and approximate (discrete) stochastic solutions. Provided the discrete solution is sufficiently close to the continuous one, their functional difference is shown to be well approximated by a simple estimate. This estimate involves the discrete solutions of the primal and associated dual problems, and the continuous adjoint solution of the dual problem. A classic surrogate of the continuous adjoint solution is proposed, resulting in an error estimate methodology requiring the resolution of the discrete primal problem and two dual problems on different approximation spaces. The dual problems to be solved being linear, the computational overhead of the error estimator is expected to be limited. In Sect. 9.3.3, we discuss the various strategies that can be subsequently used to improve the approximation in order to reduce the error. The reduction of the error can be performed by using smaller elements or by increasing the orders of the spatial and stochastic approximation spaces. As in the deterministic context, the determination of the optimal refinement strategy is an open question, which is made even more difficult and critical in the present stochastic context where the stochastic space may have many dimensions. Consequently, Sect. 9.3.4 presents some numerical tests aiming at showing the validity of the proposed dual-based error estimator in deciding which spatial/stochastic elements need priority refinement. The test problem is based on the 1D Burgers equation, with uncertainty in the viscosity and a boundary condition. Different algorithms of increasing complexity are proposed for the local refinement of the stochastic and spatial approximations, based on the dual-based error estimation.

9.3.1 Variational Formulation

9.3.1.1 Deterministic Variational Problem

We consider the standard variational problem for u on an M-dimensional domain $\mathcal{D} \subset \mathbb{R}^{\mathrm{M}}$ with homogeneous Dirichlet condition on the boundary $\partial\mathcal{D}$:

$$a(u;\varphi) = b(\varphi) \quad \forall\varphi \in \mathcal{V}, \tag{9.16}$$

to be solved for $u \in \mathcal{V}$, a suitable Hilbert space of $\mathcal{D}$. In (9.16), a is a differentiable semi-linear form and b a linear functional.

9.3.1.2 Stochastic Variational Problem

It is assumed that the mathematical model in (9.16) involves some parameters, or data, denoted by a real-valued vector d. The data may for instance consist of some physical constants involved in the model. Clearly, the solution u of the variational problem depends on d, a fact stressed by making explicit the dependence of the variational problem on d:

$$a(u;\varphi|d) = b(\varphi|d) \quad \forall\varphi \in \mathcal{V}. \tag{9.17}$$

If the actual value of the data d is not exactly known, *i.e.* if it is uncertain, it is suitable to consider d as a random quantity defined on an abstract probability space (Θ, Σ, P), Θ being the set of elementary outcomes θ, Σ the σ-algebra of the events and P a probability measure. In this context, the solution of the model is also random. In the following, we use uppercase letters to denote random quantities. Thus, the random solution U and data D are dependent stochastic quantities defined on the same probability space (Θ, Σ, P); the dependency between U and D is prescribed by the model. Uncertainty propagation and quantification thus consists in the inference of the probability law of U, given the probability law of D and the mathematical model relating the two. It is assumed that the problem is well-posed in the sense that (9.17) has almost surely a unique solution.

Denoting $\mathcal{S} = L_2(\Theta, P)$ the space of second-order random variables, we have to solve, for $U \in \mathcal{V} \otimes \mathcal{S}$,

$$A(U;\Phi|D) = B(\Phi|D) \quad \forall\Phi \in \mathcal{V} \otimes \mathcal{S}, \tag{9.18}$$

where

$$A(U;\Phi|D) \equiv \int_\Theta a(U(\theta);\Phi(\theta)|D(\theta)\,\mathrm{d}P(\theta), \tag{9.19}$$

and

$$B(\Phi|D) \equiv \int_\Theta b(\Phi(\theta)|D(\theta))\,\mathrm{d}P(\theta). \tag{9.20}$$

9.3.1.3 Probability Space

As usual, we assume that D can be parameterized using a finite number N of independent, identically distributed, real valued random variables ξ_i, defined on (Θ, Σ, P) with values in $\Xi \subset \mathbb{R}^{\mathrm{N}}$:

$$D = D(\boldsymbol{\xi}), \quad \boldsymbol{\xi} = (\xi_1, \ldots, \xi_{\mathrm{N}}) \in \Xi \subset \mathbb{R}^{\mathrm{N}}. \tag{9.21}$$

We denote by p the known probability density function of ξ_i. By virtue of the independence of the ξ_i's, the joint distribution of $\boldsymbol{\xi}$ can be expressed as:

$$p_{\boldsymbol{\xi}}(\mathbf{y}) = p_{\boldsymbol{\xi}}(y_1, \ldots, y_{\mathrm{N}}) = \prod_{i=1}^{\mathrm{N}} p(y_i). \tag{9.22}$$

Without loss of generality, we shall restrict ourself in the following to $\boldsymbol{\xi}$ having uniformly distributed components on $[-1, 1]$ and consequently we have

$$p(y) = \begin{cases} 1/2 & \text{if } y \in [-1, 1], \\ 0 & \text{otherwise.} \end{cases} \tag{9.23}$$

The variational problem can be formulated in the image probability space $(\Xi, \mathcal{B}_\Xi, P_\Xi)$, using

$$\begin{aligned} A(U; \Phi|D) &= \int_\Theta a(U(\boldsymbol{\xi}(\theta)); \Phi(\boldsymbol{\xi}(\theta))|D(\boldsymbol{\xi}(\theta))\,\mathrm{d}P(\theta) \\ &= \int_\Xi a(U(\mathbf{y}); \Phi(\mathbf{y})|D(\mathbf{y})p_{\boldsymbol{\xi}}(\mathbf{y})\,\mathrm{d}\mathbf{y} \equiv \langle a(U; \Phi|D)\rangle_\Xi, \end{aligned} \tag{9.24}$$

$$\begin{aligned} B(\Phi|D) &= \int_\Theta b(\Phi(\boldsymbol{\xi}(\theta))|D(\boldsymbol{\xi}(\theta)))\,\mathrm{d}P(\theta) \\ &= \int_\Xi b(\Phi(\mathbf{y})|D(\mathbf{y}))p_{\boldsymbol{\xi}}(\mathbf{y})\,\mathrm{d}\mathbf{y} \equiv \langle b(\Phi|D)\rangle_\Xi. \end{aligned} \tag{9.25}$$

Moreover, the stochastic functional space is now $\mathcal{S} \equiv L_2(\Xi, P_\Xi)$ and the variational problem becomes

$$A(U; \Phi|D) = B(\Phi|D) \quad \forall \Phi \in \mathcal{V} \otimes \mathcal{S}, \tag{9.26}$$

to be solved for $U \in \mathcal{U} \equiv \mathcal{V} \otimes \mathcal{S}$.

9.3.1.4 Stochastic Discretization

Following [238], we rely on piecewise orthogonal polynomials to construct the stochastic approximation space. As in Sect. 9.2, the stochastic domain Ξ is divided into a collection of Nb non-overlapping subdomains $\Xi^{(m)}$ referred to as stochastic elements (SEs). The SEs are hyper-rectangles as in Sect. 9.2.1:

$$\Xi = \bigcup_{m=1}^{\text{Nb}} \Xi^{(m)}, \quad \Xi^{(m)} = [a_1^{(m)}, b_1^{(m)}] \times \cdots \times [a_{\text{N}}^{(m)}, b_{\text{N}}^{(m)}]. \tag{9.27}$$

On $\Xi^{(m)}$, the dependence of the data and solution on $\boldsymbol{\xi}$, is expressed as a truncated Fourier-like series,

$$U(\boldsymbol{\xi} \in \Xi^{(m)}) = \sum_{k=0}^{\text{P}(m)} u_k^{(m)} \Psi_k^{(m)}(\boldsymbol{\xi}), \qquad D(\boldsymbol{\xi} \in \Xi^{(m)}) = \sum_{k=0}^{\text{P}(m)} d_k^{(m)} \Psi_k^{(m)}(\boldsymbol{\xi}), \tag{9.28}$$

where $\Psi_k^{(m)}(\boldsymbol{\xi})$ are orthogonal random polynomials in $\boldsymbol{\xi}$ and $u_k^{(m)}$, $d_k^{(m)}$ are the deterministic expansion coefficients over $\Xi^{(m)}$ of the solution and data, respectively. The orthogonality of the random polynomials is defined with regards to the expectation over the *respective* SE. Denoting $\langle \cdot \rangle_{\Xi^{(m)}}$ the expectation over the m-th SE, the orthogonality of the polynomials is expressed as:

$$\begin{aligned} \left\langle \Psi_k^{(m)} \Psi_{k'}^{(m)} \right\rangle_{\Xi^{(m)}} &= \frac{1}{|\Xi^{(m)}|} \int_{\Xi^{(m)}} \Psi_k^{(m)}(\mathbf{y}) \Psi_{k'}^{(m)}(\mathbf{y}) p_{\boldsymbol{\xi}}(\mathbf{y})\, d\mathbf{y} \\ &= \delta_{kk'} \left\langle \Psi_k^{(m)2} \right\rangle_{\Xi^{(m)}}, \end{aligned} \tag{9.29}$$

where

$$\left| \Xi^{(m)} \right| = \int_{\Xi^{(m)}} p_{\boldsymbol{\xi}}(\mathbf{y})\, d\mathbf{y}, \tag{9.30}$$

and $\delta_{kk'}$ is the Kronecker delta symbol. These polynomials vanish outside their respective support:

$$\Psi_k^{(m)}(\boldsymbol{\xi} \notin \Xi^{(m)}) = 0 \quad \forall k = 0, \ldots, \text{P}(m). \tag{9.31}$$

The number of terms $\text{P}(m)$ in (9.28) is a function of the selected stochastic expansion order $\text{No}(m)$ of the SE:

$$\text{P}(m) + 1 = \frac{(\text{No}(m) + \text{N})!}{\text{No}(m)!\text{N}!}. \tag{9.32}$$

The ξ_i being uniformly distributed, the polynomials $\Psi_k^{(m)}$ are simply rescaled and shifted multidimensional Legendre polynomials [1]. In case of a non-uniform distribution, one can rely on an expansion of uniformly-distributed random variables $\boldsymbol{\eta}$ related to $\boldsymbol{\xi}$ through the inverse distribution function of $\boldsymbol{\xi}$. In fact, relying on such transformations, the local basis for $\Xi^{(m)}$ consists of a rescaled version of $\mathcal{B}_p(\Xi^{(m)})$ of Sect. 9.2.2. The stochastic approximation space is

$$\mathcal{S}^h = \text{span}\left(\{\Psi_k^{(m)}\}, 1 \leq m \leq \text{Nb}, 0 \leq k \leq \text{P}(m) \right), \tag{9.33}$$

and the stochastic approximation can be improved by increasing the number Nb of SEs, *i.e.* by refining the partition of Ξ, and/or increasing the stochastic expansion order No(m) over some SEs.

9.3.1.5 Spatial Discretization

Consider a partition of $\mathcal{D}$ into a set of $\mathrm{N_X}$ non-overlapping finite elements (FEs) with respective support $\mathcal{D}^{(l)}$ for $l = 1, \ldots, \mathrm{N_X}$:

$$\mathcal{D} = \bigcup_{l=1}^{\mathrm{N_X}} \mathcal{D}^{(l)}. \tag{9.34}$$

The FE approximation of the continuous solution U, denoted U^h, over the element $\mathcal{D}^{(l)}$ is given by:

$$U^h(x \in \mathcal{D}^{(l)}) = \sum_{i=1}^{\mathrm{Nd}(l)} U_i^{(l)} \phi_i^{(l)}(x), \tag{9.35}$$

where Nd(l) is the number of degrees of freedom of the l-th element and $\phi_i^{(l)}$ the associated spatial shape functions. We denote q(l) the polynomial order of the shape functions over $\mathcal{D}^{(l)}$. Thus, the spatial approximation space is:

$$\mathcal{V}^h = \mathrm{span}\left(\{\phi_i^{(l)}\}, 1 \le l \le \mathrm{N_X}, 1 \le i \le \mathrm{Nd}(l)\right), \tag{9.36}$$

and the spatial approximation can be improved by refining the partition of the spatial domain $\mathcal{D}$, or increasing the spatial order q(l) of some FEs.

9.3.1.6 Approximation Space $\mathcal{U}^h$

From the stochastic and spatial approximation spaces defined above, the approximation space $\mathcal{U}^h$ of the stochastic variational problem is:

$$\mathcal{U}^h = \mathcal{V}^h \otimes \mathcal{S}^h, \tag{9.37}$$

and the solution at a point $(x, \boldsymbol{\xi})$ of $\Omega \equiv \mathcal{D} \times \Xi$ can be expressed as:

$$U(x \in \mathcal{D}^{(l)}, \boldsymbol{\xi} \in \Xi^{(m)}) = \sum_{i=1}^{\mathrm{Nd}(l)} \sum_{k=0}^{\mathrm{P}(m)} u_{i,k}^{(l,m)} \phi_i^{(l)}(x) \Psi_k^{(m)}(\boldsymbol{\xi}), \tag{9.38}$$

where the deterministic coefficient $u_{i,k}^{(l,m)}$ is the k-th uncertainty mode of the m-th SE for the i-th degree of freedom of the l-th FE.

An immediate consequence of the tensored construction of the approximation space $\mathcal{U}^h$ is that the spatial FE discretization is the same for all the stochastic elements $\Xi^{(m)}$, and conversely the stochastic discretization is the same for all spatial finite elements $\mathcal{D}^{(l)}$. This is clearly not optimal as some portions of the stochastic domain Ξ may require finer spatial discretization than others to achieve a similar accuracy. Conversely, the solution in some parts of spatial domain $\mathcal{D}$ may exhibit more complex dependences with regards to $D(\boldsymbol{\xi})$, therefore requiring a finer stochastic discretization than at other locations. However, for the tensored construction $\mathcal{U}^h$, the discrete solution can be improved through (a) refinement of the FE approximation space $\mathcal{V}^h$ uniformly over Ξ, and (b) refinement of the stochastic approximation space $\mathcal{S}^h$ uniformly over $\mathcal{D}$.

In fact, this symmetric situation can be easily relaxed: an adaptation to each SE of the spatial discretization, *i.e.* the number of elements $\mathrm{N_x}$ and/or the number of degrees of freedom of the elements Nd, causes no difficulty. This is due to the complete independence of the solution over different stochastic elements, a feature emerging from the absence of any differential operator along the uncertainty dimensions. Consequently, the adaptation of $\mathcal{V}^h$ with the SEs was actually implemented and used for the generation of the results presented below. However, to simplify the presentation of the method and the notation, this feature is not detailed here. On the other hand, using a variable stochastic approximation for different spatial FE is much more cumbersome and remains to be investigated. This adaptation would require the development of non-obvious matching conditions of the stochastic approximation across FE boundaries.

9.3.2 Dual-based a posteriori Error Estimate

9.3.2.1 *A posteriori* Error

We assume that the functionals are linear with respect to arguments placed on the right side of a semicolon. For a finite dimensional subspace $\mathcal{U}^h \subset \mathcal{U}$, the discretized solution $U^h \in \mathcal{U}^h$ is the discrete Galerkin approximation:

$$A(U^h; \Phi^h|D^h) = B(\Phi^h|D^h) \quad \forall \Phi^h \in \mathcal{U}^h. \tag{9.39}$$

Let $\mathcal{J} : \Omega \rightarrow \mathbb{R}$ be a differentiable functional of the solution. In the spirit of [12] and [11] among others, one is interested in approximating $\mathcal{J}(U)$ as closely as possible by $\mathcal{J}(U^h)$, *i.e.* to minimize the difference $\mathcal{J}(U) - \mathcal{J}(U^h)$ in some sense. We seek an expression of $\mathcal{J}(U) - \mathcal{J}(U^h)$. To this end, let us define the Lagrangian L of the continuous solution by:

$$\mathrm{L}(U; Z) \equiv \mathcal{J}(U) + B(Z|D) - A(U; Z|D), \tag{9.40}$$

where $Z \in \mathcal{U}$ is the adjoint variable of the continuous problem. The adjoint variable Z is a Lagrange multiplier of the optimization problem for the minimization of

$\mathcal{J}(U)$ under the constraints of (9.26). Formally, this minimum corresponds to the stationary points of L:

$$\frac{\partial \mathrm{L}}{\partial U} = \mathcal{J}'(U;\Phi') - A'(U;\Phi', Z|D) = 0 \quad \forall \Phi' \in \mathcal{U}, \tag{9.41}$$

$$\frac{\partial \mathrm{L}}{\partial Z} = B(\Phi|D) - A(U;\Phi|D) = 0 \quad \forall \Phi \in \mathcal{U}. \tag{9.42}$$

Equation (9.41) is the adjoint (or dual) problem, while (9.42) is the state (or primal) problem. Note that the derivatives are in the Gâteaux sense:

$$\mathcal{J}'(U;\Phi') = \lim_{\varepsilon\to 0} \frac{\mathcal{J}(U+\varepsilon\Phi') - \mathcal{J}(U)}{\varepsilon},$$

$$A'(U;\Phi', Z|D) = \lim_{\varepsilon\to 0} \frac{A(U+\varepsilon\Phi'; Z|D) - A(U; Z|D)}{\varepsilon}.$$

The discrete counterparts of the dual and primal problems are respectively given by

$$\mathcal{J}'(U^h;\Phi^{h'}) - A'(U^h;\Phi^{h'}, Z^h|D^h) = 0 \quad \forall \Phi^{h'} \in \mathcal{U}^h, \tag{9.43}$$

$$B(\Phi^h|D^h) - A(U^h;\Phi^h|D^h) = 0 \quad \forall \Phi^h \in \mathcal{U}^h. \tag{9.44}$$

Combining these results, one obtains

$$\begin{aligned} \mathrm{L}(U,Z) - \mathrm{L}(U^h, Z^h) &= \mathcal{J}(U) + B(Z) - A(U;Z) - \mathcal{J}(U^h) - B(Z^h) \\ &\quad + A(U^h; Z^h) \\ &= \mathcal{J}(U) - \mathcal{J}(U^h), \end{aligned} \tag{9.45}$$

where the dependences of A and F on D have been dropped to simplify the notations. It is seen from (9.45) that the difference in $\mathcal{J}$ for the continuous and discrete solutions is equal to the difference in their respective Lagrangians.

9.3.2.2 Posterior Error Estimation

Following [17], we now derive a more practical expression for the difference $\mathcal{J}(U) - \mathcal{J}(U^h)$. Let $K(\cdot)$ be a differentiable functional on a given functional space $\mathcal{W}$. The difference $K(v) - K(v^h)$, for v and $v' \in \mathcal{W}$, can be expressed as an integral between v and v^h of the derivative of K:

$$K(v) - K(v^h) = \int_{v^h}^{v} K'(v')\,\mathrm{d}v'. \tag{9.46}$$

The integration path can be parameterized to obtain

$$K(v) - K(v^h) = \int_0^1 K'(v^h + s(v - v^h))(v - v^h)\,\mathrm{d}s$$

$$= \int_0^1 K'(v^h + s e_v; e_v)\,\mathrm{d}s, \tag{9.47}$$

where $e_v \equiv v - v^h$. Here, use was made again of the assumption regarding the linearity of the functional forms with regard to the arguments on the right-side of the semicolon. Using $K'(v) = 0$, the right-hand side of (9.47) can be rewritten as

$$K(v) - K(v^h) = \int_0^1 K'(v^h + s e_v; e_v)\,\mathrm{d}s \\ + \frac{1}{2}\Big[K'(v^h; e_v) - K'(v^h; e_v) + K'(v; e_v)\Big]. \tag{9.48}$$

Making use of the Galerkin orthogonality and the trapezoidal rule results in:

$$K(v) - K(v^h) = \frac{1}{2}K'(v^h; e_v) + \frac{1}{2}\int_0^1 K^{(3)}(v^h + s e_v; e_v^3) s(s-1)\,\mathrm{d}s. \tag{9.49}$$

Applying this relation to the difference of the Lagrangian of the continuous and discrete solutions leads to:

$$\mathcal{J}(U) - \mathcal{J}(U^h) = \frac{1}{2}\Big[\rho(U^h, Z - \Phi'^h) + \rho^*(Z^h, U - \Phi^h)\Big] + \tilde{R}, \tag{9.50}$$

with the residuals

$$\rho(U^h, \cdot) \equiv B(\cdot) - A(U^h; \cdot), \tag{9.51}$$

$$\rho^*(Z^h, \cdot) \equiv \mathcal{J}'(U^h, \cdot) - A'(U^h; \cdot, Z^h). \tag{9.52}$$

The remainder term $\tilde{R}$ in (9.50) is given by:

$$\tilde{R} = \frac{1}{2}\int_0^1 \Big[\mathcal{J}^{(3)}(U^h + sE_U; E_U^3) - A^{(3)}(U^h + sE_U; E_U^3, Z^h + sE_Z) \\ - 3A''(U^h + sE_U; E_U^2 E_Z)\Big] s(s-1)\,\mathrm{d}s, \tag{9.53}$$

with the error terms defined as $E_U = U - U^h$ and $E_Z = Z - Z^h$. Thus $\tilde{R}$ is cubic in the error, suggesting that it can be neglected provided that the continuous and discrete solutions are sufficiently close. It is also seen that the residuals are functionals of both the primal and dual continuous solutions U and Z, such that using (9.50) to estimate $\mathcal{J}(U) - \mathcal{J}(U^h)$ would require two surrogates of U and Z even if $\tilde{R}$ is neglected. In fact, the expression can be further simplified to remove the contribution of U. Using integration by parts of $\tilde{R}$, one obtains [17]:

$$\rho^*(Z^h, U - \Phi^h) = \rho(U^h, Z - \Phi'^h) + \Delta\rho, \tag{9.54}$$

where

$$\Delta\rho = \int_0^1 \Big[A''(U^h + sE_U; E_U^2, Z^h + sE_Z) - \mathcal{J}''(U^h + sE_U; E_U^2)\Big]\,\mathrm{d}s. \tag{9.55}$$

Introducing this result into (9.50) leads to the following expression for the approximation error:

$$\mathcal{J}(U) - \mathcal{J}(U^h) = \rho(U^h, Z - \Phi'^h) + r, \tag{9.56}$$

where

$$r = \int_0^1 \left[A''(U^h + sE_U; E_U^2, Z) - \mathcal{J}''(U^h + sE_U; E_U^2) \right] s \, \mathrm{d}s. \tag{9.57}$$

The remainder term r is now quadratic in E_U and will be neglected hereafter, assuming again that the discrete solution U^h is indeed sufficiently close to U.

9.3.2.3 Methodology

At this point we have an estimate of the approximation error given by

$$\mathcal{J}(U) - \mathcal{J}(U^h) \approx B(Z - Z^h | D^h) - A(U^h; Z - Z^h | D^h), \tag{9.58}$$

where we have substituted Φ'^h by the adjoint solution of the discrete problem (9.56), as usual in *a posteriori* error methodology. To evaluate this estimate, one needs to know the solutions U^h and Z^h of the primal and dual discrete problems and the solution Z of the continuous dual problem in (9.41). However, the continuous dual problem cannot be solved as it requires the knowledge of the exact solution U. Instead, a surrogate of Z denoted $\tilde{Z}$ is used. This surrogate is classically constructed by solving a discrete dual problem on a refined finite dimensional space $\mathcal{U}^{\tilde{h}}$ containing $\mathcal{U}^h$. The methodology is thus the following. Given an approximation space $\mathcal{U}^h$ we solve the primal and dual problems in (9.44), (9.43) for U^h and $Z^h \in \mathcal{U}^h$. The refined space $\mathcal{U}^{\tilde{h}} \supset \mathcal{U}^h$ is constructed by increasing the polynomial orders of both the approximation space $\mathcal{V}^h$ and $\mathcal{S}^h$, and we solve the following dual problem for $\tilde{Z} \in \mathcal{U}^{\tilde{h}}$

$$\mathcal{J}'(U^h; \Phi) - A'(U^h; \Phi, \tilde{Z} | D^{\tilde{h}}) = 0 \quad \forall \Phi \in \mathcal{U}^{\tilde{h}}. \tag{9.59}$$

This yields the *a posteriori* error estimate given by

$$\mathcal{J}(U) - \mathcal{J}(U^h) \approx B(\tilde{Z} - Z^h | D^h) - A(U^h; \tilde{Z} - Z^h | D^h). \tag{9.60}$$

Two important remarks are made at this point. First, it is emphasized that the dual problems are linear and significantly less expensive to solve than the primal problems, even in an enriched approximation space. Second, as can be appreciated from (9.59), the adjoint solution $\tilde{Z}$ is based on a functional form A' constructed with the approximation of D on the enriched space $\mathcal{U}^{\tilde{h}}$. As a consequence, the resulting error estimate based on $\tilde{Z}$ accounts for possible error in the approximation of the uncertain data $D(\boldsymbol{\xi})$ on $\mathcal{S}^h$.

9.3.3 Refinement Procedure

9.3.3.1 Global and Local Error Estimates

The *a posteriori* error methodology described in Sect. 9.3.2 provides an estimate of $\mathcal{J}(U) - \mathcal{J}(U^h)$ according to (9.60). The global approximation error η is therefore:

$$\begin{aligned}\eta &= \left| A(U^h; \tilde{Z} - Z^h | D^h) - B(\tilde{Z} - Z^h | D^h) \right| \\ &= \left| \left\langle a(U^h; \tilde{Z} - Z^h | D^h) - b(\tilde{Z} - Z^h | D^h) \right\rangle_{\Xi} \right| \\ &\leq \sum_{m=1}^{\text{Nb}} \left| \Xi^{(m)} \right| \left| \left\langle a(U^h; \tilde{Z} - Z^h | D^h) - b(\tilde{Z} - Z^h | D^h) \right\rangle_{\Xi^{(m)}} \right|. \end{aligned} \quad (9.61)$$

Defining the local error on the element $\mathcal{D}^{(l)} \times \Xi^{(m)}$ by

$$\eta_{l,m} \equiv \left| \Xi^{(m)} \right| \left| \left\langle \int_{\mathcal{D}^{(l)}} \left[\tilde{a}(U^h; \tilde{Z} - Z^h | D^h) - \tilde{b}(\tilde{Z} - Z^h | D^h) \right] \mathrm{d}x \right\rangle_{\Xi^{(m)}} \right|, \quad (9.62)$$

where

$$\int_{\mathcal{D}} \tilde{a}(u; v|d) \, \mathrm{d}x = a(u; v|d), \qquad \int_{\mathcal{D}} \tilde{b}(v|d) \, \mathrm{d}x = b(v|d),$$

we obtain the following inequality:

$$\eta \leq \sum_{l=1}^{\mathrm{N_x}} \sum_{m=1}^{\text{Nb}} \eta_{l,m}. \quad (9.63)$$

Thus, the objective is to refine the approximation space $\mathcal{U}^h$ in order to reduce the global error η as estimated from the *a posteriori* error analysis. A popular strategy to ensure that the global error falls below a given threshold value ϵ_η is to refine the approximation such that

$$\eta_{l,m} < \frac{\epsilon_\eta}{\mathrm{N_x}\text{Nb}} = \epsilon, \quad \forall l, m \in [1, \mathrm{N_x}] \times [1, \text{Nb}]. \quad (9.64)$$

9.3.3.2 Refinement Strategies

If the criterion given in (9.64) is not satisfied for at least one SFE, the approximation space needs refinement. Different types of refinements are possible. First, from the tensored construction of the approximation space, $\mathcal{U}^h = \mathcal{V}^h \otimes \mathcal{S}^h$, it is seen that the refinement may concern the spatial or stochastic approximation spaces, or both. To distinguish these two types of refinement we shall refer to x and ξ refinement as the spatial and stochastic refinements, respectively. Second, the refinement can be based on construction of finer partitions of the domains or on increased approximation orders, hereafter referred to as h and p refinements respectively. Therefore, we can choose between four fundamental types of refinements to reduce the approximation error to satisfy (9.64), h_ξ, h_x, p_ξ, or p_x refinements, or any combination of the four.

The problem is thus to find the refinement strategy that yields the largest decay of the discretization error for the lowest computational cost. The difficulty here is that the local error estimate only provides some information about the elements (SEs and FEs) over which the approximation is insufficient. In other words, if for some l and m the local error is such that $\eta_{l,m} > \epsilon$ then we can only safely consider that the approximation error over $\Xi^{(m)} \times \mathcal{D}^{(l)}$ is too large but nothing more. Specifically, it is not possible to decide (a) between h or p refinement and (b) whether one should enrich the approximation space $\mathcal{V}^h$ or $\mathcal{S}^h$.

Difficulty (a) is a classic problem in (deterministic) hp-finite-element methods. In the deterministic context, different strategies have been proposed to support the decision regarding h or p refinement, and most of these strategies are based on trial approaches. For instance, in [100], a systematic trial of h refinement is performed. The efficiency of the h refinement is subsequently measured by comparing the resulting error reduction with its theoretical value estimated using the convergence rate of the FE scheme. If the efficiency of the h refinement is not satisfactory, a p refinement is enforced at the following refinement step. This type of trial/verification approach has not been retained here because of its numerical cost. Difficulty (b) is on the contrary specific to stochastic finite element methods and thus remains entirely to be investigated. A possible way to deal with difficulty (b) can be envisioned again by a trial approach where one would apply successively x and ξ refinements to measure the respective effectiveness in error reduction. Again, trial approaches are expected to be overly expensive in the stochastic context where the size of the discrete problems to be solved can be many times larger than for the deterministic case; better approaches, yet to be conceived, are consequently needed.

Another issue arising in the stochastic context is the potentially large dimensionality N of the stochastic domain Ξ: an isotropic h_ξ refinement, where SEs are broken into smaller ones along each dimension ξ_i, can quickly result in a prohibitively large number of SEs. This issue was already observed in [124–126] where adaptive multiwavelet approximations are used. Rather, it is desirable to gain further information on the structure of the local error $\eta_{l,m}$ in order to refine along the error's principal directions solely. Several approaches may be considered to deal with such refinement. In the context of deterministic finite element methods, several anisotropic error estimators have been rigorously derived based on higher-order information. Among others, [184] and [155] use the Hessian matrix based on Clément interpolants [35] to derive an estimate of the directional errors. Thought attractive, this method has only been derived for first-order finite elements (P1) and its extension to higher-order remains largely an open problem. This limitation precludes its use in the present context where the approximation order No is routinely larger than one.

Considering all these difficulties, it was decided to first verify the effectiveness of the dual-based *a posteriori* error estimation in indicating which elements need refinement, and to delay the question of the refinement strategy decision to a future work. Consequently, we present in the next section some numerical tests whose essential purposes are to demonstrate that the proposed error estimator indeed detects areas of Ω where the error is the most significant. Still, we perform refinements, of increasing complexity, without pretending in any way that the decision algorithms used yield optimal approximation spaces, but merely that they allow for a reduction of the global error to an arbitrary small level.

9.3.4 Application to Burgers Equation

To test the *a posteriori* error estimator, we consider the 1-D Burgers equation on the spatial domain $\mathcal{D} \in [x^-, x^+]$:

$$\begin{cases} \frac{1}{2}(u(1-u))_x - \mu u_{xx} = 0 \quad \forall x \in [x^-, x^+], \\ u(x^-) = u^-, \qquad u(x^+) = u^+. \end{cases} \tag{9.65}$$

Depending on the boundary conditions, the solution of the Burgers equation exhibits areas where $u(x)$ is nearly constant and equal to u^- (for $x \simeq x^-$) and u^+ (for $x \simeq x^+$) with a central area, the transition layer, where u quickly evolves from u^- to u^+ according to an hyperbolic tangent profile having an increasing steepness with decreasing the fluid viscosity.

9.3.4.1 Uncertainty Settings

We consider the random solution $U(x, \boldsymbol{\xi})$ of the Burgers equation which arises when the viscosity μ is uncertain and parameterized by the random vector $\boldsymbol{\xi}$: $\mu = \mu(\boldsymbol{\xi})$. As discussed above, $\boldsymbol{\xi}$ is uniformly distributed in $[-1, 1]^{\mathrm{N}}$. The number N of random variables depends on the parameterization. To ensure the existence of a solution to the stochastic problem, the parameterization is selected such that the viscosity is almost surely positive. The stochastic Burgers equation is thus:

$$\begin{cases} \frac{1}{2}\left[U(x,\boldsymbol{\xi})(1-U(x,\boldsymbol{\xi}))\right]_x - \mu(\boldsymbol{\xi})U_{xx}(x,\boldsymbol{\xi}) = 0 \quad \forall x \in [x^-, x^+], \\ U(x^-,\boldsymbol{\xi}) = u^-, \qquad U(x^+,\boldsymbol{\xi}) = u^+. \end{cases} \tag{9.66}$$

The viscosity is parameterized using $\mathrm{N} = 2$ random variables as follows

$$\mu(\boldsymbol{\xi}) = \mu_0 + \mu_1\xi_1 + \mu_2\xi_2, \quad \mu_0 > 0. \tag{9.67}$$

The expectation of the viscosity is $\langle\mu\rangle_\Xi = \mu_0$, and provided that $|\mu_1| + |\mu_2| < \mu_0$, $\mu(\boldsymbol{\xi})$ is almost surely positive. We shall set in the following $\mu_0 = 1$, $\mu_1 = 0.62$ and $\mu_2 = 0.36$. The resulting probability density function (pdf) of the random viscosity is plotted in Fig. 9.8.

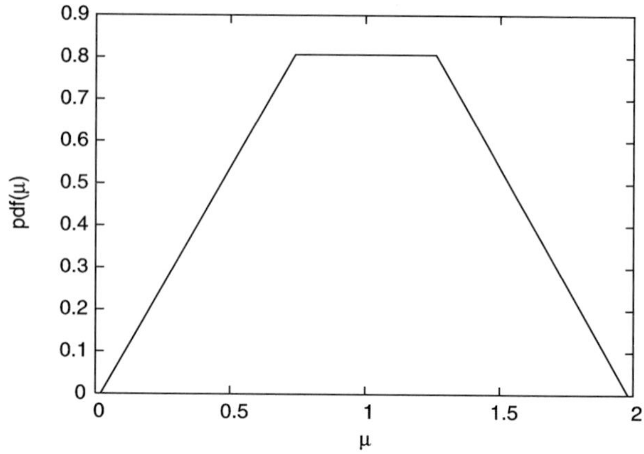

Fig. 9.8 Probability density function of the viscosity. Adapted from [147]

Finally, we set $x^- = -10$ and $x^+ = 10$ and use as boundary conditions

$$u^- = \frac{1}{2}\left[1 + \tanh\left(\frac{x^-}{4\mu_0}\right)\right] \approx 0, \qquad u^+ = \frac{1}{2}\left[1 + \tanh\left(\frac{x^+}{4\mu_0}\right)\right] \approx 1. \quad (9.68)$$

For these boundary conditions,

$$u(x) = \frac{1}{2}\left[1 + \tanh\left(\frac{x}{4\mu_0}\right)\right],$$

is the solution of the deterministic Burgers equation for $\mu = \mu_0$ [236].

9.3.4.2 Variational Problems

The variational formulation of the Burgers equation is derived. By means of integration by parts, one obtains the primal problem to be solved for $U \in \mathcal{U}$:

$$A(U;\Phi|D) - B(\Phi|D) = \left\langle \int_{\mathcal{D}} [U(1-U) - 2\mu U_x]\,\Phi_x \,\mathrm{d}x \right\rangle_{\Xi} = 0 \quad \forall \Phi \in \mathcal{U}^*, \quad (9.69)$$

where $\mathcal{U}^* = \mathcal{V}^* \otimes \mathcal{S}$ is constructed using the restriction of $\mathcal{V}$ to functions vanishing on $\partial\mathcal{D}$. For the derivation of the adjoint problem, an obvious choice is to base the *a posteriori* error estimate on the solution itself, *i.e.* using

$$\mathcal{J}(U) = \left\langle \int_{\mathcal{D}} U \,\mathrm{d}x \right\rangle_{\Xi}. \quad (9.70)$$

For this choice, we have

$$\mathcal{J}'(U;\Phi') = \lim_{\varepsilon \to 0} \frac{\mathcal{J}(U + \varepsilon\Phi') - \mathcal{J}(U)}{\varepsilon} = \left\langle \int_{\mathcal{D}} \Phi' \,\mathrm{d}x \right\rangle_{\Xi} \quad \forall \Phi' \in \mathcal{U} \quad (9.71)$$

and

$$A'(U,\Phi';Z|D) = \lim_{\varepsilon\to 0} \frac{A(U+\varepsilon\Phi';Z|D) - A(U;Z|D)}{\varepsilon} \tag{9.72}$$

$$= \left\langle \int_{\mathcal{D}} \left[(1-2U)Z_x\Phi' - 2\mu Z_x \Phi'_x\right] \mathrm{d}x \right\rangle_{\Xi}. \tag{9.73}$$

Thus, the dual problem can be expressed as

$$\left\langle \int_{\mathcal{D}} \left[(1-2U)Z_x\Phi' - 2\mu Z_x \Phi'_x + \Phi'\right] \mathrm{d}x \right\rangle_{\Xi} = 0 \quad \forall \Phi' \in \mathcal{U}, \tag{9.74}$$

for $Z \in \mathcal{U}$ and deterministic boundary conditions $Z(x^-) = Z(x^+) = 0$.

For the discretization of the primal and dual problems, we use Chebychev finite elements to construct $\mathcal{V}^h$, and Legendre polynomials (uniform distribution) for $\mathcal{S}^h$ [1]. To compute the surrogate of the exact solution of the adjoint equation, the approximation $\mathcal{U}^h$ is extended to $\mathcal{U}^{\tilde{h}}$ by increasing the orders of the Chebychev and Legendre polynomials by one unit, as explained in Sect. 9.3.2. In practice, this step is cheap due to the linearity of the dual problem, as seen from (9.74), and the resolution of the dual problem only contributes to a small fraction of the global CPU time.

A fundamental point is that primal and dual problems do not involve any operator in the stochastic directions (derivatives in ξ_i) but in the spatial direction x solely. This has the essential implication that realizations of the Burgers flow for different realizations of the viscosity are fully independent. As a result, the solution of the primal and dual problems over different SEs are uncoupled, allowing for straightforward parallelization with drastic speed-up of the computation. We took advantage of this characteristic by solving SE-wise the primal and dual problems on a Linux-cluster having 4 nodes with dual processors. Another interesting property of the stochastic decoupling between SEs is that, during the refinement process, the approximation needs only to be updated for the stochastic subdomains $\Xi^{(m)}$ that have been x or ξ refined.

9.3.4.3 Isotropic h_ξ Refinement

In a first series of tests, the spatial discretization is held fixed with $\mathrm{N_x} = 6$ Chebychev finite elements having equal size and order $\mathrm{q} = 6$. For the refinement, only h_ξ refinement is allowed, and the stochastic order is maintained to a constant value.

For the purpose of comparison, we plot in Fig. 9.9 the error in the computed mean and variance of U at the point $x = 0.52$ when the partition of Ξ is *uniformly* refined by increasing the number Nb of SEs from 2^2 to 100^2. The mean and variance

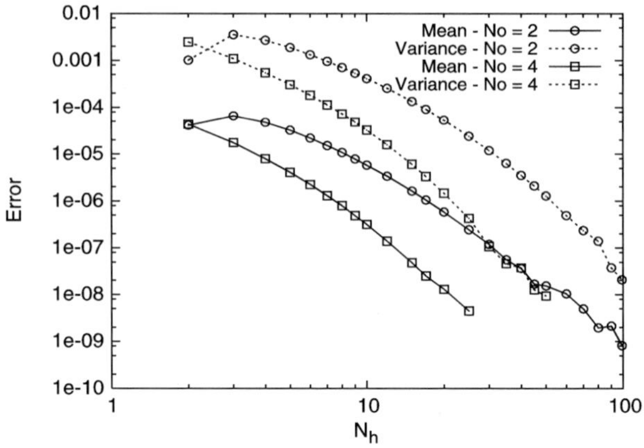

Fig. 9.9 Evolution of the errors on the computed (semi-continuous) mean and variance of the solution at $x = 0.52$ as a function of $N_h = \sqrt{\text{Nb}}$ when using uniform h_ξ refinement. Two stochastic orders No = 2 and No = 4 are reported as indicated. Adapted from [147]

are given by:

$$\left\langle U^h \right\rangle_\Xi = \sum_{m=1}^{\text{Nb}} \left|\Xi^{(m)}\right| \left\langle U^h \right\rangle_{\Xi^{(m)}},$$
$$\sigma^2(U) \equiv \left\langle \left[U^h - \left\langle U^h \right\rangle_\Xi\right]^2 \right\rangle_\Xi = \sum_{m=1}^{\text{Nb}} \left|\Xi^{(m)}\right| \left\langle \left[U^h - \left\langle U^h \right\rangle_\Xi\right]^2 \right\rangle_{\Xi^{(m)}}. \tag{9.75}$$

In this experiment, the SEs are squares with equal size. To estimate the errors, surrogates of the exact mean and variance of U were computed using $\text{N}_\text{x} = 6$, $\text{q} = 6$, $\text{Nb} = 128^2$ and No = 6. Note that these surrogates are in fact approximations of the exact mean and variance of the *semi-continuous* problem, the spatial discretization being held fixed. Consequently, it is not expected that the *a posteriori* error estimate η would drop to zero since a small but finite spatial error persists even for $\mathcal{S}^h \to \mathcal{S}$. The results in Fig. 9.9 show the convergence of the errors on the mean and variance at $x = 0.52$ of the semi-continuous solution for two stochastic orders No = 2 and No = 4. The error is seen to quickly decrease as the number of SEs increases, illustrating the convergence of the approximation. The errors on the mean and variance converge with a similar rate which is function of the stochastic order No.

However, it is known that this uniform refinement is not optimal, since some areas of Ξ may require a finer discretization than others. Thus, instead of employing a uniform refinement, we now use the *a posteriori* error estimate to identify the SEs requiring refinement. Following (9.64), an h_ξ refinement is performed on a SE $\Xi^{(m)}$ whenever $\eta_{l,m} \geq \epsilon$ for some $l \in [1, \text{N}_\text{x} = 6]$. The refinement consists of splitting $\Xi^{(m)}$ into $2^\text{N} = 4$ smaller SEs of equal size (*i.e.* isotropically). In Fig. 9.10, we plot the evolution of the errors in the mean and variance of U^h at $x = 0.52$ using this refinement scheme with No = 2. These results were generated using $\epsilon = 2 \times 10^{-5}$.

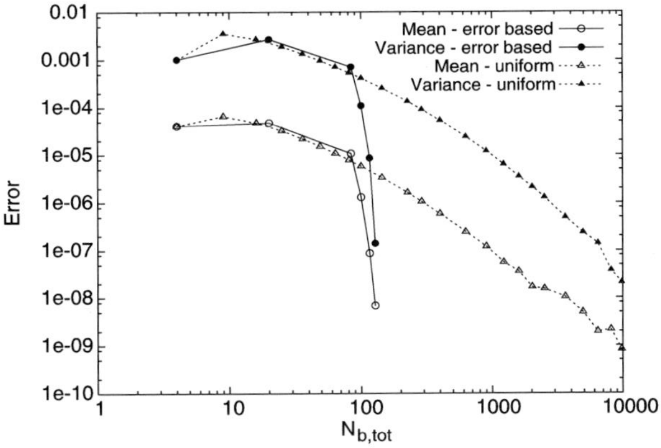

Fig. 9.10 Evolutions of the errors in the computed (semi-continuous) mean and variance of the solution at $x = 0.52$ as a function of the number of primal and dual problems solves during the isotropic h_ξ refinement and No $= 2$. Also plotted are the evolutions of the errors for the uniform refinement. Adapted from [147]

The errors are plotted as a function of the total number of dual and primal problems actually solved during the iterative refinement process. The errors for the uniform refinement previously shown in Fig. 9.9 are also reported for comparison. A dramatic improvement of the convergence of the errors on the two first moments is observed when the *a posteriori* error based refinement scheme is used, compared to the uniform refinement. Specifically, an error of $\sim 10^{-7}$ in the (semi-continuous) mean and variance is achieved at a cost of roughly 128 resolutions of the primal and dual problems when using the adaptive h_ξ refinement, whereas about 5000 primal problems have to be solved to reach a similar accuracy when using a uniform refinement. Clearly, the adaptive h_ξ refinement outperforms the uniform refinement, not only in terms of CPU-cost, but also in terms of memory requirements.

A better appreciation of the performance of the adaptive h_ξ refinement can be gained from the analysis of the data reported in Table 9.5, which provides the number Nb of SEs, the number of resolutions of primal and dual problems and the errors in the first two moments as the refinement proceeds. Starting from a partition of Ξ into 4 equal SEs, they are first all refined along the two-directions ξ_1 and ξ_2 leading to a partition involving 16 SEs. At the second iteration, all these SEs are still considered too coarse to match the prescribed accuracy and are refined again in the two stochastic directions, resulting in 64 SEs. After the third iteration, only a fraction of the SEs needs further refinement and the process eventually stops after 6 iterations with a partition of the stochastic space into 97 SEs.

In a second series of tests, the *a posteriori* error based isotropic h_ξ refinement is applied with different stochastic orders No. The refinement criterion ϵ is increased to 5×10^{-5} while other numerical parameters are kept constant ($q = 6$, $N_x = 6$). Figure 9.11 shows the resulting partition of Ξ and surface response of the solution at $x = 0.1$ for No $= 1$, 3 and 5. It is seen that to satisfy the same error criterion a lower number of FEs is needed when the stochastic order increases. Specifically, for

Table 9.5 Evolution of the SE discretization (Nb), the number of primal and dual problems solves, and the errors on mean and variance of the solution at $x = 0.52$, with adaptive h_ξ refinement using No = 2. Adapted from [147]

Iteration	Nb	Number of resolutions	Error on mean	Error on variance
1	4	4	4.1074×10^{-5}	1.0189×10^{-3}
2	16	20	4.7861×10^{-5}	2.7054×10^{-3}
3	64	84	1.0813×10^{-5}	7.1067×10^{-4}
4	76	100	1.3056×10^{-6}	1.0944×10^{-4}
5	88	116	8.7892×10^{-8}	8.5915×10^{-6}
6	97	128	6.9087×10^{-9}	1.4032×10^{-7}

No = 1, 174 SEs are needed compared to 10 for No = 5. It is also observed that the partition of Ξ is essentially refined in the lower quadrant corresponding to lower values of the viscosity. An asymmetry of the resulting partition of Ξ is also noted for No = 1, indicating different contributions of ξ_1 and ξ_2 to the uncertainty of the solution, as one may have expected from the parameterization in (9.67).

Furthermore, the surface responses in Fig. 9.11 show that the refinement of Ξ takes place in areas where the solution exhibits the steepest dependence with regards to $\boldsymbol{\xi}$, but also in areas where it is essentially unaffected by the viscosity; this is due to the fact that the refinement is based on a criterion involving all spatial locations: the solution at different spatial locations requires refinement at different places in Ξ.

9.3.4.4 Isotropic $h_{\xi,x}$ Refinement

In the previous tests, an isotropic h_ξ refinement only was applied. However, as discussed previously, the *a posteriori* error estimate incorporates both stochastic and spatial errors. In fact, it is expected that when lowering μ a finer and finer spatial FE discretization in the neighborhood of $x = 0$ is needed because the solution becomes steeper. Consequently, one may find advantages in adapting the FE discretization to $\Xi^{(m)}$. This is achieved by introducing an additional test before applying the isotropic h_ξ refinement. If the local error $\eta_{l,n}$ is greater than ϵ, the spatial discretization is first checked by computing an estimate of the spatial error $\eta^x_{l,m}$, namely using

$$\left(\eta^x_{l,m}\right)^2 = \int_{\mathcal{D}^{(l)}} \left\langle \left[U^h - \Pi^l(U^h)\right]^2 \right\rangle_{\Xi^{(m)}} \mathrm{d}x, \tag{9.76}$$

where $\Pi^l(U^h)$ is the (spatial) Clément interpolant [35] of U^h over the spatial patch defined by the union of the FEs having a common point with the element $\mathcal{D}^{(l)}$. The order of the Clément interpolant is set to $\mathrm{q}(l, m) + 1$. If this estimate of the spatial error is greater than a prescribed second threshold ϵ_x a h_x refinement is applied to the FE $\mathcal{D}^{(l)}$ (for the SE $\Xi^{(m)}$ only), consisting in its partition into two Chebychev

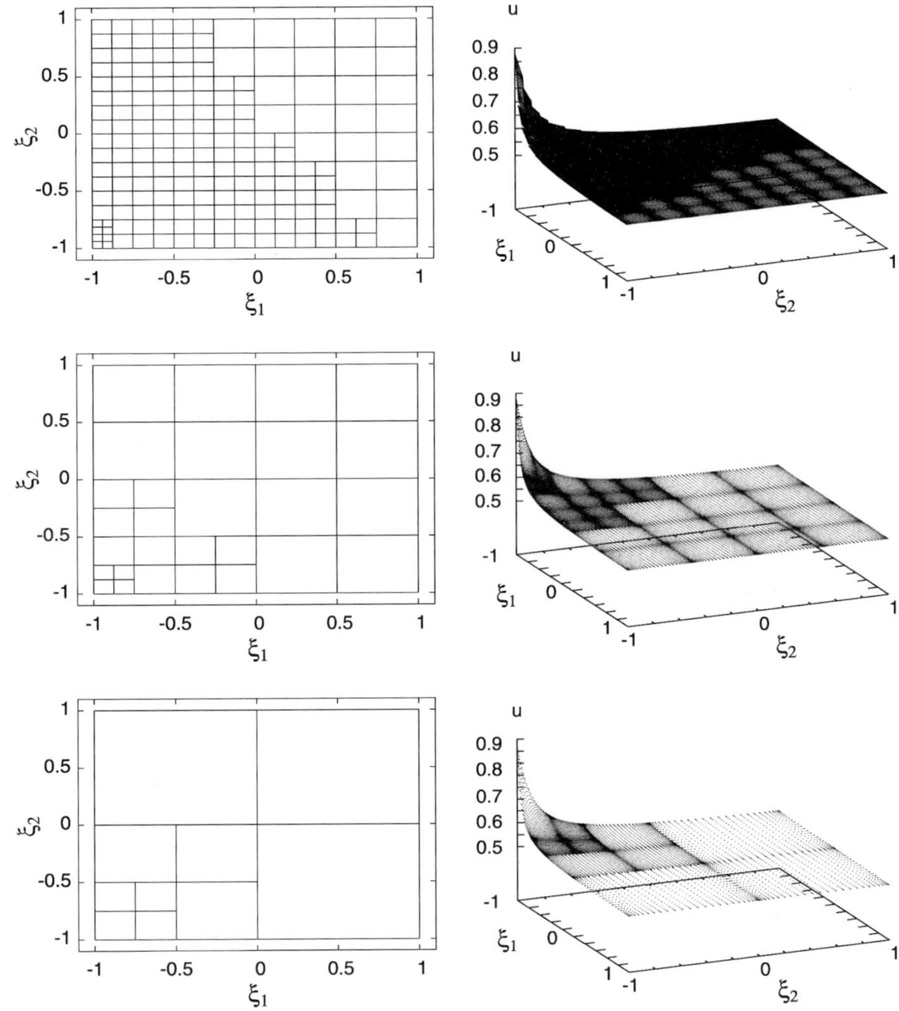

Fig. 9.11 Partition of Ξ (*left*) and surface response for $U(\boldsymbol{\xi})$ at $x = 0.1$ (*right*) at the end of the isotropic h_ξ refinement process using $\epsilon = 5.10^{-5}$. *Plots* correspond to No $= 1$, 3 and 5, arranged *from top to bottom*. Adapted from [147]

elements of equal size. On the contrary, if $\eta^x_{l,m} < \epsilon_x$ for all $l \in [1, \mathrm{Nb}(m)]$, the h_ξ refinement is applied as previously.

This strategy is applied to the test problem, with the initial discretization using $\mathrm{N_x} = 6$ identical FEs with $\mathrm{q} = 6$, over 4 equal SEs with No $= 2$ and a refinement criterion $\epsilon = 10^{-4}$. The partition of Ω at the end of the refinement process is shown in Fig. 9.12. The left plot shows the partition of Ξ and highlights again the need for refinement for the lowest values of the viscosity. The right plot shows the dependence of the refinement of the FE discretization with $\boldsymbol{\xi}$. Specifically, it is seen that h_x refinement essentially occurs for the lowest values of the viscosity (*i.e.* when the

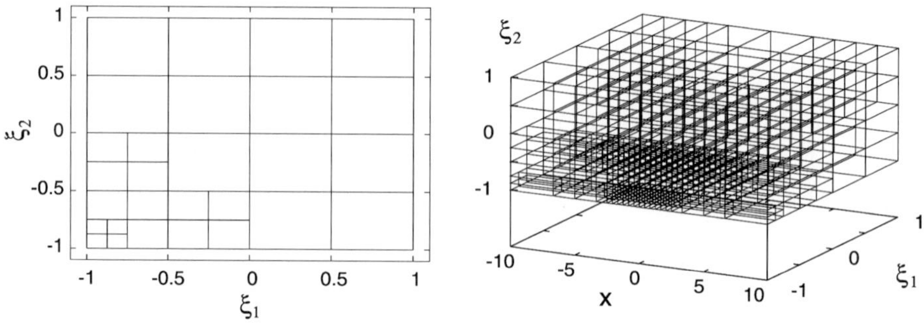

Fig. 9.12 Partition of Ξ (*left*) and Ω (*right*) after the $h_{\xi,x}$ refinement procedure. Numerical parameters are given in the text. Adapted from [147]

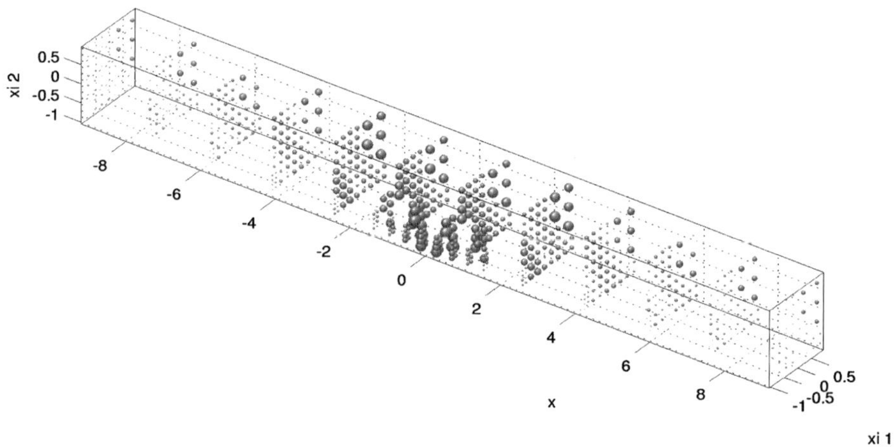

Fig. 9.13 Distribution of the local *a posteriori* error estimate $\eta_{l,m}$ after $h_{\xi,x}$ refinement. The spheres' diameter d scales as $d \sim \eta_{l,m}^{0.25}$. Adapted from [147]

solution exhibits the steepest spatial evolutions) and in the neighborhood of $x = 0$ as one may have expected.

Additional insights about the distribution of the local *a posteriori* error estimate $\eta_{l,m}$ in Ω can be gained by examining Fig. 9.13, where the local error magnitude is depicted using spheres. A large sphere corresponds to a large error $\eta_{l,m}$, with a scaling of the spheres' diameter as $d \sim \eta_{l,m}^{0.25}$. As already stated, it is seen that the maximum error occurs around $x = 0$ and that it decreases very quickly as one moves away from that location. This plot clearly exemplifies the h refinement strategy: divide elements where a large error occurs to reduce the error magnitude below the prescribed tolerance $\eta_{l,m}$.

Figure 9.14 shows the expectation and variance of the approximate solution U^h after refinement as a function of x. The plot of the expectation $\langle U^h \rangle_\Xi$ is also compared with the deterministic solution $u(x)$ for the mean viscosity $\mu_0 = 1$. This de-

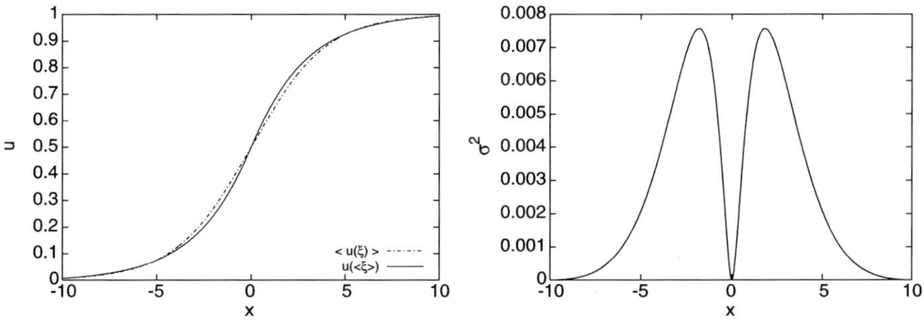

Fig. 9.14 Expectation (*left*) and variance (*right*) of the approximate solution $U^h(x, \boldsymbol{\xi})$ at the end of the $h_{\xi,x}$ refinement process. Adapted from [147]

terministic solution is given by:

$$u(x; \mu = 1) = \frac{1}{2}\left[1 + \tanh\frac{x}{4}\right]. \tag{9.77}$$

It is seen that the expected solution also has a hyperbolic tangent-like profile, which however differs from the deterministic solution. Differences arise due to the non-linearities of the Burgers equation. The right plot in Fig. 9.14 depicts the solution variance $\sigma^2(U^h)$. The boundary conditions being deterministic the variance vanishes at x^- and x^+. The uncertainty in the viscosity produces a symmetric variance with respect to $x = 0$ as it only affects the width of the profile,

$$U(x, \boldsymbol{\xi}) \approx \frac{1}{2}\left[1 + \tanh\frac{x}{4\mu(\boldsymbol{\xi})}\right]. \tag{9.78}$$

Also, due to the selected boundary conditions, we have at the center of the spatial domain $U(\boldsymbol{\xi}) = (u^- + u^+)/2 = 1/2$ almost surely, provided that $\mu(\boldsymbol{\xi}) > 0$. Therefore, the variance of U^h vanishes at $x = 0$ as shown in Fig. 9.14.

The probability density functions of U^h, together with the solution's quantiles, are reported in Fig. 9.15 as function of x. The quantiles are defined as the level $u(Q)$, for $Q \in]0, 1[$, such that the probability of $U^h(x) < u$ is equal to Q. The pdf curves show dramatic changes with x. For $x = x^-$ the pdf is a Dirac of unit mass (no uncertainty); as x increases the pdf evolves from a distribution having a sharp lower tail to one having a long lower tail. At $x = 0$ it is again a Dirac delta (no uncertainty). For x increasing further to x^+ the opposite evolution is observed (due to point symmetry around $x = 0$). Note that the distribution of the solution is bounded since U almost surely $\in [u^-, 1/2]$ for $x \le 0$ and U almost surely $\in [1/2, u^+]$ for $x \ge 0$. The quantiles reflect the complexity of the distribution with important changes in the spacing between quantiles changes as x varies.

To further illustrate the need of refinement to properly capture the solution distribution, we plot in Fig. 9.16 the pdf of U^h at $x = 0.52$ as the $h_{\xi,x}$ refinement proceeds. The left plot shows the pdf in linear-log scales to appreciate the improvement in the tails of the distribution, whereas the right plot in linear-linear scales

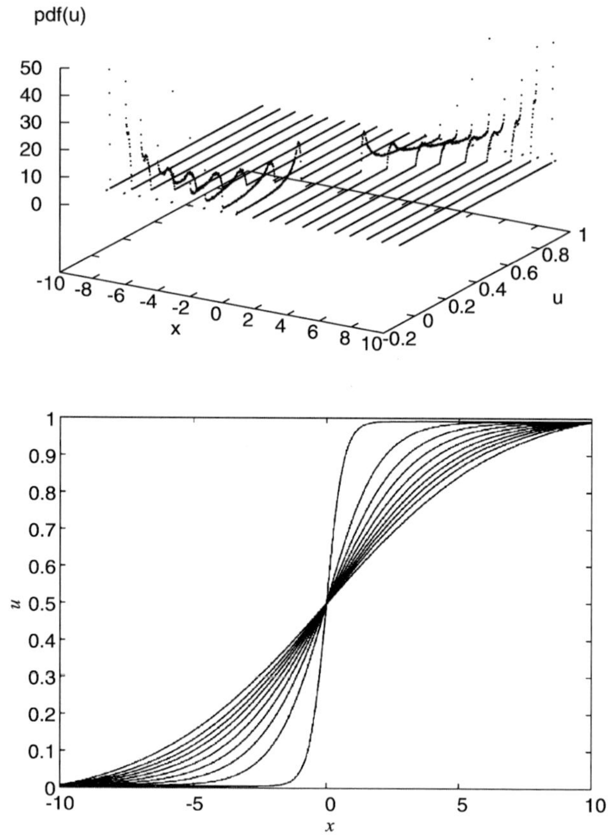

Fig. 9.15 *Top*: pdf of the approximate solution U^h as a function of x at the end of the $h_{\xi,x}$ refinement. The pdf-axis is truncated for clarity. *Bottom*: quantiles $u(Q)$ of the solution, as a function of x, for $Q = 0.05$ to 0.95 with constant increment $\Delta Q = 0.1$. Adapted from [147]

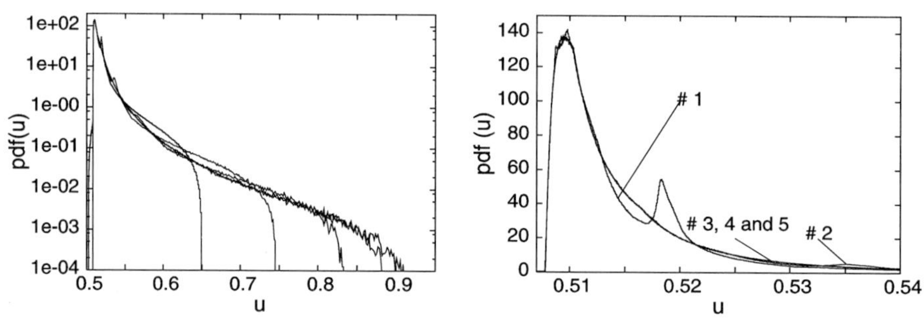

Fig. 9.16 Probability density function at $x = 0.52$ for different steps of the $h_{\xi,x}$ refinement process. Linear-log plot (*left*) and linear-linear plot = (*right*). Adapted from [147]

shows the improvement in the high density region. It is seen that during the first iterations the pdf exhibits under-estimated right-tails and some spurious oscillations, which are due to discontinuities of the approximate solution across SEs boundaries.

9.3.4.5 Anisotropic h/q Refinement

In the previous tests, an isotropic h refinement was used in the stochastic domain. As a result, each refined SE is split into 2^{N} SEs. For large N this simple procedure quickly results in a prohibitively large number of SEs. Instead, it is generally advantageous to split $\Xi^{(m)}$ only along the stochastic directions yielding the largest error reduction. Obviously, the *a posteriori* error estimate does not provide enough information to decide along which directions $\Xi^{(m)}$ should be split: an anisotropic error estimator is necessary for this purpose. In the absence of such estimator, we rely on a criterion, inspired from [125, 238], which is based on the relative contributions of each stochastic directions to the local variance. The local variance is defined as

$$\sigma^2_{\Xi^{(m)}}(U) = \left\langle \left[U - \langle U \rangle_{\Xi^{(m)}}\right]^2 \right\rangle_{\Xi^{(m)}}. \tag{9.79}$$

Since the stochastic expansion of U over $\Xi^{(m)}$ is given by:

$$U(\xi \in \Xi^{(m)}) = \sum_{k=0}^{\mathrm{P}(m)} u_k^{(m)} \Psi_k^{(m)}(\xi),$$

and because by convention $\Psi_k^{(m)} = 1$ for $k = 0$ (*i.e.* mode 0 is the mean mode), the local variance can be expressed as:

$$\sigma^2_{\Xi^{(m)}}(U) = \sum_{k=1}^{\mathrm{P}(m)} \left(u_k^{(m)}\right)^2 \left\langle \Psi_k^{(m)2} \right\rangle_{\Xi^{(m)}}, \tag{9.80}$$

and so we define

$$\sigma^2_{\Xi^{(m)} \times \mathcal{D}^{(l)}}(U) = \sum_{k=1}^{\mathrm{P}(m)} \left\langle \Psi_k^{(m)2} \right\rangle_{\Xi^{(m)}} \int_{\mathcal{D}^{(l)}} \left(u_k^{(m)}(x)\right)^2 \mathrm{d}x. \tag{9.81}$$

It is seen that the integral of the local variance on the FE $\mathcal{D}^{(l)}$ is a weighted sum of the integral of the squared stochastic expansion coefficients over the FE. The idea is thus to define, for each direction $i = 1, \ldots, \mathrm{N}$, the contribution of the polynomial of degree $\mathrm{No}(m)$ in ξ_i to this variance integrated on $\mathcal{D}^{(l)}$. This contribution is denoted $\sigma^2_{l,m}(U; i, \mathrm{No}(m))$. Using the respective contributions of each direction, it is decided that $\Xi^{(m)}$ has to be split along the i-th stochastic direction if the following test is satisfied for at least one FE:

$$\frac{\sigma^2_{l,m}(U; i, \mathrm{No}(m))}{\sum_{i=1}^{\mathrm{N}} \sigma^2_{l,m}(U; i, \mathrm{No}(m))} \geq \epsilon_2, \tag{9.82}$$

where $0 < \epsilon_2 < 1$ is an additional threshold parameter. The left-hand side of the inequality can be interpreted as a surrogate of the indicator based on the one-dimensional details along dimension i in the multiwavelet context (see (9.13) in

Sect. 9.2.3). If none of the stochastic directions satisfies the previous test, it is on the contrary decided to increment by one unit the stochastic expansion order No(m) over $\Xi^{(m)}$.

The anisotropic h/p refinement strategy now follows the algorithm:

1. Solve the primal and dual problems for the current approximation space $\mathbf{V}^h$; get U^h and Z^h.
2. Solve the adjoint problem in the enriched space $\mathbf{V}^{\tilde{h}}$; get $\tilde{Z}$.
3. Compute the local error $\eta_{l,m}$ from (9.62) for $m = 1, \ldots, \mathrm{Nb}$, $l = 1, \ldots, \mathrm{N_x}(m)$. If $\eta_{l,m} < \epsilon$ for $m = 1, \ldots, \mathrm{Nb}$, $l = 1, \ldots, \mathrm{N_x}(m)$, then end computation.
4. For $m = 1, \ldots, \mathrm{Nb}$ and $l = 1, \ldots, \mathrm{N_x}(m)$
   If $\eta_{l,m} > \epsilon$:
   a. Compute the estimate of the spatial error $\eta^x_{l,m}$ using (9.76).
   b. If $\eta^x_{l,m} > \epsilon_x$, mark element for h_x refinement.
   c. If the element has not been marked for h_x refinement,
      a) Compute the directional variances.
      b) For $i = 1, \ldots, \mathrm{N}$ if the directional variance is greater than ϵ_2 then mark element $\Xi^{(m)}$ for h_ξ refinement in direction i.
5. For $m = 1, \ldots, \mathrm{Nb}$: if $\Xi^{(m)}$ has not been marked for some h_ξ refinement, and none of the elements $\Xi^{(m)} \times \mathcal{D}^{(l)}$, $l = 1, \ldots, \mathrm{N_x}(m)$, are marked for h_x refinement but there exists at least one $l \in [1, \mathrm{N_x}(m)]$ such that $\eta_{l,m} > \epsilon$, then increase No(m) by one.
6. Construct the refined approximation space and restart from 1.

This refinement scheme has been successfully applied to the test problem, with $\mu_1 = 0.82$ and $\mu_2 = 0.16$. The viscosity parameterization was changed to increase the contribution of the first direction compared to the second to the solution uncertainty. Note that the pdf of μ is affected by this change of the parameterization, but the uncertainty range is kept constant. For illustration purposes, we present in Fig. 9.17 an example of the partition of the stochastic space into SEs with variable stochastic expansion orders. The initial discretization involves $\mathrm{Nb} = 4$ equal SEs with $\mathrm{No} = 2$. At the first iteration, all SEs were split isotropically, the expansion order being kept constant. At the second iteration, the SEs with boundary at $\xi_1 = -1$ were further refined but in the ξ_1 direction only. For the following iterations, no further h_ξ refinement was required while some SEs still have a significant estimated error: this led to an increase in the stochastic expansion order No(m). Again, the final expansion order is the greatest for the SEs with $\xi_1 = -1$ and/or $\xi_2 = -1$ boundaries (where viscosity is small), and is the lowest for the SE having boundary $\xi_1 = 1$ and $\xi_2 = 1$ where No has been kept constant.

Tests for $\mathrm{N} = 3$**:**
To conclude this series of tests, an additional uncertainty source is considered by taking the left boundary condition as random, U^-. The random boundary condition is assumed independent of viscosity, and consequently parameterized using an additional random variable ξ_3. The complete uncertainty settings are:

$$\mu(\boldsymbol{\xi}) = 1 + 0.5\xi_1 + 0.05\xi_2, \qquad U^-(\boldsymbol{\xi}) = u_0^- + u'\xi_3, \tag{9.83}$$

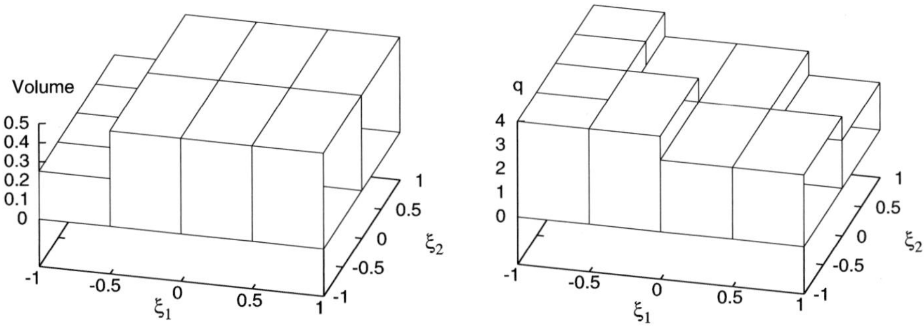

Fig. 9.17 Volume $|\Xi^{(m)}|$ and corresponding stochastic expansion order No over the partition of Ξ. The anisotropic h/q refinement scheme is used. Adapted from [147]

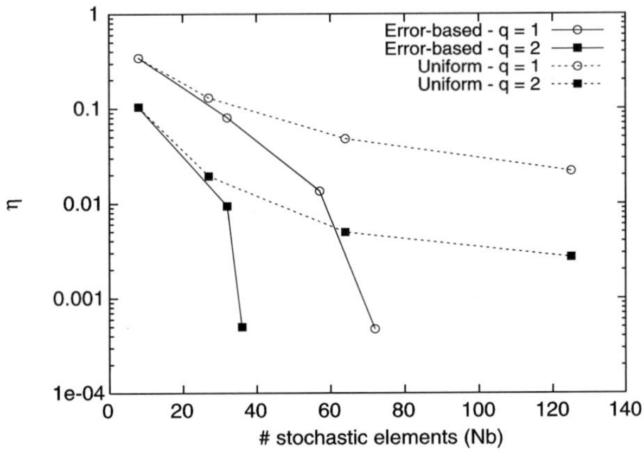

Fig. 9.18 Reduction of the *a posteriori* error estimate η with the number Nb of stochastic elements involved in the partition of Ξ. Plotted are the results for the anisotropic h_ξ refinement procedure (labeled Error-based) and uniform refinement, using No = 1 and 2 as indicated. Adapted from [147]

with u_0^- given by (9.68) and $u' = 5 \times 10^{-4}$. This low value of u' is selected as it is known that small perturbations of the boundary condition leads to $O(1)$ changes in the solution of the Burgers equation [248]. This is due to the "supersensitivity" of the transition layer location with the boundary condition: the low variability in U^- will result in large variability of the solution but essentially around the center of the spatial domain and not in the neighborhood of x^- where the solution variability is low. This problem is thus well suited to test the effectiveness of the *a posteriori* error methodology in providing correct local error estimators. Moreover, as the sensitivity of the solution with regards to U^- increases when the viscosity is lowered, a finer partition of Ξ is expected for low values of ξ_1, whereas the contribution of ξ_2 will be weaker, as can be appreciated from (9.83).

The spatial discretization ($\mathrm{N_x} = 20$, $\mathrm{q} = 6$) and stochastic orders No being held fixed, we proceed with the *a posteriori* error based anisotropic h_ξ refinement scheme described above. The target precision is set to $\epsilon_\eta = 0.001$. In Fig. 9.18 we show the reduction of the *a posteriori* error η during the refinement process for No = 1 and 2. The evolution of the error estimate for a uniform refinement of the stochas-

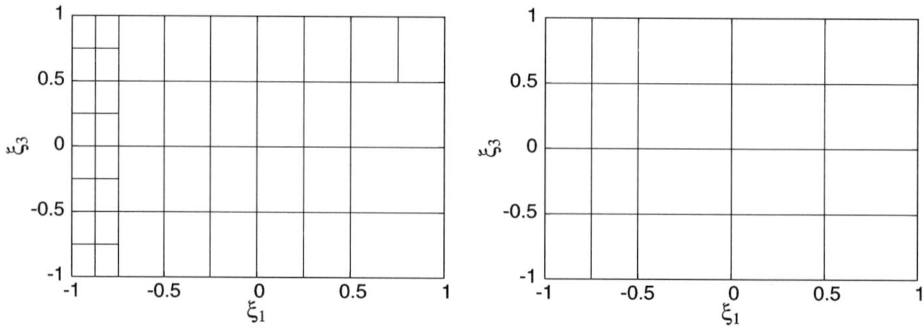

Fig. 9.19 Partition of Ξ at the end of the anisotropic h_ξ refinement process in a plane corresponding to constant ξ_2. *Left*: No = 1; *right*: No = 2. Adapted from [147]

tic space is also reported for comparison. Because the stochastic space now has 3 dimensions, the increase in number of SEs for the uniform refinement is seen to be dramatically large even for a low reduction of the *a posteriori* error. On the contrary, using the local error estimate to guide the refinement process is seen to significantly improve the error reduction with the number of SEs. It is also remarked that the anisotropic refinement requires 3 iterations to achieve the prescribed precision for No = 1, whereas only 2 iterations are needed for No = 2.

Figure 9.19 depicts the partition of the stochastic space at the end of the h_ξ refinement process. The initial partition uses $\mathrm{Nb} = 2^{\mathrm{N}} = 8$ identical SEs. Note that the anisotropic h_ξ refinement scheme never requires refinement along the second dimension ξ_2: the plots of Fig. 9.19 show the partition of Ξ in a plane where ξ_2 is constant. The independence of the partition with regards to ξ_2 denotes the capability of the proposed scheme to detect the weak influence of ξ_2 on the solution. On the other hand, it is seen that for fixed ξ_2 and ξ_3 a finer division of Ξ along the first direction is necessary when ξ_1 decreases, because of the steeper behavior of the solution when the viscosity decreases. In contrast, for fixed ξ_1 and ξ_2 the partition is uniform along the third direction, but is finer for low viscosity and No = 1, as one may have anticipated from the behavior of the Burgers solution.

Finally, we show in Fig. 9.20 the variance of the stochastic solution along the spatial domain, for the two stochastic orders No = 1 and No = 2, at the end of the anisotropic refinement process. The effect of the uncertain boundary condition on the solution variance can be appreciated through comparison with the results reported in Fig. 9.14. Specifically, the variance of the solution at the center of the spatial domain is now different from zero. It is seen that even though both orders leads to similar estimated error, small but noticeable differences are visible in the spatial distribution of the variance. These differences in the predicted variance can be better appreciated from the right plot in Fig. 9.20 where the differences for No = 1 and No = 2 are shown.

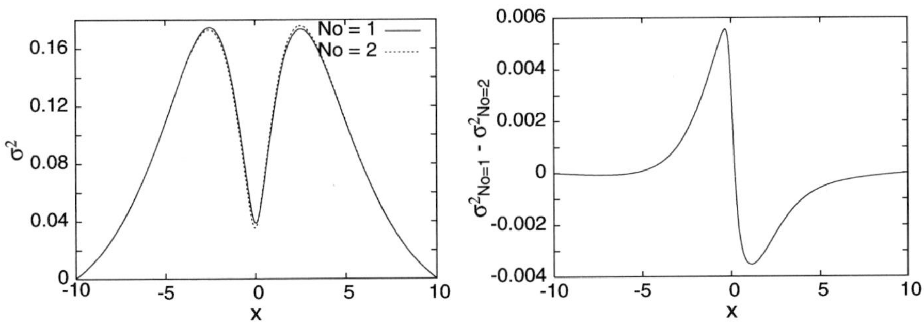

Fig. 9.20 *Left*: comparison of the variances in $U(\xi)$ as a function of x for stochastic expansion orders No $= 1$ and No $= 2$ at the end of the anisotropic h_ξ refinement process. *Right*: difference of the two variances predicted for No $= 1$ and No $= 2$. Adapted from [147]

9.4 Generalized Spectral Decomposition

Another approach to minimizing the computational cost of intrusive computations is to seek the approximate solution in a reduced space. It is remarked that such reduction approach should not be opposed or understood as an alternative to the adaptive methods mentioned above, but would actually further improve their efficiency since adaptive techniques require the resolution of large, though local, Galerkin problems. The main idea of reduced approximations is to take advantage of the structure of the full approximation space, which results from the tensor product of the deterministic and stochastic approximation spaces: if one is able to appropriately reduce the deterministic or stochastic approximation spaces, to a low dimensional subspace, the size of the Galerkin problem to be solved is drastically reduced as well. Of course, the determination of a low dimensional subspace that still accurately captures the essential features of the solution is not immediate since the solution is unknown. In [58], the Galerkin problem is first solved on a coarse deterministic mesh to provide a coarse estimate of the solution, which is then decomposed into its principal components through a KL expansion. The first random coefficients of the KL expansion are then used as a reduced stochastic basis in the Galerkin problem considered now on a fine deterministic mesh. Alternatively, in [151], a Neumann expansion of the operator is used to obtain an estimate of the covariance operator of the solution. The dominant eigenspace of the approximate covariance operator is then considered as the reduced deterministic (spatial) subspace to be used subsequently in the Galerkin procedure. In fact, as for the first approach, this can be interpreted as a coarse *a priori* KL expansion of the solution. These two approaches have demonstrated their effectiveness in reducing the size of the Galerkin problem solved *in fine*. However, the second approach, based on Neumann expansion, focused on linear problems, and the extension of the first approach to highly nonlinear problems, such as for instance the Navier-Stokes equations, seems difficult due to limitations in the possible deterministic coarsening: the reduced basis may simply miss important features of the nonlinear solution. In this section, we investigate the extension of the so-called Generalized Spectral Decomposition (GSD) method which does not require one to provide a reduced basis (*a priori* or determined by alternative means), but which instead yields by itself the "optimal" reduced basis.

The Generalized Spectral Decomposition (GSD) method consists in seeking an optimal decomposition of the solution u to a stochastic problem under the form $\sum_{i=1}^{M} U_i \lambda_i$, where the U_i are deterministic functions whereas the λ_i are random variables. In this context, the λ_i (resp. of U_i) are regarded as a reduced basis of random variables (resp. of deterministic functions). Optimal decompositions could be easily defined if the solution u were known. Such a decomposition can for instance be obtained by a KL expansion (or classical spectral decomposition) of u, which is the optimal decomposition with respect to a classical inner product. The GSD method relies on defining an optimality criterion for the decomposition which is based on the equation(s) satisfied by the solution but not on the solution itself. The construction of the decomposition therefore does not require knowledge of solution *a priori* or the availability or construction of a surrogate (approximation on coarser mesh or low order Neumann expansion) as pointed previously. The GSD method was first proposed in [168] in the context of linear stochastic problems. In the case of linear symmetric elliptic coercive problems, by defining an optimal decomposition with respect to the underlying optimization problem, the functions U_i (resp. λ_i) were shown to be solutions of an eigen-like problem. Ad-hoc algorithms, inspired by power method for classical eigenproblems, have been proposed in [168] for the resolution of the eigenproblems; improved algorithms and in-depth analysis of the GSD method for a wider class of linear problems can be found in [167]. The main advantage of these algorithms is to separate the resolution of a few deterministic problems and a few reduced stochastic problems (*i.e.* using a reduced basis of deterministic functions). These algorithms lead to significant computational savings when compared to classical resolution techniques of stochastic Galerkin equations. A first attempt for extending the GSD method to nonlinear problems has been investigated in [169]: algorithms derived for the linear case were simply applied to subsequent linear stochastic problems arising from a classical nonlinear iterative solver. Reduced bases generated at each iteration were stored, sorted and re-used for subsequent iterations. In this section, we outline a "true" extension of the GSD to nonlinear problems, where we directly construct an optimal decomposition of the solution with regard to the initial nonlinear problem.

To illustrate this development, we introduce in Sect. 9.4.1 a general formulation of nonlinear stochastic problems and the associated stochastic Galerkin schemes. In Sect. 9.4.2, we first outline the application of the GSD to linear problems, and then present its extension to the nonlinear case. In particular, we provide some basic mathematical considerations which motivate this extension. The GSD is interpreted as the solution of an eigen-like problem and two ad-hoc algorithms are presented for building the decomposition. These algorithms are inspired from the ones proposed in [168] in the context of linear stochastic problems. Then, the GSD method is applied to two nonlinear models: the steady viscous Burgers equation (Sect. 9.4.4) and a stationary diffusion equation (Sect. 9.4.5).

9.4.1 Variational Formulation

We adopt a probabilistic representation of uncertainties in an abstract probability space (Θ, Σ, P). We consider nonlinear problems having the following semi-variational form:
Given an elementary event θ, find $u(\theta) \in \mathcal{V}$ such that we have almost surely

$$b(u(\theta), v\,\theta) = l(v\,\theta) \quad \forall v \in \mathcal{V}, \tag{9.84}$$

where $\mathcal{V}$ is a given vector space, eventually of finite dimension, b and l are semi-linear and linear forms, respectively. The forms b and l may depend on the elementary event θ. Here, we consider that $\mathcal{V}$ does not depend on the elementary event. (This could be the case when considering partial differential equations defined on random domains [170, 172].) On the stochastic level, we introduce a suitable function space $\mathcal{S}$ for random variables taking values in $\mathbb{R}$. The full variational formulation of the problem is expressed as:
Find $u \in \mathcal{V} \otimes \mathcal{S}$ such that

$$B(u, v) = L(v) \quad \forall v \in \mathcal{V} \otimes \mathcal{S}, \tag{9.85}$$

where the semi-linear and linear forms B and L are respectively given by:

$$B(u, v) = \int_{\Theta} b(u(\theta), v(\theta); \theta)\, \mathrm{d}P(\theta) := \mathrm{E}\,[b(u, v; \cdot)], \tag{9.86}$$

$$L(v) = \int_{\Theta} l(v(\theta); \theta)\, \mathrm{d}P(\theta) := \mathrm{E}\,[l(v; \cdot)]. \tag{9.87}$$

Here, $\mathrm{E}\,[\cdot]$ denotes the expectation.

9.4.1.1 Stochastic Discretization

Focusing on parametric uncertainties, the semilinear form B and linear form L are parametrized using a finite set of N real continuous random variables $\boldsymbol{\xi}$ with known probability law $P_{\boldsymbol{\xi}}$. By the Doob-Dynkin's lemma [175], the solution of problem (9.84) can be written in terms of $\boldsymbol{\xi}$, *i.e.* $u(\theta) \equiv u(\boldsymbol{\xi})$. The stochastic problem can then be reformulated in the N-dimensional image probability space $(\Xi, \mathcal{B}_{\Xi}, P_{\boldsymbol{\xi}})$, where $\Xi \subset \mathbb{R}^N$ denotes the range of $\boldsymbol{\xi}$. The expectation operator has the following expression in the image probability space:

$$\mathrm{E}\big[f(\cdot)\big] = \int_{\Theta} f(\boldsymbol{\xi}(\theta))\, \mathrm{d}P(\theta) = \int_{\Xi} f(\mathbf{y})\, \mathrm{d}P_{\boldsymbol{\xi}}(\mathbf{y}). \tag{9.88}$$

Since we are interested in finding an approximate stochastic solution of (9.84), the function space $\mathcal{S}$ is considered as a finite dimensional subspace of $L_2(\Xi, P_{\boldsymbol{\xi}})$, the space of real, second-order random variables defined on Ξ. Different types of approximation are available at the stochastic level: continuous polynomial expansions [90, 218, 246], piecewise polynomial expansion [50], and multiwavelets [124, 125]. At this point, it is emphasized that the method proposed in this paper is independent of the type of stochastic approximation used.

Remark The choice of a suitable function space $\mathcal{S}$ is a non-trivial question in the infinite dimensional case. Several interpretations of stochastic partial differential equations (SPDE) are generally possible, *e.g.* by introducing the concept of Wick product between random fields, leading to well posed problems and then to different possible solutions [18, 102, 224]. These mathematical considerations are beyond the scope of the present discussion. For the nonlinear problems dealt with below, where a classical interpretation of products between random fields is used [8, 151], a possible choice could consist in classical Banach spaces $L^p(\Xi, P_{\boldsymbol{\xi}}) \subset L_2(\Xi, P_{\boldsymbol{\xi}})$, $2 \leqslant p < \infty$. Usual approximation spaces being contained and dense in these Banach spaces, this ensures the consistency of the approximation.

In the following, we will mainly use the initial probability space $(\Theta, \mathcal{B}, P)$. One must keep in mind that at any moment, the elementary event $\theta \in \Theta$ can be replaced by $\boldsymbol{\xi} \in \Xi$ in any expression.

9.4.2 General Spectral Decomposition

The GSD method consists in seeking an approximate low-order decomposition of the solution to problem (9.85):

$$u_M(\theta) = \sum_{i=1}^{M} U_i \lambda_i(\theta), \tag{9.89}$$

where $U_i \in \mathcal{V}$ are deterministic functions whereas $\lambda_i \in \mathcal{S}$ are random variables (*i.e.* real-valued functions of the elementary random event). The λ_i (resp. of U_i) can be regarded as a reduced basis of random variables (resp. of deterministic functions). In this section, we will see how optimal reduced basis can be thought as solutions of eigen-like problems. Starting from this interpretation, we will outline two simple and efficient algorithms for building the generalized spectral decomposition.

9.4.2.1 Definition of an Optimal Pair (U, λ)

First, let us explain how to define an optimal pair $(U, \lambda) \in \mathcal{V} \times \mathcal{S}$. It is remarked that if U was known and fixed, the following Galerkin orthogonality criterion would lead to a suitable definition for λ:

$$B(\lambda U, \beta U) = L(\beta U) \quad \forall \beta \in \mathcal{S}. \tag{9.90}$$

Alternatively, if λ was known and fixed, the following Galerkin orthogonality criterion would suitably define U:

$$B(\lambda U, \lambda V) = L(\lambda V) \quad \forall V \in \mathcal{V}. \tag{9.91}$$

As a shorthand notation, we denote by $\lambda = f(U)$ the solution of (9.90) and by $U = F(\lambda)$ the solution of (9.91). It should be clear that a natural definition of an optimal pair (U, λ) consists of satisfying simultaneously (9.90) and (9.91). The problem can then be expressed as:
Find $\lambda \in \mathcal{S}$ and $U \in \mathcal{V}$ such that

$$U = F(\lambda) \quad \text{and} \quad \lambda = f(U). \tag{9.92}$$

The problem can be formulated on U as follows: find $U \in \mathcal{V}$ such that

$$U = F \circ f(U) := T(U), \tag{9.93}$$

where mapping T is a homogeneous mapping of degree 1:

$$T(\alpha U) = \alpha T(U) \quad \forall \alpha \in \mathbb{R}^*. \tag{9.94}$$

This property comes from properties of f and F, which are both homogeneous mappings of degree (-1):

$$\forall \alpha \in \mathbb{R}^*, \quad f(\alpha U) = \alpha^{-1} f(U), \qquad F(\alpha\lambda) = \alpha^{-1} F(\lambda). \tag{9.95}$$

The homogeneity property of T allows us to interpret (9.93) as an eigen-like problem where the solution U is regarded as a generalized eigenfunction. By analogy with classical eigenproblems, each eigenfunction is associated with a unitary eigenvalue. The question is then: how to define the best generalized eigenfunction among all possible generalized eigenfunctions? A natural answer is: the best U is the one which maximizes the norm $\|Uf(U)\|$ of the approximate solution $Uf(U)$, *i.e.* the one giving the highest contribution to the generalized spectral decomposition. In order to provide a more classical formulation of an eigen-problem, we now rewrite the approximation as $\alpha Uf(U)/\|Uf(U)\|$, with $\alpha \in \mathbb{R}^+$. The problem is then to find a pair $(U, \alpha) \in \mathcal{V} \times \mathbb{R}^+$ such that α is maximum and such that the following Galerkin orthogonality criterion is still satisfied:

$$\alpha U = F(f(U)/\|Uf(U)\|) = \|Uf(U)\| T(U) := \tilde{T}(U). \tag{9.96}$$

The mapping $\sigma : U \in \mathcal{V} \mapsto \|Uf(U)\| \in \mathbb{R}^+$ is a homogeneous mapping of degree 0. Then, mapping $\tilde{T}$, which is a simple rescaling of T, is still homogeneous of degree 1, so that (9.96) can be interpreted as an eigen-like problem on $\tilde{T}$: find $(U, \alpha) \in \mathcal{V} \times \mathbb{R}^+$ such that

$$\tilde{T}(U) = \alpha U. \tag{9.97}$$

U is a generalized eigenfunction of $\tilde{T}$ if and only if it is a generalized eigenfunction of T. A generalized eigenfunction is associated with a generalized eigenvalue

$\alpha = \sigma(U)$ of mapping $\tilde{T}$. The best $U \in \mathcal{V}$ then appears to be the generalized eigenfunction associated with the dominant generalized eigenvalue of $\tilde{T}$.

Remark In the case where B is a bounded elliptic coercive bilinear form, it is proved in [168] that the dominant generalized eigenfunction U is the one that minimizes the error $(u - Uf(U))$ with respect to the norm induced by B.

Remark The reasoning above can be made on a problem formulated on λ:
Find $(\lambda, \alpha) \in \mathcal{S} \times \mathbb{R}^+$ such that

$$\tilde{T}^*(\lambda) = \alpha\lambda, \tag{9.98}$$

where $\tilde{T}^*(\lambda) = \sigma^*(\lambda) f \circ F(\lambda)$, with $\sigma^*(\lambda) = \|F(\lambda)\lambda\|$. We can easily show that if U is a generalized eigenfunction of $\tilde{T}$, then $\lambda = f(U)$ is a generalized eigenfunction of $\tilde{T}^*$, associated with the generalized eigenvalue $\sigma^*(\lambda) = \sigma(f(U))$. Problems on U and λ are completely equivalent. In the discussion below, we focus on the problem in terms of U.

9.4.2.2 A Progressive Definition of the Decomposition

Following the observations above, we now outline the progressive construction of the generalized spectral decomposition defined in (9.89). The pairs (U_i, λ_i) are defined successively. Let us assume that u_M is known, and denote $(U, \lambda) \in \mathcal{V} \otimes \mathcal{S}$ the next pair to be defined. A natural definition of (U, λ) still consists of satisfying the following two Galerkin orthogonality criteria:

$$B(u_M + \lambda U, \beta U) = L(\beta U) \quad \forall \beta \in \mathcal{S}, \tag{9.99}$$

$$B(u_M + \lambda U, \lambda V) = L(\lambda V) \quad \forall V \in \mathcal{V}. \tag{9.100}$$

As a shorthand notation, we denote by $\lambda = f_M(U)$ the solution of (9.99) and by $U = F_M(\lambda)$ the solution of (9.100). This problem can still be formulated in terms of U as follows:
Find $U \in \mathcal{V}$ such that

$$U = F_M \circ f_M(U) := T_M(U), \tag{9.101}$$

where the mapping T_M is a homogeneous mapping of degree 1. (9.101) can still be interpreted as an eigen-like problem. In fact, by analogy with classical eigenproblems, the operator T_M can be interpreted as a "deflation" of the initial operator T.

Introducing $\sigma_M(U) = \|U f_M(U)\|$ allows to reformulate problem (9.101) as an eigen-like problem on the mapping $\tilde{T}_M = \sigma_M(U) T_M(U)$:
Find the dominant generalized pair $(U, \alpha) \in \mathcal{V} \times \mathbb{R}^+$, satisfying:

$$\tilde{T}_M(U) = \alpha U, \tag{9.102}$$

where $\alpha = \sigma_M(U)$ is the generalized eigenvalue of $\tilde{T}_M$ associated with the generalized eigenfunction U.

Finally, denoting by $(U_i, \sigma_{i-1}(U_i))$ the dominant eigenpair of operator $\tilde{T}_{i-1}$, the generalized decomposition of order M is then defined as

$$u_M = \sum_{i=1}^{M} U_i f_{i-1}(U_i) = \sum_{i=1}^{M} \sigma_{i-1}(U_i) U_i f_{i-1}(U_i) / \| U_i f_{i-1}(U_i) \|, \tag{9.103}$$

where for consistency, we let $u_0 = 0$.

9.4.2.3 Algorithms for Building the Decomposition

With the previous definitions, optimal pairs (U_i, λ_i) appear to be the dominant eigenfunctions of successive eigen-like problems. The following algorithms, initially proposed in [168] for linear stochastic problems, are here extended to the nonlinear case. In the following, we denote $W_M = (U_1, \ldots, U_M) \in (\mathcal{V})^M$, $\Lambda_M = (\lambda_1, \ldots, \lambda_M) \in (\mathcal{S})^M$ and

$$u_M(\theta) := W_M \cdot \Lambda_M(\theta). \tag{9.104}$$

Basic power-type method: Algorithm 1
In order to find the dominant eigenpair $(U, \sigma_M(U))$ of (9.102), we use a power-type algorithm. It consists of building the series $U^{(k+1)} = \tilde{T}_M(U^{(k)})$, or equivalently $U^{(k+1)} = \gamma^{(k)} \tilde{T}_M(U^{(k)})$, where $\gamma^{(k)} \in \mathbb{R}$ is a rescaling factor. We emphasize that the rescaling factor has no influence on the convergence of this series, due to homogeneity property of mapping $\tilde{T}_M$ (inherited from those of f_M and F_M). This strategy leads to algorithm 1, which can be interpreted as a power-type algorithm with deflation for building the whole decomposition.

Algorithm 1 *Power-type algorithm*

for $i = 1, \ldots, M$ ***do***
    Initialize $\lambda \in \mathcal{S}$
    for $k = 1, \ldots, k_{max}$ ***do***
        $U := F_{i-1}(\lambda)$
        $U := U / \|U\|_{\mathcal{V}}$ *(normalization)*
        $\lambda = f_{i-1}(U)$
        Check convergence on $\sigma_{i-1}(U)$ *(tolerance* ϵ_s*)*
    end for
    $W_i := (W_{i-1}, U)$
    $\Lambda_i := (\Lambda_{i-1}, \lambda)$
    Check convergence
end for

The main advantage of this algorithm is that it only requires the resolution of problems $\lambda = f(U)$ and $U = F(\lambda)$ which respectively represent a simple nonlinear equation on λ and a nonlinear deterministic problem.

It is well known that for classical eigenproblems the power method does not necessarily converge or can exhibit a very slow convergence rate. This is the case when the dominant eigenvalue is of multiplicity greater than one or when dominant eigenvalues are very close. However, a convergence criterion based on eigenfunction U is not well suited to our problem. In fact, a pertinent evaluation of convergence should be based on the eigenvalue, which in our case corresponds to the contribution $\sigma_{i-1}(U)$ of a pair $(U, f_{i-1}(U))$ to the generalized spectral decomposition. In the case of multiplicity greater than one, a convergence of the eigenvalue indicates that the current iterate U should be a good candidate for maximizing the contribution to the generalized decomposition. When dominant eigenvalues are very close, a slow convergence rate can be observed on the eigenvalue when approaching the upper spectrum. However, close eigenvalues are associated with eigenfunctions which have similar contributions to the decomposition. Therefore, any of these eigenfunctions seems to be a rather good choice, the rest of the upper spectrum being explored by subsequent "deflations" of the operator. The above remarks indicate that a relatively coarse convergence criterion (tolerance ϵ_s) can be used for the power iterates:

$$|\sigma_{i-1}(U^{(k)}) - \sigma_{i-1}(U^{(k-1)})| \leqslant \epsilon_s \sigma_{i-1}(U^{(k)}). \tag{9.105}$$

This will be illustrated in numerical examples.

Remark A natural choice for the norm $\|U\lambda\|$ on $\mathcal{V} \otimes \mathcal{S}$ consists of taking a tensorization of norms defined on $\mathcal{V}$ and $\mathcal{S}$. The contribution of $Uf(U)$ can then be simply written $\|Uf(U)\| = \|U\|_{\mathcal{V}} \|f(U)\|_{\mathcal{S}}$. In Algorithm 1, U being normalized, the evaluation of $\sigma_{i-1}(U)$ (step (7)) only requires the evaluation of $\|\lambda\|_{\mathcal{S}}$.

Remark For computational and analytical purposes, one may want to perform an orthonormalization of the decomposition. This orthonormalization can concern the deterministic basis W_M or the stochastic basis Λ_M. In both cases, it involves a non-singular $M \times M$ matrix R such that the linear transformation yields $W_M \leftarrow W_M \cdot R$ (resp. $\Lambda_M \leftarrow \Lambda_M \cdot R$) for the orthonormalization of W_M (resp. Λ_M). To maintain the validity of the decomposition, the inverse transformation R^{-1} has also to be applied to the complementary basis, *i.e.* $\Lambda_M \leftarrow \Lambda_M \cdot R^{-1}$ (resp. $W_M \leftarrow W_M \cdot R^{-1}$).

Improved power-type method: Algorithm 2

A possible improvement of Algorithm 1 consists of updating the reduced random basis Λ_M every time a new pair is computed, while keeping unchanged the deterministic basis W_M. We denote $\mathcal{V}_M = \text{span}\{U_i, i = 1, \dots, M\} \subset \mathcal{V}$ the subspace spanned by W_M; on this subspace, (9.85) becomes:
Find $u_M \in \mathcal{V}_M \otimes \mathcal{S}$ such that

$$B(u_M, v_M) = L(v_M) \quad \forall v_M \in \mathcal{V}_M \otimes \mathcal{S}. \tag{9.106}$$

This problem is equivalent to finding $\Lambda_M \in (\mathcal{S})^M$ such that

$$B(W_M \cdot \Lambda_M, W_M \cdot \Lambda_M^*) = L(W_M \cdot \Lambda_M^*) \quad \forall \Lambda_M^* \in (\mathcal{S})^M. \tag{9.107}$$

We denote by $\Lambda_M = f_0(W_M)$ the solution to (9.107), which represents a set of M coupled nonlinear stochastic equations. The improved algorithm including stochastic basis updates can be summarized as:

Algorithm 2 *Power-type algorithm with updating of the random basis*
1: ***for*** $M = 1, \ldots, M_{max}$ ***do***
2: *Do steps* 2 *to* 10 *of algorithm* 1
3: *Orthonormalize* W_M (*optional*)
4: *Update* $\Lambda_M = f_0(W_M)$
5: *Check convergence*
6: ***end for***

In the very particular case where $b(\cdot,\cdot)$ is bilinear and deterministic, it can be shown that the updating does not modify the decomposition [167]. This can be explained by the fact that dominant eigenfunctions of successive operators $\tilde{T}_M$ are optimal regarding the initial problem, *i.e.* they are dominant eigenfunctions of the initial operator $\tilde{T} = \tilde{T}_0$. In the general case, this property is not verified and so this updating can lead to a significant improvement of the accuracy of the decomposition. This will be illustrated in numerical examples.

Remark The orthonormalization step (3) of Algorithm 2 is actually optional, as it does not affect the reduced spaces generated. Still, for numerical and analytical purposes, it is often preferred to work with orthonormal functions.

9.4.3 Extension to Affine Spaces

In many situations, *e.g.* when dealing with non-homogeneous boundary conditions, the solution u is to be sought in an affine space, with an associated vector space denoted $\mathcal{V} \otimes \mathcal{S}$. Let u_0 be a particular function of the affine space. The variational problem (9.85) becomes:
Find $u = u_0 + \tilde{u}$, with $\tilde{u} \in \mathcal{V} \otimes \mathcal{S}$, such that

$$B(u_0 + \tilde{u}, v) = L(v) \quad \forall \mathcal{V} \otimes \mathcal{S}. \tag{9.108}$$

Denoting $\tilde{u}_M = W_M \cdot \Lambda_M$ and extending the definition of u_M to

$$u_M = u_0 + \tilde{u}_M = u_0 + W_M \cdot \Lambda_M, \tag{9.109}$$

it is seen that Algorithms 1 and 2 apply immediately for the construction of the generalized spectral decomposition $\tilde{u}_M$ of $\tilde{u}$. This procedure is used in the next section, which discussed the application of the proposed iterative methods to the Burgers equation.

9.4.4 Application to Burgers Equation

We consider the steady stochastic Burgers equation on the spatial domain $\Omega = (-1, 1)$, with random (but uniform) viscosity $\mu \in L_2(\Theta, P)$. The stochastic solution,

$$u : (x, \theta) \in \Omega \times \Theta \mapsto u(x, \theta) \in \mathbb{R}, \tag{9.110}$$

satisfies almost surely

$$u\frac{\partial u}{\partial x} - \mu\frac{\partial^2 u}{\partial x^2} = 0, \quad \forall x \in \Omega. \tag{9.111}$$

We assume deterministic boundary conditions:

$$u(-1, \theta) = 1, \qquad u(1, \theta) = -1 \quad \text{(a.s.)}, \tag{9.112}$$

and further assume that $\mu(\theta) \geqslant \alpha > 0$ almost surely to ensure a physically meaningful problem. With this condition, we have almost surely $-1 \leq u \leq 1$, such that $u(x, \cdot) \in L_2(\Theta, P)$.

9.4.4.1 Variational Formulation

We introduce the function space:

$$\mathcal{U} = \{v \in H^1(\Omega); v(-1) = 1, v(1) = -1\}. \tag{9.113}$$

The space $\mathcal{U}$ is affine, and we denote $\mathcal{V}$ the corresponding vector space:

$$\mathcal{V} = \{v \in H^1(\Omega); v(-1) = 0, v(1) = 0\}. \tag{9.114}$$

The stochastic solution $u(x, \theta)$ is sought in the tensor product space $\mathcal{U} \otimes \mathcal{S}$. It is solution of the variational problem (9.108) with

$$b(u, v; \theta) = \int_\Omega \left(\mu(\theta)\frac{\partial u}{\partial x}\frac{\partial v}{\partial x} + u\frac{\partial u}{\partial x}v \right) \mathrm{d}x, \tag{9.115}$$

$$l(v; \theta) = 0. \tag{9.116}$$

Note that the variational formulation above implicitly assumes that $\mathcal{S} \subset L_2(\Theta, P)$ is finite dimensional.

To detail the methodology, we write

$$b(u, v; \theta) = \mu(\theta)a(u, v) + n(u, u, v), \tag{9.117}$$

where a and n are bilinear and trilinear forms defined as:

$$a(u, v) = \int_\Omega \frac{\partial u}{\partial x}\frac{\partial v}{\partial x}\, \mathrm{d}x, \tag{9.118}$$

and

$$n(u, v, w) = \int_{\Omega} u \frac{\partial v}{\partial x} w \, dx, \tag{9.119}$$

respectively. It is observed that the forms a and n have no explicit dependence with regards to the elementary event θ. Generalization of the methodology to situations where these forms depend on θ is, however, immediate.

The boundary conditions being deterministic, an obvious choice for $u_0 \in \mathcal{U}$ is $u_0(x, \theta) = -x$. Then, to simplify the notations, we define $\lambda_0 = 1$ and $U_0 = u_0$ such that the approximate solution u_M is given by:

$$u_M = u_0 + \sum_{i=1}^{M} \lambda_i U_i = \sum_{i=0}^{M} \lambda_i U_i. \tag{9.120}$$

9.4.4.2 Implementation of Algorithms 1 and 2

Algorithms 1 and 2 can now be applied to perform the generalized spectral decomposition of the solution. We now detail the main ingredients of the algorithms, namely steps (4) and (6) of Algorithm 1, and the update step of Algorithm 2.

Resolution of $U = F_M(\lambda)$
To compute $U = F_M(\lambda)$, one has to solve (9.100), which is equivalent to solving for U the following deterministic problem (recall that λ is given):

$$B_M(\lambda U, \lambda V) = L_M(\lambda V) \quad \forall V \in \mathcal{V}, \tag{9.121}$$

where $\forall u, v \in \mathcal{V} \otimes \mathcal{S}$,

$$B_M(u, v) = B(u_M + u, v) - B(u_M, v), \tag{9.122}$$

$$L_M(v) = L(v) - B(u_M, v). \tag{9.123}$$

After some elementary manipulations, it is easy to show that

$$B_M(\lambda U, \lambda V) = \mathrm{E}\left[\lambda^2 \mu\right] a(U, V) + \mathrm{E}\left[\lambda^3\right] n(U, U, V) + \sum_{i=0}^{M} \mathrm{E}\left[\lambda_i \lambda^2\right] \left[n(U_i, U, V) + n(U, U_i, V)\right], \tag{9.124}$$

$$L_M(\lambda V) = -\sum_{i=0}^{M} \mathrm{E}[\mu \lambda_i \lambda] a(U_i, V) - \sum_{i,j=0}^{M} \mathrm{E}\left[\lambda \lambda_i \lambda_j\right] n(U_i, U_j, V). \tag{9.125}$$

Therefore, one can recast the equation on U in the formal way:

$$\tilde{\mu} a(U, V) + n(U, U, V) + n(\tilde{U}, U, V) + n(U, \tilde{U}, V)$$

$$= -a(\breve{U}, V) - n(1, \hat{Z}, V), \quad \forall V \in \mathcal{V}, \tag{9.126}$$

where

$$\tilde{\mu} = \frac{\mathrm{E}[\lambda^2 \mu]}{\mathrm{E}[\lambda^3]}, \qquad \tilde{U} = \sum_{i=0}^{M} \frac{\mathrm{E}[\lambda_i \lambda^2]}{\mathrm{E}[\lambda^3]} U_i, \tag{9.127}$$

$$\breve{U} = \sum_{i=0}^{M} \frac{\mathrm{E}[\mu \lambda_i \lambda]}{\mathrm{E}[\lambda^3]} U_i, \qquad \hat{Z} = \frac{1}{2} \sum_{i,j=0}^{M} \frac{\mathrm{E}[\lambda_i \lambda_j \lambda]}{\mathrm{E}[\lambda^3]} U_i U_j. \tag{9.128}$$

Equation (9.126) shows that U is the solution of a nonlinear deterministic problem, involving a quadratic term ($n(U, U, V)$) which reflects the nonlinearity of the original Burgers equation. In fact, the resulting problem for U has the same structure as the weak form of the deterministic Burgers equations, with some additional (linear) terms expressing the coupling of U with u_M (due to the nonlinearity) and a right-hand side accounting for the residual for $u = u_M$. As a result, a standard nonlinear solver can be used to solve this equation, *e.g.* one can re-use a deterministic steady Burgers solver with minor adaptations.

At first thought, (9.126) suggests that a robust nonlinear solver is needed for its resolution, since *a priori* the effective viscosity $\tilde{\mu}$ may become negative and experience changes by orders of magnitudes in the course of the iterative process. However, one can always make use of the homogeneity property

$$\frac{U}{\alpha} = F_M(\alpha\lambda), \quad \forall \alpha \in \mathbb{R}^*, \tag{9.129}$$

to rescale the problem and thus satisfy the requirements of the solver if need be. Note that (9.129) together with (9.126) also indicate that the nature of the nonlinear deterministic problems to be solved is preserved along the course of the iterations. For instance, the effective viscosity goes to zero as $|\lambda| \to \infty$ but the problem does not degenerate to a hyperbolic one since the right-hand-side also goes to zero and U satisfies homogeneous boundary conditions.

Resolution of $\lambda = f_M(U)$
The random variable $\lambda \in \mathcal{S}$ is solution of the variational problem:

$$B_M(\lambda U, \beta U) = L_M(\beta U) \quad \forall \beta \in \mathcal{S}. \tag{9.130}$$

After some manipulations, this equation is found to be equivalent to:

$$\begin{aligned}
&\mathrm{E}\left[\beta\lambda^2\right] n(U, U, U) + \mathrm{E}\left[\beta\mu\lambda\right] a(U, U) \\
&\quad + \sum_{i=0}^{M} \mathrm{E}\left[\beta\lambda_i\lambda\right] \left[n(U, U_i, U) + n(U_i, U, U)\right] \\
&\quad - \sum_{i=0}^{M} \mathrm{E}\left[\beta\mu\lambda_i\right] a(U_i, U) - \sum_{i,j=0}^{M} \mathrm{E}\left[\beta\lambda_i\lambda_j\right] n(U_i, U_j, U).
\end{aligned} \tag{9.131}$$

This is a simple stochastic quadratic equation on λ; a standard nonlinear solver can be used for its resolution.

Resolution of $\Lambda_M = f_0(W_M)$
To update $\Lambda_M = (\lambda_1, \ldots, \lambda_M) \in (\mathcal{S})^M$, one has to solve:

$$B(u_0 + W_M \cdot \Lambda_M, W_M \cdot \Lambda_M^*) = L(W_M \cdot \Lambda_M^*) \quad \forall \Lambda_M^* \in (\mathcal{S})^M. \tag{9.132}$$

This equation can be split into a system of M equations:

$$\forall k \in \{1, \ldots, M\}, \quad B(u_0 + W_M \cdot \Lambda_M, U_k \beta_k) = L(U_k \beta_k) \quad \forall \beta_k \in \mathcal{S}. \tag{9.133}$$

Introducing the previously defined forms, we obtain:

$$\sum_{i=0}^{M} \mu(\theta)\lambda_i(\theta) a(U_i, U_k) + \sum_{i,j=0}^{M} \lambda_i \lambda_j n(U_i, U_j, U_k) = 0 \quad \forall k \in \{1, \ldots, M\}. \tag{9.134}$$

Again, it is seen that the updating step consists of solving a system of quadratic nonlinear equations for the $\{\lambda_i\}_{i=1}^{M}$. A standard nonlinear solver can be used for this purpose.

9.4.4.3 Spatial Discretization

Let us denote $\mathbb{P}_{\mathrm{N_x}+1}(\Omega)$ the space of polynomials of degree less or equal to $\mathrm{N_x}+1$ on Ω. We define the approximation vector space $\mathcal{V}^h$ as:

$$\mathcal{V}^h = \{v \in \mathbb{P}_{\mathrm{N_x}+1}(\Omega); v(-1) = 0, v(1) = 0\} \subset \mathcal{V}. \tag{9.135}$$

Let x_i, $i = 0, \ldots, \mathrm{N_x}+1$, be the $\mathrm{N_x}+2$ Gauss-Lobatto points [1] of the interval $[-1, 1]$, such that

$$x_0 = -1 < x_1 < \cdots < x_{\mathrm{N_x}} < x_{\mathrm{N_x}+1} = 1. \tag{9.136}$$

We denote $L_{i \in \{1,\ldots,\mathrm{N_x}\}}(x) \in \mathbb{P}_{\mathrm{N_x}+1}$, the Lagrange polynomials constructed on the Gauss-Lobatto grid:

$$L_i(x) = \prod_{\substack{j=0 \\ j \neq i}}^{\mathrm{N_x}+1} \frac{x - x_j}{x_i - x_j}. \tag{9.137}$$

These polynomials satisfy

$$L_i(x_j) = \begin{cases} 0 & \text{if } i \neq j, \\ 1 & \text{if } i = j \end{cases} \quad \forall j = 0, \ldots, \mathrm{N_x}+1, \tag{9.138}$$

and form a basis of $\mathcal{V}^h$:

$$\mathcal{V}^h = \text{span}\{L_i, i = 1, \dots, \mathrm{N_x}\}. \tag{9.139}$$

For any $v \in \mathcal{V}^h$, we have

$$v(x) = \sum_{i=1}^{\mathrm{N_x}} v^i L_i(x), \quad v^i = v(x_i). \tag{9.140}$$

The derivative of $v \in \mathcal{V}^h$ has for expression:

$$\frac{\partial v}{\partial x} = \sum_{i=1}^{\mathrm{N_x}} v^i L_i'(x), \quad L_i' \equiv \frac{\partial L_i}{\partial x}. \tag{9.141}$$

The bilinear and trilinear forms a and n are evaluated using the quadrature formula over the Gauss-Lobatto points [26]. Specifically, for $u, v \in \mathcal{V}^h$, we have

$$\begin{aligned} a(u,v) &= \int_\Omega \frac{\partial u}{\partial x}\frac{\partial v}{\partial x}\,\mathrm{d}x = \int_\Omega \left(\sum_{i=1}^{\mathrm{N_x}} u^i L_i'\right)\left(\sum_{i=1}^{\mathrm{N_x}} v^i L_i'\right)\mathrm{d}x \\ &= \sum_{i,j=1}^{\mathrm{N_x}} u^i v^j \int_\Omega L_i'(x) L_j'(x)\,\mathrm{d}x = \sum_{i,j=1}^{\mathrm{N_x}} u^i v^j a_{i,j}, \end{aligned} \tag{9.142}$$

where

$$a_{i,j} \equiv \left(\sum_{k=0}^{\mathrm{N_x}+1} L_i'(x_k) L_j'(x_k)\omega_k\right), \tag{9.143}$$

with $\omega_{k\in\{0,\dots,\mathrm{N_x}+1\}}$ the Gauss-Lobatto quadrature weights [1]. Similarly, for $u, v, w \in \mathcal{V}^h$, we have

$$\begin{aligned} n(u,v,w) &= \int_\Omega u\frac{\partial v}{\partial x} w\,\mathrm{d}x \approx \sum_{k=0}^{\mathrm{N_x}+1}\left(u(x_k)\sum_{i=0}^{\mathrm{N_x}+1} v^i L_i'(x_k) w(x_k)\right)\omega_k \\ &\approx \sum_{k=1}^{\mathrm{N_x}}\sum_{i=0}^{\mathrm{N_x}+1} n_{i,k} u^k v^i w^k, \end{aligned} \tag{9.144}$$

where $n_{i,k} \equiv L_i'(x_k)\omega_k$. The same expression holds for $u_0 \notin \mathcal{V}^h$.

9.4.4.4 Stochastic Discretization

In the results presented below, the random viscosity μ is parametrized using a set of N independent real continuous second-order random variables, $\boldsymbol{\xi} = \{\xi_1, \dots, \xi_\mathrm{N}\}$,

$$\mu(\theta) = \mu(\boldsymbol{\xi}(\theta)). \tag{9.145}$$

We denote Ξ the range of $\boldsymbol{\xi}$ and $P_{\boldsymbol{\xi}}$ the known probability law of $\boldsymbol{\xi}$. Since random variables ξ_i are independent, we have for $\mathbf{y} = (y_1, \ldots, y_N) \in \mathbb{R}^N$

$$dP_{\boldsymbol{\xi}}(\mathbf{y}) = \prod_{i=1}^{N} p_{\xi_i}(y_i)\, dy_i. \tag{9.146}$$

Let $(\Xi, \mathcal{B}_\Xi, P_{\boldsymbol{\xi}})$ be the associated probability space. The stochastic solution is then sought in the image probability space $(\Xi, \mathcal{B}_\Xi, P_{\boldsymbol{\xi}})$ instead of $(\Theta, \mathcal{B}, P)$, *i.e.* we compute $u(\boldsymbol{\xi})$. In the image probability space, the expectation operator is

$$E[f(\cdot)] = \int_\Theta f(\boldsymbol{\xi}(\theta))\, dP(\theta) = \int_\Xi f(\mathbf{y})\, dP_{\boldsymbol{\xi}}(\mathbf{y}). \tag{9.147}$$

It is clear from this relation that if $f \in L_2(\Theta, P)$ then $f \in L_2(\Xi, P_{\boldsymbol{\xi}})$, the space of second-order random variables spanned by $\boldsymbol{\xi}$. To proceed with the determination of the numerical solution, one has to construct a finite-dimensional approximation space $\mathcal{S} \subset L_2(\Xi, P_\Xi)$. Different discretizations are possible (continuous polynomial expansion, piecewise polynomial expansions, multiwavelets,...). In the following, we rely on classical Generalized Polynomial Chaos expansions, which consist in defining the stochastic space as

$$\mathcal{S} = \text{span}\{\Psi_0, \ldots, \Psi_P\}, \tag{9.148}$$

where the Ψ_i are mutually orthogonal polynomials in $\boldsymbol{\xi}$, with total degree less or equal to No. The orthogonality of the polynomials is expressed as:

$$E[\Psi_i \Psi_j] = E\left[\Psi_i^2\right]\delta_{ij}. \tag{9.149}$$

The dimension of the stochastic subspace is therefore given by

$$\dim(\mathcal{S}) = P + 1 = \frac{(N + No)!}{N!No!}, \tag{9.150}$$

and a random variable $\beta \in \mathcal{S}$ may be expanded as:

$$\beta(\boldsymbol{\xi}) = \sum_{i=0}^{P} \beta^i \Psi_i(\boldsymbol{\xi}). \tag{9.151}$$

9.4.4.5 Solvers

$U = F_M(\lambda)$

With the spatial discretization introduced previously, one has to solve for $U \in \mathcal{V}^h$ a set of N_x nonlinear equations of the form:

$$G_k(U^1, \ldots, U^{N_x}; \lambda) = 0, \quad k = 1, \ldots, N_x. \tag{9.152}$$

Here, we use a classical Newton method to solve this system.
$\lambda = f_M(U)$
Introducing the stochastic expansions of μ and of the λ_i, the coefficients λ^i of λ satisfy the following set of P + 1 nonlinear equations:

$$g_k(\lambda^0, \dots, \lambda^{\mathrm{P}}; U) = \sum_{i,j=0}^{\mathrm{P}} c_{ijk}\lambda^i\lambda^j + \sum_{i=0}^{\mathrm{P}} d_{ik}\lambda^i + e_k = 0, \quad k = 0, \dots, \mathrm{P}, \tag{9.153}$$

where

$$c_{ijk} = \mathrm{E}\left[\Psi_i\Psi_j\Psi_k\right] n(U, U, U),$$

$$d_{ik} = \sum_{j=0}^{\mathrm{P}} \mathrm{E}\left[\Psi_i\Psi_j\Psi_k\right] \left[\mu^j a(U, U) + \sum_{l=0}^{M} \lambda_l^j (n(U, U_l, U) + n(U_l, U, U)) \right],$$

$$e_k = \sum_{i,j=0}^{\mathrm{P}} \mathrm{E}\left[\Psi_i\Psi_j\Psi_k\right] \left[\mu^i \sum_{l=0}^{M} \lambda_l^j a(U_l, U) + \sum_{l,m=0}^{M} \lambda_l^i \lambda_m^j n(U_l, U_m, U) \right].$$

This system can be solved efficiently using standard techniques involving exact Jacobian computation. Here, we use the minpack subroutines [157] to solve (9.153).
$\Lambda_M = f_0(W_M)$
The stochastic expansion of Λ_M is

$$\Lambda_M = \sum_{i=0}^{\mathrm{P}} \Lambda_M^i \Psi_i. \tag{9.154}$$

Introducing this expansion in (9.134), one obtains the following system of $M \times (\mathrm{P} + 1)$ nonlinear equations:

$$g_{k,q}(\Lambda_M^0, \dots, \Lambda_M^{\mathrm{P}}; W_M) = \sum_{l,m=0}^{\mathrm{P}} \mathrm{E}\left[\Psi_l\Psi_m\Psi_q\right] \left[\sum_{i=0}^{M} \mu^l \lambda_i^m a(U_i, U_k) + \sum_{i,j=0}^{M} \lambda_i^l \lambda_j^m n(U_i, U_j, U_k) \right] = 0,$$
$$k = 1, \dots, \mathrm{M}, \; q = 0, \dots, \mathrm{P}. \tag{9.155}$$

Again, we rely on the minpack library to solve this system. It is seen that, unlike the determination of U and λ, the size of the nonlinear system of equations for the updating of Λ_M increases with M.

9.4.4.6 Results

For the purpose of analyzing convergence, we define the stochastic residual of the equation as

$$\mathcal{R}_M(x,\theta) = u_M \frac{\partial u_M}{\partial x} - \mu \frac{\partial^2 u_M}{\partial x^2} \tag{9.156}$$

and the corresponding L_2-norm

$$\|\mathcal{R}_M\|^2 = \int_\Omega \|\mathcal{R}_M(x,\cdot)\|^2_{L_2(\Xi,P_\xi)}\,\mathrm{d}x = \int_\Omega \mathrm{E}\left[\mathcal{R}_M(x,\cdot)^2\right]\mathrm{d}x. \tag{9.157}$$

It is observed that this norm accounts for the errors due to stochastic and spatial discretizations. As a result, when $(M, \dim(\mathcal{S})) \to \infty$, this error is not expected to go to zero but to level off to a finite value corresponding to the spatial discretization error. However, thanks to the spectral finite element approximation in space, the errors in the following numerical tests are dominated by the stochastic error due to $\dim(\mathcal{S}) < \infty$. In fact, we are at present more interested by the analysis of the convergence with M of u_M toward the discrete exact solution on $\mathcal{V}^h \otimes \mathcal{S}$, and the comparison of the convergence rates of the two algorithms, than in the absolute error. For this purpose, we define the stochastic residual $R_M(x,\theta)$ as the orthogonal projection of $\mathcal{R}_M(x,\theta)$ on $\mathcal{S}$:

$$\mathcal{R}_M(x,\theta) = R_M(x,\theta) + R_M^{\perp}(x,\theta), \tag{9.158}$$

such that

$$R_M(x,\cdot) \in \mathcal{S}, \qquad \mathrm{E}\left[R_M^{\perp}(x,\cdot)\beta\right] = 0, \quad \forall \beta \in \mathcal{S}. \tag{9.159}$$

In other words, $R_M(x,\cdot)$ is the classical Galerkin residual on $\mathcal{S}$,

$$R_M(x,\theta) = \sum_{k=0}^{\mathrm{P}} R_M^k(x)\Psi_k(\theta),$$

where

$$\begin{aligned}\mathrm{E}[\Psi_k\Psi_k]R_M^k(x) &= \mathrm{E}[\mathcal{R}_M(x,\cdot)\Psi_k(\cdot)]\\ &= \sum_{i,j=0}^{M} \mathrm{E}\left[\lambda_i\lambda_j\Psi_k\right]U_i\frac{\partial U_j}{\partial x} - \sum_{i=0}^{M}\mathrm{E}[\mu\lambda_i\Psi_k]\frac{\partial^2 U_i}{\partial x^2}.\end{aligned}$$

Its L_2-norm is

$$\|R_M\|^2 = \int_\Omega \left[\sum_{k=0}^{\mathrm{P}}\left(R_M^k(x)\right)^2 \mathrm{E}[\Psi_k\Psi_k]\right]\mathrm{d}x. \tag{9.160}$$

It is seen that $\|R_M\|$, though containing a contribution of the spatial discretization error deemed negligible, essentially measures the reduced basis approximation error (*i.e.* by substituting the "exact" discrete solution $u^h \in \mathcal{V}^h \otimes \mathcal{S}$ by $u_M = W_M \cdot \Lambda_M$ in the equations). Consequently, we shall refer to $\mathcal{R}_M$ as the equation residual and to R_M as the reduction residual.

Convergence analysis: To analyze the convergence of the GSD algorithms, we consider the following random viscosity setting:

$$\mu(\boldsymbol{\xi}) = \mu^0 + \sum_{i=1}^{\mathrm{N}} \mu' \xi_i, \tag{9.161}$$

with all ξ_i being uniformly distributed on $(-1,1)$, leading to $\Xi = (-1,1)^{\mathrm{N}}$. To ensure the positivity of the viscosity, we must have $\mu^0 > \mathrm{N}|\mu'|$. We set $\mu' = c\mu^0/\mathrm{N}$, with $|c| < 1$. For this parametrization, the variance of the viscosity is

$$\mathrm{E}\left[(\mu - \mu^0)^2\right] = \frac{\mathrm{N}}{3}(\mu')^2 = \frac{c^2}{3\mathrm{N}}(\mu^0)^2. \tag{9.162}$$

Accordingly, the density of μ depends on N, and μ experiences less variability as N increases. For the discretization of the stochastic space $\mathcal{S}$, we use multidimensional Legendre polynomials. The mean viscosity is set to $\mu^0 = 0.2$ and $c = 0.85$.

In a first series of tests, we set $\mathrm{N} = 4$ and $\mathrm{No} = 6$, so $\dim(\mathcal{S}) = 210$, whereas for the spatial discretization $\dim(\mathcal{V}^h) = \mathrm{N_x} = 200$ is used. This spatial discretization allows for accurate deterministic solutions for any realization $\mu(\boldsymbol{\xi})$, $\boldsymbol{\xi} \in \Xi$. If the stochastic solution were to be sought in the full approximation space $\mathcal{V}^h \otimes \mathcal{S}$, the size of the nonlinear problem to be solved would be $\dim(\mathcal{V}^h) \times \dim(\mathcal{S}) = 42000$. In contrast, the reduced basis solution $W_M \cdot \Lambda_M$ has for dimension $M \times (\dim(\mathcal{V}^h) + \dim(\mathcal{S})) = 410M$.

In Fig. 9.21, we compare the convergence of Algorithms 1 and 2, as measured by the two residual norms $\|\mathcal{R}_M\|$ and $\|R_M\|$, with the size M of the reduced basis (left plot) and with the total number of iterations performed on $U = F_M(\lambda)$ and $\lambda = f_M(U)$ (right plot). The stopping criterion used is $\epsilon_s = 10^{-3}$. Focusing first on R_M, we can conclude that both algorithms converge to the discrete solution on $\mathcal{V}^h \otimes \mathcal{S}$ with exponential rate as the dimension M of the reduced basis increases. However, Algorithm 2 is more effective in reducing R_M, compared to Algorithm 1. Specifically, the exponential convergence rates for $\|R_M\|$ are ~ 1.2 and ~ 0.3 for Algorithms 2 and 1, respectively. Also, the norms $\|\mathcal{R}_M\|$ are seen to decrease with the same rate as $\|R_M\|$. Due to the higher convergence rate of Algorithm 2, the corresponding values of $\|\mathcal{R}_M\|$ quickly saturate to a finite value (the discretization error) within just 5 iterations. For Algorithm 1, the norm of $\mathcal{R}_M$ does not reach its asymptotic value for $M \leq 10$, reflecting the slower convergence of the solution in $\mathcal{V}^h \otimes \mathcal{S}$.

Inspection of the right plot of Fig. 9.21 shows that Algorithm 2 requires less iterations on problems $U = F_M(\lambda)$ and $\lambda = f_M(U)$ to yield the next term of the decomposition. Specifically, Algorithm 2 requires 3 to 4 iterations to reach the stopping

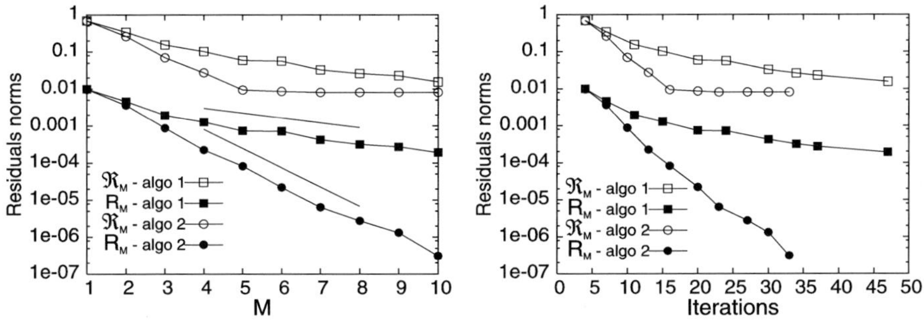

Fig. 9.21 Convergence of R_M (*closed symbols*) and $\mathcal{R}_M$ (*open symbols*) for Algorithms 1 (*squares*) and 2 (*circles*). The *left plot* depicts the residual norms as a function of the reduced basis dimension M, whereas the right plot displays the residual norms as a function of the total (cumulative) number of power-type iterations for the computation of successive pairs (U, λ). Also reported on the left using *solid lines* are fits of $\|R_M\|$ with $\sim \exp(-1.2M)$ and $\sim \exp(-0.3M)$. Adapted from [171]

criterion, whereas 3 to 8 iterations are required by Algorithm 1. This difference is essentially explained by the updating of Λ_M. Indeed, when the orthonormalization of W_M in Algorithm 2 is disregarded, the convergence of the resulting decomposition and number of iterations to yield the pair (U, λ) is unchanged (not shown). This confirms the claim made earlier that the orthonormalization of W_M is optional. The lower number of iterations and faster convergence of the residuals for Algorithm 2 does not imply a lower computational cost, since the resolution of $U = F_M(\lambda)$ is inexpensive for the 1D Burgers equation. In fact, Algorithm 2 requires a significantly larger computational time for this problem, as most of the CPU time is spent solving the stochastic update problem $\Lambda_M = f_M(W_M)$. This conclusion will not hold in general for larger problems (*e.g.* Navier-Stokes flows), where the resolution of the deterministic problems dominates the overall CPU time. Also, computational times are not the only concern and one may prefer to spend more time computing the reduced modes, to achieve a better reduced basis approximation in order to lower memory requirements, especially for problems involving large spatial approximation spaces.

To understand the higher efficiency of Algorithm 2, we compare in Fig. 9.22 the first 8 reduced modes $U_i(x)$ computed using the two algorithms. Only half of the domain is shown as the reduced modes are odd functions of x, because of the symmetry of the problem. The comparison clearly shows that Algorithm 2 yields a deterministic reduced basis $W_{M=8}$ with a higher frequency content than for that of Algorithm 1. This is explained by the improvement brought by the updating of Λ_M. In fact, because the updating procedure cancels the equation residual in the subspace span$\{W_M\} \otimes \mathcal{S}$, the following deterministic mode U constructed will be essentially orthogonal to W_M. On the contrary, Algorithm 1 solves the equations in the subspace span$\{W_M\} \otimes \mathcal{S}$ only approximatively (*i.e.* $\Lambda_M \neq f_0(W_M)$), leading to a delayed exploration of the deterministic space $\mathcal{V}^h$. This point is further illustrated in Fig. 9.23, which depicts the second moment of the residual, $\mathrm{E}[\mathcal{R}_M(x, \cdot)^2]$, for

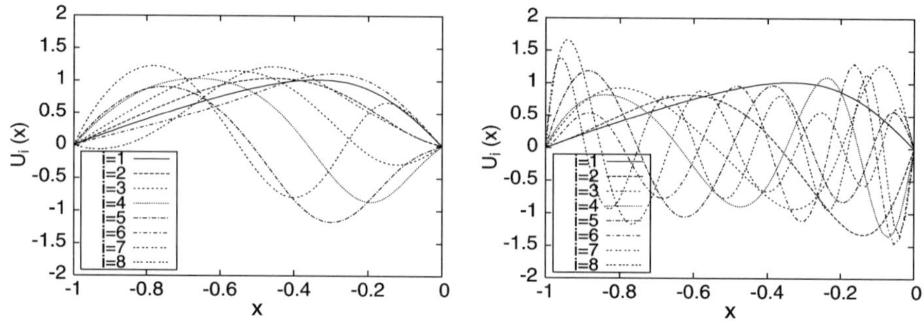

Fig. 9.22 Comparison of the first 8 reduced modes U_i obtained with Algorithms 1 (*left*) and 2 without orthonormalization of W_M (*right*). Adapted from [171]

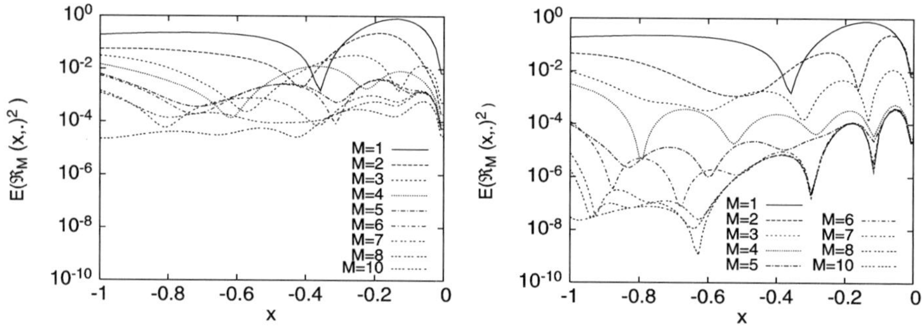

Fig. 9.23 Evolution of the second moment of the equation residual, $\mathrm{E}[\mathcal{R}_M(x,\cdot)^2]$, for different M; *left*: Algorithm 1; *right*: Algorithm 2. Adapted from [171]

different values of M. The results for Algorithm 2 highlight the efficiency of the GSD in capturing the full discrete solution on $\mathcal{V}^h \otimes \mathcal{S}$ in just few modes and indicates that the stochastic discretization mostly affects the residual in the area where the solution exhibits the steepest gradients, *i.e.* where the uncertainty has highest impact on the solution.

It is also remarked that even though the norm of the residual provides a measure of how well the reduced basis approximation satisfies the Burgers equation, it is not a direct measure of the error on the solution. Specifically, the somewhat large magnitude of $\|\mathcal{R}_M\|$ does not imply that the error ϵ_M on the solution is as high. The L_2-error of the stochastic solution can be measured using the following norm:

$$\|\epsilon_{\mathrm{M}}\|^2 = \int_\Omega \|u_M(x,\cdot) - u(x,\cdot)\|^2_{L_2(\Xi,P_\xi)}\,\mathrm{d}x, \tag{9.163}$$

where u_M is the GSD solution and u the exact stochastic solution. The exact solution being unknown, one has to rely on approximate expression for $\|\epsilon_M\|$. Here, using the fact that the stochastic error dominates the spatial error, we use a standard Monte-Carlo (MC) method to estimate the solution error. We denote $u^d(x;\boldsymbol{\xi}) \in \mathcal{V}^h$ the

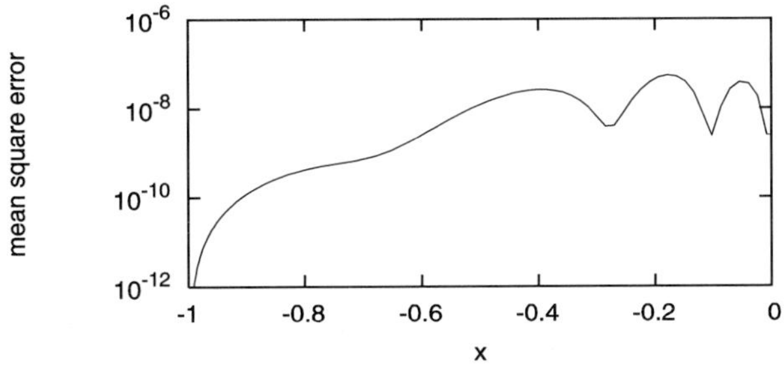

Fig. 9.24 MC estimate of the local mean square error on the solution, $\mathrm{E}[(u_M - u)^2]$ obtained with Algorithm 2 with $M = 10$. Adapted from [171]

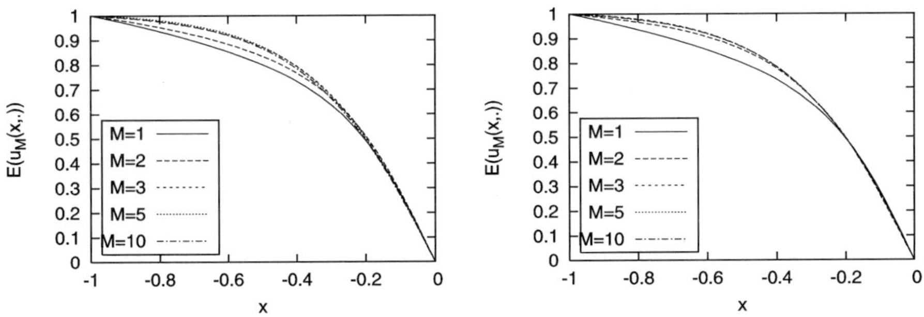

Fig. 9.25 Spatial distribution of $\mathrm{E}[u_M]$ for different values M; *left*: Algorithm 1; *right*: Algorithm 2. Adapted from [171]

deterministic solution of the Burgers equation for the viscosity realization $\mu(\boldsymbol{\xi})$. We then rely on a uniform random sampling of Ξ, with m sampling points, to construct the stochastic estimate of the local mean square error:

$$\|u_M(x,\cdot) - u(x,\cdot)\|^2_{L_2(\Xi,P_{\boldsymbol{\xi}})} \approx \frac{1}{m}\sum_{i=1}^{m}\left(u_M(x,\boldsymbol{\xi}(\theta_i)) - u^d(x;\boldsymbol{\xi}(\theta_i))\right)^2. \quad (9.164)$$

Using a set of $m = 10000$ samples we obtain the estimate $\|\epsilon_{M=10}\| = (1.55 \pm 0.1)10^{-4}$ for Algorithm 2, showing that the reduced solution u_M is indeed much more accurate than suggested by the norm of the equation residual. As for the equation residual, we provide in Fig. 9.24 the spatial distribution for the mean square error on the solution, for the MC estimate given in (9.164) using $m = 10000$ MC samples.

For a better appreciation of the convergence of the solution on the reduced basis, we plot in Figs. 9.25 and 9.26 the evolutions of the mean and standard deviation, respectively $\mathrm{E}[u_M]$ and Std-dev(u_M), for different M. Again, only half of the domain is shown, the mean and standard deviation being an odd function and an even function of x, respectively. Figure 9.25 shows a fast convergence of the mean for

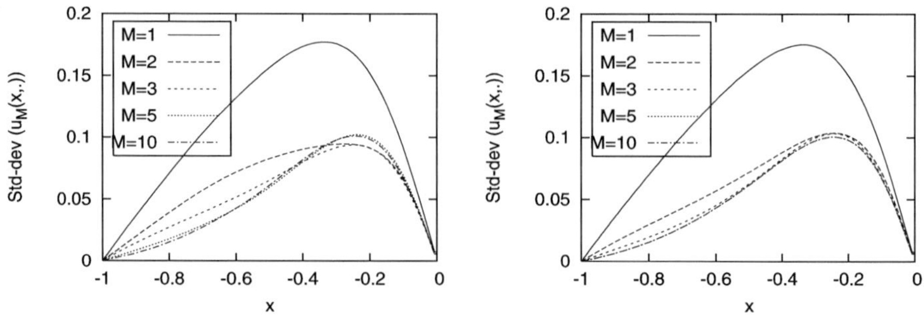

Fig. 9.26 Spatial distribution of Std-dev(u_M) for different values M; *left*: Algorithm 1; *right*: Algorithm 2. Adapted from [171]

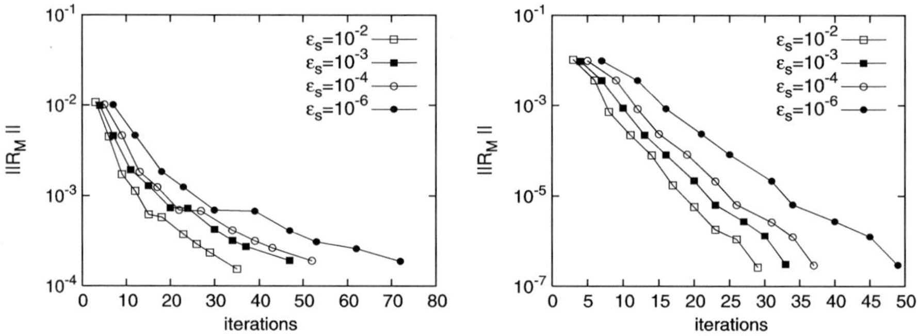

Fig. 9.27 Evolution of the reduction residual with the number of iterations for different stopping criteria ϵ_s as indicated; *left*: Algorithm 1, *right*: Algorithm 2. Adapted from [171]

the two algorithms: curves are essentially indistinguishable for $M \geq 3$. The standard deviation (Fig. 9.26) also exhibits a fast convergence, which is clearly higher for Algorithm 2 then for Algorithm 1.

Robustness of the algorithms: We now investigate the robustness of the method with regards to stochastic discretization and numerical parameters.

Impact of ϵ_s*:* The selection of an appropriate value for ϵ_s in an important issue as slow convergence was observed in some computations. Questions also arise regarding the accuracy of the computed pairs (U, λ) needed to construct an appropriate reduced basis (see discussion in Sect. 9.4.2.3). This aspect is numerically investigated by considering gradually less stringent stopping criteria ϵ_s and monitoring the convergence of $\|R_M\|$. These experiments are reported in Fig. 9.27, use the same viscosity settings and discretization parameters specified above. We test the behavior of both algorithms for $\epsilon_s = 10^{-2,-3,-4,-6}$. It is seen that for both algorithms, the selection of ϵ_s on the range tested has virtually no effect on the convergence of the decomposition, but the computations become more demanding as ϵ_s decreases. Similar experiences for other viscosity settings (see below) demonstrate that usu-

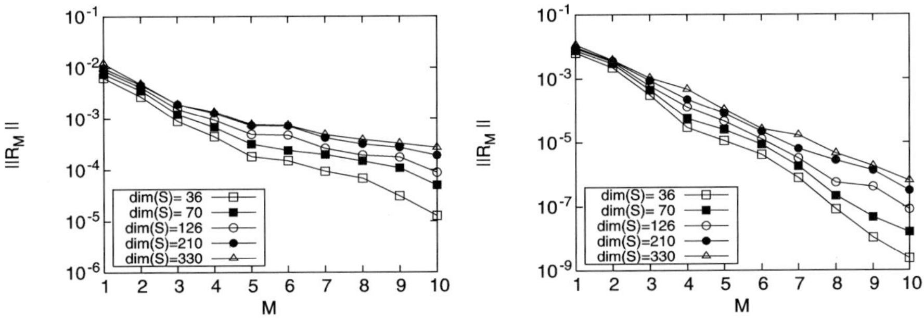

Fig. 9.28 R_M versus M for fixed N = 4 and different orders No; *left*: Algorithm 1, *right*: Algorithm 2. Adapted from [171]

ally little is gained by performing more than 3 to 4 iterations to determine the pair (U, λ).

Impact of stochastic polynomial order: We now vary the polynomial order No $= 3, \ldots, 7$ of the stochastic approximation space $\mathcal{S}$, while holding N = 4 fixed. Other parameters are the same as specified above. These experiments can be viewed as a refinement of the stochastic discretization, since dim($\mathcal{S}$) is directly related to No (see (9.150)). We monitor the convergence of the two GSD algorithms with M for different No. Results are reported in Fig 9.28. The plots show that the convergence of the algorithms becomes slower as No increases. This is not surprising since increasing No allows to capture more variability in the solution so that more modes are needed to achieve the same level of accuracy in reduction. Still, one can observe that the convergence rates tend to level off, denoting the convergence of the stochastic approximation as No increases. In fact, these results essentially highlight the need of a high polynomial order to obtain an accurate solution for the viscosity settings used. This is consistent with the decrease in the asymptotic value of the equation residual norm as No increases, as shown in Fig. 9.29. Conversely, these computations demonstrate the robustness and stability of the power type algorithms in constructing approximations on under-resolved stochastic space $\mathcal{S}$.

Impact of the stochastic dimensionality: We now wish to compare the efficiencies of the algorithms when the dimension of $\mathcal{S}$ varies with dimensionality N of the problem. Since the random viscosity, as previously parameterized, has decreasing variability when N increases, we need a different parameterization for a fair comparison. The viscosity distribution is now assumed log-normal, with median value $\overline{\mu}$ and coefficient of variation $C_{LN} > 1$. This implies that the probability of having $\mu(\theta) \in]\overline{\mu}/C_{LN}, \overline{\mu}C_{LN}[$ is equal to 0.99. Consequently, μ can be parameterized using a normalized normal random variable ζ as:

$$\mu = \exp\left[\overline{\zeta} + \sigma_\zeta \zeta\right], \quad \overline{\zeta} = \ln \overline{\mu}, \sigma_\zeta = \frac{\ln C_{LN}}{2.95}. \tag{9.165}$$

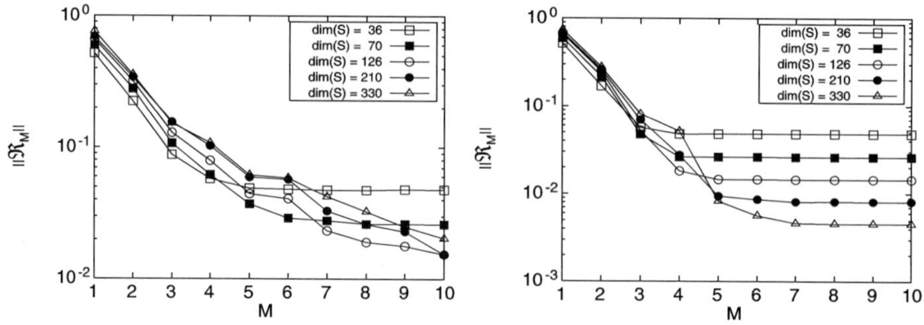

Fig. 9.29 $\mathcal{R}_M$ versus M for fixed N = 4 and different orders No; *left*: Algorithm 1, *right*: Algorithm 2. Adapted from [171]

The random variable ζ can in turns be decomposed as the sum of N independent normalized random variables ξ_i as follows:

$$\zeta = \frac{1}{\sqrt{\mathrm{N}}} \sum_{i=1}^{\mathrm{N}} \xi_i, \quad \xi_i \sim N(0,1). \tag{9.166}$$

This leads to:

$$\mu(\boldsymbol{\xi}) = \overline{\mu} \exp\left[\frac{\ln C_{LN}}{\sqrt{\mathrm{N}}} \sum_{i=1}^{\mathrm{N}} \xi_i \right], \quad \xi_i \sim N(0,1). \tag{9.167}$$

It is stressed that for this parameterization the distribution of μ is the same for any $\mathrm{N} \geq 1$. Indeed, μ keeps a log-normal distribution with constant median and coefficient of variation for any N. However, changing N implies that the stochastic solution is sought in function space $L_2(\Xi, P_{\boldsymbol{\xi}})$ with variable dimensionality for Ξ. This parametrization of μ is in fact designed to investigate the efficiency of the GSD for the same problem but considered on probability spaces with increasing dimensionalities. Specifically, we use the Hermite Polynomial Chaos system as a basis of $\mathcal{S}$, so for fixed PC order No the dimension of $\mathcal{S}$ increases with N as given by (9.150). However, the PC solution for $\mathrm{N} > 1$ involves many hidden symmetries, and we expect the GSD algorithms to "detect" these structures and to construct effective reduced basis.

We set $\overline{\mu} = 0.3$, $C_{LN} = 3$ and No = 6. The projection of μ on $\mathcal{S}$ can be determined analytically, or numerically computed by solving a stochastic ODE [52]. We compute the GSD of the solutions for $\mathrm{N} = 2, \ldots, 5$ using the two algorithms with $\epsilon_s = 10^{-2}$. Results are reported in Fig. 9.30 which shows the norms of residuals $\mathcal{R}_M$ and R_M as a function of M. The plots show that the convergence of the two algorithms is indeed essentially unaffected by the dimension of Ξ.

Robustness with regards to input variability: We now investigate the robustness of the power-type algorithms with regards to the variability in μ. We rely on the

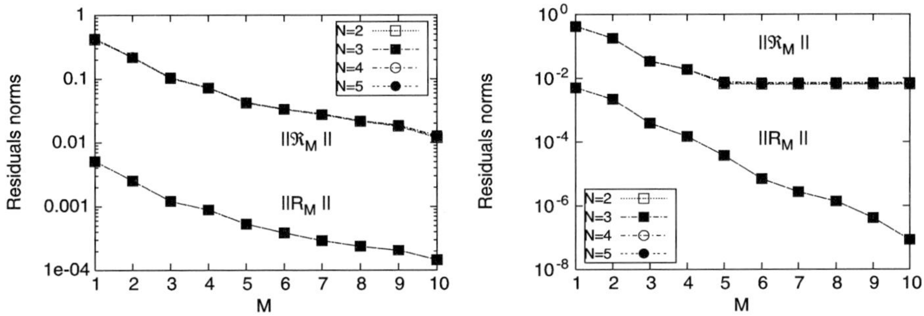

Fig. 9.30 $\mathcal{R}_M$ and R_M versus M, for different values N as indicated; *left*: Algorithm 1, *right*: Algorithm 2. Adapted from [171]

previous log-normal parameterization of the viscosity, with $\mathrm{N} = 3$ and $\mathrm{No} = 6$ ($\dim(\mathcal{S}) = 84$). We first fix $\overline{\mu} = 0.3$ and we vary C_{LN} in the range $[1.5, 4]$. We then fix $C_{LN} = 2.5$ and we vary $\overline{\mu}$ in the range $[0.1, 0.4]$. Results are presented for Algorithm 2 only, since similar trends are observed with Algorithm 1.

Figure 9.31 shows the reduced basis approximation $u_{M=10}(x)$ for all the computations, using the classical mean value $\pm$ 3 standard deviation bars representation (even though this representation is not ideal as the solution is clearly non-Gaussian). The plots of the left column correspond to $\overline{\mu} = 0.3$ and increasing coefficient of variability C_{LN} (from top to bottom). They show the increasing variability of the solution with C_{LN} while the solution mean is roughly unaffected. On the contrary, the plots of the right column corresponding to $C_{LN} = 2.5$ and increasing $\overline{\mu}$ (from top to bottom), show a large impact of the median value of the viscosity on the mean of the solution, together with a non-trivial evolution of the solution variability. Specifically, although the variance of the log-normal viscosity is fixed, the peak variance in the solution increases as $\overline{\mu}$ decreases. This complex dependence of the solution with regards to the viscosity distribution underscores the strong nonlinear character of Burger's equation.

We now analyze of the behavior of the residuals $\|R_M\|$, plotted in Fig. 9.32. Focusing first on the curves for fixed $\overline{\mu}$ (left plot of Fig. 9.32), it is first observed that the magnitude of the residual increases with C_{LN}, as one may have expected. For the two lowest values of C_{LN} the convergence rates are found roughly equal, whereas slower rates are reported for $C_{LN} = 3$ and 4. This trend can be explained by the increasing level of variability in the solutions for large COV, which demands more spectral modes to approximate the solution. Note that we have checked that $\dim(\mathcal{S})$ (*i.e.* No) was sufficiently large to account for all the variability in the solution, when $C_{LN} = 4$, by performing a computation with $\mathrm{No} = 8$, without significant change in the solution. For fixed $C_{LN} = 2.5$ (right plot of Fig. 9.32), a degradation of the convergence rate is observed as $\overline{\mu}$ decreases, together with an increase the magnitude of the residuals. This can be explained by the increasing variability in the solution as seen from Fig. 9.31, and by the more complex dependence on μ of the spatial structure of the solution as $\overline{\mu}$ decreases.

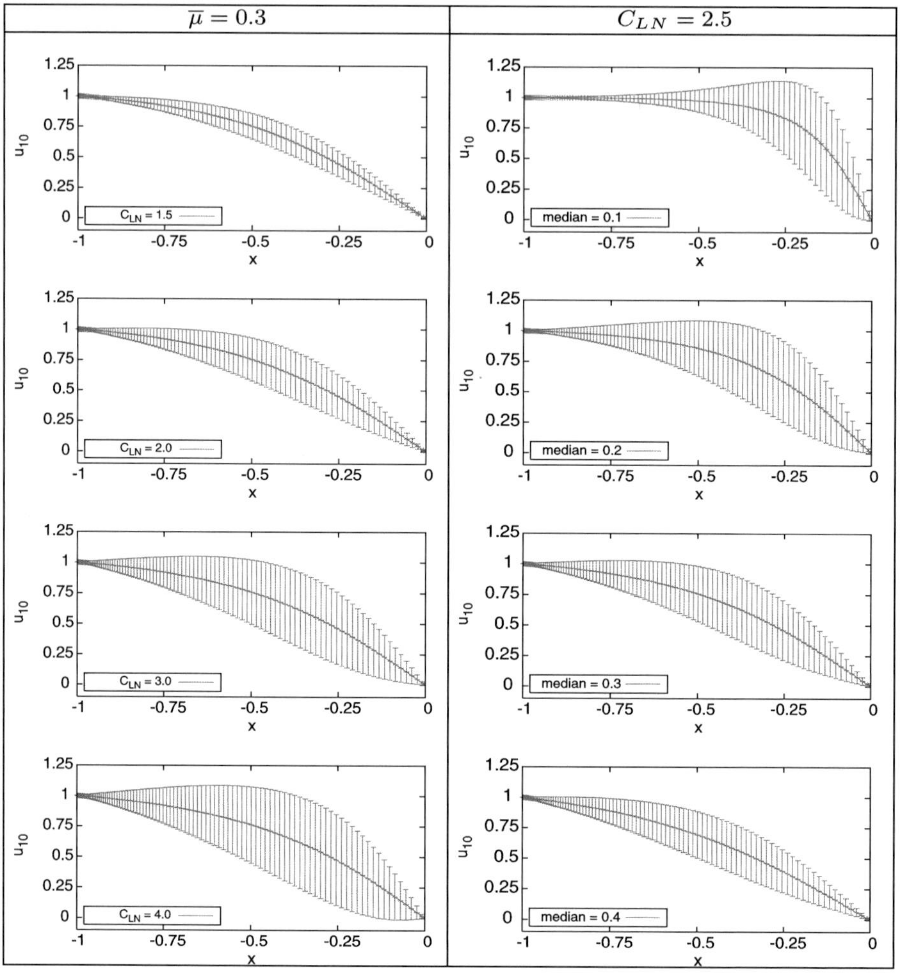

Fig. 9.31 *Left*: Mean and ± 3 standard deviation bars representation of the reduced solutions $u_{M=10}$ for $\overline{\mu} = 0.3$ and different values of C_{LN} as indicated. *Right*: Mean and ± 3 standard deviation bars representation of the reduced solutions $u_{M=10}$ for $C_{LN} = 2.5$ and different values of $\overline{\mu}$ as indicated. Computations with Algorithm 2, No $= 6$ and N $= 3$ ($\dim(\mathcal{S}) = 84$). Adapted from [171]

Convergence of probability density functions: To complete the present analysis, we examine the efficiency of GSD in light of the convergence of the pdf of u_M as M increases. To this end, we set $\overline{\mu} = 0.3$ and $C_{LN} = 3$. The parameterization of the random viscosity uses N $= 5$ with an expansion order No $= 5$, such that the dimension of the stochastic approximation space is $\dim(\mathcal{S}) = 252$. The reduced solution u_M is computed using Algorithm 2 with stopping criteria $\epsilon_s = 0.01$. We estimate the pdf of $u_M(x, \boldsymbol{\xi})$, from a Monte-Carlo sampling of Ξ. For each sample $\boldsymbol{\xi}^{(i)}$ we reconstruct the corresponding solution $u_M(x, \boldsymbol{\xi}^{(i)})$ from:

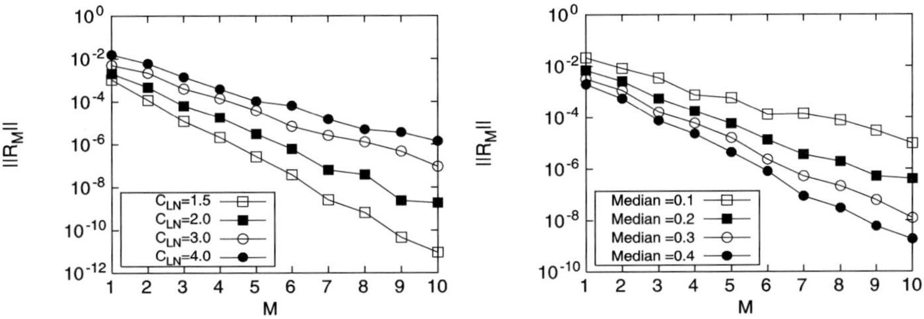

Fig. 9.32 R_M versus M for $\overline{\mu} = 3$ and different C_{LN} (*left*) and $C_{LN} = 2.5$ and different $\overline{\mu}$ (*right*). Computations with Algorithm 2, No = 6 and N = 3 ($\dim(\mathcal{S}) = 84$). Adapted from [171]

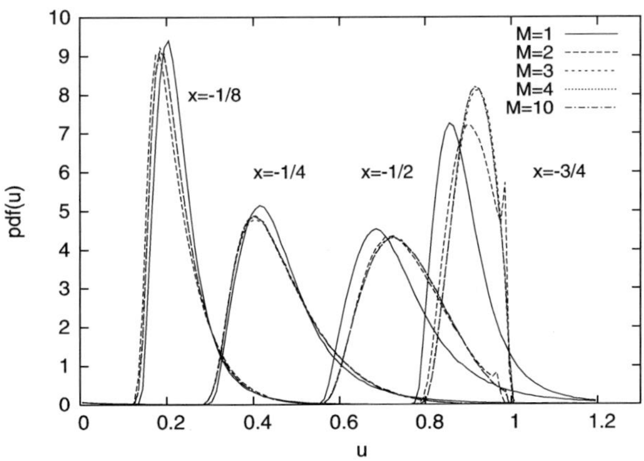

Fig. 9.33 Probability density functions of u_M at selected points for different values of M, and a stochastic viscosity with $\overline{\mu} = 0.3$ and $C_{LN} = 3$. Computations with Algorithm 2 using $\epsilon_s = 0.01$, N = 5, and No = 5 ($\dim(\mathcal{S}) = 252$). Adapted from [171]

$$u_M(x, \boldsymbol{\xi}^{(i)}) = \sum_{l=0}^{M} U_l(x)\lambda_l\left(\boldsymbol{\xi}^{(i)}\right) = \sum_{l=0}^{M} U_l(x) \sum_{k=0}^{\mathrm{P}} \lambda_l^k \Psi_k\left(\boldsymbol{\xi}^{(i)}\right). \tag{9.168}$$

These samples are then used to estimate the pdfs of u_M at selected points, $x = -1/8$, $-1/4$, $-1/2$ and $-3/4$. Since the reconstruction of the samples has a low computational cost, we use 10^6 samples to estimate the pdfs.

Figure 9.33 shows the computed pdfs at the four points for different dimensions M of the reduced basis. It is seen that for $M = 1$, the reduced approximation provides poor estimates of the pdfs, especially at $x \approx -3/4$ and $x \approx -1/2$ where the probabilities of having $u > 1$ are significant. For $M = 2$, we already obtain better estimates of the pdfs, except for the point closest to the boundary, where $M = 3$ is necessary to achieve a smooth pdf. Increasing M further leads to no significant changes in the pdfs. These results are consistent with the previous observations on the convergence of the mean and standard deviation.

To gain further confidence in the accuracy of the reduced basis approximation, we compare in Fig. 9.34 the pdfs for $u_{M=10}$ with pdfs constructed from the clas-

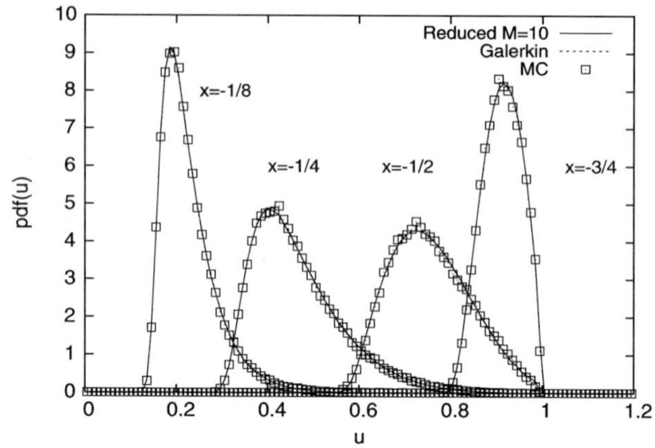

Fig. 9.34 Probability density functions of u at selected points as indicated, for the reduced approximation $u_{M=10}$, the Galerkin solution, and a Monte-Carlo simulation. Stochastic viscosity with $\overline{\mu}=0.3$ and $C_{LN}=3$. The stochastic approximation space has N $=5$, No $=5$ ($\dim(\mathcal{S})=252$) for the Galerkin and reduced solutions. Direct sampling of the log-normal distribution of μ is used in the Monte-Carlo simulation. Adapted from [171]

sical Galerkin polynomial chaos solution on $\mathcal{S}$ and a Monte-Carlo simulation. The Galerkin solution is computed using an exact Newton solver, yielding a quadratic convergence rate: it can be considered as the exact Galerkin solution on $\mathcal{S}$. The Monte-Carlo simulation is based on a direct sampling of the log-normal viscosity distribution (and not of Ξ). Only 10^4 Monte-Carlo samples are used to estimate the pdfs, due to its computational cost, while the pdfs for the Galerkin solution are generated using 10^6 samples, as for the reduced approximation. It is seen that the reduced approximation with only $M=10$ modes leads to essentially the same pdfs as the full Galerkin solution which involves 252 modes. Also, the pdfs are in close agreement with the Monte-Carlo solution, with only small differences caused by the lower sampling used.

9.4.5 Application to a Nonlinear Stationary Diffusion Equation

In this section, we apply the GSD method to a nonlinear stationary diffusion equation with a cubic nonlinearity for which the mathematical framework can be found in [151]. Specifically, we consider the L-shaped domain $\Omega \subset \mathbb{R}^2$ illustrated in Fig. 9.35: $\Omega = ((0,1)\times(0,2)) \cup ((1,2)\times(1,2))$.

Homogeneous Dirichlet boundary conditions are applied on a region Γ_1 of the boundary. A normal flux g is imposed on another region Γ_2. On the remainder of the boundary, denoted by Γ_0, a zero flux condition is used. A volumetric source f is imposed on $\Omega_1 = (1,2)\times(1,2)$.

The stochastic solution,

$$u : (x,\theta) \in \Omega \times \Theta \mapsto u(x,\theta) \in \mathbb{R}, \tag{9.169}$$

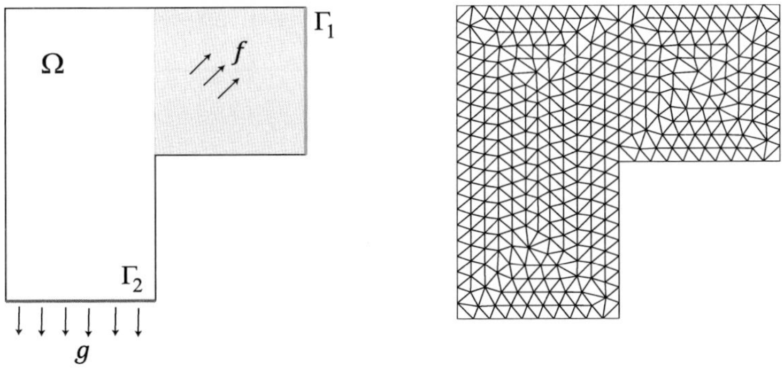

Fig. 9.35 Diffusion problem: geometry, boundary conditions and sources (*left*) and finite element mesh (*right*). Adapted from [171]

obeys almost surely

$$-\nabla \cdot ((\kappa_0 + \kappa_1 u^2)\nabla u) = \begin{cases} 0 & \text{on } \Omega \backslash \Omega_1, \\ f & \text{on } \Omega_1, \end{cases} \tag{9.170}$$

$$-(\kappa_0 + \kappa_1 u^2)\frac{\partial u}{\partial n} = \begin{cases} 0 & \text{on } \Gamma_0, \\ g & \text{on } \Gamma_2, \end{cases} \tag{9.171}$$

$$u = 0 \quad \text{on } \Gamma_1, \tag{9.172}$$

where κ_0 and κ_1 are conductivity parameters. We consider that conductivity parameters and source terms are uniform in space. They are thus modeled with real-valued random variables. The variational formulation for this problem consists of (9.85) with:

$$b(u, v; \theta) = \int_\Omega (\kappa_0(\theta) + \kappa_1(\theta)u^2)\nabla u \cdot \nabla v \,\mathrm{d}x, \tag{9.173}$$

$$l(v; \theta) = \int_{\Omega_1} f(\theta)\, v \,\mathrm{d}x + \int_{\Gamma_2} g(\theta)\, v \,\mathrm{d}s. \tag{9.174}$$

Generalization of the methodology to situations where the conductivity or source terms are discretized stochastic fields is immediate.

9.4.5.1 Application of GSD Algorithms

We now detail the main ingredients of the GSD algorithms, namely steps (4) and (6) of Algorithm 1, and the update step of Algorithm 2. To this end, we write

$$b(u, v; \theta) = \kappa_0(\theta)a(u, v) + \kappa_1(\theta)n(u^2, u, v), \tag{9.175}$$

$$l(v; \theta) = f(\theta)l_1(v) + g(\theta)l_2(v), \tag{9.176}$$

where a and n are bilinear and trilinear forms, respectively defined as:

$$a(u, v) = \int_\Omega \nabla u \cdot \nabla v \, \mathrm{d}x, \tag{9.177}$$

$$n(w, u, v) = \int_\Omega w \nabla u \cdot \nabla v \, \mathrm{d}x. \tag{9.178}$$

Resolution of $U = F_M(\lambda)$: To compute $U = F_M(\lambda)$, one has to solve the following deterministic problem:

$$B_M(\lambda U, \lambda V) = L_M(\lambda V) \quad \forall V \in \mathcal{V}, \tag{9.179}$$

where $\forall u, v \in \mathcal{V} \otimes \mathcal{S}$,

$$B_M(u, v) = B(u_M + u, v) - B(u_M, v), \tag{9.180}$$

$$L_M(v) = L(v) - B(u_M, v). \tag{9.181}$$

After some manipulations, one obtains for the left-hand side:

$$\begin{aligned} B_M(\lambda U, \lambda V) = \tilde{\kappa}_0 a(U, V) + \tilde{\kappa}_1 n(U^2, U, V) + n(\tilde{U}, U^2, V) \\ + n(U^2, \tilde{U}, V) + n(\overline{Z}, U, V) + n(U, \overline{Z}, V), \end{aligned} \tag{9.182}$$

where

$$\tilde{\kappa}_0 = \mathrm{E}[\kappa_0 \lambda\lambda], \qquad \tilde{\kappa}_1 = \mathrm{E}[\kappa_1 \lambda\lambda\lambda\lambda], \tag{9.183}$$

$$\tilde{U} = \sum_{i=1}^{M} \mathrm{E}[\kappa_1 \lambda\lambda\lambda\lambda_i] U_i, \tag{9.184}$$

$$\overline{Z} = \sum_{i,j=1}^{M} \mathrm{E}\big[\kappa_1 \lambda\lambda\lambda_i\lambda_j\big] U_i U_j. \tag{9.185}$$

We observe that the left-hand side contains the classical linear and cubic terms with deterministic parameters $\tilde{\kappa}_0$ and $\tilde{\kappa}_1$, and additional linear and quadratic terms.

For the right-hand side, one obtains the following expression:

$$L_M(\lambda V) = \tilde{f} l_1(V) + \tilde{g} l_2(V) - a(\breve{U}, V) - n(1, \hat{Z}, V), \tag{9.186}$$

where

$$\tilde{f} = \mathrm{E}\big[f\lambda\big], \qquad \tilde{g} = \mathrm{E}\big[g\lambda\big], \tag{9.187}$$

$$\breve{U} = \sum_{i=1}^{M} \mathrm{E}[\kappa_0 \lambda\lambda_i] U_i, \tag{9.188}$$

$$\hat{Z} = \frac{1}{3} \sum_{i,j,k=1}^{M} \mathrm{E}\big[\kappa_1 \lambda \lambda_i \lambda_j \lambda_k\big] U_i U_j U_k. \tag{9.189}$$

In the computations, this deterministic problem is solved using a classical Newton algorithm. Of course, various equivalent notations could have been introduced for writing left and right-hand sides of the deterministic problem. The above choice, introducing functions $\overline{Z}$ and $\hat{Z}$, allows obtaining a compact writing, without summation on spectral modes. When introducing an approximation at the spatial level (*e.g.* finite element approximation), pre-computing an approximation of functions $\overline{Z}$ and $\hat{Z}$ allows us to reduce the number of operations to be performed. This can lead to significant savings, but results in an approximation in the evaluation of left-hand and right-hand sides, and consequently in the computed solution.

Resolution of $\lambda = f_M(U)$: The random variable $\lambda \in \mathcal{S}$ is solution of the variational problem:

$$B_M(\lambda U, \beta U) = L_M(\beta U) \quad \forall \beta \in \mathcal{S}. \tag{9.190}$$

After some manipulations, this equation is found to be equivalent to:

$$\mathrm{E}\Big[\beta(\alpha^{(1)}\lambda + \alpha^{(2)}\lambda\lambda + \alpha^{(3)}\lambda\lambda\lambda)\Big] = \mathrm{E}\big[\beta\delta\big], \tag{9.191}$$

where

$$\alpha^{(1)} = \kappa_0 a(U,U) + \sum_{i,j=1}^{M} \kappa_1 \lambda_i \lambda_j \big[n(U_i U_j, U, U) + 2n(U_i U, U_j, U)\big], \tag{9.192}$$

$$\alpha^{(2)} = \sum_{i=1}^{M} \kappa_1 \lambda_i \Big[2n(U_i U, U, U) + n(U^2, U_i, U)\Big], \tag{9.193}$$

$$\alpha^{(3)} = \sum_{i,j=1}^{M} \kappa_1 \lambda_i \lambda_j \big[n(U_i U_j, U, U) + 2n(U_i U, U_j, U)\big], \tag{9.194}$$

$$\delta = f l_1(U) + g l_2(U) - \sum_{i=1}^{M} \kappa_0 \lambda_i a(U_i, U)$$
$$- \sum_{i,j,k=1}^{M} \frac{1}{3} \kappa_1 \lambda_i \lambda_j \lambda_k n(1, U_i U_j U_k, U). \tag{9.195}$$

This nonlinear equation is solved with a classical Newton algorithm.

Resolution of $\Lambda_M = f_0(W_M)$: To update the random variables $\Lambda_M = (\lambda_1, \ldots, \lambda_M) \in (\mathcal{S})^M$, one has to solve:

$$B(W_M \cdot \Lambda_M, W_M \cdot \Lambda_M^*) = L(W_M \cdot \Lambda_M^*) \quad \forall \Lambda_M^* \in (\mathcal{S})^M. \tag{9.196}$$

This equation can be split into a system of M equations:

$$\forall k \in \{1,\ldots,M\}, \quad B(W_M \cdot \Lambda_M, U_k\beta_k) = L(U_k\beta_k) \quad \forall \beta_k \in \mathcal{S}. \tag{9.197}$$

Introducing the previously defined forms, we get: $\forall k \in \{1,\ldots,M\}$,

$$\sum_{i=1}^{M} \kappa_0 a(U_i, U_k)\lambda_i + \sum_{i,j,l=1}^{M} \kappa_1 n(U_i, U_j, U_l, U_k)\lambda_i\lambda_j\lambda_l = f l_1(U_k) + g l_2(U_k). \tag{9.198}$$

This is a set of M coupled stochastic equations with a polynomial nonlinearity. This system is solved with a classical Newton algorithm.

9.4.5.2 Results

Discretization: At the stochastic level, we consider that κ_0, κ_1, f and g are random variables parametrized as follows:

$$\kappa_0 = \mu_{\kappa_0}(1 + c_{\kappa_0}\sqrt{3}\xi_1),$$
$$\kappa_1 = \mu_{\kappa_1}(1 + c_{\kappa_1}\sqrt{3}\xi_2),$$
$$f = \mu_f(1 + c_f\sqrt{3}\xi_3),$$
$$g = \mu_g(1 + c_g\sqrt{3}\xi_4)$$

where the ξ_i are four independent random variables, uniformly distributed on $(-1, 1)$. The parameters $\mu_{(\cdot)}$ and $c_{(\cdot)}$ respectively correspond to the means and coefficients of variations of the random variables. We thus work in the associated 4-dimensional image probability space $(\Xi, \mathcal{B}_\Xi, P_\xi)$, where $\Xi = (-1, 1)^4$, and use the same methodology as in Sect. 9.4.4 for defining an approximation space $\mathcal{S} \subset L_2(\Xi, P_\xi)$ based on a generalized polynomial chaos basis (multidimensional Legendre polynomials). No is the order of the polynomial chaos.

Meanwhile, the spatial discretization relies on a classical finite element approximation space $\mathcal{V}^h \subset \mathcal{V}$ associated with a triangular mesh (see Fig. 9.35).

Reference solution and error indicator: The reference Galerkin approximate solution $u^h \in \mathcal{V}^h \otimes \mathcal{S}$ satisfied:

$$B(u^h, v^h) = L(v^h), \quad \forall v^h \in \mathcal{V}^h \otimes \mathcal{S}. \tag{9.199}$$

To obtain this reference solution, the nonlinear system of equations associated with (9.199) is solved using a classical modified Newton method with a very high precision (see below for details on the reference solver).

In order to analyze the convergence of the GSD, we introduce an error indicator based on the residual of the discretized problem (9.199). This error indicator evaluates an error between the truncated GSD and the reference approximate solution

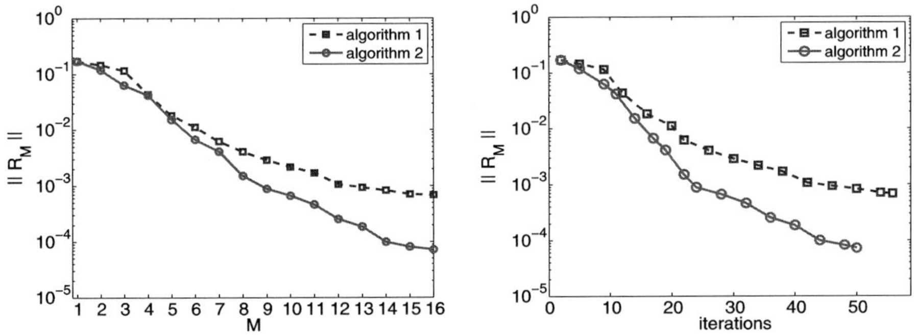

Fig. 9.36 Left: R_M versus M (*left*) for Algorithms 1 (*squares*) and 2 (*circles*). *Right*: R_M versus total (*cumulative*) number of power-type iterations for the computation of successive pairs (U, λ). Adapted from [171]

u^h but not the error due to spatial and stochastic approximations. A given function $v \in \mathcal{V}^h \otimes \mathcal{S}$ is associated with a vector $\mathbf{v} \in \mathbb{R}^{N_x} \otimes \mathcal{S}$. We denote by $R_M \in \mathcal{V}^h \otimes \mathcal{S}$ the reduction residual associated with $u_M \in \mathcal{V}^h \otimes \mathcal{S}$ and by $\mathbf{R}_M \in \mathbb{R}^{N_x} \otimes \mathcal{S}$ the associated discrete residual, defined as follows: $\forall v \in \mathcal{V}^h \otimes \mathcal{S}$, associated with $\mathbf{v} \in \mathbb{R}^{N_x} \otimes \mathcal{S}$,

$$\mathrm{E}\left[\mathbf{v}^T \mathbf{R}_M\right] = L(v) - B(u_M, v). \tag{9.200}$$

An error indicator is then simply defined by the L_2-norm of the discrete residual, defined by

$$\|\mathbf{R}_M\|^2 = \mathrm{E}\left[\mathbf{R}_M^T \mathbf{R}_M\right] \equiv \|R_M\|^2. \tag{9.201}$$

In the following, we will implicitly use a normalized error criterion $\|R_M\| \leftarrow \|R_M\|/\|R_0\|$, where R_0 denotes the right-hand side of the initial nonlinear problem.

Convergence analysis: To analyze the convergence of the GSD algorithms, we choose the following parameters for defining the basic random variables:

$$\mu_{\kappa_0} = 3, \qquad \mu_{\kappa_1} = 1.5, \qquad \mu_f = 6, \qquad \mu_g = 2.25,$$
$$c_{\kappa_0} = 0.2, \qquad c_{\kappa_1} = 0.2, \qquad c_f = 0.2, \qquad c_g = 0.2.$$

The basis of function space $\mathcal{S}$ is composed by multidimensional Legendre polynomials up to degree 5 (No = 5), so that $\dim(\mathcal{S}) = \frac{(4+\mathrm{No})!}{4!\mathrm{No}!} = 126$. For the spatial finite element discretization, we have $\dim(\mathcal{V}^h) = 368$. If the stochastic solution was to be found in the full approximation space $\mathcal{V}^h \otimes \mathcal{S}$, the size of the nonlinear problem to be solved would be $\dim(\mathcal{V}^h) \times \dim(\mathcal{S}) = 46368$. In contrast, the reduced basis solution $W_M \cdot \Lambda_M$ has dimension $M \times (\dim(\mathcal{V}^h) + \dim(\mathcal{S})) = 494M$.

In Fig. 9.36, we compare the convergence of Algorithms 1 and 2 with the size M of the reduced basis (left plot) and with the total number of power-type iterations performed for the computation of successive couples (U, λ) (right plot). The

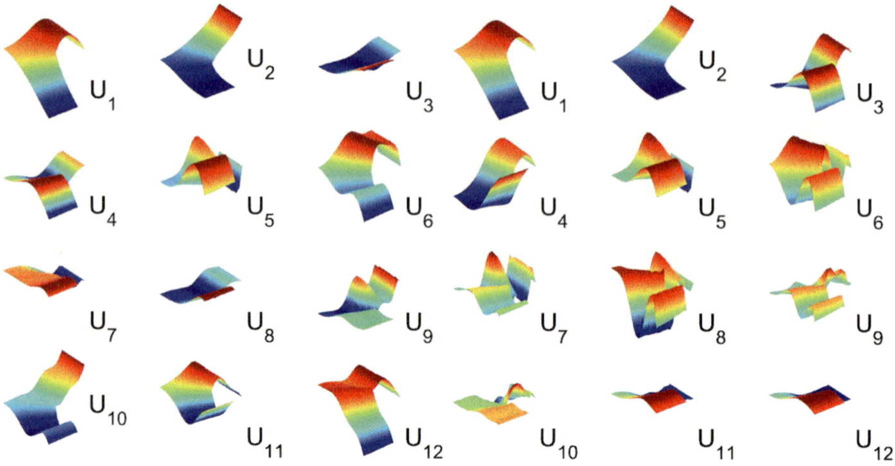

Fig. 9.37 First 12 reduced modes with Algorithms 1 (*left*) and 2 (*right*). Adapted from [171]

stopping criteria for power iterations is here $\epsilon_s = 10^{-2}$. For both algorithms, the residuals decay rapidly as the dimension M of the reduced basis increases. Algorithm 2 is more effective in reducing R_M than Algorithm 1. Although Algorithm 2 requires less power iterations, as shown in Fig. 9.36, both algorithms require similar computational costs on this particular example. Indeed, the faster convergence of Algorithm 2 is balanced with computational efforts needed for the updating of random variables. This conclusion may not hold in general for large spatial approximation spaces.

Note that in this example, we observe a quasi-exponential convergence rate for small M and a decrease of this residual decay rate for larger M. In fact, this is not due to a lack of robustness of the GSD method. It is related to the spectral content of the solution of this 2-dimensional problem. A classical spectral decomposition of the reference solution would reveal the same behavior.

We compare in Fig. 9.37 the first 12 deterministic functions U_i computed using the two algorithms. It is seen that Algorithm 2 yields a deterministic reduced basis with a higher frequency content than for that of Algorithm 1. In particular, we observe that the last modes obtained by Algorithm 2 are essentially orthogonal to the first ones. This is further illustrated in Fig. 9.38, which shows the second moment of the equation residual, $\mathrm{E}[R_M^2]$, for different M. This figure also highlights the efficiency of the GSD in capturing the full discrete solution on $\mathcal{V}^h \otimes \mathcal{S}$ in just few modes, and indicates that the stochastic discretization mostly affects the equation residual in the area where the solution exhibits the steepest gradients, *i.e.* where the uncertainty has highest impact on the solution.

Even though the equation residual norm provides a measure of the quality of the approximate solution, it is not a direct measure of the error on the solution. In Fig. 9.39, we plot the convergence curves of both algorithms with respect to the residual norm and also with respect to the L_2-norm on the solution. We observe that the error on the solution is significantly lower than the error based on the residual.

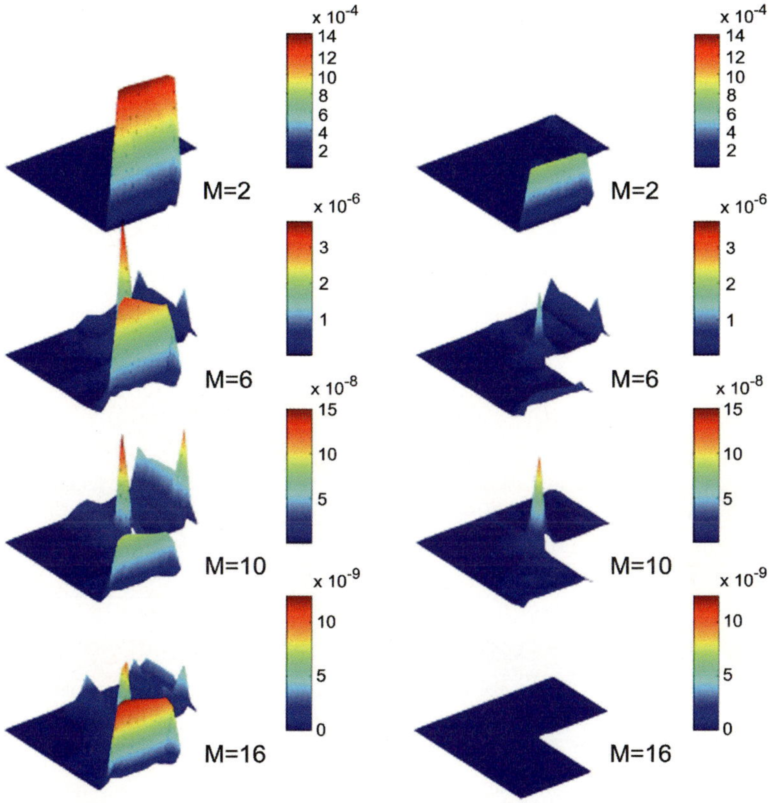

Fig. 9.38 Distribution of the second moment of the residual, $\mathrm{E}[R_M^2]$, for different M; *left*: Algorithm 1, *right*: Algorithm 2. Adapted from [171]

For a better appreciation of the convergence of the GSD, we plot in Fig. 9.40 the distributions of the relative errors in mean ε_{mean} and standard deviation ε_{Std} for different M:

$$\varepsilon_{mean} = \frac{|\mathrm{E}[u_M] - \mathrm{E}\left[u^h\right]|}{\sup(|\mathrm{E}[u^h]|)},$$

$$\varepsilon_{Std} = \frac{|\mathrm{Std}(u_M) - \mathrm{Std}(u^h)|}{\sup(Std(u^h))}.$$

We observe a very fast convergence of the GSD decomposition with both algorithms, with a faster convergence of Algorithm 2. With only $M = 4$ modes, the relative error on these first two moments is less than 10^{-3}. In Fig. 9.41, we plot the pdfs of the solution at two different points. We observe that the approximate pdfs and reference pdf are essentially indistinguishable for $M \geq 5$. We also observe the superiority of Algorithm 2, which yields more accurate pdfs with lower values of M.

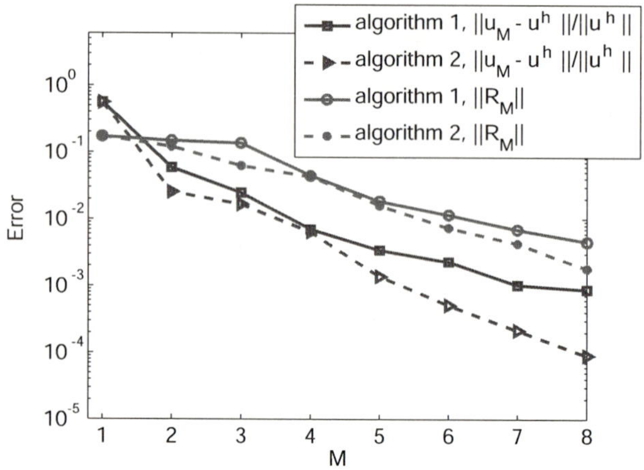

Fig. 9.39 Curves of $\|R_M\|$ and $\|u_M - u^h\|/\|u^h\|$ versus M for Algorithm 1 (*solid lines*) and Algorithm 2 (*dashed lines*). Adapted from [171]

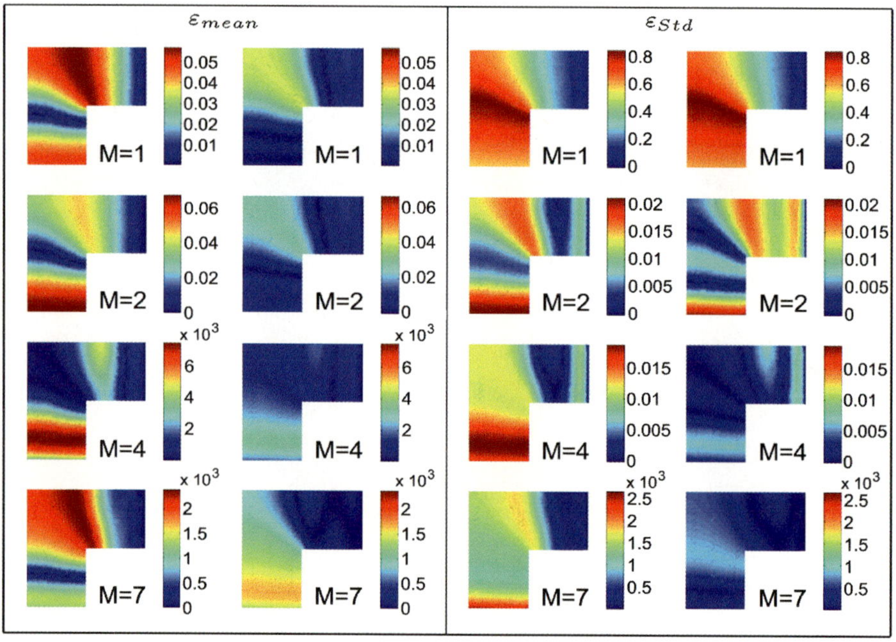

Fig. 9.40 Distributions of ε_{mean} and ε_{Std} for different M, using Algorithm 1 (*first* and *third columns*) and Algorithm 2 (*second* and *fourth columns*). Adapted from [171]

Robustness of the algorithms: We now investigate the robustness of the method with regards to stochastic discretization and numerical parameters.

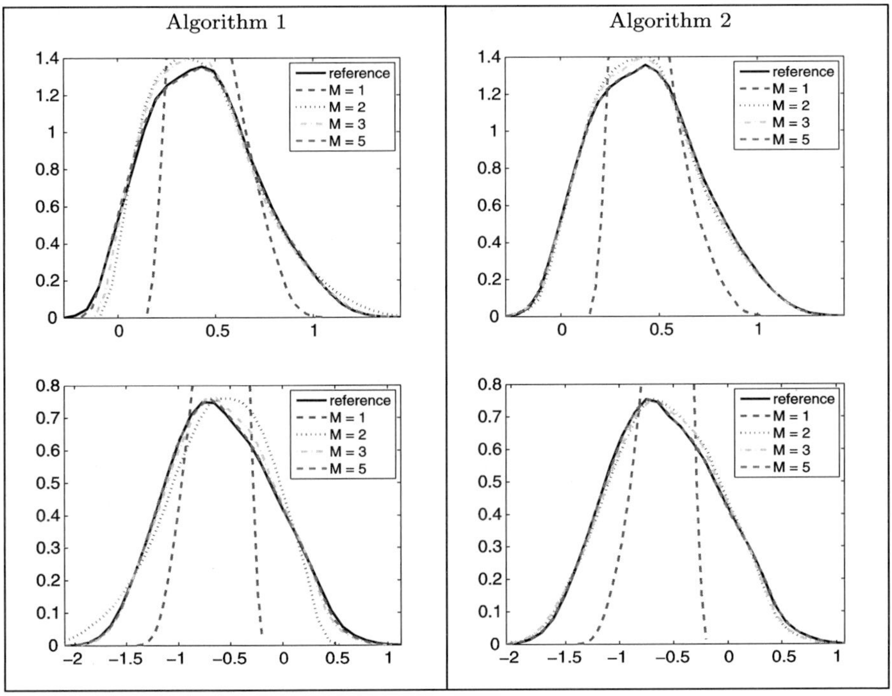

Fig. 9.41 Probability density functions of u_M for different M at points $P_1 = (1.5, 1.5)$ (*top row*) and $P_2 = (0.5, 0.1)$ (*bottom row*). Curves generated using Algorithms 1 and 2. Adapted from [171]

Impact of ϵ_s*:* We first evaluate the impact of the stopping criterion ϵ_s. To this end, we here consider gradually less stringent stopping criteria ϵ_s, and monitor the convergence of R_M. Specifically, the same probabilistic setting and discretization parameters specified above are used together with $\epsilon_s = \{5.10^{-1}, 10^{-1}, 10^{-2}, 10^{-3}\}$. Results of these experiments are reported in Figs. 9.42 and 9.43. It is seen that for both algorithms, the selection of ϵ_s in the range considered has virtually no effect on the convergence of the decomposition, which however becomes computationally more demanding as ϵ_s decreases. In practice, it is not necessary to perform more than 3 or 4 power iterations to build a new pair (U, λ); this is consistent with experiences in Sect. 9.4.4.

Impact of stochastic polynomial order: Figure 9.44 shows the dependence of the residual R_M on M for different polynomials orders. Curves are generated using algorithms 1 2, with algorithm No = 4, 5, and 6 ($\dim(\mathcal{S}) = 70$, 126, and 210, respectively). The results indicate that the polynomial order has a very low influence on the convergence. For this example, this can be explained by the fact that the error induced by the approximation at the stochastic level is lower than the error induced by the truncation of the GSD.

Impact of the input variability: To test the robustness of GSD algorithms with respect to the input variability, we first vary the coefficients of variations $c_{(\cdot)}$ of all

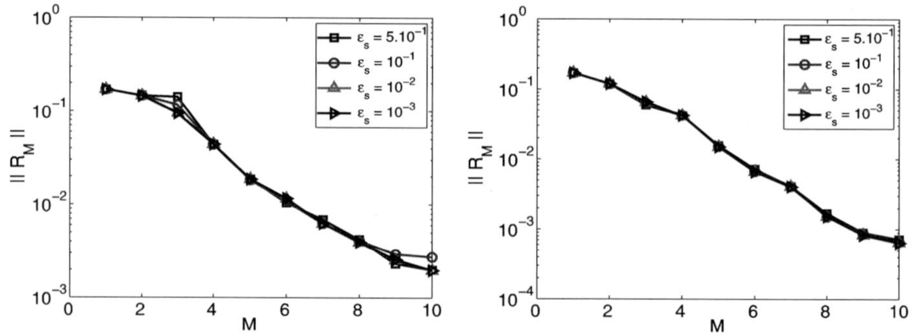

Fig. 9.42 R_M versus M for Algorithms 1 (*left*) and 2 (*right*). Different values of ϵ_s are used, as indicated. Adapted from [171]

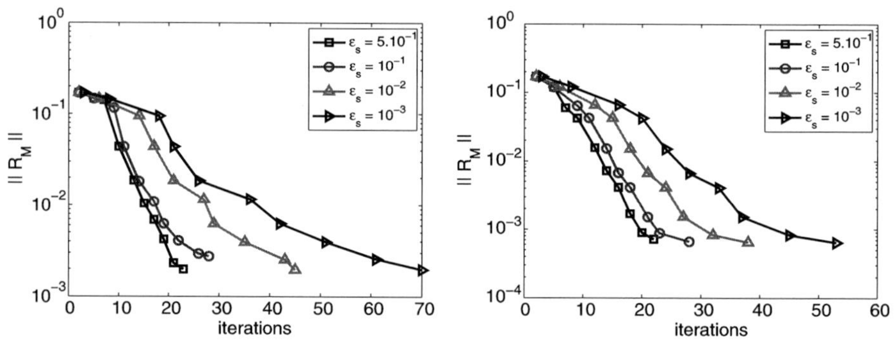

Fig. 9.43 R_M versus number of power-type iterations for Algorithms 1 (*left*) and 2 (*right*). Different values of ϵ_s are used, as indicated. Adapted from [171]

random variables at the same time. Plotted in Fig. 9.45 are curves of R_M versus M for Algorithms 1 and 2, generated using different coefficients of variation: $c_{(\cdot)} = 0.1, 0.2, 0.3$. As expected, the convergence rate decreases as the coefficient of variation increases. However, the monotonic convergence observed in the results illustrates the robustness of GSD algorithms for a wide range of input variability.

Next, we investigate the impact of the nonlinearity by varying the mean μ_{κ_1} of parameter κ_1, letting all the coefficients of variation equal to $c_{(\cdot)} = 0.2$. Figure 9.46 shows the dependence of the residual on M for Algorithms 1 and 2 for different $\mu_{\kappa_1} = 1.5, 0.5, 0.1, 0.01, 0$. We first observe that the convergence rate decreases as the magnitude of the nonlinear term increases. This can be explained by the fact that the nonlinearity induces a more complex solution, which requires more spectral modes to be correctly captured.

For the case $\mu_{\kappa_1} = 0$, corresponding to the limiting case where the equation is linear, we observe that both algorithms capture (to machine precision) the exact discrete solution in only 2 modes. This is expected since it is clear on this example that only two modes are required to exactly represent the solution of the linear problem.

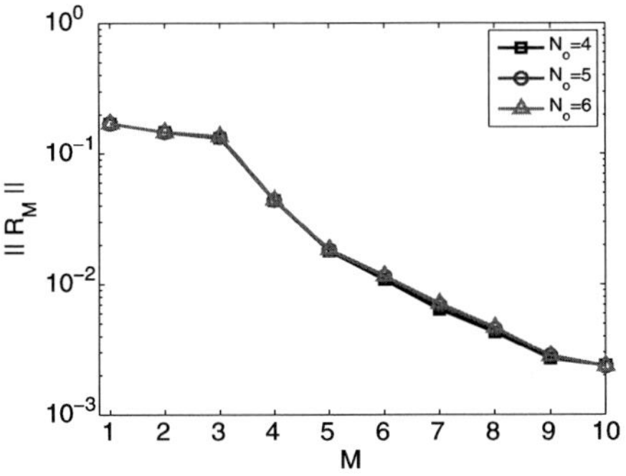

Fig. 9.44 R_M versus M for Algorithms 1 (*left*) and 2 (*right*). Curves are generated for No = 4, 5, and 6, as indicated. Adapted from [171]

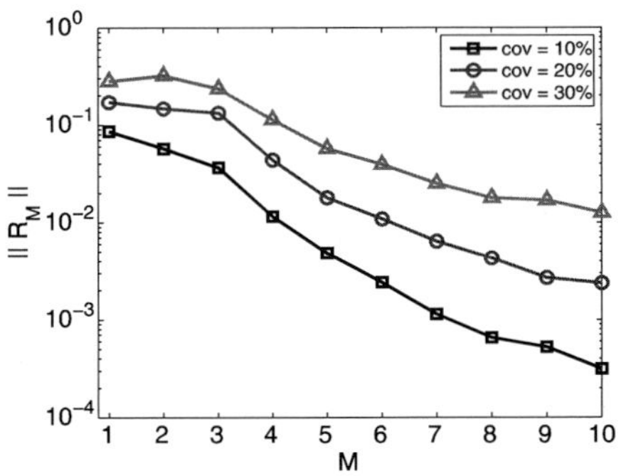

Fig. 9.45 R_M versus M for Algorithms 1 (*left*) and 2 (*right*). Curves are generated for different coefficients of variation (COV), as indicated. Adapted from [171]

Indeed, the two deterministic functions U_1 and U_2 which satisfy

$$a(U_1, V) = l_1(V) \quad \text{and} \quad a(U_2, V) = l_2(V), \quad \forall V \in \mathcal{V}^h,$$

yield an exact decomposition when associated to the ad-hoc random variables. In fact, every pair of deterministic functions in the span of these functions yields an exact decomposition. This example shows that in this particular case, the GSD algorithms yield these ideal decompositions automatically.

Computation times: The efficiency GSD method is analyzed here comparing by computation times with those of a classical modified Newton algorithm applied to the reference Galerkin system (9.199). A classical Newton method consists in the following iterations: starting from $u^{h,(0)} = 0$, iterates $u^{h,(i)} = 0$ are defined by

$$B'(u^{h,(i+1)}, v^h; u^{h,(i)}) = L(v^h) - B(u^{h,(i)}, v^h) \quad \forall v^h \in \mathcal{V}^h \otimes \mathcal{S} \tag{9.202}$$

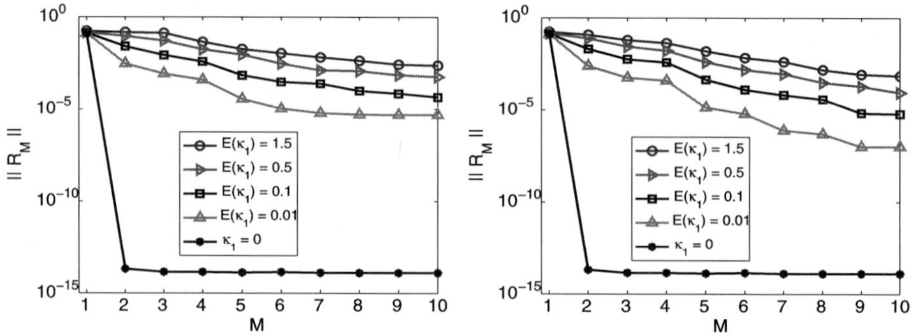

Fig. 9.46 R_M versus M for Algorithms 1 (*left*) and 2 (*right*). Curves are generated for different values of $\mathrm{E}[\kappa_1]$ as indicated. Adapted from [171]

where $B'(\cdot,\cdot;u)$ is the Gateaux derivative of semilinear form B evaluated at u:

$$\begin{aligned} B'(w,v;u) &= \lim_{\epsilon\to 0}\frac{1}{\epsilon}\left(B(u+\epsilon w,v)-B(u,v)\right)\\ &= \mathrm{E}\Big[\kappa_0 a(w,v)+\kappa_1(2n(wu,u,v)+n(u^2,w,v))\Big]. \end{aligned} \tag{9.203}$$

In order to reduce computation times of this reference solver, we use the following modification of iteration (9.202):

$$\begin{aligned} &\tilde{B}'(u^{h,(i+1)},v^h;\mathrm{E}[u^{h,(i)}]) = L(v^h) - B(u^{h,(i)},v^h) \quad \forall v^h \in \mathcal{V}^h\otimes\mathcal{S},\\ &\tilde{B}'(w,v;u) := \mathrm{E}\Big[\mu_{\kappa_0} a(w,v)+\mu_{\kappa_1}(2n(wu,u,v)+n(u^2,w,v))\Big], \end{aligned} \tag{9.204}$$

where $\tilde{B}'$ is a simple approximation of B obtained by replacing random parameters κ_0 and κ_1 by their respective mean values. Moreover, $\tilde{B}'$ is evaluated at $\mathrm{E}[u^{h,(i)}]$ instead of $u^{h,(i)}$. With these approximations, iteration (9.204) corresponds to a stochastic problem with a random right-hand side and a deterministic operator. The computation cost of this reference solver is then essentially due to the computation of the residual (right-hand side).

For the present example and moderate input variability, the modified Newton algorithm has good convergence properties. Note that for large variability of the input data, the efficiency of the modified Newton method deteriorates. A better approximation of $B'(\cdot,\cdot;u^{h,(i)})$ should be provided to keep good convergence properties. The robustness and efficiency of GSD algorithms are less affected by this increase in the input variability, as seen above.

Figure 9.47 shows the evolution of the residual norm with respect to computational time for the reference solver and for GSD algorithms. We clearly observe a computational gain with GSD algorithms (factor ≈ 6). We also observe that GSD Algorithms 1 and 2 lead to similar computational times. In fact, the computational time required by the updating step in Algorithm 2 is balanced by the fact that Al-

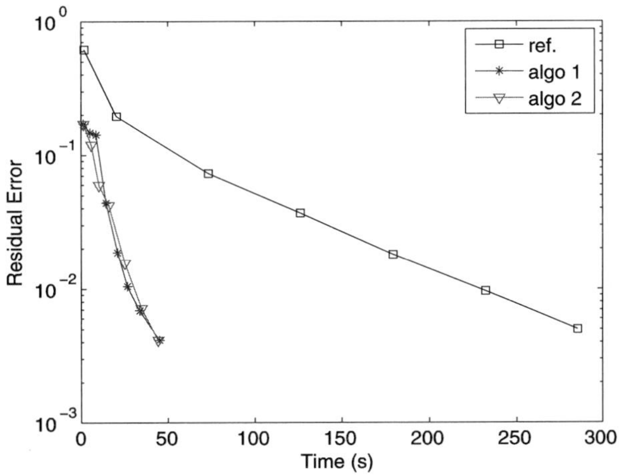

Fig. 9.47 Residual error versus computation time for reference solver, and for GSD algorithms using $\epsilon_s = 10^{-1}$ (reference discretization). Adapted from [171]

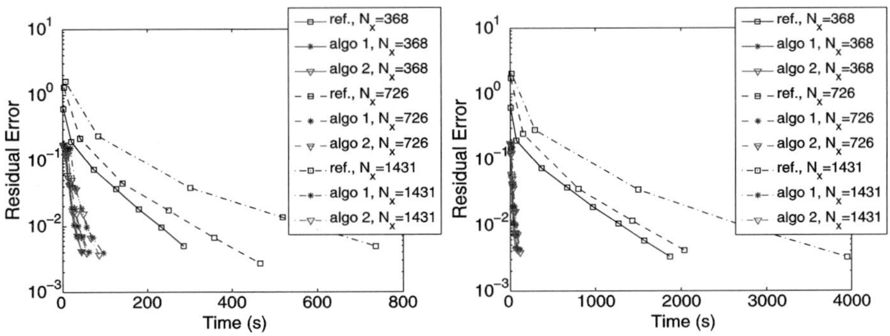

Fig. 9.48 Residual error versus computation time for the reference solver and GSD Algorithms 1 and 2 for different $\mathrm{N_x}$; *left*: No = 4, *right*: No = 5. Adapted from [171]

gorithm 2 needs for a lower-order of decomposition than Algorithm 1 for the same accuracy.

To further analyze computational costs, we consider the influence of the dimensions P and $\mathrm{N_x}$ of stochastic and deterministic approximation spaces. Specifically, we consider four finite element meshes corresponding respectively to $\mathrm{N_x} = 178$, 368, 726 and 1431. We also consider different polynomial chaos degrees No = 3, 4, 5 and 6, respectively corresponding to $P = 34$, 69, 125 and 209.

Figures 9.48 and 9.49 show curves of residual norm versus computation time for different $\mathrm{N_x}$ and No. We observe that when the dimension of approximation space increases, the efficiency of the reference solver rapidly deteriorates. GSD algorithms are far less affected by this increase of the dimension of approximation spaces.

Figure 9.50 shows the gain in computational time with respect to $\mathrm{N_x} \times P$ (for different stochastic and spatial discretization levels). The gain is computed by dividing the computational time required by GSD to reach a relative residual error of 5×10^{-2} with the computational time required by the reference solver to reach

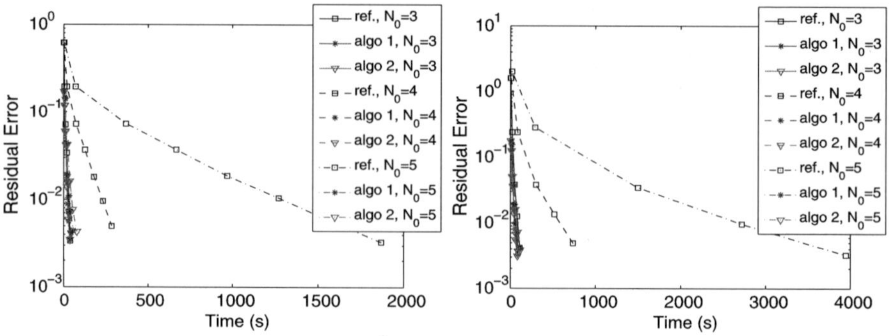

Fig. 9.49 Residual error versus computation time for the reference solver and GSD Algorithms 1 and 2 for different No; *left*: $\mathrm{N_x} = 368$, *right*: $\mathrm{N_x} = 1431$. Adapted from [171]

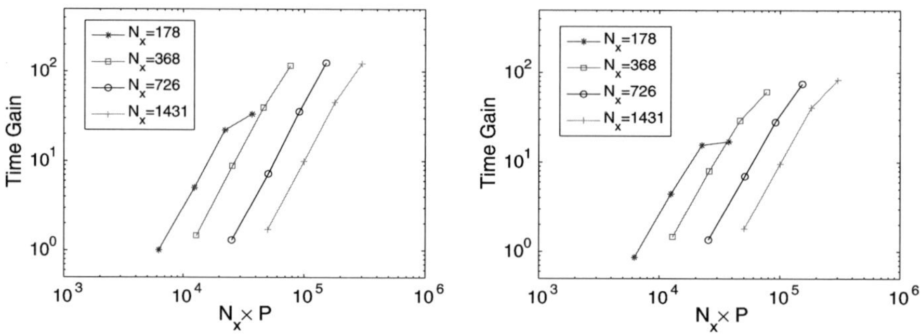

Fig. 9.50 Time gain factor versus $\mathrm{N_x} \times P$; *left*: GSD Algorithm 1, *right*: GSD Algorithm 2. Adapted from [171]

the same accuracy level. This level is sufficient to obtain accurate approximations of moments, pdfs, etc. It corresponds to the computation of 4 or 5 GSD modes. We clearly observe that GSD algorithms lead to computational savings, which increase with the dimension of approximation spaces. GSD Algorithms 1 and 2 lead to similar computational savings. For the finest discretizations, reduction of computational times by a factor of up to 100 are observed in the results.

9.5 Closing Remarks

In this chapter, we have explored different strategies towards adaptive PC computations. The first strategy is based on local refinement of a MW expansion, based on computing the "energy" content in the highest resolution levels, and locally refining the expansion where this measure exceeds a prescribed threshold. Implementation of this adaptive refinement strategy is illustrated through computations of a stochastic Rayleigh-Bénard problem. The case of a stochastic Rayleigh number is considered that uniformly distributed in a finite interval containing the critical value. Simulations indicate that the adaptive scheme can effectively handle such complex situations, as the refinement is localized in the neighborhood of the bifurcation. In particular, the present experience suggests that in situations requiring a high level of local refinement, lower-order adaptive expansions may be preferable to higher-order ones, as they are likely to result in more efficient predictions. On the other hand, the analysis also indicates that for problems with a large number of stochastic dimensions, the improvements of the local refinement strategy may not be sufficient to overcome the added complexity of multidimensional problems.

A second refinement strategy, based on adaptive block-partitioning scheme of the space of random data, is considered in order to overcome potential limitations in multi-dimensional problems. On each block, the solution is expanded in terms of a MW expansion consisting of smooth global functions and 1D details. The blocks are refined by division along individual dimensions whenever the contribution to the local variance of the corresponding 1D details exceeds a prescribed threshold. The solution is then recomputed on the newly generated blocks. Implementation of the resulting adaptive scheme is illustrated based on computations of a surface-kinetic problem having stochastic initial conditions and rate constants. Analysis of the behavior of the scheme reveals that the refinement is naturally concentrated in areas of steep variation of the solution with respect to the random data, and that errors decay rapidly as the number of blocks increases, and that the rate of decay increases with increasing order. For the present setup, the computations indicate that when the desired level of accuracy is not very large, lower-order expansions may prove more efficient than higher-order expansions.

The experiences in Sects. 9.1 and 9.2 demonstrate that adaptive refinement provides an attractive means for tackling complex, multidimensional stochastic problems. Specifically, it offers the possibility of constructing efficient and robust schemes that are able to effectively tackle situations exhibiting steep or discontinuous dependence on random data. The computations also highlight several areas where substantial enhancement may be achieved. These include the development of more elaborate strategies in which the order of the expansion is increased or reduced simultaneously with "spatial" refinement, and the construction of more efficient refinement criteria that reduces the overheads of the local analysis.

The third alternative strategy is based on dual-based *a posteriori* error analysis methodology. Fundamentally, this approach offers an alternative to the first two strategies which relied on error indicators based on the spectrum of the local stochastic expansion. The *a posteriori* error estimation involves the resolution of a linear stochastic dual problem, whose computational cost is small compared to the primal (nonlinear) problem. Numerical tests on the uncertain Burgers equation illustrate the effectiveness of the methodology in providing relevant error estimates that can be localized in the spatial and stochastic domain. The principal limitation of the methodology as presented above concerns the lack of resulting information regarding the structure of the estimated error. Specifically, the error estimator does not allow for the discrimination between the relative contributions of the stochastic directions to the overall error. We believe this is the most severe limitation in view of anisotropic refinement of the stochastic approximation space required to treat problems with high dimensional uncertainty germs.

Of course, several potential improvements of the *a posteriori* refinement schemes presented in Sect. 9.3 can be readily identified. These include the derivation of rigorous and efficient anisotropic error estimators for high-order approximation schemes, as well as extension of the methodology to transient systems.

Finally, an extension was presented of the Generalized Spectral Decomposition method [167, 168] to the resolution of nonlinear stochastic problems. The main features of the method is the approximation of the solution on reduced bases, automatically generated by the algorithms, with significant reduction of the computational requirements compared to the classical Galerkin projection schemes, and the independence of the methodology with regard to the type of stochastic discretization used.

The extended GSD is applied to nonlinear model problems, consisting of the steady Burgers equation and a steady nonlinear diffusion equation. Two power-type algorithms are developed which lead to solution methods consisting of the resolution of a series of decoupled deterministic problems and low dimensional stochastic problems. The deterministic problems inherit the properties and dimension of the initial deterministic problem, and require only slight adaptations of available deterministic codes.

Numerical experiments were used to demonstrate the effectiveness of the proposed algorithms in yielding reduced decompositions that approximate the stochastic solution with a small number of modes. For the second power algorithm, the convergence of the reduced approximation is essentially governed by the actual spectrum of the stochastic solution, and not by the dimension of the approximation space, as one may have anticipated from theoretical considerations. The power algorithm is less efficient than power-update algorithm in terms of accuracy for an equal number of modes in decomposition, but is computationally less expensive and simpler. The general superiority of one algorithm over the other is, however, problem dependent, as additional considerations may intervene, including computational complexity, the relative computational times for the deterministic and stochastic (update) problems, and memory requirements. Nonetheless, a common character of the two algorithms is their ability to yield the successive modes of the decomposition in only a few resolutions of the deterministic problem, thus resulting in large computational savings compared to a conventional stochastic Galerkin method.

Similar to other strategies explored above, improvements and extensions of the GSD strategy can be readily identified. These include the implementation of alternative algorithms for the construction of the decomposition modes using advanced subspace techniques (*e.g.* Arnoldi [167]) in order to drastically decrease the number of deterministic and reduced stochastic problems to be solved, as well as extensions to nonlinear transient problems.

Chapter 10
Epilogue

In the preceding chapters theoretical developments were presented of uncertainty propagation and quantification methods. The fundamental approach generally adopted in this monograph is based on the development of spectral representations of the input data and model solution, and on exploiting these representations in the construction of numerical methods and computational algorithms for the solution and analysis of stochastic systems. Application of these methods was illustrated through elementary examples as well as more involved problems from incompressible and weakly compressible flow. In conclusion, we provide a brief discussion of selected topics, which we attempt to group into the following (somewhat connected) categories: (a) potential extensions of methodologies outlined in previous chapters, (b) open problems, and (c) new desirable capabilities.

10.1 Extensions and Generalizations

Adaptive Methods: As discussed in Chap. 9, adaptive methods offer the promise of substantially reducing the overheads of PC computations, enabling the user to address complex behavior such as steep dependence of the solution on random data or bifurcations, and consequently mitigate the so-called "curse of dimensionality". Adaptive PC computations are quite recent; not surprisingly, refinement strategies attempted so far have net yet reached the same degree of sophistication as adaptive methods used in the solution of deterministic PDEs. Specific areas that may bring substantial benefits to adaptive PC computations include more elaborate algorithms that accommodate simultaneous refinement and coarsening of the PC representations. Additional benefits may derive from casting these elaborate refinement/coarsening strategies in more general representation framework, e.g. one accommodating change of measure or generalized basis representations.

Inverse Problems: The application of PC methods to inverse problems has been recently suggested, particularly through combinations with Bayesian inference methods [143, 144]. By providing a quantitative framework for accounting for observation errors, model uncertainties, and prior information, these methods provide

O.P. Le Maître, O.M. Knio, *Spectral Methods for Uncertainty Quantification*, Scientific Computation,
DOI 10.1007/978-90-481-3520-2_10, © Springer Science+Business Media B.V. 2010

a suitable approach to inverse problems, such as parameter estimation or data assimilation. The application of PC based methods offers the promise of substantial acceleration of Bayesian inference methods, especially in situations were forward model evaluations are costly, and consequently of substantial amplification of their scope. Further advances may also be expected by bring to bear adaptive refinement and model reduction concepts.

Change of Measure: Throughout this monograph, spectral representations were sought in a well-defined probability space in which the random data and stochastic solution are assumed to belong. The measure associated with this probability space was then exploited in the definition of random variables and orthogonal basis consisting of polynomials of these random variables. Galerkin projections exploiting the orthogonality of the basis were then used in the construction of various PC methods.

The framework above can be extended or generalized in several directions, of which we point to the follow two. The first relates to new ideas of basis enrichment [91], which essentially amounts to relaxing strict orthogonality requirement by introducing into the PC basis functions that are close to the stochastic solution. The form of these functions may be either known *a priori* or adapted during the computations. Though this approach involves additional overheads due to non-vanishing correlations between elements of the enriched basis, it offers the advantage of efficiently capturing the stochastic solution, especially when elements of the enriched basis are judiciously chosen or adapted.

A second direction that may be worthwhile to consider is based on adopting an extended mathematical framework in which the measure (and the associated probability space, and the PC basis) are adapted during the computations. Clearly, this strategy will also require numerical schemes and utility codes that implement such transformations. Despite these overheads, it appears to us that such approaches may be highly beneficial, for instance in situations where the statistical properties of the solution differ substantially at different times during the computation.

10.2 Open Problems

Structure and Properties of Galerkin Systems: Throughout this monograph, we have dealt extensively with the Galerkin approximation of stochastic PDEs in conjunction with PC representations. Various methods were also presented for the solution of coupled systems of equations that govern the behavior of the expansion modes. In addition, the suitability of these approaches has been demonstrated in a number of different settings, many involving complex dynamics.

On the other hand, one of the areas where new knowledge and results are clearly needed and would be tremendously beneficial concerns the mathematical structure and mathematical properties of the Galerkin PC systems, and their relationship with respect to the continuous stochastic PDE and when relevant its deterministic counterpart. Of particular interest would be new results concerning the stability and mathematical conditioning of Galerkin systems associated with the discretization of time-dependent stochastic PDEs.

Hyperbolic Systems: These are a prominent example of relevant equation systems for which one would like to obtain general results concerning the mathematical structure of their stochastic counterparts, and to construct numerical schemes with well-established properties. On the mathematical side, a goal would be to establish a fairly general result regarding the classification of stochastic hyperbolic systems discretized using a PC expansion. While progress in this area has been recently made for specific systems using Legendre chaos [229] (see also [79, 189]), little is known about general systems of conservation laws, including equations of motion for compressible ideal gas. On the computational side, a central issue is the construction of discretization schemes that can ensure certain properties of the stochastic solution. A particular goal concerns schemes that can ensure positivity of the solution. Such stochastic schemes would have a wide range of additional applications, including systems of reaction-diffusion equations and dynamical systems. Of course, generalization of upwind discretizations may not be the only means to achieve such capabilities. Other avenues may include transformation-based approaches, constrained optimization, etc.

Error Control: As discussed in the preceding chapter, the application of adaptive methods requires the definition of refinement criteria, and these may have a substantial effect on prediction accuracy and efficiency. In many of the cases treated in the previous chapter, in many cases the construction of these refinement criteria followed primarily heuristic approaches. Though the suitability of these approaches could be demonstrated in the examples selected, the need still exists for more robust means for controlling errors. The *a posteriori* methodology discussed above is an example of a promising approach. But the application of such methods in the context of stochastic models is still in its infancy, and many fundamental questions remain concerning the scope and capabilities that may result from their applications.

Of course, the need for estimating and controlling errors is not unique to adaptive methods; it also arises in the application of sampling or generally non-intrusive approaches. Consequently, substantial impact may result from the development and testing of new capabilities for estimating and controlling data and stochastic discretization errors.

Rapidly Broadening Spectra: One of the well-known hurdles facing PC expansions concerns the time-decorrelation of independent realizations in complex systems, including turbulent flows and systems exhibiting shock formation [31, 44, 154]. Similar difficulties also arise in deceptively simple stochastic systems exhibiting limit cycle oscillations [19]; an example consists of a linear spring-mass systems exhibiting random frequency. In this situation, straightforward application of PC expansions may lead to the approximation of sinusoidal functions of a random variable in polynomials of the same variable. While such representation may be suitable at small times, the number of terms needed in the expansion grows rapidly as time increases. Clearly, extended computations or simulations of limit cycles are not possible with such an approach. This underscores the difficulty of selecting a PC basis that is suitable for both the random input and the stochastic solution. Much needed are robust means to overcome such difficulties. Change of measure, and basis enrichment concepts outlined in Sect. 10.1 may provide possible approaches towards this capability.

Generalized/Multiscale UQ: Another set of challenges, concerns situations where one is only interested in assessing the impact of uncertainty on specific observables or selected components of the solution. These are in many ways akin to the problems just mentioned. For instance, in problems admitting limit cycle oscillations and in turbulent flows, one may only be interested in stochastic amplitudes and frequencies, but not in relating the phase of different stochastic realizations. Though these difficulties may be avoided by non-intrusive approaches, the development of computational PC methods enabling such "projections" or transformations appears to be a worthy endeavor.

A related set of issues naturally arise in complex settings involving phenomena characterized by disparate length and/or time scales. A variety of approaches have been developed to address the resulting (deterministic) multiscale problems, including asymptotic methods, computational perturbation techniques and recently hybrid or mixed discretization methods. The difficulties encountered in the deterministic setting may be severely compounded when the multiscale problem becomes subject to uncertainty. Addressing the corresponding challenge will consequently require development and optimization of new algorithms that simultaneously enable quantification of uncertainties associated with various phenomena while mitigating the stiffness of the multiscale problem. A potential avenue towards such capability may consist in judicious adaptation of UQ methods with deterministic multiscale techniques.

Representation of Input Uncertainty: Finally, we provide a few brief observations concerning the representation of input uncertainties. This is evidently a topic of enormous importance—one which also transcends the scope of the present monograph. Consequently, we simply offer the following observations.

In the preceding applications, numerous examples were provided of various sources of uncertainty. These included uncertainties in initial conditions, boundary conditions, and/or model parameters. In all cases, the representation of the input uncertainties was (relatively) straightforward, in many cases based on associating an uncertain parameter with a suitable random variable. This suited rather well the applications at hand, and enabled us to focus on various other issues, namely concerning method implementation and performance.

With the continuing trends towards more elaborate physical models, the UQ exercise is driven to contend with increasingly larger data sets and, consequently, germs of large dimension. Of course, the impact that UQ schemes can bring to such situations is in large part conditioned on a suitable representation of the uncertainty in the model inputs. Clearly, straightforward generalization of the simple approach above to complex or elaborate models would result in the definition of probability spaces having excessively large dimension. Naturally, one could mitigate this problem by establishing the appropriate correlations between dependent random variable, but this may still represent an especially delicate task in the case when data gathered from different experiments, observations and/or simulations.

10.3 New Capabilities

Automatic Evaluation of PC Transformations: The development of flexible and robust computational libraries for accurate and efficient evaluation of PC transformation can be regarded as an essential tool in the construction of stochastic PC-based stochastic codes. Briefly, these libraries have enabled efficient transformation of deterministic codes into stochastic codes. This transformation, however, requires user intervention, primarily to replace deterministic operations with the corresponding functional calls into the software libraries that implement various linear and nonlinear PC operations or mappings. An interesting concept worth pursuing consists of an "automating" transformation, in which deterministic operations would be replaced by stochastic counterparts at essentially compilation or during run time. The development of software tools that would enable such key capabilities appears to be present a key opportunity that would benefit and accelerate a wide range of investigations.

PC Transformations: The implementation of PC methods has greatly benefited from the development of software libraries that implement various PC transformations, including moment formulas, nonlinear mappings, etc... An example of such a software library is the "UQ toolkit," which includes many of the transformations, Galerkin and pseudo-spectral evaluations, moment and quadrature formulas discussed in Chap. 2 and Appendices B and C. The generalization of the capabilities afforded by such libraries would be tremendously beneficial, particularly to accommodate larger collections of basis function expansions, nonlinear PC transformations, and eventually infrastructure for change of measure methodologies.

Elliptic Solvers: The application of PC methods to elliptic PDEs of the form $\nabla \cdot (\sigma \nabla u) = f$, with random diffusivity σ and forcing f, has been the subject of numerous studies. In particular, several techniques were developed for the solution of coupled systems that result from the discretization of this equations. These include direct solvers, fast Fourier-based methods, as well as iterative solvers. In addition to the development of solution methods, theoretical results have also been established that characterize the stochastic solution as function of given or assumed properties of random diffusivity and random forcing. In light of these developments, a highly beneficial endeavor would be the construction of a software library that combines various solution schemes and error control methods for this class of stochastic PDEs.

Appendix A
Essential Elements of Probability Theory and Random Processes

Probability theory is a mathematical construction to describe phenomena whose outcomes occur non-deterministically. Depending on the context, the probability of the outcome in a random phenomenon can be interpreted either as its long-run frequency of occurrence or as a measure of its subjective uncertainty. These two points of view correspond to the frequentist and Bayesian interpretations of probabilities. This distinction is not necessary here, and we restrict the presentation to materials used in this monograph. In particular, we only introduce notions of real-valued random variables. Further elements of the probability theory can be found in standard references; see for instance [78, 104, 137, 181].

This appendix follows the exposition in Chaps. 1 and 2 of the book *Stochastic calculus: applications in science and engineering* by M. Grigoriu [94]. The discussion below distills essential elements from this reference.

A.1 Probability Theory

A.1.1 Measurable Space

We consider a random phenomenon or experiment. We denote Θ the set of all possible outcomes of the experiment. The set Θ is called the **sample space**, while θ denotes an element of Θ. The sample space may have a finite or infinite (countable or uncountable) number of elements.

Let Σ be an non-empty collection of subsets of Θ. The collection Σ is called a **σ-field on** Θ if

- $\emptyset \in \Sigma$
- $A \in \Sigma \Rightarrow \overline{A} \in \Sigma$
- $A_{i \in I} \in \Sigma \Rightarrow \bigcup_{i \in I} A_i \in \Sigma$

where $\overline{A}$ is the complement of A in Θ and I is a countable set. An element $A \in \Sigma$ is called an event or Σ-measurable subset of Θ; it is a collection of outcomes. The pair

O.P. Le Maître, O.M. Knio, *Spectral Methods for Uncertainty Quantification*, Scientific Computation,
DOI 10.1007/978-90-481-3520-2,

(Θ, Σ) is called a **measurable space**. A **Borel** σ-field is generated by the collection of all open sets of a topological set. The members of a Borel σ-field are called Borel sets. An important example of Borel σ-field is $\mathcal{B}(\mathbb{R})$, the collection of all open intervals of $\mathbb{R}$. The definition of $\mathcal{B}(\mathbb{R})$ extends to intervals; for instance $\mathcal{B}([a,b])$ denotes the Borel σ-field on the interval $[a,b]$. Denoting $\sigma(\mathcal{A})$ the smaller σ-field generated by a collection $\mathcal{A}$ of subsets of Θ, we have

$$\mathcal{B}(\mathbb{R}) = \sigma\left((a,b), -\infty \le a \le b \le +\infty\right).$$

A.1.2 *Probability Measure*

A set function $\mu : \Sigma \mapsto [0, \infty]$ is a **measure** on Σ if it is countably additive:

$$\mu\left(\bigcup_{i=1}^{\infty} A_i\right) = \sum_{i=1}^{\infty} \mu(A_i), \quad A_i \in \Sigma, A_i \cap A_{i \ne j} = \emptyset.$$

The triple (Θ, Σ, μ) is called a **measure space**. A **finite measure** is a measure such that $\mu(\Theta) < \infty$. A finite measure P such that

- $P : \Sigma \mapsto [0, 1]$
- $P(\Theta) = 1$

is called a **probability measure** or simply a probability.

A.1.3 *Probability Space*

The triple (Θ, Σ, P) is called a **probability space**. A probability space (Θ, Σ, P) such that for $A, B \in \Sigma$ with $A \subset B$ and $P(B) = 0 \Rightarrow P(A) = 0$ is said complete. A probability space can always be completed so we shall implicitly assume complete probability spaces. We then have the following properties of a probability space (Θ, Σ, P) for $A, B \in \Sigma$:

- $P(A) \le P(B)$ for $A \subseteq B$,
- $P(\overline{A}) = 1 - P(A)$,
- $P(A \cup B) = P(A) + P(B) - P(A \cap B)$.

Consider the probability spaces $(\Theta_k, \Sigma_k, P_k)$ for $k = 1, \dots, n$. From this collection, we define the **product probability space** (Θ, Σ, P) as

$$\Theta = \Theta_1 \times \cdots \times \Theta_n,$$

$$\Sigma = \Sigma_1 \times \cdots \times \Sigma_n,$$

$$P = P_1 \times \cdots \times P_n.$$

This definition also holds for $n = \infty$. Whenever, the spaces $(\Theta_k, \Sigma_k, P_k)$ are all identical, the product sample space, σ-field and probability measure will be denoted Θ^n, Σ^n and P^n respectively.

Let (Θ, Σ, P) be a probability space and let $B \in \Sigma$ be a given event such that $P(B) > 0$. We define a new probability measure, called the **probability conditional on** B as

$$P(A \in \Sigma | B) \equiv \frac{P(A \cap B)}{P(B)}.$$

From this definition, if the set of events $A_i \in \Sigma$ is a partition of the sample space (*i.e.* if $A_i \cap A_{j \neq i} = \emptyset$ and $\bigcup_i A_i = \Theta$), then

$$P(B) = \sum_i P(B \cap A_i) = \sum_i P(B|A_i) P(A_i),$$

(**law of total probability**) and

$$P(A_j|B) = \frac{P(A_j)P(B|A_j)}{P(B)} = \frac{P(A_j)P(B|A_j)}{\sum_i P(A_i)P(B|A_i)}, \qquad P(B) > 0$$

(**Bayes formula**).

A.2 Measurable Functions

A.2.1 Induced Probability

Consider two measurable spaces (Θ, Σ_Θ) and $(\Theta', \Sigma_{\Theta'})$ and a function $h : \Theta \mapsto \Theta'$ with domain Θ and range Θ'. The function h is said **measurable from** (Θ, Σ_Θ) **to** $(\Theta', \Sigma_{\Theta'})$ if for any event B in $\Sigma_{\Theta'}$

$$h^{-1}(B) \equiv \{\theta : h(\theta) \in B\} \in \Sigma_\Theta.$$

If h is measurable from (Θ, Σ_Θ) to $(\Theta', \Sigma_{\Theta'})$ and $(\Theta, \Sigma_\Theta, P)$ is a probability then $Q : \Sigma_{\Theta'} \mapsto [0, 1]$ defined by

$$Q(B \in \Sigma_{\Theta'}) \equiv P(h^{-1}(B))$$

is a probability measure on $(\Theta', \Sigma_{\Theta'})$ called the **probability induced** by h, or simply the distribution of h.

A.2.2 Random Variables

Consider a probability space (Θ, Σ, P) and a function $X : \Theta \mapsto \mathbb{R}$ measurable from (Θ, Σ) to $(\mathbb{R}, \mathcal{B})$. Then, X is a $\mathbb{R}$**-valued random variable**, sometimes denoted

$X(\theta)$. The **distribution** of X is the probability measure induced by the mapping $X : \Theta \mapsto \mathbb{R}$ on the measurable space $(\mathbb{R}, \mathcal{B})$, defined by

$$Q(B) = P(X^{-1}(B)), \quad B \in \mathcal{B}.$$

This definition implies that X is a random variable if and only if $X^{-1}((-\infty, x]) \in \Sigma, \forall x \in \mathbb{R}$. In other words, a $\mathbb{R}$-valued random variable is a function which maps the sample space Θ to $\mathbb{R}$.

The previous definition of a random variable can be extended to $\mathbb{R}^d$-valued functions $\mathbf{X}$ measurable from (Θ, Σ) to $(\mathbb{R}^d, \mathcal{B}^d)$. If all the coordinates of $\mathbf{X}$ are random variables, then $\mathbf{X}$ is called a **random vector**.

A.2.3 Measurable Transformations

This monograph is concerned with numerical techniques for the characterization of the output of physical models involving random input. We are then dealing with transformations of random variables. Denoting $\mathbf{X}$ the $\mathbb{R}^d$-valued random vector representing the model input, we are for instance interested in a model output $Y = g(\mathbf{X})$ where $g : \mathbb{R}^d \mapsto \mathbb{R}$. The model output in then defined by the mapping

$$(\Theta, \Sigma) \overset{\mathbf{X}}{\mapsto} (\mathbb{R}^d, \mathcal{B}^d) \overset{g}{\mapsto} (\mathbb{R}, \mathcal{B}).$$

It can be shown that if $\mathbf{X}$ and g are measurable functions from (Θ, Σ) to $(\mathbb{R}^d, \mathcal{B}^d)$ and from $(\mathbb{R}^d, \mathcal{B}^d)$ to $(\mathbb{R}, \mathcal{B})$, respectively, then the composed mapping $g \circ \mathbf{X} : (\Theta, \Sigma) \mapsto (\mathbb{R}, \mathcal{B})$ is measurable. As a result, the model output is a $\mathbb{R}$-valued random variable.

A.3 Integration and Expectation Operators

A.3.1 Integrability

For $X : (\Theta, \Sigma) \mapsto (\mathbb{R}, \mathcal{B})$, the integral of X with respect to P over the event $A \in \Sigma$ is

$$\int_{\Theta} I_A(\theta) X(\theta) \mathrm{d}P(\theta) = \int_A X(\theta) \mathrm{d}P(\theta),$$

where I_A is the indicator function of A. If this integral exists and is finite, X is said to be P**-integrable over A**. A random vector $\mathbf{X}$ is P-integrable over A, if all of its components are individually P-integrable over A.

Let X and Y be two random variables defined on a probability space (Θ, Σ, P). We have the following properties:

- For $A \in \Sigma$,

$$\int_A |X| \, \mathrm{d}P < \infty \Rightarrow \int_A X \, \mathrm{d}P \quad \text{is finite.}$$

- If X, Y are P-integrable over $A \in \Sigma$, the random variable $aX + bY$ is P-integrable over A and (linearity):

$$\int_A (aX + bY) \, \mathrm{d}P = a \int_A X \, \mathrm{d}P + b \int_A Y \, \mathrm{d}P.$$

- If X is P-integrable over Θ and $\{A_i \in \Sigma\}$ is a partition of Θ, then

$$\int_{\cup_i A_i} X \, \mathrm{d}P = \sum_i \int_{A_i} X \, \mathrm{d}P.$$

- If X is P-integrable on Θ, then X is finite a.s. (almost surely), *i.e.* $B = \{\theta : X(\theta) = \pm\infty\} \in \Sigma$ is such that $P(B) = 0$.
- If $X \geq 0$ a.s., then

$$\int_A X \, \mathrm{d}P \geq 0.$$

- If $Y \leq X$ a.s. then

$$\int_A Y \, \mathrm{d}P \leq \int_A X \, \mathrm{d}P.$$

A.3.2 Expectation

The particular case of $A = \Theta$ for the integration domain corresponds to the expectation operator which will be denoted $\mathrm{E}[\cdot]$:

$$\mathrm{E}[X] \equiv \int_\Theta X(\theta) \, \mathrm{d}P(\theta).$$

Provided it exists and is finite, the expectation has the following properties. Let X and Y be two $\mathbb{R}$-valued random variables defined on a probability space (Θ, Σ, P) and P-integrable over Θ. Then

- Linearity of the expectation:

$$\mathrm{E}[aX + bY] = a\mathrm{E}[X] + b\mathrm{E}[Y].$$

- $X \geq 0$ a.s. $\Rightarrow \mathrm{E}[X] \geq 0$.
- $Y \leq X$ a.s. $\Rightarrow \mathrm{E}[Y] \leq \mathrm{E}[X]$.
- $|\mathrm{E}[X]| \leq \mathrm{E}[|X|]$.

A.3.3 L_2 Space

Consider a probability space (Θ, Σ, P). For $q \geq 1$, we denote $L_q(\Theta, \Sigma, P)$ the collection of $\mathbb{R}$-valued random variables X defined on (Θ, Σ, P) such that

$$\mathrm{E}\left[|X|^q\right] < \infty.$$

The case $q = 2$, corresponding to the L_2-space is of particular importance in this book as it possesses some essential properties.

First, L_2 **is a vector space**. Indeed, for $X \in L_2, \lambda \in \mathbb{R} \Rightarrow \lambda X \in L_2$, since $\mathrm{E}[(\lambda X)^2] = \lambda^2 \mathrm{E}[X^2] < \infty$. It remains to show that for $X, Y \in L_2$, $X + Y \in L_2$, so $\mathrm{E}[(X+Y)^2] = \mathrm{E}[X^2] + \mathrm{E}[Y^2] + 2\mathrm{E}[XY] < \infty$. To prove the last inequality, we have to show that $\mathrm{E}[XY]$ is finite; to this end consider $X \neq Y$ such that

$$p(\lambda) = \mathrm{E}\left[(X + \lambda Y)^2\right] = \mathrm{E}\left[X^2\right] + \lambda^2 \mathrm{E}\left[Y^2\right] + \lambda 2\mathrm{E}\left[XY\right],$$

has no real root; this leads to $\mathrm{E}[XY]^2 - \mathrm{E}[X^2]\mathrm{E}[Y^2] \leq 0$ or equivalently

$$|\mathrm{E}\left[XY\right]| \leq \mathrm{E}\left[X^2\right]^{1/2} \mathrm{E}\left[Y^2\right]^{1/2} \quad \text{for } X, Y \in L_2,$$

which is known as the **Cauchy-Schwarz** inequality.

Second, the expectation $\mathrm{E}\left[XY\right]$ defines an **inner product** on L_2, denoted $\langle X, Y\rangle$ with the associated L_2 **norm** $\|X\|_{L_2} = \mathrm{E}[X^2]^{1/2}$. Indeed, $\langle \cdot, \cdot \rangle$ is an inner product since

$$\begin{aligned}
&\langle 0, X\rangle = 0, \quad X \in L_2,\\
&\langle X, X\rangle > 0, \quad X \in L_2, X \neq 0,\\
&\langle X, Y\rangle = \langle Y, X\rangle, \quad X, Y \in L_2,\\
&\langle X + Y, Z\rangle = \langle X, Z\rangle + \langle Y, Z\rangle, \quad X, Y, Z \in L_2,\\
&\langle \lambda X, Y\rangle = \lambda\langle X, Y\rangle, \quad X, Y \in L_2, \lambda \in \mathbb{R},
\end{aligned}$$

and $d : L_2 \times L_2 \mapsto [0, +\infty)$ defined by $d(X, Y) = \|X - Y\|_{L_2}$ is a metric on L_2 since

$$\begin{aligned}
&d(X, Y) = 0 \quad \text{iif } X = Y \text{ a.s.},\\
&d(X, Y) = d(Y, X), \quad X, Y \in L_2,\\
&d(X, Y) \leq d(X, Z) + d(Z, Y), \quad \forall X, Y, Z \in L_2.
\end{aligned}$$

Finally, L_2 equipped with the inner product $\langle \cdot, \cdot \rangle$ and the L_2-norm is a **Hilbert space**. As a result, if $X \in L_2(\Theta, \Sigma, P)$ and Σ' is a sub-σ-field of Σ, then there is a unique random variable $X' \in L_2(\Theta, \Sigma', P)$ such that

$$\|X - X'\|_{L_2} = \min\left\{\|X - Z\|_{L_2} : Z \in L_2(\Theta, \Sigma', P)\right\},$$

and

$$\langle X - X', Z\rangle = 0, \quad \forall Z \in L_2(\Theta, \Sigma', P).$$

The random variable X' is the **orthogonal projection** of X on $L_2(\Theta, \Sigma', P)$, or best mean square estimator of X, and has the smallest mean square error of all members of $L_2(\Theta, \Sigma', P)$.

A.4 Random Variables

We recall (see A.2.2) that a $\mathbb{R}$-valued random variable defined on a probability space (Θ, Σ, P) is a measurable function from (Θ, Σ) to $(\mathbb{R}, \mathcal{B}(\mathbb{R}))$.

A.4.1 Distribution Function of a Random Variable

The **cumulative distribution function**, or simply distribution function, of a random variable X defined on a probability space (Θ, Σ, P) is defined by

$$F_X(x) = P\left(X^{-1}((-\infty, x])\right) = P(\{\theta : X(\theta) \le x\}) = P(X \le x).$$

The distribution function is right-continuous, increasing, with range [0, 1]. In addition,

$$\lim_{x\to+\infty} F_X(x) = 1, \qquad \lim_{x\to-\infty} F_X(x) = 0,$$
$$P(a < x \le b) = F_X(b) - F_X(a) \ge 0,$$
$$P(a \le x < b) = F_X(b) - F_X(a) + P(x = a) - P(x = b).$$

The distribution F_X of a random variable X has only a countable number of jump discontinuities and is continuous at $x \in \mathbb{R}$ if and only if $P(X = x) = 0$.

A.4.2 Density Function of a Random Variable

If F_X is absolutely continuous in $\mathbb{R}$, there is an integrable function f called the probability density function, or density function, such that

$$F_X(b) - F_X(a) = \int_a^b f_X(x)\,\mathrm{d}x, \quad a \le b.$$

A density function has the essential properties:

$$f_X(x) = F'_X(x) \quad \text{so} \quad \int_{-\infty}^{x} f_X(y)\,\mathrm{d}y = F_X(x),$$

$$f_X(x) \geq 0,$$

$$\int_{-\infty}^{+\infty} f_X(x)\,\mathrm{d}x = 1.$$

Let X be a random variable defined on a probability space (Θ, Σ, P), and consider a measurable function $g : (\mathbb{R}, \mathcal{B}) \mapsto (\mathbb{R}, \mathcal{B})$. Then, $Y = g \circ X$ is a random variable defined on (Θ, Σ, P) and

$$\begin{aligned} \mathrm{E}[Y] &= \int_{\Theta} Y(\theta)\;\mathrm{d}P(\theta) = \int_{\Theta} g\left(X(\theta)\right)\;\mathrm{d}P(\theta) \\ &= \int_{\mathbb{R}} g(x)\,\mathrm{d}Q(x) = \int_{\mathbb{R}} g(x)\,\mathrm{d}F_X(x) = \int_{\mathbb{R}} g(x) f_X(x)\,\mathrm{d}x. \end{aligned}$$

This relation provides an expression of the expectation of Y involving the probability density function of X. In this book, we heavily rely on such expression of expectation operator in terms of density functions.

A.4.3 Moments of a Random Variable

Let X be a $\mathbb{R}$-valued random variable defined on a probability space (Θ, Σ, P), and $Y = g(X) = X^r$ for $r \geq 1$. Since $g(x)$ is continuous, it is Borel-measurable and therefore $Y = X^r$ is a random variable. The expectation of Y is called the **moment of order** r of X, and is denoted $m_r(X)$:

$$m_r(X) = \mathrm{E}\left[X^r\right] = \int_{-\infty}^{+\infty} x^r\,\mathrm{d}F_X(x) = \int_{-\infty}^{+\infty} x^r f_X(x)\,\mathrm{d}x.$$

If $X \in L_r(\Theta, \Sigma, P)$, then $m_r(X)$ exists and is finite. If instead we consider $Y = g(X) = (X - m_1(X))^r$, then the expectation of Y is the **central moment of order** r.

A.4.4 Convergence of Random Variables

Let X be a $\mathbb{R}$-valued random variable and $X_{n\geq 1}$ be a sequence of random variables defined on a probability space (Θ, Σ, P). The convergence of the sequence X_n to X depends on the way we measure $X - X_n$. Alternatives are:

- **Almost sure** convergence,

$$X_n \xrightarrow{\text{a.s.}} X \quad \text{if } \lim_{n\to\infty} X_n(\theta) = X(\theta),\ \forall\theta \in \Theta \setminus N,\ P(N) = 0.$$

- Convergence in **probability**,

$$X_n \xrightarrow{\text{pr}} X \quad \text{if } \lim_{n\to\infty} P(|X_n(\theta) - X(\theta)|) > \epsilon) = 0, \forall\epsilon > 0.$$

- Convergence in **distribution**,

$$X_n \xrightarrow{\text{d}} X \quad \text{if } \lim_{n\to\infty} F_{X_n}(x) = F_X(x), \forall x \in \mathbb{R}.$$

- Convergence in L_p,

$$X_n \xrightarrow{\text{m.p.}} X \quad \text{if } \lim_{n\to\infty} \mathrm{E}\left[|X_n - X|^p\right] = 0.$$

We will mostly use the mean square convergence, *i.e.* convergence in L_2, and denote it as $X_n \xrightarrow{\text{m.s.}} X$.

A.5 Random Vectors

Consider a probability space (Θ, Σ, P); the measurable function

$$\mathbf{X} : (\Theta, \Sigma) \mapsto (\mathbb{R}^d, \mathcal{B}^d), \quad d > 1,$$

defines a random vector in $\mathbb{R}^d$ or $\mathbb{R}^d$-valued random variable, with the induced probability measure

$$Q(B) = P(\mathbf{X} \in B) = P(\mathbf{X}^{-1}(B)).$$

A.5.1 Joint Distribution and Density Functions

The **joint distribution function** of **X** is the direct extension of the definition of the random variable distribution function:

$$F_{\mathbf{X}}(\mathbf{x}) = P\left(\bigcap_{i=1}^{d} \{X_i \le x_i\}\right), \quad \mathbf{x} = (x_1, \ldots, x_d) \in \mathbb{R}^d.$$

The joint distribution function of a random vector **X** has the following properties:

$$\lim_{x_k \to -\infty} F_{\mathbf{X}}(\mathbf{x}) = 0, \quad k = 1, \ldots, d,$$

$$x_k \mapsto F_{\mathbf{X}}(\mathbf{x}) \quad \text{is increasing, } k = 1, \ldots, d,$$

$$x_k \mapsto F_{\mathbf{X}}(\mathbf{x}) \quad \text{is right-continuous, } k = 1, \ldots, d.$$

In addition,

$$\lim_{x_k \to +\infty} F_{\mathbf{X}}(\mathbf{x}) = F_{\mathbf{X}_{|k}}(\mathbf{x}_{|k}), \qquad 1 \leq k \leq d,$$

is the joint distribution of the $\mathbb{R}^{d-1}$-valued random vector

$$\mathbf{X}_{|k} = (X_1, \ldots, X_{k-1}, X_{k+1}, \ldots, X_d).$$

The **joint density function** of the random vector $\mathbf{X}$, if it exists, is given by:

$$f_{\mathbf{X}}(\mathbf{x}) = \frac{\partial^d F_{\mathbf{X}}(\mathbf{x})}{\partial x_1 \cdots \partial x_d}.$$

The probability that the random vector takes values within the domain $(x_1, x_1 + dx_1] \times (x_d, x_d + dx_d]$, for given infinitesimal vector $\mathbf{d}x = (dx_1, \ldots, dx_d)$ is related to the joint density by

$$P\left(\bigcap_{i=1}^{d} \{X_i \in (x_i, x_i + dx_i]\}\right) \approx f_{\mathbf{X}}(\mathbf{x}) \|\mathbf{d}x\|.$$

From the joint distribution and density, one can derive expressions for the joint density or distribution of a subset of coordinates of $\mathbf{X}$. Such reduction is known as **marginalization**. For instance, the joint distribution of $\mathbf{X}_{|k}$ is

$$F_{\mathbf{X}_{|k}}(\mathbf{x}_{|k}) = F_{\mathbf{X}}(x_1, \ldots, x_{k-1}, \infty, x_{k+1}, \ldots, x_d),$$

whereas

$$f_{\mathbf{X}_{|k}}(\mathbf{x}_{|k}) = \int_{-\infty}^{+\infty} f_{\mathbf{X}}(x_1, \ldots, x_d)\, \mathrm{d}x_k.$$

In particular, $F_{X_i}(x_i) = F_{\mathbf{X}}(\infty, \ldots, \infty, x_i, \infty, \ldots, \infty)$ is the **marginal distribution** of X_i and

$$f_{X_i}(x_i) = \int_{\mathbb{R}^{d-1}} f_{\mathbf{X}}(\mathbf{x})\, \mathrm{d}\mathbf{x}_{|i}$$

is the **marginal density** of X_i.

Another useful expression concerns the **conditional density** of a subset of coordinates of $\mathbf{X}$. Consider for instance that the coordinates of $\mathbf{X}$ have been split into two subsets corresponding to two vectors $\mathbf{X}_1 \in \mathbb{R}^{0<d_1<d}$ and $\mathbf{X}_2 \in \mathbb{R}^{d-d_1}$. Then the joint density of $\mathbf{X}_1$ conditioned by $\mathbf{X}_2 = \mathbf{x}_2$ is

$$f_{\mathbf{X}_1|\mathbf{X}_2}(\mathbf{x}_1, \mathbf{x}_2) = \frac{f_{\mathbf{X}}(\mathbf{x}_1, \mathbf{x}_2)}{f_{\mathbf{X}_2}(\mathbf{x}_2)}.$$

The joint density of transformed random vectors can also be derived. Consider a one to one mapping $\mathbf{h} : \mathbf{x} \in \mathbb{R}^d \mapsto \mathbf{y} = \mathbf{h}(\mathbf{x}) \in \mathbb{R}^d$, and define the random vector $Y = \mathbf{h}(\mathbf{X})$. The joint densities $f_{\mathbf{Y}}$ and $f_{\mathbf{X}}$ are related by

$$f_{\mathbf{Y}}(\mathbf{y}) = f_{\mathbf{X}}\left(\mathbf{h}^{-1}(\mathbf{y})\right)\left|\nabla_{\mathbf{y}}\mathbf{h}^{-1}\right|,$$

where $\mathbf{h}^{-1}$ is the inverse mapping and $\nabla_{\mathbf{y}}\mathbf{h}^{-1}$ its Jacobian matrix. This relation can be intuitively derived from the principle of probability conservation. Indeed, denoting $D_{\mathbf{y}}$ a neighborhood of $\mathbf{y}$ and $D_{\mathbf{x}}$ its image by the inverse transformation $\mathbf{h}^{-1}$, the equality $P(\mathbf{Y} \in D_{\mathbf{y}}) = P(\mathbf{X} \in D_{\mathbf{x}})$ leads to

$$\begin{aligned} P(\mathbf{X} \in D_{\mathbf{x}}) &= \int_{D_{\mathbf{x}}} f_{\mathbf{X}}(\eta)\,\mathrm{d}\eta = \int_{D_{\mathbf{y}}} f_{\mathbf{X}}(\mathbf{h}^{-1}(\xi))\left|\nabla_{\xi}\mathbf{h}^{-1}\right|\mathrm{d}\xi \\ &= \int_{D_{\mathbf{y}}} f_{\mathbf{Y}}(\xi)\,\mathrm{d}\xi = P(\mathbf{Y} \in D_{\mathbf{y}}). \end{aligned}$$

A.5.2 Independence of Random Variables

Consider a set of random variables X_i, $i \in I$, defined on a probability space (Θ, Σ, P). These random variables are **independent** if and only if

$$P(X_i \le x_i, i \in J) = \prod_{i \in J} P(X_i \le x_i), \quad \forall J \subset I, x_i \in \mathbb{R}.$$

For finite index set $I = \{1, \dots, d\}$, the collection of random variables can be seen as a random vector $\mathbf{X}$ of $\mathbb{R}^d$, and the independence implies a factorized structure for the joint distribution $F_{\mathbf{X}}$,

$$F_{\mathbf{X}}(x_1, \dots, x_d) = \prod_{i=1}^{d} F_{X_i}(x_i),$$

and joint density $f_{\mathbf{X}}$,

$$f_{\mathbf{X}}(x_1, \dots, x_d) = \prod_{i=1}^{d} f_{X_i}(x_i).$$

A.5.3 Moments of a Random Vector

Consider a random vector $\mathbf{X}^d$ and an integer multi-index $\mathbf{l} = (l_1, \ldots, l_{\mathrm{d}}) \in \mathbb{N}^d$. We define $|\mathbf{l}| = \sum_i l_i$, and the function

$$g_{\mathbf{l}} : \mathbf{x} \in \mathbb{R}^d \mapsto g(\mathbf{x}) = \prod_{i=1}^{i=d} x_i^{l_i}. \tag{A.1}$$

The function $g_{\mathbf{l}}(\mathbf{X})$ is a $\mathbb{R}$-valued random variable. If $\mathbf{X} \in L_{|\mathbf{l}|}$, *i.e.* for each of the vector coordinate $X_i \in L_{|\mathbf{l}|}$, then the moments of order $|\mathbf{l}|$ of $\mathbf{X}$ are given by

$$m_{\mathbf{l}}(\mathbf{X}) = \mathrm{E}\left[g_{\mathbf{l}}(\mathbf{X})\right]. \tag{A.2}$$

If the coordinates of $\mathbf{X}$ are mutually independent, the moments $m_{\mathbf{l}}$ can be expressed as:

$$m_{\mathbf{l}}(\mathbf{X}) = \mathrm{E}\left[\prod_{i=1}^{i=d} X_i^{l_i}\right] = \prod_{i=1}^{i=d} \mathrm{E}\left[X_i^{l_i}\right] = \prod_{i=1}^{i=d} m_{l_i}(X_i). \tag{A.3}$$

Of particular importance are the mean μ_i of X_i which corresponds to the first-order moment with $l_j = \delta_{ij}$, and the correlation r_{ij} of (X_i, X_j) which is the second-order moment with all $l_k = 0$ except $l_i = l_j = 1$. Combining these first and second moments, we obtain the covariance c_{ij} of (X_i, X_j),

$$c_{ij} = \mathrm{E}\left[(X_i - \mu_i)(X_j - \mu_j)\right] = r_{ij} - \mu_i \mu_j, \tag{A.4}$$

and the variance σ_i^2 of X_i:

$$\sigma_i^2 = \mathrm{E}\left[(X_i - \mu_i)^2\right] = r_{ii} - \mu_i^2 \geq 0. \tag{A.5}$$

If $c_{ij} = 0$ for $i \neq j$, then X_i and X_j are said to be **uncorrelated**. If $r_{ij} = 0$ for $i \neq j$, X_i and X_j are said to be **orthogonal**. The first and second-order moments of a random vector can be summarized using the mean vector $\mathbf{m}$ and correlation $\mathbf{r}$ or covariance $\mathbf{c}$ matrices:

$$\begin{aligned} \mathbf{m} &= \mathrm{E}[\mathbf{X}], \\ \mathbf{r} &= [r_{ij}] = \mathrm{E}\left[\mathbf{X}\mathbf{X}^t\right], \\ \mathbf{c} &= [c_{ij}] = \mathrm{E}\left[(\mathbf{X} - \mathbf{m})(\mathbf{X} - \mathbf{m})^t\right]. \end{aligned} \tag{A.6}$$

Note that if X_i and X_j are independent, we have

$$c_{ij} = \mathrm{E}\left[(X_i - \mu_i)(X_j - \mu_j)\right] = \mathrm{E}[X_i - \mu_i]\mathrm{E}\left[X_j - \mu_j\right] = 0. \tag{A.7}$$

In other words, independent random variables are uncorrelated. However, the converse is usually not true: uncorrelated random variables are not independent in general. Also, random vectors with mutually uncorrelated coordinates have diagonal correlation matrices.

A.5.4 Gaussian Vector

The case of Gaussian vector is of particular importance in probability theory and in Polynomial Chaos expansions on Hermite bases. A Gaussian vector of $\mathbb{R}^d$ with mean $\mathbf{m}$ and covariance matrix $\mathbf{c}$ has for joint density function:

$$f_{\mathbf{X}}(\mathbf{x}) = \frac{1}{(2\pi)^{d/2}\sqrt{|\mathbf{c}|}} \exp\left[-\frac{1}{2}(\mathbf{x}-\mathbf{m})^t \mathbf{c}^{-1} (\mathbf{x}-\mathbf{m}) \right], \tag{A.8}$$

where $|\mathbf{c}|$ is the determinant of the correlation matrix. A Gaussian vector is then completely specified by its second-order properties and we shall write

$$\mathbf{X} \sim \mathrm{N}(\mathbf{m}, \mathbf{c}). \tag{A.9}$$

An important property of Gaussian vector is that if $\mathbf{c}$ is diagonal (its coordinate are uncorrelated) then the X_i are mutually independent. Indeed,

$$\begin{aligned} f_{\mathbf{X}}(\mathbf{x}) &= \frac{1}{(2\pi)^{d/2}\sqrt{|\mathbf{c}|}} \exp\left[-\frac{1}{2}\sum_{i=1}^{d} \sigma_i^{-1}(x_i - \mu_i)^2 \right] \\ &= \prod_{i=1}^{d} \frac{1}{\sqrt{2\pi\sigma_i^2}} \exp\left[-\frac{1}{2\sigma_i^2}(x_i - \mu_i)^2 \right], \end{aligned} \tag{A.10}$$

because $c_{ii} = \sigma_i^2$ and $|\mathbf{c}| = \prod c_{ii}$ for diagonal matrices. Also, if $\mathbf{X}$ and $\mathbf{Y}$ are two $\mathbb{R}^d$-valued Gaussian vectors, then $(\alpha\mathbf{X} + \beta\mathbf{Y})$ is a Gaussian random vector.

A.6 Stochastic Processes

A.6.1 Motivation and Basic Definitions

In this monograph, we are interested in the computation of the stochastic solutions of equations, essentially ODEs or PDEs. Therefore these solutions are functions of time and/or space. Then, the notion of random variable and random vector need be extended to incorporate a dependence on time or/and space coordinates.

Consider a function $\mathbf{X} : T \times \Theta \mapsto \mathbb{R}^d$ with arguments –time– $t \in T$ (a real interval) and –event– $\theta \in \Theta$. If $\mathbf{X}(t)$ is a $\mathbb{R}^d$-valued random vector on the probability space (Θ, Σ, P) for all $t \in T$, then $\mathbf{X}$ is called a $\mathbb{R}^d$-valued **stochastic process** or vector stochastic process. The function $\mathbf{X}(\cdot, \theta) : t \in T \mapsto \mathbb{R}^d$ for given $\theta \in \Theta$ is called a **sample path** or simply a **realization** of $\mathbf{X}$. Conversely, the definition shows that $\mathbf{X}(t, \cdot) : \theta \in \Theta \mapsto \mathbb{R}^d$ is a random vector on (Θ, Σ, P).

The definition can be extended to random vectors indexed by a spatial coordinate. For instance, consider $\mathbf{y} \in \mathcal{D} \subset \mathbb{R}^n$, then $\mathbf{X} : \mathcal{D} \times \Theta \mapsto \mathbf{X} \in \mathbb{R}^d$ is called a $\mathbb{R}^d$-valued

stochastic field if $\mathbf{X}(\mathbf{y},\cdot)$ is a random vector for each $\mathbf{y}\in\mathcal{D}$. The function $\mathbf{X}(\cdot,\theta)$ is called a realization of the stochastic field.

The two definitions can be combined and results in **space-time stochastic processes**:

$$\mathbf{X}: T\times\mathcal{D}\times\Theta\mapsto\mathbb{R}^d, \quad \mathbf{X}(t,\mathbf{y},\cdot)\in(\Theta,\Sigma,P)\ \forall t\in T, \mathbf{y}\in\mathcal{D}. \tag{A.11}$$

There are noticeable differences between stochastic processes and stochastic fields, arising from oriented nature of time. However, these differences will not appear in the present monograph, and we shall call stochastic processes, or random process, any type of random variable or vector indexed by a deterministic set of coordinates in $\mathcal{D}$. We shall often rely on the loose terminology of random process. Consequently, we shall denote $\mathcal{D}$ the domain of indexation of the stochastic process, making no distinction between space and time coordinates in $\mathbf{y}\in\mathbb{R}^s$.

In the following, we restrict ourself to stochastic processes $\mathbf{X}:\mathcal{D}\times\Theta\mapsto\mathbb{R}^d$ defined on a probability space (Θ,Σ,P), with $\mathcal{D}\subset\mathbb{R}$, *i.e.* $n=1$. The definitions and concepts can be extended to stochastic fields $n\geq 1$, subject to appropriate resolution of some technical difficulties and notation issues.

A.6.2 Properties of Stochastic Processes

If $\mathbf{X}$ is such that

$$\lim_{s\to t} P\left(\|\mathbf{X}(s)-\mathbf{X}(t)\|\geq\epsilon\right) \quad \forall\epsilon>0, \tag{A.12}$$

and any $t\in\mathcal{D}$, the process is said **continuous in probability**. If for given integer $p\geq 1$,

$$\lim_{s\to t}\mathrm{E}\left[\|\mathbf{X}(s)-\mathbf{X}(t)\|^p\right]=0, \quad \forall t\in\mathcal{D}, \tag{A.13}$$

the process is said **continuous in the p-th mean**, and **continuous in the mean square sense** in the case of $p=2$. If

$$P\left(\left\{\theta:\lim_{s\to t}\|\mathbf{X}(s,\theta)-\mathbf{X}(t,\theta)\|\neq 0\right\}\right)=0, \quad \forall t\in\mathcal{D}, \tag{A.14}$$

the process is said **almost surely continuous**.

In the previous definitions of continuity, the norm $\|\cdot\|$ is the classical Euclidean norm for vectors. The continuity can also be restricted to particular $t\in\mathcal{D}$ or subintervals.

A.6.2.1 Finite Dimensional Distributions and Densities

Let $m \geq 1$ be an integer, and a set of m distinct points t_i in $\mathcal{D}$. The **finite dimensional distribution (fdd) of order** m of X is defined by:

$$F_m\left(\mathbf{x}^{(1)}, \ldots, \mathbf{x}^{(m)}; t_1, \ldots, t_m\right) = P\left(\cap_{i=1}^{m}\{\mathbf{X}(t_i) \in \times_{k=1}^{d}(-\infty, x_k^{(i)}]\}\right), \quad \text{(A.15)}$$

where $\mathbf{x}^{(i)} = (x_1^{(i)}, \ldots, x_d^{(i)}) \in \mathbb{R}^d$. Thus, the finite dimensional distributions, F_m, are probability measures induced by $\mathbf{X}$ on the measurable space $(\mathbb{R}^{m\times d}, \mathcal{B}^{m\times d})$. The first-order fdd of $\mathbf{X}$ at t is called the marginal distribution. It is in fact the distribution of the random vector $\mathbf{X}(t, \cdot)$. For $d = 1$, the m-th order fdd's are

$$F_m(x^{(1)}, \ldots, x^{(m)}; t_1, \ldots, t_m) = P\left(X(t_1) \leq x^{(1)}, \ldots, X(t_m) \leq x^{(m)}\right), \quad \text{(A.16)}$$

from which the **finite dimensional densities** of X can be obtained by differentiation:

$$f_m\left(x^{(1)}, \ldots, x^{(m)}, t_1, \ldots, t_m\right) = \frac{\partial^m}{\partial x^{(1)} \cdots \partial x^{(m)}} F_m(x^{(1)}, \ldots, x^{(m)}; t_1, \ldots, t_m). \quad \text{(A.17)}$$

The marginal density of X at time t corresponds to the first-order fdd $f_1(x; t)$.

The $\mathbb{R}^d$-valued stochastic process is said **stationary in the strict sense** if all its fdd's are invariant under arbitrary shift τ of all the $t_i \to t_i + \tau$. If $\mathbf{X}$ is stationary, the marginal distribution of $\mathbf{X}$ is t-invariant.

A stochastic process $\mathbf{X}$ is a **Gaussian process** if all its fdd's are Gaussian.

A.6.3 Second Moment Properties

A **second-order stochastic process X** has all its components $X_i \in L_2(\Theta, \Sigma, P)$ for all $t \in \mathcal{D}$. For a second-order random process, we define

- the mean function: $\overline{\mathbf{x}}(t) = \mathrm{E}[\mathbf{X}]$
- the correlation function $\mathbf{r}(t, s) = \mathrm{E}[\mathbf{X}(t, \cdot)\mathbf{X}(s, \cdot)^t]$
- the covariance function $\mathbf{c}(t, s) = \mathrm{E}[(\mathbf{X}(t, \cdot) - \overline{\mathbf{x}}(t))(\mathbf{X}(s, \cdot) - \overline{\mathbf{x}}(s))^t]$

The **second moment properties** of a stochastic process are given by the couple $(\overline{\mathbf{x}}, \mathbf{r})$ or $(\overline{\mathbf{x}}, \mathbf{c})$. For a stationary process, the mean function $\overline{\mathbf{x}}$ is constant and the correlation and covariance functions depend only on the lag $\tau = t - s$. Conversely, a process with constant mean function and correlation and covariance functions of the form $\mathbf{r}(t, s) = \mathbf{r}(t - s)$ and $\mathbf{c}(t, s) = \mathbf{c}(t - s)$, is said **weakly stationary**. A weakly stationary process is non-necessarily strictly stationary.

The correlation function has the following remarkable properties:

- $r_{ii}(t, s) = r_{ii}(s, t)$. If the process is weakly stationary $r_{ii}(\tau) = r_{ii}(-\tau)$ where $\tau = t - s$.

- $\|r_{ii}(t,s)\|^2 \le r_{ii}(t,t)r_{ii}(s,s)$. If the process is weakly stationary, $|r_{ii}(\tau)| \le r_{ii}(0)$.
- $r_{ii}(t,s)$ is positive definite.
- $\mathbf{X}$ is mean square continuous at t' if and only if $r_{ii}(t,s)$ is continuous at $t = s = t'$ for $1 \le i \le d$.
- For $i \ne j$, $r_{ij}(t,s) = r_{ji}(s,t)$. If the process is weakly stationary, $r_{ij}(\tau) = r_{ji}(-\tau)$.
- $|r_{ij}(s,t)|^2 \le r_{ii}(t,t)r_{jj}(s,s)$. If the process is weakly stationary, $|r_{ij}(\tau)|^2 \le r_{ii}(0)r_{jj}(0)$.

Appendix B
Orthogonal Polynomials

By a system of orthogonal polynomials, we denote a set $\{P_n(x)\}_{n=1}^{\infty}$, where $P_n(x)$ is a polynomial of degree n. The polynomials are deemed orthogonal in the sense that the inner product

$$\langle P_n P_m \rangle$$

vanishes whenever $n \neq m$.

The variable x may be either continuous or discrete. In the former case, the inner product is defined as:

$$\langle P_n P_m \rangle = \int_a^b P_n(x) P_m(x) w(x) \, dx \tag{B.1}$$

whereas in the latter it is given by:

$$\langle P_n P_m \rangle = \sum_{i=1}^{M} P_n(x_i) P_m(x_i) w(x_i). \tag{B.2}$$

In both cases, $w(x)$ is a positive weight function. Note that the limits of integration a and b in (B.1), and the limit M in (B.2) may be either finite or infinite.

We are specifically interested in orthogonal families of polynomials that form a basis of the function space $L_w^2 = \{f : \langle ff \rangle < \infty\}$. Within this general setting, we further focus our attention on families for which expansions of the form:

$$f(x) = \sum_{n=1}^{N} a_n P_n(x)$$

result, for suitable functions f, in spectra that decay exponentially as $N \to \infty$.

In this appendix, we provide a brief outline of families of orthogonal polynomials that are frequently used in PC expansions.

O.P. Le Maître, O.M. Knio, *Spectral Methods for Uncertainty Quantification,*
Scientific Computation,
DOI 10.1007/978-90-481-3520-2, © Springer Science+Business Media B.V. 2010

B.1 Classical Families of Continuous Orthogonal Polynomials

In this section, we shall focus on the case of "continuous" polynomials defined for on the interval $a \le x \le b$. We shall denote by h_n the square of the L^2 norm of P_n, i.e.

$$h_n = \langle P_n P_n \rangle = \alpha \int_a^b P_n^2(x) w(x)\, \mathrm{d}x. \tag{B.3}$$

Note that for applications to PC expansions, it is convenient to adopt the convention that the space L^2_w has measure one, or alternatively that the norm of the indicator function of the entire space is equal to one, i.e.

$$\int_a^b w(x)\, \mathrm{d}x = 1. \tag{B.4}$$

This amounts to a constant scaling of weights functions that are normally utilized in the definition of orthogonal polynomials.

We also recall [1] that orthogonal polynomials: (a) satisfy the differential equation:

$$g_2(x) P_n'' + g_1(x) P' + a_n P_n = 0 \tag{B.5}$$

where g_1 and g_2 are independent of n and the a_n's are constants that depend on n only; and (b) can be generated using Rodrigues' formula:

$$P_n(x) = \frac{1}{e_n w(x)} \frac{\mathrm{d}^n}{\mathrm{d}x^n} \left\{ w(x) \left[g(x) \right]^n \right\} \tag{B.6}$$

where g is a polynomial in x that is independent of n and the e_n's are arbitrary normalization factors that depend on n only. Note that Rodrigues' formula immediately shows that a constant multiplicative scaling of w has no impact on the corresponding representation or normalization.

B.1.1 Legendre Polynomials

The Legendre polynomials, $\{Le_n(x), n = 0, 1, \dots\}$, are an orthogonal basis of $L^2_w[-1, 1]$ with respect to the weight function $w(x) = 1/2$ for all $x \in [-1, 1]$. The are typically normalized so that $Le_n(1) = 1$, in which case they are given by [26]:

$$Le_n(x) = \frac{1}{2^n} \sum_{l=0}^{[n/2]} (-1)^l \binom{n}{l} \binom{2n-2l}{n} x^{n-2l}. \tag{B.7}$$

Here $[n/2]$ denotes the integral part of $n/2$. The polynomials are even when n is even and odd when n is odd.

The Legendre polynomials satisfy the recurrence relation:

$$Le_{n+1}(x) = \frac{2n+1}{n+1} x Le_n(x) - \frac{n}{n+1} Le_{n-1}(x) \tag{B.8}$$

with $Le_0(x) = 1$ and $Le_1(x) = x$. The first seven Legendre polynomials are given by [1, 26]:

$$\begin{aligned}
Le_0(x) &= 1, \\
Le_1(x) &= x, \\
Le_2(x) &= \frac{1}{2}\left(3x^2 - 1\right), \\
Le_3(x) &= \frac{1}{2}\left(5x^3 - 3x\right), \\
Le_4(x) &= \frac{1}{8}\left(35x^4 - 30x^2 + 3\right), \\
Le_5(x) &= \frac{1}{8}\left(63x^5 - 70x^3 + 15x\right), \\
Le_6(x) &= \frac{1}{16}\left(2313x^6 - 315x^4 + 105x^2 5\right).
\end{aligned}$$

With the definition of w and the normalization convention specified above, the square norms of the Legendre polynomials are given by [1]:

$$h_n = \langle Le_n, Le_n \rangle = \int_1^1 Le_n^2(x) w(x)\, \mathrm{d}x = \frac{1}{2n+1}. \tag{B.9}$$

They satisfy (B.5) with [1]:

$$g_2(x) = 1 - x^2, \quad g_1(x) = -2x, \quad \text{and} \quad a_n = n(n+1), \tag{B.10}$$

and (B.6) with:

$$g(x) = 1 - x^2, \quad \text{and} \quad e_n = (-1)^n 2^n n!. \tag{B.11}$$

B.1.2 Hermite Polynomials

There are two widespread definitions of the Hermite polynomials, according to whether the weight functions are given by:

$$w(x) = \frac{1}{\sqrt{2\pi}} \exp\left(-\frac{x^2}{2}\right) \tag{B.12}$$

or

$$w(x) = \frac{1}{\sqrt{\pi}} \exp\left(-x^2\right). \tag{B.13}$$

In both cases, x is defined over the real line, i.e. $a = -\infty$ and $b = \infty$.

To distinguish between the two families, we shall use the conventional notation $He_n(x)$ to indicate that (B.12) is used, and $H_n(x)$ to indicate that (B.13) is adopted. Note that in PC expansions, we rely exclusively on He_n, but for completeness and computational convenience (see Appendix C) we address both cases.

For both definitions of w, the Hermite polynomials are conventionally normalized so that the factors appearing in Rodrigues' formula (B.6) are given by [1]:

$$e_n = (-1)^n. \tag{B.14}$$

The functions $g(x)$ appearing in (B.6) are also identical, $g(x) = 1$. It follows that the Hermite families are defined according to [111]:

$$He_n(x) = \frac{1}{(-1)^n \exp\left(-x^2/2\right)} \frac{\mathrm{d}^n}{\mathrm{d}x^n} \left[\exp\left(-x^2/2\right)\right], \tag{B.15}$$

$$He(x) = \frac{1}{(-1)^n \exp\left(-x^2\right)} \frac{\mathrm{d}^n}{\mathrm{d}x^n} \left[\exp\left(-x^2\right)\right]. \tag{B.16}$$

They have the following explicit representations:

$$He_n(x) = n! \sum_{m=0}^{[n/2]} (-1)^m \frac{1}{m! 2^m (n-2m)!} x^{n-2m}, \tag{B.17}$$

$$H_n(x) = n! \sum_{m=0}^{[n/2]} (-1)^m \frac{1}{m!(n-2m)!} (2x)^{n-2m}. \tag{B.18}$$

The polynomials $He_n(x)$ obey (B.5) with:

$$g_2(x) = 1, \quad g_1(x) = -x, \quad \text{and} \quad a_n = n, \tag{B.19}$$

and the recursion relation:

$$He_{n+1}(x) = x He_n(x) - n He_{n-1}(x). \tag{B.20}$$

With the standardization above, the square norms of the Hermite polynomials He_n are given by:

$$h_n = \langle He_n, He_n \rangle = \frac{1}{\sqrt{2\pi}} \int_{-\infty}^{\infty} [He_n(x)]^2 \exp\left(-x^2/2\right) \mathrm{d}x = n!. \tag{B.21}$$

The first seven polynomials $He_n(x)$ are given by:

$$He_0(x) = 1, \tag{B.22}$$

$$He_1(x) = x, \tag{B.23}$$

$$He_2(x) = x^2 - 1, \tag{B.24}$$

$$He_3(x) = x^3 - 3x, \tag{B.25}$$

$$He_4(x) = x^4 - 6x^2 + 3, \tag{B.26}$$

$$He_5(x) = x^5 - 10x^3 + 15x, \tag{B.27}$$

$$He_6(x) = x^6 - 15x^4 + 45x^2 - 15. \tag{B.28}$$

The polynomials $H_n(x)$ obey (B.5) with:

$$g_2(x) = 1, \quad g_1(x) = -2x, \quad \text{and} \quad a_n = 2n, \tag{B.29}$$

and the recursion relation:

$$H_{n+1}(x) = 2xH_n(x) - 2nH_{n-1}(x). \tag{B.30}$$

With the standardization above, the square norms of the Hermite polynomials H_n are given by:

$$h_n = \langle H_n, H_n \rangle = \frac{1}{\sqrt{\pi}} \int_{-\infty}^{\infty} [H_n(x)]^2 \exp\left(-x^2\right) \mathrm{d}x = 2^n n!. \tag{B.31}$$

The first seven polynomials $He_n(x)$ are given by:

$$H_0(x) = 1, \tag{B.32}$$

$$H_1(x) = 2x, \tag{B.33}$$

$$H_2(x) = 4x^2 - 2, \tag{B.34}$$

$$H_3(x) = 8x^3 - 12x, \tag{B.35}$$

$$H_4(x) = 16x^4 - 48x^2 + 12, \tag{B.36}$$

$$H_5(x) = 32x^5 - 160x^3 + 120x, \tag{B.37}$$

$$H_6(x) = 64x^6 - 480x^4 + 720x^2 - 120. \tag{B.38}$$

B.1.3 Laguerre Polynomials

The Laguerre polynomials, $\{L_n(x), n = 0, 1, \ldots\}$, are an orthogonal basis of $L_w^2[0, \infty]$ with respect to the weight function $w(x) = \exp(-x)$. They are conventionally normalized so that the factors appearing in Rodrigues' formula (B.6) are given by [1]:

$$e_n = n!. \tag{B.39}$$

The function g appearing in (B.6) is $g(x) = x$. Accordingly, the polynomials La are defined by:

$$La(x) = \frac{1}{n!\exp(-x)} \frac{\mathrm{d}^n}{\mathrm{d}x^n} \left(\exp(-x)x^n\right), \tag{B.40}$$

and admit the following explicit representation:

$$La_(x) = \sum_{m=o}^{n} (-1)^m \frac{n!}{(n-m)!(m!)^2} x^m. \tag{B.41}$$

The polynomials $La_n(x)$ obey (B.5) with:

$$g_2(x) = x, \quad g_1(x) = 1 - x, \quad \text{and} \quad a_n = n, \tag{B.42}$$

and the recursion relation:

$$La_{n+1}(x) = \frac{1}{n+1} \left[(2n+1-x)La_n(x) - nLa_{n-1}(x)\right]. \tag{B.43}$$

With the standardization above, the Laguerre polynomials La_n form an orthonormal basis of $L^2_w[0, \infty)$, i.e.

$$\langle La_n, La_m \rangle = \int_0^\infty La_n(x) La_m(x) \exp(-x)\, \mathrm{d}x = \delta_{nm}. \tag{B.44}$$

In particular, the square norm $h_n = 1$. The first six Laguerre polynomials are:

$$La_0(x) = 1, \tag{B.45}$$

$$La_1(x) = -x + 1, \tag{B.46}$$

$$La_2(x) = \frac{1}{2}(x^2 - 4x + 2), \tag{B.47}$$

$$La_3(x) = \frac{1}{6}(-x^3 + 9x^2 - 18x + 6), \tag{B.48}$$

$$La_4(x) = \frac{1}{24}(x^4 - 16x^3 + 72x^2 - 96x + 24), \tag{B.49}$$

$$La_5(x) = \frac{1}{120}(-x^5 + 25x^4 - 200x^3 + 600x^2 - 600x + 120). \tag{B.50}$$

B.2 Gauss Quadrature

In the application of PC methods, one needs in particular to evaluate moments of PC expansions. As further discussed in Appendix C, this task can be generally boiled down to evaluating inner products of one-dimensional polynomials. This section discusses evaluation of these inner products by means of Gauss quadratures which, due to their high order accuracy, provide efficient means of computing the corresponding integrals.

We focus our attention to the Legendre, Hermite and Laguerre families discussed in the previous section. Accordingly, we are interested in evaluating integrals of polynomial functions defined over the intervals $[-1, 1]$, $(-\infty, \infty)$, and $[0, \infty)$.

B.2.1 Gauss-Legendre Quadrature

The Gauss-Legendre quadrature is based on the following formula:

$$\int_{-1}^{1} f(x)\,\mathrm{d}x = \sum_{i=1}^{nq} w_i f(x_i) + R_{nq}, \tag{B.51}$$

where

$$R_{nq} = \frac{2^{2nq+1}(nq!)^4}{(2nq+1)[(2nq)!]^3} f^{(2nq)}(\xi), \quad -1 < \xi < 1. \tag{B.52}$$

Here, nq is the number of collocation points, x_i denote the coordinate of the i-th collocation point, and w_i the corresponding weight. R_n denotes the remainder of the quadrature; it indicates that the formula is exact if the order of the polynomials No $< 2nq$.

The coordinates x_i are the zeros of $Le_{nq}(x)$, and the weights are given by:

$$w_i = \frac{2}{(1-x_i^2)[Le'_{nq}(x_i)]^2}. \tag{B.53}$$

Explicit formulas for the Gauss-Legendre quadrature nodes are not known, and so they must be evaluated numerically.

The GAUSSQ routine available from netlib (http://netlib.org) can be used to determine the weights and abscissas appearing in (B.51). Table B.1 provides results using this routine for selected values of nq.

B.2.2 Gauss-Hermite Quadratures

Gauss-Hermite quadratures can be based on the following formula (see [1] p. 890):

$$\int_{-\infty}^{\infty} f(x)\exp(-x^2)\,\mathrm{d}x = \sum_{i=1}^{nq} w_i f(x_i) + R_{nq} \tag{B.54}$$

where nq is the number of collocation points, x_i denote the quadrature nodes and w_i are the corresponding weights. x_i is the i-th zero of the polynomial H_{nq}, and the

Table B.1 Nodes and weights for Gauss-Legendre quadrature for different values of nq

$nq = 5$		$nq = 10$		$nq = 15$	
x_i	w_i	x_i	w_i	x_i	w_i
−0.906180	0.236927E+00	−0.973907	0.666713E−01	−0.987993	0.307532E−01
−0.538469	0.478629E+00	−0.865063	0.149451E+00	−0.937273	0.703660E−01
0.000000	0.568889E+00	−0.679410	0.219086E+00	−0.848207	0.107159E+00
0.538469	0.478629E+00	−0.433395	0.269267E+00	−0.724418	0.139571E+00
0.906180	0.236927E+00	−0.148874	0.295524E+00	−0.570972	0.166269E+00
–	–	0.148874	0.295524E+00	−0.394151	0.186161E+00
–	–	0.433395	0.269267E+00	−0.201194	0.198431E+00
–	–	0.679410	0.219086E+00	0.000000	0.202578E+00
–	–	0.865063	0.149451E+00	0.201194	0.198431E+00
–	–	0.973907	0.666713E−01	0.394151	0.186161E+00
–	–	–	–	0.570972	0.166269E+00
–	–	–	–	0.724418	0.139571E+00
–	–	–	–	0.848207	0.107159E+00
–	–	–	–	0.937273	0.703660E−01
–	–	–	–	0.987993	0.307532E−01

weight w_i is given by:

$$w_i = \frac{2^{nq-1} nq! \sqrt{\pi}}{nq^2 [H_{nq-1}(x_i)]^2}. \tag{B.55}$$

The remainder in the quadrature (B.54) is given by:

$$R_n q = \frac{nq! \sqrt{\pi}}{2^n q (2nq)!} f^{(2nq)}(\zeta) \tag{B.56}$$

for some finite ζ. Thus, the formula (B.54) is exact if the polynomial $f(x)$ has degree smaller than $2nq$.

The GAUSSQ routine available from netlib (http://netlib.org) can be used to determine the weights and nodes appearing in (B.54). Table B.2 provides results using this routine for selected values of nq.

Note, however, the application of PC methods typically rely on the Hermite family He_n, and so the quadrature in (B.54) does not coincide with to the measure in the corresponding Hilbert space, which can be expressed as:

$$\langle f \rangle = \frac{1}{\sqrt{2\pi}} \int_{-\infty}^{\infty} f(x) \exp(-x^2/2)\, dx. \tag{B.57}$$

Table B.2 Gauss nodes and weights for the quadrature in (B.54) for different values of nq

$nq=5$		$nq=10$		$nq=15$	
x_i	w_i	x_i	w_i	x_i	w_i
−2.020183	0.199532E−01	−3.436159	0.764043E−05	−4.499991	0.152248E−08
−0.958572	0.393619E+00	−2.532732	0.134365E−02	−3.669950	0.105912E−05
0.000000	0.945309E+00	−1.756684	0.338744E−01	−2.967167	0.100004E−03
0.958572	0.393619E+00	−1.036611	0.240139E+00	−2.325732	0.277807E−02
2.020183	0.199532E−01	−0.342901	0.610863E+00	−1.719993	0.307800E−01
–	–	0.342901	0.610863E+00	−1.136116	0.158489E+00
–	–	1.036611	0.240139E+00	−0.565070	0.412029E+00
–	–	1.756684	0.338744E−01	0.000000	0.564100E+00
–	–	2.532732	0.134365E−02	0.565070	0.412029E+00
–	–	3.436159	0.764043E−05	1.136116	0.158489E+00
–	–	–	–	1.719993	0.307800E−01
–	–	–	–	2.325732	0.277807E−02
–	–	–	–	2.967167	0.100004E−03
–	–	–	–	3.669950	0.105912E−05
–	–	–	–	4.499991	0.152248E−08

In order to make use of the results above, one can apply the following change of variables:

$$\langle f\rangle = \frac{1}{\sqrt{\pi}}\int_{-\infty}^{\infty} g(x)\exp(-x^2)\,\mathrm{d}x \tag{B.58}$$

where $g(x) = f(\sqrt{2}x)$. Consequently, application of the previous quadrature to the above integral immediately results in:

$$\langle f\rangle \simeq \frac{1}{\sqrt{\pi}}\sum_{i=1}^{nq} w_i g(x_i) \tag{B.59}$$

or alternatively,

$$\langle f\rangle \simeq \sum_{i=1}^{nq} w_i' f(x_i') \tag{B.60}$$

where $w_i' \equiv w_i/\sqrt{\pi}$ and $x_i' \equiv \sqrt{2}x_i$. Equation (B.60) can now be readily used to evaluate moments and products of Hermite polynomials, He_n.

Table B.3 provides values of x_i' and w_i' for selected values of nq.

Note that our earlier observation that the quadrature in (B.54) is exact for polynomials of degree less than $2nq$ still applies to the quadrature in (B.60). In particular,

Table B.3 Gauss nodes and weights for the quadrature in (B.60) for different values of nq

$nq = 5$		$nq = 10$		$nq = 15$	
x'_i	w'_i	x'_i	w'_i	x'_i	w'_i
−2.856970	0.112574E−01	−4.859463	0.431065E−05	−6.363948	0.858965E−09
−1.355626	0.222076E+00	−3.581823	0.758071E−03	−5.190094	0.597542E−06
0.000000	0.533333E+00	−2.484326	0.191116E−01	−4.196208	0.564215E−04
1.355626	0.222076E+00	−1.465989	0.135484E+00	−3.289082	0.156736E−02
2.856970	0.112574E−01	−0.484936	0.344642E+00	−2.432437	0.173658E−01
–	–	0.484936	0.344642E+00	−1.606710	0.894178E−01
–	–	1.465989	0.135484E+00	−0.799129	0.232462E+00
–	–	2.484326	0.191116E−01	0.000000	0.318260E+00
–	–	3.581823	0.758071E−03	0.799129	0.232462E+00
–	–	4.859463	0.431065E−05	1.606710	0.894178E−01
–	–	–	–	2.432437	0.173658E−01
–	–	–	–	3.289082	0.156736E−02
–	–	–	–	4.196208	0.564215E−04
–	–	–	–	5.190094	0.597542E−06
–	–	–	–	6.363948	0.858965E−09

when sufficient quadrature points are provided, (B.60) reproduces the correct norm:

$$\left\langle He_m^2 \right\rangle = m!. \tag{B.61}$$

B.2.3 Gauss-Laguerre Quadrature

The Gauss-Laguerre quadrature is based on the following formula:

$$\int_0^{\infty} f(x)\exp(-x)\,\mathrm{d}x = \sum_{i=1}^{nq} w_i f(x_i) + R_{nq}, \tag{B.62}$$

where

$$R_{nq} = \frac{(nq!)^2}{(2nq)!} f^{(2nq)}(\xi), \quad 0 < \xi < \infty. \tag{B.63}$$

As before, nq is the number of collocation points, while x_i and w_i respectively denote the nodes and weights. From (B.63), it follows immediately that the quadrature (B.62) is exact when the order of the polynomial, f, is smaller than $2nq$.

Table B.4 Nodes and weights for Gauss-Laguerre quadrature for different values of nq

$nq = 5$		$nq = 10$		$nq = 15$	
x_i	w_i	x_i	w_i	x_i	w_i
0.263560	0.521756E+00	0.137793	0.308441E+00	0.093308	0.218235E+00
1.413403	0.398667E+00	0.729455	0.401120E+00	0.492692	0.342210E+00
3.596426	0.759424E−01	1.808343	0.218068E+00	1.215595	0.263028E+00
7.085810	0.361176E−02	3.401434	0.620875E−01	2.269950	0.126426E+00
12.640801	0.233700E−04	5.552496	0.950152E−02	3.667623	0.402069E−01
–	–	8.330153	0.753008E−03	5.425337	0.856388E−02
–	–	11.843786	0.282592E−04	7.565916	0.121244E−02
–	–	16.279258	0.424931E−06	10.120229	0.111674E−03
–	–	21.996586	0.183956E−08	13.130282	0.645993E−05
–	–	29.920697	0.991183E−12	16.654408	0.222632E−06
–	–	–	–	20.776479	0.422743E−08
–	–	–	–	25.623894	0.392190E−10
–	–	–	–	31.407519	0.145652E−12
–	–	–	–	38.530683	0.148303E−15
–	–	–	–	48.026086	0.160059E−19

The coordinates x_i are the zeros of $La_{nq}(x)$, and the weights are given by:

$$w_i = \frac{(nq!)^2 x_i}{(nq+1)^2 [La_{nq+1}(x_i)]^2}. \tag{B.64}$$

The GAUSSQ routine available from netlib (http://netlib.org) can be used to determine the weights and abscissas appearing in (B.62). Table B.4 provides results using this routine for selected values of nq.

B.3 Askey Scheme

As discussed by Xiu and Karniadakis [249], the above families of orthogonal polynomials are members of the so-called Askey scheme of polynomials [6]. The schemes classifies continuous and discrete hypergeometric orthogonal polynomials that obey certain difference and difference equations, and identifies limit relationship between them [117, 203]. The interesting aspect of the Askey family is that the weighting functions associated with some its members corresponds to frequently used probability distributions.

Specifically, in addition to the orthogonal polynomial systems discussed above, the Askey family includes the Jacobi polynomials, which in turn include the Legendre family, as well as the Charlier, Meixner, Krawtchouk and Hahn families of discrete orthogonal polynomials. The weighting functions associated with the Jacobi polynomials correspond to the beta distribution, whereas the weighting functions of the Charlier, Meixner, Krawtchouk and Hahn families respectively correspond to the Poisson, negative binormal, binormal, and hypergeometric distributions.

For brevity, and since these additional families are not used in the examples provided in the present volume, we restrict ourselves to providing definitions of corresponding polynomials, and outlining the orthogonality relationships that they obey. Implementation of these families into the corresponding Wiener chaos can be readily performed once the orthogonality relationships are exploited to define a suitable inner product, specifically in a similar fashion to the development above.

B.3.1 Jacobi Polynomials

The Jacobi polynomials $P_n^{(\alpha,\beta)}(x)$, $n = 0, 1, \ldots,$ are an orthogonal family of polynomials over $[-1, 1]$ with respect to the weight function

$$w(x) = (1-x)^{\alpha}(1+x)^{\beta}.$$

The are typically normalized so that

$$P_n^{(\alpha,\beta)}(1) = \binom{n+\alpha}{n}$$

in which case they are given by [1]:

$$P_n^{(\alpha,\beta)}(x) = \frac{1}{2^n}\sum_{m=0}^{n}\binom{n+\alpha}{m}\binom{n+\beta}{n-m}(x-1)^{n-m}(x+1)^m. \tag{B.65}$$

The Jacobi polynomials satisfy (B.5) with [1]:

$$g_2(x) = 1-x^2, \quad g_1(x) = \beta-\alpha-(\alpha+\beta+2)x, \quad \text{and} \quad a_n = n(n+\alpha+\beta+1), \tag{B.66}$$

the Rodrigues formula (B.6) with:

$$g(x) = 1-x^2, \quad \text{and} \quad e_n = (-1)^n 2^n n!, \tag{B.67}$$

the orthogonality condition:

$$\int_{-1}^{1} P_n^{(\alpha,\beta)}(x)P_m^{(\alpha,\beta)}(x)w(x)\,\mathrm{d}x = \frac{2^{\alpha}+\beta+1}{2n+\alpha+\beta+1}\,\frac{\Gamma(n+\alpha+1)\Gamma(n+\beta+1)}{n!\Gamma(n+\alpha+\beta+1)}\delta_{nm}, \tag{B.68}$$

and the recurrence relation:

$$2(n+1)(n+\alpha+\beta+1)(2n+\alpha+\beta)P_{n+1}^{(\alpha,\beta)}(x)$$

$$= \left[(2n+\alpha+\beta+1)(\alpha^2-\beta^2)+(2n+\alpha+\beta)_3 x\right] P_n^{(\alpha,\beta)}(x)$$
$$- 2(n+\alpha)(n+\beta)(2n+\alpha+beta+n) P_{n-1}^{(\alpha,\beta)}(x) \quad \text{(B.69)}$$

where $(m)_n = m(m+1)\cdots(m+n-1)$ denotes the Pochhammer symbol.

Note that the orthogonality condition (B.68) may be readily used to rescale $w(x)$ so that the resulting space $L_w^2[-1,1]$ has measure 1, and that the GAUSSQ routine also provides nodes and weights for Gauss-Jacobi quadratures.

B.3.2 Discrete Polynomials

The Charlier, Meixner, Krawtchouk, and Hahn polynomials are respectively denoted by $C_n(x,a)$, $M_n(x;\beta,c)$, $K_n(x;p,N)$, and $Q_n(x;\alpha,\beta,N)$. In all cases, x is discrete variable. For the Hahn and Krawtchouk polynomials, x is defined over the set $\{0,1,\ldots,N\}$, whereas for the Charlier and Meixner polynomials x is defined over the integers $0,1,\ldots$.

As discussed in [117], the Charlier, Meixner, Krawtchouk, and Hahn polynomials may be defined in terms of the hypergeometric series:

$${}_\tau F_s\left(\begin{matrix} a_1,\ldots,a_\tau \\ b1,\ldots,b_s \end{matrix}\middle| z\right) \equiv \sum_{k=0}^{\infty} \frac{(a_1)_k\ldots(a_\tau)_k}{(b_1)_k\ldots(b_s)_k}\frac{z^k}{k!}. \quad \text{(B.70)}$$

Relevant definitions are summarized Table B.5

The first two polynomials are:

$$C_0(x,a) = 1, \quad \text{(B.71)}$$
$$C_1(x,a) = 1 - \frac{x}{a}, \quad \text{(B.72)}$$
$$M_0(x;\beta,c) = 1, \quad \text{(B.73)}$$

Table B.5 Definition of Charlier, Meixner, Krawtchouk, and Hahn polynomials in terms of hypergeometric series

Polynomial	Definition
Charlier	${}_2F_0\left(\begin{matrix} -n,-x \\ -- \end{matrix}\middle| -\frac{1}{a}\right)$
Meixner	${}_2F_1\left(\begin{matrix} -n,-x \\ \beta \end{matrix}\middle| 1-\frac{1}{c}\right)$
Krawtchouk	${}_2F_1\left(\begin{matrix} -n,-x \\ -N \end{matrix}\middle| \frac{1}{p}\right)$
Hahn	${}_3F_2\left(\begin{matrix} -n,n+\alpha+\beta+1,-x \\ \alpha+1,-N \end{matrix}\middle| 1\right)$

Table B.6 Weights and square norms of the Charlier, Meixner, Krawtchouk, and Hahn polynomials

Polynomial	$w(x)$	h_n
Charlier	$\frac{a^x}{x!}$	$a^{-n}\exp(-a)n!$
Meixner	$\frac{(\beta)_x c^x}{x!}$	$\frac{c^{-n}n!}{(\beta)_n(1-c)^\beta}$
Krawtchouk	$\binom{N}{x} p^x(1-p)^{N-x}$	$\frac{(1)^{-n}n!}{(-N)_n}\left(\frac{1-p}{p}\right)^n$
Hahn	$\binom{\alpha+x}{x}\binom{\beta+N-x}{N-x}$	$\frac{(-1)^n(n+\alpha+\beta+1)_{N+1}(\beta+1)_n n!}{(2n+\alpha+\beta+1)(\alpha+1)_n(-N)_n N!}$

$$M_1(x;\beta,c) = 1 + \frac{x}{\beta}\left(1-\frac{1}{c}\right), \tag{B.74}$$

$$K_0(x;p,N) = 1, \tag{B.75}$$

$$K_1(x;p,N) = 1 - \frac{x}{Np}, \tag{B.76}$$

$$Q_0(x;\alpha,\beta,N) = 1, \tag{B.77}$$

$$Q_1(x;\alpha,\beta,N) = 1 - \frac{(\alpha+\beta+2)x}{(\alpha+1)N}. \tag{B.78}$$

With the definitions above, these families of discrete polynomials satisfy the discrete orthogonality relation:

$$\sum_i f_n(x_i) f_m(x_i) w(x_i) = h_n \delta_{nm} \tag{B.79}$$

where the index i ranges over the entire domain of the discrete variable x. Relevant formulas for the weights, w, and square norms, h_n, are given in Table B.6.

The discrete polynomials satisfy the Rodrigues-type formula:

$$w(x) f_n(x) = r_n \nabla^n \left[g(x)\right] \tag{B.80}$$

where $w(x)$ is the weight function, r_n is a normalizing function independent of x, g is a polynomial in x, and ∇ denotes the backward difference operator $\nabla f(x) = f(x) - f(x-1)$. Relevant formulas for the factors r and g are given in Table B.7.

For computational purposes, polynomial evaluations may be conveniently performed on the basis of recursion relations. Relevant formulas for the Charlier, Meixner, Krawtchouk and Hahn polynomials are given by:

$$aC_{n+1} = (n+a-x)C_n - nC_n{-}1, \tag{B.81}$$

$$c(n+\beta)M_{n+1} = \left[n + (n+\beta)_c - (c-1)x\right] M_n - nM_{n-1}, \tag{B.82}$$

Table B.7 Factors r and g in (B.80) for the Charlier, Meixner, Krawtchouk, and Hahn polynomials

Polynomial	r_n	$g(x)$
Charlier	1	$\frac{a^x}{x!}$
Meixner	1	$\frac{(\beta+n)_x c^x}{x!}$
Krawtchouk	$(1-p)^{-N}$	$\binom{N-n}{x}\left(\frac{p}{1-p}\right)^x$
Hahn	$\frac{(-1)^n(\beta+1)_n}{(-N)_n}$	$\binom{\alpha+n+x}{x}\binom{\beta+N-x}{N-n-x}$

$$p(N-n)K_{n+1} = \left[p(N-n)+n(1-p)-x\right]K_n - n(1-p)K_{n-1}, \quad \text{(B.83)}$$

$$A_n Q_{n+1}(x) = (A_n + D_n - x)Q_n(x) - D_n Q_{n-1}(x) \quad \text{(B.84)}$$

where $C_n \equiv C_n(x;a)$, $M_n \equiv M_n(x;\beta,c)$, $K_n \equiv K_n(x;p,N)$, $Q_n(x) \equiv Q_n(x : \alpha,\beta,N)$, and

$$A_n = \frac{(n+\alpha+\beta+1)(n+\alpha+1)(N-n)}{(2n+\alpha+\beta+1)(2n+\alpha+\beta+2)},$$

$$D_n = \frac{n(n+\alpha+\beta+N+1)(n+\beta)}{(2n+\alpha+\beta)(2n+\alpha+\beta+1)}.$$

We finally note that, with the standardizations above, the weighting functions of Charlier, Meixner, Krawtchouk, and Hahn polynomials correspond, up to a scaling factor independent of x, to the probability density functions of the Poisson, negative binormal, binormal, and hypergeometric distributions. The scaling factors can be readily obtained from the orthogonality expressions, and can be used to rescale the standard forms of w to define a suitable inner product, namely one for which the measure of the corresponding ℓ_2 space is unity.

Appendix C
Implementation of Product and Moment Formulas

In the development of computer codes implementing PC expansions, one is faced with the problem of evaluating moments of classical orthogonal polynomials, or more often of multidimensional generalizations of these polynomials. This Appendix outlines the main ingredient of a general approach that has proven to be quite useful in the construction of utility libraries that are capable of performing fundamental operations and nonlinear transformations of PC representations of random variables.

C.1 One-Dimensional Polynomials

For most situations, it is quite advantageous for users to compute and store polynomials appearing in PC expansions. As mentioned in Chapter 2, these expansions involve orthogonal polynomials that are typically defined in terms of series expansions. For computational purposes, however, it is more convenient to evaluate the polynomials recursion relations. As discussed in Appendix B, in the Hermite case, one has

$$\begin{aligned} \psi_0(x) &= 1, \\ \psi_1(x) &= x, \\ \psi_n(x) &= x\psi_{n-1}(x) - (n-1)\psi_{n-2}(x), \quad n = 2, 3, \ldots \end{aligned} \tag{C.1}$$

and similar recursion relations exist for orthogonal polynomials that are member of the Askey family. The exploitation of these relationships is recommended whenever possible.

O.P. Le Maître, O.M. Knio, *Spectral Methods for Uncertainty Quantification*, Scientific Computation,
DOI 10.1007/978-90-481-3520-2, © Springer Science+Business Media B.V. 2010

C.1.1 Moments of One-Dimensional Polynomials

We are interested in evaluating second-order moments of the form

$$\langle f \rangle \quad \text{where } f(x) \equiv \psi_i(x)\psi_j(x), \tag{C.2}$$

and more generally in higher-order moments which may be expressed as:

$$\langle f \rangle \quad \text{where } f(x) \equiv \prod_{m=1}^{M} \psi_{i_m}(x). \tag{C.3}$$

Such moments may be evaluated on the basis of analytical expressions. Alternatively as discussed in Appendix B, Gauss quadratures may be readily applied, and these yield exact values so long as a sufficient number of collocation points is provided, namely to ensure a vanishing remainder.

C.2 Multidimensional PC Basis

As discussed in Chap. 2, in the multidimensional case the PC basis functions are defined from an incomplete tensorization of the corresponding 1D basis functions. As a result, each member of the multidimensional PC basis can written as a product of 1D polynomials in the appropriate random variable. In order to specify (and hence immediately evaluate) the elements of the multidimensional basis, it is advantageous to associate with each basis function, Ψ_i a multi-index $\boldsymbol{\alpha}^i = (\alpha_1^i, \alpha_2^i, \ldots, \alpha_{\mathrm{N}}^i)$, where α_k^i denotes the order of the 1D polynomial in ξ_k. Once these multi-indices are available, evaluation of the multidimensional basis function is straightforward. We illustrate this for the case of the N-dimensional Hermite polynomials. We denote the latter by Ψ_k, in order to distinguish them from, ψ_j, their 1D counterparts.

By definition of the multi-indices, the N-dimensional polynomial chaoses can be obtained from:

$$\Psi_k(\xi_1, \xi_2, \ldots, \xi_{\mathrm{N}}) = \prod_{i=1}^{\mathrm{N}} \psi_{\alpha_i^k}(\xi_i). \tag{C.4}$$

Thus, knowledge of the multi-indices, together with the 1D basis functions, provides a very efficient and convenient means of evaluating the N-dimensional PCs.

C.2.1 Multi-Index Construction

To complete this construction, one needs to define the multi-indices $\boldsymbol{\alpha}^i$, $i = 0, \ldots, \mathrm{P}$. While multiple definitions are possible, we have adopted the same indexing scheme used in [90]. Using this indexing scheme, the multi-indices are defined recursively for arbitrary order, No, and dimension, N, as described in the following pseudo-code:

1. For the 0-th order polynomial, set:

$$\alpha_i^0 = 0, \quad i = 1, \ldots, \mathrm{N}.$$

Note that if No = 0, then we set P = 0 and the process is interrupted.
2. For the N first-order polynomials, set:

$$\alpha_j^i = \delta_{ij} \quad \text{for } 1 \le i, j \le \mathrm{N}.$$

Note that if No = 1, then we set P = N and the process is interrupted.
3. Set P = N
4. Set $p_i(1) = 1$, $i = 1, \ldots, \mathrm{N}$
5. For $k = 2, \ldots, \mathrm{No}$:

- Set $L = \mathrm{P}$
- Set $p_i(k) = \sum_{m=i}^{\mathrm{N}} p_m(k-1)$, $i = 1, \ldots, \mathrm{N}$
- For $j = 1, \ldots, \mathrm{N}$:
  + For $m = L - p_j(k) + 1, \ldots, L$:
    · Set P = P + 1
    · Set $\boldsymbol{\alpha}_i^P = \boldsymbol{\alpha}_i^m$, $i = 1, \ldots, \mathrm{N}$
    · Set $\boldsymbol{\alpha}_j^P = \boldsymbol{\alpha}_j^P + 1$

In this algorithm, for a specific k, L denotes the number of terms that have been constructed so far with order less than k. The terms of order k are then constructed from the terms of order $(k-1)$ by increasing the order of the factors in those terms belonging to a specific dimension, one at the time. For example, for the three-dimensional stochastic case, there are 6 terms of order 2. The first 6 terms of order 3 are constructed by taking all 6 terms of order 2, and increasing the multi-index corresponding to the order of the first stochastic dimension by 1. The next 3 terms of order 3 are constructed by taking the last 3 terms of order 2, and increasing the order of the contribution of the second stochastic dimension. The last term is obtained by increasing the multi-index corresponding to the third dimension with 1 in the last term of order 2. Investigation of the process shows that the terms at the current order, in which the order of the contribution of dimension j will be increased to get terms at the next order, are the terms that were generated at the current order from the previous order by increasing the order of dimension j. This bookkeeping is done by the $p_i(k)$.

Note that the procedure above also provides the number, P, of N-dimensional Polynomial Chaoses of order ≤ No.

C.2.2 Moments of Multidimensional Polynomials

We are now interested in generalizing the results of the previous section to the N-dimensional case. To this end, we exploit the tensor product construction of the multidimensional inner product, and again rely on the multi-indices $\boldsymbol{\alpha}^i$ to express the moments of N-dimensional Polynomial Chaoses in terms of products of 1-dimensional moments. For instance, the second-order moment $\langle \Psi_i \Psi_j \rangle$ can be expressed as:

$$\langle \Psi_i \Psi_j \rangle_{\mathrm{N}} = \prod_{k=1}^{\mathrm{N}} \left\langle \psi_{\alpha_k^i} \psi_{\alpha_k^j} \right\rangle_1 \tag{C.5}$$

where we have used subscripts to distinguish between the inner products in 1D and N-dimensional space. Similarly, for a moment of order M we have

$$\left\langle \prod_{m=1}^{M} \Psi_{i_m} \right\rangle_{\mathrm{N}} = \prod_{k=1}^{\mathrm{N}} \left\langle \prod_{m=1}^{M} \psi_{\alpha_k^{i_m}} \right\rangle_1 . \tag{C.6}$$

The advantage of this product factorization is that in practice only a few one-dimensional moments need to be actually computed, and this results in order-of-magnitude reduction in the effort required to form the required N-dimensional moments. It also greatly facilitates the computation of moments involving mixed bases, namely those involving tensor products of different families of orthogonal polynomials.

C.2.3 Implementation Details

We conclude with a brief outline of an attractive approach for the implementation of PC computations, especially in the multidimensional case. This approach is based on the establishment of a code infrastructure in which the multi-indices defined in Sect. C.2.1 are computed in a pre-processing step, together with the necessary multidimensional moments, namely those appearing in the definition of multiplication tensors used to evaluate products and other nonlinear transformations (see Sect. 4.5). As can be anticipated, it is generally advantageous to compute and store these tensors in a pre-processing step. One should particularly seek to take advantage of the sparse nature of these tensors, namely by storing (and using) only the non-vanishing entries of these tensors. Once properly defined in this compact format, these "transformation" tensors can be used as a basis for the construction of utility libraries that implement various transformations of PC representations.

The approach outlined above in fact forms the basis of the UQ toolkit,[1] which constitutes one incarnation of a PC-utility-library software infrastructure.

[1] B.J. Debusschere *et al.*, http://www.sandia.gov/UQToolkit/.

References

1. Abramowitz, M., Stegun, I.: Handbook of Mathematical Functions. Dover, New York (1970)
2. Agarwal, R., Wong, P.: Error Inequalities in Polynomial Interpolation and Their Applications vol. 262. Kluwer, Dordrecht (1993)
3. Alpert, B.: A class of bases in L_2 for the sparse representation of integral operators. J. Math. Anal. **24**, 246–262 (1993)
4. Alpert, B., Beylkin, G., Gines, D., Vozovoi, L.: Adaptive solution of partial differential equations in multiwavelet bases. J. Comput. Phys. **182**, 149–190 (2002)
5. Andreev, V., Pliss, N., Righetti, P.: Computer simulation of affinity capillary electrophoresis. Electrophoresis **23**(6), 889–895 (2002)
6. Askey, R., Wilson, J.: Some basic hypergeometric polynomials that generalize Jacobi polynomials. Mem. Am. Math. Soc. **54**, 1–55 (1985)
7. Babuška, I., Nobile, F., Tempone, R.: A stochastic collocation method for elliptic partial differential equations with random input data. SIAM J. Numer. Anal. **45**(3), 1005–1034 (2007)
8. Babuška, I., Tempone, R., Zouraris, G.E.: Solving elliptic boundary value problems with uncertain coefficients by the finite element method: the stochastic formulation. Comput. Methods Appl. Mech. Eng. **194**, 1251–1294 (2005)
9. Babuska, I., Chatzipantelidis, P.: On solving elliptic stochastic partial differential equations. Comput. Methods Appl. Mech. Eng. **191**, 4093–4122 (2002)
10. Babuska, I., Liu, K.M.: On solving stochastic initial-value differential equations. Technical report, TICAM Report 02–17, The University of Texas at Austin (2002)
11. Babuška, I., Miller, A.: A feedback finite element method with a posteriori error estimation. Comput. Methods Appl. Mech. Eng. **61**, 1–40 (1987)
12. Babuška, I., Rheinboldt, W.: A posteriori error estimates for the finite element method. Int. J. Numer. Methods Eng. **12**, 1597–1615 (1978)
13. Bachman, G., Narici, L.: Functional Analysis. Dover, New York (2000)
14. Batchelor, G.: An Introduction to Fluid Dynamics. Cambridge University Press, Cambridge (1985)
15. Beale, J.: On the accuracy of vortex methods at large times. In: Computational Fluid Dynamics and Reacting Gas Flows. Springer, Berlin (1988)
16. Beaudouin, A., Huberson, S., Rivoalen, E.: Simulation of anisotropic diffusion by means of a diffusion velocity method. J. Comput. Phys. **186**, 122–135 (2003)
17. Becker, R., Rannacher, R.: An optimal control approach to a posteriori error estimation in finite element methods. Acta Numer. **10**, 1–102 (2001)
18. Benth, F.E., Gjerde, J.: Convergence rates for finite element approximations of stochastic partial differential equations. Stoch. Stoch. Rep. **63**(3–4), 313–326 (1998)
19. Beran, P., Pettit, C., Millman, D.: Uncertainty quantification of limit-cycle oscillations. J. Comput. Phys. **217**, 217–247 (2006)

O.P. Le Maître, O.M. Knio, *Spectral Methods for Uncertainty Quantification,* Scientific Computation,
DOI 10.1007/978-90-481-3520-2,

20. Berveiller, M., Sudret, B., Lemaire, M.: Stochastic finite element: a non intrusive approach by regression. Eur. J. Comput. Mech. **15**, 81–92 (2006)
21. Bielewicz, E., Górski, J.: Shells with random geometric imperfections simulation-based approach. Int. J. Non-Linear Mech. **37**, 777–784 (2002)
22. Bier, M., Palusinski, O., Mosher, R., Saville, D.: Electrophoresis: Mathematical modeling and computer simulation. Science **219**(4590), 1281–1287 (1983)
23. Blatman, G., Sudret, B.: Sparse polynomial chaos expansions and adaptive stochastic finite elements using a regression approach. C. R. Méc. **336**, 518–523 (2008)
24. Boyd, J.: Chebishev and Fourier Spectral Methods. Dover, New York (2001)
25. Cameron, R., Martin, W.: The orthogonal development of nonlinear functionals in series of Fourier-Hermite functionals. Ann. Math. **48**, 385–392 (1947)
26. Canuto, C., Hussaini, M., Quarteroni, A., Zang, T.: Spectral Methods in Fluids Dynamics. Springer Series in Computational Physics. Springer, Berlin (1990)
27. Chenoweth, D., Paolucci, S.: Natural convection in an enclosed vertical layer with large horizontal temperature differences. J. Fluid Mech. **169**, 173–210 (1986)
28. Choquin, J., Lucquin-Desreux, B.: Accuracy of a deterministic particle method for Navier-Stokes equations. J. Numer. Methods Fluids **8**, 1439–1458 (1988)
29. Chorin, A.: A numerical method for solving incompressible viscous flow problems. J. Comput. Phys. **2**, 12–26 (1967)
30. Chorin, A.: Numerical study of slightly viscous flow. J. Fluid Mech. **57**, 785–796 (1973)
31. Chorin, A.: Gaussian fields and random flow. J. Fluid Mech. **63**, 21–32 (1974)
32. Christiansen, I.: Numerical simulation of hydrodynamics by the method of point vortex. J. Comput. Phys. **13**(3), 363–379 (1973)
33. Christon, M., Gresho, P., Sutton, S.: Computational predictability of natural convection flows in enclosure. In: Bathe, K. (ed.) Computational Fluid and Solid Mechanics. Proceedings of First MIT Conference on Computational Fluid and Solid Mechanics, pp. 1465–1468. Elsevier, Amsterdam (2001)
34. Christopher, D.: Numerical prediction of natural convection flows in a tall enclosure. In: Bathe, K. (ed.) Computational Fluid and Solid Mechanics, pp. 1469–1471. Elsevier, Amsterdam (2001)
35. Clément, P.: Approximation by finite element functions using local regularization. Rev. Fr. Autom. Inf. Rech. Opér., Sér. Rouge Anal. Numér. **9**(R-2), 77–84 (1975)
36. Clenshaw, C., Curtis, A.: A method for numerical integration on an automatic computer. Numer. Math. **2**, 197–205 (1960)
37. Comini, G., Manzan, M., Nonino, C., Saro, O.: Finite element solutions for natural convection in a tall rectangular cavity. In: Bathe, K. (ed.) Computational Fluid and Solid Mechanics, pp. 1472–1476. Elsevier, Amsterdam (2001)
38. Cools, R., Maerten, B.: A hybrid subdivision strategy for adaptive integration routines. J. Univ. Comput. Sci. **4**(5), 486–500 (1998)
39. Cormick, S.M.: Multigrids Methods. SIAM Books, Philadelphia (1987)
40. Cottet, G.: Large time behavior of deterministic particle approximations to the Navier-Stokes equations. Math. Comput. **56**, 45–59 (1991)
41. Cottet, G., Koumoutsakos, P.: Vortex Methods Theory and Practice. Cambridge University Press, Cambridge (2000)
42. Cottet, G., Mas-Gallic, S.: A particle method to solve the Navier-Stokes system. Numer. Math. **57**, 805–827 (1990)
43. Courant, D., Hilbert, C.: Method of Mathematical Physics. Interscience, New York (1943)
44. Crow, S., Canavan, G.: Relationship between a Wiener-Hermite expansion and an energy cascade. J. Fluid Mech. **41**, 387–403 (1970)
45. Das, S., Ghanem, R., Spall, J.: Asymptotic sampling distribution for Polynomial Chaos representation, from data: a maximum entropy and Fisher information approach. SIAM J. Sci. Comput. **30**(5), 2207–2234 (2007)
46. Daube, O.: An influence matrix technique for the resolution of 2d Navier-Stokes equation in velocity-vorticity formulation. In: Int. Conf. Num. Meth. in Laminar and Turbulent Flows (1986)

47. Daubechies, I.: Ten Lectures on Wavelets. SIAM Books, Philadelphia (1992)
48. De Vahl Davis, G., Jones, I.: Natural convection in a square cavity: a comparison exercise. Int. J. Numer. Methods Fluids **3**, 227–248 (1983)
49. De Valh Davis, G., Jones, I.: Natural convection in a square cavity: a comparison exercise. Int. J. Numer. Methods Fluids **3**, 227–248 (1983)
50. Deb, M., Babuska, I., Oden, J.: Solution of stochastic partial differential equations using Galerkin finite element techniques. Comput. Methods Appl. Mech. Eng. **190**, 6359–6372 (2001)
51. Debusschere, B., Najm, H., Matta, A., Knio, O., Ghanem, R., Le Maître, O.: Protein labeling reactions in electrochemical microchannel flow: numerical prediction and uncertainty propagation. Phys. Fluids **15**(8), 2238–2250 (2003)
52. Debusschere, B., Najm, H., Pébray, P., Knio, O., Ghanem, R., Le Maître, O.: Numerical challenges in the use of polynomial chaos representations for stochastic processes. J. Sci. Comput. **26**(2), 698–719 (2004)
53. Degond, P., Mas-Gallic, S.: The weighted particle method for convection diffusion equations, part 1: the case of an isotropic viscosity. Math. Comput. **33**(53), 485–507 (1989)
54. Degond, P., Mustieles, F.J.: A deterministic approximation of diffusion using particles. J. Sci. Stat. Comput. **1**(2), 293–310 (1990)
55. Desceliers, C., Ghanem, R., Soize, C.: Maximum likelihood estimation of stochastic chaos representations from experimental data. Int. J. Numer. Methods Eng. **66**(6), 978–1001 (2006)
56. Desceliers, C., Soize, C., Ghanem, R.: Identification of chaos representations of elastic properties of random media using experimental vibration tests. Comput. Mech. **39**(6), 831–838 (2007)
57. Ditlevsen, O., Tarp-Johansen, N.: Choice of input fields in stochastic finite-elements. Probab. Eng. Mech. **14**, 63–72 (1999)
58. Doostan, A., Ghanem, R., Red-Horse, J.: Stochastic model reduction for chaos representations. Comput. Methods Appl. Mech. Eng. **196**, 3951–3966 (2007)
59. Drazin, P., Reid, W.: Hydrodynamic Stability. Cambridge University Press, Cambridge (1982)
60. Edwards, W., Tuckerman, L., Friesner, R., Sorensen, D.: Krylov methods for the incompressible Navier-Stokes equations. J. Comput. Phys. **110**(1), 82–102 (1994)
61. Efron, B., Tibshirani, R.: An Introduction to the Bootstrap. Monographs on Statistics and Applied Probability, vol. 57. Chapman & Hall/CRC Press, London (1994)
62. Eldred, M.: Optimization strategies for complex engineering applications. Sandia Technical Report SAND98-0340, Sandia National Laboratories (1998)
63. Eldred, M., Giunta, A., Wojkiewicz, S., van Bloemen Waanders, B., Hart, W., Alleva, M.: Dakota, a multilevel parallel object-oriented framework for design optimization, parameter estimation, sensitivity analysis, and uncertainty quantification. version 3.0 reference manual. Sandia Technical Report SAND02-XXXX (in preparation), Sandia National Laboratories (2002). http://endo.sandia.gov/DAKOTA/papers/Dakota_hardcopy.pdf
64. Eldredge, J.: Numerical simulation of the fluid dynamics of 2D rigid body motion with the vortex particle method. J. Comput. Phys. **221**, 626–648 (2007)
65. Eldredge, J., Leonard, A.: A general deterministic treatment of derivatives in particles methods. J. Comput. Phys. **180**, 686–709 (2002)
66. Ermakov, S., Jacobson, S., Ramsey, J.: Computer simulations of electrokinetic transport in microfabricated channel structures. Anal. Chem. **70**, 4494–4504 (1998)
67. Ermakov, S., Jacobson, S., Ramsey, J.: Computer simulations of electrokinetic injection techniques in microfluidic devices. Anal. Chem. **72**, 3512–3517 (2000)
68. Ermakov, S., Righetti, P.: Computer simulation for capillary zone electrophoresis, a quantitative approach. J. Chromatogr. A **667**, 257–270 (1994)
69. Ermakov, S., Zhukov, M., Capelli, L., Righetti, P.: Artifactual peak splitting in capillary electrophoresis, 2: defocussing phenomena for ampholytes. Anal. Chem. **67**, 2957–2965 (1995)
70. Ern, A., Guermond, J.L.: Theory and Practice of Finite Elements vol. 159. Springer, Berlin (2004)

71. Ernst, O., Powell, C., Silvester, D., Ullmann, E.: Efficient solvers for a linear stochastic Galerkin mixed formulation of diffusion problems with random data. SIAM J. Sci. Comput. **31**(2), 1424–1447 (2009)
72. Nobile, F.R.T., Webster, C.: An anisotropic sparse grid stochastic collocation method for partial differential equations with random input data. SIAM J. Numer. Anal. **46**(5), 2411–2442 (2008)
73. Nobile, F.R.T., Webster, C.: A sparse grid stochastic collocation method for partial differential equations with random input data. SIAM J. Numer. Anal. **46**(5), 2309–2345 (2008)
74. Fletcher, C.: Computational Techniques for Fluid Dynamics, vol. 1. Springer, Berlin (1988)
75. Frauenfelder, P., Schwab, C., Todor, R.A.: Finite elements for elliptic problems with stochastic coefficients. Comput. Methods Appl. Mech. Eng. **194**(2–5), 205–228 (2005)
76. Furnival, D., Elman, H., Powell, C.: H(div) preconditioning for a mixte finite element formulation of the stochastic diffusion problem. Technical Report 2008-66, MIMS EPrint (2008)
77. Ganapathysubramanian, B., Zabaras, N.: Sparse grid collocation schemes for stochastic natural convection problems. J. Comput. Phys. **225**, 652–685 (2007)
78. Gardiner, C.: Handbook of Stochastic Methods: for Physics, Chemistry and the Natural Sciences. Springer series in Synergetics. Springer, Berlin (2004)
79. Ge, L., Cheung, K., Kobayashi, M.: Stochastic solution for uncertainty propagation in nonlinear shallow-water equations. J. Hydraul. Eng. **134**, 1732–1743 (2008)
80. Gebhart, B., Jaluria, Y., Mahajan, R., Sammakia, B.: Buoyancy-induced Flows and Transport. Hemisphere, Washington (1988)
81. Gentleman, W.: Implementing the Clenshaw-Curtis quadrature, I: methodology and experience. Commun. ACM **15**(5), 337–342 (1972)
82. Gentleman, W.: Implementing the Clenshaw-Curtis quadrature, II: computing the cosine transform. Commun. ACM **15**(5), 343–346 (1972)
83. Genz, A., Keister, B.: Fully symmetric interpolatory rules for multiple integrals over infinite regions with Gaussian weight. J. Comput. Appl. Math. **71**, 299–309 (1996)
84. Gerstner, T., Griebel, M.: Numerical integration using sparse grids. Numer. Algorithms **18**, 209–232 (1998)
85. Ghanem, R.: Probabilistic characterization of transport in heterogeneous media. Comput. Methods Appl. Mech. Eng. **158**, 199–220 (1998)
86. Ghanem, R.: The nonlinear Gaussian spectrum of lognormal stochastic processes and variables. ASME J. Appl. Mech. **66**, 964–973 (1999)
87. Ghanem, R., Dham, S.: Stochastic finite element analysis for multiphase flow in heterogeneous porous media. Transp. Porous Media **32**, 239–262 (1998)
88. Ghanem, R., Kruger, R.: Numerical solution of spectral stochastic finite element systems. Comput. Methods Appl. Mech. Eng. **129**, 289–303 (1996)
89. Ghanem, R., Spanos, P.: A spectral stochastic finite element formulation for reliability analysis. J. Eng. Mech. ASCE **117**, 2351–2372 (1991)
90. Ghanem, R., Spanos, P.: Stochastic Finite Elements: A Spectral Approach, 2nd edn. Dover, New York (2002)
91. Ghosh, D., Ghanem, R.: Stochastic convergence acceleration through basis enrichment of polynomial chaos expansions. Int. J. Numer. Methods Eng. **73**, 162–184 (2008)
92. Goluh, G., Van Loan, C.: Matrix Computations, 3rd edn. Johns Hopkins University Press, Baltimore (1996)
93. Gresho, P., Sutton, S.: 8:1 thermal cavity problem. In: Bathe, K. (ed.) Computational Fluid and Solid Mechanics, pp. 1482–1485. Elsevier, Amsterdam (2001)
94. Grigoriu, M.: Stochastic Calculus, Applications in Sciences and Engineering. Birkhäuser, Basel (2002)
95. Groce, G., Favero, M.: Simulation of natural convection flow in enclosures by an unstaggered grid finite volume algorithm. In: Bathe, K. (ed.) Computational Fluid and Solid Mechanics, pp. 1477–1481. Elsevier, Amsterdam (2001)
96. Grudmann, H., Waubke, H.: Non-linear stochastic dynamics of systems with random properties, a spectral approach combined with statistical linearization. Int. J. Non-Linear Mech. **31**(5), 619–630 (1997)

97. Hairer, E., Nørsett, S., Wanner, G.: Solving Ordinary Differential Equations I, Nonstiff Problems, 2nd edn. Springer, Berlin (1992)
98. Halton, J.: On the efficiency of certain quasi-random sequences of points in evaluating multidimensional integrals. Numer. Math. **2**(1), 84–90 (1960)
99. Hardin, R., Sloane, N.: A new approach to the construction of optimal designs. J. Stat. Plann. Inference **37**, 339–369 (1993)
100. Heuveline, V., Rannacher, R.: Duality-base adaptivity in the hp-finite element method. J. Numer. Math. **12**(2), 95–113 (2003)
101. Hien, T., Kleiber, M.: Stochastic finite element modeling in linear transient heat transfer. Comput. Methods Appl. Mech. Eng. **144**, 111–124 (1997)
102. Holden, H., Oksendal, B., Uboe, J., Zhang, T.: Stochastic Partial Differential Equations. Birkhäuser, Basel (1996)
103. Hortmann, M., Peric, M., Scheuerer, G.: Finite volume multigrid prediction of laminar natural convection: Benchmark solutions. Int. J. Numer. Methods Fluids **11**, 189–207 (1990)
104. Jacod, J., Protter, P.: Probability Essentials. Universitext/Springer, Berlin (1991)
105. Johnston, H., Krasny, R.: Computational predictability of natural convection flows in enclosures: A benchmark problem. In: Bathe, K. (ed.) Computational Fluid and Solid Mechanics, pp. 1486–1489. Elsevier, Amsterdam (2001)
106. Kac, M., Siegert, A.: An explicit representation of a stationary Gaussian process. Ann. Math. Stat. **18**, 438–442 (1947)
107. Kaminski, M., Hien, T.: Stochastic finite element modeling of transient heat transfer in layered composites. Int. Commun. Heat Mass Transfer **26**(6), 801–810 (1999)
108. Karhunen, K.: Uber lineare Methoden in der Wahrscheinlichkeitsrechnung. Am. Acad. Sci., Fennicade, Ser. A, I **37**, 3–79 (1947)
109. Keese, A.: Numerical solution of systems with stochastic uncertainties: a general purpose framework for stochastic finite elements. PhD thesis, Tech. Univ. Braunschweigh (2004)
110. Keese, A., Matthies, H.: Numerical methods and Smolyak quadrature for nonlinear stochastic partial differential equations. Technical report, Institute of Scientific Computing TU Braunschweig Brunswick (2003)
111. Keryszig, E.: Advanced Engineering Mathematics. Wiley, New York (1979)
112. Kim, J., Moin, P.: Application of a fractional-step method to incompressible Navier-Stokes equations. J. Comput. Phys. **59**, 308–323 (1985)
113. Kim, S.E., Choudhury, D.: Numerical investigation of laminar natural convection flow inside a tall cavity using a finite volume based Navier-Stokes solver. In: Bathe, K. (ed.) Computational Fluid and Solid Mechanics, pp. 1490–1492. Elsevier, Amsterdam (2001)
114. Knio, O., Ghanem, R.: Polynomial chaos product and moment formulas: A user utility. Technical report, Johns Hopkins University (2001)
115. Knio, O., Le Maître, O.: Uncertainty propagation in CFD using polynomial chaos decomposition. Fluid Dyn. Res. **38**, 616–640 (2006)
116. Knio, O., Najm, H., Wyckoff, P.: A semi-implicit numerical scheme for reacting flow, II: stiff, operator-split formulation. J. Comput. Phys. **154**, 428–467 (1999)
117. Koekoek, R., Swarttouw, R.: The Askey-scheme of hypergeometric orthogonal polynomials and its q-analogue. Technical report 98-17, Department of Technical Mathematics and Informatics, Delft University of Technology (1998). http://fa.its.tudelft.nl/~koekoek/askey
118. Koumoutsakos, P., Leonard, A., Pépin, P.: Boundary conditions for viscous vortex methods. J. Comput. Phys. **113**, 52–61 (1994)
119. Le Maître, O.: A Newton method for the resolution of steady stochastic Navier-Stokes equations. Comput. Fluids (2007, submitted)
120. Le Maître, O.: Polynomial chaos expansion of a Lagrangian model for the flow around an airfoil. C. R. Acad. Sci. **334**(11), 693–698 (2006)
121. Le Maître, O., Knio, O.: A stochastic particle-mesh scheme for uncertainty propagation in vortical flows. J. Comput. Phys. **226**, 645–671 (2007)
122. Le Maître, O., Knio, O., Debusschere, B., Najm, H., Ghanem, R.: A multigrid solver for two-dimensional stochastic diffusion equations. Comput. Methods Appl. Mech. Eng. **192**(41–42), 4723–4744 (2003)

123. Le Maître, O., Knio, O., Najm, H., Ghanem, R.: A stochastic projection method for fluid flow. I. Basic formulation. J. Comput. Phys. **173**, 481–511 (2001)
124. Le Maître, O., Knio, O., Najm, H., Ghanem, R.: Uncertainty propagation using Wiener-Haar expansions. J. Comput. Phys. **197**(1), 28–57 (2004)
125. Le Maître, O., Najm, H., Ghanem, R., Knio, O.: Multi-resolution analysis of Wiener-type uncertainty propagation schemes. J. Comput. Phys. **197**(2), 502–531 (2004)
126. Le Maître, O., Najm, H., Pébay, P., Ghanem, R., Knio, O.: Multi-resolution-analysis scheme for uncertainty quantification in chemical systems. SIAM J. Sci. Comput. **29**(2), 864–889 (2007)
127. Le Maître, O., Reagan, M., Debusschere, B., Najm, H., Ghanem, R., Knio, O.: Natural convection in a closed cavity under stochastic, non-Boussinesq conditions. SIAM J. Sci. Comput. **26**(2), 375–394 (2004)
128. Le Maître, O., Reagan, M., Najm, H., Ghanem, R., Knio, O.: A stochastic projection method for fluid flow. II. Random process. J. Comput. Phys. **181**, 9–44 (2002)
129. Le Maître, O., Scanlan, R., Knio, O.: Numerical estimation of the flutter derivatives of a NACA airfoil by means of Navier-Stokes simulation. J. Fluids Struct. **17**, 1–28 (2003)
130. Le Quéré, P.: Accurate solution to the square thermally driven cavity at high Rayleigh number. Comput. Fluids **20**(1), 29–41 (1991)
131. Le Quéré, P., de Roquefort, T.: Computation of natural convection in two-dimensional cavities with Chebyshev polynomials. J. Comput. Phys. **57**, 210–228 (1985)
132. Le Quéré, P., Masson, R., Perrot, P.: A Chebyshev collocation algorithm for 2d non-Boussinesq convection. J. Comput. Phys. **103**, 320–334 (1992)
133. Leonard, A.: Vortex methods for flow simulations. J. Comput. Phys. **37**, 289–335 (1980)
134. Liu, J.: Monte Carlo Strategies in Scientific Computing. Springer, Berlin (2001)
135. Liu, N., Hu, B., Yu, Z.W.: Stochastic finite element method for random temperature in concrete structures. Int. J. Solids Struct. **38**, 6965–6983 (2001)
136. Loève, M.: Fonctions aléatoires du second ordre. In: Lévy, P. (ed.) Processus Stochastique et Mouvement Brownien. Gauthier Villars, Paris (1948)
137. Loève, M.: Probability Theory. Springer, Berlin (1977)
138. Lumley, J.: Stochastic Tools in Turbulence. Academic Press, New York (1970)
139. Ma, X., Zabaras, N.: An adaptive hierarchical sparse grid collocation algorithm for the solution of stochastic differential equations. J. Comput. Phys. (2009, in press)
140. Majda, A., Sethian, J.: The derivation and numerical solution of the equations for zero Mach number combustion. Comb. Sci. Technol. **42**, 185–205 (1985)
141. Makeev, A., Maroudas, D., Kevrekidis, I.: Coarse stability and bifurcation analysis using stochastic simulators: kinetic Monte Carlo examples. J. Chem. Phys. **116**, 10083–10091 (2002)
142. Mao, Q., Pawliszyn, J., Thormann, W.: Dynamics of capillary isoelectric focusing in the absence of fluid flow: high-resolution computer simulation and experimental validation with whole column optical imaging. Anal. Chem. **72**, 5493–5502 (2000)
143. Marzouk, Y., Najm, H.: Dimensionality reduction and polynomial chaos acceleration of Bayesian inference in inverse problems. J. Comput. Phys. **228**, 1862–1902 (2009)
144. Marzouk, Y., Najm, H., Rahn, L.: Stochastic spectral methods for efficient Bayesian solution of inverse problems. J. Comput. Phys. **224**, 560–586 (2007)
145. Mathelin, L., Hussaini, M.: A stochastic collocation algorithm for uncertainty analysis. Technical Report NASA/CR-2003-212153, NASA Langley Research Center (2003)
146. Mathelin, L., Hussaini, M., Zang, T.: Stochastic approaches to uncertainty quantification in CFD simulations. Numer. Algorithms **38**(1), 209–236 (2005)
147. Mathelin, L., Le Maître, O.: Dual based a posteriori error estimation for stochastic finite element methods. Commun. Appl. Math. Comput. Sci. **2**(1), 83–115 (2007)
148. Mathelin, L., Zang, T., Bataille, F.: Uncertainty propagation for turbulent, compressible nozzle flow using stochastic methods. AIAA J. **42**(8), 1669–1676 (2004)
149. Matthies, H., Brenner, C., Bucher, C., Soares, C.: Uncertainties in probabilistic numerical analysis of structures and solids—stochastic finite-elements. Struct. Saf. **19**(3), 283–336 (1997)

150. Matthies, H., Bucher, C.: Finite element for stochastic media problems. Comput. Methods Appl. Mech. Eng. **168**, 3–17 (1999)
151. Matthies, H.G., Keese, A.: Galerkin methods for linear and nonlinear elliptic stochastic partial differential equations. Comput. Methods Appl. Mech. Eng. **194**(12–16), 1295–1331 (2005)
152. McGuffin, V., Tavares, M.: Computer assisted optimization of separations in capillary zone electrophoresis. Anal. Chem. **69**, 152–164 (1997)
153. McKay, M., Conover, W., Beckman, R.: A comparison of three methods for selecting values of input variables in the analysis of output from a computer code. Technometrics **21**, 239–245 (1979)
154. Meecham, W., Jeng, D.: Use of the Wiener-Hermite expansion for nearly normal turbulence. J. Fluid Mech. **32**, 225 (1968)
155. Micheletti, S., Perotto, S., Picasso, M.: Stabilized finite elements on anisotropic meshes: a priori error estimates for the advection-diffusion and the stokes problems. SIAM J. Numer. Anal. **41**(3), 1131–1162 (2003)
156. de Montigny, P., Stobaugh, J., Givens, R., Carlson, R., Srinivasachar, K., Sternson, L., Higuchi, T.: Naphtalene-2,3-dicarboxaldehyde/cyanide ion: a rationally designed fluorogenic reagent for primary amines. Anal. Chem. **59**, 1096–1101 (1987)
157. Moré, J.J., Garbow, B.S., Hillstrom, K.E.: User guide for MINPACK-1. Technical Report ANL-80-74, Argone National Lab (1980)
158. Morokoff, W., Caflisch, R.: Quasi-Monte Carlo integration. J. Comput. Phys. **122**(2), 218–230 (1995)
159. Mosher, R., Dewey, D., Thormann, W., Saville, D., Bier, M.: Computer simulation and experimental validation of the electrophoretic behavior of proteins. Anal. Chem. **61**, 362–366 (1989)
160. Mosher, R., Gebauer, P., Thormann, W.: Computer simulation and experimental validation of the electrophoretic behavior of proteins. III. use of titration data predicted by the protein's amino acid composition. J. Chromatogr. **638**, 155–164 (1993)
161. Nair, P., Keane, A.: Stochastic reduced basis methods. AIAA J. **40**, 1653–1664 (2002)
162. Najm, H.: A conservative low Mach number projection method for reacting flow modeling. In: Chan, S. (ed.) Transport Phenomena in Combustion, vol. 2, pp. 921–932. Taylor and Francis, Washington (1996)
163. Najm, H.: Uncertainty quantification and polynomial chaos techniques in computational fluid dynamics. Annu. Rev. Fluid Mech. **41**, 35–52 (2009)
164. Najm, H., Wyckoff, P., Knio, O.: A semi-implicit numerical scheme for reacting flow, I: stiff chemistry. J. Comput. Phys. **143**(2), 381–402 (1998)
165. Nicola, B., Verlinden, B., Beuselinck, A., Jancsok, P., Quenon, V., Scheerlinck, N., Verbosen, P., de Baerdemaeker, J.: Propagation of stochastic temperature fluctuations in refrigerated fruits. Int. J. Refrig. **222**, 81–90 (1999)
166. Nobile, F., Tempone, R.: Analysis and implementation issues for the numerical approximation of parabolic equations with random coefficients. Technical Report 22/2008, MOX (2008)
167. Nouy, A.: Generalized spectral decomposition method for solving stochastic finite element equations: invariant subspace problem and dedicated algorithms. Comput. Methods Appl. Mech. Eng. (2007, submitted)
168. Nouy, A.: A generalized spectral decomposition technique to solve a class of linear stochastic partial differential equations. Comput. Methods Appl. Methods Eng. **196**(45–48), 4521–4537 (2007)
169. Nouy, A.: Méthode de construction de bases spectrales généralisées pour l'approximation de problèmes stochastiques. Méc. Ind. **8**(3), 283–288 (2007)
170. Nouy, A., Clément, A., Schoefs, F., Moës, N.: An extended stochastic finite element method for solving stochastic partial differential equations on random domains. Comput. Methods Appl. Mech. Eng. (2007, submitted)
171. Nouy, A., Le Maître, O.: Generalized spectral decomposition for stochastic nonlinear problems. J. Comput. Phys. **228**, 202–235 (2009)

172. Nouy, A., Schoefs, F., Moës, N.: X-SFEM, a computational technique based on X-FEM to deal with random shapes. Eur. J. Comput. Mech. **16**, 277–293 (2007)
173. Novak, E., Ritter, K.: Simple cubature formulas with high polynomial exactness. Constr. Approx. **15**, 499–522 (1999)
174. Ogden, R.T.: Essentials Wavelets for Statistical Applications and Data Analysis. Birkhäuser, Basel (1997)
175. Oksendal, B.: Stochastic Differential Equations. An Introduction with Applications, 5th edn. Springer, Berlin (1998)
176. Paillere, H., Le Quéré, P.: Modelling and simulation of natural convection flows with large temperature differences: a benchmark problem for low Mach number solvers. In: 12th Seminar "Computational Fluid Dynamics" CEA/Nuclear Reactor Division (2000)
177. Paillere, H., Viozat, C., Kumbaro, A., Toumi, I.: Comparison of low Mach number models for natural convection problems. Heat Mass Transfer **36**, 567–573 (2000)
178. Palusinski, O., Graham, A., Mosher, R., Bier, M., Saville, D.: Theory of electrophoretic separations, part II: construction of a numerical simulation scheme and its applications. AIChE J. **32**(2), 215–223 (1986)
179. Pan, T.W., Glowinski, R.: A projection/wave-like equation method for natural convection flows in enclosures. In: Bathe, K. (ed.) Computational Fluid and Solid Mechanics, pp. 1493–1496. Elsevier, Amsterdam (2001)
180. Panton, R.: Incompressible Flow. Wiley, New York (1984)
181. Papoulis, A., Pillai, S.: Probability, Random Variables and Stochastic Processes. McGraw-Hill, New York (2002)
182. Patankar, N., Hu, H.: Numerical simulation of electroosmotic flow. Anal. Chem. **70**, 1870–1881 (1998)
183. Pellissetti, M., Ghanem, R.: Iterative solution of systems of linear equations arising in the context of stochastic finite elements. Adv. Eng. Softw. **31**, 607–616 (2000)
184. Peraire, J., Peiró, J., Morgan, K.: Adaptive remeshing for three-dimensional compressible flow computations. J. Comput. Phys. **103**(2), 269–285 (1992)
185. Pernice, M., Walker, H.: NITSOL: A Newton iterative solver for nonlinear systems. SIAM J. Sci. Comput. **19**(1), 302–318 (1998)
186. Petras, K.: On the Smolyak cubature error for analytic functions. Adv. Comput. Math. **12**, 71–93 (2000)
187. Petras, K.: Fast calculation in the Smolyak algorithm. Numer. Algorithms **26**, 93–109 (2001)
188. Peyret, R.: Spectral Methods for Incompressible Viscous Flow. Springer, Berlin (2002)
189. Poëtte, G., Després, B., Lucor, D.: Uncertainty quantification for systems of conservation laws. J. Comput. Phys. **228**, 2443–2467 (2009)
190. Powell, C.: Robust preconditioning for second-order elliptic PDEs with random field coefficients. Technical Report 2006.419, MIMS Eprint (2006)
191. Powell, C., Elman, H.: Block diagonal preconditioning for spectral stochastic finite element systems. IMA J. Numer. Anal. **29**(2), 350–375 (2009)
192. Probstein, R.: Physicochemical Hydrodynamics: An Introduction, 2nd edn. Wiley, New York (1994)
193. Pulkelsheim, F.: Optimal Design of Experiments. Classics in Applied Mathematics, vol. 50. SIAM, Philadelphia (2006)
194. Raviart, P.: An Analysis of Particles Methods. Lecture Notes in Math., vol. 1127, pp. 243–324. Springer, Berlin (1985)
195. Reagan, M., Najm, H., Debusschere, B., Maître, O.L., Knio, O., Ghanem, R.: Spectral stochastic uncertainty quantification in chemical systems. Combust. Theory Model. **8**, 607–632 (2004)
196. Rivoalen, E., Huberson, S.: Numerical simulation of axisymmetric flows by means of a particle method. J. Comput. Phys. **152**, 1–31 (1999)
197. Roberts, G., Rhodes, P., Snyder, R.: Dispersion effects in capillary zone electrophoresis. J. Chromatogr. **480**, 35–67 (1989)
198. Rosenhead, L.: The formation of vortices from a surface of discontinuity. Proc. R. Soc. **134** (1931)

199. Saad, Y.: Iterative Methods for Sparse Linear Systems, 2nd edn. SIAM, Philadelphia (2003)
200. Saad, Y., Schultz, M.: GMRES: a generalized minimal residual algorithm for solving nonsymmetric linear system. J. Sci. Stat. Comput. **7**, 856–869 (1986)
201. Salinger, A., Lehoucq, R., Pawlowski, R., Shadid, J.: Understanding the 8:1 cavity problem via scalable stability analysis algorithms. In: Bathe, K. (ed.) Computational Fluid and Solid Mechanics, pp. 1497–1500. Elsevier, Amsterdam (2001)
202. Scales, P., Grieser, F., Healy, T., White, L., Chan, D.: Electrokinetics of the silica-solution interface: a flat plate streaming potential study. Langmuir **8**, 965–974 (1992)
203. Schoutens, W.: Stochastic Processes and Orthogonal Polynomials. Springer, Berlin (2000)
204. Schuëller, G.: Computational stochastic mechanics—recent advances. Comput. Struct. **79**, 2225–2234 (2001)
205. Shafer-Nielsen, C.: A computer model for time-based simulation of electrophoresis systems with freely defined initial and boundary conditions. Electrophoresis **16**, 1369–1376 (1995)
206. Shen, W.Z., Loc, T.P.: Simulation of 2d external viscous flow by means of a domain decomposition method using an influence matrix technique. Int. J. Numer. Methods Fluids **220**, 1111 (1995)
207. Sleijpen, G., Fokkema, D.: BICGSTAB(L) for linear equations involving unsymmetric matrices with complex spectrum. Electon. Trans. Numer. Anal. **1**, 11–32 (1993)
208. Slepian, D., Pollak, H.: Prolate spheroidal wave functions; Fourier analysis and uncertainty, I. Bell Syst. Tech. J., 3–63 (1961)
209. Slepian, D., Pollak, H.: Prolate spheroidal wave functions; Fourier analysis and uncertainty, IV: extension to many dimensions; generalizes prolate spheroidal functions. Bell Syst. Tech. J., 009–3057 (1964)
210. Sluzalec, A.: Random heat flow with phase change. Int. J. Heat Mass Transfer **43**, 2303–2312 (2000)
211. Sluzalec, A.: Stochastic finite element analysis of two-dimensional eddy current problems. Appl. Math. Model. **24**, 401–406 (2000)
212. Smolyak, S.: Quadrature and interpolation formulas for tensor products of certain classes of functions. Dokl. Akad. Nauk SSSR **4**, 240–243 (1963)
213. Sobol, I.: Uniformly distributed sequences with an additional uniform property. USSR Comput. Math. Math. Phys. **16**, 236–242 (1976)
214. Soize, C.: The Fokker-Planck Equation for Stochastic Dynamical Systems ans Its Explicit Steady Solutions. World Scientific, Singapore (1994)
215. Soize, C.: A comprehensive overview of a non-parametric probabilistic approach of model uncertainty for predictive models in structural dynamics. J. Sound Vib. **288**, 623–652 (2005)
216. Soize, C.: Random matrix theory for modeling uncertainty in computational mechanics. Comput. Methods Appl. Mech. Eng. **194**(12–16), 1333–1366 (2005)
217. Soize, C.: Construction of probability distributions in high dimension using the maximum entropy principle: Applications to stochastic processes, random fields and random matrices. Int. J. Numer. Methods Eng. **76**(10), 1583–1611 (2008)
218. Soize, C., Ghanem, R.: Physical systems with random uncertainties: chaos representations with arbitrary probability measure. J. Sci. Comput. **26**(2), 395–410 (2004)
219. Soize, C., Ghanem, R.: Reduced chaos decomposition with random coefficients of vector-valued random variables and random fields. Comput. Methods Appl. Mech. Eng. **198**(21–26), 1926–1934 (2009)
220. Stoer, J., Bulirsch, R.: Introduction to Numerical Analysis, 2nd edn. Texts in Applied Mathematics, vol. 12. Springer, Berlin (1993)
221. Strang, G.: Introduction to Applied Mathematics. Wellesley-Cambridge Press, Wellesley (1986)
222. Stroud, A., Secrest, D.: Gaussian Quadrature Formula. Prentice-Hall, New York (1966)
223. Suslov, S., Paolucci, S.: A Petrov-Galerkin method for flows in cavities. In: Bathe, K. (ed.) Computational Fluid and Solid Mechanics, pp. 1501–1504. Elsevier, Amsterdam (2001)
224. Theting, T.: Solving Wick-stochastic boundary value problems using a finite element method. Stoch. Stoch. Rep. **70**, 241–270 (2000)

225. Thormann, W., Mosher, R.: Theory of electrophoretic transport and separations: the study of electrophoretic boundaries and fundamental separation principles by computer simulation. In: Chrambach, A., Dunn, M., Radola, B. (eds.) Advances in Electrophoresis, vol. 2, pp. 45–108. VCH, Weinheim (1988)
226. Thormann, W., Zhang, C.X., Caslavska, J., Gebauer, P., Mosher, R.: Modeling of the impact of ionic strength on the electroosmotic flow in capillary electrophoresis with uniform and discontinuous buffer systems. Anal. Chem. **70**, 549–562 (1998)
227. Trefethen, L.: Is Gauss quadrature better than Clenshaw-Curtis? SIAM Rev. **50**(1), 67–87 (2008)
228. Trottenberg, U., Oosterlee, C., Schüller, A.: Multigrid. Academic Press, New York (2001)
229. Tryoen, J.: Méthode intrusive de modélisation d'incertitudes dans les systèmes hyperboliques de lois de conservation. MS thesis, Université Paris VI (2008)
230. Tuckerman, L.: Steady-state solving via stokes preconditioning; recursion relations of elliptic operators. In: 11th Int. Conf. on Numer. Meth. in Fluid Dynamics. Springer, Berlin (1989)
231. Tuckerman, L., Huepe, C., Brachet, M.: Numerical methods for bifurcation problems. In: Descalzi, O., Martinez, J., Rica, S. (eds.) Instabilities and Non-equilibrium Structures IX, pp. 75–86. Kluwer Academic, Dordrecht (2004)
232. Vanel, J., Peyret, R., Bontoux, P.: A pseudo-spectral solution of vorticity-stream-function equations using the influence matrix technique. Numer. Methods Fluid Dyn. **II**, 463–475 (1986)
233. Waldvogel, J.: Fast construction of the Fejér and Clenshaw-Curtis quadrature rules. BIT Numer. Math. **43**(1), 001–018 (2003)
234. Walnut, D.F.: An Introduction to Wavelets Analysis. Applied and Numerical Harmonic Analysis. Birkhäuser, Basel (2002)
235. Walter, G.: Wavelets and Other Orthogonal Systems with Applications. CRC Press, Boca Raton (1994)
236. Walters, R., Huyse, L.: Uncertainty analysis for fluid mechanics with applications. Technical report, NASA CR-2002-211449 (2002)
237. Wan, Q.H.: Effect of electroosmotic flow on the electrical conductivity of packed capillary columns. J. Phys. Chem. B **101**(24), 4860–4862 (1997)
238. Wan, X., Karniadakis, G.: An adaptive multi-element generalized polynomial chaos method for stochastic differential equations. J. Comput. Phys. **209**, 617–642 (2005)
239. Webster, C.: Sparse grid collocation techniques for the numerical solution of partial differential equations with random input data. PhD thesis, Florida State University (2007)
240. Westerberg, K.: Thermally driven flow in a cavity using the Galerkin finite element method. In: Bathe, K. (ed.) Computational Fluid and Solid Mechanics, pp. 1505–1508. Elsevier, Amsterdam (2001)
241. Wiener, S.: The homogeneous chaos. Am. J. Math. **60**, 897–936 (1938)
242. Wojtkiewicz, S., Eldred, M., Field, R., Urbina, A., Red-Horse, J.: A toolkit for uncertainty quantification in large computational engineering models. In: Proceedings of the 42nd AIAA/ASME/ASCE/AHS/ASC Structures, Structural Dynamics, AIAA Paper 2001-1455 (2001)
243. Xin, S., Le Quéré, P.: An extended Chebyshev pseudo-spectral contribution to cpncfe benchmark. In: Bathe, K. (ed.) Computational Fluid and Solid Mechanics, pp. 1509–1513. Elsevier, Amsterdam (2001)
244. Xiu, D., Hesthaven, J.: High-order collocation methods for differential equations with random inputs. SIAM J. Sci. Comput. **27**(3), 1118–1139 (2005)
245. Xiu, D., Karniadakis, G.: Modeling uncertainty in steady state diffusion problems via generalized polynomial chaos. Comput. Methods Appl. Mech. Eng. **191**, 4927–4948 (2002)
246. Xiu, D., Karniadakis, G.: The Wiener-Askey polynomial chaos for stochastic differential equations. SIAM J. Sci. Comput. **24**, 619–644 (2002)
247. Xiu, D., Karniadakis, G.: Modeling uncertainty in flow simulations via generalized polynomial chaos. J. Comput. Phys. **187**, 137–167 (2003)
248. Xiu, D., Karniadakis, G.: Supersensitivity due to uncertain boundary conditions. Int. J. Numer. Methods Eng. **61**(12), 2114–2138 (2004)

249. Xiu, D., Lucor, D., Su, C., Karniadakis, G.: Stochastic modeling of flow structure interactions using generalized polynomial chaos. J. Fluids Eng. **124**, 51–59 (2002)
250. Xiu, D., Shen, J.: Efficient stochastic Galerkin methods for random diffusion equations. J. Comput. Phys. **228**, 266–281 (2009)
251. Youla, D.: The solution of the homogeneous Wiener-Hopf integral equation occurring in the expansion of second-order stationary random functions. IRE Tans. Inf. Theory, 18–193 (1957)

Index

O.P. Le Maître, O.M. Knio, *Spectral Methods for Uncertainty Quantification*,
Scientific Computation,
DOI 10.1007/978-90-481-3520-2, © Springer Science+Business Media B.V. 2010